Birkhäuser

Operator Theory: Advances and Applications

Volume 294

Founded in 1979 by Israel Gohberg

Yurij M. Berezansky • Mykola E. Dudkin

Jacobi Matrices and the Moment Problem

Birkhäuser

Yurij M. Berezansky
Institute of Mathematics
National Academy of Sciences of Ukraine
Kyiv, Ukraine

Mykola E. Dudkin
Igor Sikorsky Kyiv Polytechnic Institute
National Technical University of Ukraine
Kyiv, Ukraine

Translated by
Mykola E. Dudkin

ISSN 0255-0156 ISSN 2296-4878 (electronic)
Operator Theory: Advances and Applications
ISBN 978-3-031-46386-0 ISBN 978-3-031-46387-7 (eBook)
https://doi.org/10.1007/978-3-031-46387-7

Translation from the Ukrainian language edition: "Якобієві Матриці і проблема моментів" by Yurij M. Berezansky and Mykola E. Dudkin, © National Academy of Science of Ukraine 2019. Published by National Academy of Science of Ukraine. All Rights Reserved.

This book is published under the imprint Birkhäuser, www.birkhauser-science.com by the registered company Springer Nature Switzerland AG
The registered company address is: Gewerbestrasse 11, 6330 Cham, Switzerland

Paper in this product is recyclable.

Preface

The classical Hamburger moment problem has become the source and the basis for the development of various theories. Among them, there are the theory of extensions of symmetric operators, Jacobi matrices, Padé approximations, the Nevanlinna-Pick interpolation, orthogonal polynomials of the first and second order with respect to a given measure, the corresponding Weyl function, and others. On its own, the moment problem has today acquired numerous generalizations, and therefore, corresponding theories possess natural generalizations. The first famous researchers of the moment problem are Hamburger H., Stiltjes T., Chebyshev P. L., and Markov A. A.

At the beginning, the monograph deals with the usual Jacobi matrix that naturally arise by the consideration of the classical moment problem [2, 3]. Further we consider Jacobi-type matrices corresponding to the trigonometric moment problem [62, 63], the complex moment problem [37], and the real two-dimensional moment problem [87]. However, such matrices have a multi-diagonal structure.

If we put such matrices into blocks of a certain size, then they become a block tri-diagonal structure and, hence, we can apply the widely developed Berezansky theory of generalized eigenfunctions expansion for a set of commuting corresponding operators [21, 23].

In this monograph, we do not focus so much on demonstrating the form of matrices corresponding to some moment problem, as we give a sufficiently rigorous solution of the corresponding direct and inverse spectral problems. The inverse spectral problem refers to the construction of block tri-diagonal matrices of a certain structure according to a given measure, which has all corresponding moments in some functional space; the direct problem is to restore the measure in the "sense of the Parseval equality" using given matrices of a certain structure.

The material of the book will be useful for specialists in operator theory, functional analysis, theoretical physicists, graduate students, and engineers who deal with the problems of coupled pendulums. The first section of the monograph contains a number of a necessary information concerning unbounded operators, the generalized eigenvectors expansion theory, the theory of extensions of densely defined symmetric operators, etc.

The material of individual sections of this monograph was presented in special courses for students of Taras Shevchenko National University, Kyiv, National Technical University of Ukraine "Igor Sikorsky Kyiv Polytechnic Institute."

Kyiv, Ukraine

Yurij M. Berezansky
Mykola E. Dudkin

Contents

1 Introduction 1

2 Some Aspects of the Spectral Theory of Unbounded Operators 17
- 2.1 Preliminary Information About Unbounded Operators 17
- 2.2 Extensions of Hermitian Operators to Self-AdJoint Operators 27
- 2.3 The Spectral Decomposition of Bounded Operators 30
- 2.4 The Spectral Decomposition of Unbounded Operators 41
- 2.5 Representations of One-Parameter Unitary Groups 43
- 2.6 Self-Adjointness Criteria 49
- 2.7 Rigged Spaces 63
- 2.8 Tensor Products 76
- 2.9 Representations of Continuous Multi-Linear Forms 81
- 2.10 Semi-Bounded Bilinear Forms 89
- 2.11 The Generalized Eigenvectors Expansion 93
- 2.12 A General Case with a Quasi-Scalar Product 116
- Bibliographical Notes 118

3 Jacobi Matrices and the Classical Moment Problem 121
- 3.1 Difference Operators, Jacobi Matrices, and Self-Adjointness Conditions 121
- 3.2 The Generalized Eigenvectors Expansion and the Fourier Transform Corresponding to Jacobi Matrices 126
- 3.3 The Inverse Problem 136
- 3.4 Further Spectral Analysis of the Difference Operator 138
- 3.5 The Classical Moment Problem 148
- 3.6 Some Other Generalizations of the Moment Problem 164
- 3.7 Connections with the Theory of Jacobi Matrices 177
- Bibliographical Notes 180

4 The Strong Moment Problem 183
- 4.1 Preliminaries to the Strong Moment Problem 184
- 4.2 The Solution of the Strong Moment Problem 187

4.3 The Orthogonalization Procedure and the Construction of a Tri-Diagonal Block Matrix ... 191
4.4 Direct and Inverse Spectral Problems Corresponding to Tri-Diagonal Block Jacobi-Laurent Matrices Generating Self-Adjoint Operators ... 204
4.5 Considerations of Hermitian Block Jacobi-Laurent Type Matrices ... 214
4.6 The Connection Between the Strong Moment Problem and the Spectral Theory of Jacobi-Laurent Matrices ... 224
4.7 Two Addition Facts ... 227
4.8 The Inner Structure of the Jacobi-Laurent Matrix ... 231
Bibliographical Notes ... 244

5 Block Jacobi Type Matrices in the Complex Moment Problem ... 245
5.1 Construction of the Tri-Diagonal Block Jacobi Type Matrix of a Bounded Normal Operator ... 245
5.2 Direct and Inverse Spectral Problems for the Tri-Diagonal Block Jacobi Type Matrices of Bounded Normal Operators ... 257
5.3 Normality Conditions of Block Jacobi Type Matrices ... 267
5.4 The Solution of the Complex Moment Problem ... 275
5.5 The Weyl Function and Polynomials of the Second Kind in the Complex Moment Problem ... 285
Bibliographical Notes ... 287

6 Unitary Block Jacobi Type Matrices and the Trigonometric Moment Problem ... 289
6.1 Introduction to the Unitary Block Jacobi Type Matrices ... 289
6.2 Construction of the Tri-Diagonal Block Jacobi Type Matrices of a Unitary Operator ... 291
6.3 Direct and Inverse Spectral Problems for the Tri-Diagonal Block Jacobi Type Matrices of Unitary Operators ... 301
6.4 The Detailed Internal Structure of the Block Matrix of the Unitary Operator ... 310
6.5 The Solution of the Trigonometric Moment Problem ... 316
Bibliographical Notes ... 323

7 Block Jacobi Type Matrices and the Complex Moment Problem in the Exponential Form ... 325
7.1 Construction of the Tri-Diagonal Block Matrix Corresponding to the Complex Moment Problem in the Exponential Form ... 325
7.2 Direct and Inverse Spectral Problems for Tri-Diagonal Block Jacobi Type Matrices Corresponding to the Complex Moment Problem in the Exponential Form ... 345

7.3 Conditions of Unitarity and Commutativity of Matrices Corresponding to the Complex Moment Problem in the Exponential Form ... 356
7.4 The Solution of the Complex Moment Problem in the Exponential Form ... 361
Bibliographical Notes ... 370

8 Block Jacobi Type Matrices and the Two Dimensional Real Moment Problem ... 371
8.1 Construction of the Tri-Diagonal Block Jacobi Type Matrices Corresponding to the Two Dimensional Real Moment Problem ... 371
8.2 Direct and Inverse Spectral Problems for the Tri-Diagonal Block Jacobi Type Matrices Corresponding to the Two-Dimensional Real Moment Problem ... 386
8.3 An Analogue of the Weyl Function Corresponding to the Two-Dimensional Real Moment Problem ... 397
8.4 The Solution of the Two-Dimensional Real Moment Problem ... 399
8.5 Examples of Matrices Corresponding to the Two-Dimensional Real Moment Problem ... 407
Bibliographical Notes ... 411

9 Applications of the Spectral Theory of Jacobi Matrices and Their Generalizations to the Integration of Nonlinear Equations ... 413
9.1 The Integration of the Toda Chain on the Semi-axis Using the Spectral Theory of Jacobi Matrices ... 414
9.2 The Doubly Infinite Toda Chain and Its Equivalent Lax Equation for Block Jacobi Type Matrices ... 424
9.3 The Spectral Theory of Corresponding to Toda Chains Block Jacobi Type Matrices ... 430
9.4 Equations for the Weyl Function and Spectral Matrices ... 446
9.5 The Basic Statement and Applications to Hamiltonian Systems ... 458
Bibliographical Notes ... 464

References ... 467

Subject Index ... 483

Notation Index ... 487

Chapter 1
Introduction

For a better understanding of the content of the monograph, we will first outline the main content of the second chapter: direct and inverse spectral problems for classical Jacobi matrices and orthogonal polynomials on the real axis $\mathbb{R}$ (see, for example, [4, 21]). In this classical theory, the symmetric (Hermitian) Jacobi matrix

$$J = \begin{bmatrix} b_0 & a_0 & 0 & 0 & 0 & \cdots \\ a_0 & b_1 & a_1 & 0 & 0 & \cdots \\ 0 & a_1 & b_2 & a_2 & 0 & \cdots \\ \vdots & \vdots & \vdots & \vdots & \vdots & \ddots \end{bmatrix}, \tag{1.0.1}$$

$$b_n \in \mathbb{R},\ a_n > 0, \quad n \in \mathbb{N}_0 = \{0, 1, 2, \ldots\}$$

is investigated in the space l_2 of sequences $f = (f_n)_{n=0}^{\infty}$.

This matrix defines an operator J in l_2 on finite sequences $f \in l_{\text{fin}} \subset l_2$, so that the operator is Hermitian with equal defect numbers and, hence, has self-adjoint expansions in l_2. The closure $\tilde{J}$ of the operator J is the self-adjoint operator under certain conditions on J (for example, $\sum\limits_{n=0}^{\infty} a_n^{-1} = \infty$),

The direct spectral problem, i.e., the eigenfunction expansion for $\tilde{J}$ (for simplicity, we will assume that $\tilde{J}$ is self-adjoint) is considered in the following way. We introduce the sequence of polynomials $P(\lambda) = (P_n(\lambda))_{n=0}^{\infty}$, $\forall \lambda \in \mathbb{R}$ as a solution of the equation $JP(\lambda) = \lambda P(\lambda)$ with initial conditions $P_0(\lambda) = 1$, $P_{-1}(\lambda) = 0$, namely

$$\begin{aligned} &a_{n-1}P_{n-1}(\lambda) + b_n P_n(\lambda) + a_n P_{n+1}(\lambda) = \lambda P_n(\lambda), \\ &P_0(\lambda) = 1, \quad P_{-1}(\lambda) = 0, \quad \forall n \in \mathbb{N}_0. \end{aligned} \tag{1.0.2}$$

Y. M. Berezansky, M. E. Dudkin, *Jacobi Matrices and the Moment Problem*,
Operator Theory: Advances and Applications 294,
https://doi.org/10.1007/978-3-031-46387-7_1

This recurrence relation has a solution: it is only necessary to proceed step by step, starting with $P_0(\lambda)$ and $P_{-1}(\lambda) = 0$. It is possible, since $a_n > 0$.

The sequens $P(\lambda)$ of polynomials belongs to $l = \mathbb{C}^\infty\ \forall\lambda$ (more exactly, to the real part of l) and it is a generalized eigenvector for $\tilde{J}$ with the corresponding eigenvalue λ (we use a certain rigging of l_2). The appropriate Fourier transform "$\widehat{\ }$" with generalized eigenvectors of the operator $\tilde{J}$ has a form

$$l_2 \supset l_{\text{fin}} \ni f = (f_n)_{n=0}^\infty \longmapsto \hat{f}(\lambda) = \sum_{n=0}^\infty f_n P_n(\lambda) \in L_2(\mathbb{R}, d\rho(\lambda)) = L_2. \tag{1.0.3}$$

The last transformation is an isometrical operator (after taking the extension by continuity) from the whole space l_2 into the whole space L_2. Herewith, the image of $\tilde{J}$ is the multiplication by λ operator in the space L_2. The polynomials $P_n(\lambda)$ are orthogonal with respect to the measure $d\rho(\lambda)$.

The inverse problem in this classical case is the following. Let us have a probability Borel measure $d\rho(\lambda)$ on $\mathbb{R}$, that has all its moments,

$$s_n = \int_{\mathbb{R}} \lambda^n \, d\rho(\lambda), \quad n \in \mathbb{N}_0 \tag{1.0.4}$$

(and the support $d\rho(\lambda)$ contains an infinite set on a finite interval). The question is whether it is possible to recover the corresponding Jacobi matrix J such that the initial measure $d\rho(\lambda)$ would be equal to the spectral measure for $\tilde{J}$? And how to obtain such a reconstruction?

The answer is very simple: it is necessary to take the sequence of functions in L_2 (according to (1.0.4))

$$1, \lambda, \lambda^2, \ldots \tag{1.0.5}$$

that are linearly independent due to the condition on the support of $d\rho(\lambda)$ and apply the Gram-Schmidt orthogonalization procedure. As a result we get a sequence of orthonormal polynomials that form a basis in L_2

$$P_0(\lambda) = 1,\ P_1(\lambda),\ P_2(\lambda), \ldots. \tag{1.0.6}$$

Then the matrix J is reconstructed by formulas

$$a_n = \int_{\mathbb{R}} \lambda P_n(\lambda) P_{n+1}(\lambda)\, d\rho(\lambda), \quad b_n = \int_{\mathbb{R}} \lambda (P_n(\lambda))^2 \, d\rho(\lambda), \quad n \in \mathbb{N}_0. \tag{1.0.7}$$

The above mentioned connection between Jacobi matrices, the classical moment problem and orthogonal polynomials is very fruitful for studying these objects. Many mathematicians worked in this direction, but it is necessary single out the corresponding results of Krein M. G. [128, 185, 186] and Achiezer N. I. [2, 3].

The main question of further chapters is the following: in what way it is necessary to proceed for the obtaining a generalization of the above mentioned classical theory of orthogonal polynomials on the case of the complex plane $\mathbb{C}$ (or on some subset of $\mathbb{C}$, for example, on the unite circle $\mathbb{T} \subset \mathbb{C}$), on the real plane $\mathbb{R}$, etc.? Roughly spiking, it is necessary instead of one self-adjoint operator in l_2 to consider a normal (unitary) operator or the couple of commuting operators that act in some space like l_2.

More exactly, in fourth chapter, for example, instead of the space $l_2 = \mathbb{C}\oplus\mathbb{C}\oplus\cdots$ we take

$$\mathbf{l}_2 = \mathcal{H}_0 \oplus \mathcal{H}_1 \oplus \mathcal{H}_2 \oplus \cdots, \quad \text{where} \quad \mathcal{H}_n = \mathbb{C}^{n+1}, \tag{1.0.8}$$

and instead of the scalar matrix (1.0.1) we take Jacobi type block matrix with elements a_n, b_n and c_n, that are finite dimensional operators (matrices), acting between spaces $\mathcal{H}_n$ in (1.0.8), namely:

$$J = \begin{bmatrix} b_0 & c_0 & 0 & 0 & 0 & \cdots \\ a_0 & b_1 & c_1 & 0 & 0 & \cdots \\ 0 & a_1 & b_2 & c_2 & 0 & \cdots \\ \vdots & \vdots & \vdots & \vdots & \vdots & \ddots \end{bmatrix}, \quad \begin{array}{l} a_n : \mathcal{H}_n \longrightarrow \mathcal{H}_{n+1}, \\ b_n : \mathcal{H}_n \longrightarrow \mathcal{H}_n, \\ c_n : \mathcal{H}_{n+1} \longrightarrow \mathcal{H}_n, \quad n \in \mathbb{N}_0. \end{array} \tag{1.0.9}$$

Such a matrix (1.0.9) induces the operator J in $\mathbf{l}_2$ defined on finite vectors $\mathbf{l}_{\text{fin}} \subset \mathbf{l}_2$. For simplicity, *we will demand everywhere in the sequel, that the norm of elements a_n, b_n and c_n are uniformly bounded and, therefore, the operator J is bounded in $\mathbf{l}_2$.*

The essential condition $a_n > 0$ in (1.0.1) has now a long form

$$a_n = \underbrace{\begin{bmatrix} a_{n;0,0} & * & * & \dots & * \\ 0 & a_{n;1,1} & * & \dots & * \\ \vdots & \vdots & \vdots & \ddots & \vdots \\ 0 & 0 & 0 & \dots & a_{n;n,n} \\ 0 & 0 & 0 & \dots & 0 \end{bmatrix}}_{n+1} \left.\vphantom{\begin{bmatrix} a \\ 0 \\ \vdots \\ 0 \\ 0 \end{bmatrix}}\right\} n+2,$$

$$c_n = \underbrace{\begin{bmatrix} c_{n;0,0} & c_{n;0,1} & 0 & \dots & 0 & 0 \\ * & * & c_{n;1,2} & \dots & 0 & 0 \\ \vdots & \vdots & \vdots & \ddots & \vdots & \vdots \\ * & * & * & \dots & c_{n;n-1,n} & 0 \\ * & * & * & \dots & * & c_{n;n,n+1} \end{bmatrix}}_{n+2} \Bigg\} n+1,$$

$$a_{n;0,0}, a_{n;1,1}, \dots, a_{n;n,n} > 0, \quad c_{n;0,1}, c_{n;1,2}, \dots, c_{n;n,n+1} > 0, \quad n \in \mathbb{N}_0. \tag{1.0.10}$$

Under some simple condition on a_n, b_n and c_n, $n \in \mathbb{N}_n$, the matrix J is a formally normal operator, i.e., $JJ^+ = J^+J$, where J^+ denotes the matrix adjoint to J. Therefore, (due to the fact that J is bounded) the closure $\tilde{J}$ is the bounded operator in $\mathbf{l}_2$.

Let $z \in \mathbb{C}$ belong to the spectrum of $\tilde{J}$ and $P(z) = (P_n(z))_{n=0}^{\infty}$ be the corresponding generalized eigenvector $\tilde{J}$. Here $P_n(z) \in \mathcal{H}_n$ is a vector-valued polynomial with respect to variables $z, \bar{z}$ of degree n, i.e., it is some linear combination of $z^j \bar{z}^k$, $0 \le j + k \le n$. According to the generalized eigenfunctions expansion theorem, it is some solution of two equations of type (1.0.2) (but with matrix coefficients)

$$JP(z) = zP(z), \quad J^+P(z) = \bar{z}P(z). \tag{1.0.11}$$

The corresponding Fourier transform "$\widehat{\ }$" for the operator $\tilde{J}$ has a form

$$\mathbf{l}_2 \supset \mathbf{l}_{\text{fin}} \ni f = (f_n)_{n=0}^{\infty} \longmapsto \hat{f}(z) = \sum_{n=0}^{\infty} (f_n, P_n(z))_{\mathcal{H}_n} \in L_2(\mathbb{C}, d\rho(z)) = L_2, \tag{1.0.12}$$

where $d\rho(z)$ is the spectral measure of $\tilde{J}$, and it has a compact support. The mapping (1.0.12) is the isometrical operator from whole $\mathbf{l}_2$ into whoule L_2. The polynomials $P_n(z)$ are orthonormal with respect to the measure $d\rho(z)$ and form a basis in the space L_2. It is conveniens to denote these polynomials by

$$(\overline{Q_{n;0}(z)}, \overline{Q_{n;1}(z)}, \dots, \overline{Q_{n;n}(z)}) = (P_{n;0}(z), P_{n;1}(z), \dots, P_{n;n}(z)) = P_n(z).$$

So, the described above result is called the direct spectral problem for the matrix J of the form (1.0.9) with conditions on elements (1.0.10).

The inverse spectral problem is now formulated as follows. Let us have a probability Borel measure $d\rho(z)$ with a compact support on $\mathbb{C}$; assume that all complex moments

$$s_{m,n} = \int_{\mathbb{C}} z^m \bar{z}^n \, d\rho(z), \qquad m, n \in \mathbb{N}_0, \tag{1.0.13}$$

exist and the support of $d\rho(z)$ must be such that all functions $z^j \bar{z}^k$, $j, k \in \mathbb{N}_0$ (belonging to L_2, see (1.0.12)), are linearly independent in this space (for example, the support contains some open set of $\mathbb{C}$). It is necessary to construct a Jacobi type matrix (1.0.9) satisfying conditions (1.0.10) so that the spectral measure of the normal operator $\tilde{J}$ would be equal to the initial measure.

As in the classical case, it is necessary to apply the standard Gram-Schmidt orthogonalization procedure to the sequence of functions in L_2

$$(z^j \bar{z}^k)_{j,k=0}^{\infty} \tag{1.0.14}$$

(instead of (1.0.5)). But the sequence (1.0.14) has two indices, and therefore, it is necessary to choose a convenient global (linear) order for (1.0.14). We order it in the following way:

$$z^0\bar{z}^0 = 1; \quad z^1\bar{z}^0, z^0\bar{z}^1; \quad z^2\bar{z}^0, z^1\bar{z}^1, z^0\bar{z}^2; \quad \dots; \\ z^n\bar{z}^0, z^{n-1}\bar{z}^1, \dots, z^0\bar{z}^n; \quad \dots. \tag{1.0.15}$$

After such an orthogonalization, we get a sequence of polynomials

$$P_n(z) = (P_{n;0}(z), P_{n;1}(z), \dots, P_{n;n}(z)), \quad n \in \mathbb{N}_0,$$

and the matrix (1.0.9) with (1.0.10) is reconstructed by using the formulas like (1.0.7).

The above mentioned results are given in Chap. 4. All the necessary information about the projection spectral theorem is presented in the first section. Note also, that the theory of Jacobi block matrices, which are Hermitian or self-adjoint operators acting in the space $l_2(\mathcal{H}) = \mathcal{H} \oplus \mathcal{H} \oplus \cdots$, where $\mathcal{H}$ some Hilbert space, investigated in [187] for the case $\dim\mathcal{H} < \infty$ and in [17, 21] for the case $\dim\mathcal{H} \leq \infty$. The family of commuting self-adjoint operators acting on the Fock space can be seen, for example, in [30]. In particular, the Fock space has the form (1.0.8) with $\mathcal{H}_n$, which are n-particle in finite-dimensional Hilbert spaces for $n > 0$.

It would be interesting to develop the spectral theory of block matrices J of type (1.0.9) and (1.0.10) in the space $\mathbf{l}_2$ (1.0.8) in the case of unbounded normal operators $\tilde{J}$. In connection with this, a number of questions arise. What conditions on the elements of matrices J guarantee, that the operator $\tilde{J}$ is normal? In what terms can the normal extension J in $\mathbf{l}_2$ be described as in case of classical Jacobi matrices?

Let us compare the latest results with the results of the fifth chapter.

Briefly speaking, it is necessary to apply the previously explained theory in the case when the matrix J of the form (1.0.9) with elements from (1.0.10) is a unitary operator. But there is an important point here: the spectrum $\tilde{J}$ lies on the unit circle $\mathbb{T} \subset \mathbb{C}$, therefore, the functions (1.0.15) are linearly dependent on this set, because

$$z^{j+n}\bar{z}^{k+n} = z^j\bar{z}^k, \quad j, k \in \mathbb{N}_0, \quad z \in \mathbb{T}, \quad \forall n \in \mathbb{N}_0.$$

Now it is necessary to take only functions from (1.0.15), for which $j \cdot k = 0$, namely

$$\begin{aligned} &z^0\bar{z}^0 = 1; \\ &z^1\bar{z}^0 = z^1,\ z^0\bar{z}^1 = \bar{z}^1; \\ &z^2\bar{z}^0 = z^2,\ z^0\bar{z}^2 = \bar{z}^2;\ \ldots; \\ &z^n\bar{z}^0 = z^n,\ z^0\bar{z}^n = \bar{z}^n;\ \ldots, \qquad z \in \mathbb{T}. \end{aligned} \tag{1.0.16}$$

Note that, it is possible to get the classical Jacobi matrices in a similar way assuming $\bar{z} = z$, i.e., $z \in \mathbb{R}$.

The support of $d\rho(z)$ must be an infinite set on $\mathbb{T}$, hence, functions (1.0.16) are linear independent and form a total set in $L_2(\mathbb{T}, d\rho(z))$.

The global order of orthogonalization of the sequence (1.0.16) is still the same as in (1.0.15). But now "the diagonal" contains only two elements $n = 1, 2, \ldots$ (instead of $n + 1$ as before). Due to this, it is convenient to consider our operator $\tilde{J}$ on the subspace $\mathbf{l}_{2,u}$ of the space $\mathbf{l}_2$, namely

$$\mathbf{l}_{2,u} = \mathcal{H}_0 \oplus \mathcal{H}_1 \oplus \mathcal{H}_2 \oplus \cdots, \quad \text{where} \quad \mathcal{H}_0 = \mathbb{C}, \mathcal{H}_1 = \mathcal{H}_2 = \cdots = \mathbb{C}^2. \tag{1.0.17}$$

The corresponding block matrices (1.0.9) act in the space (1.0.17), so blocks have action as operators, namely,

$$\begin{aligned} a_0 &: \ \mathbb{C} \longrightarrow \mathbb{C}^2, \\ b_0 &: \ \mathbb{C} \longrightarrow \mathbb{C}, \\ c_0 &: \ \mathbb{C}^2 \longrightarrow \mathbb{C}, \\ a_n, b_n, c_n &: \ \mathbb{C}^2 \longrightarrow \mathbb{C}^2, \qquad n \in \mathbb{N} = \{1, 2, ...\}. \end{aligned} \tag{1.0.18}$$

Conditions $a_n > 0$ from (1.0.1) and from (1.0.10) have the form

$$a_0 = \underbrace{\begin{bmatrix} a_{0;0,0} \\ 0 \end{bmatrix}}_{1} \Big\} 2, \quad b_0 = \underbrace{\begin{bmatrix} b_{0;0,0} \end{bmatrix}}_{1} \Big\} 1, \quad c_0 = \underbrace{\begin{bmatrix} c_{0;0,0} & c_{0;0,1} \end{bmatrix}}_{2} \Big\} 1;$$

$$a_n = \underbrace{\begin{bmatrix} a_{n;0,0} & a_{n;0,1} \\ 0 & 0 \end{bmatrix}}_{2} \Big\} 2, \quad c_n = \underbrace{\begin{bmatrix} 0 & 0 \\ c_{n;1,0} & c_{n;1,1} \end{bmatrix}}_{2} \Big\} 2; \tag{1.0.19}$$

$$a_{0;0,0}, c_{0;0,1}, a_{n;0,0}, c_{n;1,1} > 0, \quad n \in \mathbb{N}.$$

All results of the previous chapter hold true for the unitary operators $\tilde{J}$, which act in the space $\mathbf{l}_{2,u}$ (1.0.17) and are defined by block matrices of the Jacobi type (1.0.9) with conditions (1.0.19). Namely, the direct spectral problem leads to a Fourier transform of type (1.0.12) between the spaces $\mathbf{l}_{2,u}$ and $L_2(\mathbb{T}, d\rho(z))$. The inverse spectral problem is similar to the one described in the previous chapter, where the orthogonalization is used for the sequence (1.0.16). This problem is related to the trigonometric moment problem.

It should be noted that in the fourth chapter, Jacobi matrices of the normal bounded operator J are considered, when the functions (1.0.15) are linearly independent in the space $L_2(\mathbb{C}, d\rho(z))$ constructed by the spectral measure $d\rho(z)$ of the bounded normal operator $N = \tilde{J}$. In terms of the operator $N = \tilde{J}$, this condition means that if for some coefficients $c_{j,k} \in \mathbb{C}$ and $n \in \mathbb{N}$

$$\sum_{j,k=0}^{n} c_{j,k} N^j N^{*k} = 0, \tag{1.0.20}$$

then $c_{j,k} = 0, \forall j, k \in \{0, 1, \ldots, n\}$.

It is not difficult to understand that the last condition is equivalent to linear independence of functions in (1.0.15). So, let $dE(z)$ be the expansion of the identity for N, then (1.0.20) can be rewritten

$$\int_{\mathbb{C}} \left| \sum_{j,k=0}^{n} c_{j,k} z^j \bar{z}^k \right|^2 d(E(z)f, f)_{\mathbf{l}_2} = 0, \quad \forall f \in \mathbf{l}_2.$$

By using the boundedness of the support of $E(\alpha)$, we find that the last equality means, that $\sum\limits_{j,k=0}^{n} c_{j,k} z^j \bar{z}^k \in L_2(\mathbb{C}, d\rho(z))$ and is equal to zero. By assumption, we have for all $c_{j,k} = 0$, that is, functions (1.0.15) are linearly independent in $L_2(\mathbb{C}, d\rho(z))$. The converse statement is also clear.

It follows from the previous reasoning that if the normal operator N satisfies the condition of the type $NN^* = 1$ or $N = N^*$, then its matrix is necessarily a unitary or self-adjoint operator in the corresponding basis $\mathbf{l}_2$.

Let us discuss the third chapter. Each moment problem is associated with the generalized eigenvector expansion theory, that is applicable to research in various cases. In this approach, first of all, the representation of moments is proved, using the theorem about generalized eigenvector expansion, which is applied to the corresponding operators. For such vectors, a simple equation is given that depends on the corresponding moment problem: the solution of this equation gives the form of the representation. The corresponding Parseval equality actually gives the representation of moments.

After that, we consider the connection of block tri-diagonal Jacobi type matrices with a spectral measure equivalent to the measure of the moments representation. The corresponding spectral theory of such matrices gives additional information about the mentioned problem.

This approach allows us to explore in addition classical, trigonometric, complex, as well as a strong moment problem from a single point of view.

This chapter is devoted to the demonstration such an approach to the study of the strong moment problem. The complete list of the theory of the strong moment problem provided here is based on the spectral theory of self-adjoint operators regardless of the previous and subsequent chapters.

It is important to note that, for the first time, the similar ideas to these appeared in studies of the representation of positive definite functions, moment problems and others in works of Krein M. G. (1946–1948, [180, 185]). He constructed a Hilbert space based on a positive definite kernel and considered an operator in this space related to the investigated problem. He applied this method to directional functionals, which he invented at that time. Berezansky Y. M. (1956) in [19] applied the method of generalized eigenvectors expansion to these operator, which gave, in particular, a positive result for the moment problem.

The strong moment problem is formulated as follows. For a given sequence $s = (s_n)_{n=-\infty}^{\infty}$ of real numbers $s_n \in \mathbb{R}$, it is need to find the measure $d\rho(\lambda)$ defined on the Borel σ-algebra $\mathfrak{B}(\mathbb{R})$, so that these numbers s_n are its moments, i.e.,

$$s_n = \int_{\mathbb{R}} \lambda^n \, d\rho(\lambda), \quad n \in \mathbb{Z} := \{\ldots, -1, 0, 1, \ldots\}. \tag{1.0.21}$$

Recall that, if the representation (1.0.21) holds true only for $n \in \mathbb{N}_0 := \{0, 1, 2, \ldots\} = \{0\} \cup \mathbb{N}$, then we have the classic moment problem [4, 21], the answer to which is well known: $(s_n)_{n=0}^{\infty}$ is the moment sequence if and only if for an arbitrary finite sequence $(f_n)_{n=0}^{\infty}$ of complex numbers $f_n \in \mathbb{C}$, the inequality

$$\sum_{j,k=0}^{\infty} s_{j+k} f_j \bar{f}_k \geq 0 \tag{1.0.22}$$

holds true. In other words, the matrix $(s_{j+k})_{j,k=0}^{\infty}$ must be positive definite, i.e., non-negative.

It follows from the definition, every strong moment sequence is classical on $\mathbb{N}_0 \subset \mathbb{Z}$. So, it is understandable, if we have the representation (1.0.21) for $n \in \mathbb{N}_0$, then it can be expanded to $n \in \mathbb{Z}_- := \{\ldots, -2, -1\}$, if the measure $d\rho(\lambda)$ is "small" (vanishes) in a neighborhood of the point 0: every integral for $n \in \mathbb{Z}$ must exist.

This situation corresponds to a certain reality, but this problem is not a simple one, since 1983–1984, many works have been devoted to the study of a strong moment problem. We do not provide a relevant list, but only a review in [142] and some works closer to our research.

It is important to note that the representation (1.0.21) holds true if and only if the condition like (1.0.22) holds true for an arbitrary finite set $(f_n)_{n=-\infty}^{\infty}$ with the sum from $-\infty$ to ∞.

This result was published in 1984 in [143], but we would like to note, that the representation (1.0.21) and its equivalence to positive definiteness (1.0.22) were published in 1965 by Berezansky in [21] (see also [20]), as a special case of a more general theorem. We also note, that the strong moment problem appeared for the first time in the article by Nudel'man [230].

Both, in the classical moment problem and in the strong case, similar questions arise: under what conditions is the representation (1.0.21) unique, and if we have a case of non-unity, then how to describe all measures $d\rho(\lambda)$ with given moments $s = (s_n)_{n=-\infty}^{\infty}$. Therefore, the so-called Laurent polynomials, that is, finite linear combinations λ^n, $n \in \mathbb{Z}$, are essential by this problem. Important questions are the form of an analog of the Jacobi matrix associated with s, the spectral theory of such matrices, and others.

Some of these investigations were explored in works cited in [142] and [228]. But the approach to the corresponding problem was analytical, often without applications of the corresponding natural tool—the spectral theory of operators. It is important to say that the application to this problem of the generalized eigenvectors expansion theory and the corresponding results for Jacobi matrices and positive definite kernels ([21], Sections 5, 7, 8) together give a clear, transparent picture, similarly to the classical moment problem. Unfortunately, previous authors used other ways. Works [137, 261, 267] had a great influence on our research (note, that these works also do not use the generalized eigenvectors expansion theory).

In [137] we observe the Jacobi-Laurent matrix with Laurent polynomials corresponding to the strong moment problem. The operator generated by such a matrix was considered and some of its spectral properties were investigated here. Papers [261, 267] generalize some results from [75, 142] of the strong moment problem for the matrix case. The author systematically uses the theory of operators, and it was his work that gave impetus to writing our work. We note recent works [78, 79] containing questions related to the strong moment problem, which influenced on our construction.

The structure of the chapter is as follows. First of all, we recall the main result about the generalized eigenvectors expansion, which is important for what follows.

That gives a possibility to prove a theorem about the representation (1.0.21) and some conditions under which the measure is unique.

Sections 3.3 and 3.4 are devoted to the construction and the study of Jacobi type block matrices associated with the representation (1.0.21). It is important to note that our Jacobi-Laurent matrix is block tri-diagonal instead of five-diagonal considered by previous authors. The use of block matrices is often, unlike the usual ones, very convenient in the relevant situation and is used for trigonometric and complex problems. To consider such an approach, it is important to keep in mind the results of the article [269]. The convenience of block matrices gives the simpler way for a finding of relationships between objects similarly considering for classical Jacobi matrices.

Our Jacobi-Laurent matrix J is symmetric and has an algebraic inverse J^{-1}, which is also block tri-diagonal with appropriate properties. Here we see the following tasks: to describe such a matrix J internally, similarly to the case of a five-diagonal unitary matrix (that is, a tri-diagonal block matrix). In the unitary case, such a description was provided by Verblunsky S. (see the monograph [263, 264]). A corresponding description is provided in the last section.

The chapter presents also the spectral theory of Jacobi-Laurent block matrices including the corresponding direct and inverse spectral problems. The construction is analogous to that applied to classical Jacobi matrices (Chap. 2) and uses the generalized eigenvectors expansions theory. First, the Jacobi-Laurent matrix J is considered in the general case when J defines only a Hermitian (symmetric) operator, that is not self-adjoint. The relevant theory constructed further is similarly to the case of the classical Jacobi matrix.

We also note that we do not touch upon the consideration of self-adjoint extensions of the operator generated by J leaving the space. Thus we consider *only orthogonal spectral measures*.

Note that, for ease reading, the theory is often presented first, not in the most general case. So, first we consider the simpler case of the bounded operator generated by the Jacobi-Laurent matrix, that is, the case of the measure $d\rho(\lambda)$ with a bounded support. This is below extended to the case of an arbitrary support, but the set of functions $\mathbb{R} \ni \lambda \longmapsto \lambda^m$, $m \in \mathbb{Z}$ must be total in the space $L_2(\mathbb{R}, d\rho(\lambda))$ (in particular, such a measure is quite exotic).

The last eighth chapter is devoted to the study of semi-infinite and double-sided infinite Toda chain.

Recall that the classical double-sided infinite Toda chain [282] is a system of nonlinear difference-differential equations

$$\begin{aligned}\dot{\alpha}_n(t) &= \frac{1}{2}\alpha_n(t)(\beta_{n+1}(t) - \beta_n(t)),\\ \dot{\beta}_n(t) &= \alpha_n^2(t) - \alpha_{n-1}^2(t), \quad n \in \mathbb{Z}, \quad t \in [0, T), \quad T > 0.\end{aligned} \tag{1.0.23}$$

Here, the unknowns $\alpha_n(t)$, $\beta_n(t)$ are real continuously differentiable functions and "$\cdot$" $= \frac{d}{dt}$. For (1.0.23), we can set the Cauchy problem: a solution $\alpha_n(t)$, $\beta_n(t)$, $n \in \mathbb{Z}$, $t \in [0, T)$ must be found for the given initial data $\alpha_n(0)$, $\beta_n(0)$, $n \in \mathbb{Z}$.

Note that this system is really a Hamiltonian system describing the dynamics of a chain of particles $q_n(t)$, $n \in \mathbb{Z}$ along the whole line with an exponential interaction. Let us explain that $\alpha_n(t)$, $\beta_n(t)$ are some coordinates of $q_n(t)$ and $\dot{q}_n(t)$ (Flaschka variables [103, 104]). The study of the Cauchy problem for (1.0.23) is a significant problem, that is discussed in a wide range of physical literature. The references contains articles that are important for us.

Essential for us results, obtained in [147, 219], describe the case of a finite number of equations (1.0.23), where $\mathbb{Z}$ is replaced with a finite set $\{0, \ldots, N\}$.

For the semi-infinite case, when only $\mathbb{N}_0 = \{0, 1, 2, \ldots\}$, is taken instead of the set $\mathbb{Z}$, the obtained results are close to the classical method of integrating the Cauchy problem for the KdV equation on $(x, t) \in [0, \infty) \times [0, T)$ in the sense of the inverse spectral problem for the Sturm-Liouville equation on $x \in [0, \infty)$. In our case, instead of the Sturm-Liouville equation we take its difference analog, or more precisely, the classical Jacobi matrix. Corresponding results are published in [24, 27–29]. Now the unknowns $\alpha_n(t)$ are considered positive and functions $\alpha_n(t)$, $\beta_n(t)$ are bounded uniformly with respect to $n \in \mathbb{N}_0$. Let us explain the main idea of this approach. We associate with the equation (1.0.23) the classical Jacobi matrix $J(t)$ with the main diagonal $(\beta_n(t))_{n=0}^{\infty}$ and two adjacent identical diagonals $(\alpha_n(t))_{n=0}^{\infty}$. The matrix $J(t)$ acts in the space $\ell_2 = \mathbb{C}^1 \oplus \mathbb{C}^1 \oplus \ldots$ (that is, ℓ_2 on $\mathbb{N}_0$) and defines the bounded self-adjoint operator $\mathbf{J}(t)$ (its boundedness follows from our assumption: we consider only bounded solutions of (1.0.23)).

An important point: the evolution in time $t \in [0, T)$ of the matrix $J(t)$ is complicated and given by the equation (1.0.23), but the evolution of the spectral measure $d\rho(\lambda; t)$ is simple, that is

$$d\rho(\lambda; t) = e^{\lambda t} d\rho(\lambda; 0), \quad \lambda \in \mathbb{R}, \quad t \in [0, T). \tag{1.0.24}$$

This fact characterizes the Toda systems. Note that, it follows from (1.0.24), that the spectrum of the operator $\mathbf{J}(t)$ does not depend on t. Let us also explain that using (1.0.23) it is possible to find for $\rho(\lambda; t)$ a simple differential equation with respect to the variable t and, hence, (1.0.24) is its solution.

Now the finding of a solution of the Cauchy problem is quite simple: using the initial data $\alpha_0(0)$, $\beta_0(0)$, $n \in \mathbb{N}_0$, that is, the matrix $J(0)$, we find its spectral measure $d\rho(\lambda; 0)$. Further, using (1.0.24), it is possible to obtain the spectral measure $d\rho(\lambda; t)$ of the matrix $J(t)$. This knowledge provides information about $J(t)$ according to classical formulas (it is necessary to orthogonalize the sequence $1, \lambda, \lambda^2, \ldots$, etc., that is, to solve the inverse spectral problem for the Jacobi matrix). Thus we get the solution of the Cauchy problem.

Some publications have generalizations of the approach to the semi-infinite case. Thus, solutions of the Toda chain, which is not necessarily bounded (that is, the Hermitian operator $\mathbf{J}(t)$ is generally unbounded), are investigated in [277].

A chain more general than (1.0.23), when the formula (1.0.24) is more complicated, is considered in [41, 49, 217, 218, 259] ("non-isospectral" case when spectrum $\mathbf{J}(t)$ changes with time).

The situation of matrix (operator non-Abelian) equations of type (1.0.23) is considered in [46, 106]; see also [60, 61]. Now, the corresponding operators $\mathbf{J}(t)$ acting in the space $\mathbb{C}^d \oplus \mathbb{C}^d \oplus \ldots$, where $d > 1$ is the dimension of the matrix-valued solution (or the orthogonal sum of fixed Hilbert spaces in which belongs our solution). The spectral theory of the corresponding Jacobi-type matrices is developed in [21, 36, 37, 187]. A new class of the Toda solutions is presented in [119].

Let us pass to the Toda doubly infinite (double-sided infinite) chains. Note that there is no direct application of the classical theory of Jacobi matrices in this case, since n is taken from $\mathbb{Z}$ and elements of the Jacobi matrix $(a_{m,n})_{m,n=0}^{\infty}$ is indexed by m, n. The situation with such Toda chains is much more complicated than in the case of $\mathbb{N}_0$. First, we note the significant results that are directly related to our approach.

In a classic work [282], Toda chains applied to the integration of the Cauchy problem for (1.0.23) the inverse scattering problem for the difference analog of the Sturm-Liouville equation on the whole axis $x \in \mathbb{R}$ with the basic operator for which a discrete analog of the Laplacian is taken. This approach made it possible to find a set of appropriate solutions. Later works in this direction are: [103, 104, 205].

In the paper [191], the case of periodic solutions is studied using the theta-function (see the previous paper [147]). In papers [99, 281], searches for solutions of the Cauchy problem for (1.0.23) of the difference analog of the scattering theory for the Schrödinger equation with a periodic potential are given; there considered also the case when the potential goes to different constants at $n \to +\infty$ and $n \to -\infty$. This approach was generalized in the above works to find solutions related to the more general than periodic case. Namely, the finite-band potential corresponding to the basic operator was studied. Let us explain, that the application of the scattering theory to find solutions of (1.0.23) is based on the fact that if the basic potential is changed according to (1.0.23), then the scattering indicators is changed in the same way (similarly to (1.0.24) in our approach) .

Significant results related to the search for solutions of the Cauchy problem for doubly infinite equations (1.0.23) obtained in [50, 300]. In particular, a solution is found that strongly vanishes for $|n| \to \infty$.

In general, the problem of integrating (1.0.23) with arbitrary conditions on the matrix structure is open. In the work [42], authors try to find a general solution to this problem (1.0.23), using the regarding of the equation (1.0.23) as a 2×2-matrix in the space $(\mathbb{C}^2 \oplus \mathbb{C}^2 \oplus \ldots)$. Thus, matrix-valued Toda systems are obtained, but the non-commutativity of matrices adds difficulties to the differential equation with the matrix measure $\rho(\lambda; t)$.

This approach uses a standard doubling method: a doubly infinite vector $\xi = (\ldots, \xi_{-1}, \xi_0, \xi_1, \ldots)$, $\xi_n \in \mathbb{C}^1$ is regarded as a vector

$$(x_0, x_1, \ldots) \in \mathbb{C}^2 \oplus \mathbb{C}^2 \oplus \ldots, \qquad x_n = (\xi_n, \xi_{-n-1}) \in \mathbb{C}^2,$$

namely $\xi \leftrightarrow (x_0 = (\xi_0, \xi_{-1}), x_1 = (\xi_1, \xi_{-2}), \ldots)$.

But, one can do a more convenient otherwise doubling:

$$\xi = (\dots, \xi_{-1}, \xi_0, \xi_1, \dots) \leftrightarrow (x_0 = \xi_0, x_1 = (\xi_1, \xi_{-1}), x_2 = (\xi_2, \xi_{-2}), \dots). \tag{1.0.25}$$

With such doubling, the vector ξ turns into a vector from the space

$$\mathbb{C}^1 \oplus \mathbb{C}^2 \oplus \mathbb{C}^2 \oplus \cdots . \tag{1.0.26}$$

After our doubling, the Toda chain possessed to equations in unknowns with values in the space (1.0.26). To apply the approach [24, 27–29] in our case, it was necessary to develop the spectral theory of Jacobi type matrices in the space (1.0.26). Note that this theory is an essential for doubly infinite Jacobi matrices in the space as l_2 for $\mathbb{Z}$.

It is possible to generalize the relevant results published in [33]. Note that the spectral theory of such Jacobi matrices is not standard, its development is related to works [32, 141].

It must be said that the previously discussed theory of self-adjoint matrices of the Jacobi type in the space (1.0.26) can be developed due to the theory of generalized eigenvectors expansion for the self-adjoint operator. This approach was first stated in works [18, 111]. Our researches also uses results from monographs [21, 43, 44, 47, 48].

The presented chapter is devoted to a full statement of the above results from [33] (the latter work also contains many other facts, but we only touch on the results related to Eq. (1.0.23)). It should also be said that some of the results from [24, 27–29, 33] are proved concisely. Full detailed proofs of the corresponding results are given in the chapter.

We start with a semi-bounded Toda chain and give a mathematically complete list of results. Note, that some examples in finding the solution of the corresponding Cauchy problem, which are in [27, 29], have not been added.

The first section is devoted to the proof of the writing the Toda chain (1.0.23) as a Lax equation with some matrix coefficients $A(t)$ in the space (1.0.26); properties of $A(t)$ are important. The direct and inverse spectral problems for block matrices defined in the space (1.0.26) are considered in Sect. 8.3. Note that the construction of such a theory is specific since the first space in the sum (1.0.26) is $\mathbb{C}^1$ and the others are $\mathbb{C}^2$ spaces. Now the spectral measure $d\rho(\lambda)$ is a 2×2-matrix. The corresponding inverse problem is more complicated than for the space $(\mathbb{C}^2 \oplus \mathbb{C}^2 \oplus \cdots)$.

At the end, the differential equation of functions with the variable t is written for elements of the spectral matrix $d\rho(\lambda; t) = (d\rho_{\alpha,\beta}(\lambda; t))_{\alpha,\beta=0}^{1}$ of the corresponding operator $\mathbf{J}(t)$ acting in the space (1.0.26). Now, given $\rho_{0,1}(\lambda; t) = \rho_{1,0}(\lambda; t)$, our differential equation is a first order linear system of equations with continuous real functions $r_{0,0}(\lambda; t), r_{0,1}(\lambda; t), r_{1,1}(\lambda; t)$, which are derivatives of $d\rho_{\alpha,\beta}(\lambda; t)$ by the scalar measure $d\sigma(\lambda)$, namely,

$$r_{\alpha,\beta}(\lambda; t) = \frac{d\rho_{\alpha,\beta}(\lambda; t)}{d\sigma(\lambda)}.$$

Here $t \in [0, T)$, λ is a parameter, the coefficients of the system simply depend on $\alpha_0(t)$, $\beta_0(t)$ and λ, but unfortunately the system is not integrated. This is the main difference between the semi-infinite and doubly-infinite cases for the Toda chain: in the first case, the written general solution corresponding to the (one-dimensional) differential equation is the formula (1.0.24); in the second case, only the procedure for finding solutions is written out, that is, only the linearization of the problem is obtained, but not the solution.

Sections 8.4 and 8.5 contain main theorems that give the procedure for finding the solution of the Cauchy problem for (1.0.23). These subsections also contain remarks on the obtained systems of differential equations.

The Hamiltonian system in the form of a second-order equation, associated with (1.0.23), has the form

$$\ddot{x}_n(t) = e^{x_{n-1}(t)-x_n(t)} - e^{x_n(t)-x_{n+1}(t)}, \quad n \in \mathbb{Z}, \quad t \in [0, T) \tag{1.0.27}$$

The relationship between (1.0.27) and (1.0.23) is given by the Flaschka transformation of variables

$$\alpha_n(t) = e^{1/2(x_n(t)-x_{n+1}(t))}, \quad \beta_n(t) = -\dot{x}_n(t).$$

The Cauchy problem for (1.0.27) with the given initial data $x_n(0), \dot{x}_n(0)$, $n \in \mathbb{Z}$ upon the Flaschka transformation turns into our Cauchy problem for (1.0.23), and Theorem 9.5.1 provides an opportunity for researches.

Let us pass to a brief general overview of what is presented in the book.

The first chapter, in addition to the foundations of the theory of unbounded self-adjoint and normal operators, contains new aspects of the self-adjointness criteria. Namely: the spectral expansion of the unbounded self-adjoint operator, the normal operator, and the family of commutative operators in general; elements of the theory of extensions of a symmetric densely defined operator to a self-adjoint one (without leaving the space); a number of quasi-analytic self-adjointness criteria; and the theory of generalized eigenvectors expansion (in simple cases). Some of the material of the chapter is mostly taken from [47, 48], Chapters 12–15.

The second chapter, already discussed above, is a summary of Chapter 7 from [21]. It concerns the connections of the classical moment problem with the corresponding Jacobi matrix, the operator corresponding to it and the self-adjointness, polynomials of the first and second kind, the Weyl function.

The third chapter is a generalization of the second one for the case of a strong moment problem.

The fourth chapter generalizes the second one to the case of a complex power moment problem.

The material of the fifth chapter can be considered a partial case to the fourth, since it is devoted to the trigonometric moment problem as a partial case of the complex power moment problem (which was discussed above). However, the corresponding block matrices of Jacobi type have a significant difference.

In the sixth chapter, a version of the moment problem is considered, which generalizes the complex moment problem. This chapter deals with the complex moment problem in the exponential form. If one block matrix of the normal operator corresponds to the complex moment problem in power form (and its adjoint is automatically considered), then in the case of this chapter two block matrices appear, one of them generates a positive self-adjoint (unbounded) operator and the second generate the unitary one. And operators corresponding to matrices are commutative.

The seventh chapter is also a generalization of the second one. In this chapter, the real power two-dimensional moment problem and the corresponding Jacobi block matrices are considered.

The last eighth chapter is devoted to the integration of the classical Toda chain and its generalization generated by the study of block matrices.

Chapter 2
Some Aspects of the Spectral Theory of Unbounded Operators

We hope that the reader is well acquainted with basic courses of mathematical and functional analysis, in particular with the theory of linear bounded operators in Hilbert spaces. Therefore, in this chapter we will recall only the basic concepts regarding unbounded self-adjoint operators, without which the book reading would be difficult.

2.1 Preliminary Information About Unbounded Operators

Let $\mathcal{H}$ denotes a Hilbert space. Consider a linear set $\mathfrak{D}(A) \subseteq \mathcal{H}$. We will define a mapping on this set $\mathfrak{D}(A) \ni f \mapsto Af \in \mathcal{H}$ (namely, $A(\lambda f + \mu g) = \lambda Af + \mu Ag$ for all $f, g \in \mathfrak{D}(A)$ and, in general, complex numbers $\lambda, \mu \in \mathbb{C}$). *Such a mapping is called a linear (generally possibly unbounded) operator with a domain* $\mathfrak{D}(A)$ *(sometimes we use the designations* dom (A)*). Two linear operators A and B are equivalent, if their domains and actions coincide, i.e.,* $A = B$, *if* $\mathfrak{D}(A) = \mathfrak{D}(B)$ *and* $Af = Bf$, $f \in \mathfrak{D}(A)$. In general, the domain is not dense in $\mathcal{H}$. Its density or non-density will be specified each time.

Let A and B be two operators such that $\mathfrak{D}(A) \subseteq \mathfrak{D}(B)$ and $Af = Bf$, $f \in \mathfrak{D}(A)$. *In this case, the operator B is called the extension of the operator A, and, accordingly, A is called the restriction of the operator B and is denoted by* $A \subseteq B$ *and* $B \restriction \mathfrak{D}(A) = A$. By $\mathbb{I}$ we denote the unit (identity) operator in $\mathcal{H}$, i.e., $\mathbb{I}f = f$, $f \in \mathcal{H}$.

Y. M. Berezansky, M. E. Dudkin, *Jacobi Matrices and the Moment Problem*, Operator Theory: Advances and Applications 294,
https://doi.org/10.1007/978-3-031-46387-7_2

When defining an action on linear operators, it is necessary to follow the domains of the corresponding operators. Therefore, if A and B are two operators acting in $\mathcal{H}$ and $\lambda \in \mathbb{C}$, then it is assumed

(i) $(\lambda A)f = \lambda(Af), \quad f \in \mathfrak{D}(\lambda A) = \mathfrak{D}(A)$;
(ii) $(A + B)f = Af + Bf, \quad f \in \mathfrak{D}(A + B) = \mathfrak{D}(A) \cap \mathfrak{D}(B)$;
(iii) $(AB)f = A(Bf), \quad f \in \mathfrak{D}(AB) = \{f \in \mathfrak{D}(B) \mid Bf \mid \mathfrak{D}(A)\}$.

If $\mathfrak{D}(A + B)$ and $\mathfrak{D}(AB)$, possibly, are non-dense in $\mathcal{H}$, despite the density $\mathfrak{D}(A)$ and $\mathfrak{D}(B)$, we will, additionally, require density $\mathfrak{D}(A + B)$ and $\mathfrak{D}(AB)$ in corresponding problems.

The range of the operator A is called the linear set of vectors Af, where $f \in \mathfrak{D}(A)$, and is denoted by $\mathfrak{R}(A)$ (or sometimes $\operatorname{ran}(A)$). If A defines a one-to-one correspondence between $\mathfrak{D}(A)$ and $\mathfrak{R}(A)$, then there exists an inverse operator A^{-1} defined by the expression $A^{-1}(Af) = f$ $(f \in \mathfrak{D}(A))$. Therefore, $\mathfrak{D}(A^{-1}) = \mathfrak{R}(A)$ and $\mathfrak{R}(A^{-1}) = \mathfrak{D}(A)$. The criterion for the existence of a one-to-one correspondence: the kernel of the operator A is zero, i.e, $\mathrm{Ker} A = \{f \in \mathfrak{D}(A) \mid Af = 0\} = 0$. The operator defined now is called algebraically inverse, since by the inverse operator we regard A^{-1}, if $\mathfrak{D}(A^{-1}) = \mathfrak{R}(A) = H$ and A^{-1} is bounded.

Graphs of Operators The graph of an operator A in $\mathcal{H}$ is introduced in the usual way. Consider the orthogonal sum $\mathcal{H} \oplus \mathcal{H}$ for pairs $\langle f, g\rangle$, where $f, g \in \mathcal{H}$. Linear operations for such pairs are defined coordinate-wise, and the scalar product is given by the equality:

$$(\langle f_1, g_1\rangle, \langle f_2, g_2\rangle)_{\mathcal{H}\oplus\mathcal{H}} = (f_1, f_2)_{\mathcal{H}} + (g_1, g_2)_{\mathcal{H}}, \quad f_1, f_2, g_1, g_2 \in \mathcal{H}. \tag{2.1.1}$$

A set

$$\Gamma_A = \big\{\langle f, Af\rangle \in \mathcal{H} \oplus \mathcal{H} \mid f \in \mathfrak{D}(A)\big\}. \tag{2.1.2}$$

is called the graph Γ_A of the operator A.

To simplify the study of operators by using their graphs, it is convenient to introduce the following two additional isometric operators acting in $\mathcal{H} \oplus \mathcal{H}$:

$$\begin{aligned} \mathcal{H} \oplus \mathcal{H} \ni \langle f, g\rangle &\mapsto U\langle f, g\rangle = \langle g, f\rangle \in \mathcal{H} \oplus \mathcal{H}, \\ \mathcal{H} \oplus \mathcal{H} \ni \langle f, g\rangle &\mapsto O\langle f, g\rangle = \langle -g, f\rangle \in \mathcal{H} \oplus \mathcal{H}. \end{aligned} \tag{2.1.3}$$

By virtue (2.1.3), $\mathfrak{R}(U) = \mathcal{H} \oplus \mathcal{H}$ and $\mathfrak{R}(O) = \mathcal{H} \oplus \mathcal{H}$; therefore, U and O are unitary. Definition (2.1.3) immediately implies that

$$U^2 = \mathbb{I}, \quad O^2 = -\mathbb{I}, \quad OU = -UO. \tag{2.1.4}$$

The following general theorem holds true for the operator graph.

Theorem 2.1.1 *Let A be a linear operator, generally with a non-densely domain. In order that the algebraically inverse operator A^{-1} exists, it is necessary and sufficient that a set $U\Gamma_A$ be the graph of a certain operator. Furthermore,*

$$\Gamma_{A^{-1}} = U\Gamma_A. \tag{2.1.5}$$

Closed Operators We recall the three equivalent definitions of a closed operator A acting in $\mathcal{H}$.

(1) *An operator A is closed if its graph Γ_A is closed in $\mathcal{H} \oplus \mathcal{H}$.*
(2) *An operator A is closed if, for any sequence of vectors $(f_n)_{n=1}^{\infty} \subseteq \mathfrak{D}(A)$, the fact that $f_n \to f \in \mathcal{H}$ and $Af_n \to g \in \mathcal{H}$ as $n \to \infty$ implies $f \in \mathfrak{D}(A)$ and $Af = g$.*
(3) In the domain $\mathfrak{D}(A)$ of an operator A, we introduce *a graph scalar product*

$$(f, g)_{\Gamma_A} = (f, g)_{\mathcal{H}} + (Af, Ag)_{\mathcal{H}}, \quad f, g \in \mathfrak{D}(A). \tag{2.1.6}$$

The operator A is closed if $\mathfrak{D}(A)$ is a complete space with respect to the graph scalar product.

Note that the norm corresponding to (2.1.6) is called the *norm of the graph*. For this norm, we have

$$\|f\|_{\Gamma_A}^2 = \|f\|_{\mathcal{H}}^2 + \|Af\|_{\mathcal{H}}^2, \quad f \in \mathfrak{D}(A). \tag{2.1.7}$$

Theorem 2.1.2 *Three given above definitions of closed operators are equivalent.*

Closable Operators If an operator A acting in $\mathcal{H}$ is not closed, at first sight, it is always possible to construct its closure, i.e., to extend A to a closed operator $\tilde{A}$ by adding to its domain $\mathfrak{D}(A)$ all vectors $f \in \mathcal{H}$ for which one can find a sequence $(f_n)_{n=1}^{\infty} \subseteq \mathfrak{D}(A)$ such that $f_n \to f$ as $n \to \infty$ and there exists $\lim_{n\to\infty} Af_n = g$. On the added vectors f, we naturally set $\tilde{A}f = g$. However, this procedure is, generally speaking, incorrect: $\tilde{A}f$ may depend on a sequence $(f_n)_{n=1}^{\infty}$ that approximates f.

(1) *An operator A admits a closure $\tilde{A}$ (or is closable), if the outlined above procedure is correct, namely, for all $f \in \mathcal{H}$ that admits a required approximating sequence $(f_n)_{n=1}^{\infty}$.*
(2) *An operator A is closable, if the fact*

$$\lim_{n\to\infty} f_n = 0 \quad \text{and} \quad \lim_{n\to\infty} Af_n = h \in \mathcal{H}$$

for a sequence $(f_n)_{n=1}^{\infty} \subset \mathfrak{D}(A)$ implies $h = 0$.
(3) *An operator A is closable, if the closure $\tilde{\Gamma}_A$ of its graph is the graph of some operator.*

Theorem 2.1.3 *Three given above definitions of closable operators are equivalent.*

The Adjoint Operator Let A be an operator densely defined on $\mathfrak{D}(A)$ in $\mathcal{H}$. Consider a vector $g \in \mathcal{H}$, for which on can find a vector $g^* \in \mathcal{H}$ such that

$$(Af, g)_{\mathcal{H}} = (f, g^*)_{\mathcal{H}}, \quad \forall f \in \mathfrak{D}(A). \tag{2.1.8}$$

From the linearity of the scalar product, it follows that the set of all vectors g of this type (we denote it by $\mathfrak{D}(A^*)$) forms a linear set and

$$(\lambda g + \mu h)^* = \lambda g^* + \mu h^*, \quad g, h \in \mathfrak{D}(A^*),\ \lambda, \mu \in \mathbb{C}.$$

The vector g^* is uniquely defined for a given g. Indeed, suppose that, along with (2.1.8), there exists another representation $(Af, g)_{\mathcal{H}} = (f, g^{*'})_{\mathcal{H}}$ $(f \in \mathfrak{D}(A))$. Then $(f, g^* - g^{*'})_{\mathcal{H}} = 0$, $f \in \mathfrak{D}(A)$ and we get $g^{*'} = g^*$ since $\mathfrak{D}(A)$ is dense in $\mathcal{H}$.

For $g \in \mathfrak{D}(A^*)$ we put $A^*g = g^*$. The operator, defined in this way, is linear and is called *adjoint* to A. In this case, we can write

$$(Af, g)_{\mathcal{H}} = (f, A^*g)_{\mathcal{H}}, \ f \in \mathfrak{D}(A), \ g \in \mathfrak{D}(A^*),$$

and moreover $\mathfrak{D}(A^*)$ consists all g such that (2.1.8) holds true (the last expression has also a short form, namely, $g \in \mathfrak{D}(A^*) \iff$ the functional $\mathfrak{D}(A) \ni f \mapsto l_g(f) = (Af, g)_{\mathcal{H}} \in \mathbb{C}$ is continuous).

The main properties of the adjoint operator are collected by the theorem.

Theorem 2.1.4 *Let A be a densely defined operator in the space $\mathcal{H}$ and A^* be its adjoint. Then*

(1) *the operator A^* is closed;*
(2) *if A admits a closure, then $(\tilde{A})^* = A^*$ (the operation of closure does not affect for the adjoint operator);*
(3) *assume, that $(\mathfrak{R}(A))^\sim = \mathcal{H}$ and the operator A^{-1} algebraically inverse to A exists, then the operator $(A^*)^{-1}$ exists and*

$$(A^{-1})^* = (A^*)^{-1}; \tag{2.1.9}$$

(4) *if B is another densely defined operator in $\mathcal{H}$, B^* is its adjoint and $B \supseteq A$, then*

$$B \supseteq A \Rightarrow A^* \supseteq B^*; \tag{2.1.10}$$

(5) *let A and B be operators from the previous points such that the set $\mathfrak{D}(A + B)$ is dense in $\mathcal{H}$, then*

$$(A + B)^* \supseteq A^* + B^*; \tag{2.1.11}$$

(6) *let A and B be operators from the previous points such that $\mathfrak{D}(BA)$ is dense in $\mathcal{H}$, then*

$$(BA)^* \supseteq A^*B^*. \tag{2.1.12}$$

Note that if B is a bounded operator, then (2.1.11) and (2.1.12) are equalities, i.e., $(A + B)^* = A^* + B^*$ and $(BA)^*A^*B^*$.

Theorem 2.1.5 (On the Second Adjoint Operator) *Let A be an operator densely defined in $\mathcal{H}$. Assume, that A admits a closure. Then the second adjoint operator $(A^*)^*$ exists and satisfies the equality*

$$(A^*)^* = \tilde{A}. \tag{2.1.13}$$

Conversely, assume that A has a dense domain in $\mathcal{H}$ and the operator $(A^*)^*$ exists, then A admits a closure and (2.1.13) *holds true.*

The following theorem is also often useful.

Theorem 2.1.6 (Banach on Closed Graph) *Let A be a closed operator defined in the whole Hilbert space $\mathcal{H}$, i.e., $\mathfrak{D}(A) = \mathcal{H}$, then A is bounded.*

Deficient Subspaces of Operators Consider an operator A acting in a Hilbert space $\mathcal{H}$ with a domain $\mathfrak{D}(A)$ (which may be not dense in $\mathcal{H}$, and may be $\mathfrak{D}(A) = \mathcal{H}$, i.e., A is bounded).

A point $z \in \mathbb{C}$ is called a point of regular type for the operator A if there exists a number $c_z > 0$ such that

$$\|(A - z\mathbb{I})f\|_{\mathcal{H}} \geq c_z\|f\|_{\mathcal{H}}, \quad f \in \mathfrak{D}(A). \tag{2.1.14}$$

Denote $(A - z\mathbb{I})f = g$. Then (2.1.14) means that the inverse operator

$$\mathfrak{R}(A - z\mathbb{I}) \ni g \mapsto (A - z\mathbb{I})^{-1}g \in \mathfrak{D}(A) \subseteq \mathcal{H}$$

exists and is continuous. *A point z of regular type is regular if $\mathfrak{R}(A - z\mathbb{I}) = \mathcal{H}$.* Thus, the notation of points of regular type generalizes the notation of regular points.

Let us note a number of properties of points of regular type.

(1) *The set of points of regular type is open for a given operator A.*
(2) *Let A be closed operator in $\mathcal{H}$ and $z \in \mathbb{C}$ is a point of regular type, then $\mathfrak{R}(A - z\mathbb{I})$ is a subspace (i.e., the set $\mathfrak{R}(A - z\mathbb{I})$ is closed). Conversely, let z be a point of regular type and let $\mathfrak{R}(A - z\mathbb{I})$ be a subspace in $\mathcal{H}$, then A is closed.*
(3) *Assume that an operator A admits a closure $\tilde{A}$. Then each point z of regular type of the operator A is also a point of regular type of the operator $\tilde{A}$. Furthermore,*

$$\mathfrak{R}(\tilde{A} - z\mathbb{I}) = (\mathfrak{R}(A - z\mathbb{I}))^{\sim}, \tag{2.1.15}$$

where "$\sim$" denotes the closure.

Let $z \in \mathbb{C}$ be a point of regular type of the operator A. *The subspace*

$$N_z = \mathcal{H} \ominus (\mathfrak{R}(A - z\mathbb{I})) = (\mathfrak{R}(A - z\mathbb{I}))^{\perp}$$

is called the deficient subspace of the operator A corresponding to z.Hence, we can write a decomposition

$$\mathcal{H} = (\mathfrak{R}(A - z\mathbb{I}))^{\sim} \oplus N_z. \tag{2.1.16}$$

If the operator A admits a closure or is closed, then, according to (3) and (2), the decomposition (2.1.16) can be rewritten in the form

$$\mathcal{H} = (\mathfrak{R}(\tilde{A} - z\mathbb{I})) \oplus N_z \quad \text{or} \quad \mathcal{H} = \mathfrak{R}(A - z\mathbb{I}) \oplus N_z, \tag{2.1.17}$$

respectively.

We can describe deficient subspaces in a different way. *We call φ an eigenvector of the operator B with a domain $\mathfrak{D}(B)$, if $0 \neq \varphi \in \mathfrak{D}(B)$ and $B\varphi = \lambda\varphi$ with some $\lambda \in \mathbb{C}$, which is called the eigenvalue corresponding to the eigenvector φ.*

The set $\Phi(\lambda)$, which consist of 0 and all eigenvectors corresponding to the same eigenvalue λ, is linear. It is clear that if B is closed, then the set $\Phi(\lambda)$ is also closed and is called the eigensubspace corresponding to λ.

(4) *Assume that the domain of A is dense in $\mathcal{H}$ and, therefore, A^* exists. Then the deficient subspace N_z coincides with the eigensubspace of the operator A^* corresponding to the eigenvalue $\bar{z}$.*

Defect Numbers Consider an operator A which domain $\mathfrak{D}(A)$ may be not dense in $\mathcal{H}$. Let z be its point of regular type. It is obviously that deficient subspaces N_z, corresponding to different z, are different. A remarkable property of these spaces is that the dimension $\dim N_z$ is invariant under changes of z ($\dim N_z$ is a natural number if N_z is finite-dimensional or ∞). It is described below in the theorem more precisely.

Theorem 2.1.7 (Krasnoselsky-Krein) *Let A be a closed operator in a Hilbert space $\mathcal{H}$. Then $n_z = \dim N_z$ is invariant under the changes of z within a connected component of the set of points z of regular type for the operator A. Thus, every component G of this sort can be associated with a fixed number n_z, where $z \in G$. This number is called a defect number of the operator A (in the component G).*

Hermitian and Self-Adjoint Operators Recall that the bounded operator A in the Hilbert space $\mathcal{H}$ is called self-adjoint if

$$(Af, g)_{\mathcal{H}} = (f, Ag)_{\mathcal{H}}, \quad f, g \in \mathcal{H},$$

namely $A^* = A$. The situation is more complicated in the case of an unbounded operator.

An operator A, densely defined in the Hilbert space $\mathcal{H}$, is called Hermitian if

$$(Af, g)_{\mathcal{H}} = (f, Ag)_{\mathcal{H}}, \quad f, g \in \mathfrak{D}(A). \tag{2.1.18}$$

An operator A, densely defined in $\mathcal{H}$, is called self-adjoint, if

$$A^* = A. \tag{2.1.19}$$

Let A be Hermitian, i.e., (2.1.18) holds true. This equality means that g belongs to $\mathfrak{D}(A^*)$ and $A^*g = Ag$, i.e., $A \subseteq A^*$. Thus, for an operator A with a dense domain, the fact that it is Hermitian is equivalent to the inclusion

$$A \subseteq A^*, \tag{2.1.20}$$

and its self-adjointness is equivalent to the equality (2.1.19).

Hermitian operator A always admits a closure. It follows from the inclusion (2.1.20) and from the fact that A^* is closable. Its closure $\tilde{A}$ is also the Hermitian operator.

Tn operator A is called essentially self-adjoint, if its closure $\tilde{A}$ is self-adjoint.

Lemma 2.1.8 *Each point $z \in \mathbb{C} \setminus \mathbb{R}$ is point of regular type for an arbitrary Hermitian operator.*

Criterion of Self-Adjointness The following theorems are used to find out the self-adjointness of an operator.

Theorem 2.1.9 *A closed Hermitian operator is self-adjoint if and only if its defect numbers are equal to zero, i.e., $m = n = 0$. In other words, an Hermitian operator A is self-adjoint if the equalities*

$$\mathfrak{R}(A - z_1\mathbb{I}) = \mathcal{H} \quad \text{and} \quad \mathfrak{R}(A - z_2\mathbb{I}) = \mathcal{H} \tag{2.1.21}$$

hold true for some $z_1, z_2 \in \mathbb{C}$, where $Im\, z_2 < 0$ and $Im\, z_1 > 0$. Conversely, if A is a self-adjoint operator, then the equalities (2.1.21) *hold true for any z_1 and z_2 of the indicated type.*

A pair of numbers (m, n) will be called deficiency indices of the corresponding operator.

Corollary 2.1.10 *Let A be an Hermitian operator (in general it is non-closed). This operator is essentially self-adjoint provided that its defect numbers are equal zero.*

Semi-Bounded Operators Assume that an Hermitian operator A is such that at least one its point of regular type lies in the real axis. Then the set of all points of regular type of the operator A is connectable and, therefore, its defect numbers are the same, i.e., $m = n$.

Semi-bounded operators constitute an important class of Hermitian operators with this property.

Let A be an operator with a dense domain $\mathfrak{D}(A)$ in $\mathcal{H}$. Assume that there exists a number $\alpha \in \mathbb{R}$ such that

$$(Af, f)_{\mathcal{H}} \geq \alpha \|f\|^2_{\mathcal{H}}, \quad f \in \mathfrak{D}(A). \tag{2.1.22}$$

The operator A is called semi-bounded below, and the number α is called its vertex.

The operator semi-bounded above are defined similarly. If an operator A is semi-bounded below, then $(-A)$ is semi-bounded above. *A semi-bounded operator with a vertex $\alpha = 0$ is called non-negative.*

If A is semi-bounded and admits a closure $\tilde{A}$, then it is evident that $\tilde{A}$ is also semi-bounded and has the same vertex.

Lemma 2.1.11 *Let A be a semi-bounded operator with a vertex $\alpha \in \mathbb{R}$. Any $z \in \mathbb{R} \setminus [\alpha, +\infty)$ is a point of regular type for this operator.*

Theorem 2.1.12 *Let A be a closed semi-bounded operator wit a vertex $\alpha \in \mathbb{R}$. It has a (mutually) equal defect numbers. If*

$$\mathfrak{R}(A - z\mathbb{I}) = \mathcal{H} \tag{2.1.23}$$

for some $z \in \mathbb{C} \setminus [\alpha, +\infty)$, then the operator is self-adjoint.

Let we have $A = A^*$. Recall that *a point $z \in \mathbb{C}$ is called a regular point of the operator A, if the inverse operator $R_z = (A - z\mathbb{I})^{-1}$ exists and is bounded and defined in the whole $\mathcal{H}$ (this operator is called the resolvent of A)*. Thus, the point z is regular, if it is of regular type and $\mathfrak{R}(A - z\mathbb{I}) = \mathcal{H}$. As in the case of bounded operators, one can easily prove that the set of regular points is open in $\mathbb{C}$ and the Hilbert identity (equality)

$$R_z - R_\zeta = (z - \zeta)R_z R_\zeta. \tag{2.1.24}$$

holds true for any two regular points z and ζ.

The spectrum $S(A)$ of a self-adjoint operator A if defined as the complement to the set of its regular points in $\mathbb{C}$.

Thus, the spectrum $S(A)$ is closed, generally speaking, unbounded set on the real axis. If, in addition, the operator A is semi-bounded with a vertex α, then $S(A) \subseteq [\alpha, +\infty)$ (see. Lemmas 2.1.8 and 2.1.11).

Isometric and Unitary Operators In general, *an operator U, acting from the subspace $\mathfrak{D}(U) \subseteq \mathcal{H}_1$ to the subspace $\mathfrak{R}(U) \subseteq \mathcal{H}_2$, is called isometric if*

$$(Uf, Ug)_{\mathcal{H}_2} = (f, g)_{\mathcal{H}_1}, \quad f, g \in \mathfrak{D}(U). \tag{2.1.25}$$

Since, as before, all operators considered below act in a Hilbert space $\mathcal{H}$, then we say that *an operator U, acting from the subspace $\mathfrak{D}(U) \subseteq \mathcal{H}$ to the subspace $\mathfrak{R}(U) \subseteq \mathcal{H}$, is called isometric if*

$$(Uf, Ug)_{\mathcal{H}} = (f, g)_{\mathcal{H}}, \quad f, g \in \mathfrak{D}(U). \tag{2.1.26}$$

This operator is called unitary if, in addition, $\mathfrak{D}(U) = \mathfrak{R}(U) = \mathcal{H}$.

Note that the isometric operator U is necessarily continuous. Therefore, it is always assumed that $\mathfrak{D}(U)$ and $\mathfrak{R}(U)$ are subspaces and the operator U is closed.

Let us describe the set containing only points of regular type for an isometric operator.

Lemma 2.1.13 *Every $z \in \mathbb{C}$, $|z| \neq 1$ is a point of regular type for an isometric operator.*

Thus, the structure of the set of regular type points for isometric operators characterized by the presence of two connected component $\{z \in \mathbb{C} \mid |z| > 1\}$ and $\{z \in \mathbb{C} \mid |z| < 1\}$. According to Theorem 2.1.7, their defect numbers are m and n, respectively and (m, n) is the deficiency index of the operator U.

The theorem below is the similar to Theorem 2.1.9.

Theorem 2.1.14 *An isometric operator U in the Hilbert space $\mathcal{H}$ is unitary iff its defect numbers are equal to zero, i.e., $m = n = 0$.*

Cayley Transforms Let $\mathcal{H}$ be a Hilbert space and A be a closed Hermitian operator in $\mathcal{H}$. The domain $\mathfrak{D}(A)$ of A may be not dense in $\mathcal{H}$. We fixe $z \in \mathbb{C}$ such that $\operatorname{Im} z > 0$.

Consider a vector $g \in \mathfrak{R}(A - z\mathbb{I})$ of the form $g = (A - z\mathbb{I})f$, where $f \in \mathfrak{D}(A)$. We construct a mapping $g \mapsto (A - \bar{z}\mathbb{I})f = Ug$. This mapping is correct since f is uniquely defined for a given g. It is also clear that U is a linear operator with a domain $\mathfrak{R}(A - z\mathbb{I})$ and a range $\mathfrak{R}(A - \bar{z}\mathbb{I})$. Hence, we have

$$g = (A - z\mathbb{I})f, \quad Ug = (A - \bar{z}\mathbb{I})f, \quad f \in \mathfrak{D}(A); \tag{2.1.27}$$

$$\mathfrak{D}(U) = \mathfrak{R}(A - z\mathbb{I}), \quad \text{and} \quad \mathfrak{R}(U) = \mathfrak{R}(A - \bar{z}\mathbb{I}). \tag{2.1.28}$$

It is clear that (2.1.27) can be written in a more concise form

$$Ug = (A - \bar{z}\mathbb{I})(A - z\mathbb{I})^{-1}g. \tag{2.1.29}$$

The operator U *is called the Cayley transform of an operator* A. It has the following simple properties:

(1) *The the Cayley transform of a closed Hermitian operator is an isometric operator.*

(2) *Let* $m(A), n(A)$ *and* $m(U), n(U)$ *be the defect numbers of the operators* A *and* U*, respectively. Then*

$$m(A) = m(U) \quad and \quad n(A) = n(U). \tag{2.1.30}$$

(3) *The the Cayley transform of a self-adjoint operator is a unitary operator.*

(4) *Let* $B \supseteq A$ *be the closed Hermitian extension of an operator* A*. Then its Cayley transform* V *is an isometric extension of the operator* U*.*

Inverse Cayley Transform To write the inverse Cayley transform, let us express the operator A in terms of U using (2.1.27). Subtracting the first equality in (2.1.27) from the second one, we get

$$(U - \mathbb{I})g = (z - \bar{z})f, \quad f \in \mathfrak{D}(A). \tag{2.1.31}$$

Further, by multiplying the second equality in (2.1.27) by z $(\bar{z})$ and subtracting again, we obtain

$$(zU - \bar{z}\mathbb{I})g = (z - \bar{z})Af, \quad f \in \mathfrak{D}(A). \tag{2.1.32}$$

Equalities (2.1.31) and (2.1.32) can be rewritten in the form

$$f = \frac{1}{z - \bar{z}}(U - \mathbb{I})g, \quad Af = \frac{1}{z - \bar{z}}(zU - \bar{z}\mathbb{I})g. \tag{2.1.33}$$

Assume now that an isometric operator U acting in $\mathcal{H}$ exists. Then, for vectors f of the form $f = (z - \bar{z})^{-1}(U - \mathbb{I})g$ where $g \in \mathfrak{D}(U)$, we define an operator A by setting $Af = (z - \bar{z})^{-1}(zU - \bar{z}\mathbb{I})g$. This definition is correct and the operator A is correct and it follows to a linear operator A if

$$\operatorname{Ker}(U - \mathbb{I}) = \{0\}. \tag{2.1.34}$$

Suppose the condition (2.1.34) holds true. The operator A constructed as a result is called the inverse Cayley transform of the operator U, herewith

$$\mathfrak{D}(A) = \mathfrak{R}(U - \mathbb{I}), \quad \mathfrak{R}(A) = \mathfrak{R}(zU - \bar{z}\mathbb{I}). \tag{2.1.35}$$

Formally, the operator A can be expressed in terms of U:

$$Af = (zU - \bar{z}\mathbb{I})(U - \mathbb{I})^{-1} f. \tag{2.1.36}$$

Let us recall some properties of the inverse Cayley transform.

(5) *The inverse Cayley transform of the isometric operator is a closed Hermitian operator.*
(6) *Defect numbers of operators U and A satisfy equalities* $m(A) = m(U), n(A) = n(U)$.
(7) *The inverse Cayley transform of the unitary operator is the self-adjoint operator, if it is, additionally, known that* $\mathfrak{D}(A) = \mathfrak{R}(U - \mathbb{I})$ *is dense in* $\mathcal{H}$.

Lemma 2.1.15 *If* $\mathfrak{R}(U - \mathbb{I})$ *is dense in* $\mathcal{H}$, *then* $\operatorname{Ker}(U - \mathbb{I}) = \{0\}$.

It is now convenient to formulate the property 4 in the form.

(8) *Let* $V \supseteq U$ *be the isometric extension of an isometric operator* U *such that* $\mathfrak{R}(U - \mathbb{I})$ *is dense in* $\mathcal{H}$. *Then the operator* V *possesses the inverse Cayley transform* B, *which is the closed Hermitian extension of a closed Hermitian operator* A.
(9) *Let us construct the inverse Cayley transform of an operator* U, *which is, in turn, the Cayley transform of a given operator* A. *As a result, we obtain A, i.e.,* $A \mapsto U \mapsto A$. *similarly,* $U \mapsto A \mapsto U$.

2.2 Extensions of Hermitian Operators to Self-AdJoint Operators

Let us recall the main statements of the theory of the extension of a densely defined symmetric (Hermitian) operator to a self-adjoint one without leaving the space.

Below, we assume that the defect numbers m, n of an operator acting in a Hilbert space $\mathcal{H}$ has the value $0, 1, \ldots,$ or ∞. The last is really so sins $\mathcal{H}$ is separable.

Theorem 2.2.1 *Let* U *be an isometric operator in* $\mathcal{H}$ *with a domain* $\mathfrak{D}(U)$, *a range* $\mathfrak{R}(U)$, *and the defect numbers* $m = \dim(\mathcal{H} \ominus \mathfrak{D}(U)) > 0$ *and* $n = \dim(\mathcal{H} \ominus \mathfrak{R}(U)) > 0$. *Fix a number* $k \leq \min(m, n)$, *choose two k-dimensional subspaces* $F \subseteq \mathcal{H} \ominus \mathfrak{D}(U)$, $G \subseteq \mathcal{H} \ominus \mathfrak{R}(U)$, *and construct an isometric operator* W *acting from whole* F *into whole* G, *i.e.,* $\mathfrak{D}(W) = F$ *and* $\mathfrak{R}(W) = G$. *Then, the orthogonal sum*

$$V = U \oplus W, \quad \mathfrak{D}(V) = \mathfrak{D}(U) \oplus \mathfrak{D}(W), \quad \mathfrak{R}(V) = \mathfrak{R}(U) \oplus \mathfrak{R}(W)$$

is an isometric extension of the operator U. All possible isometric extensions of this operator can be obtained by using the same procedure for all possible k, F, G, and W.

The last theorem has simple but important corollaries.

Corollary 2.2.2 *If at least one defect number (m or n) of an isometric operator U is equal to zero, then this operator has no no non-trivial isometric extensions.*

Corollary 2.2.3 *In order that the operator U admit unitary extensions, it is necessary and sufficient that its defect numbers are equal to each other, that is $m{=}n$. To construct a unitary extension, one must set $F = \mathcal{H} \ominus \mathfrak{D}(U)$ and $G = \mathcal{H} \ominus \mathfrak{R}(U)$, and take an isometric operator W with $\mathfrak{D}(W) = F$ and $\mathfrak{R}(W) = G$.*

If one of the defect numbers m or n is zero, and the other is positive, then U is called the maximal operator.

We now consider the case of Hermitian operators. Let A be a closed Hermitian operator with dense domain and let $B \supseteq A$ be its closed Hermitian extension. Since $B^* \subseteq A^*$ and $A \subseteq A^*$, then we get the chain:

$$A \subseteq B \subseteq B^* \subseteq A^*. \tag{2.2.1}$$

If B is a self-adjoint operator, then the relation (2.2.1) turns into:

$$A \subseteq B \subseteq A^*. \tag{2.2.2}$$

Let us clarify the condition under which, for a given operator A, one can construct an operator B satisfying (2.2.1) or (2.2.2) and describe the set of such B.

Theorem 2.2.4 *Let A be a closed Hermitian operator with a dense domain in the separable Hilbert space $\mathcal{H}$ and defect numbers m, n. In order that A admits non-trivial closed Hermitian extension $B \supseteq A$, it is necessary and sufficient that its defect numbers m and n be positive. In order that A admits a self-adjoint extension $B = B^* \supseteq A$, it is necessary and sufficient that its defect numbers m and n be equal to each other, i.e. $m = n$.*

These extensions are constructed as follows. Fix a point $z \in \mathbb{C}$ such that $\operatorname{Im} z > 0$. In the defect subspaces $N_z = \mathcal{H} \ominus \mathfrak{R}(A - z\mathbb{I})$ and $N_{\bar{z}} = \mathcal{H} \ominus \mathfrak{R}(A - \bar{z}\mathbb{I})$ whose dimensions are $m > 0$ and $n > 0$, respectively, we chose subspaces $F \subseteq N_z$ and $G \subseteq N_{\bar{z}}$ of the same dimensions and construct an isometric operator W that maps the whole $F = \mathfrak{D}(W)$ into whole $G = \mathfrak{R}(W)$. Let U be the Cayley transform of the operator A, $\mathfrak{D}(U) = \mathfrak{R}(A - z\mathbb{I})$, and $\mathfrak{R}(U) = \mathfrak{R}(A - \bar{z}\mathbb{I})$. Consider an isometric operator

$$V = U \oplus W, \quad \mathfrak{D}(V) = \mathfrak{D}(U) \oplus \mathfrak{D}(W), \quad \mathfrak{R}(V) = \mathfrak{R}(U) \oplus \mathfrak{R}(W).$$

The inverse Cayley transform B of the operator V is a closed Hermitian extension of the operator A. By taking all possible F, G, and W (for a fixed z), we obtain the set of all closed Hermitian extensions B of the operator A.

If $m = n$, then, in particular, we can set $F = N_z$ and $G = N_{\bar{z}}$. In this case, the operator $B \supseteq A$ is a self-adjoint extension of A. By takin all possible W, we obtain the set of all self-adjoint extensions B of the operator A.

Von Neumann Formulas The following theory belongs to Von Neumann J.

(1) Consider a formula that describes the action of the operator A^*. Firs, it is necessary to recall that a linear set $L \subseteq \mathcal{H}$ is called the direct sum of linear sets $L_1, \dots, L_n \subseteq \mathcal{H}$, if, for any $f \in L$, one can write a representation $f = f_1 + \dots + f_n$, where $f_j \in L_j$, and this representation is unique (in other words, $0 = f_1 + \dots + f_n \Rightarrow f_1 = \dots = f_n = 0$). Denote this direct sum

$$L = L_1 \dotplus L_2 \dotplus \dots \dotplus L_n. \tag{2.2.3}$$

Let A be a closed Hermitian operator with a dense domain in $\mathcal{H}$ and let $z \in \mathbb{C} \setminus \mathbb{R}$ be fixed. Then

$$\mathfrak{D}(A^*) = \mathfrak{D}(A) \dotplus N_z \dotplus N_{\bar{z}}. \tag{2.2.4}$$

Thus, according to (2.2.4), any $g \in \mathfrak{D}(A^*)$ admits a unique decomposition

$$g = f + h_z + h_{\bar{z}} \tag{2.2.5}$$

$$f = f(g) \in \mathfrak{D}(A), \quad h_z = h_z(g) \in N_z, \quad h_{\bar{z}} = h_{\bar{z}}(g) \in N_{\bar{z}}.$$

The result of the actin of the operator A^* upon vector (2.2.5) can now be determined in a very simple way. Indeed, since $A^* \supseteq A$, it follows from (4) that

$$A^* g = Af + \bar{z} h_z + z h_{\bar{z}}. \tag{2.2.6}$$

(2) Let us describe the action of a closed Hermitian extension B of an operator A, by using the decomposition (2.2.4). Fix $z \in \mathbb{C}$ such that $\operatorname{Im} z > 0$. Let W be the operator associated with the extension B according to Theorem 2.2.4, $\mathfrak{D}(W) = F \subseteq N_z$, and $\mathfrak{R}(W) = G \subseteq N_{\bar{z}}$.

The set $\mathfrak{D}(B)$ admits a decomposition

$$\mathfrak{D}(B) = \mathfrak{D}(A) \dotplus (W - \mathbb{I})F, \tag{2.2.7}$$

i.e., for all $g \in \mathfrak{D}(B) \subseteq \mathfrak{D}(A^*)$, the decomposition (2.2.5) *has a form*

$$g = f - h_z + W h_z, \ f \in \mathfrak{D}(A), \ h_z \in F \subseteq N_z, \ W h_z \in WF \subseteq N_{\bar{z}}. \tag{2.2.8}$$

Since $B \subseteq A^*$, the action B upon the vector g is is defined by (2.2.6), *namely*

$$Bg = A^*G = Af - zh_z + zWh_z. \tag{2.2.9}$$

We will recall another important theorem of Von Neumann J.

Theorem 2.2.5 *Let A be closed densely defined operator in $\mathcal{H}$. Then the operator A^*A is self-adjoint and non-negative.*

2.3 The Spectral Decomposition of Bounded Operators

We begin by the describing the resolution of the identity for a bounded operator in order to use it to get the resolution of the identity of an unbounded one. The resolution of the identity is a projector-valued measure. Before constructing the spectral integral, we give a number of statements.

Lemma 2.3.1 *Let P_{G_1} and P_{G_2} be projectors onto subspaces $G_1, G_2 \subseteq \mathcal{H}$, respectively. The sum $P = P_{G_1} + P_{G_2}$ is projector if an only if $G_1 \perp G_2$, i.e., $P_{G_1} P_{G_2} = 0$ (or, that is equivalent $P_{G_2} P_{G_1} = 0$). In this case, P is projector onto $G_1 \oplus G_2 : P = P_{G_1 \oplus G_2}$.*

We now proceed to the definition of a general resolution of the identity. Let R be an abstract space and let $\mathfrak{R}$ be σ-algebra of its subsets. In other words, we define a measurable space $\langle R, \mathfrak{R}\rangle$. In particular, is also considered a Hilbert space $\mathcal{H}$.

An operator-valued function $\mathfrak{R} \ni \alpha \mapsto E(\alpha) \in \mathcal{L}(\mathcal{H})$ is called a resolution of the identity (on R), provided that:

(a) *for any $\alpha \in \mathfrak{R}$, $E(\alpha)$ is a projector in $\mathcal{H}$; $E(\emptyset) = 0$, and $E(R) = \mathbb{I}$.*
(b) *the function E is countably-additive, i.e., for an arbitrary sequence $(\alpha_j)_{j=1}^{\infty}$ that consists of disjoint set from $\mathfrak{R}$, the equality*

$$E\left(\bigcup_{j=1}^{\infty} \alpha_j\right) = \sum_{j=1}^{\infty} E(\alpha_j), \tag{2.3.1}$$

holds true, where the series converges in the sense of the strong convergence of operators (i.e., on every vector $f \in \mathcal{H}$ with respect the norm of $\mathcal{H}$).

Theorem 2.3.2 *The resolution of the identity possesses the property of orthogonality, namely*

$$E(\alpha)E(\beta) = E(\alpha \cap \beta), \quad \alpha, \beta \in \mathfrak{R}. \tag{2.3.2}$$

Corollary 2.3.3 *It follows from* (2.3.2) *that operators $E(\alpha)$ $(\alpha \in \mathfrak{R})$ are commuting.*

Let $\mathfrak{R} \ni \alpha \mapsto E(\alpha)$ be the resolution of the identity. We fix $f \in \mathcal{H}$. Then the set function

$$\mathfrak{R} \ni \alpha \mapsto \rho_{f,f}(\alpha) = \big(E(\alpha)f, f\big)_{\mathcal{H}} = \|E(\alpha)f\|^2_{\mathcal{H}} \geq 0, \tag{2.3.3}$$

is, clearly, a non-negative finite measure on $\mathfrak{R}$ (in (2.3.3), we have used the fact that $\big(E(\alpha)\big)^2 = E(\alpha)$, $\big(E(\alpha)\big)^* = E(\alpha)$). For fixed $f, g \in \mathcal{H}$, the function of sets

$$\mathfrak{R} \ni \alpha \mapsto \rho_{f,g}(\alpha) = \big(E(\alpha)f, g\big)_{\mathcal{H}} \in \mathbb{C} \tag{2.3.4}$$

is a complex-valued measure (a charge) on $\mathfrak{R}$. It is a linear combination of measures (2.3.3) by virtue of the polarization identity.

For a scalar measure, the following result holds true.

Theorem 2.3.4 *Let $(\alpha_n)_{n=1}^{\infty}$, $(\beta_n)_{n=1}^{\infty}$ be decreasing and increasing sequences of sets $\alpha_n, \beta_n \in \mathfrak{R} : \alpha_1 \supseteq \alpha_2 \supseteq \dots, \beta_1 \subseteq \beta_2 \subseteq \dots$. Then*

$$\lim_{n\to\infty} E(\alpha_n) = E(\bigcap_{n=1}^{\infty} \alpha_n) \quad \text{and} \quad \lim_{n\to\infty} E(\beta_n) = E(\bigcup_{n=1}^{\infty} \beta_n). \tag{2.3.5}$$

in the sense of strong convergence in $\mathcal{H}$,

Theorem on Extension We recall one more result about the resolution of the identity. As in the case of scalar measure, by the construction of the resolution of the identity, it is often convenient to define it on a certain algebra of sets, and then apply the theorem about an extension.

Let R be a space and let $\mathfrak{R}$ be an *algebra* of subsets R. As above, the function $\mathfrak{R} \ni \alpha \mapsto E(\alpha) \in \mathcal{L}(\mathcal{H})$ is called a resolution of the identity provided that conditions (a) and (b) are satisfied with an additional requirement that $\cup_{j=1}^{\infty}\alpha_j \in \mathfrak{R}$ in (b). It is clear that Theorem 2.3.2 remain true in the case where $\mathfrak{R}$ is an algebra.

Theorem 2.3.5 *Let E be a resolution of the identity on the algebra $\mathfrak{R}$. Then there exists its extension to the resolution of the identity E_σ on its σ-hull $\mathfrak{R}_\sigma$, i.e., the restriction $E_\sigma \upharpoonright \mathfrak{R} = E$. The restriction of the identity E_σ is uniquely defined for given E.*

The Construction of Spectral Integrals Suppose, that the measurable space $\langle R, \mathfrak{R}\rangle$ is given or E is given on R. Elements from R are defined by $\lambda, \mu, \dots$.

Recall that a function $R \ni \lambda \mapsto F(\lambda) \in \mathbb{C}$ is called a simple function (or a step function) if it is a linear combination of indicators $\chi_\alpha(\lambda)$ of sets $\alpha \in \mathfrak{R}$. A collection of all simple functions is denoted by $S(R, \mathfrak{R}) = S$. The set S is an algebra with respect to ordinary summation and multiplication and contains all constants.

It is clear that every simple functions can be represented in the form of a linear combination of indicators of disjoint sets, i.e.,

$$F(\lambda) = \sum_{k=1}^{n} F_k \chi_{\alpha_k}(\lambda), \quad F_k \in \mathbb{C};\ \alpha_k \cap \alpha_j = \emptyset,\ k \neq j;\ \lambda \in R. \tag{2.3.6}$$

By definition, we set

$$\int_R F(\lambda)\, dE(\lambda) = \int_R \Big(\sum_{k=1}^{n} F_k \chi_{\alpha_k}(\lambda)\Big)\, dE(\lambda) = \sum_{k=1}^{n} F_k E(\alpha_k) \in \mathcal{L}(\mathcal{H}). \tag{2.3.7}$$

It is clear that according to (2.3.6), it can be assumed that the sets α_k form the decomposition of the set $R : \cup_{k=1}^{n} \alpha_k = R$. The representation (2.3.6) of a simple function F is, obviously, not unique because sets α_k are not defined uniquely for a given F (they can be subdivided into smaller sets). Nevertheless, integrals (2.3.7) is defined correctly and does not depend on the representation (2.3.6). The constructed integral possesses a series of simple properties.

(1) *The linearity*, i.e.,

$$\int_R \big(aF(\lambda)+bG(\lambda)\big)\, dE(\lambda) = a \int_R F(\lambda)\, dE(\lambda) + b \int_R G(\lambda)\, dE(\lambda) \tag{2.3.8}$$

$$a, b \in \mathbb{C};\ F, G \in S.$$

(2) The unusual property of *the multiplicativety of an integral*

$$\int_R F(\lambda)\, dE(\lambda) \cdot \int_R G(\lambda)\, dE(\lambda) = \int_R F(\lambda)G(\lambda)\, dE(\lambda), \quad F, G \in S \tag{2.3.9}$$

that yields the property of *commutativity* of any two integrals.

(3) For any $F \in S$, we have

$$\Big(\int_R F(\lambda)\, dE(\lambda)\Big)^* = \int_R \overline{F(\lambda)}\, dE(\lambda). \tag{2.3.10}$$

(4) For any $F \in S$ and arbitrary $f, g \in \mathcal{H}$, we have

$$\Big(\Big(\int_R F(\lambda)\, dE(\lambda)\Big) f, g\Big)_{\mathcal{H}} = \int_R F(\lambda)\, d(E(\lambda) f, g)_{\mathcal{H}}, \tag{2.3.11}$$

where $d(E(\lambda) f, g)_{\mathcal{H}}$ denotes the integration with respect to the charge (2.3.4).

(5) For any $F \in S$ and some $f \in \mathcal{H}$, we have

$$\left\| \Big(\int_R F(\lambda)\, dE(\lambda) \Big) f \right\|_{\mathcal{H}}^2 = \int_R |F(\lambda)|^2\, d(E(\lambda) f, f)_{\mathcal{H}}. \tag{2.3.12}$$

(6) For any $F \in S$, we have

$$\left\| \int_R F(\lambda)\, dE(\lambda) \right\| \le \sup \big\{ |F(\lambda)| \mid \lambda \in R \big\}. \tag{2.3.13}$$

Integrals of Bounded Functions Consider the set $L_\infty(R, \mathfrak{R}) = L_\infty$ of all bounded measurable functions connected with a measurable space $\langle R, \mathfrak{R} \rangle$. Just as S, this set is an algebra with respect to the standard algebraic operations.

By definition, we set

$$\mathcal{L}(\mathcal{H}) \ni \int_R F(\lambda)\, dE(\lambda) = \lim_{n \to \infty} \int_R F_n(\lambda)\, dE(\lambda), \quad \forall F \in L_\infty, \tag{2.3.14}$$

where the limit is regarded in the operator norm.

(7) *Integrals of bounded measurable functions $F, G \in L_\infty$ also possess properties (1)–(6).*

Integrals of Unbounded Functions Consider a set $L_0 := L_0(R, \mathfrak{R}, E)$ of functions of the form $R \ni \lambda \mapsto F(\lambda) \in \mathbb{C} \cup \{\infty\}$, measurable with respect to $\mathfrak{R}$ and almost everywhere finite with respect to an operator-valued measure E. Clearly, the last statement means that

$$E\big(\{ \lambda \in R \mid |F(\lambda)| = \infty \} \big) = 0. \tag{2.3.15}$$

Just as S and L_∞, by virtue of the additivity of the measure E, the set L_0 allows us to form an algebra with ordinary operations of summation and multiplication of functions (and with standard formal rules for operations with ∞).

Lemma 2.3.6 *Let $F \in L_0$. Then the set*

$$\mathfrak{D}_F = \left\{ f \in \mathcal{H} \,\middle|\, \int_R |F(\lambda)|^2\, d\big(E(\lambda) f, f\big)_{\mathcal{H}} < \infty \right\} \tag{2.3.16}$$

is linear and everywhere dense in $\mathcal{H}$.

We define now the corresponding integral. As usual, for $F \in L_0$ and $N \ge 0$, we denote by F_N its cutoff function, i.e., the bounded function of the form $F_N(\lambda) =$

$F(\lambda)$ for $\lambda \in \{\lambda \in R \mid |F(\lambda)| \le N\}$ and $F_N(\lambda) = N$ for all other $\lambda \in R$. By definition, for $f \in \mathfrak{D}_F$, we set

$$\mathcal{I}_F f = \int_R F(\lambda)\, dE(\lambda) f = \lim_{N\to\infty} \int_R F_N(\lambda)\, dE(\lambda) f \tag{2.3.17}$$

in the sense of convergence in $\mathcal{H}$.

This integral $\mathcal{I}_F$ (2.3.17) *exists, generally speaking, as un unbounded operator with a dense domain $\mathfrak{D}(\mathcal{I}_F) = \mathfrak{D}_F$ in $\mathcal{H}$ defined in* (2.3.16).

Let us describe the properties of the integral $\mathcal{I}_F$. First, note that the property (6) is meaningless.

(8) *The integral of unbounded functions $F \in L_0$ possesses properties* (4) *and* (5) *with $f \in \mathfrak{D}(\mathcal{I}_F)$ in* (2.3.11) *and* (2.3.12).

Other Properties of Spectral Integral The generalization of properties (1)–(3) appears to be more complicated.

Theorem 2.3.7 *Let $F \in L_0$. Then the operator $\mathcal{I}_F$ of the form* (2.3.17) *with a dense domain $\mathfrak{D}(\mathcal{I}_F) = \mathfrak{D}_F$ is closed. The equality*

$$(\mathcal{I}_F)^* = \mathcal{I}_{\bar{F}}, \qquad \mathfrak{D}\big((\mathcal{I}_F)^*\big) = \mathfrak{D}_F,$$

holds true, i.e., the property (3) *admits the generalization*

$$\left(\int_R F(\lambda)\, dE(\lambda)\right)^* = \int_R \overline{F(\lambda)}\, dE(\lambda), \qquad F \in L_0. \tag{2.3.18}$$

Theorem 2.3.8 *Let $F, G \in L_0$ and $a, b \in \mathbb{C}$. Then the following equalities, which generalize properties* (1) *and* (2),

$$\int_R \big(aF(\lambda) + bG(\lambda)\big)\, dE(\lambda) = \left(a \int_R F(\lambda)\, dE(\lambda) + b \int_R G(\lambda)\, dE(\lambda)\right)^{\sim}, \tag{2.3.19}$$

$$\int_R F(\lambda)G(\lambda)\, dE(\lambda) = \left(\int_R F(\lambda)\, dE(\lambda) \int_R G(\lambda)\, dE(\lambda)\right)^{\sim} \tag{2.3.20}$$

are true.

An operator A acting in the Hilbert space $\mathcal{H}$ is called normal if it is densely defined, closed and commutes with adjoint operator, i.e.,

$$A\, A^* = A^*\, A. \tag{2.3.21}$$

It is clear that self-adjoint and unitary operators are normal. *For all* $F \in L_0$, *operator* (2.3.17) *is normal.*

The Representation of a Resolution of the Identity by Change of Variables Let us explain, how to replace the variable in the spectral integral and construct the Cartesian product of resolutions of the identity.

Let $\langle R, \mathfrak{R}\rangle$ be a measurable space, R' be another space of points $\lambda',, \mu', \ldots,$ and $R \ni \lambda \to \varphi(\lambda) \in R'$ be a fixed one-to-one mapping of R into R'. Without loss of generality, we can assume that $\varphi(R) = R'$. Given $\mathfrak{R}$ and φ, we can define on R' (in the standard way) σ-algebra $\mathfrak{R}'$ that contains of all sets $\alpha' \subseteq R'$ whose preimage $\varphi^{-1}(\alpha')$ lies in $\mathfrak{R}$. It can be shown that $\mathfrak{R}'$ is a σ-algebra. Thus we have constructed the measurable space $\langle R', \mathfrak{R}'\rangle$.

Let E be a resolution of the identity given on $\langle R, \mathfrak{R}\rangle$. We now describe the standard method for constructing the image of E, i.e., the resolution of the identity E' in $\langle R', \mathfrak{R}'\rangle$. Thus, we set

$$\mathfrak{R}' \ni \alpha' \mapsto E'(\alpha') = E\big(\varphi^{-1}(\alpha')\big). \tag{2.3.22}$$

It is not difficalt to show that the operator-valued measure (2.3.22) is a resolution of the identity, i.e., satisfies the requirements (a) and (b).

The following theorem gives the rule of the change of variables in spectral integral.

Theorem 2.3.9 *Let* $L_0(R', \mathfrak{R}', E')$ *be a collection of complex-valued functions defined on* R', *measurable with respect to* $\mathfrak{R}'$ *and finite* E'*-almost everywhere. If* $F' \in L_0(R', \mathfrak{R}', E')$, *then* $F' \circ \varphi \in L_0(R, \mathfrak{R}, E)$ *and formulas about the change of variables*

$$\int_R F'\big(\varphi(\lambda)\big)\, dE(\lambda) = \int_{R'} F'(\lambda')\, dE'(\lambda') = \mathcal{I}_{F'}; \tag{2.3.23}$$

$$\begin{aligned} \mathfrak{D}(\mathcal{I}_{F'}) &= \Big\{ f \in \mathcal{H} \,\Big|\, \int_R |F'\big(\varphi(\lambda)\big)|^2\, d\big(E(\lambda) f, f\big)_{\mathcal{H}} < \infty \Big\} \\ &= \Big\{ f \in \mathcal{H} \,\Big|\, \int_{R'} |F'(\lambda')|^2\, d\big(E'(\lambda') f, f\big)_{\mathcal{H}} < \infty \Big\} \end{aligned} \tag{2.3.24}$$

holds true.

The Product of Resolutions of the Identity Let us describe the direct product of resolutions of the identity.

Let $\langle R_1, \mathfrak{R}_1\rangle$, $\langle R_2, \mathfrak{R}_2\rangle$ be two measurable spaces with resolutions of the identity E_1 and E_2, respectively. The values of theses resolutions of the identity are projectors in the same Hilbert space $\mathcal{H}$. Suppose that E_1 and E_2 commute, i.e., the operators $E_1(\alpha_1)$ and $E_2(\alpha_2)$ commute for all $\alpha_1 \in \mathfrak{R}_1$ and $\alpha_2 \in \mathfrak{R}_2$.

As in the case of scalar measures, one can construct the direct product E of resolution of the identity E_1 and E_2 on the space $R = R_1 \times R_2$. More precisely, denote by $\mathfrak{R}$ the direct product $\mathfrak{R}_1 \times \mathfrak{R}_2$ of the σ-algebras $\mathfrak{R}_1$ and $\mathfrak{R}_2$ consisting all subsets $R_1 \times R_2$ that belong to σ-span of all possible rectangles $\alpha_1 \times \alpha_2$, $\alpha_1 \in \mathfrak{R}_1, \alpha_2 \in \mathfrak{R}_2$. It is necessary to construct a resolution of the identity E in the measurable space $\langle R, \mathfrak{R}\rangle$ such that

$$E(\alpha_1 \times \alpha_2) = E_1(\alpha_1)E_2(\alpha_2), \quad \alpha_1 \in \mathfrak{R}_1,\ \alpha_2 \in \mathfrak{R}_2, \tag{2.3.25}$$

i.e., the measure E of a rectangle is the product of the corresponding measures of its sides. Note that, in view of the fact that $E_1(\alpha_1)$ and $E_2(\alpha_2)$ commute, then the operator in (2.3.25) is projector.

This E is called the direct product of E_1 and E_2 and it is denoted by $E = E_1 \times E_2$.

We arrive at a quite unexpected result that, unlike the scalar case, the direct product E does not always exist. Nevertheless, in "proper" cases, it exists.

Thus, let R be a complete metric separable space and let $\mathfrak{R} = \mathfrak{B}(R)$ be the σ-algebra of its Borel subsets.

The resolution of the identity defined on $\mathfrak{B}(R)$ is called the Borel resolution of the identity.

Theorem 2.3.10 *Let E_1 and E_2 be two commuting Borel resolutions of the identity R_1 and R_2, respectively. Then condition* (2.3.25) *determines a unique resolution of the identity $E = E_1 \times E_2$ defined on $\mathfrak{R} = \mathfrak{B}\ (R_1 \times R_2)$.*

The given result remains correct for an arbitrary finite number of resolutions of the identity, and also has a generalization to an infinite number of resolutions of the identity.

The Spectral Decomposition of a Bounded Self-Adjoint Operator Let us start with the construction of the spectral integral for the bounded operator.

Consider a bounded operator A acting in a Hilbert space $\mathcal{H}$. Assume that its spectrum $S(A)$ is a (closed) set in a finite intervale $(a, b) \subset \mathbb{R}$. Denote its resolvent by $R_z = (A - z\mathbb{I})^{-1}$, $z \in \mathbb{C} \setminus S(A)$.

Let $F(z)$ be an analytic function defined in a certain (complex) neighborhood of the interval $[a, b]$. The collection of all functions of this sort is denoted by $\mathcal{A}\big([a, b]\big)$. For any closed Jordan contour γ that encloses the spectrum $S(A)$ and lies in the domain of the function F, one can write the integral

$$F(A) = -\frac{1}{2\pi i}\oint\limits_{\gamma} F(z)R_z\, dz. \tag{2.3.26}$$

As we know, this integral is independent of the choice of the Jordan contour γ. The mapping

$$\mathcal{A}\big([a,b]\big) \ni F \mapsto F(A) \in \mathcal{L}(\mathcal{H}) \tag{2.3.27}$$

is the homomorphism of the algebra of functions $\mathcal{A}\big([a,b]\big)$ to the algebra $\mathcal{L}(\mathcal{H})$ of bounded operators in $\mathcal{H}$ that transforms the function $F(z) \equiv 1$ to the identity operator and the function $F(z) = z$ to the operator A.

Recall that $\mathcal{A}\big([a,b]\big)$ is an algebra with respect to the standard algebraic operations

$$(F+G)(z) = F(z) + G(z), \quad (FG)(z) = F(z)G(z), \quad F, G \in \mathcal{A}\big([a,b]\big), \tag{2.3.28}$$

where the functions $F+G$ and FG are defined on the intersection of the domains of the functions F and G (note that $\mathcal{A}\big([a,b]\big)$ also contains constants).

Lemma 2.3.11 *The formula,*

$$\begin{aligned} F(A) &= \lim_{\varepsilon \to +0} \frac{1}{2\pi i} \int_a^b F(\lambda)(R_{\lambda+i\varepsilon} - R_{\lambda-i\varepsilon})\, d\lambda \\ &= \lim_{\varepsilon \to +0} \frac{\varepsilon}{\pi} \int_a^b F(\lambda) R_{\lambda+i\varepsilon} R_{\lambda-i\varepsilon}\, d\lambda, \quad F \in \mathcal{A}\big([a,b]\big), \end{aligned} \tag{2.3.29}$$

where the limit is regarded in the operator norm, holds true.

Now let us construct the resolution of the identity of the self-adjoint operator.

Theorem 2.3.12 *Let A be a bounded self-adjoint operator. Then one can define a resolution of the identity E on the σ-algebra $\mathfrak{B}(\mathbb{R})$ of Borel subsets of the real axis such that the spectral representation*

$$A = \int_{\mathbb{R}} \lambda\, dE(\lambda) \tag{2.3.30}$$

is true.

By proving this theorem, the lemma below is used, which is important in its own right. Let A be self-adjoin. Then

$$R_z^* = \big((A - z\mathbb{I})^{-1}\big)^* = \big(A^* - \bar{z}\mathbb{I}\big)^{-1} = R_{\bar{z}}$$

and relation (2.3.29) can be rewritten in the form

$$F(A) = \lim_{\varepsilon \to +0} \frac{\varepsilon}{\pi} \int_a^b F(\lambda) R^*_{\lambda+i\varepsilon} R_{\lambda+i\varepsilon} \, d\lambda, \quad F \in \mathcal{A}\big([a,b]\big). \tag{2.3.31}$$

Lemma 2.3.13 *Assume that $F \in \mathcal{A}\big([a,b]\big)$ takes values on the real axis (the algebra of such F is denoted by $\mathcal{A}_{Re}\big([a,b]\big)$). The operator $F(A)$ is self-adjoint and admits the estimate*

$$\big|(F(A)f, g)_{\mathcal{H}}\big| \le \max\{|F(\lambda)| \mid \lambda \in [a,b]\} \, \|f\|_{\mathcal{H}} \|g\|_{\mathcal{H}}, \tag{2.3.32}$$
$$f, g \in \mathcal{H}; \quad F \in \mathcal{A}_{Re}\big([a,b]\big).$$

Functions of Operators and Their Spectrum An analytic function of the form (2.3.26) of a self -adjoint operator $A \in \mathcal{L}(\mathcal{H})$ admit a representation as a spectral integral

$$F(A) = \int_{\mathbb{R}} F(\lambda) \, dE(\lambda), \quad F \in \mathcal{A}([a,b]). \tag{2.3.33}$$

In (2.3.33), it is assumed that $S(A) \subset (a,b)$; therefore, E concentrated on (a,b)).

The spectral representation (2.3.30) enables us to generalize the definition of a function of an operator A from the class of analytic functions to the class $L_0 = L_0(\mathbb{R}, \mathfrak{B}(\mathbb{R}), E)$ of functions of the form $\mathbb{R} \ni \lambda \mapsto F(\lambda) \in \mathbb{C} \cup \{\infty\}$ measurable with respect to $\mathfrak{B}(\mathbb{R})$ and almost everywhere finite with respect to E. Namely, we set

$$L_0(\mathbb{R}, \mathfrak{B}(\mathbb{R}), E) \ni F \mapsto F(A) = \int_{\mathbb{R}} F(\lambda) \, dE(\lambda), \tag{2.3.34}$$

$$\mathfrak{D}(F(A)) = \left\{ f \in \mathcal{H} \;\middle|\; \int_{\mathbb{R}} |F(\lambda)|^2 \, d(E(\lambda)f, f)_{\mathcal{H}} < \infty \right\}.$$

As in the case of a scalar measure, *the support of a general resolution of identity* E on the measurable space $\langle \mathbb{R}, \mathfrak{B}(\mathbb{R}) \rangle$ is regarded as the intersection of all closed sets $\varphi \subseteq \mathbb{R}$ of full measure E, i.e., such that $E(\varphi) = \mathbb{I}$. This support is denoted $\operatorname{supp} E$. It is closed and, hence, belongs to $\mathfrak{B}(\mathbb{R})$. It is easy to see that

$$E(\operatorname{supp} E) = \mathbb{I}. \tag{2.3.35}$$

Also note that *if o is an open set on the axis $\mathbb{R}$ such that $o \cap \operatorname{supp} E \ne \emptyset$, then* $E(o) \ne 0$.

Theorem 2.3.14 *The spectrum of a self-adjoint operator* $A \in \mathcal{L}(\mathcal{H})$ *coincides with the support of its resolution of the identity, namely* $S(A) = \operatorname{supp} E$. *Thus, the integral over* $\mathbb{R}$ *in relations* (2.3.30), (2.3.33) *and* (2.3.34) *can be replaced with the integrals over* $S(A)$.

The resolution of the identity E is uniquely defined for a given self-adjoint operator $A \in \mathcal{L}(\mathcal{H})$. It is called the resolution of the identity of the operator A.

We remark that, if R is a topological space with a countable basis of neighborhoods and $\mathfrak{R}$ is the σ-algebra $\mathfrak{B}(R)$ of its Borel subsets, then the properties of the support, in particular (2.3.35), remain true.

The Spectral Decomposition of Unitary Operators For a unitary operator, spectral decomposition is constructed as for self-adjoint operator with change of the real axis on the unit circle.

Let U be a unitary operator in $\mathcal{H}$ and let $S(U)$ be its spectrum. It is well known that $S(U)$ is a closed set on the unit circle $\mathbb{T} = \{z \in \mathbb{C} \mid |z| = 1\}$. Denote by $\mathcal{A}(\mathbb{T})$ the collection of all functions analytic in a certain neighborhood of $\mathbb{T}$. Jus as $\mathcal{A}([a, b])$, the class $\mathcal{A}(\mathbb{T})$ is an algebra with respect to the standard algebraic operations.

Relations (2.3.26)–(2.3.28) remain true with A replaced with U, $\mathcal{A}([a, b])$ by $\mathcal{A}(\mathbb{T})$, and γ by a contour that encloses the unit circle and is bypassed in the required direction. For example, $\gamma = \gamma_1 \cup \gamma_2$, where $\gamma_1 (\gamma_2)$ is a circle whose radius is greater then one (less than one) which is bypassed counterclockwise (clockwise).

The role of the transition $\mathbb{C} \ni z \mapsto \bar{z} \in \mathbb{C}$ is now played by the reflection with respect to the unit circle: $\mathbb{C} \setminus \{0\} \ni z \mapsto z^* = \bar{z}^{-1} \in \mathbb{C} \setminus \{0\}$.

Lemma 2.3.15 *Let* $A_{Re}(\mathbb{T})$ *be the class of functions from* $A(\mathbb{T})$ *that take real values in* $\mathbb{T}$. *Then the operator* $F(U)$ *is self-adjoint and satisfies the estimate*

$$|(F(U)f, g)_{\mathcal{H}}| \le \max\{|F(\lambda)| \mid \lambda \in \mathbb{T}\} \, \|f\|_{\mathcal{H}} \, \|g\|_{\mathcal{H}}, \qquad (2.3.36)$$
$$F \in \mathcal{A}_{Re}(\mathbb{T}); \quad f, g \in \mathcal{H}.$$

The following theorem is similar to Theorems 2.3.12 and 2.3.14.

Theorem 2.3.16 *Let* U *be a unitaty operator. Then, on the* σ*-algebra* $\mathfrak{B}(\mathbb{T})$ *of the Borel subsets of the unit circle* $\mathbb{T} = \{z \in \mathbb{C} \mid |z| = 1\}$, *the resolution of the identity* E *of the operator* U *is defined so that the representation*

$$U = \int_{\mathbb{T}} \lambda \, dE(\lambda) = \int_0^{2\pi} e^{i\varphi} \, dE(e^{i\varphi}) \qquad (2.3.37)$$

holds true.

In (2.3.37), $\mathbb{T}$ can be replaced with the spectrum $S(U)$ of the operator U. For every function $F(z)$ analytic in the neighborhood of $S(U)$, the equality

$$F(U) = \int_{S(U)} F(\lambda)\, dE(\lambda), \tag{2.3.38}$$

holds true, where the operator $F(U)$ is constructed as a contour integral of the type (2.3.26). The resolution of the identity E from (2.3.37) is defined uniquely.

As in the case of self-adjoint operator, the representation (2.3.37) gives the possibility to generalize the notion of a function of an operator U for arbitrary functions from the class $L_0\big(\mathbb{T}, \mathfrak{B}(\mathbb{T}), E\big)$, of E-almost everywhere finite functions defined on $\mathbb{T}$ and measurable with respect to $\mathfrak{B}(\mathbb{T})$. This generalization is given by the formula

$$L_0\,(\mathbb{T}, \mathfrak{B}(\mathbb{T}), E) \ni F \mapsto F(U) = \int_{\mathbb{T}} F(\lambda)\, dE(\lambda), \tag{2.3.39}$$

$$\mathfrak{D}\,(F(U)) = \left\{ f \in \mathcal{H} \,\middle|\, \int_{\mathbb{T}} |F(\lambda)|^2\, d(E(\lambda)f, f)_{\mathcal{H}} < \infty \right\}.$$

The Spectral Theorem for a Normal Operator Let us construct the spectral decomposition for a bounded normal operator A. The corresponding resolution of the identity will be constructed as a direct product of the resolutions of the identity of two commuting self-adjoint operators associated with A.

Theorem 2.3.17 *Let A_1 and A_2 be two self-adjoint bounded operators. In order that their resolutions of the identity be commuting, it is necessary and sufficient that their resolvents $R_{z_1}(A_1)$ and $R_{z_2}(A_2)$ commute for some fixed z_1 and z_2 regular for the operators A_1 and A_2, respectively.*

Consider a bounded normal operator $\mathcal{H}$, i.e., an operator $A \in \mathcal{L}(\mathcal{H})$ such that $A^*A = AA^*$.

Theorem 2.3.18 *Let A be a bounded normal operator. Then, on the σ-algebra $\mathfrak{B}(\mathbb{C})$ of Borel subsets of the complex plane, one can construct the resolution of the identity E of the operator A satisfying the representation*

$$A = \int_{\mathbb{C}} \lambda\, dE(\lambda). \tag{2.3.40}$$

In (2.3.40), $\mathbb{C}$ can be replaced with the spectrum $S(A)$ of the operator A. Any function $F(z)$ analytic in a neighborhood of $S(A)$ satisfies the equality

$$F(A) = \int_{S(A)} F(\lambda)\, dE(\lambda), \tag{2.3.41}$$

where the operator $F(A)$ *is constructed as a contour integral of the type* (2.3.26). *The resolution of the identity* E *from* (2.3.40) *is determined uniquely.*

2.4 The Spectral Decomposition of Unbounded Operators

Now let us generalize the previous material to the case of an unbounded operator. We consider first a self-adjoint operator A. The idea is as follows. To construct the resolution of the identity E of the operator A, we first find a bounded (self-adjoint or unitary) operator for which the resolution of the identity has already been constructed; its proper representation would give the required E.

Theorem 2.4.1 *Let* A *be an arbitrary self-adjoint operator. Then the resolution of the identity* E *of the operator* A *is defined on the* σ*-algebra* $\mathfrak{B}(\mathbb{R})$ *of Borel subsets of the real axis and the representation*

$$A = \int_{\mathbb{R}} \lambda \, dE(\lambda), \quad \mathfrak{D}(A) = \left\{ f \in \mathcal{H} \,\middle|\, \int_{\mathbb{R}} \lambda^2 \, d(E(\lambda)f, f)_{\mathcal{H}} < \infty \right\} \tag{2.4.1}$$

holds true.

In (2.4.1), $\mathbb{R}$ *can be replaced with the spectrum* $S(A)$ *of the operator* A. *The resolution of the identity* E *in* (2.4.1) *is defined uniquely.*

As before, using the spectral integral, it is possible to construct the theory of functions of the unbounded self-adjoint operator A. Thus, for $F \in L_0(\mathbb{R}, \mathfrak{B}(\mathbb{R}), E)$, where E is the resolution of the identity that corresponds to A, we construct the operator

$$F(A) = \int_{\mathbb{R}} F(\lambda) \, dE(\lambda), \tag{2.4.2}$$

$$\mathfrak{D}\big(F(A)\big) = \left\{ f \in \mathcal{H} \,\middle|\, \int_{\mathbb{R}} |F(\lambda)|^2 \, d\big(E(\lambda)f, f\big)_{\mathcal{H}} < \infty \right\}.$$

The properties of the mapping $L_0\big(\mathbb{R}, \mathfrak{B}(\mathbb{R}), R\big) \ni F \mapsto F(A)$ are described previously.

Let us mention some important functions of a self-adjoint operator A (operators in (2.4.5)–(2.4.7) are, generally speaking, unbounded).

(1) *A resolution of the identity*. For the indicator χ_α of a set $\alpha \in \mathfrak{B}(\mathbb{R})$, we have

$$E(\alpha) = \chi_\alpha(A) = \int_{\mathbb{R}} \chi_\alpha(\lambda) \, dE(\lambda). \tag{2.4.3}$$

(2) *A resolvent*. Let $z \notin S(A)$, then the representation

$$R_z = \left(A - z\mathbb{I}\right)^{-1} = \int\limits_{S(A)} \frac{1}{\lambda - z}\, dE(\lambda) \tag{2.4.4}$$

holds true.

(3) *The square root of a non-negative operator A*.

$$\sqrt{A} = \int\limits_{S(A)} \sqrt{\lambda}\, dE(\lambda) = \int\limits_{0}^{\infty} \sqrt{\lambda}\, dE(\lambda) \geq 0; \quad (\sqrt{A})^2 = A. \tag{2.4.5}$$

(4) *The absolute value of an operator*:

$$|A| = \int\limits_{S(A)} |\lambda|\, dE(\lambda) = \int\limits_{\mathbb{R}} |\lambda|\, dE(\lambda) \geq 0. \tag{2.4.6}$$

(5) *The exponent*. For any $z \in \mathbb{C}$,

$$\exp(zA) = e^{zA} = \int\limits_{S(A)} e^{zA}\, dE(\lambda). \tag{2.4.7}$$

The Stone Formula There are more items related to the resolution of the identity E of the self-adjoint operator A. Thus, we present a usual expression for E in terms of the resolvent of A.

Theorem 2.4.2 *Let E be a resolution of the identity of a self-adjoint operator and let R_z be its resolvent. Then, the formula*

$$\frac{1}{2}\left(E(\delta) + E(\tilde{\delta})\right) = \lim_{\varepsilon \to +\infty} \frac{1}{2\pi i} \int\limits_{\delta + i\varepsilon} (R_z - R_{\bar{z}})\, dz \tag{2.4.8}$$

holds true in the sense of the strong convergence for every open finite interval $\delta \subset \mathbb{R}$.

Theorem 2.4.3 *For general self-adjoint operators A_1 and A_2, Theorem* 2.3.17 *holds true in the same formulation.*

As in the case of scalar measures, for a resolution of the identity defined on Borel subsets of the real axis, we often use (instead of an operator-valued measure) a non-decreasing operator-valued function that is also called a resolution of the identity.

The operator-valued function $\mathbb{R} \ni \lambda \mapsto E_\lambda$ *whose values are projectors in a fixed Hilbert space* $\mathcal{H}$ *is called a resolution of the identity if it is satisfies conditions*

(a) *monotonicity:* $\forall \lambda, \mu \in \mathbb{R},\ \lambda < \mu \Rightarrow E_\lambda \leq E_\mu$;
(b) *completeness:* $\lim\limits_{\lambda \to -\infty} E_\lambda = 0$ *and* $\lim\limits_{\lambda \to +\infty} E_\lambda = \mathbb{I}$ *in the sense of strong convergence;*
(c) *left-continuity:* $\lim\limits_{\lambda \to \mu - 0} E_\lambda = E_\mu$ *in the sense of the strong convergence.*

Theorem 2.4.4 *Let E be a resolution of the identity (a measure) given on* $\mathfrak{B}(\mathbb{R})$*. Then*

$$\mathbb{R} \ni \lambda \mapsto E_\lambda = E\big((-\infty, \lambda)\big) \tag{2.4.9}$$

is resolution of the identity (a function). Conversely, for a given resolution of the identity E_λ*, one can construct a resolution of the identity E on* $\mathfrak{B}(\mathbb{R})$ *such that* E_λ *and E satisfy* (2.4.9).

The Case of Normal Operators Recall that a closed densely defined operator A is called *normal*, if $A^*A = AA^*$ for the case of a bounded operator and commute their resolutions of the identity for the case of the unbounded operator.

Theorem 2.4.5 *Let A be an arbitrary normal operator. Then, on the* σ*-algebra* $\mathfrak{B}(\mathbb{C})$ *of Borel subsets of the complex plane, one can define a resolution of the identity E of the operator A such that the spectral representation*

$$A = \int\limits_{\mathbb{C}} \lambda\, dE(\lambda), \quad \mathfrak{D}(A) = \left\{ f \in \mathcal{H} \,\middle|\, \int\limits_{\mathbb{C}} |\lambda|^2\, d(E(\lambda)f, f)_{\mathcal{H}} < \infty \right\} \tag{2.4.10}$$

holds true.

In (2.4.10)*,* $\mathbb{C}$ *can be replaced with the spectrum* $S(A)$ *of the operator A. The resolution of the identity E from* (2.4.10) *is defined uniquely.*

2.5 Representations of One-Parameter Unitary Groups

Let A be a self-adjoint operator acting in a Hilbert space $\mathcal{H}$ and let E be its resolution of the identity. For a given function $\mathbb{R} \times \mathbb{R} \ni (\lambda, t) \mapsto e^{it\lambda} \in \mathbb{C}$, we construct the operator-valued function

$$\mathbb{R} \ni t \mapsto U(t) = \int\limits_{\mathbb{R}} e^{it\lambda}\, dE(\lambda) = e^{itA} \in \mathcal{L}(H) \tag{2.5.1}$$

according to (2.4.7).

It follows from the properties of spectral integral that the operator $U(t)$ is unitary for all $t \in \mathbb{R}$ and satisfies the equality

$$U(t+s) = U(t)U(s), \quad t, s \in \mathbb{R}. \tag{2.5.2}$$

The function, defined in (2.5.1), is strongly continuous, i.e., $\forall f \in \mathcal{H}$, $\forall t \in \mathbb{R}$: $U(s)f \to U(t)f$ as $s \to t$. To prove this fact, we use relation (2.3.12) and obtain

$$\|U(s)f - U(t)f\|_{\mathcal{H}}^2 = \left\| \int_{\mathbb{R}} \left(e^{is\lambda} - e^{it\lambda}\right) dE(\lambda) f \right\|_{\mathcal{H}}^2$$
$$= \int_{\mathbb{R}} \left|e^{is\lambda} - e^{it\lambda}\right|^2 d(E(\lambda)f, f)_{\mathcal{H}} \underset{s \to t}{\to} 0.$$

Moreover, $U(t)$ *is strongly continuous*, i.e., for all $f \in \mathfrak{D}(A)$ and $t \in \mathbb{R}$, the strong derivative

$$U'(t)f = \lim_{h \to 0} \frac{1}{h}\left(U(t+h) - U(t)\right) f = iU(t)Af \tag{2.5.3}$$

exists and is a continuous vector function.

Indeed, by virtue of relations (2.5.1), (2.4.7), and (2.3.12), we obtain

$$\left\| iU(t)Af - \frac{1}{h}(U(t+h) - U(t))f \right\|_{\mathcal{H}}^2$$
$$= \left\| \int_{\mathbb{R}} \left(ie^{it\lambda}\lambda - \frac{1}{h}\left(e^{i(t+h)\lambda} - e^{it\lambda}\right)\right) dE(\lambda) f \right\|_{\mathcal{H}}^2$$
$$= \int_{\mathbb{R}} \left|i\lambda - \frac{1}{h}\left(e^{ih\lambda} - 1\right)\right|^2 d(E(\lambda)f, f)_{\mathcal{H}},$$

where the integrand, on the right-hand side of this equality, vanishes as $h \to 0$ for all $\lambda \in \mathbb{R}$ and is uniformly bounded with respect to $\mathcal{H}$ by the function $c\lambda^2$. For $f \in \mathfrak{D}(A)$, by virtue of the Lebesgue theorem, we can pass to the limit under the integral sign and get (2.5.3). The continuity of the derivative follows from the inclusion $f \in \mathfrak{D}(A)$.

Relations (2.5.1) and (2.5.3) have the following interpretation: Consider an operator-differential equation

$$u'(t) = iAu(t), \quad t \in \mathbb{R}, \tag{2.5.4}$$

where $\mathbb{R} \ni t \mapsto u(t) \in \mathcal{H}$ is required solution. The function u is assumed to be strongly continuously differentiable and $u(t) \in \mathfrak{D}(A)$ for all t. Such solution are called strong. The strong solution of the Eq. (2.5.3) satisfying the initial condition $u(0) = u_0 \in \mathfrak{D}(A)$, i.e., the solution of the corresponding Cauchy problem exists and is given by the formula

$$u(t) = U(t)u_0, \quad t \in \mathbb{R}. \tag{2.5.5}$$

It is not difficult to verify that this Cauchy problem has a unique solution. Another view of these formulas is as follows: *We say that a function* $\mathbb{R} \ni t \mapsto U(t)$, *whose value are unitary operators in* $\mathcal{H}$, *satisfying the relation* (2.5.2), *defines a one-parameter unitary group. Or, in other words, there is given the unitary representation of the group* $\mathbb{R}$.

Thus, (2.5.1) is an example of a one-parameter unitary group which is, in addition, strongly continuous. The theorem below say that the formula (2.5.1) gives the general form of these groups.

Theorem 2.5.1 (Stone) *A strongly continuous one-parameter unitary group* $U(t)$, ($t \in \mathbb{R}$) *always admits the representation* (2.5.1) *with a certain resolution of the identity* E *uniquely determined for a given group. The corresponding operator* A *is called the infinitesimal generator of this group.*

Proof Let us construct a linear set $\mathfrak{D} \subseteq H$ dense in $\mathcal{H}$, and such that the vector function

$$\mathbb{R} \ni t \mapsto U(t)f \in \mathcal{H} \tag{2.5.6}$$

is strongly continuously differentiable for all $f \in \mathfrak{D}$ (the presented below construction is a particular case of the construction of the so-called Gårding domain).

For $F \in C_0^\infty(\mathbb{R})$ and $g \in \mathcal{H}$, we consider a vector of the form

$$g_F = \int_{\mathbb{R}} F(s)U(s)g\, ds \in \mathcal{H}, \tag{2.5.7}$$

where the integral is regarded as the limit of Riemann integral sums in $\mathcal{H}$. By virtue of the continuity of $F(\cdot)$ and $U(\cdot)f$ and finiteness of $F(\cdot)$, the standard arguments enable us to conclude that integral (2.5.7) exists and possesses natural properties of Riemann integrals. Thus, the set $\mathfrak{D}$ be chosen as the collection of linear combinations of vectors (2.5.7) with $F \in C_0^\infty(\mathbb{R})$ and $g \in \mathcal{H}$.

The set $\mathfrak{D}$ is dense in $\mathcal{H}$. To prove this, we consider a vector $\mathcal{H} \ni h \perp \mathfrak{D}$. Multiplying scalar (2.5.7) on this vector, we obtain

$$0 = (g_F, h)_{\mathcal{H}} = \int_{\mathbb{R}} F(s)(U(s)g, h)_{\mathcal{H}}\, ds, \quad \forall F \in C_0^\infty(\mathbb{R}).$$

In view of the arbitrariness of F in this equality and the continuity of the function $\mathbb{R} \ni s \mapsto \big(U(s)g, h\big)_{\mathcal{H}} \in \mathbb{C}$ (which follows from our assumptions), we conclude that $\forall s \in \mathbb{R} : (U(s)g, h)_{\mathcal{H}} = 0,\ g \in \mathcal{H}$. In particular, by setting $s = 0$ and using the equality $U(0) = \mathbb{I}$ (a consequence of (2.5.2)), we obtain $(g, h)_{\mathcal{H}} = 0, \forall g \in \mathcal{H}$ $\Rightarrow h = 0$. Therefore, $\mathfrak{D}$ is dense in $\mathcal{H}$.

To prove that the function (2.5.6) is strongly continuously differentiable, it suffices to consider f of the form (2.5.7). By using (2.5.2) and changing variables in the integral, we get

$$
\begin{aligned}
\frac{1}{h}\left(U(t+h)g_F - U(t)g_F\right) &= \frac{1}{h}\int_{\mathbb{R}} F(s)\left(U(t+h)U(s) - U(t)U(s)\right) g\, ds \\
&= \frac{1}{h}\int_{\mathbb{R}} F(s)\left(U(t+h+s) - U(t+s)\right) g\, ds \\
&= \int_{\mathbb{R}} \frac{1}{h}\left(F(s-t-h) - F(s-t)\right) U(s) g\, ds \\
&\underset{h\to 0}{\to} -\int_{\mathbb{R}} F'(s-t)U(s)g\, ds = g_{-F'(\cdot - t)} \\
&= U'(t)g_F, \quad \forall t, h \in \mathbb{R}.
\end{aligned}
\tag{2.5.8}
$$

The limit transition in (2.5.8) can be easily justified by the following simple estimate (established by limit transition from integral sums)

$$
\left\| \int_{\mathbb{R}} G(s)U(s)g\, ds \right\|_{\mathcal{H}} \le \int_{\mathbb{R}} \|U(s)g\|_{\mathcal{H}}\, |G(s)|\, ds, \quad G \in C_0(\mathbb{R}, \ g \in \mathcal{H}.
\tag{2.5.9}
$$

As follows, from(2.5.9), the vector function on the right-hand side of (2.5.8) is strongly continuous. Note that $U'(t)g_F$ has the form (2.5.7) for all $t \in \mathbb{R}$.

On $\mathfrak{D}$, we introduce the operator A in the space $\mathcal{H}$ by the formula

$$
\mathfrak{D} = \mathfrak{D}(A) \ni f \mapsto Af = \frac{1}{i}U'(0)f \in \mathfrak{D} \subseteq \mathcal{H}.
\tag{2.5.10}
$$

Note that $\mathfrak{D}$ is invariant under the action of A. By applying the operator $U(t)$ to (2.5.7), using (2.5.2), and changing the variables in the integral, we establish that $\mathfrak{D}$ is also invariant under the action of $U(t)$ for all $t \in \mathbb{R}$.

The introduced above operator A is Hermitian. To prove this, it suffices to show that $(Ag_F, h_G)_{\mathcal{H}} = (g_F, Ah_G)_{\mathcal{H}}$, where $F, G \in C_0^\infty(\mathbb{R})$ and $g, h \in \mathcal{H}$. According to (2.5.8), we have

$$
\begin{aligned}
(Ag_F, h_G)_{\mathcal{H}} &= \lim_{h\to 0} \big(\frac{1}{ih}(U(h) - \mathbb{I})g_F, h_G\big)_{\mathcal{H}} \\
&= \lim_{h\to 0} \big(g_F, -\frac{1}{ih}(U(-h) - \mathbb{I})h_G\big)_{\mathcal{H}} = (g_F, Ah_G)_{\mathcal{H}}.
\end{aligned}
$$

Furthermore, we can prove that A is essentially self-adjoint. Let $z \in \mathbb{C} \setminus \mathbb{R}$ and $\varphi \in \mathfrak{D}(A^*)$ be such that $A^*\varphi = z\varphi$; it is necessary to prove that $\varphi = 0$. To do this, we first show that

$$
U'(t)g_F = iAU(t)g_F \quad F \in C_0^\infty(\mathbb{R}, \quad g \in \mathcal{H},\ t \in \mathbb{R}. \tag{2.5.11}
$$

This equality will be proved if we notice that, in this case, $U(t)g_F$ also has the form (2.5.7) (this vector is equal to $g_{F(\cdot - t)}$) and then compute $AU(t)g_F$ according to (2.5.10).

In view of (2.5.11), we obtain, for any $F \in C_0^\infty(\mathbb{R})$ and $g \in \mathcal{H}$

$$
\begin{aligned}
\frac{d}{dt}(U(t)g_F, \varphi)_{\mathcal{H}} &= (U'(t)g_F, \varphi)_{\mathcal{H}} = i(AU(t)g_F, \varphi)_{\mathcal{H}} \\
&= i(U(t)g_F, A^*\varphi)_{\mathcal{H}} = i\bar{z}(U(t)g_F, \varphi)_{\mathcal{H}}, \quad t \in \mathbb{R}.
\end{aligned}
$$

Hence, the complex-valued bounded function $a(t) = (U(t)g_F, \varphi)_{\mathcal{H}}$ satisfies the equation $a' = i\bar{z}a$. Therefore, $a(t) = e^{i\bar{z}t}a(0)$, $t \in \mathbb{R}$, and, by virtue of the condition $\operatorname{Im} z \neq 0$, it may be bounded only if $a(0) = 0$. Hence $(g_F, \varphi)_{\mathcal{H}} = a(0) = 0$, $F \in C_0^\infty(\mathbb{R}, g \in \mathcal{H})$, i.e., $\varphi \perp \mathfrak{D} \Rightarrow \varphi = 0$.

Thus, we have proved that the operator $\tilde{A}$ is self-adjoint. In terms of this operator, according to formula (2.5.1), we can now construct the operator-valued function $\mathbb{R} \ni t \mapsto V(t) = e^{it\tilde{A}} \in \mathcal{L}(H)$. It remains to show that $V(t) = U(t)$, $t \in \mathbb{R}$.

According to (2.5.4) and (2.5.5), $v(t) = V(t)v_0$, $v_0 \in \mathfrak{D}(\tilde{A})$ is a strong solution of the Cauchy problem

$$
v'(t) = i\tilde{A}v(t), \quad t \in \mathbb{R},\ v(0) = v_0. \tag{2.5.12}
$$

We set $v_0 = g_F$ for some $F \in C_0^\infty(\mathbb{R})$ and $g \in \mathcal{H}$. In view of (2.5.11), the function $u(t) = U(t)g_F$ is also a strong solution of the problem (2.5.12) (in (2.5.11)), A can be replaced with $\tilde{A}$, since $U(t)g_F \in \mathfrak{D}$). Further in Theorem 2.6.1, it will be shown that strong solutions of the Eq. (2.5.12) with self-adjoint $\tilde{A}$ that correspond to the same initial data coincide. Therefore, $V(t)g_F = U(t)g_F$ and, by virtue of the denseness of $\mathfrak{D}$ in $\mathcal{H}$, we get $V(t) = U(t)$, $t \in \mathbb{R}$. (It is clear that this coincidence can be proved even without using the general Theorem 2.6.1)

In order to prove that E is uniquely determined by(2.5.1), we note that

$$(U(t)f, g)_{\mathcal{H}} = \int_{\mathbb{R}} e^{it\lambda}\, d(E(\lambda)f, g)_{\mathcal{H}}, \quad t \in \mathbb{R}, \quad \forall f, g \in \mathcal{H}, \tag{2.5.13}$$

i.e., the function on the left-hand side of (2.5.13) is the Fourier-Stieltjes transform of the charge $\mathfrak{B}(\mathbb{R}) \mapsto (E(\alpha)f, g)_{\mathcal{H}} \in \mathbb{C}$, and it is well known that a charge is uniquely determined by its Fourier transform. This proves that E is uniquely determined. □

We remark that the strong continuity of a one-parameter unitary group is equivalent to its weak continuity, i.e., to the continuity of the function $\mathbb{R} \ni t \mapsto (U(t)f, g)_{\mathcal{H}} \in \mathbb{C}$ for all $f, g \in \mathcal{H}$.

Indeed, it suffices to prove that weak continuity implies strong continuity. For $f \in \mathcal{H}$ we have

$$\begin{aligned}\|U(t)f - U(s)f\|_{\mathcal{H}}^2 &= ((U(t) - U(s))f, (U(t) - U(s))f)_{\mathcal{H}} \\ &= ((U(t) - U(s))^*(U(t) - U(s))f, f)_{\mathcal{H}} \\ &= ((2 \cdot \mathbb{I} - U(t-s) - U(s-t))f, f)_{\mathcal{H}} \underset{s \to t}{\to} 0.\end{aligned}$$

Using the last remark, Theorem 2.5.1 can be reformulated.

A function $\mathbb{R} \ni t \mapsto k(t) \in \mathbb{C}$ is called *positive definite* if it satisfies the inequality

$$\sum_{j,l=1}^{n} k(t_j - t_l)\xi_j \bar{\xi}_l \geq 0 \tag{2.5.14}$$

for all $t_1, \dots, t_n \in \mathbb{R},\ \xi_1, \dots, \xi_n \in \mathbb{C}$, and $n \in \mathbb{N}$.

The well-known Bochner-Khinchin theorem (see [Shi1]) asserts that a continuous positive definite function admits a representation

$$k(t) = \int_{\mathbb{R}} e^{i\lambda t}\, d\sigma(\lambda), \quad t \in \mathbb{R}, \tag{2.5.15}$$

where $\mathfrak{B}(\mathbb{R}) \ni \alpha \mapsto \sigma(\alpha) \geq 0$ is a finite measure uniquely defined in terms of this function. Conversely, each function of the form (2.5.15) is positive definite. By comparing

$$(U(t)f, g)_{\mathcal{H}} = \int_{\mathbb{R}} e^{it\lambda}\, d(E(\lambda)f, g)_{\mathcal{H}}, \quad t \in \mathbb{R},\ \forall f, g \in \mathcal{H}$$

with (2.5.15), we conclude that the function

$$\mathbb{R} \ni t \mapsto k(t) = (U(t)f, f)_{\mathcal{H}}, \quad f \in \mathcal{H} \tag{2.5.16}$$

is positive definite.

2.6 Self-Adjointness Criteria

By the study of the operator differential Eq. (2.5.4), we have already shown that relations (2.5.5) and (2.5.1) give a solution of the corresponding Cauchy problem. Here, we somewhat generalize these arguments.

Consider a densely defined operator B acting in a Hilbert space $\mathcal{H}$ and let $I \subseteq \mathbb{R}$ be a finite or infinite closed, open, or half-open interval and $r \in \mathbb{N}$ (in fact, we restrict ourselves to the cases where $r = 1, 2$).

A vector-function $I \ni t \mapsto u(t) \in \mathcal{H}$ is called a strong solution of the equation

$$\left(\frac{d^r u}{dt^r}\right)(t) + Bu(t) = 0 \quad t \in I \tag{2.6.1}$$

on I, if it is r times strongly continuously differentiable (i.e., has r strong derivatives on I, the last of which is continuous), $u(t) \in \mathfrak{D}(B)$ for all $t \in I$ and it satisfies the Eq. (2.6.1).

An r times strongly continuously differentiable vector function $I \ni t \mapsto u(t) \in \mathcal{H}$ is a strong solution of the equation

$$\left(\frac{d^r u}{dt^r}\right)(t) + B^* u(t) = 0, \quad t \in I \tag{2.6.2}$$

if and only if the "weak" equality

$$\left(\left(\frac{d^r u}{dt^r}\right)(t), f\right)_{\mathcal{H}} + \big(u(t), Bf\big)_{\mathcal{H}} = 0, \quad f \in \mathfrak{D}(B),\ t \in I \tag{2.6.3}$$

hods true. This statement immediately follows from the definition of adjoint operator, since (2.6.3) implies that $u(t) \in \mathfrak{D}(B^*)$ for all $t \in I$ in view of inclusion $u^{(r)}(t) \in \mathcal{H}$.

We say that *the Cauchy problem for Eq.* (2.6.1) *on* $I = [0, b)$, $(0 < b \leq \infty)$ *is uniquely solvable in the strong sense if each strong solution of this equation on* $[0, b)$ *such that* $u(0) = \cdots = u^{(r-1)}(0) = 0$ *vanishes for all* $t \in (0, b)$.

If the Cauchy problem is uniquely solvable on $[0, b)$ for some $b > 0$, then it is uniquely solvable on $[0, \infty)$.

Indeed, let $[0, \infty) \ni t \mapsto u(t) \in \mathcal{H}$ be a strong solution of Eq. (2.6.1) on $[0, \infty)$ such that $u(0) = \cdots = u^{(r-1)}(0) = 0$. In view of the assumed unique solvability on

$[0, b)$, we have $u(t) = 0$ for $t \in (0, b)$ and, in particular, $u(t) = 0$ in a neighborhood of the point $c = b/2$; therefore $u(c) = \cdots = u^{(r-1)}(c) = 0$. The function $[0, \infty) \ni t \mapsto u_1(t) = u(t + c)$ is a strong solution of (2.6.1) on $[0, \infty)$ such that $u_1(0) = u(c) = 0, \ldots, u_1^{r-1}(0) = u^{r-1}(c) = 0$ and, hence, $u_1(t) = 0$ for $t \in (0, b)$. By repeating the same reasoning, we can show that the function

$$[0, \infty) \ni t \mapsto u_2(t) = u_1(t + c) = u(t + 2c)$$

also vanishes for $t \in (0, b)$. Then we construct the function $u_3(t)$, etc. As a final result, we conclude that $u(t) = 0$ $(t \in [0, \infty))$. □

If the operator B in (2.6.1) has the form $B = \zeta A$, where $\zeta \in \mathbb{C}$ is a fixed number and A is a self-adjoint operator, then, under the corresponding restriction imposed on the initial conditions, the Cauchy problem for (2.6.1) is solvable, and one can write a representation of this solution in terms of the resolution of the identity E for the operator A.

Let us show that the self-adjointness of operators is closely related to the uniqueness of strong solutions of the Cauchy problems for the corresponding evolutionary equations.

The Schrödinger Criterion of Self-Adjointness The Schrödinger criterion of self-adjointness is formulated in the theorem.

Theorem 2.6.1 *Let A be an Hermitian operator acting in the separable Hilbert space $\mathcal{H}$. For its essential self-adjointness, it is necessary that the Cauchy problems for both equations*

$$\left(\frac{du}{dt}\right)(t) \pm (iA^*)u(t) = 0, \quad t \in [0, b) \tag{2.6.4}$$

be strongly uniquely solvable on $[0, b)$ for all $b \in (0, \infty]$ and it is sufficient that these problems be uniquely solvable for some b.

Proof The proof is split into several steps.

I. Let us first establish sufficiency under the assumption that A has equal defect numbers. Assume the contrary, namely, let $\tilde{A}$ be not self-adjoint. Then A has two different self-adjoint extensions A_1 and A_2 in $\mathcal{H}$. Let E_1 and E_2 be corresponding resolutions of the identity. For every $g \in \mathfrak{D}(A) \subseteq \mathfrak{D}(A_1)$ the integral $\int_{\mathbb{R}} \lambda^2 \, d\big(E_1(\lambda)g, g\big)_{\mathcal{H}}$ is convergent. Therefore, the vector-function

$$[0, \infty) \ni t \mapsto u_1(t) = \int_{\mathbb{R}} e^{i\lambda t} \, dE_1(\lambda)g \tag{2.6.5}$$

is strongly continuously differentiable and

$$u_1'(t) = i \int_{\mathbb{R}} \lambda e^{i\lambda t}\, dE_1(\lambda)g.$$

It is easy to see that it is a strong solution of Eq. (2.6.4) with the sign "+" on $[0, \infty)$. Indeed, it is necessary to check the corresponding weak equality (2.6.3), which now has the form

$$\left(\left(\frac{du_1}{dt}\right)(t), f\right)_{\mathcal{H}} + (u_1(t), iAf)_{\mathcal{H}} = 0, \quad f \in \mathfrak{D}(A), \quad t \in [0, \infty)).$$

Since

$$\begin{aligned} d(E_1(\lambda)g, Af)_{\mathcal{H}} &= d(E_1(\lambda)g, A_1 f)_{\mathcal{H}} \\ &= d\Bigg(\int_{-\infty}^{\lambda} \mu\, d(E_1(\mu)g, f)_{\mathcal{H}}\Bigg) = \lambda d(E_1(\lambda)g, f)_{\mathcal{H}}, \end{aligned}$$

then we have

$$\begin{aligned} &\left(\left(\frac{du_1}{dt}\right)(t), f\right)_{\mathcal{H}} + (u_1(t), iAf)_{\mathcal{H}} \\ &\qquad = i \int_{\mathbb{R}} \lambda e^{i\lambda t}\, d(E_1(\lambda)g, f)_{\mathcal{H}} - i \int_{\mathbb{R}} e^{i\lambda t} d(E_1(\lambda)g, Af)_{\mathcal{H}} = 0, \\ &\qquad f \in \mathfrak{D}(A);\ t \in [0, \infty), \end{aligned}$$

i.e., the required relation is satisfied.

□

Similarly, the function $u_2(t)$ constructed according to (2.6.5) for given E_2 is a strong solution of the same equation; $u_1(0) = u_2(0) = g$. Thus, $u(t) = u_1(t) - u_2(t)$ is also a strong solution of the Eq. (2.6.4) with the sign "+" on $[0, \infty)$ such that $u(0) = 0$. By virtue of the condition of the theorem, the problem is uniquely solvable on $[0, \infty)$. Therefore, $u(t) = 0$ for $t \in [0, \infty)$, whence

$$\int_{\mathbb{R}} e^{i\lambda t}\, d((E_1(\lambda) - E_2(\lambda))g, h)_{\mathcal{H}} = 0, \quad g \in \mathfrak{D}(A),\ h \in \mathcal{H},\ t \in [0, \infty). \tag{2.6.6}$$

Consider the Eq. (2.6.4) with the sign "−". By repeating the arguments presented above with $e^{i\lambda t}$ replaced with $e^{-i\lambda t}$ in (2.6.5), we arrive at the relation that differs from (2.6.6) by the same change. Therefore, if we introduce the charge

$$\omega(\alpha) = ((E_1(\alpha) - E_2(\alpha))g, h)_{\mathcal{H}}, \quad \alpha \in \mathfrak{B}(\mathbb{R}),$$

then, according to (2.6.6) and the indicated modification of this relation, we have $\int_{\mathbb{R}} e^{i\lambda t}\, d\omega(\lambda) = 0$, for any $t \in \mathbb{R}$. Taking into account the theorem on uniqueness of the Fourier-Stieltjes transform of a charge already applied, we conclude that $\omega = 0$, i.e., $((E_1(\alpha) - E_2(\alpha))g, h)_{\mathcal{H}} = 0$, $\alpha \in \mathfrak{B}(\mathbb{R})$. Since $g \in \mathfrak{D}(A)$ and $h \in \mathcal{H}$ are arbitrary, this implies that $E_1 = E_2$, and we arrive at a contradiction.

II. In the case of an operator A with distinct defect numbers, we use the lemma.

Lemma 2.6.2 *Consider the orthogonal sum of spaces $\mathcal{H} \oplus \mathcal{H}$ with vectors $f = \langle f_1, f_2\rangle$, $f_1, f_2 \in \mathcal{H}$ and an operator C with a dense domain $\mathfrak{D}(C) = \mathfrak{D}(A) \oplus \mathfrak{D}(A)$, acting in this space according to the formula $Cf = \langle Af_1, -Af_2\rangle$, $f \in \mathfrak{D}(C)$, and the equality*

$$\left(\frac{du}{dt}\right)(t) + (iC)^* u(t) = 0, \quad t \in [0, b), \quad b \in (0, \infty], \tag{2.6.7}$$

for the vector functions with values in $\mathcal{H} \oplus \mathcal{H}$.

It is stated that if the Cauchy problem for both Eq. (2.6.4) *is uniquely solvable in the strong sense on* $[0, b)$, *then the Cauchy problem for Eq.* (2.6.7) *is also uniquely solvable in the sense of strong solutions, and vice versa.*

Proof Let

$$[0, b) \ni t \mapsto u(t) = \langle u_1(t), u_2(t)\rangle \in \mathcal{H} \oplus \mathcal{H}$$

be a strong solution of the Cauchy problem for Eq. (2.6.7). Since

$$C^* f = \langle A^* f_1, -A^* f_2\rangle, \quad f \in \mathfrak{D}(C^*) = \mathfrak{D}(A^*) \oplus \mathfrak{D}(A^*),$$

then functions

$$[0, b) \ni t \mapsto u_1(t) \in \mathcal{H}, \text{ and } [0, b) \ni t \mapsto u_2(t) \in \mathcal{H}$$

are strong solutions of the Eq. (2.6.4) with signs "+" and "−", respectively. In view of the fact that, by assumption, strong solutions of the Cauchy problem for (2.6.4) are unique on $[0, b)$, this implies the required uniqueness for (2.6.7). The converse statement is deduced similarly. □

III. Let us prove the sufficiency for an operator A with deficiency index (m, n). As in Lemma 2.6.2, we construct the operator C. It is easy to verify that the deficiency index of this operator is equal to $(m+n, m+n)$. By virtue 2.6.2 the Cauchy problem for (2.6.7) is uniquely solvable on $[0, b)$. By applying this lemma to the case where A is replaced with $-A$, we conclude that this uniqueness is preserved for the Eq. (2.6.7), in which "+" is replaced with "−". In view of the fact that the defect numbers of the operator C coincide and are equal to $m+n$, the reasoning used in step I is applicable in this case, and we conclude that C is essentially self-adjoint. But then $m+n=0$, whence $m=n=0$, i.e., A is also essentially self-adjoint.

IV. To prove necessity, we first establish a general lemma that, in our case, reflects the Holmgren principle in the theory of partial differential equations.

Lemma 2.6.3 *Consider Eq.* (2.6.2) *on* $[0, b)$, $b \in (0, \infty)$. *Assume that there exists a set* Φ *dense in* $\mathcal{H}$ *such that the Cauchy problem*

$$\begin{gathered}\left(\frac{d^r\varphi}{dt^r}\right)(t)+(-1)^r B\varphi(t)=0, \quad t\in[0,T];\\ \varphi(T)=\varphi_0,\ \ldots,\ \varphi^{(r-1)}(T)=\varphi_{r-1}\end{gathered} \tag{2.6.8}$$

has a strong solution for all $T \in (0, b)$ *and* $\varphi_0, \ldots \varphi_{r-1} \in \Phi$.

Then the Cauchy problem for (2.6.2) *is uniquely solvable on* $[0, b)$ *in the sense of strong solutions.*

Proof Let us prove Lemma 2.6.2, e.g., in the case of $r=2$. One can easily verify the following formula of integration by parts. Let $[0, T] \ni t \mapsto \alpha(t), \beta(t) \in \mathcal{H}$ be twice strongly continuously differentiable vector functions. Then

$$\begin{aligned}\int_0^T (\alpha''(t), \beta(t))_{\mathcal{H}}\, dt &= \int_0^T (\alpha(t), \beta''(t))_{\mathcal{H}}\, dt\\ &\quad + [(\alpha'(t), \beta(t))_{\mathcal{H}} - (\alpha(t), \beta'(t))_{\mathcal{H}}]\Big|_0^T.\end{aligned} \tag{2.6.9}$$

Let $u(t)$ be a strong solution of the Cauchy problem for the Eq. (2.6.2) with $r=2$ on $[0, b)$ such that $u((0)=u'(0)=0$ and let $\varphi(t)$ be a strong solution mentioned in the formulation of the lemma. By using (2.6.9), we obtain

$$\int_0^T \left((u''(t), \varphi(t))_{\mathcal{H}} - (u(t), \varphi''(t))_{\mathcal{H}}\right) dt = (u'(T), \varphi_0 r)_{\mathcal{H}} - (u(T), \varphi_1)_{\mathcal{H}}. \tag{2.6.10}$$

Note that $\varphi(s) \in \mathfrak{D}(B)$ for every $s \in [0, T]$. Therefore, according to equality (2.6.3) with $f=\varphi(s)$, we can write

$$(u''(t), \varphi(s))_{\mathcal{H}} + (u(t), B\varphi(s))_{\mathcal{H}} = 0, \quad t \in [0, b).$$

We set now $t = s$ and then replace s with t. This yields

$$(u''(t), \varphi(t))_{\mathcal{H}} = -(u(t), B\varphi(t))_{\mathcal{H}}, \quad t \in [0, T].$$

By virtue of (2.6.8) with $r = 2$, we have

$$(u(t), \varphi''(t))_{\mathcal{H}} = -(u(t), B\varphi(t))_{\mathcal{H}}, \quad t \in [0, T].$$

These two equalities imply that the expression on the left-hand side of (2.6.10) vanishes. Consequently,

$$(u'(T), \varphi_0)_{\mathcal{H}} - (u(T), \varphi_1)_{\mathcal{H}} = 0, \quad \varphi_0, \varphi_1 \in \Phi.$$

Hence, it follows from the denseness of Φ in $\mathcal{H}$ that $u(T) = u'(T) = 0$. Since $T \in (0, b)$ is arbitrary, this yields the required assertion.

In the case where $r = 1$ the reasoning is similar, one should only use the following formula for integration by parts:

$$\int_0^T (\alpha'(t), \beta(t))_{\mathcal{H}} dt = -\int_0^T (\alpha(t), \beta'(t))_{\mathcal{H}}\, dt + [(\alpha(t), \beta(t))_{\mathcal{H}}]\Big|_0^T, \tag{2.6.11}$$

which holds true for continuously differentiable vector functions $[0, T), \ni t \mapsto \alpha(t), \beta(t) \in \mathcal{H}$. In the case where r is arbitrary, one must iterate relation (2.6.11) r times (note that relation (2.6.10) is, in fact, the relation (2.6.11), iterated twice). □

V. Let us prove necessity. Let $\tilde{A}$ be self-adjoint and let E be its resolution of the identity. We apply Lemma 2.6.3, setting $r = 1$, $B = (iA)^* = -i\tilde{A}$, $\Phi = \cup_{n=1}^{\infty} E((-n, n))H$. A strong solution of the Cauchy problem (2.6.8), which now has the form

$$\varphi'(t) + i\tilde{A}\varphi(t) = 0, \quad t \in [0, T]), \ \varphi(T) = \varphi_0,$$

exists and is equal to

$$\varphi(t) = \int_{\mathbb{R}} e^{-i\lambda(t-T)}\, dE(\lambda)\varphi_0, \quad t \in [0, T], \tag{2.6.12}$$

(since $\varphi_0 \in \Phi$, the integration in (2.6.12) is, in fact, carried out over a finite interval and, therefore, the function $[0, T] \ni t \mapsto \varphi(t)$ is continuously differentiable; it is clear that it solves the problem under consideration). Thus, by virtue of this lemma, the Eq. (2.6.4) with the sign "+" is uniquely solvable on $[0, b)$. The Eq. (2.6.4) with sign "−" is investigated similarly. Finally, we conclude that $B = -(iA)^* = i\tilde{A}$. □

The Hyperbolic Criterion of Self-Adjointness The "hyperbolic" criterion of self-adjointness is formulated in the form of two theorems.

Theorem 2.6.4 *Let A be an Hermitian operator acting in $\mathcal{H}$. For its essential self-adjointness, it is necessary that the Cauchy problem for the equation*

$$\left(\frac{d^2u}{dt^2}\right)(t) + A^*u(t) = 0, \quad t \in [0, b) \tag{2.6.13}$$

be uniquely solvable on $[0, b)$ *for all* $b \in (0, \infty]$ *(in the sense of strong solutions) and it is sufficient that A be semi-bounded below and that the indicated Cauchy problem be uniquely solvable in the same sense for some* $b > 0$.

Proof

Sufficiency. Suppose that $\tilde{A}$ is not self-adjoint. Then A has two different self-adjoint extensions A_1 and A_2 in $\mathcal{H}$ bounded below by a number $c > -\infty$ Let E_1 and E_2 be the corresponding resolutions of the identity. For every $g \in \mathfrak{D}(A) \subseteq \mathfrak{D}(A_1)$, the integral $\int_{\mathbb{R}} \lambda^2\, d(E_1(\lambda)g, g)_{\mathcal{H}}$ is convergent and, therefore, the vector function

$$[0, \infty) \ni t \mapsto u_1(t) = \int_c^\infty \cos(\sqrt{\lambda}t)\, dE_1(\lambda)g \tag{2.6.14}$$

is twice strongly continuously differentiable. As in the proof of Theorem 2.6.1, one can easily show that it is a strong solution of the Eq. (2.6.13) on $[0, \infty)$. For this purpose, one must check the validity of the corresponding weak equality of the form (2.6.3). In addition, we have, $u_1(0) = g$ and $u_1'(0) = 0$. Similarly, by changing E_1 by E_2 in (2.6.14), we construct the function $u_2(t)$. The difference $u(t) = u_1(t) - u_2(t)$ is also a strong solution of Eq. (2.6.13) on $[0, \infty)$ such that $u(0) = u'(0) = 0$. In view of the assumed uniqueness of strong solutions of the Cauchy problem, $u(t) = 0$ for $t \geq 0$. Multiplying this equality by $h \in \mathcal{H}$, we obtain

$$\int_c^\infty \cos(\sqrt{\lambda}t)\, d((E_1(\lambda) - E_2(\lambda))g, h)_{\mathcal{H}} = 0, \quad t \in [0, \infty).$$

Since the charge ω is uniquely determined in terms of its cosine Fourier-Stieltjes transform, then this makes it possible to assert $E_1 = E_2$, which leads to a contradiction.

Necessity. Let $\tilde{A}$ be self-adjoint and let E be its resolution of the identity. Let us apply Lemma 2.6.3. setting $r = 2$, $B = A^* = \tilde{A}$, and $\Phi = \cup_{n=1}^{\infty} E((-n, n))\mathcal{H}$. A strong solution of the Cauchy problem (2.6.8), which now has the form

$$\varphi''(t) + \tilde{A}\varphi(t) = 0, \quad t \in [0, T]),\ \varphi(T) = \varphi_0,\ \varphi'(T) = \varphi_1,$$

exists and is equal to

$$\varphi(t) = \int_{\mathbb{R}} \cos(\sqrt{\lambda}(t-T))\, dE(\lambda)\varphi_0 + \int_{\mathbb{R}} 1/\sqrt{\lambda}\sin(\sqrt{\lambda}(t-T))\, dE(\lambda)\varphi_1,$$

where, as in (2.6.12), the integration is, in fact, carried out over a finite segment. Therefore, according to Lemma 2.6.3, we conclude that the Cauchy problem for (2.6.13) is uniquely solvable on $[0, b)$ for all $b \in (0, \infty]$ in the sense of strong solutions.

□

It is convenient to use this theorem in a simple combination with Lemma 2.6.3.

Theorem 2.6.5 *Let A be an Hermitian operator acting in $\mathcal{H}$ and semi-bounded below. Assume that there exists a linear set $\Phi \subseteq \mathcal{H}$ dense in $\mathcal{H}$ and such that the Cauchy problem*

$$\left(\frac{d^2\varphi}{dt^2}\right)(t) + A\varphi(t) = 0, \quad t \in [0, T],\ \varphi(T) = \varphi_0,\ \varphi'(T) = \varphi_1 \tag{2.6.15}$$

has a strong solution for some $b > 0$ and all $T \in (0, b)$ and $\varphi_0, \varphi_1 \in \Phi$. Then the operator A is essentially self-adjoint.

Proof By virtue of Lemma 2.6.3, it follows from the condition of the theorem that the Cauchy problem for the Eq. (2.6.13) has a unique strong solution on $[0, b)$. But then, according to Theorem 2.6.4, the operator $\tilde{A}$ is self-adjoint. □

The Parabolic Criterion of Self-Adjointness

Theorem 2.6.6 *Let A be an Hermitian operator acting in $\mathcal{H}$. For its essential self-adjointness, it is necessary that the Cauchy problem for the equation*

$$\left(\frac{du}{dt}\right)(t) + A^*u(t) = 0, \quad t \in [0, \infty) \tag{2.6.16}$$

be uniquely solvable in the sense of strong solutions. For an operator semi-bounded below, this is also a sufficient condition.

Proof

Necessity. As in Theorems 2.6.1 and 2.6.4, it is proved by using Lemma 2.6.3 with $r = 1$, $B = \tilde{A}$, and $\Phi = \cup_{n=1}^{\infty} E((-n, n))\mathcal{H}$, where E is the resolution of the identity of $\tilde{A}$. A strong solution of the corresponding Cauchy problem which now has the form

$$\varphi'(t) - \tilde{A}\varphi(t) = 0, \quad t \in [0, T],\ \varphi(T) = \varphi_0 \in \Phi,$$

exists and is equal to

$$\varphi(t) = \int_{\mathbb{R}} e^{\lambda(t-T)}\, dE(\lambda)\varphi_0, \quad t \in [0, T].$$

Sufficiency. It is established as in Theorems 2.6.1 and 2.6.4. Suppose that $\tilde{A}$ is not self-adjoint. Let A_1 and A_2 be two different self-adjoint extensions of A bounded below by a number $c > -\infty$ and let E_1 and E_2 be the corresponding resolutions of the identity. The vector function

$$[0, \infty) \ni t \mapsto u_1(t) = \int_c^\infty e^{-\lambda t}\, dE_1(\lambda)g, \quad g \in \mathfrak{D}(A) \subseteq \mathfrak{D}(A_1) \tag{2.6.17}$$

is strongly continuously differentiable and $u_1(t) \in \mathfrak{D}(A_1) \subseteq \mathfrak{D}(A^*)$. The derivative $u_1'(t)$ is expressed by integral (2.6.17) with the factor $-\lambda$ before $e^{-\lambda t}$. The expression $A^*u_1(t) = A_1u_1(t)$ also has the same form. Thus, (2.6.17) is a strong solution of Eq. (2.6.16) with $u_1(0) = g$. Further, by the same procedure, we construct $u_2(t)$ in terms of E_2 and consider the difference $u(t) = u_1(t) - u_2(t)$. For this difference, we have $u(0) = 0$ and, therefore, in view of the assumed uniqueness of strong solutions, $u(t) = 0$, whence

$$\int_c^\infty e^{-\lambda t}\, d((E_1(\lambda) - E_2(\lambda))g, h)_{\mathcal{H}} = 0, \quad t \in [0, \infty). \tag{2.6.18}$$

This relation means that the Laplace-Stieltjes transform of the charge appearing in (2.6.18) is equal to zero, but then the charge is also identically equal to zero. This gives the equality $E_1 = E_2$, which leads to a contradiction.

□

The Quasi-Analytic Criterion of Self-Adjointness First, we recall some facts from the theory of quasi-analytic functions. Let $[a, b] \subset \mathbb{R}$ be a finite segment and let $(m_n)_{n=1}^\infty$ be a fixed sequence of positive numbers.

The class $C\{m_n\}$ is defined as the linear set of all functions $f \in C^\infty([a, b])$,

$$|(D^n f)(t)| \leq K_f^n m_n, \quad t \in [a, b],\ n \in \mathbb{N}, \tag{2.6.19}$$

where K_f is a constant that depends on f.

As is known, the class of analytic functions defined on $[a, b]$ is characterized by the estimates (2.6.19) with $m_n = n!$. It is clear that the class $C\{n!\}$ is characterized by the following property. If $f \in C\{n!\}$ is such that $(D^n f)(t_0) = 0$ for all $n \in \mathbb{N}$ and $f(t_0) = 0$ at a fixed point $t_0 \in [a, b]$, then $f(t) = 0$ for $t \in [a, b]$. In order to generalize this situation, we introduce the following definition.

The class $C\{m_n\}$ is called quasi-analytic if the fact that a function $f \in C\{m_n\}$ satisfies the equalities $(D^n f)(t_0) = 0$, $n \in \mathbb{N}$ and $f(t_0) = 0$ at a fixed point $t_0 \in [a, b]$ implies $f(t) = 0$, $t \in [a, b]$.

We will recall the well-known Denjoy-Càrleman theorem. *The class $C\{m_n\}$ is quasi-analytic if and only if*

$$\sum_{n=1}^{\infty} \left(\inf \left\{ m_k^{1/k} \mid k \geq n \right\} \right)^{-1} = \infty. \tag{2.6.20}$$

For example, the class $C\{n^{pn}\}$ is quasi-analytic if and only if $p \leq 1$.

Let $\mathcal{H}$ be a Hilbert space and let A be an Hermitian operator in it. *A vector $\varphi \in \mathcal{H}$ is called quasi-analytic (with respect to A), if $\varphi \in \cap_{n=1}^{\infty} \mathfrak{D}(A^n)$ and the class $C\{\|A^n \varphi\|_{\mathcal{H}}\}$ is quasi-analytic.*

Lemma 2.6.7 *A vector $\varphi \in \cap_{n=1}^{\infty} \mathfrak{D}(A^n)$ is quasi-analytic if and only if*

$$\sum_{n=1}^{\infty} \|A^n \varphi\|_{\mathcal{H}}^{-1/n} = \infty. \tag{2.6.21}$$

Proof It is clear that $C\{\|A^n \varphi\|_{\mathcal{H}}\} = C\{\|A^n(\lambda\varphi)\|_{\mathcal{H}}\}$, where $\lambda > 0$ is fixed. This implies that it suffices to verify the lemma for a vector φ such that $\|\varphi\|_{\mathcal{H}} = 1$. For a vector of this sort, the sequence

$$\left(\|A^n \varphi\|_{\mathcal{H}}^{1/n} \right)_{n=1}^{\infty} \tag{2.6.22}$$

is non-decreasing. Indeed,

$$\|A\varphi\|_{\mathcal{H}}^2 = (A\varphi, A\varphi)_{\mathcal{H}} = (A^2\varphi, \varphi)_{\mathcal{H}} \leq \|A^2\varphi\|_{\mathcal{H}} \|\varphi\|_{\mathcal{H}},$$

i.e., $\|A\varphi\|_{\mathcal{H}} \leq \|A^2\varphi\|_{\mathcal{H}}^{1/2}$.

Assume that the inequality $\|A^n\varphi\|_{\mathcal{H}}^{1/n} \leq \|A^{n+1}\varphi\|_{\mathcal{H}}^{1/(n+1)}$ is already proved and prove that $\|A^{n+1}\varphi\|_{\mathcal{H}}^{1/(n+1)} \leq \|A^{n+2}\varphi\|_{\mathcal{H}}^{1/(n+2)}$, $n \in \mathbb{N}$. In view of the assumed inequality, we get

$$\begin{aligned} \|A^{n+1}\varphi\|_{\mathcal{H}}^2 &= (A^{n+1}\varphi, A^{n+1}\varphi)_{\mathcal{H}} = (A^{n+2}\varphi, A^n\varphi)_{\mathcal{H}} \\ &\leq \|A^{n+2}\varphi\|_{\mathcal{H}} \, \|A^n\varphi\|_{\mathcal{H}} \leq \|A^{n+2}\varphi\|_{\mathcal{H}} \, \|A^{n+1}\varphi\|_{\mathcal{H}}^{n/(n+1)}, \end{aligned}$$

whence $\|A^{n+1}\varphi\|_{\mathcal{H}}^{1+1/(n+1)} \leq \|A^{n+2}\varphi\|_{\mathcal{H}}$. Thus, (2.6.22) is a non-decreasing sequence.

Let us apply the Denjoy-Càrleman criterion to the class $C\{\|A^n\varphi\|_{\mathcal{H}}\}$, where $\|\varphi\|_{\mathcal{H}} = 1$. Since (2.6.22) is a non-decreasing sequence, we have

$$\inf \left\{ \|A^k\varphi\|_{\mathcal{H}}^{1/k} \,\middle|\, k \geq n \right\} = \|A^n\varphi\|_{\mathcal{H}}^{1/n}.$$

Therefore, the condition (2.6.20) on the quasi-analyticity of this class, i.e., for the quasi-analyticity of the vector φ, can be rewritten in the form (2.6.21). □

Theorem 2.6.8 *Let A be Hermitian operator acting in $\mathcal{H}$. It is self-adjoint if and only if $\mathcal{H}$ contains total set of quasi-analytic vectors.*

Proof In one direction, this statement is trivial. Indeed, let A be self-adjoint. Then it suffices to prove the quasi-analyticity of each vector φ of the form $\varphi = E((a, b))f$, where E is a resolution of the identity that corresponds to A, $a, b \in \mathbb{R}$, $(a < b)$, and $f \in \mathcal{H}$. It is obvious that $\varphi \in \cap_{n=1}^{\infty}\mathfrak{D}(A^n)$. Further, we have:

$$\|A^n\varphi\|_{\mathcal{H}}^2 = \int_a^b \lambda^{2n}\, d(E(\lambda)f, f)_{\mathcal{H}} \le c^{2n}\|f\|_{\mathcal{H}}^2, \quad c = \max(|a|, |b|),\ n \in \mathbb{N}.$$

Therefore, the series (2.6.21) is divergent and, according to Lemma 2.6.7, the vector φ is quasi-analytic.

Suppose that A has a total set M of quasi-analytic vectors φ. Since A is closed, it suffices to prove its essential self-adjointness or, according to Theorem 2.6.1, the uniqueness of strong solutions of the Cauchy problem for the Eq. (2.6.4) with $b = \infty$. Let $u(t)$ be a strong solution of the problem

$$\left(\frac{du}{dt}\right)(t) - (\zeta A)^* u(t) = 0, \quad t \in [0, \infty),\ u(0) = 0, \tag{2.6.23}$$

where $\zeta = \pm i$. It suffices to establish that $u(t) = 0$ for $t \in [0, T]$ with any $T > 0$.

The "weak" equality (2.6.3) for (2.6.23) with a quasi-analytic vector $f = \varphi \in \cap_{n=1}^{\infty}\mathfrak{D}(A^n)$ gives

$$\frac{d}{dt}(u(t), \varphi)_{\mathcal{H}} = \left(\left(\frac{du}{dt}\right)(t), \varphi\right)_{\mathcal{H}} = (u(t), (\zeta A)\varphi)_{\mathcal{H}}, \quad t \in [0, T].$$

But $(\zeta A)\varphi \in \cap_{n=1}^{\infty}\mathfrak{D}(A^n)$ and, therefore,

$$\frac{d}{dt}(u(t), (\zeta A)\varphi)_{\mathcal{H}} = (u(t), (\zeta A)^2\varphi)_{\mathcal{H}}, \quad t \in [0, T].$$

And we continue in such a way. This implies that $(u(t), \varphi)_{\mathcal{H}} \in C^{\infty}([0, T])$ and

$$D^n(u(t), \varphi)_{\mathcal{H}} = D^{n-1}(u(t), (\zeta A)\varphi)_{\mathcal{H}} = \cdots = (u(t), (\zeta A)^n\varphi)_{\mathcal{H}}, \\ t \in [0, T],\ n \in \mathbb{Z}_+. \tag{2.6.24}$$

Since the values of $u(t)$ on $[0, T]$ are bounded, it follows from (2.6.24) that

$$|D^n(u(t), \varphi)_{\mathcal{H}}| \le c\|(\zeta A)^n\varphi\|_{\mathcal{H}} = c\|A^n\varphi\|_{\mathcal{H}}, \quad t \in [0, T],\ n \in \mathbb{Z}_+,$$

i.e., the scalar function $[0, T] \ni t \mapsto f(t) = (u(t), \varphi)_{\mathcal{H}}$ belongs to the class $C\{\|A^n\varphi\|_{\mathcal{H}}\}$. Equalities (2.6.24) and $u(0) = 0$ imply that $(D^n f)(0) = 0$, $n \in \mathbb{Z}_+$. Therefore, by virtue of the quasi-analyticity of $C\{\|A^n\varphi\|_{\mathcal{H}}\}$, we get the equality $f(t) = (u(t), \varphi)_{\mathcal{H}} = 0$, $t \in [0, T]$. Since the set M of vectors φ is total, we have $u(t) = 0$, $t \in [0, T]$. □

Other Criteria of Self-Adjointness As above, let A be an Hermitian operator acting in a Hilbert space $\mathcal{H}$. *A vector $\varphi \in \mathcal{H}$ is called analytic (with respect to A) if φ belongs to $\bigcap_{n=1}^{\infty} \mathfrak{D}(A^n)$* and the series

$$\sum_{n=0}^{\infty} \frac{\|A^n\varphi\|_{\mathcal{H}}}{n!} z^n \tag{2.6.25}$$

has a non-zero radius of convergence. This vector is called entire if this radius is equal to infinity.

It is clear that each analytic vector is quasi-analytic but the converse statements is not true. Note that, *if A is self-adjoint, then it possesses a total set of entire vectors.* Indeed, every vector $\varphi = E((a, b))f$, satisfying the estimate $\|A^n\varphi\|_{\mathcal{H}} \le c^n \|\varphi\|_{\mathcal{H}}$, $n \in \mathbb{N}$, is entire.

Let us formulate a theorem that makes Theorem 2.6.8 more precise in the case of operators semi-bounded from below.

Let A be an Hermitian operator acting in $\mathcal{H}$. *A vector $\varphi \in \mathcal{H}$ is called the Stieltjes vector (with respect to A), if $\varphi \in \bigcap_{n=1}^{\infty} \mathfrak{D}(A^n)$ and the class $C\{\|A^n\varphi\|_{\mathcal{H}}^{1/2}\}$ is quasi-analytic, or, in other words,*

$$\sum_{n=1}^{\infty} \|A^n\varphi\|_{\mathcal{H}}^{-1/(2n)} = \infty. \tag{2.6.26}$$

The equivalence of the fact that a vector belongs to the class of Stieltjes vectors and condition (2.6.26) follows from the fact that the sequence $(\|A^n\varphi\|_{\mathcal{H}}^{1/(2n)})_{n=1}^{\infty}$, where $\|\varphi\|_{\mathcal{H}} = 1$, is non-decreasing together with (2.6.22). To complete the proof, one should use the Denjoy-Càrleman criterion.

It is clear that every quasi-analytic vector is a Stieltjes vector but not vice versa. Thus, if we denote the sets of all entire, analytic, quasi-analytic, and Stieltjes vectors with respect to the operator A by $\mathcal{E}(A)$, $\mathcal{A}(A)$, $\mathcal{Q}(A)$, and $\mathcal{S}(A)$, respectively, then we get the inclusions

$$\mathcal{E}(A) \subseteq \mathcal{A}(A) \subseteq \mathcal{Q}(A) \subseteq \mathcal{S}(A). \tag{2.6.27}$$

Theorem 2.6.9 *Let A be a closed Hermitian operator semi-bounded below. If $\mathcal{H}$ contains a total set that consists of Stieltjes vectors, then A is self-adjoint.*

The converse statement is evident by virtue of the already proved assertions and (2.6.27).

Proof According to Theorem 2.6.4, it suffices to prove the uniqueness of strong solutions of the Cauchy problem for Eq. (2.6.13) for $b = \infty$. Let $u(t)$ be a strong solution of this problem such that $u(0) = u'(0) = 0$. Let us show that $u(t) = 0, t \in [0, T])$ for all $T > 0$. Assume that M is a total set of Stieltjes vectors φ appearing in the condition of the theorem. We set $[0, T] \ni t \mapsto f(t) = (u(t), \varphi)_{\mathcal{H}}$, where $\varphi \in M$. It follows from the relation (2.6.3), written for (2.6.13) that $f \in C^2([0, T])$ and

$$\left(\frac{d^2 f}{dt^2}\right)(t) = -(u(t), A\varphi)_{\mathcal{H}}, \quad t \in [0, T].$$

Since $A\varphi \in \cap_{n=1}^{\infty} \mathfrak{D}(A^n)$, by the same reasoning, we conclude that

$$(u(t), A\varphi)_{\mathcal{H}} \in C^2([0, T]),$$

$$\frac{d^2}{dt^2}(u(t), A\varphi)_{\mathcal{H}} = -(u(t), A^2\varphi)_{\mathcal{H}}, \quad t \in [0, T], \quad \text{etc.}$$

As a result, we get $f \in C^{\infty}([0, T])$ and

$$\begin{aligned}(D^{2k} f)(t) &= D^{2k}(u(t), \varphi)_{\mathcal{H}} = -D^{2(k-1)}(u(t), A\varphi)_{\mathcal{H}} = \dots \\ &= (-1)^k (u(t), A^k\varphi)_{\mathcal{H}}, \quad t \in [0, T],\ k \in \mathbb{Z}_+.\end{aligned} \tag{2.6.28}$$

One can also deduce a similar equality for odd derivatives. Indeed, by differentiating (2.6.28), we obtain

$$(D^{2k+1} f)(t) = (-1)^k (u'(t), A^k\varphi)_{\mathcal{H}}, \quad t \in [0, T],\ k \in \mathbb{Z}_+. \tag{2.6.29}$$

Values of $u(t)$ and $u'(t)$ on $[0, T]$ are bounded; therefore, it follows from (2.6.28), (2.6.29), the fact that the operator is Hermitian, and the Cauchy-Bunyakovsky inequality that

$$|(D^{2k} f)(t)|, |(D^{2k+1} f)(t)| \le c\|A^k\varphi\|_{\mathcal{H}} \le c\|\varphi\|_{\mathcal{H}}^{1/2} \|A^{2k}\varphi\|^{1/2},$$

$$t \in [0, T],\ k \in \mathbb{Z}_+,$$

i.e., $f \in C\{m_n\}$, where $(m_n)_{n=1}^{\infty}$ is a sequence of numbers

$$\|\varphi\|_{\mathcal{H}}^{1/2},\ \|\varphi\|_{\mathcal{H}}^{1/2},\ \|A\varphi\|_{\mathcal{H}}^{1/2},\ \|A\varphi\|_{\mathcal{H}}^{1/2},\ \|A^2\varphi\|_{\mathcal{H}}^{1/2},\ \|A^2\varphi\|_{\mathcal{H}}^{1/2}, \dots.$$

The class $C\{m_n\}$ will be not changed if we normalize φ. But then, as already mentioned, the sequence $(\|A^n\varphi\|_{\mathcal{H}}^{1/(2n)})_{n=1}^{\infty}$ is non-decreasing. This means that the Denjoy-Càrleman condition (2.6.20) on the sequence, under considerations, can be written in the form (2.6.26). Hence, the class $C\{m_n\}$ is quasi-analytic.

Furthermore, relations (2.6.28), (2.6.29), and the condition $u(0) = u'(0) = 0$ imply that $(D^n f)(0) = 0$, $n \in \mathbb{Z}_+$. Therefore, $(u(t), \varphi)_{\mathcal{H}} = f(t) = 0$, $t \in [0, T]$. In view of the fact that M is total, we conclude that $u(t) = 0$ $(t \in [0, T])$. □

Commutativity Criteria of Operators Here we consider another application of quasi-analytic vectors. Let A_1 and A_2 be two bounded self-adjoint operators. To commute mutually the operators $A_1 A_2 = A_2 A_1$, it is necessary and sufficient to commute mutually their resolutions of identity E_1 and E_2. In what follows, we exclude the word "mutually" by considerations of the commutativity, because we do not intend otherwise. In the case of unbounded A_1 and A_2, their formal commutativity, that is, commutativity on a dense set of vectors from $\mathcal{H}$, does not lead to the commutativity of E_1 and E_2. Additional conditions under which the commutativity of E_1 and E_2 are satisfied are established by the following theorem.

Theorem 2.6.10 *Let A_1 and A_2 be two Hermitian operators with domains $\mathfrak{D}(A_1)$ and $\mathfrak{D}(A_2)$ in $\mathcal{H}$, and let $\mathfrak{D}$ be a linear set such that $\mathfrak{D} \subseteq \mathfrak{D}(A_1) \cap \mathfrak{D}(A_2)$. Suppose that our operators commute on $\mathfrak{D}$, i.e., $A_1\mathfrak{D} \subseteq \mathfrak{D}(A_2)$, $A_2\mathfrak{D} \subseteq \mathfrak{D}(A_1)$, and $A_1 A_2 = A_2 A_1 f$, $f \in \mathfrak{D}$.*

Let in addition A_1, A_2, and the restriction $A_1 \restriction ((A_2 - z\mathbb{I})\mathfrak{D})$ (for some non-real z) have total set of quasi-analytic vectors. Then the operators A_1 and A_2 are essentially self-adjoint both and their resolutions of the identity commute.

Proof By virtue of Theorem 2.6.8, operators $\tilde{A}_1$ and $\tilde{A}_2$ are self-adjoint both. Let $R_z(\tilde{A}_1)$ and $R_z(\tilde{A}_2)$ be their resolvents, respectively. By Theorem 2.3.17, to prove the commutativity of resolutions of identity $\tilde{A}_1$ and $\tilde{A}_2$, it is enough to check the commutativity of $R_z(\tilde{A}_1)$ and $R_z(\tilde{A}_2)$. For $f \in \mathfrak{D}$, due to the commutativity of A_1 and A_2, we obtain

$$\begin{aligned} R_z(\tilde{A}_1)R_z(\tilde{A}_2)(A_1 - z\mathbb{I})(A_2 - z\mathbb{I})f \\ &= R_z(\tilde{A}_1)R_z(\tilde{A}_2)(A_2 - z\mathbb{I})(A_1 - z\mathbb{I})f \\ &= f = R_z(\tilde{A}_2)R_z(\tilde{A}_1)(A_1 - z\mathbb{I})(A_2 - z\mathbb{I})f. \end{aligned}$$

Thus, to prove that $R_z(\tilde{A}_1)$ commutes with $R_z(\tilde{A}_2)$, it suffices to show that $(A_1 - z\mathbb{I})(A_2 - z\mathbb{I})\mathfrak{D}$ is dense in $\mathcal{H}$. But this set coincides with the range $\mathfrak{R}(A_1 \restriction ((A_2 - z\mathbb{I})\mathfrak{D}) - z\mathbb{I})$ and the operator $A_1 \restriction ((A_2 - z\mathbb{I})\mathfrak{D})$ is essentially self-adjoint by the same Theorem 2.6.8. Therefore, the specified range is dense in $\mathcal{H}$. □

The following theorem from the work of [225] is also used.

Theorem 2.6.11 *Let A_1 and A_2 be two Hermitian operators defined on $\mathfrak{D}(A_1)$ and $\mathfrak{D}(A_2)$ in a Hilbert space $\mathcal{H}$, and let $\mathfrak{D}$ be a linear set dense in $\mathcal{H}$ such that $\mathfrak{D}$ is contained in domains of operators A_1, A_2, A_1^2, $A_1 A_2$, $A_2 A_1$, and A_2^2, so that $A_1 A_2 f = A_2 A_1 f$, for all $f \in \mathfrak{D}$.*

If the restriction $A_1^2 + A_2^2$ on $\mathfrak{D}$ is essentially self-adjoint operator, then A_1 and A_2 are essentially self-adjoint and their closures commute in the strong resolvent sense.

The Self-Adjointness of Perturbed Operators Ahead, it is found out when the Hermitian operator is self-adjoint. Now we will find out the self-adjointness of $A + B$, when A is self-adjoint and B is Hermitian.

Let A be a self-adjoint operator in a Hilbert space $\mathcal{H}$ and let B be a bounded self-adjoint operator. Then the operator $A + B$ is self-adjoint.

However, in the case of an unbounded perturbation B, the situation becomes much more complicated. First, we introduce the following definition

Consider operators A and B acting in $\mathcal{H}$ and such that $\mathfrak{D}(B) \supseteq \mathfrak{D}(A)$. We say that B is *subordinated* to A if it satisfies the inequality

$$\|Bf\|_{\mathcal{H}} \leq p\,\|Af\|_{\mathcal{H}} + q\|f\|_{\mathcal{H}}, \quad f \in \mathfrak{D}(A) \tag{2.6.30}$$

with constants $p, q \geq 0$ (the constants of the subordination). If, for any $p > 0$, there exists $q = q(p)$ such that inequality (2.6.30) is satisfied, then the operator B is called *infinitely small* as compared to A.

Theorem 2.6.12 (Rellich-Kato) *Consider a self-adjoint operator A and an Hermitian operator B acting in $\mathcal{H}$ and such that $\mathfrak{D}(B) \supseteq \mathfrak{D}(A)$. If B is subordinated to A with a constant of the subordination $p \in [0, 1)$, then the operator $A + B$, : $(\mathfrak{D}(A + B) = \mathfrak{D}(A))$ is self-adjoint.*

2.7 Rigged Spaces

Let $\mathcal{H}_0$ be a Hilbert space with a scalar product $(\cdot\,,\cdot)_{\mathcal{H}_0}$ and a norm $\|\cdot\|_{\mathcal{H}_0}$, and let f and g be its elements. Assume that a linear set $\mathcal{H}_+$, which is also a Hilbert space with respect to a new scalar product $(\cdot\,,\cdot)_{\mathcal{H}_+}$, is dense in $\mathcal{H}_0$ and let $\|\cdot\|_{\mathcal{H}_+}$ be a norm in $\mathcal{H}_+$ such that

$$\|u\|_{\mathcal{H}_0} \leq \|u\|_{\mathcal{H}_+}, \quad u \in \mathcal{H}_+. \tag{2.7.1}$$

We denote by u, v elements of $\mathcal{H}_+$ and this space will be called positive.

Every element $f \in \mathcal{H}_0$ generates an antilinear continuous functional l_f in $\mathcal{H}_+$ by the formula

$$l_f(u) = (f, u)_{\mathcal{H}_0}, \quad u \in \mathcal{H}_+; \tag{2.7.2}$$

its continuity follows from the estimate

$$|l_f(u)| = |(f, u)_{\mathcal{H}_0}| \leq \|f\|_{\mathcal{H}_0}\|u\|_{\mathcal{H}_0} \leq \|f\|_{\mathcal{H}_0}\|u\|_{\mathcal{H}_+}.$$

We now introduce a new norm $\|\cdot\|_{\mathcal{H}_-}$ in $\mathcal{H}_0$ by taking the norm of the functional l_f that corresponds to an element f as the norm of this element, i.e.,

$$\|f\|_{\mathcal{H}_-} = \|l_f\| = \sup\left\{ \frac{|(f, u)_{\mathcal{H}_0}|}{\|u\|_{\mathcal{H}_+}} \middle| u \in \mathcal{H}_+ \right\}. \tag{2.7.3}$$

We must check that if $\|f\|_{\mathcal{H}_-} = 0$, then $f = 0$; all the other properties of norm are evident. If $\|f\|_{\mathcal{H}_-} = 0$, then we have $(f, u)_{\mathcal{H}_0} = 0$ for all $u \in \mathcal{H}_+$. Since $\mathcal{H}_+$ is dense in $\mathcal{H}_0$, this yields $f = 0$.

By completing the space $\mathcal{H}_0$ with respect to the norm (2.7.3) we obtain linear normed space denoted by $\mathcal{H}_-$, which is called *a space with negative norm or negative space*. Thus, we have constructed the chain of spaces

$$\mathcal{H}_- \supseteq \mathcal{H}_0 \supseteq \mathcal{H}_+ \tag{2.7.4}$$

with negative, zero, and positive norm, respectively. Their elements are denoted by $\alpha, \beta, \ldots \in \mathcal{H}_-$, $f, g, \ldots \in \mathcal{H}_0$, and $u, v, \ldots \in \mathcal{H}_+$. Sometimes they are called generalized, ordinary, and smooth vectors, respectively. We say that (2.7.4) is the (Hilbert) rigging (chain) of the spaces $\mathcal{H}_0$, $\mathcal{H}_+$, and $\mathcal{H}_-$. We also say the space $\mathcal{H}_0$ is equipped by the spaces $\mathcal{H}_+$ and $\mathcal{H}_-$.

Since $\mathcal{H}_0 \ni f \mapsto l_f \in (\mathcal{H}_+)'$ is a one-to-one linear mapping, it is easy to show that $\mathcal{H}_-$ can be regarded as a subset of the dual space of antilinear functionals over $\mathcal{H}_+$, namely, $\mathcal{H}_- \subseteq (\mathcal{H}_+)'$. Therefore, the expression $\alpha(u)$ is meaningful. As in the case of the form $(\alpha, u)_{L_2(\mathbb{R}^N)}$ in the Sobolev–Schwartz theory of generalized functions, we denote this expression by $(\alpha, u)_{\mathcal{H}_0} = \overline{(u, \alpha)}_{\mathcal{H}_0}$.

A bilinear form

$$\mathcal{H}_- \times \mathcal{H}_+ \ni \langle \alpha, u \rangle \mapsto (\alpha, u)_{\mathcal{H}_0} \in \mathbb{C} \tag{2.7.5}$$

is an extension of the form

$$\mathcal{H}_+ \times \mathcal{H}_+ \ni \langle v, u \rangle \mapsto (v, u)_{\mathcal{H}_0} \in \mathbb{C}$$

to $\mathcal{H}_- \times \mathcal{H}_+$ by continuity. Clearly, the Cauchy-Bunyakowsky inequality admits the generalization

$$|(\alpha, u)_{\mathcal{H}_0}| \leq \|\alpha\|_{\mathcal{H}_-} \|u\|_{\mathcal{H}_+}, \quad \alpha \in \mathcal{H}_-, \ u \in \mathcal{H}_+. \tag{2.7.6}$$

Let us prove several simple but important facts.

Theorem 2.7.1 *The negative space $\mathcal{H}_-$ is a Hilbert space.*

In further studies, the construction of the scalar product of $\mathcal{H}_-$ will be important.

Proof Let us construct a scalar product in $\mathcal{H}_-$. Consider a bilinear form

$$\mathcal{H}_0 \times \mathcal{H}_+ \ni \langle f, u\rangle \mapsto b(f, u) = (f, u)_{\mathcal{H}_0} \in \mathbb{C}. \tag{2.7.7}$$

It is continuous. Indeed,

$$|b(f, u)| \leq \|f\|_{\mathcal{H}_0}\|u\|_{\mathcal{H}_0} \leq \|f\|_{\mathcal{H}_0}\|u\|_{\mathcal{H}_+}.$$

Therefore, it is representable in the form

$$b(f, u) = (f, Au)_{\mathcal{H}_0} = (A^* f, u)_{\mathcal{H}_+},$$

where $A : \mathcal{H}_+ \to \mathcal{H}_0$, $A^* : \mathcal{H}_0 \to \mathcal{H}_+$ are mutually adjoint continuous operators. According to (2.7.7), the operator A is equal to the imbedding operator $O : \mathcal{H}_+ \to \mathcal{H}_0$. Denote $I = O^* : \mathcal{H}_0 \to \mathcal{H}_+$. Thus,

$$(f, u)_{\mathcal{H}_0} = (f, Ou)_{\mathcal{H}_0} = (If, u)_{\mathcal{H}_+}, \quad f \in \mathcal{H}_0,\ u \in \mathcal{H}_+. \tag{2.7.8}$$

In $\mathcal{H}_0$ we introduce a quasi-scalar product

$$(f, g)_{\mathcal{H}_-} = (If, Ig)_{\mathcal{H}_+} = (f, Ig)_{\mathcal{H}_0} = (If, g)_{\mathcal{H}_0}, \quad f, g \in \mathcal{H}_0. \tag{2.7.9}$$

According to (2.7.3), (2.7.8), and (2.7.9), we have

$$\begin{aligned}\|f\|_{\mathcal{H}_-} &= \sup\left\{ \frac{|(f, u)_{\mathcal{H}_0}|}{\|u\|_{\mathcal{H}_+}} \,\middle|\, u \in \mathcal{H}_+ \right\} = \sup\left\{ \frac{|(If, u)_{\mathcal{H}_+}|}{\|u\|_{\mathcal{H}_+}} \,\middle|\, u \in \mathcal{H}_+ \right\} \\ &= \|If\|_{\mathcal{H}_+} = \sqrt{(f, f)_{\mathcal{H}_-}}, \quad f \in \mathcal{H}_0.\end{aligned}$$

Since $\|\cdot\|_{\mathcal{H}_-}$ is a norm in $\mathcal{H}_0$, then the relation (2.7.9) actually determines not just a quasi-scalar but a scalar product. As a result of completion, this scalar product becomes meaningful not only in $\mathcal{H}_0$ but also in $\mathcal{H}_-$ and $\mathcal{H}_-$ is transformed into a Hilbert space. □

Hence, $\mathcal{H}_-$ is now equipped with the scalar product

$$(\alpha, \beta)_{\mathcal{H}_-}, \quad \|\alpha\|_{\mathcal{H}_-} = \sqrt{(\alpha, \alpha)_{\mathcal{H}_-}}, \quad \alpha, \beta \in \mathcal{H}_-.$$

Since $\|I\| = \|O\| = 1$, then $\|f\|_{\mathcal{H}_-} = \|If\|_{\mathcal{H}_+} \leq \|f\|_{\mathcal{H}_0}$, $f \in \mathcal{H}_0$.

The first equality in (2.7.9) indicates that I is an isometric operator acting from $\mathcal{H}_-$ into $\mathcal{H}_+$ defined on a dense set in the space $\mathcal{H}_-$. Its extension by continuity is an isometric operator $\mathbf{I} : \mathcal{H}_- \to \mathcal{H}_+$ acting from the whole $\mathcal{H}_-$ into $\mathcal{H}_+$; $I = \mathbf{I} \upharpoonright \mathcal{H}_0$.

It is easy to see that $\mathbf{I}$ is *an isometry between the whole $\mathcal{H}_-$ and the whole $\mathcal{H}_+$* (i.e., $\mathfrak{R}(\mathbf{I}) = \mathcal{H}_+$).

Thus, $\mathfrak{R}(\mathbf{I})$ is dense in $\mathcal{H}_+$. If $u \in \mathcal{H}_+$ such that $u \perp \mathfrak{R}(\mathbf{I})$ in $\mathcal{H}_+$, then, by virtue of (2.7.8), for any $f \in \mathcal{H}_0$, we have

$$0 = (\mathbf{I}f, u)_{\mathcal{H}_+} = (If, u)_{\mathcal{H}_+} = (f, u)_{\mathcal{H}_0},$$

whence we conclude that $u = 0$. Moreover, $\mathfrak{R}(\mathbf{I})$ is closed in $\mathcal{H}_+$. Therefore, $\mathfrak{R}(\mathbf{I}) = \mathcal{H}_+$.

Further, we have

$$(\alpha, u)_{\mathcal{H}_0} = (\mathbf{I}\alpha, u)_{\mathcal{H}_+}, \quad \alpha \in \mathcal{H}_-, \ u \in \mathcal{H}_+. \tag{2.7.10}$$

Indeed, assume that $\mathcal{H}_0 \ni f_n \to \alpha$ as $n \to \infty$ in $\mathcal{H}_-$. Then, by virtue of (2.7.6), (2.7.8), and the continuity of $\mathbf{I}$, we obtain

$$(\alpha, u)_{\mathcal{H}_0} = \lim_{n\to\infty} (f_n, u)_{\mathcal{H}_0} = \lim_{n\to\infty} (\mathbf{I}f_n, u)_{\mathcal{H}_+} = (\mathbf{I}\alpha, u)_{\mathcal{H}_+}.$$

Theorem 2.7.2 *The equality*

$$\mathcal{H}_- = (\mathcal{H}_+)',$$

holds true, i.e., the negative space can be regarded as the space of antilinear functionals dual to the positive space.

Proof It suffices to show that every functional $l \in (\mathcal{H}_+)'$ can be represented in the form $l(u) = (\alpha, u)_{\mathcal{H}_0}$, $u \in \mathcal{H}_+$ with some $\alpha \in \mathcal{H}_-$. According to the Riesz theorem, there exists $\alpha \in \mathcal{H}_+$ such that $l(u) = (a, u)_{\mathcal{H}_+}$, $u \in \mathcal{H}_+$. We set $\alpha = \mathbf{I}^{-1}a \in \mathcal{H}_-$. Since $\mathfrak{R}(\mathbf{I}) = \mathcal{H}_+$, in view of (2.7.10), we obtain

$$l(u) = (a, u)_{\mathcal{H}_+} = (\mathbf{I}\cdot\mathbf{I}^{-1}a, u)_{\mathcal{H}_+} = (\mathbf{I}\alpha, u)_{\mathcal{H}_+} = (\alpha, u)_{\mathcal{H}_0}, \quad u \in \mathcal{H}_+.$$

□

We emphasize that the identification of the dual space $(\mathcal{H}_+)'$ with $\mathcal{H}_+$ (the classical identification) and with $\mathcal{H}_-$ depends on the form of a functional $l \in (\mathcal{H}_+)'$, $\forall u \in \mathcal{H}_+$. It can be written in the form $l\,(u) = (a, u)_{\mathcal{H}_+}$ or in the form $l(u) = (\alpha, u)_{\mathcal{H}_0}$, $\alpha \in \mathcal{H}_-$, $a \in \mathcal{H}_+$. In the second case, we say that the spaces $\mathcal{H}_+$ and $\mathcal{H}_-$ are coupled by a form $(\cdot, \cdot)_{\mathcal{H}_0}$ of the type (2.7.5).

Note that all results discussed above admit a natural generalization to the case of real spaces $\mathcal{H}_0$, $\mathcal{H}_+$, and $\mathcal{H}_-$.

The operator $\mathbf{I}$ can be decomposed into a product of two operators.

Theorem 2.7.3 *The isometry* $\mathbf{I} : \mathcal{H}_- \to \mathcal{H}_+$ *admits a decomposition into a product of two isometries, namely*

$$\mathbf{I} = J\mathbf{J}, \tag{2.7.11}$$

$$\mathbf{J} : \mathcal{H}_- \to \mathcal{H}_0, \quad J : \mathcal{H}_0 \to \mathcal{H}_+, \quad OJ = \mathbf{J} \restriction \mathcal{H}_0.$$

Proof Let $A = OI : \mathcal{H}_0 \to \mathcal{H}_0$. This operator is bounded and non-negative. Indeed, according to (2.7.8), we have

$$(Af, f)_{\mathcal{H}_0} = (OIf, f)_{\mathcal{H}_0} = (If, If)_{\mathcal{H}_+} \geq 0, \quad f \in \mathcal{H}_0.$$

We set $B = \sqrt{A} : \mathcal{H}_0 \to \mathcal{H}_0$ and consider this operator as acting from a dense set $\mathcal{H}_0$ in the space $\mathcal{H}_-$. In this case, it is isometric,

$$(Bf, Bg)_{\mathcal{H}_0} = (B^2 f, g)_{\mathcal{H}_0} = (If, Ig)_{\mathcal{H}_+} = (f, g)_{\mathcal{H}_-}, \quad f, g \in \mathcal{H}_0$$

and its extension by continuity is an isometric operator $\mathbf{J} : \mathcal{H}_- \to \mathcal{H}_0$.

Let us prove that $\mathfrak{R}(\mathbf{J}) = \mathcal{H}_0$. It suffices to show that the fact that $f \perp \mathfrak{R}(\mathbf{J})$ in $\mathcal{H}_0$ implies that $f = 0$. For an arbitrary $g \in \mathcal{H}_0$, we have

$$0 = (\mathbf{J}g, f)_{\mathcal{H}_0} = (Bg, f)_{\mathcal{H}_0} = (g, Bf)_{\mathcal{H}_0},$$

whence $Bf = 0$. Then $OIf = B^2 f = 0$, i.e., $If = 0$ and, hence, $f = 0$.

We now show that $\mathfrak{R}(B) \subseteq \mathcal{H}_+$. It suffices to prove the equality

$$B\mathbf{J} = O\mathbf{I}. \tag{2.7.12}$$

For $f \in \mathcal{H}_0$, we have $B\mathbf{J}f = B^2 f = OIf = O\mathbf{I}f$. Since $\mathcal{H}_0$ is dense in $\mathcal{H}_-$ as a set and the operator $\mathbf{J} : \mathcal{H}_- \to \mathcal{H}_0$, $B : \mathcal{H}_0 \to \mathcal{H}_0$, and $O\mathbf{I} : \mathcal{H}_- \to \mathcal{H}_0$ are continuous, then this yields (2.7.12).

Denote by J the operator B regarded as an operator from $\mathcal{H}_0$ into $\mathcal{H}_+$. Then (2.7.12) implies $J\mathbf{J} = \mathbf{I}$, and it remains to prove that J is an isometric operator and $\mathfrak{R}(J) = \mathcal{H}_+$. But this follows from the equality $J = \mathbf{I}\mathbf{J}^{-1}$ because $\mathbf{J}^{-1} : \mathcal{H}_0 \to \mathcal{H}_-$ and $\mathbf{I} : \mathcal{H}_- \to \mathcal{H}_+$ are isometries. □

Let us write the most important relations connected with the rigging of the Hilbert space $\mathcal{H}_0$ by the spaces $\mathcal{H}_+$ and $\mathcal{H}_-$ constructed above (Fig. 2.1). Thus, we have

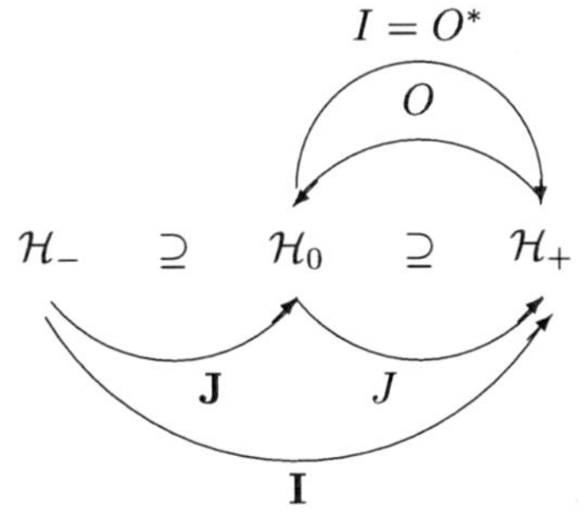

Fig. 2.1 Hilbert equipment diagram

$$
\begin{aligned}
& I = O^* \\
& \mathcal{H}_- \supseteq \mathcal{H}_0 \supseteq \mathcal{H}_+ \\
& \|\cdot\|_{\mathcal{H}_-} \leq \|\cdot\|_{\mathcal{H}_0} \leq \|\cdot\|_{\mathcal{H}_+}, \\
& OJ = \mathbf{J} \restriction \mathcal{H}_0, \quad \mathbf{I} = J\mathbf{J}, \\
& (\alpha, u)_{\mathcal{H}_0} = (\mathbf{I}\alpha, u)_{\mathcal{H}_+} = (\alpha, \mathbf{I}^{-1}u)_{\mathcal{H}_-}, \\
& (\mathbf{I}\alpha, \beta)_{\mathcal{H}_0} = (\alpha, \mathbf{I}\beta)_{\mathcal{H}_0}, \quad (\mathbf{J}\alpha, f)_{\mathcal{H}_0} = (\alpha, Jf)_{\mathcal{H}_0}, \\
& \alpha, \beta \in \mathcal{H}_-; \quad f \in \mathcal{H}_0; \quad u \in \mathcal{H}_+
\end{aligned}
\tag{2.7.13}
$$

(the last two equalities in (2.7.13) are obtained by setting, respectively $u = \mathbf{I}\beta$ and $u = Jf$).

Note that, instead of (2.7.1) one can require the validity of the inequality

$$
\|u\|_{\mathcal{H}_0} \leq c\|u\|_{\mathcal{H}_+}, \quad u \in \mathcal{H}_+
$$

with some $c > 0$; by a simple renormalization of $\mathcal{H}_+$, this case is reduced to (2.7.1).

Let us introduce an important definition.

The rigging (2.7.4) *is called quasi-nuclear if the imbedding operator $O : \mathcal{H}_+ \to \mathcal{H}_0$ is quasi-nuclear, i.e., a Hilbert-Schmidt operator.*

Operators in Chain In this subsection, we present another procedure for constructing chain (2.7.4) and generalize the notion of adjointness. Relations (2.7.13) imply that

$$
\begin{aligned}
(u, v)_{\mathcal{H}_+} &= (J^{-1}u, J^{-1}v)_{\mathcal{H}_0}, \quad u, v \in \mathcal{H}_+, \\
(\alpha, \beta)_{\mathcal{H}_-} &= (\mathbf{J}\alpha, \mathbf{J}\beta)_{\mathcal{H}_0}, \quad \alpha, \beta \in \mathcal{H}_-.
\end{aligned}
\tag{2.7.14}
$$

These equalities indicate that it is possible to construct riggings of the space $\mathcal{H}_0$ in terms of a certain operator similar to J or J^{-1} acting between this spaces and (that is more convenient).

Consider a closed operator D acting in $\mathcal{H}_0$ with a dense domain $\mathfrak{D}(D)$ in $\mathcal{H}_0$ and such that

$$
\|Du\|_{\mathcal{H}_0} \geq \|u\|_{\mathcal{H}_0}, \quad u \in \mathfrak{D}(D). \tag{2.7.15}
$$

By virtue of the closeness of D, the set $\mathfrak{D}(D)$ is a Hilbert space with the scalar product

$$
(u, v)_{\mathcal{H}_+} = (Du, Dv)_{\mathcal{H}_0}, \quad u, v \in \mathfrak{D}(D). \tag{2.7.16}
$$

This set is regarded as a positive space $\mathcal{H}_+$ and we construct the corresponding Hilbert space $\mathcal{H}_-$ dual to $\mathcal{H}_+$ with respect to $\mathcal{H}_0$ according to the procedure described above.

Let J be an isometric operator corresponding to the chain $\mathcal{H}_- \supseteq \mathcal{H}_0 \supseteq \mathcal{H}_+$ just constructed and let $(J^{-1})_{\mathcal{H}_0}$ be the operator $J^{-1} : \mathcal{H}_+ \to \mathcal{H}_0$ regarded as an operator in $\mathcal{H}_0$ with the domain $\mathcal{H}_+$. It follows from the proof of Theorem 2.7.3 that $(J^{-1})_{\mathcal{H}_0} = B^{-1}$ and, therefore, $(J^{-1})_{\mathcal{H}_0}$ is a positive self-adjoint operator in $\mathcal{H}_0$. By comparing(2.7.14) with (2.7.16), we conclude that $\|(J^{-1})_{\mathcal{H}_0} f\|_{\mathcal{H}_0} = \|Df\|_{\mathcal{H}_0}$, $f \in \mathcal{H}_+$, i.e., operators $(J^{-1})_{\mathcal{H}_0}$ and D are metrically equal. This immediately implies that

$$(J^{-1})_{\mathcal{H}_0} = \sqrt{D^* D}. \tag{2.7.17}$$

The relation (2.7.17) for the operators D and J becomes much simpler if the operator D is self-adjoint. In this case $(J^{-1})_{\mathcal{H}_0} = D$. It follows from (2.7.14) that $(f, g)_{\mathcal{H}_-} = (D^{-1} f, D^{-1} g)_{\mathcal{H}_0}$, $f, g \in \mathcal{H}_0$. The negative space $\mathcal{H}_-$ is obtained as the completion of $\mathcal{H}_0$ with respect to the last scalar product.

Let us generalize the concept of adjointness for continuous operators acting between the spaces of chain (2.7.4) ("adjointness with respect to $\mathcal{H}_0$"). Let we have $A \in \mathcal{L}(\mathcal{H}_+, \mathcal{H}_-)$. Then the operator $A^+ \in \mathcal{L}(\mathcal{H}_+, \mathcal{H}_-)$ is adjoint to A with respect to $\mathcal{H}_0$ and it is defined by the equality

$$(Au, v)_{\mathcal{H}_0} = (u, A^+ v)_{\mathcal{H}_0}, \quad u, v \in \mathcal{H}_+. \tag{2.7.18}$$

It is easy to see that the operator A^+ exists and can be expressed in terms of the ordinary adjoint operator $A^* \in \mathcal{L}(\mathcal{H}_-, \mathcal{H}_+)$. Thus, by using the relation $(\alpha, u)_{\mathcal{H}_0} = (\alpha, \mathbf{I}^{-1} u)_{\mathcal{H}_-}$ $(\alpha \in \mathcal{H}_-, u \in \mathcal{H}_+)$, which follows from (2.7.10), we obtain

$$(Au, v)_{\mathcal{H}_0} = (Au, \mathbf{I}^{-1} v)_{\mathcal{H}_-} = (u, A^* \mathbf{I}^{-1} v)_{\mathcal{H}_+} = (u, \mathbf{I}^{-1} A^* \mathbf{I}^{-1} v)_{\mathcal{H}_0}$$
$$(u, v \in \mathcal{H}_+),$$

i.e.,

$$A^+ = \mathbf{I}^{-1} A^* \mathbf{I}^{-1}, \quad A^* : \mathcal{H}_- \to \mathcal{H}_+. \tag{2.7.19}$$

The concept of adjointness with respect to $\mathcal{H}_0$ can also be introduced for operators acting between the other spaces of chain (2.7.4). Thus, let $A \in \mathcal{L}(\mathcal{H}_+, \mathcal{H}_0)$. Then $A^+ \in \mathcal{L}(\mathcal{H}_0, \mathcal{H}_-)$ is defined by the following equality (similar to (2.7.18)):

$$(Au, f)_{\mathcal{H}_0} = (u, A^+ f)_{\mathcal{H}_0}, \quad u \in \mathcal{H}_+, \ f \in \mathcal{H}_0. \tag{2.7.20}$$

It is easy to see that the operator A^+ exists and satisfies the equality $A^+ = \mathbf{I}^{-1} A^*$, where $A^* \in \mathcal{L}(\mathcal{H}_0, \mathcal{H}_+)$. Indeed, by virtue of (2.7.10), we have

$$(Au, f)_{\mathcal{H}_0} = (u, A^* f)_{\mathcal{H}_+} = (u, \mathbf{I}^{-1} A^* f)_{\mathcal{H}_0}, \quad u \in \mathcal{H}_+, \ f \in \mathcal{H}_0.$$

By using the definitions similar to (2.7.18) and (2.7.20), we can write $\mathbf{I}^+ = \mathbf{I}$, $\mathbf{J}^+ = J$, and $J^+ = \mathbf{J}$.

For operators $A \in \mathcal{L}(\mathcal{H}_+, \mathcal{H}_-)$, the concept of self-adjointness admits the following natural generalization. *We say that A is self-adjoint if $(u, Av)_{\mathcal{H}_0} = (Au, v)_{\mathcal{H}_0}$ for all $u, v \in \mathcal{H}_+$, that is $A^+ = A$. Similarly, an operator $A \in \mathcal{L}(\mathcal{H}_+, \mathcal{H}_-)$ is called non-negative if $(Au, u)_{\mathcal{H}_0} \geq 0$, $u \in \mathcal{H}_+$.*

Non-negative operators are certainly self-adjoint.

An ordinary bounded self-adjoint operator A in $\mathcal{H}_0$ regarded as an operator acting from $\mathcal{H}_+$ to $\mathcal{H}_-$ is obviously self-adjoint in the generalized sense. Generally speaking, an operator $A \in \mathcal{L}(\mathcal{H}_+, \mathcal{H}_-)$ is self-adjoint in the generalized sense if and only if the operator $\mathbf{I}A \in \mathcal{L}(\mathcal{H}_+, \mathcal{H}_+)$ is self-adjoint in $\mathcal{H}_+$ (or $A\mathbf{I}^{-1}$ is self-adjoint in $\mathcal{H}_-$). This follows from the equality (2.7.13)

$$(\mathbf{I}Au, v)_{\mathcal{H}_+} = (Au, v)_{\mathcal{H}_0} = (u, Av)_{\mathcal{H}_0} = (u, \mathbf{I}Av)_{\mathcal{H}_+}, \quad u, v \in \mathcal{H}_+.$$

The operator $A \in \mathcal{L}(\mathcal{H}_+, \mathcal{H}_-)$ is non-negative if $\mathbf{I}A \in \mathcal{L}(\mathcal{H}_+, \mathcal{H}_+)$ is non-negative in $\mathcal{H}_+$.

Rigging of Hilbert Spaces by Linear Topological Spaces We now proceed to the construction a chain of the type (2.7.4), where the roles of $\mathcal{H}_+$ and $\mathcal{H}_-$ are played a linear topological space and its dual space, respectively. First, we recall some facts from the theory of linear topological spaces

Let R be an abstract set. It becomes a topological space if we indicate a family (collection) Σ of its subsets $U, V, \cdots \subseteq R$, which are called *neighborhoods*. Let $x \in R$ and U is a neighborhood from Σ which contains x. Then we say that U is a neighborhood of the point x, and this is denoted by $U = U(x)$. The notion of a neighborhood plays an important role in the theory if the family Σ satisfies the following two axioms.

(a) Every point $x \in R$ belongs to a certain neighborhood and, moreover, for all $x, y \in R$ such that $x \neq y$, one can indicate disjoint neighborhoods $U(x)$ and $U(y)$ of these points (this is the Hausdorff separation axiom; we consider only spaces satisfying this axiom);
(b) For every $x \in R$ and any two its neighborhoods $U(x)$ and $V(x)$, there exists a neighborhood of this point $W(x)$ such that $W(x) \subseteq U(x) \cap V(x)$.

We consider only such topological spaces that satisfy the given axioms.

In topological spaces, one can introduce the notions similarly to well-known topological notions from analysis on the real axis.

Thus, for any $\alpha \subseteq R$ *its closure $\tilde{\alpha}$ is defined as a set that consists of all points $x \in R$ such that for every $U(x)$, the intersection $U(x) \cap \alpha$ is non-empty.* It is clear, that $\tilde{\alpha} \supseteq \alpha$.

A set $\alpha \subseteq R$ is called closed if $\tilde{\alpha} = \alpha$ and open if its complement $R \setminus \alpha$ is closed.

(1) *Every neighborhood $U \in \Sigma$ is an open set*, i.e., its complement $R \setminus U$ is closed.

(2) *A set $\alpha \subseteq R$ is open if and only if each point of this set belongs to α together with its neighborhood.*

The choice of Σ is not unambiguous. The following statement describes cases, when there are other equivalent systems.

(3) *In order for Σ and Σ' to be equivalent, it is necessary and sufficient that, $\forall x \in R$ and $\forall U(x) \in \Sigma, \exists U'(x) \in \Sigma', U'(x) \subseteq U(x)$ and the same relation holds true with Σ replaced with Σ' and Σ' replaced with Σ.*

In topological spaces, one can introduce the notion of convergence. Namely, a sequence $(x_n)_{n=1}^{\infty}$ of points $x_n \in R$ is called convergent to a point $x \in R$ if, for any $U(x) \in \Sigma$, there exists $N = N(U(x))$ such that $x_n \in U(x)$ whenever $n > N$.

One more note. If $R_1 = R_{1,\Sigma_1}$ and $R_2 = R_{2,\Sigma_2}$ are topological spaces and $R = R_1 \times R_2$ is their direct product, which consists of the points $x = \langle x_1, x_2 \rangle$, where $x_1 \in R_1$ and $x_2 \in R_2$, then this direct product can always be topologized by the family Σ that consists of all rectangles $U = U_1 \times U_2$ such that $U_1 \in \Sigma_1$ and $U_2 \in \Sigma_2$ (it is easy to show that Σ satisfies axioms (a) and (b) of topological space).

Projective Limits of Spaces We now introduce the notion of linear topological space. Let Φ be a linear space over the field $\mathbb{C}$ of complex numbers (for definiteness), which is, at the same time, a topology space with a base of neighborhoods $\Sigma = \{U(\varphi) | \varphi \in \Phi\}$.

If the linear structure of Φ is consistent, in a certain sense, with the topology, then Φ is called a linear topological space. The consistence is regarded in the sense of the validity of the following axiom.

(c) *Mappings*

$$\Phi \times \Phi \ni \langle \varphi, \psi \rangle \mapsto \varphi + \psi \in \Phi, \text{ and } \mathbb{C} \times \Phi \ni \langle \lambda, \varphi \rangle \mapsto \lambda\varphi \in \Phi$$

are continuous. In other words, operations of summation and multiplication by a scalar are continuous.

It is clear that the continuity of these mapping should be regarded as follows. For the first mapping, for any $\varphi, \psi \in \Phi$ and $U(\varphi + \psi)$, one can choose $U(\varphi)$ and $U(\psi)$ such that $U(\varphi) + U(\psi) \subseteq U(\varphi + \psi)$; for the second mapping, for any $\lambda \in \mathbb{C}$, $\varphi \in \Phi$, and $U(\lambda\varphi)$, one can choose $U(\lambda)$ and $U(\varphi)$ such that $U(\lambda)U(\varphi) \subseteq U(\lambda\varphi)$ (where $U(\lambda)$ denotes a neighborhood of a point λ in the ordinary topology of $\mathbb{C}$;

$$\forall \alpha, \beta \subseteq \Phi, \quad \alpha + \beta \stackrel{\text{def}}{=} \{\varphi + \psi \mid \varphi \in \alpha, \psi \in \beta\};$$
$$\forall \alpha \subseteq \mathbb{C}, \forall \beta \subseteq \Phi, \quad \alpha\beta = \{\lambda\varphi | \lambda \in \alpha, \varphi \in \beta\}.$$

Thus, the base of neighborhoods of a linear topological space Φ must satisfy axioms (a)–(c). Due to the fact that Φ is equipped with linear structure, it is often convenient to introduce first a family Σ_0 of subsets $U, V, \ldots$ of Φ, which are regarded as neighborhoods of the origin, then construct sets

$$\{\varphi \in \Phi | \ \varphi - \varphi_1 \in U\}, \quad U \in \Sigma_0, \ \varphi_1 \in \Phi \tag{2.7.21}$$

shifted by vectors $\varphi_1 \in \Phi$, and, finally, take all possible sets of the form (2.7.21) as a base of neighborhoods Σ (as in the case of a linear normed space, where we first consider open spheres centered at the origin and then realize all possible shifts of these spheres). Axioms (a)–(c) can be easily reformulated in terms of the family Σ_0.

Clearly, a linear normed space is an example of a linear topological space with open spheres with arbitrary centers and radius regarded as base neighborhoods.

Consider a family of Banach spaces $(B_\tau)_{\tau\in T}$ parameterized by elements of an arbitrary indexing set T. Assume that the set $\Phi = \bigcap_{\tau\in T} B_\tau$ is dense in each B_τ and the family $(B_\tau)_{\tau\in T}$ is directed by imbedding operators, i.e.,

$$\forall\tau', \tau'' \in T,\ \exists\tau''' \in T, \quad B_{\tau'''} \subseteq B_{\tau'}, \quad \text{and} \quad B_{\tau'''} \subseteq B_{\tau''}, \tag{2.7.22}$$

where all imbedding operators are dense and continuous. In Φ, we introduce the projective topology with respect to the families of Banach spaces $(B_\tau)_{\tau\in T}$ and natural imbedding operators $O_\tau \Phi \to B_\tau$. By definition, the family of all possible sets

$$U(\varphi_1; \tau; \varepsilon) = \left\{\varphi \in \Phi \mid \|\varphi - \varphi_1\|_{B_\tau} < \varepsilon\right\}, \quad \varphi_1 \in \Phi,\ \tau \in T,\ \varepsilon > 0. \tag{2.7.23}$$

is a system Σ, i.e., the base of neighborhoods in this topology.

Thus, the base of neighborhoods (2.7.23) is intersections of the set Φ with open spheres in the spaces B_τ (with arbitrary index τ, radius, and centers). The system (2.7.23) is obtained according to (2.7.21) by the shifts of the corresponding spheres centered at the origin by $\varphi_1 \in \Phi$.

The system (2.7.23) *satisfies axioms* (a)–(c) *of the linear topological space.*

The space $\Phi = \bigcap_{\tau\in T} B_\tau$ equipped with the topology generated by the base of neighborhoods (2.7.23) *is called the projective limit of Banach spaces B_τ and denoted by*

$$\Phi = \operatorname{pr}\lim_{\tau\in T} B_\tau. \tag{2.7.24}$$

The projective limit introduced above is sometimes called *reduced*—this emphasizes the fact that Φ is dense in each B_τ. Note that the convergence of a sequence $(\varphi_n)_{n=1}^\infty$ of vectors $\varphi_n \in \Phi$ to $\varphi \in \Phi$ in terms of the space (2.7.24) means that $\|\varphi_n - \varphi\|_{B_\tau} \to 0$ as $n \to \infty$ for all $\tau \in T$.

If T is countable, then the projective limit (2.7.24) *is called a countably normed space.*

In the case $T = \mathbb{Z}_+$, it is also denoted by $\Phi = \operatorname{pr}\lim_{n\to\infty} B_n$. In this case, we often encounter a situation where norms of spaces B_n are monotone, i.e.,

$$B_0 \supseteq B_1 \supseteq \ldots, \quad \|\varphi\|_{B_0} \le \|\varphi\|_{B_1} \le \ldots, \quad \varphi \in \Phi = \bigcap_{n=0}^{\infty} B_n \tag{2.7.25}$$

and the condition on the directionality by imbedding is clearly satisfied. However, it is possible to show that the general case of countably normed spaces can be reduced to the monotone case (2.7.25) by a proper renormalization of $B_\tau (\tau \in T)$.

If $B_\tau = \mathcal{H}_\tau$ *are Hilbert spaces, then* (2.7.24) *is called a projective limit of Hilbert spaces. If* T *is countable,* Φ *is called a countably Hilbert space.*

In the class of projective limits of Hilbert spaces, one can select an important subclass of nuclear spaces that are extensively used in the book.

The projective limit of Hilbert spaces $\Phi = \operatorname{pr}\lim_{\tau \in T} \mathcal{H}_\tau$ *is called nuclear if, for any* $\tau \in T$*, one can find* $\tau' \in T$ *such that* $\mathcal{H}_{\tau'} \subseteq \mathcal{H}_\tau$ *and the imbedding operator* $\mathcal{H}_{\tau'} \to \mathcal{H}_\tau$ *is quasi-nuclear.*

Riggings Constructed by Using Projective Limits Let Φ be some linear topological space over the field $\mathbb{C}$. *An antilinear continuous functional* l *over* Φ *is defined as a continuous mapping* $\Phi \ni \varphi \mapsto l(\varphi) \in \mathbb{C}$ *satisfying the antilinearity condition*

$$l(\lambda\varphi + \mu\psi) = \bar{\lambda} l(\varphi) + \bar{\mu} l(\psi), \qquad \varphi, \psi \in \Phi; \ \lambda, \mu \in \mathbb{C}.$$

A collection of all functionals of this sort (for a fixed Φ) obviously forms a linear space over $\mathbb{C}$ with the ordinary summation of functions and multiplication by a scalar.

As above, this space is called dual to Φ *and denoted by* Φ'. Note that, it is now convenient to consider an antilinear functional instead of linear one.

There are many methods for introducing topology in the dual space Φ'. Here, we study only the case of weak topology.

The weak convergence in Φ' *is regarded in a standard way, i.e., a sequence* $(l_n)_{n=1}^\infty$ *of functionals* $l_n \in \Phi'$ *is called weakly convergent to* $l \in \Phi'$ *if* $l_n(\varphi) \to l(\varphi)$ *as* $n \to \infty$ *for any* $\varphi \in \Phi$.

The weak topology in Φ' is generated by a system that is the base of neighborhoods in Φ' of the form

$$U(l_1; \varphi_1, \ldots, \varphi_n) = \big\{ l \in \Phi' \mid |l(\varphi_k) - l_1(\varphi_k)| < 1; \ k = 1, \ldots, n \big\}, \qquad (2.7.26)$$
$$l_1 \in \Phi'; \ \varphi_k \in \Phi, \ k = 1, \ldots, n; \ n \in \mathbb{N}.$$

The convergence with respect to this topology is equivalent to the weak convergence introduced above.

The fact that Φ' equipped with the system of the base of neighborhoods (2.7.26), satisfies axioms (a)–(c) of linear topological space can be verified by repeating the reasoning used in the same situation for Banach spaces.

The following simple but important theorem clarifies the structure of Φ' in the case where Φ is a projective limit.

Theorem 2.7.4 (Schwartz) *Assume that Φ is a projective limit of Banach spaces, i.e., $\Phi = \operatorname{pr}\lim_{\tau\in T} B_\tau$. Then*

$$\Phi' = \bigcup_{\tau\in T} B'_\tau \tag{2.7.27}$$

and this equality must be interpreted as follows. For all $l \in \Phi'$, there exists $\tau \in T$ such that l can be extended by continuity from Φ on B_τ to an element of B'_τ and, vice versa, if $l \in B'_\tau$ for some $\tau \in T$, then $l \restriction \Phi \in \Phi'$.

The representation (2.7.27) enables one to introduce in Φ', parallel with the weak topology, so called *topology of inductive limit of spaces*. Thus, relation (2.7.22) yields the following property of the directionality of the family $(B'_\tau)_{\tau\in T}$:

$$\forall\tau', \tau'' \in T, \ \exists\tau''' \in T, \quad B'_{\tau'''} \supseteq B'_{\tau'}, \quad \text{and} \quad B'_{\tau'''} \supseteq B'_{\tau''}, \tag{2.7.28}$$

where imbedding operators are dense and continuous. By using (2.7.28), one can easily prove that the collection of all possible sets of the form

$$U\big(l_1; \varepsilon(\cdot)\big) = \bigcup_{\tau\in T} \left\{l \in \Phi' \mid l - l_1 \in B'_\tau, \quad \|l - l_1\|_{B'_\tau} < \varepsilon(\tau)\right\}, \tag{2.7.29}$$

$$l_1 \in \Phi'; \ T \ni \tau \mapsto \varepsilon(\tau) > 0$$

forms a system of the base of neighborhoods in Φ'. In other words, for $l_1 = 0$, all possible unions of open spheres with positive radius centered at the origin form systems of the base of neighborhoods at the origin in the spaces B'_τ $(\tau \in T)$. Then these neighborhoods are shifted by arbitrary $l_1 \in \Phi'$.

In fact, we have given the general definition of the *inductive limit of Banach spaces*. Thus, let $(B_\tau)_{\tau\in T}$ be a given family of Banach spaces with the following property of directedness of the type (2.7.28): for any $\tau', \tau'' \in T$ there exits $\tau''' \in T$ such that $B_{\tau'''} \supseteq B_{\tau'}$ and $B_{\tau'''} \supseteq B_{\tau''}$.

In this case, the union $\bigcup_{\tau\in T} B_\tau = \Psi$ can be equipped with a natural linear structure. Indeed, let

$$\varphi, \psi \in \bigcup_{\tau\in T} B_\tau \Rightarrow \exists\tau', \tau'' \in T, \ \varphi \in B_{\tau'}, \ \psi \in B_{\tau'} \Rightarrow \exists\tau''' \in T, \ \varphi, \psi \in B_{\tau'''}.$$

Therefore, the expression $\lambda\varphi + \mu\psi$ is meaningful as a vector in $B_{\tau'''}$ and, hence, in Ψ.

A linear space Ψ is called the inductive limit of the spaces B_τ (the notation: $\Psi = \operatorname{ind}\lim_{\tau\in T} B_\tau$) if it is equipped with a system of base of neighborhoods of the form (2.7.29), *where Φ' and B'_τ are replaced with Ψ and B_τ respectively.*

After a brief survey of simple properties of linear topological spaces, we now proceed to the construction of the rigging of a Hilbert space by linear topological spaces.

Let $\mathcal{H}_0$ be a Hilbert space with elements $f, g, \ldots$ and Φ be a linear topological space, with elements $\varphi, \psi, \ldots$, densely and continuously imbedded in $\mathcal{H}_0$. Every element $f \in \mathcal{H}_0$ generates an antilinear continuous functional in Φ by the formula $l_f(\varphi) = (f, \varphi)_{\mathcal{H}_0}$, $\varphi \in \Phi$. By identifying f with l_f (this is possible in view of the fact that Φ is dense in $\mathcal{H}_0$), we arrive at the imbedding of $\mathcal{H}_0$ in the space Φ' of antilinear continuous functionals on Φ. It is clear that if Φ' is equipped with the weak topology, then the imbedding $H_0 \to \Phi'$ is continuous. Thus, we have constructed a chain that generalizes (2.7.4), namely,

$$\Phi' \supseteq \mathcal{H}_0 \supseteq \Phi. \tag{2.7.30}$$

We also say that (2.7.30) is the rigging of the space $\mathcal{H}_0$ by the spaces Φ and Φ' or, in other words, that we have defined the pairing of the spaces Φ and Φ' by the space $\mathcal{H}_0$. If Φ is a nuclear space, then the rigging (2.7.30) is called nuclear.

In what follows, we consider only riggings of the form (2.7.30), where Φ is a projective limit of Hilbert spaces, i.e.,

$$\Phi = \operatorname{pr}\lim_{\tau \in T} \mathcal{H}_\tau. \tag{2.7.31}$$

In addition, we suppose that each $\mathcal{H}_\tau$ is densely and continuously imbedded into $\mathcal{H}_0$ and, moreover, $\|\varphi\|_{\mathcal{H}_0} \leq \|\varphi\|_{\mathcal{H}_\tau}, \varphi \in \mathcal{H}_\tau, \tau \in T$, where T is assumed to contain the index "0" (in fact, this is not a restriction because any general situation can always be reduced to this case).

Thus, for any $\tau \in T$, one can construct a chain of the form (2.7.4)

$$\mathcal{H}_{-\tau} \supseteq \mathcal{H}_0 \supseteq \mathcal{H}_\tau, \tag{2.7.32}$$

where $\mathcal{H}_\tau$ is a positive space and $\mathcal{H}_{-\tau}$ is the corresponding negative space. Since $\mathcal{H}_{-\tau} = (\mathcal{H}_\tau)'$, (see Theorem 2.7.2), then the equality (2.7.27) implies that

$$\Phi' = \bigcup_{\tau \in T} \mathcal{H}_{-\tau}. \tag{2.7.33}$$

This construction takes a fairly simple form in the case where Φ is a countably Hilbert space and the norms of the spaces are monotone (see (2.7.25)). In this case, we arrive at the chain

$$\Phi' = \bigcup_{n=1}^{\infty} \mathcal{H}_{-n} \supseteq \ldots \supseteq \mathcal{H}_{-2} \supseteq \mathcal{H}_{-1} \supseteq \mathcal{H}_0 \supseteq \mathcal{H}_1 \supseteq \mathcal{H}_2 \supseteq \ldots \supseteq \bigcap_{n=1}^{\infty} \mathcal{H}_n = \Phi, \tag{2.7.34}$$

where each space $\mathcal{H}_{m+1}$ is densely imbedded in $\mathcal{H}_m$ and

$$\|\varphi\|_{\mathcal{H}_m} \le \|\varphi\|_{\mathcal{H}_{m+1}}, \quad \varphi \in \mathcal{H}_{m+1}; \; m \in \mathbb{Z}.$$

Note that the last results also hold for real spaces.

And one more lemma plays an important role in following chapters.

Lemma 2.7.5 *Suppose we have two riggings*

$$\mathcal{H}_- \supset \mathcal{H} \supset \mathcal{H}_+, \quad \mathcal{F}_- \supset \mathcal{F} \supset \mathcal{F}_+ = \mathcal{H}_+ \tag{2.7.35}$$

with equal positive spaces. Then there exists a unitary operator $U : \mathcal{H}_- \to \mathcal{F}_-$, $U\mathcal{H}_- = \mathcal{F}_-$ such that

$$(U\xi, f)_{\mathcal{F}} = (\xi, f)_{\mathcal{H}}, \quad \xi \in \mathcal{H}_-, \quad f \in \mathcal{H}_+ = \mathcal{F}_+. \tag{2.7.36}$$

This operator can be given by expression $U = \mathbf{I}_{\mathcal{F}}^{-1}\mathbf{I}_{\mathcal{H}}$, where $\mathbf{I}_{\mathcal{F}}$ and $\mathbf{I}_{\mathcal{H}}$ are standard two isometric isomorphisms in corresponding chains $\mathbf{I}_{\mathcal{F}}\mathcal{F}_- = \mathcal{F}_+$, $\mathbf{I}_{\mathcal{H}}\mathcal{H}_- = \mathcal{H}_+$.

Proof This is simple. So, standard operators $\mathbf{I}_{\mathcal{H}}$: $\mathcal{H}_- \longmapsto \mathcal{H}_+$, $\mathbf{I}_{\mathcal{F}}$: $\mathcal{F}_- \longmapsto \mathcal{F}_+$ are two isometric isomorphisms in the corresponding spaces. For such operators we have $\forall \alpha \in \mathcal{H}_-$, $f \in \mathcal{H}_+$

$$(\alpha, f)_{\mathcal{H}} = (\mathbf{I}_{\mathcal{H}}\alpha, f)_{\mathcal{H}_+} = (\alpha, \mathbf{I}_{\mathcal{H}}^{-1} f)_{\mathcal{H}_-}, \quad (\mathbf{I}_{\mathcal{H}}\alpha, \beta)_{\mathcal{H}} = (\alpha, \mathbf{I}_{\mathcal{H}}\beta)_{\mathcal{H}}$$

and the similar equality for the second chain (2.7.35). Using these equalities, we get

$$\begin{aligned}(U\xi, f)_{\mathcal{F}} &= (\mathbf{I}_{\mathcal{F}}^{-1}\mathbf{I}_{\mathcal{H}}\xi, f)_{\mathcal{F}} = (\mathbf{I}_{\mathcal{H}}\xi, f)_{\mathcal{F}_+} \\ &= (\mathbf{I}_{\mathcal{H}}\xi, f)_{\mathcal{H}_+} = (\xi, f)_{\mathcal{H}}, \quad \xi \in \mathcal{H}_-, \quad f \in \mathcal{H}_+ = \mathcal{F}_+.\end{aligned}$$

□

2.8 Tensor Products

To consider the multi-dimensional moment problem, the concept of tensor product will be needed. For simplicity, we consider only tensor products of Hilbert spaces.

Tensor Products of Spaces Let $(\mathcal{H}_k)_{k=1}^n$ be a finite sequence of separable Hilbert spaces and let and $(e_j^{(k)})_{j=0}^\infty$ be an orthonormal basis in $\mathcal{H}_k$. Consider a formal product

$$e_\alpha = e_{\alpha_1}^{(1)} \otimes \cdots \otimes e_{\alpha_n}^{(n)}, \tag{2.8.1}$$

where $\alpha = (\alpha_1, \dots, \alpha_n) \in \mathbb{Z}_+^n = \mathbb{Z}_+ \times \cdots \times \mathbb{Z}_+$ (n times), i.e., we consider the ordered sequence $(e_{\alpha_1}^{(1)}, \dots, e_{\alpha_n}^{(n)})$ and construct a Hilbert space spanned by the formal vectors (2.8.1), which are assumed to be an orthonormal basis of this space. The separable Hilbert space, thus constructed, is called the tensor product of the spaces $\mathcal{H}_1, \dots, \mathcal{H}_n$ and is denoted by $\mathcal{H}_1 \otimes \cdots \otimes \mathcal{H}_n = \otimes_{k=1}^n \mathcal{H}_k$. Its vectors have the form

$$f = \sum_{\alpha \in \mathbb{Z}_+^n} f_\alpha e_\alpha, \quad f_\alpha \in \mathbb{C}, \quad \|f\|^2_{\otimes_{k=1}^n \mathcal{H}_k} = \sum_{\alpha \in \mathbb{Z}_+^n} |f_\alpha|^2 < \infty, \tag{2.8.2}$$

$$(f, g)_{\otimes_{k=1}^n \mathcal{H}_k} = \sum_{\alpha \in \mathbb{Z}_+^n} f_\alpha \bar{g}_\alpha, \quad g = \sum_{\alpha \in \mathbb{Z}_n^+} g_\alpha e_\alpha \in \bigotimes_{k=1}^n \mathcal{H}_k.$$

Let $f^{(k)} = \sum_{j=0}^\infty f_j^{(k)} e_j^{(k)} \in \mathcal{H}_k, k = 1, 2, \dots, n$ be some vectors. By definition,

$$f = f^{(1)} \otimes \cdots \otimes f^{(n)} = \sum_{\alpha \in \mathbb{Z}_+^n} f_{\alpha_1}^{(1)} \dots f_{\alpha_n}^{(n)} e_\alpha. \tag{2.8.3}$$

The coefficients $f_\alpha = f_{\alpha_1}^{(1)} \dots f_{\alpha_n}^{(n)}$ of decomposition (2.8.3) satisfy condition (2.8.2). Therefore, vector (2.8.3) belongs to $\otimes_{k=1}^n \mathcal{H}_k$ in addition,

$$\|f\|_{\otimes_{k=1}^n \mathcal{H}_k} = \prod_{k=1}^n \|f_k\|_{\mathcal{H}_k}. \tag{2.8.4}$$

Clearly, the function

$$\mathcal{H}_1 \oplus \cdots \oplus \mathcal{H}_n \ni \langle f^{(1)}, \dots, f^{(n)} \rangle \mapsto f^{(1)} \otimes \cdots \otimes f^{(n)} \in \bigotimes_{k=1}^n \mathcal{H}_k$$

is linear in each argument and the linear span L of vectors (2.8.3) is dense in $\otimes_{k=1}^n \mathcal{H}_k$. This linear span is called an algebraic (non-completed) tensor product of the spaces $\mathcal{H}_1, \dots, \mathcal{H}_n$ and is denoted by a. $\otimes_{k=1}^n \mathcal{H}_k$. If L_k is a linear set in $\mathcal{H}_k$, $k = 1, 2, \dots, n$), then, by analogy,

$$\text{a.} \bigotimes_{k=1}^n L_k = \text{l.s.}\{f^{(1)} \otimes \cdots \otimes f^{(n)} \mid f^{(k)} \in L_k, \quad k = 1, 2, \dots, n\},$$

$$\bigotimes_{k=1}^n L_k = \text{c.l.s.}\{f^{(1)} \otimes \cdots \otimes f^{(n)} \mid f^{(k)} \in L_k, \quad k = 1, 2, \dots, n\}.$$

This definition of tensor product depends, clearly, on the choice of an orthonormal basis $\big(e_j^{(k)}\big)_{j=0}^\infty$ in each multiplier $\mathcal{H}_k$. However, it is easy to show that, by changing the basis, one always arrives at a tensor product which is isomorphic to the original one with preservation of the structure.

In fact, for the case of two Hilbert spaces $\mathcal{H}_1$ and $\mathcal{H}_2$, the concept of tensor product introduced above has the following meaning. We consider the linear span L of the formal products $f^{(1)} \otimes f^{(2)}$ and suppose that

$$\begin{aligned} (f^{(1)} + g^{(1)}) \otimes f^{(2)} &= f^{(1)} \otimes f^{(2)} + g^{(1)} \otimes f^{(2)}, \\ f^{(1)} \otimes (f^{(2)} + g^{(2)}) &= f^{(1)} \otimes f^{(2)} + f^{(1)} \otimes g^{(2)}, \\ (\lambda f(1)) \otimes f^{(2)} &= \lambda (f^{(1)} \otimes f^{(2)}), \\ f^{(1)} \otimes (\lambda f^{(2)}) &= \lambda (f^{(1)} \otimes f^{(2)}) \end{aligned} \quad , \tag{2.8.5}$$

$$f^{(1)}, g^1 \in \mathcal{H}_1, \; f^{(2)}, g^{(2)} \in \mathcal{H}_2, \; \lambda \in \mathbb{C}.$$

In other words, the linear space L is factorized by its linear subspace spanned by all possible vectors representable as differences between right-hand and left-hand sides of equalities (2.8.5).

Then L is equipped with a scalar product. For vectors of the form $f^{(1)} \otimes f^{(2)}$, it is defined by the formula

$$(f^{(1)} \otimes f^{(2)}, g^{(1)} \otimes g^{(2)})_{\mathcal{H}_1 \otimes \mathcal{H}_2} = (f^{(1)}, g^{(1)})_{\mathcal{H}_1} \, (f^{(2)}, g^{(2)})_{\mathcal{H}_2},$$
$$f^{(1)}, g^{(1)} \in \mathcal{H}_1, \quad f^{(2)}, g^{(2)} \in \mathcal{H}_2$$

and then bilinearly extended to other elements of the factorized space L.

Tensor Products of Operators We will present the definition of the tensor product of bounded operators.

Theorem 2.8.1 *Let $(\mathcal{H}_k)_{k=1}^n$ and $(G_k)_{k=1}^n$ be a finite sequence of separable Hilbert spaces and let $(A_k)_{k=1}^n$ be a sequence of operators $A_k \in \mathcal{L}(\mathcal{H}_k, G_k)$. The tensor product $A_1 \otimes \cdots \otimes A_n = \otimes_{k=1}^n A_k$ is defined by the formula*

$$(\bigotimes_{k=1}^n A_k) f = (\bigotimes_{k=1}^n A_k) \, (\sum_{\alpha \in \mathbb{Z}_+^n} f_\alpha e_\alpha) = \sum_{\alpha \in \mathbb{Z}_+^n} f_\alpha \big(A_1 e_{\alpha_1}^{(1)}\big) \otimes \cdots \otimes \big(A_n e_{\alpha_n}^{(n)}\big), \tag{2.8.6}$$

$$f \in \bigotimes_{k=1}^n \mathcal{H}_k.$$

It is stated that the series on the right-hand side of (2.8.6) *is weakly convergent in $\otimes_{k=1}^n G_k$ and defines the operator*

$$\bigotimes_{k=1}^n A_k \in \mathcal{L}\big(\otimes_{k=1}^n \mathcal{H}_k, \otimes_{k=1}^n G_k \big).$$

Furthermore,

$$\| \overset{n}{\underset{k=1}{\otimes}} A_k \| = \prod_{k=1}^{n} \|A_k\|. \tag{2.8.7}$$

The definition (2.8.6) yields the equality

$$\Big(\overset{n}{\underset{k=1}{\otimes}} A_k \Big) \big(f^{(1)} \otimes \cdots \otimes f^{(n)} \big) = A_1 f^{(1)} \otimes \cdots \otimes A_n f^{(n)}, \tag{2.8.8}$$

$$f^{(k)} \in \mathcal{H}_k; \ k = 1, 2, \ldots, n,$$

which uniquely determines the operator $\otimes_{k=1}^n A_k$. The mapping

$$\overset{n}{\underset{k=1}{\times}} \mathcal{L}(\mathcal{H}_k, G_k) \ni \langle A_1, \ldots, A_n \rangle \mapsto \overset{n}{\underset{k=1}{\otimes}} A_k \in \mathcal{L}\Big(\overset{n}{\underset{k=1}{\otimes}} \mathcal{H}_k, \otimes_{k=1}^n G_k \Big)$$

is linear in each variable. Note that, by using (2.8.6), we can get the relations

$$\Big(\overset{n}{\underset{k=1}{\otimes}} B_k \Big) \Big(\overset{n}{\underset{k=1}{\otimes}} A_k \Big) = \overset{n}{\underset{k=1}{\otimes}} (B_k A_k), \quad \Big(\overset{n}{\underset{k=1}{\otimes}} A_k \Big)^* = \overset{n}{\underset{k=1}{\otimes}} A_k^* \tag{2.8.9}$$

for $A_k \in \mathcal{L}(\mathcal{H}_k, G_k)$ and $B_k \in \mathcal{L}(G_k, F_k)$, $k = 1, 2, \ldots, n$.

Suppose that each A_k in Theorem 2.8.1 is a Hilbert-Schmidt operator. Then $\otimes_{k=1}^n A_k$ is also a Hilbert-Schmidt operator and

$$| \overset{n}{\underset{k=1}{\otimes}} A_k | = \prod_{k=1}^{n} | A_k |. \tag{2.8.10}$$

Corollary 2.8.2 *Let $\mathcal{H}_k \subseteq G_k$ be Hilbert spaces such that each imbedding operator $O_k : \mathcal{H}_k \to G_k$ is continuous $k = 1, 2, \ldots, n$. Then $\otimes_{k=1}^n \mathcal{H}_k \subseteq \otimes_{k=0}^n G_k$ and, for the corresponding imbedding operator, we have $O = \otimes_{k=0}^n O_k$. If the operators O_k are quasi-nuclear for all k, then the operator O is also quasi-nuclear.*

The Tensor Product of Chains Let us turn to the tensor product of chains of Hilbert spaces type (2.7.4)

$$\mathcal{H}_{-,k} \supseteq \mathcal{H}_{0,k} \supseteq \mathcal{H}_{+,k}, \quad k = 1, 2, \ldots, n. \tag{2.8.11}$$

According to Corollary 2.8.2 from Theorem 2.8.1, we have

$$\overset{n}{\underset{k=1}{\otimes}} \mathcal{H}_{-,k} \supseteq \overset{n}{\underset{k=1}{\otimes}} \mathcal{H}_{0,k} \supseteq \overset{n}{\underset{k=1}{\otimes}} \mathcal{H}_{+,k}. \tag{2.8.12}$$

Theorem 2.8.3 *The Hilbert space* $\otimes_{k=1}^{n}\mathcal{H}_{-,k}$ *can be considered as a negative space with respect to the zero* $\otimes_{k=1}^{n}\mathcal{H}_{0,k}$ *and the positive space* $\otimes_{k=1}^{n}\mathcal{H}_{+,k}$, *i.e.,* (2.8.12) *is a chain.*

The Projective Limit By using the properties of the tensor products of Hilbert spaces established above, we can investigate the same collection of problems for riggings by linear topological spaces. Consider a set of riggings of the form (2.7.30)

$$\Phi_k' \supseteq \mathcal{H}_{0,k} \supseteq \Phi_k, \quad k = 1, 2, \ldots, n, \tag{2.8.13}$$

where $\Phi_k = \operatorname{pr}\lim_{\tau_k \in T_k} \mathcal{H}_{+,\tau_k}$ is the projective limit of a family of Hilbert spaces $(\mathcal{H}_{+,\tau_k})_{\tau_k \in T_k}$, $k = 1, 2, \ldots, n$, directed by imbedding operators and satisfying the required conditions. For every multi-index $\tau = (\tau_1, \ldots, \tau_n) \in T = \times_{k=1}^{n} T_k$, we consider a collection of Hilbert riggings

$$\mathcal{H}_{-,\tau_k} \supseteq \mathcal{H}_{0,\tau_k} \supseteq \mathcal{H}_{+,\tau_k}, \quad k = 1, 2, \ldots, n, \tag{2.8.14}$$

where $\mathcal{H}_{-,\tau_k}$ is the Hilbert space dual to $\mathcal{H}_{+,\tau_k}$ with respect to $\mathcal{H}_{0,k}$. According to Theorem 2.8.2, for a fixed $\tau \in T$ the tensor product of chains (2.8.14) is also a chain

$$\bigotimes_{k=1}^{n} \mathcal{H}_{-,\tau_k} \supseteq \bigotimes_{k=1}^{n} \mathcal{H}_{0,\tau_k} \supseteq \bigotimes_{k=1}^{n} \mathcal{H}_{+,\tau_k}. \tag{2.8.15}$$

Since each family $(\mathcal{H}_{+,\tau_k})_{\tau_k \in T_k}$, $k = 1, 2, \ldots, n$ is directed, the family of Hilbert spaces $(\otimes_{k=1}^{n}\mathcal{H}_{+,\tau_k})_{\tau_k \in T_k}$ is also directed by imbedding operators following from (2.8.7). Furthermore, in addition, the set $\bigcap_{\tau \in T} \otimes_{k=1}^{n} \mathcal{H}_{+,\tau_k}$ is dense in each space of this family. According to our assumption, for any $\tau_k \in T_k$ we have $\|\cdot\|_{\mathcal{H}_{0,k}} \leq \|\cdot\|_{\mathcal{H}_{+,\tau_k}}$. Then, we conclude that

$$\forall \tau \in T, \quad \|\cdot\|_{\otimes_{k=1}^{n}\mathcal{H}_{0,k}} \leq \|\cdot\|_{\otimes_{k=1}^{n}\mathcal{H}_{+,\tau_k}}.$$

The tensor product $\otimes_{k=1}^{n}\Phi_k$ of the spaces Φ_k, $k = 1, 2, \ldots, n$), is defined as the projective limit

$$\bigotimes_{k=1}^{n} \Phi_k = \operatorname{pr}\lim_{\tau=(\tau_1,\ldots,\tau_n)\in T} \bigotimes_{k=1}^{n} \mathcal{H}_{+,\tau_k}, \quad T = \mathop{\times}_{k=1}^{n} T_k. \tag{2.8.16}$$

Hence, as the multi-index $\tau = (\tau_1, \ldots, \tau_n)$ in (2.8.15) runs over the indexing set T, we obtain the family of chains of the form (2.7.32) with $\mathcal{H}_0 = \otimes_{k=1}^{n}\mathcal{H}_{0,k}$. This enables us to apply the scheme described above to construct the chain

$$\Phi' \supseteq \mathcal{H}_0 = \bigotimes_{k=1}^{n} \mathcal{H}_{0,k} \supseteq \operatorname{pr}\lim_{\tau \in T} \bigotimes_{k=1}^{n} \mathcal{H}_{+,\tau_k} = \bigotimes_{k=1}^{n} \Phi_k = \Phi.$$

The space Φ' can be equipped by the topology of the inductive limit of negative spaces $\otimes_{k=1}^{n}\mathcal{H}_{-,\tau_k}$ of the chain (2.8.15). By definition,

$$\mathop{\otimes}_{k=1}^{n} \Phi'_k = \operatorname{ind} \lim_{\tau=(\tau_1,\ldots,\tau_n)\in T} \mathop{\otimes}_{k=1}^{n} \mathcal{H}_{-,\tau_k}. \tag{2.8.17}$$

Finally, we arrive at the chain

$$\mathop{\otimes}_{k=1}^{n} \Phi'_k \supseteq \mathop{\otimes}_{k=1}^{n} \mathcal{H}_{0,k} \supseteq \mathop{\otimes}_{k=1}^{n} \Phi_k. \tag{2.8.18}$$

Corollary 2.8.2 implies that if each Φ_k, $k = 1, 2, \ldots, n$ is a nuclear space, then the space $\otimes_{k=1}^{n}\Phi_k$ is also nuclear.

In fact, we have proved the theorem.

Theorem 2.8.4 *The tensor product of chains*

$$\Phi'_k \supseteq \mathcal{H}_{0,k} \supseteq \Phi_k, \quad k = 1, 2, \ldots, n,$$

where $\Phi_k = \operatorname{pr} \lim_{\tau_k\in T_k} \mathcal{H}_{+,\tau_k}$, *is also a chain*

$$\mathop{\otimes}_{k=1}^{n} \Phi'_k \supseteq \mathop{\otimes}_{k=1}^{n} \mathcal{H}_{0,k} \supseteq \mathop{\otimes}_{k=1}^{n} \Phi_k,$$

where spaces $\otimes_{k=1}^{n}\Phi_k$ *and* $\otimes_{k=1}^{n}\Phi'_k$ *are defined by equalities* (2.8.16), (2.8.17).

Furthermore, assume that each rigging (2.8.14) *is nuclear, then the rigging* (2.8.18), *constructed as indicated above is also nuclear.*

2.9 Representations of Continuous Multi-Linear Forms

First, we introduce the notion of a generalized kernel. Consider a collection of n chains

$$\mathcal{H}_{-,k} \supseteq \mathcal{H}_{0,k} \supseteq \mathcal{H}_{+,k}, \quad k = 1, 2, \ldots, n. \tag{2.9.1}$$

and their tensor product

$$\mathop{\otimes}_{k=1}^{n} \mathcal{H}_{-,k} \supseteq \mathop{\otimes}_{k=1}^{n} \mathcal{H}_{0,k} \mathop{\otimes}_{k=1}^{n} \mathcal{H}_{+,k}. \tag{2.9.2}$$

Elements

$$F, G, \dots \in \bigotimes_{k=1}^{n} \mathcal{H}_{0,k}, \quad U, V, \dots \in \bigotimes_{k=1}^{n} \mathcal{H}_{+,k}, \quad A, B, \dots \in \bigotimes_{k=1}^{n} \mathcal{H}_{-,k}$$

are called ordinary, smooth, and generalized kernels, respectively.

Also consider a continuous n-linear form $a(f^{(1)}, \dots, f^{(n)})$ regarded as a continuous function

$$\mathcal{H}_{0,1} \oplus \dots \oplus \mathcal{H}_{0,n} \ni \langle f^{(1)}, \dots, f^{(n)} \rangle \mapsto a(f^{(1)}, \dots, f^{(n)}) \in \mathbb{C}, \tag{2.9.3}$$

linear in each $f^{(k)}$ provided that all other variables are fixed. The continuity of (2.9.3) is equivalent to the existence of the estimate

$$|a(f^{(1)}, \dots, f^{(n)})| \le c \prod_{k=1}^{n} \|f^{(k)}\|_{\mathcal{H}_{0,k}}, \quad f^{(k)} \in \mathcal{H}_{0,k};\ k = 1, 2, \dots, n \tag{2.9.4}$$

with some constant $c > 0$. In every $\mathcal{H}_{0,k}$, we fix an orthonormal basis $(e_j^{(k)})_{j=0}^{\infty}$. Let $f^{(k)} = \sum_{\alpha_k=0}^{\infty} f_{\alpha_k} e_{\alpha_k}^{(k)}$ be the decomposition of a vector $f^{(k)} \in \mathcal{H}_{0,k}$ in this basis. We set $\alpha = (\alpha_1, \dots, \alpha_n) \in \mathbb{Z}_+^n$. In view of the continuity and multi-linearity of a, it can be represented in the form of a convergent series in its coordinates a_α and the coordinates of the vectors $f^{(k)}$

$$a(f^{(1)}, \dots, f^{(n)}) = \sum_{\alpha \in \mathbb{Z}_+^n} a_\alpha f_{\alpha_1}^{(1)} \cdots f_{\alpha_n}^{(n)}, \quad a_\alpha = a(e_{\alpha_1}^{(1)}, \dots, e_{\alpha_n}^{(n)}). \tag{2.9.5}$$

The main result of the subsection is the kernel theorem, which is based on two lemmas.

Lemma 2.9.1 *Let a be the continuous n-linear form* (2.9.3) *and $A_k \in \mathcal{L}(\mathcal{H}_{0,k})$, $k = 2, 3, \dots, n$ be Hilbert-Schmidt operators. Consider a continuous n-linear form*

$$\begin{aligned} \mathcal{H}_{0,1} \oplus \dots \oplus \mathcal{H}_{0,n} \ni \langle f^{(1)}, \dots, f^{(n)} \rangle &\mapsto b(f^{(1)}, \dots, f^{(n)}) \\ &= a(f^{(1)}, A_2 f^{(2)}, \dots, A_n f^{(n)}). \end{aligned}$$

It is stated that the coordinates $(b_\alpha)_{\alpha \in \mathbb{Z}_+^n}$ of the form b are such that

$$\sum_{\alpha \in \mathbb{Z}_+^n} |b_\alpha|^2 < \infty. \tag{2.9.6}$$

Conversely, if $A_k \in \mathcal{L}(\mathcal{H}_{0,k})$, $A_k \neq 0$, $k = 2, 3, \dots, n$ and, for any continuous form a, the coordinates of the form b satisfy condition (2.9.6), *then all A_k are Hilbert-Schmidt operators.*

Lemma 2.9.2 *Let*

$$\mathcal{H}_{0,1} \oplus \cdots \oplus \mathcal{H}_{0,n} \ni \langle f^{(1)}, \ldots, f^{(n)} \rangle \mapsto b(f^{(1)}, \ldots, f^{(n)})$$

be a continuous n-linear form. It can be represented in the form

$$b(f^{(1)}, \ldots, f^{(n)}) = (f^{(1)} \otimes \cdots \otimes f^{(n)}, K)_{\otimes_{k=1}^{n} \mathcal{H}_{0,k}},$$

where $K \in \otimes_{k=1}^{n} \mathcal{H}_{0,k}$, *if and only if its coordinates* $(b_\alpha)_{\alpha \in \mathbb{Z}_+^n}$ *satisfy the condition* (2.9.6).

Theorem 2.9.3 *Assume that chains* (2.9.1) *are such that all imbedding operators* $O_k : \mathcal{H}_{+,k} \mapsto \mathcal{H}_{0,k}$, $k = 2, 3, \ldots, n$ *are quasi-nuclear. Then every continuous n-linear form*

$$\mathcal{H}_{0,1} \oplus \cdots \oplus \mathcal{H}_{0,n} \ni \langle f^{(1)}, \ldots, f^{(n)} \rangle \mapsto a(f^{(1)}, \ldots, f^{(n)}) \in \mathbb{C}$$

can be associated with a unique generalized kernel

$$A \in \mathcal{H}_{0,k} \otimes \mathcal{H}_{-,2} \otimes \cdots \otimes \mathcal{H}_{-,n},$$

such that

$$a(f^{(1)}, u^{(2)}, \ldots, u^{(n)}) = (f^{(1)} \otimes u^{(2)} \otimes \cdots \otimes u^{(n)}, A)_{\otimes_{k=1}^{n} \mathcal{H}_{0,k}}, \tag{2.9.7}$$

$$f^{(1)} \in \mathcal{H}_{0,1}; \quad u^{(k)} \in \mathcal{H}_{+,k}; \ k = 2, 3, \ldots, n.$$

Conversely, if every form a of the indicated type admits representation (2.9.7) *with*

$$A \in \mathcal{H}_{0,1} \otimes H_{-,2} \otimes \cdots \otimes \mathcal{H}_{-,n},$$

then imbedding operators O_k, $k = 2, 3, \ldots, n$ *are quasi-nuclear.*

Corollary 2.9.4 *The statement of Theorem* 2.9.3 *can be made somewhat in "symmetric form", (if we simplify the result)*: assume that each imbedding $\mathcal{H}_{+,k} \to \mathcal{H}_{0,k}$, $k = 1, 2, \ldots, n$ in (2.9.1) is quasi-nuclear. Then every n-linear continuous form (2.9.3) admits the representation

$$a(u^{(1)}, \ldots, u^{(n)}) = (u^{(1)} \otimes \cdots \otimes u^{(n)}, \mathrm{A})_{\otimes_{k=1}^{n} \mathcal{H}_{0,k}}, \tag{2.9.8}$$

$$u^{(k)} \in \mathcal{H}_{+,k}, \ k = 1, 2, \ldots, n.$$

Moreover, the kernel $\mathrm{A} \in \otimes_{k=1}^{n} \mathcal{H}_{-,k}$ is determined uniquely in this representation.

Often Corollary 2.9.4 is used instead of Theorem 2.9.3.

Nuclear Riggings Let us now modify Theorem 2.9.3 for the case of nuclear riggings and forms defined in nuclear spaces. Consider a collection of n nuclear riggings given by (2.7.30) and (2.7.33)

$$\Phi_k' \supseteq \mathcal{H}_{0,k} \supseteq \Phi_k = \text{pr} \lim_{\tau_k \in T_k} \mathcal{H}_{+,\tau_k}, \quad k = 1, 2, \ldots, n. \tag{2.9.9}$$

According to the previous scheme (see (2.8.18)), we construct the nuclear chain

$$\bigotimes_{k=1}^{n} \Phi_k' \supseteq \bigotimes_{k=1}^{n} \mathcal{H}_{0,k} \supseteq \bigotimes_{k=1}^{n} \Phi_k \tag{2.9.10}$$

and consider, as in (2.9.3), n-linear forms

$$\bigoplus_{k=1}^{n} \Phi_k \ni \langle \varphi^{(1)}, \ldots, \varphi^{(n)} \rangle \mapsto a(\varphi^{(1)}, \ldots, \varphi^{(n)}) \in \mathbb{C}, \tag{2.9.11}$$

continuous in the direct product $\oplus_{k=1}^{n} \Phi_k$ of linear topological spaces Φ_k. Since

$$\bigotimes_{k=1}^{n} \Phi_k = \text{pr} \lim_{\tau = (\tau_1, \ldots, \tau_n) \in T} \bigotimes_{k=1}^{n} \mathcal{H}_{+,\tau_k}, \quad T = \mathop{\times}_{k=1}^{n} T_k,$$

one can easily prove that the continuity of the form (2.9.11) is equivalent to its continuity in the norm of the space $\otimes_{k=1}^{n} \mathcal{H}_{+,\tau_k}$ with certain set $\tau = (\tau_1, \ldots, \tau_n)$ and, hence, to the validity of the estimate

$$|a(\varphi^{(1)}, \ldots, \varphi^{(n)})| \leq c_\tau \prod_{k=1}^{n} \|\varphi^{(k)}\|_{\mathcal{H}_{+,\tau_k}}, \tag{2.9.12}$$

$$c_\tau > 0, \; \varphi^{(k)} \in \mathcal{H}_{+,\tau_k}; \; k = 1, 2, \ldots, n.$$

Theorem 2.9.5 *For the nuclear riggings* (2.9.9)*, every continuous n-linear form* (2.9.11) *admits the representation*

$$a(\varphi^{(1)}, \ldots, \varphi^{(n)}) = (\varphi^{(1)} \otimes \cdots \otimes \varphi^{(n)}, A)_{\otimes_{k=1}^{n} \mathcal{H}_{0,k}}, \tag{2.9.13}$$

$$\varphi^{(k)} \in \Phi_k, \; k = 1, 2, \ldots, n;$$

moreover, the generalized kernel $A \in \otimes_{k=1}^{n} \Phi_k'$ *in this representation is uniquely determined for given a.*

In the case where each Φ_k in (2.9.11) is a countable Hilbert space, then each multi-linear form (2.9.11) is also continuous with respect to the collection of its variables, i.e., in $\times_{k=1}^{n} \Phi_k$. Therefore, for countable Hilbert space Φ_k, Theorem 2.9.1 remains true for separately continuous forms.

Bilinear Forms By using Theorem2.9.3, one can also establish the kernel theorem for bilinear (sesquilinear) forms. Thus, consider the chain

$$\mathcal{H}_- \supseteq \mathcal{H}_0 \supseteq \mathcal{H}_+ \tag{2.9.14}$$

and assume that $\mathcal{H}_+$ is equipped with an involution, which is, at the same time, an involution in $\mathcal{H}_0$. It means that one can indicate an antilinear mapping $\mathcal{H}_+ \ni u \mapsto u^* \in \mathcal{H}_+$ acting in $\mathcal{H}_+$ such that $(u^*)^* = u$, $(u^*, v^*)_{\mathcal{H}_+} = \overline{(u, v)_{\mathcal{H}_+}}$ and $(u^*, v^*)_{\mathcal{H}_0} = \overline{(u, v)_{\mathcal{H}_0}}$ $(u, v \in \mathcal{H}_+)$.

It is easy to see that, in this case, $(u^*, v^*)_{\mathcal{H}_-} = \overline{(u, v)_{\mathcal{H}_-}}$, i.e., the operation "$*$" is also an involution in $\mathcal{H}_-$ and, consequently, can be extended by continuity to the involution in the whole $\mathcal{H}_- : \mathcal{H}_- \ni \alpha \mapsto \alpha^* \in \mathcal{H}_-$. The restriction of this mapping $\mathcal{H}_- \supseteq \mathcal{H}_0 \ni f \mapsto f^* \in \mathcal{H}_0$ is an involution in $\mathcal{H}_0$. Clearly, $(\mathbf{I}\alpha)^* = \mathbf{I}\alpha^*$, $\alpha \in \mathcal{H}_-$. If the spaces of chain (2.9.14) are equipped with this involution, then we say that (2.9.14) is a chain with the involution "$*$".

Let $a(f, g)$ be a bilinear form defined in $\mathcal{H}_0$, i.e., a continuous function $\mathcal{H}_0 \oplus \mathcal{H}_0 \ni \langle f, g\rangle \mapsto a(f, g) \in \mathbb{C}$ is linear in the first variable and antilinear in the second variable.

Theorem 2.9.6 *Assume that the imbedding $\mathcal{H}_+ \to \mathcal{H}_0$ in chain* (2.9.14) *is quasi-nuclear. Then, for every continuous bilinear form $a(f, g)$, $f, g \in \mathcal{H}_0$ one can construct a unique generalized kernel $A_a \in \mathcal{H}_0 \otimes \mathcal{H}_-$ such that*

$$a(u, g) = (A_a, g \otimes u^*)_{\mathcal{H}_0 \otimes \mathcal{H}_0}, \quad u \in \mathcal{H}_+;\ g \in \mathcal{H}_0. \tag{2.9.15}$$

Conversely, if every form a of the indicated type admits representation (2.9.15) *with $A_a \in \mathcal{H}_0 \otimes \mathcal{H}_-$, then the imbedding $\mathcal{H}_+ \to \mathcal{H}_0$ is quasi-nuclear.*

In Theorem 2.9.3 one can fix any other variable instead of the first one. Hence, in addition to (2.9.15), the form a admits the following representation

$$a(f, v) = (\mathrm{A}'_a, v \otimes f^*)_{\mathcal{H}_0 \otimes \mathcal{H}_0}, \quad f \in \mathcal{H}_0,\ v \in \mathcal{H}_+, \tag{2.9.16}$$

where the kernel $\mathrm{A}'_a \in \mathcal{H}_- \otimes \mathcal{H}_0$ is uniquely determined for a given form a. It follows from (2.9.15) and (2.9.16), that

$$a(v^*, u) = (\mathrm{A}_a, u \otimes v)_{\mathcal{H}_0 \otimes \mathcal{H}_0} = (\mathrm{A}'_a, u \otimes v)_{\mathcal{H}_0 \otimes \mathcal{H}_0}$$

for $u, v \in \mathcal{H}_+$, i.e., the restrictions of the functionals A_a and A'_a to $\mathcal{H}_+ \otimes \mathcal{H}_+$ mutually coincide. Denote this common restriction by A. Thus, under the conditions of Theorem 2.9.6, we have the representation

$$a(u, v) = \left(\mathrm{A}, v \otimes u^*\right)_{\mathcal{H}_0 \otimes \mathcal{H}_0}, \quad u, v \in \mathcal{H}_+ \tag{2.9.17}$$

with the kernel $\mathrm{A} \in \mathcal{H}_- \otimes \mathcal{H}_-$.

The assertions of Theorem 2.9.5 for the case of nuclear riggings can also be easily generalized to the case of bilinear forms. Thus, let Φ be a nuclear space with the natural involution $* \ \Phi \to \Phi$. Consider a bilinear form defined in Φ, i.e., a continuous function $\Phi \times \Phi \ni \langle \varphi, \psi \rangle \mapsto a(\varphi, \psi) \in \mathbb{C}$ linear in the first variable and antilinear in the second variable. Then $a(\varphi, \psi) = \big(\mathrm{A}, \psi \otimes \varphi^*\big)_{\mathcal{H}_0 \otimes \mathcal{H}_0}$, where $\mathrm{A} \in \Phi' \otimes \Phi'$.

Let $A \in \mathcal{L}(\mathcal{H}_0)$. Given A we construct the continuous bilinear form $a(f, g) = (Af, g)_{\mathcal{H}_0}$, $f, g \in \mathcal{H}_0$. Then, by using this form, we construct, according to (2.9.17), the kernel $\mathcal{A}_A = \mathcal{A} \in \mathcal{H}_- \otimes \mathcal{H}_-$. It is called the kernel of the operator A:

$$(Au, v)_{\mathcal{H}_0} = (\mathcal{A}, v \otimes u^*)_{\mathcal{H}_0 \otimes \mathcal{H}_0}, \quad u, v \in \mathcal{H}_+. \tag{2.9.18}$$

A Separate Kernel Theorem In order to formulate the kernel theorem for the case of multi-linear (bilinear) forms defined in the spaces $L_2(G)$, $G \subseteq \mathbb{R}^N$, $N \in \mathbb{N}$, with respect to the Lebesgue measure, one can use the Sobolev space with a quasi-nuclear imbedding in $L_2(G)$ or, in the case of nuclear riggings, the spaces $C^\infty(G)$, $\mathcal{S}(\mathbb{R}^N)$ and $\mathfrak{D}(\mathbb{R}^N)$.

Let us give an "elementary" kernel theorem.

Theorem 2.9.7 *Let $L_2(G) \oplus L_2(G) \ni \langle f, g \rangle \mapsto a(f, g) \in \mathbb{C}$ be a continuous bilinear form. Then there exists a kernel $T \in C(\mathbb{R}^N \times \mathbb{R}^N)$ such that*

$$a(u, v) = \int\limits_{G \times G} T(x, y)(\mathfrak{D}u)(y)\overline{(\mathfrak{D}v)(x)}\, dx dy \tag{2.9.19}$$

$$\mathfrak{D} = D_1 \dots D_N; \ u, v \in C_0^N(G),$$

for the smooth functions finite with respect to G.

Note that if the kernel T is smooth, then (2.9.19) has the form

$$(Af, g)_{\mathcal{H}_0} = \int\limits_{R \times R} K(x, y) f(y) \overline{g(x)}\, d(\mu \times \mu)(x, y) = (K, g \otimes f^*)_{\mathcal{H}_0 \otimes \mathcal{H}_0}$$

with $K = \mathcal{A}_A = \mathfrak{D}_x \mathfrak{D}_y T$. Generally speaking, these derivatives should be regarded in the sense of generalized functions and the last kernel is generalized. Note that one can indicate a quasi-nuclear rigging (2.9.14) of the space $\mathcal{H}_0 = L_2(G)$, which transforms (2.9.17) into (2.9.19). Also note that a relation, similar to (2.9.19), can be written for the case of multi-linear forms.

The formula (2.9.19) is connected with the following expression for the elements of the matrix $(a_{jk})_{j,k=1}^m$ of an operator A in the m-dimensional space $\mathbb{C}^m$:

$$a_{jk} = (A\delta_k, \delta_j)_{\mathbb{C}^N}, \quad j, k = 1, 2, \dots, m. \tag{2.9.20}$$

Here, $\delta_j = (\delta_{jn})_{n=1}^m$ are vectors of an orthonormal basis in $\mathbb{C}^m$. Let us clarify this assertion. On passing from $\mathbb{C}^m$ to $L_2(G)$, a role of δ_j must play δ-functions $\delta_x(\xi)$, $x \in G$, but they do not belong to $L_2(G)$, and, therefore, the relation (2.9.20) becomes meaningless. But at the same time, one can transform the basis in (2.9.20) by setting $\omega_j = \sum_{l=1}^{j} \delta_l$, $j = 1, 2, \ldots, m$, and construct the matrix $t_{jk} = (A\omega_k, \omega_j)_{\mathbb{C}^m}$, $j, k = 1, 2, \ldots, m$. This matrix enables us to present the action of the operator in a simple form. The kernel T is an analogue of the matrix $(t_{jk})_{j,k=1}^m$, e.g., if $G = (0, \infty) \subset \mathbb{R}$, then we can formally write

$$\omega_x(\xi) = \int_0^x \delta_z(\xi)\, dz, \qquad x \in (0, \infty)).$$

The generalized kernel $\mathfrak{D}_x\mathfrak{D}_y T$ is an analogue of matrix (2.9.20) and the relation (2.9.19) is similar to the formula that reconstructs the action of the operator A in terms of the matrix $(t_{jk})_{j,k=1}^m$.

Completions of a Space with Respect to Two Different Norms Let L be a linear space, let $L \ni f \mapsto \|f\|_E \geq 0$ be a norm in this space, and let E be the completion of L with respect to this norm. Recall that E consists of the classes f_E of equivalent fundamental sequences $(f_n)_{n=1}^\infty$, $f_n \in L$. The relation of equivalence $(f_n)_{n=1}^\infty \sim (g_n)_{n=1}^\infty$ means that $\|f_n - g_n\|_E \to 0$ as $n \to \infty$. The linear structure in the space E is induced by the linear operations over the sequences. If $\|\cdot\|_E$ is a Hilbert norm, i.e., if L is equipped with the scalar product $(f, g)_E$, $(f, g \in L)$ and $\|\cdot\|_E = (\cdot, \cdot)_E^{1/2}$, then E is a Hilbert space. The space L is imbedded in E by the identification of $f \in L$ with a class that contains a stationary sequence $(f, f, \ldots)$.

Now, assume that L is a linear set with two different norms, namely, $L \ni f \mapsto \|f\|_{E_1} \geq 0$ and $L \ni f \mapsto \|f\|_{E_2} \geq 0$. Let E_1 and E_2 be the corresponding completions of L. Assume that these norms are comparable in the following sense:

$$\|f\|_{E_1} \leq \|f\|_{E_2}, \quad f \in L \tag{2.9.21}$$

(it is clear that, instead of (2.9.21) one can write the inequality $\exists c > 0$, $\forall f \in L$, $\|f\|_{E_1} \leq c\|f\|_{E_2}$).

Arguing somewhat inaccurately, one can conclude that, as a result of the completion, inequality (2.9.21) yields the inclusion $E_1 \supseteq E_2$ and the inequality $\|f\|_{E_1} \leq \|f\|_{E_2}$, $f \in E_2$. However, it has been already mentioned that this is not true. Let us clarify the situation.

Assume that $(f_n)_{n=1}^\infty$, $f_n \in L$ is a fundamental sequence with respect to the norm $\|\cdot\|_{E_2}$. Then, by virtue of (2.9.21), it is also fundamental with respect to the norm $\|\cdot\|_{E_1}$. Let $(f_n)_{n=1}^\infty \in f_{E_2} \in E_2$ and $(f_n)_{n=1}^\infty \in f_{E_1} \in E_1$. Let us associate the vector f_{E_2} with the vector f_{E_1}. This mapping is well defined. Indeed, if $(g_n)_{n=1}^\infty \in f_{E_2}$, i.e., $(f_n)_{n=1}^\infty \sim (g_n)_{n=1}^\infty$ with respect to $\|\cdot\|_{E_2}$, then, by virtue of (2.9.21), the same relation of equivalence can be written for $\|\cdot\|_{E_1}$.

Thus, we have constructed the mapping

$$E_2 \ni f_{E_2} \mapsto Qf_{E_2} = f_{E_1} \in E_1.$$

It follows from the method, according to which the completions are equipped with a linear structure, that Q is linear. Indeed, by virtue of (2.9.21):

$$\|Qf_{E_2}\|_{E_1} = \|f_{E_1}\|_{E_1} = \lim_{n\to\infty} \|f_n\|_{E_1} \leq \lim_{n\to\infty} \|f_n\|_{E_2} = \|f_{E_2}\|_{E_2},$$
$$(f_n)_{n=1}^{\infty} \in f_{E_2} \in E_2.$$

Note that the restriction $Q \restriction L$ is the imbedding operator which imbeds $L \subseteq E_2$ in the set L regarded as a subset of the space E_1.

Consider a subspace

$$\text{Ker}\, Q = \{f \in E_2 |\, Qf = 0\} \subseteq E_2. \tag{2.9.22}$$

If Ker $Q = \{0\}$, then E_2 can be identified with the range $\mathfrak{R}(Q)$ and one can assume that $E_2 \subseteq E_1$ and $\|f\|_{E_1} \leq \|f\|_{E_2}$, $f \in E_2$. In the general case, these inclusion and inequality hold true for the factor-space $E_2/\text{Ker}Q \subseteq E_1$.

In the case where E_2 is a Hilbert space, instead of the factor-space, we can take the orthogonal complement $E_2 \ominus \text{Ker}\, Q$. Note that the second ultimate case (Ker $Q = E_2$) is impossible. Moreover, $L \cap \text{Ker}\, Q = \{0\}$, as follows from the fact that $Q \restriction L$, is the indicated imbedding.

Let us formulate these results as a theorem.

Theorem 2.9.8 *Let L be a linear space with two norms $\|\cdot\|_{E_1}$ and $\|\cdot\|_{E_2}$ comparable in the sense of* (1.9.21), *let E_1 and E_2 be the corresponding completions of L, and let Q be the operator introduced above. Then*

$$E_1 \supseteq E_2/Ker\, Q, \quad \|f\|_{E_1} \leq \|f\|_{E_2/Ker\, Q}, \quad f \in E_2/KerQ. \tag{2.9.23}$$

If E_2 is a Hilbert space, then $E_2 \ominus Ker\, Q$ plays a role of the factor-space in (2.9.23). *If Ker $Q = \{0\}$, then $E_1 \supseteq E_2$ and for $f \in E_2$, $\|f\|_{E_1} \leq \|f\|_{E_2}$.*

The theorem below immediately follows from the construction of the operator Q and relation (2.9.22).

Theorem 2.9.9 *The kernel is trivial, i.e., Ker $Q = \{0\}$ if and only if any sequence $(f_n)_{n=1}^{\infty}$, $f_n \in L$ fundamental with respect to the norm $\|\cdot\|_{E_2}$ and converges to zero with respect to the norm $\|\cdot\|_{E_1}$, also converges to zero with respect to the norm $\|\cdot\|_{E_2}$.*

2.10 Semi-Bounded Bilinear Forms

It is known that an arbitrary continuous bilinear form a in a Hilbert space $\mathcal{H}$ admits the representation

$$a(f, g) = (Af, g)_{\mathcal{H}}, \quad f, g \in \mathcal{H},$$

where A is a bounded operator in $\mathcal{H}$. An important role is played the similar theorem on representations in the case of forms that are not continuous.

Further, the material is based on statements about the completion of a space according to two norms.

Lemma 2.10.1 *Assume that* $\mathfrak{D}(A) = \{u \in \mathcal{H}_+ \mid \mathbf{I}^{-1}u \in \mathcal{H}_0\}$. *In* $\mathcal{H}_0$, *we consider an operator* $A = \mathbf{I}^{-1} \restriction \mathfrak{D}(A)$. *It is stated that* A *is self-adjoint and satisfies the relations*

$$(u, v)_{\mathcal{H}_+} = (Au, v)_{\mathcal{H}_0}, \quad u \in \mathfrak{D}(A), \; v \in \mathcal{H}_+, \tag{2.10.1}$$
$$(u, v)_{\mathcal{H}_+} = (\sqrt{A}u, \sqrt{A}v)_{\mathcal{H}_0}, \quad u, v \in \mathcal{H}_+ = \mathfrak{D}(\sqrt{A}).$$

Positive Forms Here, we introduce the notion of a prechain, which is closely related to the notion of the chain. In fact, the presence of a prechain is equivalent to the determination of a positive form and the existing of the close relation between the theories of bilinear forms and rigged Hilbert spaces is largely based on this fact.

Let $\mathcal{H}_0$ be a Hilbert space and let L be a linear set dense in this space with a scalar product $(f, g)_{L_+}$, $f, g \in L$ such that $\|f\|_{\mathcal{H}_0} \leq \|f\|_{L_+}$, $\|f\|_{L_+} = \sqrt{(f, f)_{L_+}}$, $f \in L$. In this case, we say that the *prechain*

$$\mathcal{H}_0 \supseteq L \tag{2.10.2}$$

is defined.

Denote by L_+ the completion of L with respect to the norm $\|\cdot\|_{L_+}$. In this case, all requirements of the scheme are satisfied for $E_1 = \mathcal{H}_0$ and $E_2 = L_+$. Let $Q : L_+ \to \mathcal{H}_0$ be the corresponding operator. According to (2.9.23), for a given prechain (2.10.2), one can construct the chain

$$\mathcal{H}_- \supseteq \mathcal{H}_0 \supseteq \mathcal{H}_+ = L_+ \ominus \operatorname{Ker} Q. \tag{2.10.3}$$

We say that the prechain (2.10.2) is *closed* if L is complete with respect to the norm $\|\cdot\|_{L_+}$ and *closable* (admitting a closure) if $\operatorname{Ker} Q = \{0\}$ (it is obvious that closeness yields a closability). In view of Theorem 2.9.9 , we can formulate the following criterion of the closability.

The prechain (2.10.2) *is closable if and only if every sequence* $(f_n)_{n=1}^{\infty} \subset L$ *fundamental with respect to the norm* $\|\cdot\|_{L_+}$ *and convergent to zero with respect to the norm* $\|\cdot\|_{\mathcal{H}_0}$ *converges to zero with respect to the norm* $\|\cdot\|_{L_+}$.

The prechain $\mathcal{H} \supseteq L_+$ constructed according to the closable prechain (2.10.1) by completing L with respect to $\|\cdot\|_{L_+}$ is called a closure of the prechain (2.10.1). Throughout this book, we consider only closed or closable prechains $\mathcal{H}_0 \supseteq L$ (in the sense of closeness $L_+ = L$). Every prechain of this sort can be extended to the chain

$$\mathcal{H}_- \supseteq \mathcal{H}_0 \supseteq \mathcal{H}_+ = L_+, \tag{2.10.4}$$

by constructing the corresponding negative space.

Now, we pass in our presentations to the concept of forms. *A function $\mathfrak{D}(a) \times \mathfrak{D}(a) \ni \langle f, g\rangle \mapsto a(f, g) \in \mathbb{C}$ linear in the first variable and antilinear in the second variable is called a bilinear (sesquilinear) form a in a Hilbert space $\mathcal{H}_0$ (where $\mathfrak{D}(a)$ is a linear set dense in $\mathcal{H}_0$ and it is the domain of the form a). The diagonal values of the functional $a(\,\cdot\,,\,\cdot\,)$ represent the quadratic form $a[\cdot]$ associated with the bilinear form under consideration, i.e., $\mathfrak{D}(a) \ni f \mapsto a[f] = a(f, f) \in \mathbb{C}$.*

For a given quadratic form, one can uniquely reconstruct the corresponding bilinear form by using the polarization identity

$$a(f, g) = \frac{1}{4}\big(a[f+g] - a[f-g] + ia[f+ig] - ia[f-ig]\big), \tag{2.10.5}$$
$$f, g \in \mathfrak{D}(a).$$

The linear operations are introduced on bilinear forms in a natural way. Thus, if a and b are two bilinear forms and, at the same time, the intersection $\mathfrak{D}(a) \cap \mathfrak{D}(b)$ is dense in $\mathcal{H}_0$, then the bilinear form $a + b$ is defined by the equality

$$(a+b)(f, g) = a(f, g) + b(f, g), \quad f, g \in \mathfrak{D}(a+b) = \mathfrak{D}(a) \cap \mathfrak{D}(b).$$

The product λa, where $\lambda \in \mathbb{C}$, is always defined. Indeed,

$$(\lambda a)(f, g) = \lambda a(f, g), \quad f, g \in \mathfrak{D}(\lambda a) = \mathfrak{D}(a).$$

For a given bilinear form a, one can always construct the adjoint bilinear form a^* according to the equality

$$a^*(f, g) = \overline{a(g, f)}, \quad f, g \in \mathfrak{D}(a^*) = \mathfrak{D}(a).$$

A bilinear form a is called Hermitian if $a^* = a$. It follows from (2.10.5) that in order for a to be Hermitian, it is necessary and sufficient that the quadratic form $a[\cdot]$ take only real values. Every bilinear form a can be expressed as a linear combination of two Hermitian forms $\operatorname{Re} a$ and $\operatorname{Im} a$, namely,

$$a = \operatorname{Re} a + i\operatorname{Im} a, \quad \operatorname{Re} a = \frac{1}{2}(a + a^*), \quad \operatorname{Im} a = \frac{1}{2i}(a - a^*).$$

A bilinear form a is called positive with vertex $\alpha > 0$ *if*

$$a(f, f) \geq \alpha \|f\|^2_{\mathcal{H}_0}, \quad f \in \mathfrak{D}(a). \tag{2.10.6}$$

This is convenient to assume that $\alpha = 1$. Positive forms are always Hermitian because $a[\cdot]$ is real-valued.

For a given positive form a in $\mathcal{H}_0$, one can naturally construct the prechain (2.10.2) by setting $L = \mathfrak{D}(a)$, $(f, g)_{L_+} = a(f, g)$, $f, g \in L$. Conversely, the prechain (2.10.2) determines the positive form

$$a(f, g) = (f, g)_{L_+}, \quad f, g \in \mathfrak{D}(a) = L.$$

The definitions introduced above for prechains can be easily reformulated for positive forms.

A positive form a is called closed if the corresponding prechain is closed. The closure $\tilde{a}$ *of a closable form a is defined by the equality* $\tilde{a}(f, g) = (f, g)_{L_+}$, *where* $f, g \in \mathfrak{D}(\tilde{a}) = L_+$, L *and* $(\cdot, \cdot)_{L_+}$ *are constructed according to* a.

Thus, to calculate $\tilde{a}(f, g)$ for $f, g \in \mathfrak{D}(\tilde{a}) \subseteq \mathcal{H}_0$, we must construct sequences $(f_n)_{n=1}^{\infty}$, $(g_m)_{m=1}^{\infty} \subset \mathfrak{D}(a)$, fundamental in the norm $(a[\cdot])^{1/2}$ and convergent in $\mathcal{H}_0$ to f and g , respectively. Then

$$\tilde{a}(f, g) = \lim_{n,m\to\infty} a(f_n, g_m).$$

Lemma 2.10.1 yields the following theorem on the representation of a positive form.

Theorem 2.10.2 *Let a be a closed positive bilinear form with vertex* $\alpha = 1$. *It is stated that there exists a self-adjoint operator* $A \geq \mathbb{I}$ *acting in the space* $\mathcal{H}_0$ *such that*

$$a(f, g) = (Af, g)_{\mathcal{H}_0}, \quad f \in \mathfrak{D}(A) \subseteq \mathfrak{D}(a), \ g \in \mathfrak{D}(a). \tag{2.10.7}$$

Its domain $\mathfrak{D}(A)$ *is dense in* $\mathfrak{D}(a)$ *with respect to the norm* $(a\,[\cdot])^{1/2}$ *and, moreover,* $\|Af\|_{\mathcal{H}_0} \geq (a[f])^{1/2}$, $f \in \mathfrak{D}(a)$. *In addition to* (2.10.7), *the form a admits the following representation in terms of the operator* $\sqrt{A}$:

$$a(f, g) = (\sqrt{A}f, \sqrt{A}g)_{\mathcal{H}_0}, \quad f, g \in \mathfrak{D}(\sqrt{A}) = \mathfrak{D}(a). \tag{2.10.8}$$

Semi-Bounded Forms Usually, the representation theory is based on semi-bounded forms. *A bilinear form a in the space* $\mathcal{H}_0$ *is called semi-bounded (from below) with vertex* $\alpha \in \mathbb{R}$ *if*

$$a(f, f) \geq \alpha \ \|f\|^2_{\mathcal{H}_0}, \quad f \in \mathfrak{D}(a). \tag{2.10.9}$$

If $\alpha = 0$, then the form a is called non-negative. For $\alpha > 0$, it is called a positive form and has been already introduced above.

It is clear that semi-bounded forms are Hermitian. Each semi-bounded form a can be associated with a positive form a_p (whose vertex is equal to one) by setting

$$a_p(f, g) = a(f, g) + (1 - \alpha)(f, g)_{\mathcal{H}_0} \quad f, g \in \mathfrak{D}(a_p) = \mathfrak{D}(a). \tag{2.10.10}$$

Definition, related to the form a, are formulated in terms of the form a_p, namely, *a is closed (closable) if a_p is closed (closable); the closure $\tilde{a}$ of a closable form a is determined, according to* (2.10.10) *by the formula*

$$\tilde{a}(f, g) = \tilde{a}_p(f, g) - (1 - \alpha)(f, g)_{\mathcal{H}_0}, \quad f, g \in \mathfrak{D}(\tilde{a}) = \mathfrak{D}(\tilde{a}_p); \tag{2.10.11}$$

this closure is a semi-bounded form with the same vertex α.

In the case of positive forms, we can act in a somewhat different manner. If a is a positive form with vertex α, then $\frac{1}{\alpha}a$ is a positive form with vertex one and the application of the definitions introduced to $\frac{1}{\alpha}a$ leads us to corresponding definitions for a (since the norms $(a_p[\,\cdot\,])^{1/2}$ and $(\frac{1}{\alpha}a\,[\,\cdot\,])^{1/2}$ are equivalent). This implies that, in the case of semi-bounded forms, a_p can also be defined by relation (2.10.10) with "1" replaced with $\varepsilon > 0$.

For semi-bounded forms, Theorem 2.10.2 takes the following form.

Theorem 2.10.3 *Let a be a closed semi-bounded bilinear form with vertex $\alpha \in \mathbb{R}$. There exists a self-adjoint operator $A \geq \alpha\mathbb{I}$ acting in the space $\mathcal{H}_0$ such that the representation* (2.10.7) *holds true. Its domain $\mathfrak{D}(A)$ is dense in $\mathfrak{D}(a)$ with respect to the norm $(a_p[\,\cdot\,])^{1/2}$. If a is non-negative, then it is also representable in the form* (2.10.8).

If the form a admits a closure $\tilde{a}$, then we can write it in the form (2.10.7) and (2.10.8), whence we get the required representations for the form a. Let us dwell upon an important procedure for constructing extensions of semi-bounded operators to self-adjoint operators, which was introduced by Friedrichs K. O. Consider an Hermitian semi-bounded operator $A \geq \alpha\mathbb{I}$, $\alpha \in \mathbb{R}$ acting in $\mathcal{H}_0$ with dense domain. It generates, in a standard way, a semi-bounded form a with vertex α

$$a(f, g) = (Af, g)_{\mathcal{H}_0}, \quad f, g \in \mathfrak{D}(a) = \mathfrak{D}(A). \tag{2.10.12}$$

The properties of this form are described by the following theorem.

Theorem 2.10.4 (Friedrichs) *The bilinear form a, defined in (2.10.12), admits the closure $\tilde{a}$ representable in the form $\tilde{a}(f, g) = (A_F f, g)_{\mathcal{H}_0}$, $f \in \mathfrak{D}(A_F) \subseteq \mathfrak{D}(\tilde{a})$, $g \in \mathfrak{D}(\tilde{a})$, where $A_F \geq \alpha\mathbb{I}$ is a self-adjoint operator in $\mathcal{H}_0$ and it is an extension of the operator A (the so-called Friedrichs extension). The operator A_F is the unique self-adjoint extension of A whose domain lies in $\mathfrak{D}(\tilde{a})$.*

The Form Sum of Operators Let A and B be, respectively, self-adjoint and Hermitian operators in the space $\mathcal{H}_0$. It is necessary to study the operator $A+B$, $\mathfrak{D}(A+B)=\mathfrak{D}(A)\cap\mathfrak{D}(B)$ and indicate conditions under which it is self-adjoint or essentially self-adjoint. For this, the form sum method is considered here.

First, we present the well-known result for forms which allows one, under certain restrictions, to make the operator $A+B$ meaningful even in the case where $\mathfrak{D}(A)\cap\mathfrak{D}(B)=\{0\}$.

Theorem 2.10.5 (KLMN) *Let a be a closed positive form and let b be an Hermitian form on $\mathfrak{D}(b)=\mathfrak{D}(a)$ such that*

$$|b(f,f)|\leq pa(f,f)+q(f,f)_{\mathcal{H}_0},\quad f\in\mathfrak{D}(a), \tag{2.10.13}$$

for some $p\in[0,1)$ and $q\in\mathbb{R}$. Then the form $a+b$, $\mathfrak{D}(a+b)=\mathfrak{D}(a)$ is semi-bounded and closed.

In the name of the theorem, the abbreviation KLMN means that the following authors contributed to the theorem: Kato T. (1955), Lax P. and Milgram A. (1954), Lions J.-L. (1961), Nelson E. (1964) [151, 196, 200, 226].

2.11 The Generalized Eigenvectors Expansion

As already noted earlier, the expansion in eigenvectors of the self-adjoint operator A in a finite-dimensional Hilbert space $\mathcal{H}$ cannot be directly generalized to the case of infinite-dimensional space because the operator A may have no eigenvectors.

We will give a simple example of a self-adjoint operator A that has no eigenvectors. Assume that $\mathcal{H}=L_2((a,b))$ with respect to Lebesgue measure. Consider the operator $(Af)(x)=xf(x)$, $f\in L_2((a,b))$, $x\in(a,b)$ which is bounded and self-adjoint and has the spectrum $S(A)=[a,b]$. The equation for the eigenvector $\varphi\in L_2((a,b))$ that corresponds to a point $\lambda\in[a,b]$ of the spectrum has the form

$$(x-\lambda)\varphi(x)=0. \tag{2.11.1}$$

On the one hand, this implies that $\varphi(x)=0$ almost everywhere, i.e., $\varphi(x)=0$ in $L_2((a,b))$ and, therefore, it is not an eigenvector in $\mathcal{H}$. On the other hand, the δ-function at the point λ, i.e., $\varphi=\delta_\lambda$, is also a formal solution of Eq. (2.11.1). As an element of a corresponding space, it differs from zero and, therefore, can be regarded as an eigenvector.

Thus, the operator A has no ordinary eigenvectors but, at the same time, it has eigenvectors which are generalized functions. It turns out that this is a general property of self-adjoint operators in a separable space $\mathcal{H}$. We show that, under certain restrictions, the spectral theorem for A, i.e., the formulas

$$\mathbb{I} = \int_{-\infty}^{\infty} dE(\lambda), \qquad A = \int_{-\infty}^{\infty} \lambda\, dE(\lambda), \tag{2.11.2}$$

can be rewritten in the form similar to that in the case of discrete spectrum where

$$\mathbb{I} = \sum_{k=1}^{\infty} P(\lambda_k), \qquad A = \sum_{k=1}^{\infty} \lambda_k P(\lambda_k).$$

Namely, formulas (2.11.2) can be rewritten in the form

$$\mathbb{I} = \int_{-\infty}^{\infty} P(\lambda)\, d\rho(\lambda), \qquad A = \int_{-\infty}^{\infty} \lambda P(\lambda)\, d\rho(\lambda), \tag{2.11.3}$$

where ρ is a measure and $P(\lambda)$ is an operator of "generalized projection", whose range consists of the generalized eigenvectors of the operator A that correspond to the eigenvalue λ.

In what follows, we will state the theorem of Radon-Nikodym type on differentiation of an operator-valued measure with respect to its trace and present the corollary concerning the differentiation of a resolution of the identity.

The Differentiation of Operator-Valued Measures Let us fix a chain

$$\mathcal{H}_- \supseteq \mathcal{H}_0 \supseteq \mathcal{H}_+, \tag{2.11.4}$$

in which all spaces are separable (clearly, it is sufficient to assume that $\mathcal{H}_+$ is separable). Recall that an operator $A : \mathcal{H}_+ \to \mathcal{H}_-$ is called non-negative if $(Au, u)_{\mathcal{H}_0} \geq 0$, $u \in \mathcal{H}_+$. By definition, the trace of a non-negative operator is equal to

$$\operatorname{Tr}(A) = \sum_{j=1}^{\infty} (Ae_j, e_j)_{\mathcal{H}_0},$$

where $(e_j)_{j=1}^{\infty}$ is an orthonormal basis in $\mathcal{H}_+$. The value $\operatorname{Tr}(A)$ does not depend on the choice of this basis. Indeed, if $\mathbf{I}$ is an isometry associated with (2.11.4), then, by virtue of the relation $(\alpha, u)_{\mathcal{H}_0} = (\mathbf{I}\alpha, u)_{\mathcal{H}_+}$ $\alpha \in \mathcal{H}_-$, $u \in \mathcal{H}_+$, we can conclude that the non-negativity of A is equivalent to the ordinary non-negativity of $\mathbf{I}A : \mathcal{H}_+ \to \mathcal{H}_+$ and $\operatorname{Tr}(A) = \operatorname{Tr}(\mathbf{I}A)$.

Assume that R is an abstract space not necessarily equipped with a topology and $\mathfrak{R}$ is a σ-algebra of sets from R. We say that a function $\mathfrak{R} \ni \alpha \mapsto \theta(\alpha)$ is *an operator-valued measure with a finite trace*, if the following conditions are fulfilled:

(a) $\theta(\alpha)$ is a non-negative operator from $\mathcal{H}_+$ to $\mathcal{H}_-$ such that $\theta(\emptyset) = 0$ and $\operatorname{Tr}(\theta(R)) < \infty$;
(b) the property of counted additivity is fulfilled, i.e., if the sets $\alpha_j \in \mathfrak{R}$, $j \in \mathbb{N}$) do not intersect each other, then

$$\theta\Big(\bigcup_{j=1}^{\infty} \alpha_j\Big) = \sum_{j=1}^{\infty} \theta(\alpha_j),$$

where the series converges in the weak sense.

It follows from the additivity and non-negativity of θ that it is monotone, i.e., if $\alpha' \subseteq \alpha''$, then $\theta(\alpha') \leq \theta(\alpha'')$. Therefore, $\theta(\alpha) \leq \theta(R)$ and $\operatorname{Tr}(\theta(\alpha)) \leq \operatorname{Tr}(\theta(R))$, $\alpha \in \mathfrak{R}$.

Let us introduce a numerical non-negative function of sets

$$\mathfrak{R} \ni \alpha \mapsto \rho(\alpha) = \operatorname{Tr}(\theta(\alpha)).$$

If $\alpha_j \in \mathfrak{R}$, $j \in \mathbb{N}$ are disjoint, then, by virtue of condition (b) and non-negativity of the terms, we have

$$\begin{aligned}
\rho\Big(\bigcup_{j=1}^{\infty} \alpha_j\Big) &= \operatorname{Tr}\Big(\theta\Big(\bigcup_{j=1}^{\infty} \alpha_j\Big)\Big) = \operatorname{Tr}\Big(\sum_{j=1}^{\infty} \theta(\alpha_j)\Big) \\
&= \sum_{k=1}^{\infty} \Big(\Big(\sum_{j=1}^{\infty} \theta(\alpha_j)\Big) e_k, e_k\Big)_{\mathcal{H}_0} \\
&= \sum_{k=1}^{\infty} \sum_{j=1}^{\infty} (\theta(\alpha_j) e_k, e_k)_{\mathcal{H}_0} \\
&= \sum_{j=1}^{\infty} \operatorname{Tr}(\theta(\alpha_j)) = \sum_{j=1}^{\infty} \rho(\alpha_j).
\end{aligned}$$

Thus, $\mathfrak{R} \ni \alpha \mapsto \rho(\alpha)$ is a numerical non-negative finite measure. The measure ρ is called *a trace measure* for θ.

Theorem 2.11.1 *An operator-valued measure θ with a finite trace can be differentiated with respect to its trace measure ρ. This means that there exists an operator-valued function*

$$Q(\lambda) : \mathcal{H}_+ \to \mathcal{H}_-, \ Q(\lambda) \geq 0, \ | Q(\lambda)| \leq \mathrm{Tr}(Q(\lambda)) = 1$$

weakly measurable with respect to $\mathfrak{R}$, defined for ρ-almost all $\lambda \in R$, and such that

$$\theta(\alpha) = \int_\alpha Q(\lambda)\, d\rho(\lambda), \quad \alpha \in \mathfrak{R} \tag{2.11.5}$$

(the integral converges in the Hilbert-Schmidt norm $| \cdot |$). The function $Q(\lambda)$ is uniquely defined to within its values on the set of measure ρ zero and is called the Radon-Nikodym derivative $(d\theta/d\rho)(\lambda) = Q(\lambda)$.

Note that the convergence of the integral (2.11.5) in the Hilbert-Schmidt norm means its convergence in the Bochner sense if $Q(\lambda)$ is regarded as a vector-function with values in the space of Hilbert-Schmidt operators from $\mathcal{H}_+$ to $\mathcal{H}_-$.

The strong measurability of $Q(\lambda)$ can be proved as follows. Let $(Q_{jk}(\lambda))_{j,k=1}^\infty$ be a matrix of $Q(\lambda)$. One can construct a sequence of matrices $(Q_{jk}(\lambda))_{j,k=1}^n$, $n \in \mathbb{N}$. The corresponding finite-dimensional operators converge to $Q(\lambda)$ for any $\lambda \in R$ in the norm $| \cdot |$. As is well known, measurable functions $Q_{jk}(\lambda)$, $j, k = 1, 2, \dots, n$ can be approximated.

Proof We fix an orthonormal basis $(e_j)_{j=1}^\infty$ in the space $\mathcal{H}_+$. The measure θ is absolutely continuous with respect to ρ, i.e., if $\rho(\alpha) = 0$, then $\theta(\alpha) = 0$, $\alpha \in \mathfrak{R}$. Indeed,

$$|(\theta(\alpha)e_j, e_k)_{\mathcal{H}_0}|^2 \leq (\theta(\alpha)e_j, e_j)_{\mathcal{H}_0}(\theta(\alpha)e_k, e_k)_{\mathcal{H}_0} \leq \rho^2(\alpha) = 0, \quad j, k \in \mathbb{N}.$$

This implies that, for fixed $u, v \in \mathcal{H}_+$, the complex-valued measure $\mathfrak{R} \ni \alpha \mapsto (\theta(\alpha)u, v)_{\mathcal{H}_0} \in \mathbb{C}$ is also absolutely continuous with respect to ρ and, according to the ordinary Radon-Nikodym theorem, we have

$$(\theta(\alpha)u, v)_{\mathcal{H}_0} = \int_\alpha q(\lambda; u, v)\, d\rho(\lambda), \quad \alpha \in \mathfrak{R}, \ u, v \in \mathcal{H}_+, \tag{2.11.6}$$

where the derivative $q(\lambda; u, v)$ is defined on the set $\beta_{u,v} \subseteq R$ of the complete measure ρ, measurable with respect to $\mathfrak{R}$, and summable; for $u = v$, it is non-negative. Denote by L a linear span of vectors $(e_j)_{j=1}^\infty$ with rational complex coefficients; $\tilde{L} = \mathcal{H}_+$. Since L is countable, the set $\bigcap_{u,v \in L} \beta_{u,v}$ is also a set of the complete measure; for λ from this set, all functions $q(\lambda; u, v)(u, v \in L)$ are defined and $q(\lambda; u, u) \geq 0$, $u \in L$.

Since derivative is uniquely defined to within its values on the set of measure zero, the bilinearity of the left-hand side of (2.11.6) with respect to u and v yields the bilinearity of $q(\lambda; u, v)$. More exactly, there exists a set of the complete measure $\beta \subseteq \bigcap_{u,v\in L} \beta_{u,v}$ such that for $\lambda \in \beta$,

$$q(\lambda; p_1u_1 + p_2u_2, r_1v_1 + r_2v_2) = p_1\bar{r}_1q(\lambda; u_1, v_1) + p_1\bar{r}_2q(\lambda; u_1, v_2) + p_2\bar{r}_1q(\lambda; u_2, v_1) + p_2\bar{r}_2q(\lambda; u_2, v_2)$$

for any $u_1, u_2, v_1, v_2 \in L$ and complex rational p_1, p_2, r_1, and r_2. To prove this, we use the bilinearity of $(\theta(\alpha)u, v)_{\mathcal{H}_0}$ and the fact that $\alpha \in \mathfrak{R}$ in (2.11.6) and conclude that this equality holds true for λ from the set $\beta_{p_1,p_2,r_1,r_2,u_1,u_2,v_1,v_2} \subseteq \bigcap_{u,v\in L} \beta_{u,v}$ of the complete measure. Then we take the (countable) intersection of all such sets as β. Furthermore, as was mentioned above, for such λ, we have $q(\lambda; u, u) \geq 0$, $u \in L$. The bilinearity and non-negativity yield, in a standard way, the Cauchy-Bunyakovsky inequality

$$|q(\lambda; u, v)|^2 \leq q(\lambda; u, u)q(\lambda; v, v), \quad \lambda \in \beta,\ u, v \in L. \tag{2.11.7}$$

By setting $u = v = e_j$ in (2.11.6), summing over $j \in \mathbb{N}$, and using the Fubini theorem, we get

$$\rho(\alpha) = \int_\alpha \Big(\sum_{j=1}^{\infty} q(\lambda; e_j, e_j)\Big)\, d\rho(\lambda), \quad \alpha \in \mathfrak{R}.$$

Hence, for almost all $\lambda \in \beta$, we have

$$\sum_{j=1}^{\infty} q(\lambda; e_j, e_j) = 1. \tag{2.11.8}$$

Reducing, if necessary, the set β, we can assume that (2.11.8) holds true for all $\lambda \in \beta$. Relations (2.11.7) and (2.11.8) yield

$$\sum_{j,k=1}^{\infty} |q(\lambda; e_j, e_j)|^2 \leq \sum_{j,k=1}^{\infty} q(\lambda; e_j, e_j)q(\lambda; e_k, e_k) = 1, \quad \lambda \in \beta. \tag{2.11.9}$$

Let us fix $\lambda \in \beta$ and denote by $A(\lambda)$ the operator in $\mathcal{H}_+$ that corresponds to the matrix $(a_{jk}(\lambda))_{j,k=1}^{\infty}$ in the basis $(e_j)_{j=1}^{\infty}$; here, $a_{jk}(\lambda) = q(\lambda; e_k, e_j)$. By virtue of (2.11.9), $A(\lambda)$ is well defined and is a Hilbert-Schmidt operator. The measurability of each

$$\beta \ni \lambda \mapsto q(\lambda; e_j, e_k), \quad j, k \in \mathbb{N}$$

implies that the operator-valued function $\beta \ni \lambda \mapsto A(\lambda)$ is weakly measurable. Let us introduce a continuous operator $Q(\lambda) = \mathbf{I}^{-1}A(\lambda) : \mathcal{H}_+ \to \mathcal{H}_-$ and show that this operator is the required one.

It follows from the measurability of $A(\lambda)$ that $\beta \ni \lambda \mapsto Q(\lambda)$ is weakly measurable. Further, for $u = \sum_{k=1}^{\infty} p_k e_k$, $v = \sum_{j=1}^{\infty} r_j e_j \in L$, we have

$$\begin{aligned}(Q(\lambda)u, v)_{\mathcal{H}_0} &= (\mathbf{I}^{-1}A(\lambda)u, v)_{\mathcal{H}_0} = (A(\lambda)u, v)_{\mathcal{H}_+} \\ &= \sum_{j,k=1}^{\infty} q(\lambda; e_k, e_j) p_k \bar{r}_j = q(\lambda; u, v).\end{aligned} \tag{2.11.10}$$

In particular, $(Q(\lambda)u, u)_{\mathcal{H}_0} = q(\lambda; u, u) \geq 0$. By passing to the limit, we find that the inequality remains valid for an arbitrary $u \in \mathcal{H}_+$, i.e., $Q(\lambda) \geq 0$. According to(2.11.8), $\mathrm{Tr}(Q(\lambda)) = \mathrm{Tr}(A(\lambda)) = 1$, $\lambda \in \beta$. Thus, $|\, Q(\lambda)| \leq 1$ and $Q(\lambda)$, $\lambda \in \beta$ is weakly measurable. Hence, there exists the integral

$$\int_{\alpha} Q(\lambda)\, d\rho(\lambda), \qquad \alpha \in \mathfrak{R}$$

convergent in the Hilbert-Schmidt norm. According to (2.11.10) and (2.11.6), for $u, v \in L$ and $\alpha \in \mathfrak{R}$, we have

$$\begin{aligned}\left(\left(\int_{\alpha} Q(\lambda)\, d\rho(\lambda)\right)u, v\right)_{\mathcal{H}_0} &= \int_{\alpha} (Q(\lambda)u, v)_{\mathcal{H}_0}\, d\rho(\lambda) \\ &= \int_{\alpha} q(\lambda; u, v)\, d\rho(\lambda) = (\theta(\alpha)u, v)_{\mathcal{H}_0},\end{aligned} \tag{2.11.11}$$

i.e., (2.11.5) holds true.

Finally, let us establish the uniqueness of $Q(\lambda)$. Assume that, parallel with $Q(\lambda)$, there is an operator-valued function $Q_1(\lambda)$ of the same type satisfying the equality

$$\int_{\alpha} Q(\lambda) d\rho(\lambda) = \int_{\alpha} Q_1(\lambda)\, d\rho(\lambda), \qquad \alpha \in \mathfrak{R}.$$

Then, for every $u, v \in L$, we have $(Q(\lambda)u, v)_{\mathcal{H}_0} = (Q_1(\lambda)u, v)_{\mathcal{H}_0}$ for λ from the set of the complete measure $\beta_{1;u,v} \subseteq R$. But then, for λ from the set of the complete measure $\beta_1 = \bigcap_{u,v \in L} \beta_{1;u,v}$, we have

$$(Q(\lambda)u, v)_{\mathcal{H}_0} = (Q_1(\lambda)u, v)_{\mathcal{H}_0}$$

for all $u, v \in L$. This and the continuity of the operators $Q(\lambda)$ and $Q_1(\lambda)$ for each fixed $\lambda \in \beta_1$ imply that $Q(\lambda) = Q_1(\lambda)$. □

In this way, we can also consider an operator-valued measure with σ-finite trace. This means that there exists a sequence $(R_k)_{k=1}^{\infty} \subset \mathfrak{R}$ such that

$$R_1 \subseteq R_2 \subseteq, \dots, \quad \bigcup_{k=1}^{\infty} R_k = R \text{ and } \operatorname{Tr}(\theta(R_k)) < \infty, \quad k \in \mathbb{N}.$$

In this case, Theorem 2.11.1 is modified as follows. One must consider a σ-finite trace measure instead of a finite one and state that the representation (2.11.5) holds true for every $\mathfrak{R} \ni \alpha \subseteq R_k$ for some $k \in \mathbb{N}$. The proof remains the same.

The formulation of Theorem 2.11.1 can be made similar to the Radon-Nikodym theorem. Namely, assume that an operator-valued measure θ with σ-finite trace and a σ-finite non-negative numerical measure $\mathfrak{R} \ni \alpha \mapsto \rho(\alpha) \in [0, \infty]$ with respect to which θ is absolutely continuous (i.e., if $\rho(\alpha) = 0$ for some $\alpha \in \mathfrak{R}$, then $\theta(\alpha) = 0$). In this case, the representation (2.11.5), in which $\mathfrak{R} \ni \alpha \subseteq R_k$, $k \in \mathbb{N}$ and $Q(\lambda)$ is a weakly measurable operator-valued function defined for ρ-almost all $\lambda \in R$. Values of the function $Q(\lambda)$ are non-negative operators from $\mathcal{H}_+$ to $\mathcal{H}_-$, each having a finite trace summable with respect to ρ over R_k, $k \in \mathbb{N}$) (one should write representation (2.11.5) with a trace measure and differentiate this measure with respect to ρ).

The Differentiation of a Resolution of the Identity Assume that R is an abstract space, $\mathfrak{R}$ is a σ-algebra of its sets, and $\mathfrak{R} \ni \alpha \mapsto E(\alpha)$ is a general resolution of the identity acting in the space $\mathcal{H}_0$. As a rule, the measure E has no finite or σ-finite trace and, therefore, Theorem 2.11.1 cannot be directly applied. It is convenient for us to act as follows. Assume that the rigging (2.11.4) of the space $\mathcal{H}_0$ is given. Let $O : \mathcal{H}_+ \to \mathcal{H}_0$, $O^+ : \mathcal{H}_0 \to \mathcal{H}_-$ be the corresponding imbedding operators (by the equality

$$(f, Ou)_{\mathcal{H}_0} = (f, u)_{\mathcal{H}_0} = (O^+ f, u)_{\mathcal{H}_0}, \quad f \in \mathcal{H}_0,\ u \in \mathcal{H}_+,$$

it follows that O^+ is, in fact, adjoint to O with respect to $\mathcal{H}_0$). The function

$$\mathfrak{R} \ni \alpha \mapsto \theta(\alpha) = O^+ E(\alpha) O, \tag{2.11.12}$$

whose values are continuous operators from $\mathcal{H}_+$ to $\mathcal{H}_-$, is an operator-valued measure ($O^+ E(\alpha) O \geq 0$ because

$$(O^+ E(\alpha) O u, u)_{\mathcal{H}_0} = (E(\alpha) O u, O u)_{\mathcal{H}_0} \geq 0$$

for $u \in \mathcal{H}_+$. Recall that the rigging (2.11.4) is called quasi-nuclear if the imbedding operator O is quasi-nuclear.

Lemma 2.11.2 *If the rigging* (2.11.4) *is quasi-nuclear, then the operator-valued measure* (2.11.12) *has a finite trace.*

We note also that *if* $A : \mathcal{H}_0 \to \mathcal{H}_0$ *is the non-negative operator, then* $O^+AO : \mathcal{H}_+ \to \mathcal{H}_-$ *is also non-negative and*

$$\mathrm{Tr}(O^+AO) \leq \|A\| \mid O\mid^2. \tag{2.11.13}$$

Indeed, the inequality $O^+AO \geq 0$ has already been explained by the example of $A = E(\alpha)$. Further, let $(e_j)_{j=1}^{\infty}$ be an orthonormal basis in $\mathcal{H}^+$. Then

$$\mathrm{Tr}(O^+AO) = \sum_{j=1}^{\infty}(O^+AOe_j, e_j)_{\mathcal{H}_0} = \sum_{j=1}^{\infty}(AOe_j, Oe_j)_{\mathcal{H}_0}$$

$$\leq \|A\| \sum_{j=1}^{\infty} \|(Oe_j)\|^2_{\mathcal{H}_0} = \|A\| \mid O\mid^2.$$

Proof of Lemma 2.11.2. According to (2.11.13), we have

$$\mathrm{Tr}(\theta(R)) = \mathrm{Tr}(O^+E(R)O) \leq \mid O\mid^2 < \infty.$$

□

Let us fix the quasi-nuclear rigging (2.11.4). A non-negative finite measure $\mathfrak{R} \ni \alpha \mapsto \rho(\alpha) = \mathrm{Tr}(O^+E(\alpha)O) \in [0, \infty)$ is called the *spectral measure of resolution of the identity* E. Clearly, E and ρ are mutually absolutely continuous: for some $\alpha \in \mathfrak{R}$, the equalities $E(\alpha) = 0$ and $\rho(\alpha) = 0$ are equivalent. By applying Theorem 2.11.1 to (2.11.13) and ρ, we obtain the following assertion.

Theorem 2.11.3 *Suppose that* $\mathfrak{R} \ni \alpha \mapsto E(\alpha)$ *is a resolution of the identity acting in the space* $\mathcal{H}_0$, (2.11.4) *is a fixed quasi-nuclear rigging, and* $\mathfrak{R} \ni \alpha \mapsto \rho(\alpha) \in [0, \infty)$ *is the corresponding spectral measure. Then the following representation in the form of an integral convergent in the Hilbert-Schmidt norm*

$$O^+E(\alpha)O = \int_{\alpha} P(\lambda)\, d\rho(\lambda), \quad \alpha \in \mathfrak{R} \tag{2.11.14}$$

holds true. Here $P(\lambda) : \mathcal{H}_+ \to \mathcal{H}_-$ *is an operator-valued function weakly measurable with respect to* $\mathfrak{R}$, *defined for* ρ*-almost all* $\lambda \in R$, *and such that* $P(\lambda) \geq 0$ *and* $|P(\lambda)| \leq Tr(P(\lambda)) = 1$. *In particular,* $P(\lambda)$ *is called a generalized projection.*

In the case of resolution of the identity E of a self-adjoint operator in $\mathcal{H}_0$ with a discrete spectrum $(\lambda_j)_{j=1}^{\infty}$, the equality

$$E(\alpha) = \sum_{\lambda_j \in \alpha} P(\lambda_j), \quad \alpha \in \mathfrak{B}(\mathbb{R}),$$

holds true, in which $P(\lambda_j)$ is the projector onto the eigensubspace A of corresponding to the eigenvalue λ_j. The comparison of this formula with (2.11.14) has determined the choice of the term "generalized projector".

We can also introduce the concept of the general spectral measure corresponding to resolution of the identity E. This measure is defined as a σ-finite non-negative measure $\mathfrak{R} \ni \alpha \mapsto \rho(\alpha) \in [0, \infty]$ such that ρ and E are mutually absolutely continuous. The representation (2.11.14) remains valid for the general spectral measure except that the operator $P(\lambda)$ takes a scalar multiplier.

The Case of a Nuclear Rigging Similar results can be obtained if we use the nuclear rigging

$$\Phi' \supseteq \mathcal{H}_0 \supseteq \Phi \tag{2.11.15}$$

of the Hilbert space $\mathcal{H}_0$ instead of its quasi-nuclear rigging (2.11.4) (recall, that a rigging is called nuclear if $\Phi = \operatorname{pr\,lim}_{\tau \in T} \mathcal{H}_\tau$ is a nuclear space). More exactly, the case of the chain (2.11.15) is reduced to the chain (2.11.4).

Consider the rigging (2.11.15) and denote by O and O^+ imbedding operators $\Phi \subseteq \mathcal{H}_0$ and $\mathcal{H}_0 \subseteq \Phi'$, respectively. We also consider continuous operators $A : \Phi \to \Phi'$. These operators are called non-negative ($A \geq 0$) if $(A\varphi, \varphi)_{\mathcal{H}_0} \geq 0$, $\varphi \in \Phi$. In particular, if $\mathfrak{R} \ni \alpha \mapsto E(\alpha)$ is the resolution of the identity mentioned above, then the operator $O^+ E(\alpha) O : \Phi \to \Phi'$ is non-negative. The function $\mathfrak{R} \ni \alpha \mapsto \theta(\alpha) = O^+ E(\alpha) O$ is a measure similar to those was before but with values in $\mathcal{L}(\Phi, \Phi')$.

Theorem 2.11.4 *Let $\mathfrak{R} \ni \alpha \mapsto E(\alpha)$ and $\rho(\alpha)$ be a resolution of the identity acting in the space $\mathcal{H}_0$ and some spectral measure corresponding to it, respectively. Suppose that* (2.11.15) *is a fixed nuclear rigging. Then the representation* (2.11.14) *in the form of a weakly convergent integral is true, in which $0 \leq P(\lambda) : \Phi \mapsto \Phi'$ (a generalized projector) is an operator-valued function weakly measurable with respect to $\mathfrak{R}$ and defined for ρ-almost all $\lambda \in R$.*

Proof As was shown above, every $\mathcal{H}_\tau$, $\tau \in T$, is imbedded densely and continuously in $\mathcal{H}_0$ and $0 \in T$. We choose τ so that the imbedding $\mathcal{H}_\tau \subseteq \mathcal{H}_0$ is quasi-nuclear; this is possible because Φ is nuclear. As a result, we obtain the chain

$$\Phi' \supseteq \mathcal{H}_{-\tau} \supseteq \mathcal{H}_0 \supseteq \mathcal{H}_\tau \supseteq \Phi \tag{2.11.16}$$

with dense and continuous imbedding operators. Let us take the central part of chain (2.11.15) as (2.11.4) and introduce imbedding operators $O_1 : \mathcal{H}_\tau \to \mathcal{H}_0$ and $O_1^+ : \mathcal{H}_0 \to \mathcal{H}_{-\tau}$. Then, by virtue of Theorem 2.11.3 and in accordance with the previous remark, we get

$$O_1^+ E(\alpha) O_1 = \int\limits_{\alpha} P_1(\lambda)\, d\rho(\lambda), \quad \alpha \in \mathfrak{R}, \tag{2.11.17}$$

where $P_1(\lambda) : \mathcal{H}_\tau \to \mathcal{H}_{-\tau}$ is a corresponding generalized projector.

Consider imbedding operators $O_2 : \Phi \to \mathcal{H}_\tau$ and $O_3 : \mathcal{H}_{-\tau} \to \Phi'$. Multiplying (2.11.17) from the right and from the left by O_2 and O_3, respectively, we obtain the required equality (2.11.14) with $P(\lambda) = O_3 P_1(\lambda) O_2$. The convergence of integral (2.11.17) in the Hilbert-Schmidt norm (from $\mathcal{H}_\tau$ to $\mathcal{H}_{-\tau}$) obviously yields the weak convergence of integral (2.11.14). □

The results presented in this section are valid, e.g., for the resolution of the identity corresponding to a given self-adjoint or normal operator. In the next considerations, we examine this situation in detail.

Generalized Eigenvectors and the Projection Spectral Theorem Consider the simplest classic case of only one self-adjoint operator A acting in a rigged Hilbert space $\mathcal{H}_0$. Assume that there exists a linear topological space D densely and continuously imbedded in $\mathcal{H}_+$ such that $D \subseteq \mathfrak{D}(A)$ and the restriction $A \restriction D$ acts continuously from D to $\mathcal{H}_+$. In this case, instead of (2.11.4) we have the rigging (chain)

$$\mathcal{H}_- \supseteq \mathcal{H}_0 \supseteq \mathcal{H}_+ \supseteq D. \tag{2.11.18}$$

We say that the operator A with the mentioned properties and rigging (2.11.18) *are standardly connected (or A admits* (2.11.18)*. The chain* (2.11.18) *is called an extension of* (2.11.4). As before, (2.11.18) is quasi-nuclear by definition if the imbedding of the (separable) space $\mathcal{H}_+$ in $\mathcal{H}_0$ is quasi-nuclear.

A non-zero vector $\varphi \in \mathcal{H}_-$ is called a *generalized eigenvector* of the operator A corresponding to an eigenvalue $\lambda \in \mathbb{C}$ if

$$(\varphi, Au)_{\mathcal{H}_0} = \lambda(\varphi, u)_{\mathcal{H}_0}, \quad u \in D. \tag{2.11.19}$$

If $\varphi \in \mathfrak{D}(A)$, then the operator A in (2.11.19) can be transferred to φ and, since $u \in D$ is arbitrary, we get $A\varphi = \lambda\varphi$, i.e., φ is an ordinary eigenvector of A corresponding to the eigenvalue λ. Thus, the definition introduced above generalizes the classic concept.

The collection of eigenvalues corresponding to all possible generalized eigenvectors is called *a generalized spectrum* $g(A)$ generalized spectrum A. Clearly, this spectrum depends on the choice of rigging (2.11.18) and, generally speaking, differs from the spectrum $S(A)$ of operator A.

Example Let $\mathcal{H}_0 = L_2(\mathbb{R}, dx)$ with respect to the Lebesgue measure dx and let A be the minimal operator generated by the expression $(\mathcal{L}u)(x) = -iu'(x)$, i.e., the closure of the operator $\mathcal{H}_0 \supseteq C_0^\infty(\mathbb{R}) \ni u \mapsto iu'(x) \in \mathcal{H}_0$; this operator is self-adjoint and $S(A) = \mathbb{R}$. Assume that $\mathcal{H}_+ = L_2(\mathbb{R}, e^{x^2}dx)$ and $D = \mathfrak{D}(\mathbb{R})$. Thus, the conditions concerning the rigging (2.11.18) are satisfied and $\mathcal{H}_- = L_2(\mathbb{R}, e^{-x^2}dx)$. For every $\lambda \in \mathbb{C}$, the function $\varphi(x) = e^{i\lambda x}$, $x \in \mathbb{R}$ belongs to $\mathcal{H}_-$ and satisfies (2.11.19) Thus,

$$g(A) = \mathbb{C} \neq \mathbb{R} = S(A), \quad g(A) \supset S(A).$$

Consider the same operator A and assume that $\mathcal{H}_+ = \mathcal{H}_0$. In this case, $\mathcal{H}_- = \mathcal{H}_0$. There are no vectors $\mathcal{H}_- \ni \varphi \neq 0$ satisfying (2.11.19). Indeed, each $\varphi \in \mathfrak{D}'(\mathbb{R}) \supset \mathcal{H}_0$, satisfying (2.11.19) is a generalized solution of the equation $-\varphi' = \lambda\varphi$ and, therefore, $\varphi(x) = ce^{i\lambda x}$, $x \in \mathbb{R}$, $\mathbb{C} \ni c \neq 0$. But such a function φ can not belong to $\mathcal{H}_0 = \mathcal{H}_-$ for any $\lambda \in \mathbb{C}$. Thus, $g(A) = \emptyset \neq \mathbb{R} = S(A)$, $g(A) \subset S(A)$.

Assume that A is a self-adjoint operator acting in a separable Hilbert space $\mathcal{H}_0$ and standardly connected with the quasi-nuclear rigging (2.11.18). According to Theorem 2.11.3, its resolution of the identity E defined on the σ-algebra of Borel sets $\mathfrak{B}(\mathbb{R})$, admits the representation (2.11.14), i.e.,

$$O^+E(\alpha)O = \int_\alpha P(\lambda)\,d\rho(\lambda), \quad \alpha \in \mathfrak{B}(\mathbb{R}). \tag{2.11.20}$$

Here, by definition,

$$\mathfrak{B}(\mathbb{R}) \ni \alpha \mapsto \rho(\alpha) = \mathrm{Tr}(O^+E(\alpha)O) \in [0, \infty)$$

is the spectral measure of A and $P(\lambda) : \mathcal{H}_+ \to \mathcal{H}_-$ is an operator-valued function weakly measurable with respect to $\mathfrak{B}(\mathbb{R})$, defined for ρ-almost all $\lambda \in \mathbb{R}$ and such that $P(\lambda) \geq 0$ and $|\,P(\lambda)| \leq \mathrm{Tr}(P(\lambda)) = 1$. Since, for $B \in \mathcal{L}(\mathcal{H}_+, \mathcal{H}_-)$, $B \geq 0$, the equalities $B = 0$ and $\mathrm{Tr}(B) = 0$ are equivalent, we have $S(A) = \mathrm{supp}E = \mathrm{supp}\rho$. Therefore, we can assume that λ takes values not in $\mathbb{R}$ but in $S(A)$. Naturally, such measure connected with the resolution of the identity of the operator A is called a *general spectral measure* of this operator.

Theorem 2.11.5 (BGKM) *Let A be a self-adjoint operator acting in a separable Hilbert space $\mathcal{H}_0$ and standardly connected with the quasi-nuclear chain* (2.11.18) *in which D is separable. Then there exists an operator-valued function $P(\lambda)$ weakly measurable, defined for almost all λ from the spectrum $S(A)$ in the sense of the spectral measure ρ and such that its values are non-negative operators from the positive space $\mathcal{H}_+$ to the negative space $\mathcal{H}_-$ and $|\,P(\lambda)| \leq \mathrm{Tr}(P(\lambda)) = 1$, which* *realizes the following representation of the operator A and its resolution of the identity E:*

$$E(\alpha)u = \left(\int\limits_{\alpha} P(\lambda)\, d\rho(\lambda)\right)u, \quad \alpha \in \mathfrak{B}(\mathbb{R}),\ u \in \mathcal{H}_+,$$
$$Au = \left(\int\limits_{S(A)} \lambda P(\lambda)\, d\rho(\lambda)\right)u, \quad u \in \mathfrak{D}(A) \cap \mathcal{H}_+. \tag{2.11.21}$$

The range $\mathfrak{R}(P(\lambda)) \subseteq \mathcal{H}_-$ consists of generalized eigenvectors φ of the operator A corresponding to the eigenvalue λ, i.e.,

$$(\varphi, Au)_{\mathcal{H}_0} = \lambda(\varphi, u)_{\mathcal{H}_0}, \quad u \in D. \tag{2.11.22}$$

In the name of the theorem, the abbreviation BGKM means that the following authors contributed to the theorem: Berezansky Yu. M.(1956), Gårding L.(1954), Gelfand I. M., Kostyuchenko A. G. (1955), Kac G. I. (1958), and Maurin K. (1958) [18–21, 111, 116, 131, 132, 146, 210, 212, 213].

Thus, formulas (2.11.21) , in fact, have the form of the representation (2.11.3) with selected "projectors on the eigensubspaces" consisting of generalized eigenvectors. In this connection, *the spectral theorem for A in the form* ((2.11.21) *is called the projection theorem.*

Proof By comparing Theorem 2.11.5 with Theorem 2.11.3 in the particular case of the resolution of the identity E, corresponding to A (see (2.11.20)), corresponding to (2.11.21) follows from the representation (2.4.1) and the equality $dE(\lambda)u = P(\lambda)d\rho(\lambda)u$ for $u \in \mathcal{H}_+$, which, in fact, is the first equality in (2.11.21) (i.e., (2.11.20)), rewritten in a different form.

Let us prove that $\varphi \in \mathfrak{R}(P(\lambda))$ satisfies (2.11.22). In other words, we must establish the following relation: there exists the set $\beta \in \mathfrak{B}(\mathbb{R})$ of a complete spectral measure ρ such that for any $\lambda \in \beta$

$$(P(\lambda)u, Av)_{\mathcal{H}_0} = \lambda(P(\lambda)u, v)_{\mathcal{H}_0}, \quad u \in \mathcal{H}_+,\ v \in D. \tag{2.11.23}$$

First, note that it suffices to establish the existence of the set $\beta \in \mathfrak{B}(\mathbb{R})$ of the complete measure ρ such that for $\lambda \in \beta$ the relation (2.11.23) holds true for u and v taking values, respectively, in fixed sets $G \subseteq \mathcal{H}_+$ and $F \subseteq D$, which are dense in these spaces.

Indeed, assume that $u \in \mathcal{H}_+$, $v \in D$, $D \ni u_n \underset{n\to\infty}{\longrightarrow} u$ in $\mathcal{H}_+$, and $F \ni v_m \underset{m\to\infty}{\longrightarrow} v$ in D. By assumption, we have

$$(P(\lambda)u_n, Av_m)_{\mathcal{H}_0} = \lambda(P(\lambda)u_n, v_m)_{\mathcal{H}_0}, \quad \lambda \in \beta,\ n, m \in \mathbb{N}. \tag{2.11.24}$$

Let us pass to the limit in (2.11.24), first, as $n \to \infty$ and then as $m \to \infty$. This is possible because $P(\lambda) \in \mathcal{L}(\mathcal{H}_+, \mathcal{H}_-)$ and $A \restriction D \in \mathcal{L}(D, \mathcal{H}_+)$. As a result, we obtain the required relation (2.11.23).

We fix $u, v \in D$. By using (2.11.20), we get

$$\begin{aligned}\int_\alpha (P(\lambda)u, Av)_{\mathcal{H}_0}\, d\rho(\lambda) &= \Big(\Big(\int_\alpha P(\lambda)\, d\rho(\lambda)\Big)u, Av\Big)_{\mathcal{H}_0} \\ &= (O^+E(\alpha)Ou, Av)_{\mathcal{H}_0} = (AE(\alpha)u, v)_{\mathcal{H}_0} \qquad (2.11.25) \\ &= \Big(\Big(\int_\alpha \lambda\, dE(\lambda)\Big)u, v\Big)_{\mathcal{H}_0} = \int_\alpha \lambda\, d(O^+E(\lambda)Ou, v)_{\mathcal{H}_0}\end{aligned}$$

for all $\alpha \in \mathfrak{B}(\mathbb{R})$. Here, we have used the relation

$$AE(\alpha) = \int_\alpha \lambda\, dE(\lambda), \quad \alpha \in \mathfrak{B}(\mathbb{R}),$$

which follows from the equalities (2.4.1) and (2.3.20). Further, by replacing the differential in the last integral in (2.11.25) according to (2.11.20), we obtain

$$\int_\alpha (P(\lambda)u, Av)_{\mathcal{H}_0}\, d\rho(\lambda) = \int_\alpha \lambda(P(\lambda)u, v)_{\mathcal{H}_0}\, d\rho(\lambda), \quad \alpha \in \mathfrak{B}(\mathbb{R}).$$

Since α is arbitrary, the last relation implies that there exists a set $\beta_{u,v} \in \mathfrak{B}(\mathbb{R})$ of the complete measure ρ such that for $\lambda \in \beta_{u,v}$, the relation (2.11.23) holds true.

Assume now that L is a countable set dense in the space D and, hence, in $\mathcal{H}_+$. Consider a countable intersection $\bigcap_{u,v\in L} \beta_{u,v} = \beta$, which is a set of complete measure ρ. If $\lambda \in \beta$, relation (2.11.23) holds true for all $u, v \in L$. But then, according to the remark made at the beginning of the proof, this relation is also true for all $u \in \mathcal{H}_+$ and $v \in D$. □

Under somewhat stronger restrictions on the operator A, this theorem can be reformulated in terms of nuclear riggings of the space $\mathcal{H}_0$.

Consider a rigging of $\mathcal{H}_0$ by linear topological spaces

$$\Phi' \supseteq \mathcal{H}_0 \supseteq \Phi. \qquad (2.11.26)$$

We say that a self-adjoint operator A in $\mathcal{H}_0$ and the chain (2.11.26) are *standardly connected* (or A admits (2.11.26)) if $\Phi \subseteq \mathfrak{D}(A)$ and $A \restriction \Phi$ acts continuously in Φ. The definition of generalized eigenvectors remains, in fact, the same, i.e., equality (2.11.19) must hold for $u \in \Phi$. Furthermore, the generalized spectrum $g(A)$ is, as before, a collection of all eigenvalues corresponding to generalized eigenvectors.

Note that if A is standardly connected with (2.11.26) and $\Phi = \operatorname{pr} \lim_{\tau \in T} \mathcal{H}_\tau$, then A is also standardly connected with each chain

$$\mathcal{H}_{-\tau} \supseteq \mathcal{H}_0 \supseteq \mathcal{H}_\tau \supseteq \Phi = D \tag{2.11.27}$$

of the form (2.11.18). Let us apply Theorem 2.11.4 instead of Theorem 2.11.3 to the resolution of the identity E of the operator A and repeat the proof scheme.

Theorem 2.11.6 *Let A be a self-adjoint operator acting in a separable Hilbert space $\mathcal{H}_0$ and standardly connected with the nuclear chain* (2.11.26). *Then all statements of Theorem* 2.11.5 *remain valid with the only modification that $P(\lambda)$ acts continuously from Φ to Φ' (acts continuously from). In this case, the measure ρ is a spectral measure of the operator A.*

The Case of a Normal Operator Let us show how the previous result looks like when a self-adjoint operator is replaced with the normal operator A. First, we modify the definition of the generalized eigenvector.

Let A be a normal operator in $\mathcal{H}_0$ and let $\varphi \in \mathcal{H}_0$ be its eigenvector corresponding to an eigenvalue λ_0, in a certain neighborhood $U \subset \mathbb{C}$ of which there are no other points of the spectrum of A. Then the spectral decomposition of A has the following form

$$A = \int_{\mathbb{C}} \lambda \, dE(\lambda) = \lambda_0 P(\lambda_0) + \int_{\mathbb{C}\setminus U} \lambda \, dE(\lambda), \tag{2.11.28}$$

where E is the resolution of the identity of the operator A and $P(\lambda_0)$ is the projector to the eigensubspace corresponding to λ_0 and consisting of the vectors φ. For the adjoint operator, the relation like (2.11.28), namely

$$A^* = \int_{\mathbb{C}} \bar{\lambda} \, dE(\lambda) = \bar{\lambda}_0 P(\lambda_0) + \int_{\mathbb{C}\setminus U} \bar{\lambda} \, dE(\lambda) \tag{2.11.29}$$

holds true.

It follows from (2.11.29) that φ is also an eigenvector of the operator A^* corresponding to the eigenvalue $\bar{\lambda}_0$. In view of this, it is convenient to introduce the following definition.

The chain (2.11.18) considered above and a normal operator A acting in $\mathcal{H}_0$ are called standardly connected if $D \subseteq \mathfrak{D}(A)$ and the restrictions $A \upharpoonright D$ and $A^* \upharpoonright D$ act continuously from D to $\mathcal{H}_+$ both. The generalized eigenvector of the operator A corresponding to an eigenvalue $\lambda \in \mathbb{C}$ is defined as a vector $\varphi \in \mathcal{H}_-$ such that

$$(\varphi, A^* u)_{\mathcal{H}_0} = \lambda(\varphi, u)_{\mathcal{H}_0}, \quad (\varphi, Au)_{\mathcal{H}_0} = \bar{\lambda}(\varphi, u)_{\mathcal{H}_0}, \quad u \in D. \tag{2.11.30}$$

As above, we can conclude from (2.11.30) that if, in addition, $\varphi \in \mathfrak{D}(A)$, then $A\varphi = \lambda\varphi$ and $A^*\varphi = \bar{\lambda}\varphi$. Thus, (2.11.30) is a generalization of the concept of an eigenvector of a normal operator. Obviously, the equality (2.11.20) remains valid; however, $\alpha \in \mathfrak{B}(\mathbb{C})$. The concept of the spectral measure ρ of an operator A is introduced similarly. In the considered case, the theorem similar to Theorem 2.11.5 is also true.

Theorem 2.11.7 *Let A be a normal operator acting in a separable Hilbert space $\mathcal{H}_0$ and standardly connected with the quasi-nuclear chain* (2.11.18) *in which D is separable. Then all statements of Theorem* 2.11.5 *remain valid with the formulas* (2.11.21) *replaced with*

$$
\begin{aligned}
E(\alpha)u &= \Big(\int_{\alpha} P(\lambda)\, d\rho(\lambda)\Big)u, \quad \alpha \in \mathfrak{B}(\mathbb{C}, \ u \in \mathcal{H}_+, \\
Au &= \Big(\int_{S(A)} \lambda P(\lambda)\, d\rho(\lambda)\Big)u, \quad u \in \mathfrak{D}(A) \cap \mathcal{H}_+, \\
A^*u &= \Big(\int_{S(A)} \bar{\lambda} P(\lambda)\, d\rho(\lambda)\Big)u, \quad u \in \mathfrak{D}(A) \cap \mathcal{H}_+
\end{aligned} \tag{2.11.31}
$$

and the equality (2.11.22) *is replaced with* (2.11.30).

Proof As in the case of Theorem 2.11.5, the question is reduced to proving the following assertion. There exists a set $\beta \in \mathfrak{B}(\mathbb{C})$ of complete spectral measure such that for every $\lambda \in \beta$, we have

$$
\begin{aligned}
(P(\lambda)u, A^*v)_{\mathcal{H}_0} &= \lambda(P(\lambda)u, v)_{\mathcal{H}_0}, \\
(P(\lambda)u, Av)_{\mathcal{H}_0} &= \bar{\lambda}(P(\lambda)u, v)_{\mathcal{H}_0}, \quad u \in \mathcal{H}_+, \ v \in D.
\end{aligned} \tag{2.11.32}
$$

First, we prove the existence of the set $\beta_1 \in \mathfrak{B}(\mathbb{C})$ of complete measure ρ such that for any $\lambda \in \beta_1$, the first relation in (2.11.32) holds true. It is done in the same way as by the proving Theorem 2.11.5; it is only necessary to use the relation (2.4.10) instead of (2.4.1).

Similarly, one can show that there exists the set $\beta_2 \in \mathfrak{B}(\mathbb{C})$ of complete measure ρ such that for any $\lambda \in \beta_2$, the second relation in (2.11.32) is true (by using the representation (2.4.10) instead of the representation (2.4.1). Finally, we set $\beta = \beta_1 \cap \beta_2$. □

As in the case of self-adjoint operators, we can use the nuclear chain (2.11.26) standardly connected with a normal operator. The definitions of standard connection and generalized eigenvector are similar to the corresponding definitions presented above. The corresponding analogue of Theorem 2.11.6 can be obtained by obvious modification.

In a special case, the results of this subsection can be applied to unitary operators. Theorems corresponding to normal, unitary and others variants of operators related to the generalized eigenvectors expansions, we will present in chapters where necessary.

The Set of Commuting Operators Let us discuss one more problem concerning self-adjoint operators. We assume that unbounded self-adjoint operators $A_1, \dots, A_n$ act in a Hilbert space $\mathcal{H}_0$ and denote their resolutions of the identity by $E_1, \dots, E_n$, respectively. These operators are called *commuting* if their resolutions of the identity commute, namely, $E_j(\alpha_j)E_k(\alpha_k) = E_k(\alpha_k)E_j(\alpha_j)$, $\alpha_j, \alpha_k \in \mathfrak{B}(\mathbb{R})$; $j, k = 1, \dots, n$. The commutativity conditions of operators $A_1, \dots, A_n$ have already been clarified earlier. Recall that if these operators are bounded, then relations $A_jA_k = A_kA_j$, $j, k = 1, 2, \dots, n$ are necessary and sufficient conditions for these operators to commute. If they are unbounded, the situation is much more complicated and, for example, the fact that such conditions are satisfied on certain dense sets does not guarantee that the corresponding resolutions of the identity commute, i.e., that A_j commute.

Consider the family $A = (A_j)_{j=1}^n$ of commuting self-adjoint operators in $\mathcal{H}_0$. Let us construct the expansion of these operators in generalized joint eigenvectors with generalized eigenvalues. *A vector $0 \neq \varphi \in \mathcal{H}_0$ is called an (ordinary) joint eigenvector of the family A if $\varphi \in \mathfrak{D}(A_j)$ and $A_j\varphi = \lambda_j\varphi$ with some $\lambda_j \in \mathbb{R}$, $j = 1, 2, \dots, n$), where $\lambda = (\lambda_1, \dots, \lambda_n) \in \mathbb{R}^n$ is the eigenvalue of the family A corresponding to φ.*

In accordance with this definition, we introduce a concept of generalized joint eigenvector. Consider a chain (2.11.18) standardly connected with each A_j ($j = 1, \dots, n$). Then, by definition, $\varphi \in \mathcal{H}_-$ is a *generalized joint eigenvector of the family A corresponding to the eigenvalue $\lambda = (\lambda_1, 2, \dots, \lambda_n \in \mathbb{C}^n$ if*

$$(\varphi, A_j u)_{\mathcal{H}_0} = \lambda_j (\varphi, u)_{\mathcal{H}_0}, \quad j = 1, 2, \dots, n, \ u \in D. \tag{2.11.33}$$

The collection of all these λ forms the generalized spectrum $g(A)$ of the family.

The family A is naturally associated with so called joint resolution of the identity E, i.e., the resolution of the identity in the space $\mathbb{R}^n$ defined on a σ-algebra $\mathfrak{B}(\mathbb{R}^n)$ as a direct product $E = \times_{j=1}^n E_j$ of resolutions of the identity $E_1, \dots, E_n$. By definition, the support of the measure E is the spectrum of A,

$$S(A) = \operatorname{supp} E \subseteq \mathop{\times}_{j=1}^{n} \operatorname{supp} E_j = \mathop{\times}_{j=1}^{n} S(A_j).$$

As before, by using E and the quasi-nuclear chain (2.11.18), we construct the spectral measure of the family A, namely,

$$\mathfrak{B}(\mathbb{R}^n) \ni \alpha \mapsto \rho(\alpha) = \operatorname{Tr}(O^+ E(\alpha) O).$$

Similarly, we introduce the general spectral measure of the family A. Recall that the operators A_j, by using E, can be reconstructed

$$A_j = \int_{\mathbb{R}^n} \lambda_j \, dE(\lambda), \quad j = 1, \dots, n. \tag{2.11.34}$$

Theorem 2.11.8 *Let $A = (A_j)_{j=1}^n$ be a family of commuting self-adjoint operators acting in a separable Hilbert space $\mathcal{H}_0$, each of which is standardly connected with the quasi-nuclear chain* (2.11.18) *in which D is separable. Then all statements of Theorem* 2.11.5 *remain valid with the formulas* (2.11.21) *replaced with*

$$E(\alpha)u = \Big(\int_{\alpha} P(\lambda)\, d\rho(\lambda)\Big)u, \quad \alpha \in \mathfrak{B}(\mathbb{R}^n), \ u \in \mathcal{H}_+, \tag{2.11.35}$$

$$A_j u = \Big(\int_{S(A)} \lambda_j P(\lambda)\, d\rho(\lambda)\Big)u, \quad j = 1, 2, \dots, n, \ u \in \mathfrak{D}(A_j) \cap \mathcal{H}_+$$

and equality (2.11.22) *replaced with* (2.11.33).

Proof As in the case of Theorem 2.11.5, the problem is reduced to proving the following relation: there exists a set $\beta \in \mathfrak{B}(\mathbb{R}^n)$ of complete measure ρ such that for any $\lambda \in \beta$, we have

$$(P(\lambda)u, A_j v)_{\mathcal{H}_0} = \lambda_j (P(\lambda)u, v)_{\mathcal{H}_0}, \quad j = 1, 2, \dots, n, \ u \in \mathcal{H}_+, \ v \in D. \tag{2.11.36}$$

As in the proof of Theorem 2.11.5, we can conclude that, for any fixed $j = 1, 2, \dots, n$, there is a set $\beta_j \in \mathfrak{B}(\mathbb{R}^n)$, for which the relation (2.11.36) is satisfied. To prove Theorem 2.11.8, one must repeat the proof of Theorem 2.11.5, using the representation (2.11.34) instead of (2.4.1) instead of $\beta = \bigcap_{j=1}^n \beta_j$. □

Theorem 2.11.8 remains valid for commuting normal operators (i.e., operators whose resolutions of the identity commute). The corresponding results can be also formulated for the case where the nuclear chain (2.11.26) is used instead of the quasi-nuclear chain (2.11.18). For infinite families of commuting self-adjoint (or normal) operators, the facts presented above remain valid as well, but their formulations and proofs are more complicated).

It is worth noting that the space D from (2.11.18) may be not dense in $\mathcal{H}_+$. If D is dense in $\mathcal{H}_0$, then one can construct a chain $D' \supseteq \mathcal{H}_0 \supseteq D$, where D' is a dual space of antilinear continuous functionals on D. In a certain sense, D' contains $\mathcal{H}_-$; indeed, any $\alpha \in \mathcal{H}_-$ is also a continuous antilinear functional in D and, therefore, α can be interpreted as an element $l_\alpha \in D'$ (more exactly, l_α is identified with the class of $\beta \in \mathcal{H}_-$ such that $(\beta, u)_{\mathcal{H}_0} = (\alpha, u)_{\mathcal{H}_0}$, $u \in D$). It is clear that all results obtained above for $\varphi \in \mathcal{H}_-$, remain valid for generalized eigenvectors φ from D'.

Cyclic Vectors Recall that the spectral measure of a self-adjoint operator A in $\mathcal{H}_0$ was defined as the trace measure

$$\mathfrak{B}(\mathbb{R}) \ni \alpha \mapsto \rho(\alpha) = \operatorname{Tr}(O^+ E(\alpha) O) \in [0, \infty),$$

constructed by using the quasi-nuclear chain (2.11.18). The general spectral measure was introduced as a scalar measure ρ such that ρ and E are absolutely continuous with respect to each other

Here, we consider an important case where the role of the spectral measure ρ can be played by a measure different from a trace one.

A unit vector $\Omega \in \mathcal{H}_0$ is called a *cyclic vector* (or a *vacuum*) of an operator A if $\Omega \in \bigcap_{m=1}^{\infty} \mathfrak{D}(A^m)$ and the set of vectors $\{A^m \Omega \mid m \in \mathbb{Z}_+\}$ is total in $\mathcal{H}_0$.

Theorem 2.11.9 *Assume that a self-adjoint operator A with the resolution of the identity E has a cyclic vector* Ω. *Then a finite non-negative measure*

$$\mathfrak{B}(\mathbb{R}) \ni \alpha \mapsto \rho(\alpha) = (E(\alpha)\Omega, \Omega)_{\mathcal{H}_0} \in [0, \infty)$$

is one of spectral measures of this operator.

Proof If $E(\alpha) = 0$ for some $\alpha \in \mathfrak{B}(\mathbb{R})$, then $\rho(\alpha) = 0$. Let us show the inverse implication. Assume that

$$0 = \rho(\alpha) = (E(\alpha)\Omega, \Omega)_{\mathcal{H}_0} = \|E(\alpha)\Omega\|^2_{\mathcal{H}_0},$$

i.e., $E(\alpha)\Omega = 0$. Then, for any $m \in \mathbb{Z}_+$, we have

$$0 = A^m E(\alpha)\Omega = E(\alpha) A^m \Omega,$$

and, hence, $E(\alpha) f = 0$, where f belongs to the linear span of vectors $A^m \Omega$, $m \in \mathbb{Z}$. By assumption, this span is dense in $\mathcal{H}_0$ and, therefore, $E(\alpha) = 0$. □

In the case of normal operators, the formulation of the theorem remains the same. For a family of commuting operators $A = (A_j)_{j=1}^n$, instead of A^m, one must take the products $A_1^{m_1} \cdot \ldots \cdot A_n^{m_n}$, where $m_1, \ldots, m_n \in \mathbb{Z}_+$.

The Fourier Transform by Generalized Eigenvectors Note that initial formulas (2.11.1) and (2.11.2) of the expansion in eigenvectors of a self-adjoint operator in a finite-dimensional space differ from expressions (2.11.21) proved above. Namely, the initial formulas contain the eigenvectors while the expressions (2.11.21) contain the projectors $P(\lambda)$. Let us show that analogous expansions can be also obtained in the general case.

We consider the case of one self-adjoint operator. But the results obtained can be easily generalized to normal operators and families of commuting operators considered above. We leave formulations of corresponding results for chapters below.

Assume that a self-adjoint operator A satisfies the conditions of the projection spectral theorem 2.11.5. By virtue of the first relation in (2.11.21), we have

$$(E(\alpha)u, v)_{\mathcal{H}_0} = \int_{\alpha} (P(\lambda)u, v)_{\mathcal{H}_0} \, d\rho(\lambda), \tag{2.11.37}$$

$$u, v \in \mathcal{H}_+, \; \alpha \in \mathfrak{B}(\mathbb{R}).$$

In particular, for $\alpha = \mathbb{R}$, the equality (2.11.37) gives the decomposition of $(u, v)_{\mathcal{H}_0}$ into the integral of $(P(\lambda)u, v)_{\mathcal{H}_0}$. The last expression determines a scalar product and (2.11.37) turns into the decomposition of $\mathcal{H}_0$, which is called a *direct integral* of the corresponding Hilbert spaces.

Let us consider such a case in detail. We fix $\lambda \in \mathbb{R}$ so that $\mathrm{Tr}(P(\lambda)) = 1$. Points λ, possessing this property, form a set of complete spectral measure. Below, we assume that λ belong to this set. By virtue of the relations $P(\lambda) \geq 0$ and $| \, P(\lambda)| \leq \mathrm{Tr}(P(\lambda)) = 1$, the operator $\mathbf{J}P(\lambda)J : \mathcal{H}_0 \to \mathcal{H}_0$ is a non-negative Hilbert-Schmidt operator. Let $h_\gamma(\lambda) \in \mathcal{H}_0$ ($\gamma = 1, 2, \ldots, N(\lambda) \leq \infty$) be an orthonormal sequence of eigenvectors of the operator $\mathbf{J}P(\lambda)J$ corresponding to eigenvalues $\nu_\gamma(\lambda)$. Since the dependence of $\mathbf{J}P(\lambda)J$ on the parameter λ is weakly measurable with respect to $\mathfrak{B}(\mathbb{R})$, we can assume that $h_\gamma(\lambda)$ is weakly measurable and $\nu_\gamma(\lambda)$ is measurable ($\gamma = 1, 2, \ldots, N(\lambda)$). Then

$$\begin{aligned}(P(\lambda)Jf, Jg)_{\mathcal{H}_0} &= (\mathbf{J}P(\lambda)Jf, g)_{\mathcal{H}_0} = \sum_{\gamma=1}^{N(\lambda)} \nu_\gamma(\lambda)(f, h_\gamma(\lambda))_{\mathcal{H}_0}\overline{(g, h_\gamma(\lambda))_{\mathcal{H}_0}} \\ &= \sum_{\gamma=1}^{N(\lambda)} (Jf, \varphi_\gamma(\lambda))_{\mathcal{H}_0}\overline{(Jg, \varphi_\gamma(\lambda))_{\mathcal{H}_0}}, \quad f, g \in \mathcal{H}_0,\end{aligned} \tag{2.11.38}$$

where

$$\varphi_\gamma(\lambda) = \sqrt{\nu_\gamma(\lambda)}\mathbf{J}^{-1}h_\gamma(\lambda) = P(\lambda)((\nu_\gamma(\lambda))^{-1/2}Jh_\gamma(\lambda)) \in \mathfrak{R}(P(\lambda)) \subseteq \mathcal{H}_-.$$

The vectors $\varphi_\gamma(\lambda)$, $\lambda \in \mathbb{R}$, $\gamma = 1, 2, \ldots, N(\lambda) \leq \infty$ are individual generalized eigenvectors of the operator A.

It follows from (2.11.38) that

$$(P(\lambda)u, v)_{\mathcal{H}_0} = \sum_{\gamma=1}^{N(\lambda)} (u, \varphi_\gamma(\lambda))_{\mathcal{H}_0} \overline{(v, \varphi_\gamma(\lambda))_{\mathcal{H}_0}}, \quad u, v \in \mathcal{H}_+, \tag{2.11.39}$$

for ρ-almost all $\lambda \in \mathbb{R}$.

Denote $l_2(\infty) = l_2$ and $l_2(N) = \mathbb{C}^N$, $N < \infty$, assuming that the last space is imbedded in l_2 (all the coordinates of the vector, beginning from $N+1$, are equal to zero). *The mapping*

$$\begin{aligned} &\mathcal{H}_+ \ni u \mapsto \hat{u}(\lambda) = (\hat{u}_1(\lambda), \hat{u}_2(\lambda), \ldots) \in l_2(N(\lambda)), \\ &\hat{u}_\gamma(\lambda) = (u, \varphi_\gamma(\lambda))_{\mathcal{H}_0}, \quad \gamma = 1, 2, \ldots, N(\lambda) \end{aligned} \tag{2.11.40}$$

is called the Fourier transform corresponding to the operator A (the inclusion $\hat{u}(\lambda) \in l_2(N(\lambda))$ follows from (2.11.39) for $v = u$). The Fourier transform $\hat{u}(\lambda)$ of the vector u is defined for ρ-almost all $\lambda \in \mathbb{R}$ and each its coordinate is measurable with respect to $\mathfrak{B}(\mathbb{R})$. By inserting (2.11.39) into (2.11.37) and using (2.11.40), we obtain the Parseval equality for Fourier transforms

$$(E(\alpha)u, v)_{\mathcal{H}_0} = \int_\alpha (\hat{u}(\lambda), \hat{v}(\lambda))_{l_2(N(\lambda))}\, d\rho(\lambda), \tag{2.11.41}$$

$$\alpha \in \mathfrak{B}(\mathbb{R}), \ u, v \in \mathcal{H}_+.$$

Note one important fact.

Lemma 2.11.10 *For a fixed λ, the set $\{\hat{u}(\lambda) \mid u \in \mathcal{H}_+\}$ coincides with $l_2(N(\lambda))$ for $N(\lambda) < \infty$ and contains all finite vectors from l_2 for $N(\lambda) = \infty$.*

Proof It suffices to show that every vector $(0, \ldots, 0, 1, 0, \ldots)$ with the unit on the k-th position belongs to this set. We take $u = (\nu_k(\lambda))^{-1/2} J h_k(\lambda) \in \mathcal{H}_+$. Then

$$\begin{aligned} (u, \varphi_\gamma(\lambda)_{\mathcal{H}_0}) &= ((\nu_k(\lambda))^{-1/2} (J h_k(\lambda), \varphi_\gamma(\lambda))_{\mathcal{H}_0} \\ &= (\nu_k(\lambda))^{-1/2} (h_k(\lambda), \mathbf{J}\varphi_\gamma(\lambda))_{\mathcal{H}_0} = (h_k(\lambda), h_\gamma(\lambda))_{\mathcal{H}_0} = \delta_{k\gamma}, \\ &\gamma = 1, 2, \ldots, N(\lambda). \end{aligned}$$

□

The Direct Integral of Hilbert Spaces Consider the direct integral of Hilbert spaces $l_2(N(\lambda))$ over $\mathbb{R}$ with a measure ρ

$$L_2 = \int_{\mathbb{R}} \bigoplus l_2(N(\lambda))\, d\rho(\lambda).$$

This integral is defined as a collection of all vector functions $\mathbb{R} \ni \lambda \mapsto F(\lambda) \in l_2(N(\lambda))$, given for ρ-almost all λ, measurable with respect to $\mathfrak{B}(\mathbb{R})$ in the sense that each coordinate $F_\gamma(\lambda)$, $\gamma = 1, 2, \ldots, N(\lambda)$ is measurable, and such that

$$\int_{\mathbb{R}} \|F(\lambda)\|^2_{l_2(N(\lambda))}\, d\rho(\lambda) < \infty.$$

One can easily prove that the direct integral is a Hilbert space with the scalar product has a form

$$(F(\cdot), G(\cdot))_{L_2} = \int_{\mathbb{R}} (F(\lambda), G(\lambda))_{l_2(N(\lambda))}\, d\rho(\lambda) \tag{2.11.42}$$

$$F(\cdot), G(\cdot) \in L_2.$$

Comparing (2.11.41) and (2.11.42), we obtain that for $\alpha = \mathbb{R}$, the expression on the right (2.11.41) defines the scalar product in the direct integral, and hence the Parseval equality can be rewritten as $(u, v)_{\mathcal{H}_0} = (\hat{u}(\cdot), \hat{v}(\cdot))_{L_2}$, $u, v \in \mathcal{H}_+$. Extending this equality by continuity on the whole $\mathcal{H}_0$, we get

$$(f, g)_{\mathcal{H}_0} = \int_{\mathbb{R}} (\hat{f}(\lambda), \hat{g}(\lambda))_{l_2(N(\lambda))}\, d\rho(\lambda), \quad f, g \in \mathcal{H}_0, \tag{2.11.43}$$

where the Fourier transform $\hat{f}(\lambda)$ of the vector f is regarded as the limit of the Fourier transform (2.11.40) in the norm of the direct integral (clearly, it is no possible longer to use the last formula in (2.11.40) for $\hat{f}_\gamma(\lambda)$).

Theorem 2.11.11 *If D is a base of the operator A, i.e., if the closure of $A \restriction D$ in $\mathcal{H}_0$ coincides with A, then the Fourier transforms $\hat{u}(\lambda)$, $u \in \mathcal{H}_+$ are dense in the direct integral and, hence, the Fourier transform $\mathcal{H}_0 \ni f \mapsto \hat{f}(\lambda) \in L_2$ realizes an isomorphism between spaces $\mathcal{H}_0$ and L_2.*

Thus, in the indicated sense, $\mathcal{H}_0$ can be regarded as decomposed into the direct integral

$$\mathcal{H}_0 = \int_{\mathbb{R}} \bigoplus l_2(N(\lambda))\, d\rho(\lambda). \tag{2.11.44}$$

Lemma 2.11.12 *Under mapping* (2.11.40), *the operator $A \restriction D$ is transferred into the operator of the multiplication by λ, i.e., for ρ-almost all $\lambda \in \mathbb{R}$, and the relation*

$$\widehat{(Au)}(\lambda) = \lambda \hat{u}(\lambda), \quad u \in D \tag{2.11.45}$$

hold true.

Proof Denote by $\beta \in \mathfrak{B}(\mathbb{R})$ a set of complete spectral measure ρ such that for $\lambda \in \beta$, $\mathfrak{R}(P(\lambda))$ consists of generalized eigenvectors of the operator A corresponding to the eigenvalue λ. The existence of this set is guaranteed by Theorem 2.11.5. Let $\lambda \in \beta$. Then, for all $\gamma = 1, 2, \ldots, N(\lambda)$, we have $\varphi_\gamma(\lambda) \in \mathfrak{R}(P(\lambda))$. Therefore, according to (2.11.22) and (2.11.40), the equality

$$\begin{aligned}\widehat{(Au)}(\lambda) &= (Au, \varphi_\gamma(\lambda))_{\mathcal{H}_0} = \overline{(\varphi_\gamma(\lambda), Au)}_{\mathcal{H}_0} \\ &= \lambda\overline{(\varphi_\gamma(\lambda), u)}_{\mathcal{H}_0} = \lambda(u, \varphi_\gamma(\lambda))_{\mathcal{H}_0} = \lambda\hat{u}_\gamma(\lambda), \quad \forall u \in D\end{aligned} \tag{2.11.46}$$

holds true. This relation is the required equality (2.11.45) in the coordinate form. □

Proof of Theorem 2.11.11 Let us fix a non-real $z \in \mathbb{C}$. It should be proved that, for every $u \in \mathcal{H}_+$, we have $(\lambda - z)^{-1}\hat{u}(\lambda) \in \hat{\mathcal{H}}_0 \subset L_2$. Indeed, since $(\lambda - z)^{-1}$ regarded as a function of $\lambda \in R$ is bounded, the multiplication operator by $(\lambda - z)^{-1}$ is continuous in L_2 and, therefore, for $v \in D$, according to (2.11.45) and (2.11.43), we have

$$\begin{aligned}\|(\lambda - z)^{-1}\hat{u}(\lambda) - \hat{v}(\lambda)\|_{L_2} &\le c\|\hat{u}(\lambda) - (\lambda - z)\hat{v}(\lambda)\|_{L_2} \\ &= c\|\hat{u}(\lambda) - ((A - z\mathbb{I})v)\;(\lambda)\|_{L_2} \\ &= c\|u - (A - z\mathbb{I})v\|_{\mathcal{H}_0}.\end{aligned}$$

Since D forms the base of the self-adjoint operator A, the right-hand side of this estimate can be made as small as desired, which proves the assertion made above. □

Let $F(\lambda) \in L_2$ be orthogonal in L_2 to $\hat{\mathcal{H}}_0$.Then, in particular, for non-real $z \in \mathbb{C}$ and $u \in \mathcal{H}_+$ is chosen so that $\hat{u}(\lambda) = (1, 0, 0, \ldots)$ (this is possible, see Lemma 2.11.10)

$$\begin{aligned}0 &= \int_{\mathbb{R}} \left(F(\lambda), \frac{1}{\lambda - z}\hat{u}(\lambda)\right)_{l_2(N(\lambda))} d\rho(\lambda) \\ &= \int_{\mathbb{R}} \frac{1}{\lambda - \bar{z}} F_1(\lambda)\, d\rho(\lambda) = \int_{\mathbb{R}} \frac{1}{\lambda - \bar{z}}\, d\omega(\lambda); \\ \omega(\delta) &= \int_{\delta} F_1(\lambda)\, d\rho(\lambda), \quad \delta \in \mathfrak{B}(\mathbb{R}).\end{aligned}$$

This implies that $\omega = 0$, and, hence, $F_1(\lambda) = 0$ for ρ-almost all $\lambda \in \mathbb{R}$ (this follows from the following well-known fact: if a charge ω of a bounded variation given on $\mathfrak{B}(\mathbb{R})$ is such that $\int_{\mathbb{R}}(\lambda - z)^{-1}\, d\omega(\lambda) = 0$ for all non-real $z \in \mathbb{C}$, then $\omega = 0$.

The same reasoning can be applied to $F_2(\lambda)$ and so on. As a result, we find that $F(\lambda) = 0$ in L_2, which implies that $\hat{\mathcal{H}}_0 = L_2$. □

The obtained above result yields $N(\lambda) = \dim(\mathfrak{R}(P(\lambda)))$, i.e., $N(\lambda)$ is the "multiplicity" of the eigenvalue λ. In the case where $N(\lambda) = 1$ for ρ-almost all λ, i.e., if "spectrum is simple", then we have

$$\hat{u}(\lambda) = \hat{u}_1(\lambda) = (u, \varphi_1(\lambda))_{\mathcal{H}_0} \in \mathbb{C}, \quad u \in \mathcal{H}_+,$$

and (2.11.44) gives a decomposition of $\mathcal{H}_0$ into the direct integral of complex planes $\mathbb{C} = l_2(1)$.

Theorem 2.11.13 *Suppose that a self-adjoint operator A, satisfying conditions of the projection spectral Theorem* 2.11.5, *has a cyclic vector $\Omega \in D$ such that for any $m \in \mathbb{N}$, $A^m\Omega \in D$ and the linear span of these vectors is dense not only in $\mathcal{H}_0$, and also in $\mathcal{H}_+$. Then the spectrum of the operator A is simple.*

Proof Assume that $N(\lambda) > 1$ for a set of λ of positive measure ρ. Then there exists $\lambda \in \mathbb{R}$ such that $N(\lambda) > 1$, $\mathrm{Tr}(P(\lambda)) = 1$, and $\mathfrak{R}(P(\lambda))$ consists of generalized eigenvectors corresponding to λ. For $f \in \mathcal{H}_0$ and $g = J^{-1}\Omega$, relation (2.11.38) yields

$$\begin{aligned}(P(\lambda)Jf, \Omega)_{\mathcal{H}_0} &= \sum_{\gamma=1}^{N(\lambda)} \nu_\gamma(\lambda)(f, h_\gamma(\lambda))_{\mathcal{H}_0}\overline{(J^{-1}\Omega, h_\gamma(\lambda))_{\mathcal{H}_0}} \\ &= \sum_{\gamma=1}^{N(\lambda)} (f, h_\gamma(\lambda))_{\mathcal{H}_0}\bar{a}_\gamma.\end{aligned} \tag{2.11.47}$$

Since $(h_\gamma(\lambda))_{\gamma=1}^{N(\lambda)}$ is an orthonormal sequence in $\mathcal{H}_0$, vectors

$$b(f) = ((f, h_1(\lambda))_{\mathcal{H}_0}, (f, h_2(\lambda))_{\mathcal{H}_0}, \dots) \in l_2(N(\lambda))$$

run over $l_2(N(\lambda))$ as f runs over $\mathcal{H}_0$. This implies that, for $N(\lambda) > 1$, there exists $f_0 \in \mathcal{H}_0$ such that $b(f_0)$ is not equal to zero and is orthogonal in $l_2(N(\lambda))$ to the vector $a = (a_1, a_2, \dots)$, introduced in (2.11.47) (the vector a belongs to $l_2(N(\lambda))$, because $J^{-1}\Omega \in \mathcal{H}_0$ and the factors $\nu_\gamma(\lambda)$ are bounded). By setting $f = f_0$ in (2.11.47), we get $(P(\lambda)Jf_0, \Omega)_{\mathcal{H}_0} = 0$.

Since $P(\lambda)Jf_0$ is a generalized eigenvector of the operator A corresponding to λ, the application of the equality (2.11.19) yields

$$(P(\lambda)Jf_0, A^m\Omega)_{\mathcal{H}_0} = \lambda^m(P(\lambda)Jf_0, \Omega)_{\mathcal{H}_0} = 0, \quad m \in \mathbb{Z}_+.$$

In view of the fact that the linear span of the vectors $A^m\Omega$ is dense in $\mathcal{H}_+$, this implies $P(\lambda)Jf_0 = 0$.

By setting $f = f_0$ in (2.11.38), we get

$$0 = \sum_{\gamma=1}^{N(\lambda)} \nu_\gamma(\lambda) b_\gamma(f_0) \overline{b_\gamma(g)}, \quad g \in \mathcal{H}_0.$$

Vectors $b(g) \in l_2(N(\lambda))$ run over the whole $l_2(N(\lambda))$ as g runs over $\mathcal{H}_0$. Therefore, the last equality and relations $0 < \nu_\gamma(\lambda) \le c < \infty$ yield $b(f_0) = 0$. We arrived at a contradiction. □

In all formulas of this section, we can, clearly, replace $\mathbb{R}$ and $\mathfrak{B}(\mathbb{R})$ with the spectrum $S(A)$ of the operator A and $\mathfrak{B}(S(A))$, respectively.

2.12 A General Case with a Quasi-Scalar Product

Consider a chain of Hilbert spaces with an involution "*", namely,

$$G_- \supset G_0 \supset G_+. \tag{2.12.1}$$

This means that an involution is defined in the space $G_- \ni \xi \mapsto \xi^* \in G_-$, the restriction of which on G_0 or G_+ is an involution in the corresponding space. Let us construct the tensor square of chains (2.12.1) in the form

$$G_- \otimes G_- \supset G_0 \otimes G_0 \supset G_+ \otimes G_+ \tag{2.12.2}$$

and fix some generalized kernel $K \in G_- \otimes G_-$. A kernel is said to be positive definite if

$$(K, \varphi \otimes \varphi^*)_{G_0 \otimes G_0} \ge 0, \quad \varphi \in G_+, \tag{2.12.3}$$

that is, a continuous bilinear form

$$G_+ \oplus G_+ \ni (\varphi, \psi) \mapsto a(\varphi, \psi) = (K, \psi \otimes \varphi^*)_{G_0 \otimes G_0} \in \mathbb{C}$$

is non-negative. The kernel $K \in G_- \otimes G_-$ is called Hermitian, if the form is Hermitian, i.e.,

$$(K, \psi \otimes \varphi^*)_{G_0 \otimes G_0} = \overline{(K, \varphi \otimes \psi^*)}_{G_0 \otimes G_0}.$$

Every positive definite kernel is Hermitian.

Let us construct the Hilbert space generated by such a kernel. For the positive definite kernel K, we introduce, in general, the quasi-scalar product in G_+

$$\langle \varphi, \psi \rangle = (K, \psi \otimes \varphi^*)_{G_0 \otimes G_0} \geq 0, \quad \varphi, \psi \in G_+. \tag{2.12.4}$$

After identifying with zero those $\varphi \in G_+$ for which $\langle \varphi, \varphi \rangle = 0$, and further completion of the set of classes

$$\hat{\varphi} = \{\chi \in G_+ \mid \langle \varphi - \chi, \varphi - \chi \rangle = 0\}$$

we get the space denoted by $\mathcal{H}_K$.

In more detail, the procedure looks like this. Let

$$N = \{\varphi \in G_+ \mid \langle \varphi, \varphi \rangle = 0\}.$$

Due to the inequality

$$\langle \varphi, \varphi \rangle \leq \|K\|_{G_0 \otimes G_0} \|\varphi\|^2_{G_+}, \quad \varphi \in G_+, \tag{2.12.5}$$

N is a subspace in G_+. The class $\hat{\varphi}$, for which $\varphi \in G_+$, coincides with hyperplane $\varphi + N$. The form (2.12.4) does not change if φ and ψ run over the classes $\hat{\varphi}$ and $\hat{\psi}$ and it defines their scalar product $(\hat{\varphi}, \hat{\psi})_{\mathcal{H}_K}$. The space $\mathcal{H}_K$ is a complement of this pre-Hilbert space of classes. The described procedure is isomorphic to the following: on the orthogonal complement $G_+ \ominus N$ in G_+, the formula (2.12.4) already generates a scalar product. The completion it gives $\mathcal{H}_K$. The isomorphism is defined accordingly (before completion) between the class $\hat{\varphi}$ and the projection in G_+ of the vector φ onto $G_+ \ominus N$ (all vector projections $\varphi \in \hat{\varphi}$ coincide with each other).

Consider the case of Hilbert rigging. Let be given a chain with an involution

$$\mathcal{H}_- \supset G_- \supset G_0 \supset G_+ \supset \mathcal{H}_+, \tag{2.12.6}$$

where a Hilbert space $\mathcal{H}_+$ is topologically imbedded into G_+. Let us construct the equipment of the space $\mathcal{H}_K$ according to (2.12.6) . For this, we note that $N \cap \mathcal{H}_+$ is a subspace in $\mathcal{H}_+$ (due to continuity of the imbedding $\mathcal{H}_+ \mapsto G_+$) and (2.12.5)), and consider $\mathcal{H}_+ \ominus (N \cap \mathcal{H}_+)$ (orthogonal difference in $\mathcal{H}_+$). *Vectors of the space $\mathcal{H}_+ \ominus (N \cap \mathcal{H}_+)$, can be identified with the set of all classes $\hat{\varphi}$, where $\varphi \in \mathcal{H}_+$: each such class $\hat{\varphi}$ has only one common vector φ_N with $\mathcal{H}_+ \ominus (N \cap \mathcal{H}_+)$, i.e., a projection φ in $\mathcal{H}_+$ on $\mathcal{H}_+ \ominus (N \cap \mathcal{H}_+)$.*

Thus, let φ_1 be a such projection. Then $\varphi - \varphi_1 \in \mathcal{H}_+ \ominus (N \cap \mathcal{H}_+)$, i.e., $\varphi_1 \in \hat{\varphi}$. Let now $\varphi_1, \varphi_2 \in \hat{\varphi} \cap \mathcal{H}_+ \ominus (N \cap \mathcal{H}_+)$, then $\varphi_1, \varphi_2 \in \mathcal{H}_+$, $\varphi_1 - \varphi_2 \in N$, whence follows $\varphi_1 - \varphi_2 \in N \cap \mathcal{H}_+$; since $\varphi_1 - \varphi_2 \in \mathcal{H}_+ \ominus (N \cap \mathcal{H}_+)$, then $\varphi_1 - \varphi_2 = 0$. □

We denote by $\mathcal{H}_{+,K}$ the linear set of all classes $\hat{\varphi}$, where $\varphi \in \mathcal{H}_+$, and transform it into a Hilbert space by putting

$$(\hat{\varphi}, \hat{\psi})_{\mathcal{H}_{+,K}} = (\varphi_N, \psi_N)_{\mathcal{H}_+}, \quad \varphi, \psi \in \mathcal{H}_+. \tag{2.12.7}$$

This definition is obviously correct; the obtained space $\mathcal{H}_{+,K}$ is isometric to $\mathcal{H}_+ \ominus (N \cap \mathcal{H}_+)$.

The space $\mathcal{H}_{+,K}$ is imbedded into $\mathcal{H}_K$, moreover, the imbedding operator O can be represented as a product $O = SO_+$, where $O_+ : \mathcal{H}_+ \to G_+$ is the imbedding operator $\mathcal{H}_+$ into G_+, and $S_+ : G_+ \to \mathcal{H}_K$ is an operator which transfers the vector $\varphi \in G_+$ to the corresponding class $\hat{\varphi} \in \mathcal{H}_K$. Thus, taking into account (2.12.5)), we have

$$\|\hat{\varphi}\|_{\mathcal{H}_K} \le \|SO_+\| \|\hat{\varphi}\|_{\mathcal{H}_{+,K}} \le \|K\|^{1/2}_{G_- \otimes G_-} \|O_+\| \|\hat{\varphi}\|_{\mathcal{H}_{+,K}}, \quad \hat{\varphi} \in \mathcal{H}_{+,K}.$$

It is clear that $\mathcal{H}_{+,K}$ is imbedded in $\mathcal{H}_K$. This fact follows from the density $\mathcal{H}_+$ in G_+ and the inequality (2.12.5). Hence, it can be assumed that $\mathcal{H}_K$ is the zero space, $\mathcal{H}_{+,K}$ is a positive space, and we construct the chain

$$\mathcal{H}_{-,K} \supset \mathcal{H}_K \supset \mathcal{H}_{+,K}. \tag{2.12.8}$$

From the representation of $O = SO_+$, it follows that *the chain (2.12.8) is quasi-nuclear if the imbedding $\mathcal{H}_+ \to G_+$ is quasi-nuclear.*

In the case of a degenerate kernel, i.e., when $N = \{0\}$, the procedure (2.12.8) of the construction of a chain becomes quite simple, then $\mathcal{H}_{+,K} = \mathcal{H}_+$.

Bibliographical Notes

The first section of the book deals with the general principles of the theory, mainly of unbounded self-adjoint operators, outlined in [47, 48]. In more detail, sections of the theory of unbounded operators and their spectral properties are considered in [4, 43, 44, 54, 152, 194, 211, 236, 238, 239, 242, 247, 250]. In particular, generalized resolvents and a description of all self-adjoint extensions are given in [4]. A proof of the spectral theorem, which uses the theory of commutative Banach algebras, can be found in [211, 247, 298]. Some other methods of establishing the self-adjointness of operators are presented in [43, 44, 239]. Spectral representations of families of commutative operators and their application to the Stone theorems, harmonic analysis, and non-commutative operators are described in [43, 44, 250].

The detailed information on differential equations in Banach spaces can be found in [190]. Quasi-analytic classes of functions are studied in [206].

For a detailed description of the theory of equipped (rigged) spaces, see: [21, 43, 44, 116, 128, 203, 211]; the spaces of test and generalized functions are carefully studied in [114] in the case of a finite number of variables and in [43, 44] in the case of an infinite number of variables. Various versions of the kernel theorem can be found in [21, 23, 43, 44, 116, 211, 242]. Many results about bilinear forms are presented in [43, 44, 152, 170, 171, 239, 242].

The generalized eigenvectors expansion method is considered in detail in [21, 43, 44, 51, 116, 211]. In particular, the book [43, 44] is devoted to the generalized eigenvectors expansion for a family of commutative normal operators of any power. Direct integrals of Hilbert spaces are studied in [211, 221].

The self-adjointness of differential operators with partial derivatives is studied in [21, 239]; in the case of an infinite number of independent variables, these problems are considered in [43, 44]. Books [51, 122, 241] contain additional information on the spectral properties of differential operators with partial derivatives (including the usual differential operators). Books [4, 95, 197, 198, 207, 220, 222] devoted to the study of spectral properties of ordinary differential operators. Spectral properties of the specified operators with operator-valued coefficients are studied in [128, 203] (such operators cover certain classes of operators with partial derivatives).

The first chapter and the book as a whole did not include some areas of functional analysis. Among them: group theory [16, 52, 138, 152, 211, 239, 298]; scattering theory [4, 240]; perturbations of operators [4, 152, 170, 171]; non-self-adjoint operators [125, 248]; spaces with an indefinite metric [14, 80]; Banach algebras (normed rings) [112, 138, 202, 211, 221, 247, 298]; topological groups and their representations [16, 161, 202, 223, 237]; differential calculus in infinite-dimensional spaces and nonlinear functional analysis [43, 44, 82, 150, 163, 175, 201]; approximation methods of functional analysis [68, 150, 174]; topology [7, 10, 246, 251]; equations of mathematical physics [243, 244, 291, 292].

In general, an important information related to functional analysis, we recommend in particular [8, 9, 15, 69, 72, 94–96, 98, 123, 134, 140, 176, 184, 245, 279, 283, 284, 294]. Issues of generalized functions and related to them are covered in [109, 110, 113–116, 162, 192, 227, 293]. The main questions of the theory of extensions of symmetric operators to self-adjoint ones are presented in [58, 129, 130, 170, 193, 204]. The theory of analytic functions is in [208, 209]. For general and applied questions of mathematical analysis and the measure theory, are recommended [1, 100, 108, 133, 199, 214, 224, 225, 229, 256–258, 268, 280, 285, 286] theory of probabilities [121]; differential and integral equations [67, 149, 183, 195, 215, 216, 234, 235]. Some applied questions of functional analysis and the theory of singularly perturbed operators are in works [5, 6, 170].

Chapter 3
Jacobi Matrices and the Classical Moment Problem

This section is devoted to the presentation of the basic provisions of the theory of (ordinary numerical tri-diagonal) Jacobi matrices and the classical moment problem which is closely related to this theory.

3.1 Difference Operators, Jacobi Matrices, and Self-Adjointness Conditions

Consider the transformation defined by the expression L of the form

$$(Lu)_j = a_{j-1}u_{j-1} + b_j u_j + a_j u_{j+1}, \quad j \in \mathbb{N}_0, \tag{3.1.1}$$

where $a_j > 0$, $b_j \in \mathbb{R}$ are given coefficients, $u = (u_j)_{j=0}^{\infty} = (u_0, u_1, u_2, \ldots)$ is a sequence of numbers on which L acts, and for the convenience of writings, we put $u_{-1} := 0$. The expression (3.1.1) is the differential analogue of the Sturm-Liouville differential expression on the semi-axis $[0, \infty)$:

$$\frac{d}{dx}\left(a(x)\frac{du}{dx}\right) + q(x)u(x), \quad u(0) = 0.$$

Y. M. Berezansky, M. E. Dudkin, *Jacobi Matrices and the Moment Problem*,
Operator Theory: Advances and Applications 294,
https://doi.org/10.1007/978-3-031-46387-7_3

The correctness of the Green formula can be easily verified by simple calculations:

$$\sum_{j=k}^{l}[(Lu)_j\bar{v}_j - u_j\overline{(Lv)_j}] = a_l(u_{l+1}\bar{v}_l - u_l\bar{v}_{l+1})$$
$$- a_{k-1}(u_k\bar{v}_{k-1} - u_{k-1}\bar{v}_k), \; k, l \in \mathbb{N}_0, \; k < l, \tag{3.1.2}$$

where a dash above the expression means a complex conjugation.

We consider the operator L in the Hilbert space $l_2 = \mathcal{H}_0$, which consists of sequences $u = (u_0, u_1, u_2, \ldots)$, $u_j \in \mathbb{C}$, for which the series converges $\sum_{j=0}^{\infty} |u_j|^2 < \infty$ and the scalar product is given by the expression

$$(u, v) = \sum_{j=0}^{\infty} u_j\bar{v}_j, \quad u = (u_0, u_1, u_2, \ldots), \; v = (v_0, v_1, v_2, \ldots).$$

We denote by L_0 the operator defined by (3.1.1) on the finite vectors combined into the linear set $l_{2,0}$. Then, through L we denote the closure L_0; it is the so-called minimal operator generated by the expression (3.1.1). Therefore, in what follows, the expression (3.1.1) and the corresponding operator will be denoted by L. Moreover, the domain of the adjoint operator L^* consists of vectors $v \in l_2$ for which $Lv \in l_2$ and $(L^*v)_j = (Lv)_j$, where $v_{-1} = 0$.

In general, $L^* \supseteq L$. Self-adjointness conditions for L will be discussed in detail below. For now, we only note that the defect numbers of the operator L are equal to each other. This follows from the fact that L is real with respect to the involution $u = (u_j)_{j=0}^{\infty} \to \bar{u} = (\bar{u}_j)_{j=0}^{\infty}$.

Thus, now the action of the operator, L can be associated with the action of the matrix

$$J = \begin{bmatrix} b_0 & a_0 & 0 & 0 & 0 & \cdots \\ a_0 & b_1 & a_1 & 0 & 0 & \cdots \\ 0 & a_1 & b_2 & a_2 & 0 & \cdots \\ \vdots & \vdots & \vdots & \vdots & \vdots & \ddots \end{bmatrix}, \quad a_j > 0, \; b_j \in \mathbb{R}, \; j \in \mathbb{N}_0 \tag{3.1.3}$$

on the vector $u = (u_j)_{j=0}^{\infty} = (u_0, u_1, u_2, \ldots)$. Matrices with this structure are called *Jacobi matrices*. Note that often, the minimal operator generated by the matrix (3.1.3) will be denoted by a bold letter $\mathbf{J}$.

Consider the equation given by the matrix (3.1.3):

$$(Lu)_j = a_{j-1}u_{j-1} + b_ju_j + a_ju_{j+1} = zu_j, \quad j \in \mathbb{N}_0, \tag{3.1.4}$$

where $z \in \mathbb{C}$, and u_j are desired solutions. Considering $u_{-1} = 0$ and $u_0 = 1$, all u_j are calculated recursively. Thus, for example, $u_1 = \frac{z-b_0}{a_0}$. Let us redefine the solutions $P_j(z) := u_j$; they are polynomials depending on z. Polynomials $P_j(z)$ are called *polynomials of the first kind*. Therefore,

$$(LP(z))_j = zP_j(z), \quad j \in \mathbb{N}_0, \; P_{-1}(z) = 0, \; P_0(z) = 1. \tag{3.1.5}$$

Due to the real a_j and b_j, $j \in \mathbb{N}_0$, coefficients of polynomials $P_j(z)$, $j \in \mathbb{N}_0$ are real, because $a_j > 0$, $j \in \mathbb{N}_0$, the coefficient by z^j in $P_j(z)$, $j \in \mathbb{N}_0$ is always positive. The difference equation in (3.1.4) is a second-order equation, therefore, it has two linearly independent solutions. The second solution will be indicated below.

We denote by N_z for $\mathrm{Im} z \neq 0$ the orthogonal complement to $\mathfrak{R}(L - \bar{z}\mathbb{I})$ in $\mathcal{H}_0 = l_2$, where we recall that $\mathbb{I}$ is the identity operator and N_z is the defect subspace of the operator L corresponding to the regular point z. This subspace coincides with the set of solutions of the equation $L^* f = zf$, that is, due to the form of L^*, (it coincides with the subspace of solutions of the difference equation $(Lu)_j = zu_j$, $u_{-1} = 0$, which are included in l_2). Each solution of such an equation is represented in the form $u_j = u_0 P_j(z)$, therefore, the defect subspace is no more than one-dimensional, and it is not empty if and only if $(u_j) = (u_0 P_j(z)) \in l_2$, that is, when the series

$$\sum_{j=0}^{\infty} |P_j(z)|^2 \tag{3.1.6}$$

converges. In this way, we can formulate the theorem.

Theorem 3.1.1 *The operator L, defined using* (3.1.1) *by the matrix generated by the difference expression* (3.1.4) *in the space l_2, has defect indices* $(0, 0)$ *or* $(1, 1)$. *In the case of defect indices* $(0, 0)$, *the series* (3.1.6) *diverges for z,* $\mathrm{Im} z \neq 0$. *In the case of defect indices* $(1, 1)$, *the series* (3.1.6) *converges, and the defect subspace N_z is formed by the vector* $(P_0(z), P_1(z), \ldots)$.

The Self-Adjointness Criteria It follows from Theorem 3.1.1 that for the operator L to be self-adjoint it is necessary that the series (3.1.6) diverges as z, $\mathrm{Im} z \neq 0$, and it is sufficient that it diverges for at least one such z. In terms of coefficients a_j and b_j, $j \in \mathbb{N}_0$, it is possible to provide sufficient conditions for the convergence or divergence of the series (3.1.6), i.e., conditions for the self-adjointness or non-self-adjointness of the operator L. We will present the simplest of such criteria.

Theorem 3.1.2 *If elements a_j, b_j, $j \in \mathbb{N}_0$ of the matrix* (3.1.3) *are bounded, then the corresponding operator L generated by the expression* (3.1.1) *is bounded, and therefore also self-adjoint.*

Proof Let $a_j, \|b_j\| \leq C, j = 0, 1, \ldots$, then it follows from the triangle inequality

$$\|Lu\|_0 = \left(\sum_{j=0}^{\infty} |a_{j-1}u_{j-1} + a_j u_{j+1} + b_j u_j|^2\right)^{1/2}$$

$$\leq \left(\sum_{j=0}^{\infty} |a_{j-1}u_{j-1}|^2\right)^{1/2} + \left(\sum_{j=0}^{\infty} |a_j u_{j+1}|^2\right)^{1/2} + \left(\sum_{j=0}^{\infty} |b_j u_j|^2\right)^{1/2}$$

$$\leq C(\left(\sum_{j=0}^{\infty} |u_{j-1}|^2\right)^{1/2} + \left(\sum_{j=0}^{\infty} |u_{j+1}|^2\right)^{1/2} + \|u\|_0) \leq 3C\|u\|_0.$$

Thus, the theorem is proved. □

Theorem 3.1.3 *If elements a_j, $j \in \mathbb{N}_0$ of the matrix* (3.1.3) *generated by the expression* (3.1.6) *are such that*

$$\sum_{j=0}^{\infty} \frac{1}{a_j} = \infty, \tag{3.1.7}$$

and b_n are arbitrary, then the corresponding (3.1.3) *operator L is self-adjoint.*

Proof It suffices to check that, it follows from (3.1.7) that, the series (3.1.6) diverges for some z, $\mathrm{Im} z \neq 0$. By using the Green formula (3.1.2), we find

$$(z - \bar{z}) \sum_{j=0}^{n} P_j(z)\overline{P_j(z)}$$

$$= \left(\sum_{j=0}^{n} (L[P(z)])_j \overline{P_j(z)} - P_j(z)\overline{(L[P(z)])_j}\right)$$

$$= a_n(P_{n+1}(z)\overline{P_n(z)} - P_n(z)\overline{P_{n+1}(z)}),\ n \in \mathbb{N}_0.$$

Hence,

$$\sum_{j=0}^{n} |P_j(z)|^2 = \frac{a_n}{(z - \bar{z})}(P_{n+1}(z)\overline{P_n(z)} - P_n(z)\overline{P_{n+1}(z)}),\ n \in \mathbb{N}_0. \tag{3.1.8}$$

Since $P_0(z) = 1$, then, evaluating (3.1.8) by the modulo, we get

$$\begin{aligned}1 \leq& \frac{a_n}{|z-\bar{z}|}(|P_{n+1}(z)||P_n(z)| - |P_n(z)||P_{n+1}(z)|)\\ =& \frac{2a_n}{|z-\bar{z}|}|P_n(z)||P_{n+1}(z)|.\end{aligned}$$

Hence,

$$\frac{1}{a_n} \leq C|P_n(z)||P_{n+1}(z)|.$$

And finally,

$$\begin{aligned}\infty = \sum_{j=0}^{\infty}\frac{1}{a_n} &\leq C\sum_{j=0}^{\infty}|P_n(z)||P_{n+1}(z)|\\ &\leq C\left(\sum_{j=0}^{\infty}|P_n(z)|^2 \cdot \sum_{j=0}^{\infty}|P_{n+1}(z)|^2\right)^{1/2}\\ &< C\sum_{j=0}^{\infty}|P_n(z)|^2,\end{aligned}$$

i.e., $\sum_{j=0}^{\infty}|P_n(z)|^2 = \infty$. Thus, the theorem is proved. □

Let us prove, to some extent, the opposite assertion to the previous theorem.

Theorem 3.1.4 *If elements a_j, b_j, $j \in \mathbb{N}_0$ of the matrix* (3.1.3), *generated by the expression* (3.1.1), *are such that $|b_j| \leq C$ and starting from some number, the inequality $a_{j-1}a_{j+1} \leq a_j^2$ is valid and*

$$\sum_{j=0}^{\infty}\frac{1}{a_j} < \infty, \tag{3.1.9}$$

then the corresponding to (3.1.3) *operator L is not self-adjoint.*

Proof Let us fix z, $\mathrm{Im} z \neq 0$. In accordance with (3.1.9), to prove the inequality $\sum_{j=0}^{\infty}|P_n(z)|^2 < \infty$, and hence the theorem, it suffices to show that

$$\sqrt{a_j}|P_j(z)| \leq A < \infty, \quad j \in \mathbb{N}_0. \tag{3.1.10}$$

Assume that the inequality $\sqrt{a_j}|P_j(z)| \le A_n$, $j = 1, 2, \ldots, n$ holds true, and select a constant A_{n+1} taking into account A_n; while we consider that $n \ge n_0$, where $a_{j-1}a_{j+1} \le a_j^2$, $j = n_0, n_0 + 1, \ldots$. Since

$$P_{n+1}(z) = \frac{1}{a_n}(z - b_n)P_n(z) - \frac{a_{n-1}}{a_n}P_{n-1}(z),$$

then

$$\begin{aligned}
\sqrt{a_{n+1}}|P_{n+1}(z)| \le & \frac{\sqrt{a_{n+1}}}{a_n}|z - b_n||P_{n+1}(z)| + \frac{\sqrt{a_{n+1}a_{n-1}}}{a_n}|P_{n-1}(z)| \\
\le & C_1\frac{\sqrt{a_{n+1}a_{n-1}}}{a_n}\frac{1}{\sqrt{a_{n-1}a_n}}\sqrt{a_n}|P_n(z)| \\
& + \frac{\sqrt{a_{n+1}a_{n-1}}}{a_n}\sqrt{a_{n-1}}|P_{n-1}(z)| \le A_n\left(1 + \frac{C_1}{\sqrt{a_{n-1}a_n}}\right),
\end{aligned}$$

where we denote $C_1 = C + |z|$. Hence, we put

$$A_{n+1} = A_n\left(1 + \frac{C_1}{\sqrt{a_{n-1}a_n}}\right).$$

Thus,

$$A_n = A_{n_0}\prod_{j=n_0}^{n-1}\left(1 + \frac{C_1}{\sqrt{a_{n-1}a_n}}\right) \le A_{n_0}\prod_{j=n_0}^{\infty}\left(1 + \frac{C_1}{\sqrt{a_{n-1}a_n}}\right) = A,$$

for $n > n_0$. If the last product converges, then $A < \infty$ and (3.1.10) is proved. But the convergence is really valid, because, due to (3.1.10),

$$\sum_{j=0}^{\infty}\frac{1}{\sqrt{a_{j-1}a_j}} \le \left(\sum_{j=0}^{\infty}\frac{1}{a_{j-1}}\cdot\sum_{j=0}^{\infty}\frac{1}{a_j}\right)^{1/2} < \infty.$$

Thus, the theorem is proved. □

3.2 The Generalized Eigenvectors Expansion and the Fourier Transform Corresponding to Jacobi Matrices

In this section, we construct the generalized eigenvectors expansion for some self-adjoint extension of the operator L defined in the previous section.

Let $E(\cdot)$ be the generalized resolution of the identity of the operator L (see Sect. 2.11). Since the operator L acts in the space l_2, then according to the spectral theorem

$$E(\Delta) = \int_{\Delta} P(\lambda)\, d\varrho(\lambda), \tag{3.2.1}$$

where Δ is from Borel σ-algebra on $\mathbb{R}$, $P(\lambda)$ is the generalized projection operator acting from the space l_2 into the space, $l_2((p_j)^{-1})$ that is the space l_2 with weight $(p_j)^{-1}$, where elements of the sequence $p_j \geq 1$ such that $\sum_{j=0}^{\infty} \frac{1}{p_j} < \infty$. The operator $P(\lambda)$ is a positive defined matrix $[\Phi_{j,k}(\lambda)]_{j,k=0}^{\infty}$ with elements which ϱ-almost everywhere satisfy the condition

$$\sum_{j,k=0}^{\infty} \frac{|\Phi_{j,k}(\lambda)|^2}{p_j p_k} \leq 1. \tag{3.2.2}$$

The Representation of the Generalized Projector Recall that the generalized eigenvectors $P_j(\lambda)$ are defined in the previous section. Before the main result, we prove the lemma.

Lemma 3.2.1 *For matrix elements of the generalized projection operator and generalized eigenvectors, the representation*

$$\Phi_{j,k}(\lambda) = P_j(\lambda) P_k(\lambda) \Phi_{0,0}(\lambda), \quad j, k \in \mathbb{N}_0 \tag{3.2.3}$$

holds true.

Proof Each vector $\varphi \in \mathfrak{R}(P(\lambda))$ is a generalized eigenvector with a corresponding eigenvalue λ of the operator L. It means that $(\varphi, (L^* - \lambda\mathbb{I})u)_0 = 0$ for all vectors $u \in l_{2,0}$, or in other words $(\varphi, (L - \lambda\mathbb{I})u)_0 = 0$ for all vectors $u \in l_{2,0}$, where we put $u_{-1} = 0$. Using the Green formula (3.1.2), we get

$$0 = (\varphi, (L - \lambda\mathbb{I})u)_0 = ((L - \lambda\mathbb{I})\varphi, u)_0 - a_{-1}\varphi_{-1}\bar{u}_0,$$

i.e., $a_{-1}\varphi_{-1}\bar{u}_0 = ((L - \lambda\mathbb{I})\varphi, u)_0$. Due to the arbitrariness of u , the last expression is possible only when $((L - \lambda\mathbb{I})\varphi)_j = 0$, $j \in \mathbb{N}$ and $a_0\varphi_1 + (b_0 - \lambda)\varphi_0 = 0$. Hence, it can be assumed that $((L - \lambda\mathbb{I})\varphi)_j = 0$, $j \in \mathbb{N}_0$ and $\varphi_{-1} = 0$. Thus, using (3.1.5), we get $\varphi_j = \varphi_0 P_j(\lambda)$, $j \in \mathbb{N}_0$.

In particular, for fixed $k \in \mathbb{N}_0$, the vector

$$(\Phi_{0,k}(\lambda), \Phi_{1,k}(\lambda), \ldots) = P(\lambda)\delta_k \in \mathfrak{R}(P(\lambda)),$$

where $\delta_k = (\delta_{0,k}, \delta_{1,k}, \ldots)$. Then $\Phi_{j,k}(\lambda) = \Phi_{0,k}(\lambda)P_j(\lambda)$, $j \in \mathbb{N}_0$. Hence $\overline{\Phi_{0,k}(\lambda)} = \Phi_{k,0}(\lambda) \in \mathfrak{R}(P(\lambda))$, therefore, $\overline{\Phi_{0,k}(\lambda)} = \overline{\Phi_{0,0}(\lambda)P_k(\lambda)}$. Due to the fact that $P_k(\lambda)$ is real, we have $\Phi_{0,k}(\lambda) = \Phi_{0,0}(\lambda)P_k(\lambda)$, $k \in \mathbb{N}_0$. Substituting the last equality into the resulting expression for $\Phi_{j,k}(\lambda)$, we get (3.2.3). □

Denote $d\sigma(\lambda) = \Phi_{0,0}(\lambda)d\varrho(\lambda)$. Since $\Phi_{0,0} \geq 0$, then the expression $d\sigma(\lambda)$ is positive and, hence, it is a measure. From (3.2.1) and (3.2.3), we get

$$E(\Delta)_{j,k} = (E(\Delta)\delta_k, \delta_j)_0 = \int_{\Delta} P_j(\lambda)P_k(\lambda)\, d\sigma(\lambda), \quad j, k \in \mathbb{N}_0, \tag{3.2.4}$$

where δ_k are basis vectors in l_2: $\delta_k = (0, 0, \ldots, 0, 1, 0, \ldots)$ in which the unite is situated on the k-th position. Therefore, from (3.2.4), it follows

$$\sigma(\Delta) = (E(\Delta)\delta_0, \delta_0)_0. \tag{3.2.5}$$

In what follows we will call the measure $d\sigma(\lambda)$ spectral. It is clear that both measures $d\sigma(\lambda)$ and $d\varrho(\lambda)$ are absolutely continuous with respect to each other. It also follows from (3.2.4) that the polynomials $P_j(\lambda)$, $j \in \mathbb{N}$ form an orthonormal system with measure $d\sigma(\lambda)$:

$$\int_{\mathbb{R}} P_j(\lambda)P_k(\lambda)\, d\sigma(\lambda) = \delta_{j,k}, \quad j, k \in \mathbb{N}_0, \tag{3.2.6}$$

where $\delta_{j,k}$ is the Kronecker symbol.

Let us write the Parseval equality. To do this, we introduce the Fourier transform of the finite sequence u, by setting for $\lambda \ni \mathbb{C}$:

$$\tilde{u}(\lambda) = \sum_{j=0}^{\infty} u_j P_j(\lambda), \qquad u \in l_{2,0}. \tag{3.2.7}$$

Now (3.2.4) gives the Parseval equality

$$(E(\Delta)u, v)_0 = \int_{\Delta} \tilde{u}(\lambda)\overline{\tilde{v}(\lambda)}\, d\sigma(\lambda), \qquad u, v \in l_{2,0}. \tag{3.2.8}$$

Note that with the Fourier transform introduced in this way in (3.2.7), $\tilde{u}(\lambda)$ polynomially depends on λ. It also follows from (3.2.7) and (3.2.8) that the spectrum of the operator L in this case is unique. So, we can formulate the theorem.

Theorem 3.2.2 *Let $E(\Delta)$ be the resolution of the identity (ordinary or generalized) of one of the self-adjoint expansions of the operator L. Define the spectral measure $\sigma(\Delta) = (E(\Delta)\delta_0, \delta_0)_0$. Then, relative to this measure polynomials $P_j(\lambda)$, $j \in \mathbb{N}_0$ form an orthonormal system. The corresponding to $E(\Delta)$ the generalized* *eigenvectors expansion of the operator L is described by formulas* (3.2.7) *and* (3.2.8).

Thus, from the solutions $P_j(\lambda)$ of the difference equation $(LP(\lambda))_j = \lambda P_j(\lambda)$, $j \in \mathbb{N}_0$, $P_{-1}(\lambda) = 0$ as a function (sequences) are selected those solutions for which λ is situated on the spectrum of the corresponding extension. From (3.2.1) and (3.2.3), it follows that for σ-almost all λ, the sequence $(P_0(\lambda), P_1(\lambda), \ldots)$ satisfies the estimate

$$\sum_{j=1}^{\infty} \frac{|P_j(\lambda)|^2}{p_j} < \infty, \tag{3.2.9}$$

where $p_j \geq 0$, $\sum_{j=1}^{\infty} \frac{1}{p_j} < \infty$. The set of λ, for which (3.2.9) is valid, depends on the choice of the sequence $(p_j)_{j=0}^{\infty}$.

Proof of Theorem 3.2.2. It remains to prove only formulas (3.2.7) and (3.2.8). Let us write the action of the operator L on the vector δ_j. For an arbitrary sequence $u \in l_{2,0}$, we have

$$\begin{aligned}(u, L\delta_j)_0 =&(Lu, \delta_j)_0 = (Lu)_j \\ =&a_{j-1}u_{j-1} + b_j u_j + a_j u_{j+1} \\ =&(u, a_{j-1}\delta_{j-1} + b_j\delta_j + a_j\delta_{j+1})_0.\end{aligned}$$

Due to the arbitrariness of the choice of u we have

$$L\delta_j = a_{j-1}\delta_{j-1} + b_j\delta_j + a_j\delta_{j+1}, \ j \in \mathbb{N}_0, \tag{3.2.10}$$

where, it is also set $\delta_{-1} = 0$. The formula (3.2.10) shows that $L\delta_j$ is also a finite sequence and, therefore, it included to $\mathfrak{D}(L)$. Hence, $\delta \in \mathfrak{D}(L^n)$, $n \in \mathbb{N}_0$.

The equality (3.2.10) is considered as a recurrence relation to find δ_j by given δ_0 with $\delta_{-1} = 0$. Thus, at $j = 0$ we get

$$a_0\delta_1 + b_0\delta_0 = L\delta_0, \ \delta_{-1} = 0,$$

and hence

$$\delta_1 = \frac{1}{a_0}(L - b_0\mathbb{I})\delta_0 = P_1(L)\delta_0.$$

Then, we find $\delta_2 = P_2(L)\delta_0$ and similarly further values. Thus,

$$\delta_j = P_j(L)\delta_0, \ j \in \mathbb{N}_0. \tag{3.2.11}$$

It is now easily proved (3.2.4), namely,

$$\begin{aligned}(E(\Delta)\delta_k, \delta_l)_0 &= (E(\Delta)P_k(L)\delta_0, P_j(L)\delta_0)_0 \\ &= (P_j(L)E(\Delta)P_k(L)\delta_0, \delta_0)_0 \\ &= \int\limits_{\Delta} P_j(\lambda)P_k(\lambda)\, d(E_\lambda\delta_0, \delta_0)_0, \quad j, k, \in \mathbb{N}_0.\end{aligned}$$

□

We also note that formulas (3.2.7) and (3.2.11) are very useful.

The Generalized Fourier Transform The polynomials $P_j(\lambda)$, $j \in \mathbb{N}_0$ are orthonormal, hence, they are linearly independent in $L_2(\mathbb{R}, d\sigma(\lambda))$. Each polynomial has a power j. Therefore, they form a basis in the space of polynomials. And each polynomial is represented as a linear combination of polynomials $P_j(\lambda)$, $j \in \mathbb{N}_0$. Therefore, it follows from (3.2.7) that the Fourier transform maps all finite sequences into the set of all polynomials on λ.

Since $P_j(\lambda) \in L_2(\mathbb{R}, d\sigma(\lambda))$ (3.2.6), then each polynomial belongs to $L_2(\mathbb{R}, d\sigma(\lambda))$. In other words, the spectral measure is such that

$$\int\limits_{\mathbb{R}} \lambda^m \, d\sigma(\lambda) = c_m < \infty, \qquad m \in \mathbb{N}_0. \tag{3.2.12}$$

However, the set of all polynomials does not have to be dense in $L_2(\mathbb{R}, d\sigma(\lambda))$. The density criterion is given below.

In what follows, we will study separately the problem in which the measure $d\sigma(\lambda))$ is restored from a given sequence of numbers c_m, $m \in \mathbb{N}_0$, which are called moments of the measure, that is, we will consider the classical moment problem.

Let us point out one more important property of the spectral measure—*the set of growth points of the function* $\sigma(\lambda)$ is infinite. Indeed, if we were to assume the opposite, then this set (denoted by Λ) is finite. Let us construct a sufficiently high degree n polynomial $P(\lambda)$ with a set of zeros on Λ. Expand $P(\lambda)$ by the system $P_j(\lambda)$, $j \in \mathbb{N}_0$:

$$P(\lambda) = \sum_{j=0}^{n} c_j P_j(\lambda),$$

then

$$c_j = \int\limits_{\mathbb{R}} P(\lambda)P_j(\lambda)\, d\sigma(\lambda) = 0, \qquad j \in \mathbb{N}_0,$$

i.e., $P(\lambda) \equiv 0$, that is impossible.

The proven property can also be formulated in the form *if the polynomial* $P(\lambda)$ *such that* $\int_{\mathbb{R}} |P(\lambda)|^2 \, d\sigma(\lambda) = 0$, *then* $P(\lambda) \equiv 0$.

The correspondence $u \to \tilde{u}(\lambda)$ is one-to-one between a set of finite sequences and its Fourier image, that is, some set of polynomials of the variable λ, which are elements of the space $L_2(\mathbb{R}, d\sigma(\lambda))$. Note, the Parseval equality has a form

$$(u, v)_0 = \int_{\mathbb{R}} \tilde{u}(\lambda)\overline{\tilde{v}(\lambda)} \, d\sigma(\lambda), \tag{3.2.13}$$

which follows from (3.2.8) for $\Delta = \mathbb{R}$. The equality (3.2.13) is an isometry between parts of l_2 and $L_2(\mathbb{R}, d\sigma(\lambda))$. Extending such an isometry by continuity, we obtain an isometry between whole spaces l_2 and $L_2(\mathbb{R}, d\sigma(\lambda))$, that coincides with the closure of all polynomials in this space. We denote this closure $\widetilde{l_2}$ as the closure of the Fourier image. Let the sequence $u \in l_2$ corresponds to its image $\tilde{u} \in \widetilde{l_2}$. It is clear that $\tilde{u}(\lambda)$ is calculated for u using the same formula (3.2.7), only now, the series will be infinite and convergent in the sense of the space $L_2(\mathbb{R}, d\sigma(\lambda))$. Thus, the Fourier transform is extended to infinite sequences l_2.

The operator L by the transferring from l_2 to $\widetilde{l_2}$ turns into a multiplication operator by the independent variable λ in $L_2(\mathbb{R}, d\sigma(\lambda))$. More precisely, the operator L is defined on $l_{2,0}$; its image $\tilde{L}$ during the Fourier transform acts on the image $l_{2,0}$, that is, on $\widetilde{l_{2,0}}$ (the set of all polynomials) as a multiplication on λ. Thus, thanks to(3.1.2) we have

$$\begin{aligned}(\widetilde{Lu})(\lambda) &= \sum_{j=0}^{\infty}(Lu)_j P_j(\lambda) = \sum_{j=0}^{\infty} u_j (LP_j(\lambda))_j \\ &= \lambda \sum_{j=0}^{\infty} u_j P_j(\lambda)) = \lambda \tilde{u}(\lambda), \ u \in l_{2,0}, \ u_{-1} = 0.\end{aligned}$$

The operator L is the closure of this operator from $l_{2,0}$, so $\tilde{L}$ is equal to the closure of the multiplication operator by λ defined initially for the set of polynomials.

Theorem 3.2.3 *In order for the set of all polynomials of the variable* λ *to be dense in the space* $L_2(\mathbb{R}, d\sigma(\lambda))$, *it is necessary and sufficient that* $d\sigma(\lambda)$ *be generated by the usual resolution of the identity. (Such spectral measures are called orthogonal.)*

Proof Let $\sigma(\Delta) = (E(\Delta)\delta_0, \delta_0)$, where $E(\Delta)$, be the usual resolution of the identity. Let us show the density of polynomials in $L_2(\mathbb{R}, d\sigma(\lambda))$. Denote $A \supset L$ a self-adjoint extension of the operator L in l_2 corresponding to $E(\Delta)$. Denote also $R_z = (A - z\mathbb{I})^{-1}$, $\mathrm{Im} z \neq 0$ its resolvent. Take $u \in l_{2,0}$. Since $R_z u \in \mathfrak{D}(A) \subset l_2$, then for a fixed z, there is $v \in l_{2,0}$ such that $\|R_z u - v\|_0 < \varepsilon$ for an arbitrary ε.

When passing by isometry from the space l_2 to the space $L_2(\mathbb{R}, d\sigma(\lambda))$, we get that for each $\tilde{u}(\lambda)$, there is a polynomial $\tilde{v}(\lambda)$ such that

$$\left\| \frac{\tilde{u}(\lambda)}{\lambda - z} - \tilde{v}(\lambda) \right\|_{L_2(\mathbb{R}, d\sigma(\lambda))} < \varepsilon.$$

Given the arbitrariness of ε, the last inequality means that an arbitrary function $\frac{\tilde{u}(\lambda)}{\lambda - z}$ can be approximated by a polynomial, i.e., $\frac{\tilde{u}(\lambda)}{\lambda - z} \in \widetilde{l_2}$ for an arbitrary polynomial $\tilde{u}(\lambda)$.

Now, the proof of sufficiency is easily completed. Let there exist a vector $h(\lambda) \in L_2(\mathbb{R}, d\sigma(\lambda))$ orthogonal to $\widetilde{l_2}$. Then for an arbitrary non-real z

$$\int_{\mathbb{R}} \frac{1}{\lambda - z} h(\lambda)\, d\sigma(\lambda) = 0, \quad u = \delta_0,\ \tilde{u}(\lambda) = 1$$

holds true. By the general Stieltjes transform theorem, this means that

$$\eta(\Delta) = \int_{\Delta} h(\lambda)\, d\sigma(\lambda) = 0,$$

were Δ is from Borel σ-algebra on $\mathbb{R}$, i.e., $h(\lambda) = 0$ σ-almost everywhere. Therefore, the equality $\widetilde{l_2} = L_2(\mathbb{R}, d\sigma(\lambda))$ is proved, which was required.

Let us show the necessity. Take $\widetilde{l_2} = L_2(\mathbb{R}, d\sigma(\lambda))$. It is necessary to show that $d\sigma(\lambda))$ is generated by some self-adjoint extension A of operator L in l_2. When passing by isometry from the space l_2 to the space $L_2(\mathbb{R}, d\sigma(\lambda))$, the operator L is transferred to the operator $\tilde{L}$ of multiplication on λ, defined initially on polynomials, and then we take its closure. We denote by $\tilde{A}$ the usual multiplication operator by λ in the space $L_2(\mathbb{R}, d\sigma(\lambda))$. The operator $\tilde{A}$ is self-adjoint. $\tilde{L} \subseteq \tilde{A}$ and $\sigma(\Delta) = (\tilde{E}(\Delta)1, 1)_{L_2(\mathbb{R}, d\sigma(\lambda))}$, where $\tilde{E}(\Delta)$ is the resolution of the identity corresponding to $\tilde{A}$, and it coincides with the multiplication operator on a characteristic function. Since $L_2(\mathbb{R}, d\sigma(\lambda)) = \widetilde{l_2}$ then by inverse isometry, we go from the operator $\tilde{A}$ to A in the space l_2. Obviously, this is the self-adjoint operator which is an extension of L and which generates the measure $d\sigma(\lambda)$. □

Theorem 3.2.4 *If the spectral measure such that*

$$\int_{\mathbb{R}} \frac{\log \sigma'(\lambda)}{1 + \lambda^2}\, d\lambda > -\infty, \tag{3.2.14}$$

where $\sigma'(\lambda)$ is its derivative, then it is generated by the generalized resolution of the identity, i.e., it is not orthogonal one.

Proof Thanks to the theorem from the monograph of Akhiezer N. I. [3] (chapter 2), it is known that for the density of the linear envelope of functions $e^{i\alpha\lambda}$, $0 \le \alpha < \infty$

in $L_2(\mathbb{R}, d\sigma(\lambda))$ is necessary and sufficient that

$$\int_{\mathbb{R}} \frac{\log \sigma'(\lambda)}{1+\lambda^2}\, d\lambda = -\infty,$$

where $d\sigma(\lambda)$ is an arbitrary non-negative finite measure on the axis. According to this theorem, in connection with (3.2.14), there exists a function $0 \neq h(\lambda) \in L_2(\mathbb{R}, d\sigma(\lambda))$ such that

$$\int_{\mathbb{R}} h(\lambda) e^{i\alpha\lambda}\, d\sigma(\lambda) = 0, \ 0 \le \alpha < \infty.$$

We differentiate this integral by α and then set $\alpha = 0$ and get

$$\int_{\mathbb{R}} h(\lambda)\lambda^j\, d\sigma(\lambda) = 0, \ j \in \mathbb{N}_0.$$

This means that the polynomials are not dense in $L_2(\mathbb{R}, d\sigma(\lambda))$. And therefore, according to the previous theorem, the measure $d\sigma(\lambda)$ is generated by the generalized resolution of the identity. □

Previously, the spectral measure was defined as $\sigma(\Delta) = (E(\Delta)\delta_0, \delta_0)$. We will show that it can be, additionally, described in another way.

Theorem 3.2.5 *Let $d\sigma(\lambda)$ be a finite measure on the real axis of $\mathbb{R}$, which has all moments, that is, there exist integrals* (3.2.12). *If for arbitrary finite sequences u, v and their Fourier transforms $\tilde{u}(\lambda)$, $\tilde{v}(\lambda)$ defined in* (3.2.7), *the Parseval equality* (3.2.13) *holds true (or, that is the same, the orthogonality relation* (3.2.6) *holds true), then $d\sigma(\lambda)$ is a spectral measure, that is, there exists a generalized resolution of the identity $E(\Delta)$ constructed by the self-adjoint extension of the operator L, so that $\sigma(\Delta) = (E(\Delta)\delta_0, \delta_0)$.*

Proof is almost obvious due to Parseval equality. In the same way as before, an isometry is established between l_2 and $\widetilde{l_2} \subset L_2(\mathbb{R}, d\sigma(\lambda))$, under which the operator L is transferred to the multiplication operator $\tilde{L}$ by λ, which is initially defined on polynomials. Using this isomorphism, the problem is reduced to constructing such a self-adjoint extension $\tilde{A}$ of the operator $\tilde{L}$ that $\sigma(\Delta) = (\tilde{E}(\Delta)1, 1)_{L_2(\mathbb{R}, d\sigma(\lambda))}$, where $\tilde{E}(\Delta)$ is the the resolution of the identity of the operator $\tilde{A}$. But A can be taken as the self-adjoint multiplication operator by λ in $L_2(\mathbb{R}, d\sigma(\lambda))$. If $\widetilde{l_2} = L_2(\mathbb{R}, d\sigma(\lambda))$, then the resulting extension corresponds to the extension of L without leaving l_2. In case $\widetilde{l_2} \subset L_2(\mathbb{R}, d\sigma(\lambda))$ we have an extension with an exit from the space. □

Since the spectral measure can be determined using the Parseval equality, regardless of the theory of operators, the following corollary is meaningful.

Corollary 3.2.6 *The operator generated by the difference expression* (3.1.1) *has a unique spectral measure if and only if the operator L generated by it is self-adjoint.*

Indeed, due to the equality (3.2.8), there is a one-to-one correspondence between the spectral measures and the generalized resolution of the identity.

We say that a difference equation or the corresponding Jacobi matrix is determined or (non-determined) indeterminate if the operator L is self-adjoint or not self-adjoint.

Let us fix the difference expression (its operator L) and consider the set Σ of all measures corresponding to this expression, that is, corresponding to various self-adjoint extensions of the operator L without exit and with exit from the space. In the indeterminate case, Σ consists of more than one element. The set Σ is convex, that is, if $\sigma_1, \sigma_2 \in \Sigma$, then for $\mu_1 + \mu_2 = 1$, $\mu_1, \mu_2 \geq 0$ the measure $\sigma(\Delta) = \mu_1\sigma_1(\Delta) + \mu_2\sigma_2(\Delta)$ is also included in Σ. Indeed, if the Parseval equality holds for σ_1 and σ_2, then it also holds for σ.

Recall that the elements of this set are called extreme if they cannot be represented in the form $\sigma(\Delta) = \mu_1\sigma_1(\Delta) + \mu_2\sigma_2(\Delta)$ with some $\sigma_1, \sigma_2 \in \Sigma$, $\mu_1+\mu_2 = 1$, $\mu_1, \mu_2 > 0$ and an arbitrary Δ. In a certain case, the unique measure is extreme. Then, it will be shown that every orthogonal measure is extreme. However, not each extreme measure is orthogonal.

Theorem 3.2.7 *In order, for the set of all polynomials from λ, to be dense in the space $L_2(\mathbb{R}, d\sigma(\lambda))$, it is necessary and sufficient that $d\sigma(\lambda)$ be extreme spectral measure.*

Before proving the theorem, we consider the following lemma.

Lemma 3.2.8 *Two measures $d\sigma_1(\lambda)$ and $d\sigma_2(\lambda)$ on the real axis of $\mathbb{R}$ satisfying* (3.2.12) (*that is, they have all moments*) *belong both to the same class Σ if and only if for an arbitrary polynomial $P(\lambda)$*

$$\int\limits_{\mathbb{R}} P(\lambda)\, d\sigma_1(\lambda) = \int\limits_{\mathbb{R}} P(\lambda)\, d\sigma_2(\lambda). \tag{3.2.15}$$

Proof If there is equality (3.2.15) and, for example, the Parseval equality (3.2.13) holds true for $d\sigma_1(\lambda)$, then it also holds for $d\sigma_2(\lambda)$, since $\tilde{u}(\lambda)$ and $\tilde{v}(\lambda)$ are polynomials $u, v \in l_{2,0}$. Let (3.2.13) holds true for $d\sigma_1(\lambda)$ and $d\sigma_2(\lambda)$ with one operator L, then $P_j(\lambda)$ are orthonormal in each space $L_2(\mathbb{R}, d\sigma_1(\lambda))$ and $L_2(\mathbb{R}, d\sigma_2(\lambda))$. Expanding $P(\lambda)$ by $P_j(\lambda)$, we get

$$\int\limits_{\mathbb{R}} P(\lambda)\, d\sigma_1(\lambda) = \int\limits_{\mathbb{R}} \left(\sum_{j=0}^{n} c_j P_j(\lambda)\right) d\sigma_1(\lambda) = c_0 = \int\limits_{\mathbb{R}} P(\lambda)\, d\sigma_2(\lambda).$$

Therefore, the lemma is proved. □

Proof of Theorem 3.2.7 First, we will show that if $d\sigma(\lambda)$ is not extreme, then the set of polynomials is not dense in $L_2(\mathbb{R}, d\sigma(\lambda))$. Thus, let

$$\sigma(\Delta) = \mu_1\sigma_1(\Delta) + \mu_2\sigma_2(\Delta),\ \sigma_1, \sigma_2 \in \Sigma,\ \mu_1 + \mu_2 = 1,\ \mu_1, \mu_2 > 0.$$

Since $\sigma_j(\Delta) \leq \frac{1}{\mu_j}\sigma(\Delta)$, then $d\sigma_j(\lambda)$ is absolutely continuous with respect to $d\sigma_1(\lambda)$ and

$$\varphi_j(\lambda) = \frac{d\sigma_j(\lambda)}{d\sigma(\lambda)} \leq \mu_j, \quad j = 1, 2.$$

By Lemma 3.2.8, for an arbitrary polynomial $P(\lambda)$, we have

$$\int_{\mathbb{R}} P(\lambda)\varphi_1(\lambda)\, d\sigma(\lambda) = \int_{\mathbb{R}} P(\lambda)\, d\sigma_1(\lambda)$$
$$= \int_{\mathbb{R}} P(\lambda)\, d\sigma_2(\lambda) = \int_{\mathbb{R}} P(\lambda)\varphi_2(\lambda)\, d\sigma(\lambda),$$

that is, the bounded function $\psi(\lambda) = \varphi_1(\lambda) - \varphi_2(\lambda)$ is not equal to zero σ-almost everywhere is such that

$$\int_{\mathbb{R}} P(\lambda)\psi(\lambda)\, d\sigma(\lambda) = 0$$

for an arbitrary $P(\lambda)$. This means that on $L_1(\mathbb{R}, d\sigma(\lambda))$, there exists a non-zero functional ψ that is vanishes on all polynomials, i.e., the set of polynomials is not dense in $L_2(\mathbb{R}, d\sigma(\lambda))$. Conversely, if polynomials are not dense in $L_2(\mathbb{R}, d\sigma(\lambda))$, then there exists a function $\psi(\lambda)$ bounded by one and not equal to zero σ-almost everywhere, for which

$$\int_{\mathbb{R}} P(\lambda)\psi(\lambda)\, d\sigma(\lambda) = 0$$

for an arbitrary polynomial $P(\lambda)$. Let us construct two measures

$$d\sigma_1(\lambda) = (1 + \psi(\lambda)d\sigma(\lambda)),\ d\sigma_2(\lambda) = (1 - \psi(\lambda)d\sigma(\lambda)).$$

These measures obviously satisfy (3.2.12), that is, have all moments, and are such that

$$\int_{\mathbb{R}} P(\lambda)\, d\sigma(\lambda) = \int_{\mathbb{R}} P(\lambda)\, d\sigma_1(\lambda) = \int_{\mathbb{R}} P(\lambda)\, d\sigma_2(\lambda),$$

for an arbitrary polynomial $P(\lambda)$. According to Lemma 3.2.8, $\sigma_1\sigma_2 \in \Sigma$, where Σ is the convex set of spectral measures corresponding to the operator L. However, $\sigma_1 \neq \sigma_2$ and $\sigma = \frac{1}{2}\sigma_1 + \frac{1}{2}\sigma_2$, that is, σ is not extreme. □

3.3 The Inverse Problem

In the previous section, a direct problem was considered: the spectral representation was constructed according to the given expression L. Quite naturally, the inverse question arises as to whether it is possible to reconstruct L from the spectral data. We will show that such a reconstruction (recovery) is possible and that the spectral measure should be taken as spectral data. Moreover, it turns out that the set of all possible spectral measures is easy to describe.

Consider the difference expression:

$$a_{j-1}P_{j-1}(\lambda) + b_j P_j(\lambda) + a_j P_{j+1}(\lambda) = \lambda P_j(\lambda), \quad j \in \mathbb{N},$$

were we put as usual $P_{-1}(\lambda) := 0$ and $a_1 := 0$. Multiplying the left and right parts of the previous equality on $P_k(\lambda)$ relative to the scalar product $L_2(\mathbb{R}, d\sigma(\lambda))$, we obtain

$$a_j = \int_{\mathbb{R}} \lambda P_j(\lambda) P_{j+1}(\lambda)\, d\sigma(\lambda),\ b_j = \int_{\mathbb{R}} \lambda P_j^2(\lambda)\, d\sigma(\lambda),\ j \in \mathbb{N}_0. \tag{3.3.1}$$

Now let there be given a measure $d\sigma(\lambda)$ on the real axes so that all integrals (3.2.12) converges, that is, the measure has all moments, $\sigma(\mathbb{R}) = 1$, and $d\sigma(\lambda)$ has an infinite number of growth points. We will show that such a measure is a spectral measure of some difference expression L. To do this, consider the space $L_2(\mathbb{R}, d\sigma(\lambda))$ and the system of functions

$$1, \lambda, \lambda^2, \lambda^3, \ldots, \lambda^j, \ldots. \tag{3.3.2}$$

We orthogonalize this sequence according to the Gram-Schmidt procedure and get a system of polynomials

$$P_0(\lambda) = 1,\ P_1(\lambda),\ P_2(\lambda), \ldots,\ P_j(\lambda), \ldots \tag{3.3.3}$$

orthonormal in $L_2(\mathbb{R}, d\sigma(\lambda))$. Polynomials (3.3.3) are defined up to the sign of the leading coefficient. For uniqueness (3.3.3) according to the given sequence (3.3.2), we set the higher coefficients in $P_j(\lambda)$ to be positive. The orthogonalization process will be infinite because $d\sigma(\lambda)$ has an infinite number of growth points.

Now we construct a_j, b_j by polynomials $P_j(\lambda)$ using equalities (3.3.1), where $a_j > 0,\ j \in \mathbb{N}_0$. Indeed, $\lambda P_j(\lambda)$ is a polynomial of degree $j+1$ with a positive leading coefficient. Therefore, in the representation

$$\lambda P_j(\lambda) = c_{j+1} P_{j+1}(\lambda) + \ldots + c_0 P_0(\lambda),$$

coefficients $c_{j+1} > 0$, and hence $c_{j+1} = a_j$. We take numbers $a_j, b_j,\ j \in \mathbb{N}_0$ as coefficients of the difference expression L (3.1.1).

We will show that $P_j(\lambda)$ are polynomials of the first kind for L, i.e

$$\lambda P_j(\lambda) = a_{j-1} P_{j-1}(\lambda) + b_j P_j(\lambda) + a_j P_{j+1}(\lambda),$$
$$j \in \mathbb{N}_0,\ P_{-1}(\lambda) = 0,\ P_0(\lambda) = 1. \qquad (3.3.4)$$

To prove this proposition, it is necessary to show that in the expansion of the $j+1$-th degree polynomial $\lambda P_j(\lambda)$ by $P_0(\lambda),\ P_1(\lambda), \ldots,\ P_{j+1}(\lambda)$, coefficients at $P_0(\lambda)$, $P_1(\lambda), \ldots,\ P_{j-1}(\lambda)$ are equal to zero, i.e.,

$$\int_{\mathbb{R}} \lambda P_j(\lambda) P_k(\lambda)\, d\sigma(\lambda) = 0,$$

as $0 \le k \le j-2$. But $\lambda P_k(\lambda)$ is a polynomial maximum $(j-1)$-th degree, and hence $P_j(\lambda)$ is orthogonal to it.

Therefore, the measure $d\sigma(\lambda)$ has the property that polynomials of the first kind, corresponding to the expression L, constructed by the measure, are orthonormal with respect to this measure. This implies the Parseval equality (3.2.13) on the finite sequences u and v, that is, $d\sigma(\lambda)$ is the spectral measure of L. Hence, we prove the theorem.

Theorem 3.3.1 *Let the measure $d\sigma(\lambda)$ be given on the real axis $\mathbb{R}$, which has an infinite number of growth points, is normalized to one and has all moments, namely,*

$$\int_{\mathbb{R}} d\sigma(\lambda) = 1,\ \int_{\mathbb{R}} \lambda^n\, d\sigma(\lambda) < \infty,\ n \in \mathbb{N}. \qquad (3.3.5)$$

Then such a measure is a spectral measure of the difference expression, the coefficients of which are uniquely determined by $d\sigma(\lambda)$, using formulas (3.3.1), *where the system of polynomials* (3.2.15) *is constructed by the orthogonalization of the sequence* (3.3.2), *according to the scalar product of the space $L_1(\mathbb{R}, d\sigma(\lambda))$.*

3.4 Further Spectral Analysis of the Difference Operator

Polynomials of the Second Kind In this section, we will present the main general results from the spectral analysis of Jacobi matrices. Consider the difference equation as in (3.1.5): $(Lu)_j = zu_j,\ j \in \mathbb{N}$, but with initial data $u_0 = 0,\ u_1 = \frac{1}{a_0}$. Let $Q_1(z) = \frac{1}{a_0},\ Q_2(z),\ \dots,\ Q_j(z),\ \dots$ be the solutions of this Cauchy problem. Obviously, $Q_j(z),\ j \in \mathbb{N}$ are polynomials of the j-th degree with real coefficients. *Polynomials* $Q_0(z) = 0,\ Q_j(z),\ j \in \mathbb{N}$ *are called polynomials of the second kind generated by the expression* L. It is clear that $P(z) = (P_1(z), P_2(z), \dots)$ and $Q(z) = (Q_1(z), Q_2(z), \dots)$ are linear independent systems of solutions of the equation $(Lu)_j = zu_j,\ j \in \mathbb{N}$. Polynomials of the second kind will be needed to describe the Weyl function, which uniquely characterizes the spectral measure of the operator.

Polynomials $P_j(z)$ and $Q_j(z),\ j \in \mathbb{N}_0$ are related by

$$Q_j(z) = \int\limits_{\mathbb{R}} \frac{P_j(\lambda) - P_j(z)}{\lambda - z}\, d\sigma(\lambda),\ j \in \mathbb{N}_0. \tag{3.4.1}$$

Indeed, the sequence

$$u_j = \int\limits_{\mathbb{R}} \frac{P_j(\lambda) - P_j(z)}{\lambda - z}\, d\sigma(\lambda)$$

satisfies the equation

$$\begin{aligned}
(Lu)_j =& \int\limits_{\mathbb{R}} \frac{(LP(\lambda))_j - (LP(z))_j}{\lambda - z}\, d\sigma(\lambda) = \int\limits_{\mathbb{R}} \frac{\lambda P_j(\lambda))_j - zP_j(z)_j}{\lambda - z}\, d\sigma(\lambda)\\
=& \int\limits_{\mathbb{R}} \frac{\lambda P_j(\lambda) - zP_j(z)_j + zP_j(\lambda) - zP_j(\lambda)_j}{\lambda - z}\, d\sigma(\lambda)\\
=& z\int\limits_{\mathbb{R}} \frac{P_j(\lambda) - P_j(z)}{\lambda - z}\, d\sigma(\lambda) + \int\limits_{\mathbb{R}} P_j(\lambda)\, d\sigma(\lambda) = zu_j,\ j \in \mathbb{N},
\end{aligned}$$

where the last integral is zero due to mutual orthogonality $P_j(z),\ j \in \mathbb{N}_0$ (3.2.6), in particular $P_j(\lambda),\ j \in \mathbb{N}$ are orthogonal to $P_0(\lambda) = 1$. Moreover, $u_0 = 0$ and

$$u_1 = \int\limits_{\mathbb{R}} \frac{\left(\frac{1}{a_0}(\lambda - b_0) - \frac{1}{a_0}(z - b_0)\right)}{\lambda - z}\, d\sigma(\lambda) = \frac{1}{a_0},$$

hence, $u_j = Q_j(z),\ j \in \mathbb{N}$. Thus, (3.4.1) is proved.

Note that polynomials $Q_j(z)$, $j \in \mathbb{N}$ are also orthogonal with respect to some measure $d\hat{\sigma}(\lambda)$ on the real axis $\mathbb{R}$. Indeed, $a_0 Q_j(z)$ can be considered as polynomials of the first kind of the shifted difference expression $\hat{L}$ with coefficients $\hat{a}_j = a_{j+1}$, $\hat{b}_j = b_{j+1}$, $j = -1, 0, 1, \ldots$. Therefore, as $d\hat{\sigma}(\lambda)$ we can take the spectral measure corresponding to $\hat{L}$.

In what follows, this lemma will be useful.

Lemma 3.4.1 *Let $B_0(\lambda)$, $B_1(\lambda)$, ..., $B_j(\lambda)$, ... be an arbitrary sequence of real polynomials degrees up to $j \in \mathbb{N}_0$ respectively, orthogonal on the real axis $\mathbb{R}$ with respect to the measure $d\sigma(\lambda)$ satisfying conditions of Theorem 3.3.1. Then the roots (zeros) of such polynomials are real and different.*

Proof Let some polynomial $B_n(\lambda)$ have less than n zeros. We denote by $\lambda_1, \lambda_2, \ldots, \lambda_m$ $(m < n)$ only those of its real zeros, when passing through which $B_n(\lambda)$ changes sign. Let us construct a polynomial $P(\lambda)$ of degree m with zeros $\lambda_1, \lambda_2, \ldots, \lambda_m$, so that $\operatorname{sgn} P(\lambda) = \operatorname{sgn} B_n(\lambda)$, $\lambda \in \mathbb{R}$. Then, on the one hand, $P(\lambda)$ is represented as a linear combination of $\sum_{j=0}^{m} c_j B_j(\lambda)$ and hence

$$\int_{\mathbb{R}} P(\lambda) B_n(\lambda)\, d\sigma(\lambda) = 0,$$

and on the other hand, this equality is impossible due to the construction of $P(\lambda)$. The obtained contradiction completes the proof of the lemma. □

One Corollary Polynomials $B_j(\lambda)$ of the previous lemma satisfy the estimate

$$|B_j(x + iy')| \leq |B_j(x + iy'')|, \quad |y'| \leq |y''|, \quad j \in \mathbb{N}_0. \tag{3.4.2}$$

The inequality (3.4.2) follows from the representation

$$B_j(z) = a \prod_{\alpha=1}^{j} (z - \lambda_\alpha),$$

where λ_α are the real zeros of polynomials $B_j(z)$.

In particular, all zeros of polynomials of the first and second kind are real and the estimate (3.4.2) is carried out for them.

The Resolvent Let us pass to consider the resolvent associated with the operator L. Let A be some self-adjoint extension of the operator L, which is generally possible with exit to a wider space $\tilde{H}_0$ than l_2. We denote by $\tilde{E}_\lambda$ and $\tilde{R}_z$ the resolution of the identity and the resolvent of the operator A in $\tilde{H}$, respectively, and by E_λ and

R_z the generalized resolution of the identity and the resolvent. If P is the projection operator $\tilde{H}_0$ on l_2, then $E_\lambda = P\tilde{E}_\lambda P$ and $R_z = P\tilde{R}_z P$. Since

$$R_z = \int\limits_{\mathbb{R}} P(\lambda)\frac{1}{\lambda - z}\, dE_\lambda,$$

then due to (3.2.4), the operator matrix R_z has a form

$$R_{z;j,k} = (R_z\delta_k, \delta_j)_0 = \int\limits_{\mathbb{R}} \frac{P_j(\lambda)P_k(\lambda)}{\lambda - z}\, d\sigma(\lambda), \quad j, k \in \mathbb{N}_0, \tag{3.4.3}$$

in the basis $\delta_0, \delta_1, \ldots, \delta_j, \ldots$.

In what follows, the function, holomorphic outside the spectrum of the operator A,

$$m(z) = R_{z;0,0} = (R_z\delta_0, \delta_0)_0 = \int\limits_{\mathbb{R}} \frac{1}{\lambda - z}\, d\sigma(\lambda), \tag{3.4.4}$$

plays a significant role which is the Stieltjes transform of the spectral measure. As is well known, $\sigma(\Delta)$ is written using the inversion formula for $m(z)$. Consider the relation in the case when $d\sigma(\lambda)$ is orthogonal, that is, the relation is constructed by the usual resolution of the identity. According to (3.4.1) and (3.4.3) we have:

$$\begin{aligned}(R_z\delta_0)_j &= \int\limits_{\mathbb{R}} \frac{P_j(\lambda)}{\lambda - z}\, d\sigma(\lambda) = \int\limits_{\mathbb{R}} \frac{P_j(\lambda) - P_j(z)}{\lambda - z}\, d\sigma(\lambda) \\ &+ P_j(z)\int\limits_{\mathbb{R}} P(\lambda)\frac{1}{\lambda - z}\, dE_\lambda = Q_j(z) + m(z)P_j(z), \quad j \in \mathbb{N}_0.\end{aligned} \tag{3.4.5}$$

For the ordinary resolvent R_z, the Hilbert identity (2.1.24):

$$R_\mu - R_\lambda = (\mu - \lambda)R_\mu R_\lambda$$

holds true. Using this identity and (3.4.5) we get

$$\begin{aligned}\frac{\overline{m(\xi)} - m(z)}{\bar{\xi} - z} &= \left(\frac{R_{\bar{\xi}} - R_z}{\bar{\xi} - z}\delta_0, \delta_0\right)_0 = (R_{\bar{\xi}}R_z\delta_0, \delta_0)_0 \\ &= (R_\xi^* R_z\delta_0, \delta_0)_0 = (R_z\delta_0, R_\xi\delta_0)_0 = \sum_{j=0}^{\infty}(R_z\delta_0)_j\overline{(R_\xi\delta_0)_j} \\ &= \sum_{j=0}^{\infty}\left(Q_j(z) + m(z)P_j(z)\right)\overline{\left(Q_j(\xi) + m(\xi)P_j(\xi)\right)}, \ \xi \neq z.\end{aligned}$$

Therefore, for arbitrary ξ and z we have

$$\frac{\overline{m(\xi)} - m(z)}{\bar{\xi} - z} = \sum_{j=0}^{\infty} \left(Q_j(z) + m(z)P_j(z)\right) \overline{\left(Q_j(\xi) + m(\xi)P_j(\xi)\right)}, \ \bar{\xi} \neq z. \tag{3.4.6}$$

We put $\xi = z$ and get

$$\frac{\overline{m(z)} - m(z)}{\bar{z} - z} = \sum_{j=0}^{\infty} \left|Q_j(z) + m(z)P_j(z)\right|^2, \ \mathrm{Im} z \neq 0. \tag{3.4.7}$$

Let us write the relation (3.4.7) in the case when $d\sigma(\lambda)$ is constructed according to the generalized resolution of the identity. Now the Hilbert identity for $\tilde{R}_z$ is used, i.e., $\tilde{R}_\mu - \tilde{R}_\lambda = (\mu - \lambda)\tilde{R}_\mu \tilde{R}_\lambda$ and $\xi = z$ is also assumed. Hence

$$\begin{aligned}
\frac{\overline{m(z)} - m(z)}{\bar{z} - z} &= \left(\frac{R_{\bar{z}} - R_z}{\bar{z} - z}\delta_0, \delta_0\right)_0 = \left(\frac{\tilde{R}_{\bar{z}} - \tilde{R}_z}{\bar{z} - z}\delta_0, \delta_0\right)_{\tilde{\mathcal{H}}_0} = \\
&= (\tilde{R}_{\bar{z}}\tilde{R}_z\delta_0, \delta_0)_0 = (\tilde{R}_z^*\tilde{R}_z\delta_0, \delta_0)_0 \\
&= (\tilde{R}_z\delta_0, \tilde{R}_z\delta_0)_0 \geq (P\tilde{R}_z\delta_0, P\tilde{R}_z\delta_0)_0 \\
&= (R_z\delta_0, R_z\delta_0)_{\tilde{\mathcal{H}}_0} = \sum_{j=0}^{\infty} \left\| Q_j(z) + m(z)P_j(z) \right\|^2.
\end{aligned}$$

Therefore, if $d\sigma(\lambda)$ is orthogonal, then the relation (3.4.7) *holds true. If $d\sigma(\lambda)$ is not orthogonal, then the inequality*

$$\frac{\overline{m(z)} - m(z)}{\bar{z} - z} \geq \sum_{j=0}^{\infty} \left|Q_j(z) + m(z)P_j(z)\right|^2, \ \mathrm{Im} z \neq 0,$$

holds true.

The relation (3.4.7) sometimes makes it possible to find $m(z)$ and, therefore, $d\sigma(\lambda)$ for a given L. Indeed, let us have a determined case. Suppose that such a function $f(z)$ is found, $\mathrm{Im} z \neq 0$, that the sequence $Q_j(z) + f(z)P_j(z) \in l_2$ for each non-real z. Then $m(z) = f(z)$. Indeed, if there exists z_0 for which $m(z_0) \neq f(z_0)$, then $Q_j(z_0) + m(z)P_j(z_0) \in l_2$ and $Q_j(z_0) + f(z)P_j(z_0) \in l_2$, as well as $(m(z_0) - f(z_0))P_j(z_0) \in l_2$, i.e., $P_j(z_0) \in l_2$—which is not possible due to the definiteness of L.

The Indeterminate Case Suppose that the expression L corresponds to the indeterminate case, that is, the corresponding operator has non-trivial defect indices and, therefore, has many self-adjoint extensions (without exit from space). It turns out that the set of all spectral measures corresponding to L can be described. Let us first describe only the orthogonal ones.

For non-real z, as already said, the series $\sum_{j=0}^{\infty} |P_j(z)|^2$ converges. The stronger fact is actually correct.

Lemma 3.4.2 *In the indeterminate case, series* $\sum_{j=0}^{\infty} |P_j(z)|^2$ *and* $\sum_{j=0}^{\infty} |Q_j(z)|^2$ *both simultaneously converge uniformly in each bounded region of the complex plane.*

Proof Consider a function

$$m(z) = \int_{\mathbb{R}} \frac{1}{\lambda - z} \, d\sigma(\lambda),$$

corresponding to the usual resolution of the identity. This function $m(z)$ satisfies (3.4.7). Since the left-hand side of the last equality is continuous everywhere outside the real axis $\mathbb{R}$, then, according to the Dini theorem, the series

$$\sum_{j=0}^{\infty} |Q_j(z) + m(z)P_j(z)|^2$$

the series converges uniformly in every bounded region located at a certain distance from $\mathbb{R}$.

Let $d\sigma_1(\lambda)$ be a spectral measure for L different from $d\sigma(\lambda)$ and let $m_1(z)$ be the corresponding function (3.4.4). Since $m(z) \neq m_1(z)$, the equality $m(z) = m_1(z)$ is possible only on some sequence $z_1, z_2, \ldots, z_n, \ldots$ with boundary points only on the real axis $\mathbb{R}$. Let us show that in each bounded region G of the complex plane, which is located at a distance from the real axis $\mathbb{R}$ and from the points z_j, the series $\sum_{j=0}^{\infty} |P_j(z)|^2$ converges uniformly.

Indeed, both series

$$\sum_{j=0}^{\infty} \left|Q_j(z) + m(z)P_j(z)\right|^2, \quad \sum_{j=0}^{\infty} \left|Q_j(z) + m_1(z)P_j(z)\right|^2$$

converge in G uniformly. Then the series

$$|m(z) - m_1(z)|^2 \sum_{j=0}^{\infty} |P_j(z)|^2 = \left|Q_j(z) + m(z)P_j(z) - (Q_j(z) + m(z)P_j(z))\right|^2$$

also converges and, hence, we get the conclusion.

Due to the estimate established in (3.4.2), the series $\sum_{j=0}^{\infty} |P_j(x + iy')|^2$ is majorized by the series $\sum_{j=0}^{\infty} |P_j(x + iy'')|^2$, if $|y''| \geq |y'|$. Therefore, the uniform convergence of the series $\sum_{j=0}^{\infty} |P_j(z)|^2$ in the region G implies its uniform convergence in an arbitrary bounded region. And finally, from the uniform convergence of the series $\sum_{j=0}^{\infty} \left|Q_j(z) + m(z)P_j(z)\right|^2$ and $\sum_{j=0}^{\infty} \left|P_j(z)\right|^2$ in a limited region that is separated from the real axis $\mathbb{R}$, it follows that the series uniformly converges $\sum_{j=0}^{\infty} \left|Q_j(z)\right|^2$ in this region. Takin into account the estimate (3.4.2) we get the convergence. □

Let us fix a non-real z and consider $m(z)$ as points in the space of functions such that, differen $m(z)$ correspond to different $d\sigma(z)$ that is generated by the usual resolution of the identity. It turns out that, similar to the case of the Sturm-Liouville equation, all such points lie on some Weyl-Hamburger circle. Recall that the equation

$$a|\omega|^2 + b\bar{\omega} + \bar{b}\omega + c = 0, \quad a > 0, \ \mathrm{Im} c = 0, \ |b|^2 - ac > 0 \tag{3.4.8}$$

with the complex variable ω, forms a circle with the center in O and a radius R on the complex plane $\mathbb{C}$, which are calculated by formulas

$$O = -\frac{b}{a}, \quad R^2 = \frac{1}{a^2}(|b|^2 - ac) \tag{3.4.9}$$

After simple transformations, due to Lemma 3.4.2, the equality (3.4.7) is rewritten in the form

$$\left(\sum_{j=0}^{\infty} |P_j(z)|^2\right) |m(z)|^2 \overline{\left(\frac{1}{z - \bar{z}} + \sum_{j=0}^{\infty} \overline{P_j(z)} Q_j(z)\right) m(z)}$$
$$+ \left(\frac{1}{z - \bar{z}} + \sum_{j=0}^{\infty} \overline{P_j(z)} Q_j(z)\right) m(z) + \sum_{j=0}^{\infty} |Q_j(z)|^2 = 0. \tag{3.4.10}$$

This is the relation of the type (3.4.8). (Note that the inequality $|b|^2 - ac > 0$ is fulfilled automatically if it is known that ω exists that satisfies the Eq. (3.4.8)). Thus, all values of $m(z)$ really belong to some circle. We will show that these values fill the whole circle.

Consider the Cayley transform of the operator L : $U_z = (L - \bar{z}\mathbb{I})(L - z\mathbb{I})^{-1}$. This operator maps isometrically all $\mathfrak{R}(L - z\mathbb{I})$ to all $\mathfrak{R}(L - \bar{z}\mathbb{I})$ in $\mathcal{H}_0 = l_2$. Let us construct an isometric operator X that maps the deficiency subspace N_z into $N_{\bar{z}}$. The orthogonal sum $V_z = U_z \oplus X$ is given by the Cayley transform of the self-adjoint extension A in l_2 of the operator L. Going through all such X, we get all extensions. Recall that the defect subspace N_z is spaned by vectors $P_0(z), P_1(z), \dots, P_j(z), \dots$.

Let us decompose the vector δ_0 into orthogonal components $\mathfrak{R}(L - z\mathbb{I})$ and N_z:

$$\delta_0 = e_z + \eta_z, \quad e_z \in \mathfrak{R}(L - z\mathbb{I}), \quad \eta_z \in N_z. \tag{3.4.11}$$

Let us calculate $\|\eta_z\|_0$. We denote by ϱ the distance. Then

$$\|\eta_z\|_0^2 = \varrho^2(\delta_0, \mathfrak{R}(L - z\mathbb{I})) = \varrho^2(\delta_0, (L - z\mathbb{I})l_{2,0}).$$

The last equality holds true because the set $(L - z\mathbb{I})l_{2,0}$ is dense in $\mathfrak{R}(L - z\mathbb{I})$. Let us pass to the Fourier transform. Since $\tilde{\delta} = 1$, and Fourier image $(L - z\mathbb{I})l_{2,0}$, which we denote by Θ, consists of all possible polynomials of the form $(\lambda - z)P(\lambda)$, where $P(\lambda)$ is an arbitrary one, then

$$\begin{aligned}
\|\eta_z\|_0^2 &= \varrho^2(\delta_0, \Theta)) = \varrho^2(\tilde{\delta}_0, \Theta) \\
&= \varrho^2|_{P(z)=0}(1, P(\lambda)) = \inf_{P:P(z)=0} \|1 - P(\lambda)\|^2_{L_2(\mathbb{R}, d\rho(\lambda))} \\
&= \inf_{P:P(z)=1} \|P(\lambda)\|^2_{L_2(\mathbb{R}, d\rho(\lambda))}.
\end{aligned}$$

Let us calculate the last infimum. An arbitrary polynomial $P(\lambda)$ with a condition $P(z) = 1$ has a form

$$P(\lambda) = \sum_{j=0}^{\infty} c_j P_j(\lambda) \left(\sum_{j=0}^{\infty} c_j P_j(z) \right)^{-1},$$

where the sequence of coefficients c_j is finite. Now, by the Cauchy-Bunyakovsky inequality, we get

$$\begin{aligned}
\|P(\lambda)\|^2_{L_2(\mathbb{R}, d\rho(\lambda))} &= \int_{\mathbb{R}} \left| \frac{\sum_{j=0}^{\infty} c_j P_j(\lambda)}{\sum_{j=0}^{\infty} c_j P_j(z)} \right|^2 d\sigma(\lambda) \\
&= \frac{\sum_{j=0}^{\infty} |c_j|^2}{\left| \sum_{j=0}^{\infty} c_j P_j(z) \right|^2} \geq \frac{1}{\sum_{j=0}^{\infty} |P_j(z)|^2}.
\end{aligned}$$

On the other hand, putting $c_j = \overline{P_j(z)}$, $j = 0, 1, \dots, n$ and $c_j = 0$, $j > n+1$, we get

$$\|P(\lambda)\|^2_{L_2(\mathbb{R},d\rho(\lambda))} = \left(\sum_{j=0}^{\infty} |P_j(z)|^2\right)^{-1}.$$

Thus,

$$\|\eta_z\|_0^2 = \inf_{P(z)=1} \|P(\lambda)\|^2_{L_2(\mathbb{R},d\rho(\lambda))} = \frac{1}{\sum\limits_{j=0}^{\infty} |P_j(z)|^2}. \tag{3.4.12}$$

In particular, (3.4.12) shows that $\|\eta_z\| \neq 0$. If $\delta_0 = e_{\bar{z}} + \eta_{\bar{z}}$ is a decomposition of δ_0 in accordance with

$$l_2 = \mathfrak{R}(L - \bar{z}\mathbb{I}) \oplus N_{\bar{z}},$$

then $\|\eta_z\|_0 = \|\eta_{\bar{z}}\|_0$.

Let A^0 be some fixed self-adjoint extension in l_2 of the operator L, and let A be some other arbitrary self-adjoint extension. We denote their Cayley transform by V_z^0 and V_z, respectively. Since V_z^0 and V_z act on $\mathfrak{R}(L - z\mathbb{I})$ in the same way, then for some $\varphi \in [0, 2\pi)$, we have

$$\begin{aligned} V_z\delta_0 =& V_z e_z + V_z\eta_z = V_z^0 e_z + e^{i\varphi} V_z^0 \eta_z; \\ (V_z\delta_0, \delta_0)_0 =& (V_z^0 e_z + e^{i\varphi} V_z^0 \eta_z, e_{\bar{z}} + \eta_{\bar{z}})_0 \\ =& (V_z^0 e_z, e_{\bar{z}}) + e^{i\varphi} (V_z^0 \eta_z, \eta_{\bar{z}})_0. \end{aligned} \tag{3.4.13}$$

From the set of isometric operators, we choose one that transfers N_z into $N_{\bar{z}}$ and is defined by V_z^0 in a certain way: that is, we put $V_z^0\eta_z = \eta_{\bar{z}}$. Then (3.4.13) has the form

$$(V_z\delta_0, \delta_0)_0 = (V_z^0 e_z, e_{\bar{z}}) + e^{i\varphi}\|\eta_{\bar{z}}\|_0^2 = (V_z^0 e_z, e_{\bar{z}}) + e^{i\varphi}\|\eta_z\|_0^2.$$

However

$$V_z = (A - \bar{z}\mathbb{I})(A - z\mathbb{I})^{-1} = \mathbb{I} + (z - \bar{z})R_z,$$

therefore $(V_z\delta_0, \delta_0)_0 = 1 + (z - \bar{z})m(z)$. Thus,

$$1 + (z - \bar{z})m(z) = (V_z^0 e_z, e_{\bar{z}}) + e^{i\varphi}\|\eta_{\bar{z}}\|_0^2,$$

hence

$$m(z) = \frac{1}{z-\bar{z}}((V_z^0 e_z, e_{\bar{z}}) - 1 + e^{i\varphi}\|\eta_{\bar{z}}\|_0^2). \tag{3.4.14}$$

Taking all φ in $[0, 2\pi)$, we obtain all possible isometric operators mapping N_z to $N_{\bar{z}}$. Then operator A runs through all self-adjoint extensions of L in l_2. According to (3.4.14), the point $m(z)$ describes some circle that coincides with the circle (3.4.10). Its center and radius are calculated using the formulas (3.4.9), but for the radius, the expression, following resulting from (3.4.12) and (3.4.14), is more convenient

$$R = \frac{\|\eta_z\|_0^2}{|z-\bar{z}|} = \left(|z-\bar{z}|\sum_{j=0}^{\infty}|P_j(z)|^2\right)^{-1}.$$

Thus, we proved the first part of the following theorem.

Theorem 3.4.3 *Let the expression L corresponds to an indeterminate case. Let us fix an non-real z and consider the point*

$$m(z) = \int_{\mathbb{R}} \frac{1}{\lambda - z}\, d\sigma(\lambda),$$

where $d\sigma(\lambda)$ is an orthogonal spectral measure. Going through all such measures, we get that the point $m(z)$ describes some circle K_z, i.e., the Weyl-Hamburger circle with the center and the radius written by expressions

$$\begin{aligned} O(z) &= -\frac{1}{\sum_{j=0}^{\infty}|P_j(z)|^2}\left(\frac{1}{z-\bar{z}} + \sum_{j=0}^{\infty}\overline{P_j(z)}Q_j(z)\right), \\ R(z) &= \frac{1}{|z-\bar{z}|\sum_{j=0}^{\infty}|P_j(z)|^2}. \end{aligned} \tag{3.4.15}$$

If L is determined, then the circle (3.4.15) *degenerates into the Weyl-Hamburger point.*

Points ω of the Weyl-Hamburger circle are in one-to-one correspondence with the orthogonal spectral measures $d\sigma(\lambda)$, and therefore, due to (3.2.4)*—with the spectral resolution of the identity. Namely, each measure $d\sigma(\lambda)$ corresponds to a point*

$$\omega = m(z) = \int_{\mathbb{R}} \frac{1}{\lambda - z}\, d\sigma(\lambda).$$

Conversely, each $\omega \in K_z$ *corresponds to* $d\sigma(\lambda)$*, whose Stieltjes transform is calculated by the formula*

$$m(\zeta) = \frac{E_0(\zeta, z)\omega + E_1(\zeta, z)}{D_0(\zeta, z)\omega + D_1(\zeta, z)}, \tag{3.4.16}$$

where $E_0(\zeta, z)$, $E_1(\zeta, z)$, $D_0(\zeta, z)$, $D_1(\zeta, z)$ *are entire functions of each variable with the other fixed and are defined by series*

$$\begin{aligned} E_0(\zeta, z) =& 1 + (\zeta - z)\sum_{j=0}^{\infty} Q_j(\zeta)P_j(z), \\ E_1(\zeta, z) =& (\zeta - z)\sum_{j=0}^{\infty} Q_j(\zeta)Q_j(z), \\ D_0(\zeta, z) =& -(\zeta - z)\sum_{j=0}^{\infty} P_j(\zeta)P_j(z), \\ D_1(\zeta, z) =& 1 - (\zeta - z)\sum_{j=0}^{\infty} P_j(\zeta)Q_j(z). \end{aligned} \tag{3.4.17}$$

Proof of the second part of the theorem. Let us replace ζ with $\bar{\zeta}$ in (3.4.6) and from this equality, we get $m(\zeta)$. Thus, we have

$$m(\zeta) = \frac{E_0(\zeta, z)m(z) + E_1(\zeta, z)}{D_0(\zeta, z)m(z) + D_1(\zeta, z)}, \tag{3.4.18}$$

where $E_0(\zeta, z)$, $E_1(\zeta, z)$, $D_0(\zeta, z)$, $D_1(\zeta, z)$ have the form (3.4.17) (the convergence of series (3.4.17) and the properties of analyticity follow from Lemma 3.4.2. The relation (3.4.18) shows that if $m(z)$ is known at one point, then it is known everywhere. The statement of the theorem easy follows from the last conclusion.

□

Remark 3.4.4 The Weyl—Hamburger circle K_z is located in the upper half-plane if $\operatorname{Im} z > 0$, and in the lower one if $\operatorname{Im} z < 0$. This follows from the fact that the points $m(z)$, if we go through all the orthogonal measures, describes the whole circle K_z. At the same time

$$\operatorname{Im} m(z) = \operatorname{Im} \int_{\mathbb{R}} \frac{1}{\lambda - z}\, d\sigma(\lambda) = y \int_{\mathbb{R}} \frac{1}{(\lambda - x)^2 + y^2}\, d\sigma(\lambda), \ z = x + iy.$$

Corollary 3.4.5 *In the indeterminate case, the spectrum of the self-adjoint extensions in the space l_2 of the operator L is always discrete and has no condensation points on an arbitrary finite interval.*

It follows from (3.4.16) that $m(z)$ is a meromorphic function and, therefore, is equivalent to the given in the last corollary.

3.5 The Classical Moment Problem

The classic moment problem (the Hamburger moment problem) consists in finding out the conditions for the sequence of real numbers (s_n), $n \in \mathbb{N}_0$, $s_n \in \mathbb{R}$, which are necessary and sufficient for the existence of the representation

$$s_n = \int_{\mathbb{R}} \lambda^n \, d\rho(\lambda), \quad n \in \mathbb{N}_0,$$

where $d\rho(\lambda)$ is a Borel measure on σ-algebra of Borel sets on the real axis $\mathbb{R}$ in the usual topology. It contains questions about the uniqueness of such a measure and a description of all measures in the case of non-uniqueness, etc.

There are several approaches to researching such questions. In our opinion, the more natural is the spectral approach, which was first proposed by Krein M. G. for similarly problems. He used a spectral representation for the self-adjoint operator, which enables broad generalizations of this problem, particularly interesting and important for the mathematical physics.

This approach is closely related to the theory of Jacobi matrices, and therefore its representation in this chapter is natural (see also Chap. 1, Sects. 2.4 and 2.11).

Necessary Chains of Hilbert Spaces For conveniens, we repeat with additional explanations some material from Sect. 2.12. Consider a chain of Hilbert spaces

$$G_- \supset G_0 \supset G_+ \tag{3.5.1}$$

with an involution "$*$". In other words, an involution $G \ni \xi \mapsto \xi^* \in G_-$ is defined in the negative space G_-. The restriction of this involution to the zero space G_0 or the positive G_+ is also an involution in the corresponding space. Let us construct the tensor square of the chain (3.5.1)

$$G_- \otimes G_- \supset G_0 \otimes G_0 \supset G_+ \otimes G_+ \tag{3.5.2}$$

and fix some generalized kernel $K \in G_- \otimes G_-$. It is called positive definite if

$$\forall g \in G_+, \quad (K, \varphi \otimes \varphi^*)_{G_0 \otimes G_0} \geq 0. \tag{3.5.3}$$

A kernel $K \in G_- \otimes G_-$ is called Hermitian if the bilinear form

$$a(\varphi, \psi) = (K, \psi \otimes \varphi^*)_{G_0 \otimes G_0}, \ \varphi, \psi \in G_+$$

is Hermitian, that is, $a(\varphi, \psi) = \overline{a(\psi, \varphi)}$. It is clear that every positive definite kernel will be Hermitian.

For each positive definite kernel K, a corresponding Hilbert space H_K is constructed by default. The construction is as follows. There is introduced a quasi-scalar product

$$\langle \varphi, \psi \rangle = (K, \psi \otimes \varphi^*)_{G_0 \otimes G_0}, \quad \varphi, \psi \in G_+ \tag{3.5.4}$$

All $\varphi \in G$, for which $\langle \varphi, \varphi \rangle = 0$, are identified with zero. Consider the corresponding classes of vectors $\varphi \in G_+$, namely,

$$\hat{\varphi} := \{\chi \in G_+ \mid \langle \varphi - \chi, \varphi - \chi \rangle = 0\}. \tag{3.5.5}$$

The expression (3.5.4) will already be a non-degenerate scalar product on classes (3.5.5). The space H_K is the complement of the linear space of class (3.5.5) by the scalar product (3.5.4).

In more detail, this procedure is as follows. Take

$$N = \{\varphi \in G_+ \mid \langle \varphi, \varphi \rangle = 0\}. \tag{3.5.6}$$

Due to the inequality that follows from (3.5.4),

$$\begin{aligned} |\langle \varphi, \psi \rangle| &= |(K, \psi \otimes \varphi^*)_{G_0 \otimes G_0}| \leq \|K\|_{G_- \otimes G_-} \|\psi \otimes \varphi^*\|_{G_+ \otimes G_+} \\ &\leq \|K\|_{G_- \otimes G_-} \|\psi\|_{G_+} \|\varphi\|_{G_+}, \quad \varphi, \psi \in G_+, \end{aligned} \tag{3.5.7}$$

N will be a (linear) closed subspace in G_+. The class $\hat{\varphi}$ corresponding to $\varphi \in G_+$ coincides with the hyperplane $\varphi + N$ or with the elements of the quotient space G_+/N. The form (3.5.4) does not change when φ and ψ run through the corresponding classes $\hat{\varphi}$ and $\hat{\psi}$ and gives their scalar product $\langle \hat{\varphi}, \hat{\psi} \rangle_{H_K}$. The space H_K is the completion of this pre-Hilbert space of classes.

It is clear that the described picture is isomorphic to this one: on the orthogonal (in G_+) complement $G_+ \ominus N$, the formula (3.5.4) already gives the scalar product of the classes $\hat{\varphi}$, $\hat{\psi}$, the completion of which will be the space H_K. An isomorphism is defined by the correspondence (before completion) between the classes $\hat{\varphi}$ and the projection in G_+ of the vector $\varphi \in G_+$ onto $G_+ \ominus N$ (all projections of the vectors $\varphi \in \hat{\varphi}$ are the same).

We will use the equipment procedure of the space H_K, constructed according to the positive definite kernel $K \in G_- \otimes G_-$. Its difference from an arbitrary equipment of the space H_K is that it is constructed using the chain (3.5.1), which defines the kernel K with the help of (3.5.2).

Let us have a chain with involution

$$H_- \supset G_- \supset G_0 \supset G_+ \supset H_+, \tag{3.5.8}$$

that continues the chain (3.5.1). That is, H_+ is continuously and densely embedded in G_+ with an involution "$*$" continued from G_+ and the negative space H_- is constructed in (3.5.8) by the zero G_0.

To construct such an equipment of H_K, note that $N \ominus H_+$ is a subspace in H_+ due to the continuity of the imbedding $H_+ \hookrightarrow G_+$ and continuities (see (3.5.7))

$$\langle \varphi, \varphi \rangle \leq \|K\|_{G_- \otimes G_-} \|\varphi\|^2_{G_+}, \quad \varphi \in G_+. \tag{3.5.9}$$

Namely, $\exists C \in (0, \infty)$: $\langle \varphi, \varphi \rangle \leq C \|\varphi\|^2_{G_+}$, $\varphi \in N \cap H_+$.

Lemma 3.5.1 *Vectors of the space*

$$H_+ \ominus (N \cap H_+) \tag{3.5.10}$$

(*the orthogonal difference is taken in the space* H_+) *can be identified with the set of all classes* $\hat{\varphi}$ *where* $\varphi \in H_+$. *Each such class has only one common vector* φ_N *from* (3.5.10)—*the projection* $\varphi \in H_+$ *onto* (3.5.10).

Proof Indeed, let φ_1 be a projection. Then $\varphi - \varphi_1 \in N \cap H_+ \subset N$, i.e., $\varphi_1 \in \hat{\varphi}$. Let now $\varphi_1 - \varphi_2 \in \hat{\varphi} \cap (H_+ \ominus (N \cap H_+))$, then $\varphi_1, \varphi_2 \in H_+$, $\varphi_1 - \varphi_2 \in N$, due to $\varphi_1 - \varphi_2 \in N \cap H_+$. Since $\varphi_1 - \varphi_2 \in H_+ \ominus (N \cap H_+)$, then $\varphi_1 - \varphi_2 = 0$, i.e., $\varphi_1 = \varphi_2 = \varphi_N$. □

Let us denote by $H_{+,K}$ the linear set of all classes $\hat{\varphi}$, where $\varphi \in H_+$, and transform it into a Hilbert space by putting

$$(\hat{\varphi}, \hat{\psi})_{H_{+,K}} = (\varphi_N, \psi_N)_{H_+}, \quad \varphi, \psi \in H_+, \tag{3.5.11}$$

where φ_N, ψ_N are vectors from Lemma 3.5.1. This definition is obviously correct; the resulting space $H_{+,K}$ is isometric to $H_+ \ominus (N \cap H_+)$ and imbedded, as a set, in H_+.

The space $H_{+,K}$ is imbedded in H_K, and the imbedding operator O, as it is easy to see, has the form of the product $O = RO_+$, where $O_+ : H_+ \mapsto G_+$ is the imbedding operator H_+ in G_+ is a chain (3.5.8), and $R : G_+ \mapsto H_+$ is an operator transfers the vector $\varphi \in G_+$ corresponding to $\hat{\varphi} \in H_K$. Thus, (see (3.5.9))

$$\begin{aligned} \|\hat{\varphi}\|_{H_K} &= \|RO_+\hat{\varphi}\|_{H_+} \leq \|O_+\| \|\hat{\varphi}\|_{H_{+,K}} \\ &\leq \|K\|^{1/2}_{G_- \otimes G_-} \|O_+\| \|\hat{\varphi}\|_{H_{+,K}}, \quad \varphi \in H_{+,K}. \end{aligned} \tag{3.5.12}$$

It is clear that $H_{+,K}$ is dense in H_K: this follows from the density of H_+ in G_+ and the inequality (3.5.9).

Thus, it is possible to consider H_+ to be the zero space, and $H_{+,K}$ to be a positive space and construct a chain that equips the space H_K:

$$H_{-,K} \supset H_K \supset H_{+,K}.$$

These constructions lead to the following result.

Theorem 3.5.2 *Consider the chain of Hilbert spaces with involution "$*$" and its tensor square*

$$G_- \supset G_0 \supset G_+, \quad G_- \otimes G_- \supset G_0 \otimes G_0 \supset G_+ \otimes G_+. \tag{3.5.13}$$

We fix the generalized kernel $K \in G_- \otimes G_-$, which is considered positive definite, i.e.,

$$(K, \varphi \otimes \varphi^*)_{G_0 \otimes G_0} \geq 0, \quad \varphi \in G_+. \tag{3.5.14}$$

Consider the quasi-scalar product on G_+, namely,

$$G_+ \ni \varphi, \psi \mapsto \langle \varphi, \psi \rangle := (K, \psi \otimes \varphi^*)_{G_0 \otimes G_0} \tag{3.5.15}$$

and construct the Hilbert space H_K of classes

$$\hat{\varphi} = \{\chi \in G_+ \mid \langle \varphi - \chi, \varphi - \chi \rangle = 0\}. \tag{3.5.16}$$

It is claimed that the space H_K can be equipped in the following way. Let us have an arbitrary extension of the first chain in (3.5.13) *to a chain with the same involution "$*$" and the zero space G_0, namely,*

$$H_- \supset G_- \supset G_0 \supset G_+ \supset H_+. \tag{3.5.17}$$

Then there is an equipment

$$H_{-,K} \supset H_K \supset H_{+,K}, \tag{3.5.18}$$

where the positive space $H_{+,K}$ coincides with the space of classes $\hat{\varphi}$ and $\varphi \in H_+ \subset G_+$.

If the imbedding operator $H_+ \subset G_+$ is the Hilbert-Schmidt one, then the chain (3.5.18) *is quasi-nuclear.*

Proof is actually reduced to some components relative to the construction on the previous pages. Thus, the inequality (3.5.12) gives the continuity of the imbedding of spaces of class $H_{+,K} \hookrightarrow H_K$, which is necessary because in (3.5.18), space H_K plays the role of the zero space. The density of $H_{+,K}$ in H_K is a consequence of the density of H_+ in G_+ and the inequality (3.5.9).

The formula $O = RO_+$, for imbedding operators of classes, holds true because each class $\hat{\varphi}$ has the form of a hyperplane $\varphi + N$, where $\varphi \in H_+$, and due to Lemma 3.5.1 it is defined by its projection φ_N.

If the imbedding operator $O_+ : H_+ \mapsto G_+$ is Hilbert-Schmidt, then the operator $RO_+ : H_+ \mapsto H_K$ will be like this, since R is a continuous operator. That is, the chain (3.5.18) will be quasi-nuclear. This follows from a simple general Lemma 2.7.5. □

Remark 3.5.3 Note that if the kernel $K \in G_- \otimes G_-$ is strictly positive definite, i.e., in the inequality (3.5.14), the sign ">" stands by $\varphi \neq 0$, then (3.5.15) will be a scalar and not a quasi-scalar product, and the theorem becomes obvious: the chain (3.5.17) and the scalar product (3.5.15) generate the chain (3.5.18). It will be quasi-scalar if the imbedding operator nesting operator for $H_+ \subset G_+$ is the Hilbert-Schmidt type.

Let the chain (3.5.1) and the positive definite kernel $K \in G_- \otimes G_-$ be given. Consider an operator A acting in the space G_+ and having a dense domain $\mathfrak{D}(A)$ in G_+. When applying Theorem 3.5.2, the question arises when this operator can be "restricted" to the operator $\hat{A}$ acting in the H_K space of $\hat{\varphi}$ classes generated by N in (3.5.6). That is, when A transfers class into a class. Then it can be assumed that

$$\hat{A}\hat{\varphi} := (A\varphi)\hat{}, \quad \varphi \in \hat{\varphi}, \; \varphi \in \mathfrak{D}(A); \; \mathfrak{D}(\hat{A}) = (\mathfrak{D}(A))\hat{}. \tag{3.5.19}$$

It is easy to do in such two cases.

Lemma 3.5.4 *The operator A is Hermitian with respect to the quasi-scalar product (3.5.4), i.e., $\langle A\varphi, \psi\rangle = \langle \varphi, A\psi\rangle$, $\varphi, \psi \in \mathfrak{D}(A)$.*

The operator A is isometric with respect to (3.5.6), i.e., $\langle A\varphi, A\psi\rangle = \langle \varphi, \psi\rangle$, $\varphi, \psi \in \mathfrak{D}(A)$.

Proof Consider the case of the Hermitian operator A. In order to perform (2.7.35), it is necessary to make sure that A transfers a class into a class, and for this, it is necessary to prove: if $\varphi \in \mathfrak{D}(A)$, then from the equality $\langle \varphi, \varphi\rangle = 0$, it follows that $\langle A\varphi, A\varphi\rangle = 0$.

Let $\psi \in \mathfrak{D}(A)$, then

$$|\langle A\varphi, \varphi\rangle|^2 = |\langle \varphi, A\varphi\rangle|^2 \le \langle \varphi, \varphi\rangle\langle A\psi, A\psi\rangle, \tag{3.5.20}$$

since the Cauchy-Bunyakovsky inequality holds true for the quasi-scalar product (3.5.4).

Since $\langle \varphi, \varphi\rangle = 0$, from (3.5.20), we have that $\langle A\varphi, \psi\rangle = 0$, $\psi \in \mathfrak{D}(A)$. Due to the density of $\mathfrak{D}(A)$ in G_+, we approximate the vector $A\varphi$ in G_+ by the vectors

$\psi_n \in \mathfrak{D}(A)$. They also converge in the space H_K according to the estimate (3.5.9). Again, due to the same Cauchy-Bunyakovsky inequality, we have

$$\langle A\varphi, A\varphi\rangle = \lim_{n\to\infty} \langle A\varphi, \psi\rangle = 0,$$

and this had to be proved.

The case of the isometric operator A is obvious: if $\langle\varphi, \varphi\rangle = 0$, then $\langle A\varphi, A\varphi\rangle = 0$. □

Some Results of the Spectral Theory Let the self-adjoint operator A defined on $\mathfrak{D}(A)$ be given in the separable Hilbert space $\mathcal{H}$. Consider the equipment of the space $\mathcal{H}$:

$$\mathcal{H}_- \supset \mathcal{H} \supset \mathcal{H}_+ \supset \mathcal{D}, \tag{3.5.21}$$

where $\mathcal{H}_+$ is a Hilbert space topologically and quasi-nuclearly embedded in $\mathcal{H}$; $\mathcal{H}_-$ is the space dual to $\mathcal{H}_+$ with respect to the space $\mathcal{H}$; $\mathcal{D}$ is a linear topological space topologically imbedded into $\mathcal{H}_+$.

Recall that an operator A is said to be standardly connected with a chain (3.5.21) if $\mathcal{D} \subset \mathfrak{D}(A)$ and the restriction $A \restriction \mathcal{D}$ acts continuously from $\mathcal{D}$ to $\mathcal{H}_+$. Such a case is considered below.

We also recall that the vector $\Omega \in \mathcal{D}$ is called strong cyclic for the operator A if for all $n \in \mathbb{N}$, $\Omega \in \mathfrak{D}(A^n) \in \mathcal{D}$ and the set of all vectors $A^n\Omega$, for $n = \mathbb{N}_0$ is total in the space $\mathcal{H}_+$ (and therefore also in $\mathcal{H}$).

Assuming that a strictly cyclic vector exists, we formulate some reduced version of the projective spectral theorem (see Chap. 1, Sect. 2.11).

Theorem 3.5.5 *For a self-adjoint operator A with a strong cyclic vector in a separable Hilbert space $\mathcal{H}$, there exists a Borel measure $d\rho(\lambda)$ on the real axis, so that for ρ-almost all $\lambda \in \mathbb{R}$ there exists a generalized eigenvector $\xi_\lambda \in \mathcal{H}_-$, that is, for each $f \in \mathcal{D}$*

$$(\xi_\lambda, Af)_{\mathcal{H}} = \lambda(\xi_\lambda, f)_{\mathcal{H}}, \quad \xi_\lambda \neq 0. \tag{3.5.22}$$

The corresponding Fourier transform F acts according to the rule:

$$\mathcal{H} \supset \mathcal{H}_+ \ni f \mapsto (Ff)(\lambda) = \hat{f}(\lambda) = (f, \xi_\lambda)_{\mathcal{H}} \in L_2(\mathbb{R}, d\rho(\lambda)), \tag{3.5.23}$$

is an isometry operator (after its extension by continuity) acting from the space $\mathcal{H}$ in $L_2(\mathbb{R}, d\rho(\lambda))$. The image of the operator A under the transformation by F is the multiplication operator by the independent variable λ in $L_2(\mathbb{R}, d\rho(\lambda))$, that is, $(FA)f(\lambda) = \lambda f(\lambda)$.

We also recall that for the self-adjoint operator A defined on $\mathfrak{D}(A)$ in $\mathcal{H}$, the vector $f \in \bigcap_{n=0}^{\infty} \mathfrak{D}(A^n)$ is called quasi-analytic if the class $C\{m_n\}$, where in this case $m_n = \|A^n f\|_{\mathcal{H}}$, is quasi-analytic (the class of functions on $[a,b] \subset \mathbb{R}$ is defined by the expression

$$C(\{m_n\}) = \{g \in C^\infty([a,b])\ \exists K = K_f > 0,$$
$$|g^{(n)}(t)| \leq K^n m_n,\ t \in [a,b],\ n \in \mathbb{N}_0\},$$

hence it is quasi-analytic if

$$\sum_{n=1}^{\infty} (\inf\{m_k^{1/k} \mid n\})^{-1} = \infty).$$

The quasi-analyticity of the vector is used in the criterion of a self-adjointness and a commutativity, that is, results from Chap. 1 are used (see Sect. 2.6).

The Moment Problem As mentioned above, the classical moment problem is regarded as the problem in finding conditions on the sequence (s_n), $n \in \mathbb{N}_0$, $s_n \in \mathbb{R}$, when for which there exists a Borel measure $d\rho(x)$ on the real axis $\mathbb{R}$, so that

$$s_n = \int_{\mathbb{R}} \lambda^n \, d\rho(\lambda),\ n \in \mathbb{N}_0. \tag{3.5.24}$$

The main result of the Section is contained in the theorem.

Theorem 3.5.6 *A sequence of real numbers* $(s_n)_{n \in \mathbb{N}_0}$ *has a representation* (3.5.24) *if and only if it is positive definite, i.e.,*

$$\sum_{m,n=0}^{\infty} s_{n+m} f_m \bar{f}_n \geq 0 \tag{3.5.25}$$

on arbitrary finite sequences $(f_n)_{n=0}^{\infty}$ *of numbers* $f_n \in \mathbb{C}$.

The representation (3.5.24) *exists and the measure* $d\rho(\lambda)$ *is unique if the sequence of numbers* (s_n), $n \in \mathbb{N}_0$ *is positive definite and, additionally,*

$$\sum_{p=1}^{\infty} \frac{1}{\sqrt[2p]{s_{2p}}} = \infty. \tag{3.5.26}$$

Note that the condition (3.5.25) is necessary and sufficient for the existence of the representation (3.5.24), but there can be many representations (measures). The condition (3.5.25) together with (3.5.26) guarantees the uniqueness (of the measure) of the representation (3.5.24).

Proof We will prove the necessity of the condition (3.5.25). If the sequence (s_n), $n \in \mathbb{N}_0$ has the representation (3.5.24), then for an arbitrary finite sequence $f = (f_n)_{n=0}^{\infty}$, $f_n \in \mathbb{C}$ we have

$$\sum_{m,n=0}^{\infty} s_{n+m} f_m \bar{f}_n = \int_{\mathbb{C}} \left| \sum_{n=0}^{\infty} \lambda^n f_n \right|^2 d\rho(\lambda) \geq 0,$$

which proves the necessity.

Let us prove sufficiency. We will use the scheme outlined in Theorem 3.5.2. Thus, the role of the first of the chains (3.5.13) plays the equipment of the usual space $l_2 = G_0$ of sequences

$$f = (f_j)_{j=0}^{\infty}, \; f_j \in \mathbb{C}; \; \|f\|_{l_2}^2 = \sum_{j=0}^{\infty} |f_j|^2 < \infty.$$

Spaces G_+ and G_- will be a like l_2 space with a weight, namely, we denote by $l_2(q)$ the space l_2 with weight $q = (q_j)_{j=0}^{\infty}$, $q_j > 0$; it consists of sequences $f = (f_j)_{j=0}^{\infty}$, $f_j \in \mathbb{C}$, for which

$$\|f\|_{l_2(q)}^2 = \sum_{j=0}^{\infty} |f_j|^2 q_j < \infty, \quad (f, g)_{l_2(q)} = \sum_{j=0}^{\infty} f_j \bar{g}_j q_j. \tag{3.5.27}$$

We fix $q = (q_j)_{j=0}^{\infty}$, $q_j \geq 1$. Then the first of the chains (3.5.13) (i.e., (3.5.1)) will have the form

$$l_2(q^{-1}) \supset l_2 \supset l_2(q) \supset l_{\text{fin}}, \quad q^{-1} = (q_j^{-1})_{j=0}^{\infty}, \; q = (q_j)_{j=0}^{\infty}, \; q_j \geq 1, \tag{3.5.28}$$

where l_{fin} denotes the space of finite sequences. The involution of "$*$" is now the usual complex conjugation: $(f_j)_{j=0}^{\infty} = f \mapsto f^* = (\bar{f}_j)_{j=0}^{\infty}$. Accordingly, it is also convenient for us to change the notation of vectors in Theorem 3.5.2 from φ to f.

Let us specify the choice of weight q. The condition (3.5.25) in the theorem means that the matrix $K = (K_{jk})_{j,k=0}^{\infty}$, where $K_{jk} = s_{j+k}$, is positive definite. Then

$$|K_{jk}|^2 \leq K_{jj} K_{kk}, \text{ i.e., } |s_{j,k}|^2 \leq s_{2j} s_{2k}, \; j, k \in \mathbb{N}_0. \tag{3.5.29}$$

Let $q = (q_n)_{n=0}^{\infty}$, $q_n \geq 1$ be the sequence such that

$$C(s, q) := \sum_{n=0}^{\infty} s_{2n} q_n^{-1} < \infty. \tag{3.5.30}$$

Using (3.5.29) and (3.5.30) we get

$$
\begin{aligned}
0 &\le \sum_{j,k=0}^{\infty} s_{j+k} f_k \bar{f}_j \le \sum_{j,k=0}^{\infty} |s_{j+k}||f_k||f_j| \\
&\le \sum_{j,k=0}^{\infty} \sqrt{s_{2j}s_{2k}}|f_k||f_j| \le \sum_{j,k=0}^{\infty} \sqrt{s_{2j}s_{2k}q_j^{-1}q_k^{-1}}|f_k q_k^{1/2}||f_j q_j^{1/2}| \\
&\le \left(\sum_{j,k=0}^{\infty} s_{2j}s_{2k}q_j^{-1}q_k^{-1}\right)^{1/2} \left(\sum_{j,k=0}^{\infty} |f_k|^2 q_k |f_j|^2 q_j\right)^{1/2} \qquad (3.5.31) \\
&= \left(\sum_{j=0}^{\infty} s_{2j}q_j^{-1}\right)\left(\sum_{j=0}^{\infty} |f_j|^2 q_j\right) = C(s,q)\|f\|^2_{l_2(q)}, \quad f \in l_{\text{fin}}.
\end{aligned}
$$

Passing here to the limit from the finite sequences l_{fin} to $l_2(q)$, we obtain the chain

$$
\begin{aligned}
&l_2(q^{-1}) \supset l_2 \supset l_2(q) \supset l_{\text{fin}}, \; q = (q_n)_{n=0}^{\infty}, \; q_n \ge 1, \\
&C(s,q) = \sum_{n=0}^{\infty} \frac{s_{2n}}{q_n} < \infty.
\end{aligned} \qquad (3.5.32)
$$

We fix the weight q for which the condition (3.5.32) is fulfilled (the sequence (s_n), $n \in \mathbb{N}_0$ is considered given). Then the chain of spaces (3.5.32) plays the role of the first of chains (3.5.13).

As already mentioned, the matrix $K = (K_{jk})_{j,k=0}^{\infty}$, where $K_{jk} = s_{j+k}$, is the kernel K of discrete variable $j \in \mathbb{N}_0$; we will denote it by $S = K$. It will be positive definite, since the condition (3.5.3) is fulfilled: it coincides with (3.5.25).

The next step in the scheme of the Theorem 3.5.2 is the construction of a chain (3.5.17). Here we require that the imbedding $H_+ \hookrightarrow G_+$ must be Hilbert-Schmidt in the space $G_+ = l_2(q)$, where q satisfies the construction (3.5.32), i.e., it is imbedded in the space $l_2(p)$ quasi-nuclear. It is so for $p = (p_n)_{n=0}^{\infty}$ if

$$
\sum_{n=0}^{\infty} p_n^2 q_n^2 < \infty. \qquad (3.5.33)
$$

Thus, we take as the chain (3.5.17) the following one:

$$
(l_{\text{fin}})' \supset l_2(p^{-1}) \supset l_2(q^{-1}) = l_2 \supset l_2(q) \supset l_2(p) \supset l_{\text{fin}}, \qquad (3.5.34)
$$

where the weight q is chosen such that the condition (3.5.32) is fulfilled, and the weight p is chosen so that the condition (3.5.33) is fulfilled. Let us fix the weight $p = (p_n)_{n=0}^{\infty}$.

According to Theorem 3.5.2, the necessary chain (3.5.18) is constructed

$$H_{-,S} \supset H_S \supset H_{+,S} \supset l_{\text{fin}}, \tag{3.5.35}$$

where we put $K = S$.

Let us first prove the theorem in the case of non-degeneracy of the kernel $K = S$, i.e., when the inequality (3.5.25) contains the sign ">" for an arbitrary finite sequence $(f_n)_{n=0}^{\infty} \in l_{\text{fin}}$, which is not equal to zero (that is, for at least one n we have $f_n \neq 0$).

The proof is completely based on the Theorem 3.5.2.

In this case, the expression

$$(f, g)_S := \sum_{j,k=0}^{\infty} s_{j+k} f_k \bar{g}_j, \quad f, g \in l_{\text{fin}} \tag{3.5.36}$$

is a scalar (and not a quasi-scalar) product in the space l_{fin}. The completion of this space with respect to (3.5.36) will be denoted by H_S.

A linear space $\mathbb{C}^{\infty}$ of all sequences $f = (f_n)_{n=0}^{\infty}$, $f_n \in \mathbb{C}$, will be denoted by l. Let $\delta_n = (0, \ldots, 0, 1, 0, \ldots)$, $n \in \mathbb{N}_0$ be a δ-sequence ("1" is on the n-th position), then for each $f \in l_{\text{fin}}$, we have $f = \sum_{n=0}^{\infty} f_n \delta_n$. In this way, a chain of linear spaces

$$l \supset H_S \supset l_{\text{fin}}, \quad \forall f \in l_{\text{fin}}, \ f = (f_n)_{n=0}^{\infty} = \sum_{n=0}^{\infty} f_n \delta_n. \tag{3.5.37}$$

is constructed.

In the case of a non-degenerate kernel S, the chains (3.5.35) and (3.5.34) give the following two chains of Hilbert spaces

$$(l_2(p))_{-,S} \supset H_S \supset l_2(p) \supset l_{\text{fin}}, \tag{3.5.38}$$

$$l = (l_2)' \supset l_2(p^{-1}) \supset l_2 \supset l_2(p) \supset l_{\text{fin}}, \tag{3.5.39}$$

where the weight p satisfies the requirements (3.5.33), (3.5.30).

Consider a linear expression in the space l:

$$J = \begin{bmatrix} 0 & 0 & 0 & 0 & \cdots \\ 1 & 0 & 0 & 0 & \cdots \\ 0 & 1 & 0 & 0 & \cdots \\ 0 & 0 & 0 & 0 & \cdots \\ \cdots & \cdots & \cdots & \cdots & \ddots \end{bmatrix}; \tag{3.5.40}$$

$$J : (Jf)_k = f_{k-1}, \quad k \in \mathbb{N}_0, \quad (Jf)_0 = 0; \quad f \in l.$$

This expression is a "creation" type operator. Then the relation

$$J\delta_n = \delta_{n+1}, \; n \in \mathbb{N}_0 \tag{3.5.41}$$

holds true for δ-sequence.

Its restriction to $l_{\text{fin}} \subset l_2$ will not be a Hermitian operator in the space l_2 but it will be Hermitian in H_S, which is given by the inner product (3.5.36). Indeed, for $f, g \in l_{\text{fin}}$ we have:

$$\begin{aligned} (Jf, g)_S &= \sum_{j,k=0}^{\infty} s_{j+k}(Jf)_k \bar{g}_j = \sum_{j,k=0}^{\infty} s_{j+k} f_{k-1} \bar{g}_j \\ &= \sum_{j,k=0}^{\infty} s_{j+k+1} f_k \bar{g}_j = \sum_{j,k=0}^{\infty} s_{j+k} f_k \overline{(Jg)_j} = (f, Jg)_S, \end{aligned} \tag{3.5.42}$$

(here it is taken into account that due to the definition (3.5.40) $f_{-1} = g_{-1} = 0$).

Thus, the operator $l_{\text{fin}} \ni f \mapsto Jf \in l_{\text{fin}}$ will be Hermitian with a dense domain in the space H_S. It has the equal defect numbers because $J\bar{f} = \widehat{Jf}$, where $\bar{f}$ is the complex conjugate of $f \in l$. Let A be some of its fixed self-adjoint extension in the space H_S. Thus, we have

$$Af = Jf, \quad f \in l_{\text{fin}} \subset \mathfrak{D}(A) \subset H_S. \tag{3.5.43}$$

Let us now apply Lemma 2.7.5. As chains (2.7.35) we take (3.5.38) and (3.5.39). Let $\xi_\lambda \in (l_2(p))_{-,S}$ be the generalized eigenvector of the operator A in terms of the rigging (3.5.38). In this case, according to Theorem 3.5.5, we have

$$(\xi_\lambda, Af)_S = \lambda(\xi_\lambda, f)_S, \; \lambda \in \mathbb{R}, \; f \in l_{\text{fin}}. \tag{3.5.44}$$

Denote

$$P(\lambda) = U\xi_\lambda \in l_2(p^{-1}) \subset l, \; P(\lambda) = (P_n(\lambda))_{n=0}^{\infty}, \; P_n(\lambda) \in \mathbb{C}.$$

Using (2.7.36), the expression (3.5.44) is rewritten as

$$(P(\lambda), Af)_{l_2} = \lambda(P(\lambda), f)_{l_2}, \quad \lambda \in \mathbb{R}, \ f \in l_{\text{fin}}. \tag{3.5.45}$$

The corresponding Fourier transform has the form

$$H_S \supset l_{\text{fin}} \ni f \to (Ff)(\lambda) = \hat{f}(\lambda) = (f, P(\lambda))_{l_2} \in L_2(\mathbb{R}, d\rho(\lambda)). \tag{3.5.46}$$

Let us calculate $P(\lambda)$. The operator A is defined in (3.5.43) using the expression in (3.5.40), therefore, in (3.5.45) gives $\forall f \in l_{\text{fin}}$

$$\begin{aligned} \sum_{n=0}^{\infty} \lambda P_n(\lambda) \bar{f}_n &= \lambda(P(\lambda), f)_{l_2} = (P(\lambda), Af)_{l_2} \\ &= (P(\lambda), Jf)_{l_2} = (JP(\lambda), f)_{l_2} = \sum_{n=0}^{\infty} P_{n+1}(\lambda) \bar{f}_n. \end{aligned} \tag{3.5.47}$$

Thus,

$$\lambda P_n(\lambda) = P_{n+1}(\lambda), \quad n \in \mathbb{N}_0.$$

Without loss of generality, we assume $P_0(\lambda) = 1, \ \lambda \in \mathbb{R}$. Now the last two formulas give

$$P_n(\lambda) = \lambda^n, \ n \in \mathbb{N}_0. \tag{3.5.48}$$

Therefore, the Fourier transform (3.5.46) has the form

$$H_S \supset l_{\text{fin}} \ni f \to (Ff)(\lambda) = \hat{f}(\lambda) = \sum_{n=0}^{\infty} f_n \lambda^n \in L_2(\mathbb{R}, d\rho(\lambda)), \tag{3.5.49}$$

and the Parseval equality:

$$(f, g)_S = \int_{\mathbb{C}} \hat{f}(\lambda)\overline{\hat{g}(\lambda)}\, d\rho(\lambda), \ f, g \in l_{\text{fin}}. \tag{3.5.50}$$

To construct the Fourier transform (3.5.46) and check formulas (3.5.47)–(3.5.50) it is necessary to verify the expression of a strong cyclic vector for the operator A $\Omega = \delta_0 \in l_{\text{fin}}$ in the sense of the rigging (3.5.21). The last is really so since from (3.5.41), we have

$$A^p \Omega = J^p \delta_0 = \delta_p.$$

The Parseval equality (3.5.50) leads directly to the representation (3.5.24): according to (3.5.48), (3.5.49) $\hat{\delta}_n = \lambda^n$ and $\hat{\delta}_0 = 1$; from (3.5.36), it follows

$$s_n = (\delta_n, \delta_0)_S = (\hat{\delta}_n, \hat{\delta}_0)_{L_2(\mathbb{R}, d\rho(\lambda))} = \int_{\mathbb{R}} \lambda^n \, d\rho(\lambda), \quad n \in \mathbb{N}_0. \tag{3.5.51}$$

Therefore, the first part of the theorem in the case of non-degeneracy of the kernel $K = S$ is proved. To prove its second part, it is necessary to make sure that the condition (3.5.26) leads to the self-adjointness of the operator A in the space H_S. To do this, consider a Hermitian operator defined on a linear set invariant under its action

$$\mathcal{D} = l_{\text{fin}} = \operatorname{span}\{\delta_n \mid n \in \mathbb{N}_0\}$$

by the expression $A\delta_n = \delta_{n+1}$, and for $p \geq 1$, we have $A^p \delta_n = \delta_{n+p}$.

According to (3.5.36), the space H_S has the norm $\|f\|_S = \sqrt{(f, f)_S}$. Therefore, for each $\delta_n \in \mathcal{D}$, we have $\|A^p \delta_n\|_S^2 = \|\delta_{n+p}\|_S^2 = s_{2n+2p}$. Since

$$\sum_{p=1}^{\infty} \frac{1}{\sqrt[p]{\|A^p \delta_n\|}} = \sum_{p=1}^{\infty} \frac{1}{\sqrt[2p]{s_{2n+2p}}},$$

then it can be asserted that the quasi-analyticity of the class $\mathrm{C}\{\|A^p \delta_n\|\}$ is equivalent to the quasi-analyticity of the class $\mathrm{C}\{\sqrt{s_{2n+2p}}\}$ and according to the properties of quasi-analytic classes, this class is equivalent to $\mathrm{C}\{\sqrt{s_{2p}}\}$. Here we used Theorem 2.6.8, Lemma 2.6.7 and the formula (2.6.21).

Therefore, the Theorem 3.5.6 can be considered proved in the previous case.

Let us consider the case of the degenerate kernel $S = K$, that is, when the left expression in the inequality (3.5.25) is equal to zero for some non-zero sequence $f \in l_{\text{fin}}$. Consider the (linear) set of all such sequences. Since the inequality (3.5.29) holds true, it will be closed in the space $l_2(q)$ constructed according to (3.5.32), i.e., in the space G_+ of the chain (3.5.17).

Thus, we arrive at the general situation of the quasi-scalar product considered in Theorem 3.5.2. The expression $(f, g)_S$, i.e., (3.5.36) will now be a quasi-scalar product, but the operator J by equality (3.5.42) will be Hermitian in the quasi-scalar product (3.5.4). According to Lemma 3.5.4, this operator transforms a class into a class, and the proof of the theorem must be repeated starting from the definition (3.5.43). □

Remark 3.5.7 Let the sufficient conditions of the Theorem 3.5.6 be fulfilled and we have a representation of the moment sequence (3.5.24) with some measure $d\rho(\lambda)$ on the axis $\mathbb{R}$. Suppose that the support of this measure is larger than the set consisting of a finite number of points on the axis. Then in the inequality (3.5.25) there should be a sign ">", that is, the corresponding product (3.5.36) will not be quasi-scalar, but scalar.

Indeed, for the support α of the measure $d\rho(\lambda)$, we have a relation

$$\int_{\alpha} |F_n(\lambda)|^2 \, d\rho(\lambda) = \int_{\mathbb{R}} |F_n(\lambda)|^2 \, d\rho(\lambda) = \int_{\alpha} \left| \sum_{j=0}^{n} f_j \lambda^j \right|^2 d\rho(\lambda)$$
$$= \sum_{j,k=0}^{n} s_{j+k} f_j \bar{f}_k \geq 0, \text{ where } F_n(\lambda) = \sum_{j=0}^{n} f_j \lambda^j \tag{3.5.52}$$

following from (3.5.24) and (3.5.25).

Suppose that the product $(\,\cdot\,,\,\cdot\,)_S$ is quasi-scalar, then there exists $n \in \mathbb{N}_0$ such that $f_0, \ldots, f_n$ not all equal zero, so that the sign "$\geq$" will be changed on "$=$" in (3.5.52) and for a corresponding polynomial $F_n(\lambda)$, we obtain $F_n(\lambda) = 0$, $\lambda \in \alpha$ due to (3.5.52). But the polynomial $F_n(\lambda) = 0$ has at most n different zeros, and α has more than n points. Thus, we obtain a contradiction. □

One Generalization of the Moment Problem Let us consider a generalization of Theorem 3.5.6, which also uses the spectral theory of the operator but which is somewhat more complicated than the operator generated by the matrix (3.5.40). Namely, instead of (3.5.40), let us considered a Jacobi matrix with real elements $a_n \neq 0, b_n, c_n$

$$J = \begin{bmatrix} b_0 & c_0 & 0 & 0 & \cdots \\ a_0 & b_1 & c_1 & 0 & \cdots \\ 0 & a_1 & b_2 & c_2 & \cdots \\ 0 & 0 & a_2 & b_3 & \cdots \\ \cdots & \cdots & \cdots & \cdots & \ddots \end{bmatrix} \tag{3.5.53}$$

$$\begin{aligned} (Jf)_n &= a_{n-1} f_{n-1} + b_n f_n + c_n f_{n+1}, \quad (Jf)_0 = 0; \\ J\delta_n &= a_n \delta_{n+1} + b_n \delta_n + c_{n-1} \delta_{n-1}, \quad n \in \mathbb{N}, \\ J\delta_0 &= a_0 \delta_1 + b_0 \delta_0. \end{aligned}$$

Let $(s) = (s_{j,k})_{j,k=0}^{\infty}$, $s_{j,k} \in \mathbb{R}$ be a fixed non-negative definite matrix. Denote by S the Hilbert space, constructed by the scalar product

$$l_{\text{fin}} \ni f, g \longrightarrow (f, g)_S = \sum_{j,k=0}^{\infty} s_{j,k} f_k \bar{g}_j \in \mathbb{C}. \tag{3.5.54}$$

We assume that the matrix (3.5.53) is Hermitian with respect to the scalar product (3.5.54), i.e.,

$$(Jf, g) = (f, Jg)_S, \quad f, g \in l_{\text{fin}}. \tag{3.5.55}$$

As in Sect. 3.1, we introduce the operator $\dot{J}$ and its self-adjoint extension A in the space S. The generalized eigenvectors expansion of this operator A can be performed using the construction of Sect. 3.1. Namely, we introduce the chain (3.5.38), but in its construction, we use a sequence $q = (q_n)_{n=0}^{\infty}$ for which $\sum\limits_{n=0}^{\infty} s_{n,n} q_n^{-1} < \infty$ (instead of the condition $\sum\limits_{n=0}^{\infty} s_{2n} q_n^{-1} < \infty$), the chain (3.5.39) is as before. In this case, the vector $\Omega = \delta_0 \in l_{\text{fin}}$ is also a strong cyclic vector *because* $\forall n \in \mathbb{N}_0$ *the linear space spanned by vectors* $\delta_0, J\delta_0, \ldots, J^n\delta_0$ *coincides with the subspace* $\{f \in l_{\text{fin}} \mid f_{n+1} = f_{n+2} = \ldots = 0\}$.

To prove the statement, it is sufficient to check that for any vector

$$f = (f_0, f_1, \ldots, f_n, 0, 0, \ldots), \quad f_k \neq 0, \ k = 0, 1, \ldots, n$$

there exist numbers $x_0, x_1, \ldots, x_n \in \mathbb{C}$ such that

$$x_0\delta_0 + x_1 J\delta_0 + \cdots + x_n J^n \delta_0 = f. \tag{3.5.56}$$

From the last two formulas (3.5.53), we get that the equality (3.5.56), in the coordinate form, is a system of $n+1$ linear equations with the desired $x_0, x_1, \ldots, x_n$ with a triangular matrix having the values $1, a_0, a_0a_1, \ldots, a_0a_1 \ldots a_{n-1}$ on the main diagonal. Since all $a_n \neq 0$, the system (3.5.56) has a solution. □

Therefore, we can claim that (3.5.45) and (3.5.46) hold true. Now $d\rho(\lambda)$ is the spectral measure of the operator A and

$$P(\lambda) = (P_n(\lambda))_{n=0}^{\infty} \in l_2(p^{-1}) \subset l$$

is the corresponding generalized eigenvector wit an eigenvalue λ. The calculation as in (3.5.47) for matrix (3.5.53) gives

$$\begin{aligned} c_{n-1}P_{n-1}(\lambda) + b_n P_n(\lambda) + a_n P_{n+1}(\lambda) &= \lambda P_n(\lambda), \ n \in \mathbb{N} \\ b_0 P_0(\lambda) + a_0 P_1(\lambda) &= \lambda P_0(\lambda), \ \lambda \in \mathbb{R}. \end{aligned} \tag{3.5.57}$$

Setting $P_0(\lambda) = 1$, we conclude that $P_n(\lambda)$ *is a real polynomial in* λ *of degree* $n \in \mathbb{N}$. The Fourier transform and the Parseval equality now have a form similar to (3.5.49) and (3.5.50),

$$\begin{aligned} S \supset l_{\text{fin}} \ni f \to (Ff)(\lambda) = \hat{f}(\lambda) &= \sum_{n=0}^{\infty} f_n P_n(\lambda) \in L_2(\mathbb{R}, d\rho(\lambda)), \\ (f, g)_S &= \int_{\mathbb{R}} \hat{f}(\lambda)\overline{\hat{g}(\lambda)}\, d\rho(\lambda), \ f, g \in l_{\text{fin}}. \end{aligned} \tag{3.5.58}$$

Since $\hat{\delta}_n(\lambda) = P_n(\lambda)$, $n \in \mathbb{N}_0$, then from (3.5.58), we get a representation of the form (3.5.51):

$$s_{j,k} = (\delta_k, \delta_j)_S = (\hat{\delta}_k, \hat{\delta}_j)_{L_2(\mathbb{R}, d\rho(\lambda))} = \int_{\mathbb{R}} P_j(\lambda) P_k(\lambda)\, d\rho(\lambda), \ \ j, k \in \mathbb{N}_0. \tag{3.5.59}$$

We will look for the form of the matrix (s), which is non-negatively defined and satisfies the Eq. (3.5.55). Using the adjoint J^+, we rewrite the equality in the form

$$(Jf, g)_S = \sum_{j,k=0}^{\infty} s_{j,k} (Jf)_k \bar{g}_j = \sum_{j=0}^{\infty} \left(\sum_{j=0}^{\infty} (J^+ s_{j,\cdot})_k f_k \right) \bar{g}_j$$

$$= \sum_{j,k=0}^{\infty} (c_{k-1} s_{j,k-1} + b_k s_{j,k} + a_k s_{j,k+1}) f_k \bar{g}_j,$$

$$(f, Jg)_S = \sum_{j,k=0}^{\infty} s_{j,k} f_k \overline{(Jg)_j} = \sum_{j=0}^{\infty} f_k \left(\sum_{j=0}^{\infty} (J^+ s_{\cdot,j})_j \bar{g}_j \right)$$

$$= \sum_{j,k=0}^{\infty} (c_{j-1} s_{j-1,k} + b_j s_{j,k} + a_j s_{j+1,k}) f_k \bar{g}_j, \quad \forall f, g \in l_{\text{fin}}.$$

Using the equalities (3.5.55) and the arbitrariness of f and g, we obtain

$$\begin{aligned} c_{j-1} s_{j-1,k} + b_j s_{j,k} + a_j s_{j+1,k} &= c_{k-1} s_{j,k-1} + b_k s_{j,k} + a_k s_{j,k+1} \\ s_{-1,k} = s_{j,-1} &= 0, \quad j, k \in \mathbb{N}_0. \end{aligned} \tag{3.5.60}$$

The equality (3.5.60) is a partial difference equation with two-indexed unknowns $s_{j,k}$ in which every five points $(j-1, k)$, (j, k), $(j+1, k)$, $(j, k-1)$, $(j, k+1)$, are connected and the corresponding "exterior" coefficients a_j, a_k are non-zero.

Therefore, proceeding step by step with increasing k, we fine a solution $s_{j,k}$ for $k \leq j$ using the initial data $s_{j,0}$ and the "boundary" conditions $s_{j,-1} = 0$, $j \in \mathbb{N}_0$. Similarly, $s_{j,k}$, for $j \leq k$, are found, using initial data $s_{0,k}$ and $s_{-1,k} = 0$, $j \in \mathbb{N}_0$. Taking into account the symmetry $(s) = (s_{j,k})_{j,k=0}^{\infty}$ and coefficients of the Eq. (3.5.60) (with respect to the replacement $(j, k) \to (k, j)$), we conclude that the above procedure gives a symmetric solution $s_{j,k}$ of (3.5.60) satisfying the prescribed initial data $s_{j,0} = s_{0,j}$ and the boundary conditions $s_{j,-1} = s_{-1,j} = 0$, $j \in \mathbb{N}_0$.

Thus, we have obtained the generalization of the first part of Theorem 3.5.6.

Theorem 3.5.8 *Let J be a Jacobi matrix of the form* (3.5.53) *and let* $(s) = (s_{j,k})_{j,k=0}^{\infty}$ *be non-negative defined matrix whose elements* $s_{j,k} \in \mathbb{R}$ *are a solution of the partial differential Eq.* (3.5.60) *with the indicated zero boundary conditions. Then, for* $s_{j,k}$ *the representation*

$$s_{j,s} = \int_{\mathbb{R}} P_j(\lambda) P_k(\lambda)\, d\rho(\lambda), \quad j, k \in \mathbb{N}_0, \tag{3.5.61}$$

holds true, where $P_j(\lambda)$ *are polynomials of degree* j *and determined as a solution of recurrence relations* (3.5.57) *and* $d\rho(\lambda)$ *is finite Borel measure on the real axis* $\mathbb{R}$.

Conversely, every sequence (3.5.61) *is such a solution of the Eq.* (3.5.60).

It is only necessary to explain the last assertion. Substituting (3.5.61) into (3.5.60) and using (3.5.57), we see that the left and right-hand sides of the equalities (3.5.60) are equal to

$$s_{j,s} = \int_{\mathbb{R}} \lambda P_j(\lambda) P_k(\lambda)\, d\rho(\lambda),$$

i.e., (3.5.60) holds true.

Substituting (3.5.61) into (3.5.54) with $g = f$, we conclude $\forall f \in l_{\text{fin}}$

$$\sum_{j,k=0}^{\infty} s_{j,k} f_k \bar{f}_j = \int_{\mathbb{R}} \left| \sum_{j=0}^{\infty} f_j P_j(\lambda) \right|^2 d\rho(\lambda), \geq 0$$

i.e., the non-negative definiteness of the matrix (s) is also performed. □

3.6 Some Other Generalizations of the Moment Problem

An Algebraic Approach to the Classical Moment Problem The main problem here is now to define the convolution "$*$", conveniently related to the Eq. (3.5.60). For this, auxiliary constructions are necessary. Let us introduce the convolution "$*$" on the linear space l_{fin}, namely,

$$l_{\text{fin}} \ni f, g \longrightarrow (f * g)_n = \sum_{j+k=n} f_j g_k \in l_{\text{fin}}. \tag{3.6.1}$$

As a result, the space l_{fin} becomes a commutative algebra $\mathcal{A}$ with unit δ_0 and involution

$$l_{\text{fin}} \ni f = (f_n)_{n=0}^{\infty} \longrightarrow \bar{f} := (\bar{f}_n)_{n=0}^{\infty} \in l_{\text{fin}},$$

where $\bar{f}_n$ denotes the complex conjugation.

Every sequence $s = (s_n)_{n=0}^{\infty}$, $s_n \in \mathbb{R}$ can be interpreted as a linear functional σ on l_{fin}, i.e.,

$$l_{\text{fin}} \ni f \longrightarrow \sigma(f) := \sum_{n=0}^{\infty} s_n f_n \in \mathbb{C}. \tag{3.6.2}$$

If the sequence s is moment one, then the functional is positive (more exactly, non-negative), i.e., $\sigma(f * \bar{f}) \geq 0$, $f \in l_{\text{fin}}$. This fact follows from the equality

$$\sigma(f * \bar{g}) = \sum_{n=0}^{\infty} s_n \left(\sum_{j+k=n} f_j \bar{g}_k \right) = \sum_{j,k=0}^{\infty} f_j \bar{g}_k, \quad f, g \in l_{\text{fin}} \tag{3.6.3}$$

and the assumption (3.5.25).

Expression 3.5.36) and (3.6.3) imply that

$$(f, g)_S = \sigma(f * \bar{g}), \quad f, g \in l_{\text{fin}} = \mathcal{A}, \tag{3.6.4}$$

i.e., quasi-scalar product on the algebra $\mathcal{A}$ is given in a standard way using a positive functional and the involution (the so-called (GNS)-construction: Gelfand I. M., Naimark M .A., Segal I.).

The operator J defined in (3.5.40) has, in this algebraic approach, the form

$$S \supset \mathcal{A} = l_{\text{fin}} \ni f \longrightarrow Jf = \delta_1 * f \in l_{\text{fin}} = \mathcal{A}. \tag{3.6.5}$$

Thus, we can say that *the results of Sect. 3.1 have been obtained by applying the theory of the generalized eigenvectors expansion to the operator* (3.6.5) (*or to any its self-adjoint extension*) *acting in the Hilbert space obtained vie the* (*GNS*)-*construction* (3.6.4).

Let us combine this algebraic approach with the generalization of the moment problem described above. Let the matrix (3.5.53) be given. Then from the Eq. (3.5.57) it is easy to concluded that $P_n(\lambda)$ is degree n polynomial with real coefficients. For every fixed $n \in \mathbb{N}_0$, the transformation of the functions 1, λ, $\lambda^2, \ldots, \lambda^n$ into $P_0(\lambda) = 1$, $P_1(\lambda)$, $P_2(\lambda), \ldots, P_n(\lambda)$, $(\lambda \in \mathbb{R})$ is made by a real lower triangular matrix C having values $1, a_0, a_0a_1, \ldots, a_0a_1 \ldots a_{n-1}$ on the main diagonal,

$$P_j(\lambda) = \sum_{k=0}^{n} C_{j,k}\lambda^k, \quad j = 0, 1, 2, \ldots, n; \quad C = (C_{j,k})_{j,k=0}^{n}, \tag{3.6.6}$$

where $C_{j,k} = 0$ for $k > j$. This matrix is invertible and for the inverse matrix $D = (D_{j,k})_{j,k=0}^{n} = C^{-1}$, the equality (3.6.6) holds true with λ^j, $P_k(\lambda)$ replacing $P_j(\lambda)$, λ^k, respectively.

The relations (3.6.6) for matrices D and C shows that every real polynomial $\sum\limits_{k=0}^{n} a_k \lambda^k = F(\lambda)$ can be represented in the form

$$F(\lambda) = \sum_{j=0}^{n} b_j P_j(\lambda), \quad \lambda \in \mathbb{R} \tag{3.6.7}$$

and the coefficients b_j in (3.6.7) depend uniquely and linearly on F (namely $C^+ b = a$, $b = D^+ a$, $b = (b_0, b_1, \ldots, b_n)$, $a = (a_0, a_1, \ldots, a_n)$; "+" denotes a conjugation).

In the lineal real space $\mathcal{P}$ of all real-valued polynomials $F(\lambda), G(\lambda), \ldots, \lambda \in \mathbb{R}$, we introduce a scalar product by setting

$$(F, G)_{\mathcal{P}} = \sum_{j=0}^{\infty} f_j g_j \text{ where } F(\lambda) = \sum_{j=0}^{\infty} f_j P_j(\lambda), \;\; G(\lambda)) = \sum_{j=0}^{\infty} g_j P_j(\lambda). \tag{3.6.8}$$

The expression (3.6.8) determines a real scalar product, because a finite sequence of coordinates $(f_j)_{j=0}^{\infty}$ and polynomials $F(\lambda)$ are in one-to-one correspondence. For each $F \in \mathcal{P}$, we have

$$f_j = (F, P_j)_{\mathcal{P}}, \;\; j \in \mathbb{N}_0; \quad F(\lambda) = \sum_{j=0}^{\infty} (F, P_j)_{\mathcal{P}} P_j(\lambda), \;\; \lambda \in \mathbb{R}. \tag{3.6.9}$$

(Note, that $f_j = 0$ for j greater than the degree of F). The sequence $(P_j(\lambda))_{j=0}^{\infty}$ forms an orthogonal basis in $\mathcal{P}$.

Let us fix $l \in \mathbb{N}_0$ and denote by $q_{j,k}(l)$ a solution of the partial difference Eq. (3.5.60) with the initial data $q_{j,0}(l) = q_{0,j}(l) = \delta_{j,l}$.

Proposition 3.6.1 *The representation*

$$q_{j,k}(l) = (P_j(\lambda) P_k(\lambda), P_l(\lambda))_{\mathcal{P}}, \;\; j, k, l \in \mathbb{N}_0; \;\; q_{j,k}(l) = 0 \, for \, l > j + k, \tag{3.6.10}$$

hold true.

Indeed, let us set $s_{j,k} = q_{j,k}(l)$ in the left-hand side (3.5.60). Using the formula (3.5.57), we get $\forall j, k \in \mathbb{N}_0$,

$$c_{j-1} q_{j-1,k}(l) + b_j q_{j,k}(l) + a_j q_{j+1,k}(l) = (\lambda P_j(\lambda) P_k(\lambda), P_l(\lambda))_{\mathcal{P}}. \tag{3.6.11}$$

Similarly, for the right-hand side of (3.5.57), we get

$$c_{k-1}q_{j,k-1}(l) + b_k q_{j,k}(l) + a_k q_{j,k+1}(l) = (P_j(\lambda)\lambda P_k(\lambda), P_l(\lambda))_{\mathcal{P}}.$$

This equality together with (3.6.11) imply that $q_{j,k}$ is a solution of the Eq. (3.5.60).

The last equality in 3.6.10) holds true, because the degree of the polynomial $P_j(\lambda)P_k(\lambda)$ is equal to $j+k$.

Using the orthogonality relation for $P_j(\lambda)$ we conclude that

$$q_{j,0}(l) = (P_j(\lambda)1, P_j(\lambda))_{\mathcal{P}} = \delta_{j,l}, \quad j, l \in \mathbb{N}_0.$$

□

We can now introduce a convolution in the linear space l_{fin} by setting

$$\begin{aligned} l_{\text{fin}} \ni f, g \longrightarrow (f * g)_n &:= \sum_{j,k=0}^{\infty} f_j g_k q_{j,k}(n) \\ &= \sum_{j,k=0}^{\infty} f_j g_k (\lambda P_j(\lambda) P_k(\lambda), P_n(\lambda))_{\mathcal{P}}, \quad \forall n \in \mathbb{N}_0. \end{aligned} \tag{3.6.12}$$

The right-hand side in (3.6.12) is a finite sequence, according to the last equality in (3.6.10).

Proposition 3.6.2 *The space l_{fin} with the convolution (3.6.12) is a commutative algebra $\mathcal{A}$ with the unit δ_0 and involution*

$$\mathcal{A} \ni f = (f_n)_{n=0}^{\infty} \mapsto \bar{f} := (\bar{f}_n)_{n=0}^{\infty} \in \mathcal{A}.$$

This assertion immediately follows from the following observation. We introduce for $f \in \mathcal{A}$ its Fourier transform

$$l_{\text{fin}} = \mathcal{A} \ni f \longrightarrow \hat{f}(\lambda) = \sum_{j=0}^{\infty} f_j P_j(\lambda) \in \mathcal{P}_c, \tag{3.6.13}$$

where $\mathcal{P}_c$ is the complexification of $\mathcal{P}$ (i.e., the space of all polynomials with complex coefficients).

The mapping (3.6.13) is an isomorphism between the spaces $\mathcal{A}$ and $\mathcal{P}_c$. Under this isomorphism *the convolution "$*$" goes over to the ordinary multiplication of polynomials*. Indeed, using (3.6.13) and (3.6.9) we have

$$\begin{aligned}(f * g)\hat{}(\lambda) &= \sum_{j=0}^{\infty}(f * g)_n P_n(\lambda)\\ &= \sum_{j=0}^{\infty}\left(\sum_{j,k=0}^{\infty} f_j g_k (P_j(\lambda)P_k(\lambda),\, P_n(\lambda))_{\mathcal{P}}\right) P_n(\lambda)\\ &= \sum_{j=0}^{\infty}(\hat{f}(\lambda)\hat{g}(\lambda),\, P_n(\lambda))_{\mathcal{P}} P_n(\lambda) =\\ &\hat{f}(\lambda)\hat{g}(\lambda), \quad \lambda \in \mathbb{R},\ \forall f, g \in \mathcal{A}.\end{aligned} \tag{3.6.14}$$

Proof of Proposition 3.6.2. The space $\mathcal{P}_c$ is a commutative algebra with a standard multiplication, unit, and an involution. $\mathcal{A}$ is isomorphic to $\mathcal{P}_c$. Therefore, $\mathcal{A}$ has the corresponding properties.

Thus, the introduced convolution is connected with Jacobi matrix (3.5.53). Now we have a possibility to reformulate the results of Theorem 3.5.8 in an algebraic form, similar to the classical moment problem. □

Let the Jacobi matrix (3.5.53) be given. we construct the corresponding convolution "$*$" (3.6.12) and an algebra $\mathcal{A} = l_{\text{fin}}$. Each sequence $s = (s_n)_{n=0}^{\infty}$, $s_n \in \mathbb{R}$ according to (3.6.2), can be interpreted as a functional σ on $\mathcal{A}$.

Theorem 3.6.3 *Let $s = (s_n)_{n=0}^{\infty}$, $s_n \in \mathbb{R}$ generate a positive functional σ on the algebra $\mathcal{A}$, i.e.,*

$$\sigma(f * \bar{f}) = \sum_{j=0}^{\infty} s_n (f * \bar{f})_n = \sum_{j=0}^{\infty} s_n \left(\sum_{j,k=0}^{\infty} f_j \bar{f}_k q_{j,k}(n)\right) \geq 0, \quad f \in \mathcal{A}. \tag{3.6.15}$$

Then the representation

$$s_n = \int_{\mathbb{R}} P_n(\lambda)\, d\rho(\lambda), \quad n \in \mathbb{N}_0, \tag{3.6.16}$$

holds true, where $P_n(\lambda)$ is a solution of the Eq. (3.5.57) *and $d\rho(\lambda)$ is a Borel finite measure on $\mathbb{R}$.*

Conversely, each sequence (3.6.16) *generates a positive functional σ, i.e., the inequality* (3.6.15) *holds true.*

Proof We reduce Theorem 3.6.3 to Theorem 3.5.8. On the other hand, this reducing explains connection between constructions introduced above.

Let a sequence s satisfying condition (3.6.14) be given. Put

$$s_{j,k} = \sum_{j=0}^{\infty} s_n q_{j,k}(n), \quad j,k \in \mathbb{N}_0 \tag{3.6.17}$$

($\forall j,k$ the sum (3.6.17) is finite, according to Proposition 3.6.1). The sequence (3.6.17) is a solution of the Eq. (3.5.60) with the initial data $s_{j,0} = s_{0,j} = 0$, $j \in \mathbb{N}_0$. The matrix $s = (s_{j,k})_{j,k=0}^{\infty}$ is non-negative definite, that is for l_{fin} we have, according to (3.6.15),

$$\sum_{j=0}^{\infty} s_{j,k} f_j \bar{f}_k = \sum_{j,k=0}^{\infty} \left(\sum_{n=0}^{\infty} s_n q_{j,k}(n) \right) f_j \bar{f}_k = \sigma(f * \bar{f}) \geq 0. \tag{3.6.18}$$

Therefore, we apply Theorem 3.5.8 to the sequence (s). Then the representation (3.5.61), for $k = 0$, gives (3.6.16).

To prove the inverse assertion, we first note that

$$\sum_{j=0}^{\infty} q_{j,k}(n) P_n(\lambda) = P_j(\lambda) P_k(\lambda), \quad \lambda \in \mathbb{R}, \ \forall j,k \in \mathbb{N}_0. \tag{3.6.19}$$

Then for $f \in l_{\text{fin}}$ and s_n of the form (3.6.16), we have

$$\begin{aligned} \sum_{n=0}^{\infty} s_n \left(\sum_{j,k=0}^{\infty} f_j \bar{f}_k q_{j,k}(n) \right) &= \int_{\mathbb{R}} \sum_{j,k=0}^{\infty} f_j \bar{f}_k \left(\sum_{n=0}^{\infty} P_n(\lambda) q_{j,k}(n) \right) d\rho(\lambda) \\ &= \int_{\mathbb{R}} \left| \sum_{j=0}^{\infty} f_j P_j(\lambda) \right|^2 d\rho(\lambda) \geq 0. \end{aligned}$$

□

Remark 3.6.4 We note that teh representation (3.6.16) leads to the representation (3.5.61), since a solution $s_{j,k}$ of (3.5.60) with the initial data $s_{j,0} = s_{0,j} = s_n$ has a form (3.6.17). By using the representation (3.6.16) and the equality (3.6.19), we conclude that (3.5.61).

Remark 3.6.5 The scalar product (3.5.54), defined the space S, has $\forall f, g \in \mathcal{A}$ the form $(f, g)_S = \sigma(f * \bar{g})$, i.e., this space is constructed via the GNS-construction (see (3.6.18)).

In the case of the classical moment problem, the action of the matrix J on $f \in l_{\mathrm{fin}}$ is equal $\delta_1 * f$ (see (3.6.5)). Let us consider the corresponding situation for a general matrix (3.5.53).

Remark 3.6.6 In terms of the Fourier transform (3.6.13), the action of the matrix J is equal to the multiplication by λ

$$(Jf)\widehat{\ }(\lambda) = \lambda \hat{f}(\lambda), \quad f \in l_{\mathrm{fin}}, \quad \lambda \in \mathbb{R}. \tag{3.6.20}$$

Also, we have

$$\begin{aligned} a_n =&(\lambda P_n(\lambda), P_{n+1}(\lambda))_{\mathcal{P}}, \\ b_n =&(\lambda P_n(\lambda), P_n(\lambda))_{\mathcal{P}}, \\ c_n =&(\lambda P_{n+1}(\lambda), P_n(\lambda))_{\mathcal{P}}, \quad n \in \mathbb{N}_0. \end{aligned} \tag{3.6.21}$$

Thus, for an arbitrary Jacobi matrix (3.5.53), it is possible to write formulas as in the classical self-adjoint case, but using the scalar product (3.6.8) instead of integration with respect to the scalar measure. The proof of (3.6.20) and (3.6.21) is simple. Indeed, it is sufficient to check (3.6.20) for $f = \delta_n$, $n \in \mathbb{N}_0$. Using the last formula in (3.5.53), equality $\hat{\delta}_j(\lambda) = P_j(\lambda)$ and (3.5.57), we get

$$\begin{aligned} (J\delta_n)\widehat{\ }(\lambda) &= (a_n\delta_{n+1} + b_n\delta_n + c_{n-1}\delta_{n-1})\widehat{\ }(\lambda) \\ &= a_n P_{n+1}(\lambda) + b_n P_n(\lambda) + c_{n-1}P_{n-1}(\lambda) = \lambda P_n(\lambda) = \lambda\hat{\delta}_n(\lambda), \ \lambda \in \mathbb{R}. \end{aligned}$$

To prove (3.6.21), it is necessary to multiply (in the scalar sense, in the space $\mathcal{P}$) the equality (3.5.57) by $P_j(\lambda)$. Then (3.6.21) follows from the orthogonality of the system $(P_j(\lambda))_{j=0}^{\infty}$ in $\mathcal{P}$.

Let us return to $\delta_1 * f$.

Remark 3.6.7 The equality

$$\delta_1 * f = (a_0^{-1}J - a_0^{-1}b_0\mathbb{1})f, \quad f \in l_{\mathrm{fin}} \tag{3.6.22}$$

holds true.

It is possible to derive this formula directly using (3.6.12) and calculating $q_{1,k}(n)$ from (3.5.60), but it is more convenient to use the Fourier transform (3.6.13). Takin into account (3.6.14), the equality $\hat{\delta}_1(\lambda) = P_1(\lambda) = a_0^{-1}(\lambda - b_0)$, $\lambda \in \mathbb{R}$ (see (3.5.57)) and (3.6.20), we find that $\forall f \in l_{\mathrm{fin}}$

$$\begin{aligned} (\delta_1 * f)\widehat{\ }(\lambda) &= \hat{\delta}(\lambda)\hat{f}(\lambda) = a_0^{-1}(\lambda - b_0)\hat{f}(\lambda) \\ &= ((a_0^{-1}J - a_0^{-1}b_0\mathbb{1})f)\widehat{\ }(\lambda), \quad \lambda \in \mathbb{R}. \end{aligned}$$

This equality gives (3.6.22).

Thus the equality (3.6.22) shows that the operation $f \mapsto \delta_1 * f$ coincides with the action of the matrix J. Therefore, the proof of Theorem 3.5.8 could be based on the spectral theory of the operator $f \mapsto \delta_1 * f$ instead of J.

The Self-Adjointness of Operators and the Uniqueness of Representations Pass now dwell upon a generalization of the second part of Theorem 3.5.6 to the case when J has the form (3.5.53), that is, we will consider the problem of the uniqueness of the measure $d\rho(\lambda)$ in the representations (3.5.61) and (3.6.16). As we have already recalled, *such uniqueness is equivalent to the essential self-adjointness of the operator* $l_{\mathrm{fin}} \ni f \mapsto Jf \in l_{\mathrm{fin}}$*(or* $l_{\mathrm{fin}} \ni f \mapsto \delta_1 * f \in l_{\mathrm{fin}}$ *in the space* S)

We will start with some simple modifications of the evolution and quasi-analytic criteria of the self-adjointness and the commutativity considered in Chap. 1, Sect. 2.6.

Consider an equipment of the type (3.5.21) of a complex Hilbert space $\mathcal{H}$, namely,

$$\mathcal{H}_- \supset \mathcal{H} \supset \mathcal{H}_+. \tag{3.6.23}$$

Let B be an operator in the space $\mathcal{H}_+$ with a dense domain $\mathrm{Dom}(B)$ and let B^* be its adjoint operator (*in the space* $\mathcal{H}_+$). On the interval $[0, b)$, $b \le \infty$, we consider the differential equation

$$\left(\frac{dv}{dt}\right)(t) - B^* v(t) = 0, \quad t \in [0, b). \tag{3.6.24}$$

We say that a vector-valued function $[0, b) \ni t \mapsto v(t) \in \mathcal{H}_+$ is a strong solution of (3.6.24) if it is strongly continuously differentiable, $v(t) \in \mathrm{Dom}(B^*)$ and satisfies the equality (3.6.24). It is easy to understand that, instead of (3.6.24), we can demand that the "weak," equality

$$\left(\left(\frac{dv}{dt}\right)(t), f\right)_{\mathcal{H}_+} - (v(t), Bf)_{\mathcal{H}_+} = 0, \quad f \in \mathrm{Dom}(B),\ t \in [0, b) \tag{3.6.25}$$

holds true.

Let us consider B as an operator A acting in the space $\mathcal{H}$, that is, let $\mathrm{Dom}(A) = \mathrm{Dom}(B)$ and $Af = Bf \in \mathcal{H}_+ \subset \mathcal{H}$, $f \in \mathrm{Dom}(A)$. *Suppose, A is Hermitian in the space $\mathcal{H}$.*

Theorem 3.6.8 *For the essential self-adjointness of the operator A, it is sufficient that for some $b > 0$, both equations*

$$\left(\frac{dv}{dt}\right)(t) \pm (iB)^* v(t) = 0, \quad t \in [0, b) \tag{3.6.26}$$

have a unique strong solution of the Cauchy problem.

Proof Note first that the domain $\mathrm{Dom}(A)$ is dense in $\mathcal{H}$. Applying Theorem 2.6.1, we assert that to prove the theorem, it is sufficient to check the uniqueness of strong solutions of the Cauchy problem for both equations in the space $\mathcal{H}$,

$$\left(\frac{dv}{dt}\right)(t) \pm (iA)^* v(t) = 0, \quad t \in [0, b). \tag{3.6.27}$$

Let, for example, $[o, b) \ni t \mapsto u(t) \in \mathcal{H}$ be a strong solution of the Eq. (3.6.27) with the sign "−" and $u(0) = 0$. It is necessary to show that $u(t) = 0, t \in [o, b)$.

To this end, we introduce a continuous operator $I : \mathcal{H} \longrightarrow \mathcal{H}_+$ by the equality

$$(If, \omega)_{\mathcal{H}_+} = (f, \omega)_{\mathcal{H}}, \quad f \in \mathcal{H}, \ \omega \in \mathcal{H}_+,$$

where I denotes a canonical isometric isomorphism—one of the standard operators of the theory of rigging (see (2.7.13)). The function $[0, b) \ni t \mapsto v(t) = Iu(t) \in \mathcal{H}_+$ is strongly continuously differentiable and is a strong solution of Eq. (3.6.26) with the sign "−". Indeed, using the equality of the type (3.6.25) for $B = iA$ and the definition of I, we conclude $\forall f \in \mathrm{Dom}(A) = \mathrm{Dom}(B)$,

$$\begin{aligned} 0 &= \left(\left(\frac{du}{dt}\right)(t), f\right)_{\mathcal{H}} - (u(t), iAf)_{\mathcal{H}} = \left(I\left(\frac{du}{dt}\right)(t), f\right)_{\mathcal{H}_+} - (Iu(t), iAf)_{\mathcal{H}_+} \\ &= \left(\left(\frac{dv}{dt}\right)(t), f\right)_{\mathcal{H}_+} - (v(t), iBf)_{\mathcal{H}_+}, \quad t \in [0, \infty). \end{aligned}$$

From this equality, it follows (for a fixed t) that $v(t) \in \mathrm{Dom}((iB)^*)$ (because $\left(\frac{dv}{dt}\right)(t) \in \mathcal{H}_+$) and, therefore, $v(t)$ is a strong solution of the Eq. (3.6.26) with the sign "−", for which $v(0) = Iu(0) = 0$.

The assumption of Theorem 3.6.8 about the uniqueness of a strong solution implies that $v(t) = 0, t \in [0, b)$. But the kernel of the operator I is equal to zero, therefore $u(t) = 0, t \in [0, b)$.

Thanks to the mentioned Theorem 2.6.1, we get that A is essentially self-adjoint.

□

Remark 3.6.9 Similarly, it is also possible to modify other evolution criteria of the essential self-adjointness. Namely, it is easy to formulate results similar to Theorem 3.6.8, as a generalization of Theorems 2.6.1 and 2.6.6.

Let us present for consideration a generalization of the quasi-analytical criterion. As usual, a vector $\psi \in \mathcal{H}_+$ is called quasi-analytic (with respect to the operator $B : \mathcal{H} \longrightarrow \mathcal{H}_+$) if $\psi \in \bigcap_{n=0}^{\infty} \mathrm{Dom}(B^n)$ and the class $C\{\|B^n\psi\|_{\mathcal{H}_+}\}$ is quasi-analytic.

Theorem 3.6.10 *If for B a total set of quasi-analytic vectors in $\mathcal{H}_+$ exists, then the Hermitian operator A in the space $\mathcal{H}$ is essentially self-adjoint.*

Note, that in the case $\mathcal{H}_+ = \mathcal{H}$, we have the well-known sufficient condition of the essential self-adjointness of the operator $A = B$ (in such a case this condition is also necessary).

Proof According to Theorem 3.6.8, it is sufficient to prove the uniqueness of a strong solution of the Cauchy problem for both Eq. (3.6.26). The equality (3.6.26) with the sign "−" means (see (3.6.25)) that

$$(v'(t), f)_{\mathcal{H}_+} = (v(t), \zeta Bf)_{\mathcal{H}_+}, \quad \zeta = +i, \ f \in \operatorname{Dom}(B), t \in [0, b). \qquad (3.6.28)$$

Taking $f = \psi \in \bigcap_{n=0}^{\infty} \operatorname{Dom}(B^n)$ in (3.6.28), we conclude

$$\frac{d}{dt}(v(t), \psi)_{\mathcal{H}_+} = (v'(t), \psi)_{\mathcal{H}_+} = (v(t), (\zeta B)\psi)_{\mathcal{H}_+}.$$

Thus, we have a situation similar to the proof of Theorem 2.6.8. We can repeat the corresponding part of the proof of this theorem and conclude that the condition $v(0) = 0$ implies that $v(t) = 0$, $t \in [0, b)$, i.e., get the uniqueness of the strong solution of the Cauchy problem for (3.6.26). The case of the sign "+" in the Eq. (3.6.26) can be considered similarly.

But it is possible to reduce directly our theorem to Theorem 2.6.8. Namely, let $\psi \in \mathcal{H}_+$ be a quasi-analytic vector for the operator B. Then

$$\psi \in \bigcap_{n=0}^{\infty} \operatorname{Dom}(A^n), \quad B^n\psi = A^n\psi$$

and

$$C\{\|B^n\psi\|_{\mathcal{H}_+}\} \supset C\{\|A^n\psi\|_{\mathcal{H}}\}, \qquad (3.6.29)$$

since $\|f\|_{\mathcal{H}_+} \geq \|f\|_{\mathcal{H}}$, $f \in \mathcal{H}_+$. The inclusion (3.6.29) shows that the class $C\{\|A^n\psi\|_{\mathcal{H}}\}$ is quasi-analytic and therefore, the vector ψ is quasi-analytic for A in $\mathcal{H}$. The space $\mathcal{H}_+$ is dense (as a set) in $\mathcal{H}$, thus, A has a total set of quasi-analytic vectors and, according to mentioned Theorem 2.6.8, it is essentially self-adjoint.

□

Remark 3.6.11 By using Theorem 2.6.8, it can be easily shown that for a semi-bounded operator A from Theorem 2.6.10, for its essential self-adjointness ot os sufficient to demand that the class $C\{\|B^n\psi\|_{\mathcal{H}_+}^{1/2}\}$ must be quasi-analytic for a total set of vectors $\psi \in \bigcap_{n=0}^{\infty} \operatorname{Dom}(A^n)$.

Similarly, a variant of Theorem 2.6.10 is obtained for the semi-bounded operators A_1 and A_2. Here, it is possible to replace classes $C\{\|B_1^n\psi\|_{\mathcal{H}_+}^{1/2}\}$, $C\{\|B_2^n\psi\|_{\mathcal{H}_+}^{1/2}\}$ and

$C\{\|(B_1 \restriction (B_2 - z\mathbb{I}))^n \psi\|_{\mathcal{H}_+}^{1/2}\}$ with corresponding classes where the norm $\|\cdot\|_{\mathcal{H}_+}$ is replaced with $\|\cdot\|_{\mathcal{H}_+}^{1/2}$.

Pass now to give an application of Theorem 3.6.10 to the problem of uniqueness of the measure $d\rho(\lambda)$ in the representations (3.5.61) or (3.6.16).

Theorem 3.6.12 *The measure $d\rho(\lambda)$ in the representation* (3.5.61) (*or* (3.6.16)) *is defined by a matrix $(s) = (s_{j,k})_{j,k=0}^{\infty}$ uniquely if the class*

$$C\{A_{2n}^n \sqrt{\max_{j=0,1,\dots,n} s_{j,j}}\}, \tag{3.6.30}$$

is quasi-analytic. Here

$$A_n = \max_{j=0,1,\dots,n} \{1, |a_j|, |b_j|, |c_j|\}. \quad n \in \mathbb{N}_0 \tag{3.6.31}$$

Proof We will use Theorem 3.6.10 connected with rigging (3.6.23). Now, the space $\mathcal{H}$ in this rigging is equal to the space S as in (3.5.54), and $\mathcal{H}_+ = l_2(q)$, where

$$q = (q_n)_{n=0}^{\infty}, \quad q_n = (1+n)^{1+\varepsilon} s_{n,n}, \quad \varepsilon > 0 \tag{3.6.32}$$

(the condition $\sum\limits_{n=0}^{\infty} s_{n,n} q_n^{-1} < \infty$ is now fulfilled, therefore, $\mathcal{H} \supset \mathcal{H}_+$ (see the beginning of Section)).

We will show that the condition (3.6.30) implies that each $\delta_k = (\delta_{j,k})_{j=0}^{\infty} \in \mathcal{H}$, $k \in \mathbb{N}_0$ is a quasi-analytic vector for the operator B with the domain $\mathrm{Dom}(B) = l_{\mathrm{fin}}$ in the space $\mathcal{H}_+$. The operator B is generated by the matrix J given in (3.5.53).

The set $\{\delta_0, \delta_1, \dots, \}$ is total in $\mathcal{H}_+$, therefore, our assertion gives the essential self-adjointness of the operator $l_{\mathrm{fin}} \ni \mapsto Af := Jf \in l_{\mathrm{fin}}$ in the space $\mathcal{H} = S$.

First, we note that the structure (3.5.53) of the matrix J and the condition (3.6.31) give the following: let $f \in l_{\mathrm{fin}}$ and $|f_J| \le \Phi_j$, $j \in \mathbb{N}_0$, where $0 < \Phi_0 \le \Phi_1 \le \dots$, then

$$|(Jf)_j| = |a_{j-1} f_{j-1} + b_j f_j + c_j f_{j+1}| \le 3A_j \Phi_{j+1}, \quad j \in \mathbb{N}_0.$$

Repeating this estimate and using $3A_j\Phi_{j+1}$ as Φ_j for $(Jf)_j$, we find that $j \in \mathbb{N}_0$,

$$|(J^2 f)_j| = |(J(Jf))_j| \le 3A_j(3A_{j+1}\Phi_{j+2}) = 3^2 A_j A_{j+1} \Phi_{j+2},$$

and so on in the same way for $n = 3, 4, \dots$,

$$|(J^n f)_j| \le 3^n A_j \dots A_{j+n-1} \Phi_{j+n}) \le 3^n A_{j+n-1} \Phi_{j+n}, \quad j \in \mathbb{N}_0. \tag{3.6.33}$$

Let $k \in \mathbb{N}_0$ be fixed and $f = \delta_k$; we put $\Phi_0 = \Phi_1 = \ldots = 1$. Then $(J^n \delta_k)_j = 0$ for $j > k + n + 1$ and for (3.6.33), using (3.6.32) and (3.6.31), we have

$$\begin{aligned}
\|J^n \delta_k\|^2_{\mathcal{H}_+} &= \|J^n \delta_k\|^2_{l_2(q)} = \sum_{j=0}^{k+n+1} |(J^n \delta_k)_j|^2 q_j \\
&\le \sum_{j=0}^{k+n+1} (3^n A^n_{j+n+1})^2 (1+j)^{1+\varepsilon} s_{j,j} \\
&\le (k+n+2)(3^n A^n_{2k+2n})^2 (k+n+2)^{1+\varepsilon} \max_{j=0,1,\ldots,k+n+1} s_{j,j} \\
&\le (k+n+2)^{2+\varepsilon} 3^{2n} A^{2(k+1+n)}_{k+1+n} \max_{j=0,1,\ldots,k+1+n} s_{j,j} \\
&=: (k+n+2)^{2+\varepsilon} 3^{2n} m^2_{k+1+n}, \quad n \in \mathbb{N}.
\end{aligned} \tag{3.6.34}$$

As already mentioned in Chaps. 1 and 2, the quasi-analyticity of $C\{m_{l+m}\}$ and $C\{m_n\}$ are equivalent. Therefore, the quasi-analyticity of (3.6.30) and the estimate (3.6.34) give that each vector δ_k, $k \in \mathbb{N}_0$ is quasi-analytic for the operator B. □

If, instead of the general Theorem 3.6.10, we apply Theorem 3.6.8 and use a special form of the operator B (the form of the Jacobi matrix), then a somewhat finer statement is obtained.

Proposition 3.6.13 *The result of Theorem 3.6.12 holds true if the class*

$$C\{A^n_n \sqrt{\max_{j=0,1,\ldots,n} s_{j,j}}\} \tag{3.6.35}$$

is quasi-analytic.

Proof We use the equality (3.6.29) from the proof of Theorem 3.6.10. Let B be defined as above by the matrix J, (see (3.5.53)), let $\mathcal{H}_+ = l_2(q)$, where q is given by (3.6.32) and $\psi \in l_{\text{fin}}$. Then equalities

$$\begin{aligned}
\frac{d^n}{dt^n}(v(t), \psi)_{l_2(q)} &= \sum_{j=0}^{\infty} v_j^{(n)}(t) \bar{\psi}_j q_j = (v(t), (\zeta J)^n \psi)_{l_2(q)} \\
&= \sum_{j=0}^{\infty} v_j(t) \overline{((\zeta J)^n \psi)} q_j = \bar{\zeta}^n \sum_{j=0}^{\infty} ((J^+)^n (q v(t)))_j \bar{\psi}_j, \\
&\quad t \in [0, b), \quad \forall n \in \mathbb{N}
\end{aligned} \tag{3.6.36}$$

hold true. Here J^+ denotes the adjoint matrix of J, $qv(t) = (q_j v_j(t))_{j=0}^{\infty}$. From (3.6.36), we conclude, that for $w(t) = qv(t)$, the equality

$$w_j^{(n)}(t) = ((J^+)^n w(t))_j, \quad t \in [0, b_1),\ b_1 \in (0, b),\ j \in \mathbb{N}_0 \tag{3.6.37}$$

holds true.

The vector-valued function $[0, b_1) \ni t \mapsto v(t) \in l_2(q)$ is strongly continuous and, therefore, is bounded. Then for some $C > 0$, $\forall t \in [0, b_1)$, we have

$$\begin{gathered}\sum_{j=0}^{\infty} |w_j(t)|^2 q_j^{-1} = \sum_{j=0}^{\infty} |v_j(t)|^2 q_j^{-1} = \|v(t)\|_{l_2(q)}^2 \le C; \\ |w_j(t)| \le \sqrt{C q_j}, \quad j \in \mathbb{N}_0.\end{gathered} \tag{3.6.38}$$

For $j \in \mathbb{N}_0$, we introduce a numbers

$$\Phi_j = (C \max_{k=0,1,\dots,j} q_k)^{1/2}, \quad 0 < \Phi_0 \le \Phi_1 \le \dots.$$

Similarly to (3.6.33), from (3.6.38), we get

$$|((J^+)^n w(t))_j| \le 3^n A_{j+n-1}^n \Phi_{j+n}, \quad t \in [0, b_1],\ j \in \mathbb{N}_0,\ n \in \mathbb{N}.$$

Therefore, (3.6.37) and (3.6.32) gives for these j and n that

$$\begin{aligned}|w_j^{(n)}(t)| &\le 3^n A_{j+n-1}^n \Phi_{j+n} \le 3^n A_{j+n}^{j+n} \Phi_{j+n} \\ &= C^{1/2}(1 + j + n)^{\frac{1+\varepsilon}{2}} 3^n A_{j+n}^{j+n} \sqrt{\max_{j=0,1,\dots,j+n} s_{k,k}} =: m_{j+n},\ t \in [0, b_1].\end{aligned} \tag{3.6.39}$$

It is sufficient to prove that, under the condition (3.6.35), each our solution $v(t)$ for which $v(0) = 0$ is equal zero on $[0, b_1]$. The equality $v(0) = 0$ gives, by (3.6.37), that $w_j^{(n)}(0) = 0$, $j \in \mathbb{N}_0$, $n \in \mathbb{N}$. Fix j and consider the estimate (3.6.39). The quasi-analyticity of the class (3.6.35) gives, as before, the quasi-analyticity of the class $C\{m_n\}$, therefore, $w(0) = 0$, $t \in [0, b_1]$. And since $j \in \mathbb{N}_0$ is arbitrary, then $\forall t \in [0, b_1]$, $w(0) = 0$ and $v(0) = 0$.

The case of the sign "+" in the Eq. (3.6.26) is similar. □

Remark 3.6.14 The results of the last section can be generalized to more complicated matrices J. That is, it is possible to obtain a representation of the form (3.6.26) for matrices (3.5.53) with complex coefficients. Moreover, instead of tri-diagonal Jacobi matrices, one can take matrices with an arbitrary finite number of non-zero diagonals, that is, consider the case of difference expressions of order higher than two.

3.7 Connections with the Theory of Jacobi Matrices

The space H_K is a completion of $H_+ = l_2(p_j)$ or $l_{2,0}$ with respect to the scalar product

$$\langle u, v\rangle = \sum_{j,k=0}^{\infty} s_{j+k} u_k \bar{v}_j.$$

Herewith, it is associated with the moment problem. For this, if necessary, an identification is first performed, i.e., $\sum_{j,k=0}^{N} s_{j+k}\xi_k\bar{\xi}_j = 0$ for $(\xi_0, \ldots, \xi_N) \not\equiv 0$, which simplifies the case. *In this case, the measure will be concentrated in a finite number of points.* Therefore, the absence of degeneracy is assumed further. Also, without loss of generality, we can assume that $s_0 = 1$.

Let us carry out in H_K the orthogonalization procedure of δ-sequences $\{\delta_0, \delta_1, \ldots, \delta_k, \ldots\}$. Thus, we get a sequence of vectors $\{p_0 = \delta_0, p_1, \ldots\} \in l_{2,0}$, which is orthonormal basis in H_K. The orthogonalization procedure is unambiguous if each vector $p_n \perp \{\delta_0, \ldots, \delta_{n-1}\}$ will be taken in the form $p_n = c_0\delta_0 + \ldots + c_n\delta_n$, where all c_j are real and $c_n > 0$; that is exactly what we will do. Since the absence of degeneracy is assumed, the basis $p_0, p_1, \ldots$ is infinite. In general, it is possible to write out the usual formulas for the transition from the basis $\delta_0, \delta_1, \ldots$ to the basis $p_0, p_1, \ldots$, but they will not be used. Let us only denote the operator F in H_K, which transfers $\{\delta_0, \delta_1, \ldots\}$ into $\{p_0, p_1, \ldots\}$, i.e., $F\delta_k = p_k$, $k \in \mathbb{N}_0$. There is an operator F^{-1} on $l_{2,0} \subset H_K$, but it is not bounded, generally speaking.

Sometimes it is convenient to use another interpretation of the space H_K for the moment problem. Let us correspond each $u \in l_{2,0}$ to the polynomial $\tilde{u}(\lambda) = \sum_{j=0}^{\infty} u_j\lambda^j$ and H_K is regarded as the space of all polynomials $\tilde{u}(\lambda)$ for which a non-degenerate scalar product

$$\langle \tilde{u}(\lambda), \tilde{v}(\lambda)\rangle = \langle u, v\rangle = \sum_{j=0}^{\infty} s_{j+k} u_k \bar{v}_j, \quad u, v \in\in l_{2,0} \tag{3.7.1}$$

is defined. If an image (3.5.24) with some measure $d\sigma(\lambda)$ is already written for the sequence s_j, then the scalar product (3.7.1) will be written in the form

$$\langle \tilde{u}(\lambda), \tilde{v}(\lambda)\rangle = \int_{\mathbb{R}} \tilde{u}(\lambda)\overline{\tilde{v}(\lambda)}\, d\sigma(\lambda). \tag{3.7.2}$$

We introduce polynomials

$$P_j(\lambda) = \tilde{p}_j(\lambda), \ j \in \mathbb{N}_0; \quad P_0 = \tilde{\delta}_0(\lambda) \equiv 1,$$

which are orthonormal in the scalar product (3.7.1) or (3.7.2) and form a basis in the space H_K.

It is clear from what has been said that H_K can be interpreted as the space l_2 if the basis $\{p_0, p_1, \ldots\}$ (or $\{P_0(\lambda), P_1(\lambda), \ldots\}$) is used. In any case, for $u \in l_{2,0}$, we can write

$$u = \sum_{k=0}^{\infty} u_k \delta_k = \sum_{j=0}^{\infty} \hat{u}_j p_j,$$

hence

$$\hat{u} = \sum_{k=0}^{\infty} u_k \langle \delta_k, p_j \rangle = \sum_{k=0}^{\infty} F_{jk} u_k,$$

where

$$\| F_{jk} \|_0^\infty = \| \langle \delta_k, F \delta_j \rangle \|_0^\infty = F;$$

u_j is calculated for $\hat{u}_j$ using the formally inverse matrix F^{-1}. Since δ_k and p_k are included in $l_{2,0}$, the matrices F and F^{-1} transfer finite sequences into finite sequences.

Let us write the operator $Bu = T^+ u$, $u \in l_{2,0}$ in the basis $\{p_0, p_1, \ldots\}$. Obviously he looks like

$$\hat{u}_j \to (F T^+ F^{-1} \hat{u}), \quad (\hat{u}_0, \hat{u}_1, \ldots) \in l_{2,0}.$$

Calculate the elements of the matrix FT^+F^{-1}. It is easier to do this not directly, but with the help of such considerations. Since $T^+\delta_k = \delta_{k+1}$, then

$$\widetilde{(T^+ \delta_k)}(\lambda) = \tilde{\delta}_{k+1}(\lambda) = \lambda \tilde{\delta}_k(\lambda).$$

Thus, H_K has an interpretation as a space of polynomials with a scalar product (3.7.2), the operator B has the form of a multiplication operator by λ. But it is known that such an operator can also be considered as an action of the Jacobi matrix (see Sect. 3.3). Elements of this matrix are calculated by formulas (3.3.1), where polynomials $P_j(\lambda)$ coincide with those just introduced.

Theorem 3.7.1 *Let the non-degenerate moment sequence s_j, $j \in \mathbb{N}_0$ be given. We construct the space H_K and $K_{jk} = s_{j+k}$, $j, k \in \mathbb{N}_0$. Let us pass in this space from the basis $\delta_0, \delta_1, \ldots$ to the described orthonormal basis $p_0, p_1, \ldots$. The operator $Bu = T^+u$, $u \in l_{2,0}([0, \infty))$ in the basis $p_0, p_1, \ldots$, will be written as an action of the Jacobi matrix, the elements of which can be calculated by formulas*

$$a_j = \langle T^+ p_j, p_{j+1} \rangle = \langle \lambda P_j(\lambda), P_{j+1}(\lambda) \rangle,$$

$$b_j = \langle T^+ p_j, p_j \rangle = \langle \lambda P_j(\lambda), P_j(\lambda) \rangle, \quad j \in \mathbb{N}_0.$$

Conversely, let some Jacobi matrix (3.1.3) *be given. We will construct a system of polynomials of the first kind* $\{P_0(\lambda), P_1(\lambda), \dots\}$*; and* $d\sigma(\lambda)$ *is some of its spectral measure. Let*

$$\lambda^n = c_0^{(n)} P_0(\lambda) + \dots + c_n^{(n)} P_n(\lambda), \quad n \in \mathbb{N}_0.$$

Then

$$s_n = c_0^{(n)} = \int_{\mathbb{R}} \lambda^n \cdot 1 \, d\sigma(\lambda), \quad n \in \mathbb{N}_0$$

is a non-degenerate moment sequence. If, according to it, using the first part of the theorem, namely construct the Jacobi matrix, then we get the initial matrix.

Since the solution of the moment problem is reduced to the spectral theory of the operator B, and Sects. 3.1–3.4 describe the construction of the spectral theory of this particular operator, only written in a different basis, then all the results indicated there are transferred to the moment problem. In particular, Theorem 3.4.3 gives a description of all solutions of the moment problem, that is, all spectral measures $d\sigma(\lambda)$ of Sect. 2.4.

Finally, we will show that the Carleman criterion (3.5.26) of the definiteness of the moment problem is a consequence of the criterion (3.1.9) of self-adjointness of the difference operator. Indeed, from the construction of polynomials of the first kind $P_j(\lambda)$, it is clear that $P_j(\lambda) = \frac{1}{a_0 \cdots a_{j-1}} \lambda^j + \dots$, therefore, taking into account their orthogonality and normality with respect to each spectral measure $d\sigma(\lambda)$ and the formula (3.3.1), we obtain:

$$\begin{aligned} a_j &= \int_{\mathbb{R}} \lambda P_j(\lambda) P_{j+1}(\lambda) \, d\sigma(\lambda) \\ &= \int_{\mathbb{R}} \left(\frac{\lambda^{j+1}}{a_0 \cdots a_{j-1}} + \lambda(\dots) \right) P_{j+1}(\lambda) \, d\sigma(\lambda) = \\ &= \frac{1}{a_0 \dots a_{j-1}} \int_{\mathbb{R}} \lambda^{j+1} P_{j+1}(\lambda) \, d\sigma(\lambda) \\ &\le \frac{1}{a_0 \cdots a_{j-1}} \left(\int_{\mathbb{R}} \lambda^{2(j+1)} \, d\sigma(\lambda) \right)^{1/2} \left(\int_{\mathbb{R}} P_{j+1}^2(\lambda) \, d\sigma(\lambda) \right)^{1/2} \\ &= \frac{1}{a_0 \cdots a_{j-1}} \sqrt{s_{2(j+1)}}, \end{aligned}$$

i.e., $a_0 \cdots a_j \leq \sqrt{s_{2(j+1)}}$, $j \in \mathbb{N}_0$. Let the condition (3.5.26) be fulfilled. Then, from the obtained, it follows

$$\sum_{j=0}^{\infty} \frac{1}{\sqrt[j]{a_0 \cdots a_j}} = \infty. \tag{3.7.3}$$

If we now use the general inequality

$$\sum_{n=1}^{\infty} \sqrt[n]{q_1 \cdots q_n} \leq e \sum_{n=1}^{\infty} q_n, \quad q_n \geq 0, \quad n \in \mathbb{N}_0,$$

we get that the series $\sum_{j=0}^{\infty} \frac{1}{a_j}$ diverges. According to Theorem 3.1.3, the difference operator L is self-adjoint, that is, the moment problem is determined.

Bibliographical Notes

The main information of the chapter is taken from the seventh chapter of the book by Berezansky Yu. M. [21]. Presentations are motivated by the works of Akhiezer N. Y., Krein M. G., Krasnosel'skii M. A. [2, 188]. The main used results are published in works of Berezansky Y. M., Kondratiev Y. G., Naimark M. A., Glazman I. M. [17, 124, 179, 227, 295].

A close approach to the theory of Jacobi matrices is found in [3, 260].

Works of Akhiezer N. I. [2, 3] is the beginning and the reason for the research presented in the chapter. These studies received a modern description in works of Berezansky Yu. M. [17–23, 43, 44, 50], Kostyuchenko A. G., Sargsyan I. S. [169], Shohat J. A, Tamarkin J. D. [260]. Initial works on this issues were papers by Krein M. G., [179, 180, 185, 186].

Important applications are in works Krasnosel'skii M. A., Vainikko G. M., Zabreiko P. P., Rutitskii J. B., Stetsenko V. Ya., Lifshitz E. A., Sobolev A. V. [173, 174]. There is also information about the eigenfunactions expansion of differential and other operators by Gelfand I. M., Kostyuchenko A.G. [111]. General orthogonal functions are described by Glazman I. M., Naiman P. B. [124]. General problems of the theory of extensions of symmetric densely defined operators to self-adjoint ones are in the works of Gorbachuk M. L. and Gorbachuk V. I. [128, 129], Derkach V. A., Malamud M. M. [77, 80], Kochubei A. N. [165], Krein M. G. [177, 178, 181, 185, 187], Krein M. G., Krasnosel'skii M. A. [188], Kuzhel A. V. [193]. The application of monotonic sequences to the moment problem was studied by Nudel'man A. A. [230], Krein M.G., Nudel'man A. A. [189]. General information can be found in works of Dunford N., Schwartz J. T. [95, 236].

Generalizations, in particular, concerning Jacobi matrices related to the moment problem, can be found in works of Berezansky Yu. M. [26, 30, 31] Berezansky Yu. M., Mierzejewski D. A., Shmoish M. [45, 49], Derevyagin M. S., Derkach V. A. [78], Derkach V. A., Malamud M. M. [76], Devinatz A. [81], Gesztesy F., Simon B. [120], Manakov S. V. [205], Schmüdgen K. [255], Szafraniec F. H. [275], Wouk A. [295].

The question of the uniqueness of the moment representation is solved with the help of works Carleman T. [64], Nelson E. [225], Nussbaum A. E. [231, 232].

An interesting direction related to the moment problem is the so-called truncated moment problem. We do not touch this direction in our book and refer to the most informative works in this direction, for example, Krein M. G. [70, 182].

Chapter 4
The Strong Moment Problem

In this chapter, we present the main provisions concerning the strong moment problem. More precisely, strong classical, when studying the possibility of representing a sequence of real numbers (s_n) in the form $s_n = \int_{\mathbb{R}} \lambda^n \, d\rho(\lambda)$ for all integers n, i.e., $n \in \mathbb{Z}$.

It is known that various formulations of the moment problem are investigated using the method of the generalized eigenvectors expansion. With this approach, first of all, it is necessary to represent the moment sequence by applying the theory of the generalized eigenvectors expansion for the corresponding operators. For such vectors, we write a simple equation corresponding to the problem under some consideration: the solution of such an equation gives the form of the representation. The corresponding Parseval equality gives immediately the desired representation.

After that, we pass from the moment problem to a block tri-diagonal matrices (Jacobi type). The spectral measure of such matrix gives the measure of the moment representation.

The corresponding spectral theory of such a matrix provides additional information about the corresponding considered moment problem.

The structure of the chapter is as follows. In the first section, we recall the results on the generalized eigenvectors expansion, which are necessary for what follows. In Sect. 4.2 we prove the main representation theorem and some conditions for the uniqueness of the existence of a measure.

Sections 4.3 and 4.4 are devoted to the construction and study of block matrices related to the strong moment problem. It should be noted that our J matrices of the Jacobi-Laurent type have a tri-diagonal block structure. The main result of the subsections are Theorems 4.3.5 and 4.3.11.

Our Jacobi-Laurent matrix J is symmetric and has an algebraic inverse J^{-1} which is also block tri-diagonal with appropriate properties. An additional problem is the internal description of such matrices J and J^{-1}.

Y. M. Berezansky, M. E. Dudkin, *Jacobi Matrices and the Moment Problem*,
Operator Theory: Advances and Applications 294,
https://doi.org/10.1007/978-3-031-46387-7_4

Section 4.3 presents the theory of Jacobi-Laurent block matrices including direct and inverse spectral problems. The construction is the same as for the classical Jacobi matrices and uses the generalized eigenvectors expansion. In Sect. 4.4, the Jacobi-Laurent matrices J are considered in the case when J is generated only by the Hermitian operator, which is not self-adjoint. We construct a corresponding theory similar to the case of classical Jacobi matrices. But we do not present the theory of the description of all self-adjoint extension in the original Hilbert space and we do not give a complete such description, because the construction is very similar to the case of classical Jacobi matrices, which is described in the second chapter.

Section 4.5 is devoted to the construction of s_n, $n \in \mathbb{Z}$, or the initial measure $d\rho(\lambda)$ of the Jacobi-Laurent matrices J, and an establishing connection of this measure with the spectral measure of the self-adjoint operator generated by J. From this section it becomes clear why the theory of self-adjoint extensions of Jacobi-Laurent matrices provides a description of all solutions of the strong moment problem.

4.1 Preliminaries to the Strong Moment Problem

Let $\mathcal{H}$ be a separable Hilbert space and let A be a self-adjoint operator defined on $\mathrm{Dom}(A)$ in $\mathcal{H}$. Consider an equipment of the space $\mathcal{H}$:

$$\mathcal{H}_- \supset \mathcal{H} \supset \mathcal{H}_+ \supset \mathcal{D}, \tag{4.1.1}$$

such that $\mathcal{H}_+$ is a Hilbert space topologically and quasi-nuclear embedded into $\mathcal{H}$ (we recall that topologically means densely and continuously; quasi-nuclear means that the imbedding operator is of the Hilbert-Schmidt type); $\mathcal{H}_-$ is a space dual to $\mathcal{H}_+$ with respect to the space $\mathcal{H}$; $\mathcal{D}$ is a linear, topological space, topologically embedded into $\mathcal{H}_+$.

Recall also that the operator A is called standardly connected with the chain (4.1.1) if $\mathcal{D} \subset \mathrm{Dom}(A)$ and the restriction $A \upharpoonright \mathcal{D}$ acts from $\mathcal{D}$ into $\mathcal{H}_+$ continuously (see for details Chap. 2).

Let us formulate a shortened version of the projection spectral theorem about generalized eigenvectors expansion, which is convenient for considering the strong moment problem.

Theorem 4.1.1 *Let A be a self-adjoint operator defined on a separable Hilbert space $\mathcal{H}$ and standardly connected with the chain* (4.1.1)*, where $\mathcal{D}$ is also separable. Then there exists an operator-valued function $\Phi(\lambda)$ and a bounded Borel (general) spectral measure $d\sigma(\lambda)$ such that $\Phi(\lambda)$ is weakly measurable and defined for almost all λ on the spectrum $s(A)$ of the operator A in the sense of the spectral measure $d\sigma(\lambda)$ and takes values as non-negative operators acting from $\mathcal{H}_+$ into $\mathcal{H}_-$, and for every λ its Hilbert-Schmidt norm satisfies the equality $|\Phi(\lambda)| \leq \mathrm{Tr}(\Phi(\lambda)) = 1$.*

Here "Tr" denotes the trace of a corresponding operator. The function $\Phi(\lambda)$ *and the measure* $d\sigma(\lambda)$ *give a representation of the expansion of the identity* E *of* A*, namely,*

$$E(\alpha)f = \Big(\int\limits_{\Delta} \Phi(\lambda)\, d\sigma(\lambda)\Big) f, \quad \alpha \in \mathfrak{B}(\mathbb{R}), \quad f \in \mathcal{H}_+ \tag{4.1.2}$$

and for the operator A

$$Af = \Big(\int\limits_{s(A)} \lambda\Phi(\lambda)\, d\sigma(\lambda)\Big) f, \quad f \in \mathrm{Dom}(A) \cap \mathcal{H}_+. \tag{4.1.3}$$

The set of values (range) $\mathrm{Ran}(\Phi(\lambda)) \subset \mathcal{H}_-$ *consists of generalized eigenvectors* $\varphi(\lambda) \in \mathcal{H}_-$ *of the operator* A *with corresponding eigenvalues* λ*, i.e.,*

$$(\varphi(\lambda), Af)_{\mathcal{H}} = \lambda(\varphi(\lambda), f)_{\mathcal{H}}, \quad \lambda \in \mathbb{R}, \quad f \in \mathcal{D}; \quad \varphi(\lambda) \neq 0. \tag{4.1.4}$$

In general case, for the operator A appearing in Theorem 4.1.1, it is possible to construct the generalized eigenvectors expansion of each $f \in \mathcal{H}$ according to the operator A Fourier transform. But in general situation the dimension of vector, by the corresponding Fourier transform, $\hat{f}(\lambda)$ depends on "multiplicity" of eigenvalue λ and corresponding formulas are small effective. But in some special case of operators A the terms of Fourier transform is very convenient and replaces effectively formulas (4.1.2) and (4.1.3).

Just such a case appeared in Chap. 3, when the existing of a cyclic vector helped the situation (see Theorem 3.5.6). Now there will be a similar situation. Let us introduce a corresponding definition.

Let A be some self-adjoint operator in $\mathcal{H}$, vector $q \in \mathcal{H}$ is called cyclic if $q \in \mathrm{Dom}(A^n)$, $n \in \mathbb{N}$. Let for A, algebraically inverse operator A^{-1} exists (i.e., for $f \in \mathrm{Dom}(A)$), $A^{-1}Af = f$, $\mathrm{Dom}(A^{-1}) = \mathrm{Ran}(A)$). The cyclic vector q, is called a double cyclic if $q \in \mathrm{Dom}(A^n)$, $n \in \mathbb{Z}$.

The following theorem is required by the prove of Theorem 4.2.1.

Theorem 4.1.2 *Let* A *be a self-adjoint operator for which the conditions of Theorem 4.1.1 are satisfied. Assume that for this operator the cyclic vector* q *exists (or is doubly cyclic, in this case we assume that* $A^{-1}f \in \mathcal{D}$ *if* $f \in \mathcal{D}$ *and the algebraic inverse* A^{-1} *is defined on* $\mathcal{D}$*), for which* $\forall n \in \mathbb{N}_0$ *(or* $\forall n \in \mathbb{Z}$*)* $A^n q \in \mathcal{D}$ *and the set of such vectors is total in* $\mathcal{D}$*. Then the spectrum* A *is simple and for each* $\lambda \in \sigma(A)$ *the corresponding generalized eigenvector* $\varphi(\lambda) \in \mathcal{H}_-$ *exists and the Fourier transform* F *can be written*

$$\mathcal{D} \ni f \longmapsto (Ff)(\lambda) = \hat{f}(\lambda) := (f, \varphi(\lambda))_{\mathcal{H}} \in \mathbb{C}. \tag{4.1.5}$$

Instead of (4.1.2), (4.1.3) *we have representations equivalent to them*

$$(f, g)_{\mathcal{H}} = \int_{\mathbb{R}} f(\lambda)\overline{g(\lambda)}\, d\sigma(\lambda), \quad (\widehat{Af})(\lambda) = \lambda \hat{f}(\lambda), \quad \lambda \in s(A),\ \forall f, g \in \mathcal{D}. \tag{4.1.6}$$

Due to the extension by continuity, (4.1.5) *is possible to extend on all* $f \in \mathcal{H}$, *then* $\hat{f}(\lambda) \in L_2(\mathbb{R}, d\sigma(\lambda))$. *The first equality in* (4.1.6) *can be extended on* $f, g \in \mathcal{H}$ *and it is the Parseval equality. The second equality can be extended on* $f \in \text{Dom}(A)$ *and shows that our operator is unitary equivalent to the multiplication operator by* λ *in the space* $L_2(\mathbb{R}, d\sigma(\lambda))$, *defined by* $\hat{f}(\lambda)$, $f \in \text{Dom}(A)$.

Proof This theorem in the case of cyclic vector is proved in Chap. 2. In the case of double cyclic vector q it is necessary to repeat proof of such a theorem. Only one place is necessary to explain in this repeating. Namely, let $\varphi(\lambda) := P(\lambda)Jf_0 \in \mathcal{H}_-$ is the vector from the proof of Theorems 2.11.9 and 2.11.11. We have in such a situation $(\varphi(\lambda), q)_{\mathcal{H}} = 0$ (previously q was denoted by Ω).

To prove the theorem, it is necessary to show that $\varphi(\lambda) = 0$. This vector is the generalized eigenvector with the eigenvalue λ, i.e.,

$$(\varphi(\lambda), Af)_{\mathcal{H}} = \lambda(\varphi(\lambda), f)_{\mathcal{H}}, \quad f \in \mathcal{D}.$$

Therefore, $\forall f \in \mathcal{D}$

$$(\varphi(\lambda), f)_{\mathcal{H}} = (\varphi(\lambda), A(A^{-1}f))_{\mathcal{H}} = \lambda(\varphi(\lambda), A^{-1}f)_{\mathcal{H}}, \tag{4.1.7}$$

since by conditions of the theorem $A^{-1}f \in \mathcal{D}$. From (4.1.7), we conclude that $\lambda \neq 0$: if $\lambda = 0$, then $(\varphi(\lambda), f)_{\mathcal{H}} = 0$, $f \in \mathcal{D}$, i.e., $\varphi(\lambda) = 0$ and the proof is finished.

Thus, let $\lambda \neq 0$. By using (4.1.7), we get

$$(\varphi(\lambda), A^{-1}f)_{\mathcal{H}} = \lambda^{-1}(\varphi(\lambda), f)_{\mathcal{H}}, \quad f \in \mathcal{D}. \tag{4.1.8}$$

By the iteration of (4.1.8) (note that $A^{-1}f \in \mathcal{D}$), we have

$$(\varphi(\lambda), A^{-n}f)_{\mathcal{H}} = \lambda^{-n}(\varphi(\lambda), f)_{\mathcal{H}}, \quad f \in \mathcal{D}, \quad n \in \mathbb{N}. \tag{4.1.9}$$

For non-negative powers of A we evidently have

$$(\varphi(\lambda), A^n q)_{\mathcal{H}} = \lambda^n(\varphi(\lambda), f)_{\mathcal{H}} = 0, \quad n \in \mathbb{N}_0. \tag{4.1.10}$$

Taking in (4.1.9) $f = q$ we conclude that (4.1.10) is fulfilled for $n \in \mathbb{Z}$. But on conditions of the theorem, the set $\{A^n q, n \in \mathbb{Z}\}$ is total in $\mathcal{D}$. Therefore, in the case of double cyclic vector we have (4.1.10) for $n \in \mathbb{Z}$ and, hence, $\varphi(\lambda) = 0$. □

Then, some self-adjointness conditions related to the concept of a quasi-analytic vector are used. Let us recall these results. Therefore, for the Hermitian operator A defined on $\mathrm{Dom}(A)$ in $\mathcal{H}$, the vector $f \in \bigcap_{n=1}^{\infty} \mathrm{Dom}(A^n)$ is quasi-analytic if

$$\sum_{n=1}^{\infty} \frac{1}{\sqrt[n]{\|A^n f\|_{\mathcal{H}}}} = \infty.$$

4.2 The Solution of the Strong Moment Problem

The solution of the strong moment problem is given by the following theorem.

Theorem 4.2.1 *The given sequence of real numbers $s = (s_n)$, $n \in \mathbb{Z}$, $s_n \in \mathbb{R}$ admits a representation*

$$s_n = \int_{\mathbb{R}} \lambda^n \, d\rho(\lambda), \quad n \in \mathbb{Z} \tag{4.2.1}$$

with some Borel measure $d\rho(\lambda)$ if and only if it is positive defined, i.e.,

$$\sum_{j,k \in \mathbb{Z}} s_{j+k} f_j \bar{f}_k \geq 0 \tag{4.2.2}$$

for every finite sequences of complex numbers (f_j), $j \in \mathbb{Z}$, $f_j \in \mathbb{C}$.

The measure in the representation (4.2.1) *is unique if, additionally, to* (4.2.2), *we have*

$$\sum_{n=1}^{\infty} \frac{1}{\sqrt[2n]{s_{2n}}} = \infty. \tag{4.2.3}$$

Proof The Necessity of the condition (4.2.2) is obvious. Indeed, if the sequence s has representation (4.2.1), then for an arbitrary finite sequence $f = (f_k)_{k \in \mathbb{Z}}$, $f_k \in \mathbb{C}$, we have

$$\sum_{j,k \in \mathbb{Z}} s_{j+k} f_j \bar{f}_k = \int_{\mathbb{R}} \left| \sum_{n \in \mathbb{Z}} \lambda^n f_n \right|^2 d\rho(\lambda) \geq 0. \tag{4.2.4}$$

Denote by l a linear space $\mathbb{C}^{\infty}$ of sequences $f = (f_j)$, $j \in \mathbb{Z}$, $f_j \in \mathbb{C}$, and by l_{fin} we denote its linear subspace consisting of finite sequences $f = (f_j)$, $j \in \mathbb{Z}$, i.e., the sequences such that $f_j \neq 0$ only for finite numbers j. Let δ_m, $m \in \mathbb{Z}$, be the δ-sequence, i.e., $\delta_m = (0, \ldots, \underset{m\text{-th position}}{0, 1, 0, 0}, \ldots)$. Then each vector $f \in l_{\text{fin}}$ has the representation $f = \sum_{m\in\mathbb{Z}} f_m \delta_m$.

Let us consider the linear operator J on the form

$$(Jf)_j = f_{j-1}, \quad j \in \mathbb{Z}; \quad \text{Dom}(J) = l_{\text{fin}}. \tag{4.2.5}$$

The operator J is the "creation" type operator. For the δ-sequence we get

$$J\delta_m = \delta_{m+1}, \quad m \in \mathbb{Z}. \tag{4.2.6}$$

The operator J is Hermitian with respect to the (quasi)scalar product

$$(f, g)_S = \sum_{j,k\in\mathbb{Z}} s_{j+k} f_j \bar{g}_k, \quad f, g \in l_{\text{fin}}. \tag{4.2.7}$$

Indeed, as in the case of (3.5.42), we have

$$\begin{aligned}(Jf, g)_S &= \sum_{j,k\in\mathbb{Z}} s_{j+k} (Jf)_j \bar{g}_k = \sum_{j,k\in\mathbb{Z}} s_{j+k} f_{j-1} \bar{g}_k \\ &= \sum_{j,k\in\mathbb{Z}} s_{j+k+1} f_j \bar{g}_k = \sum_{j,k\in\mathbb{Z}} s_{j+k} f_j \bar{g}_{k-1} \\ &= \sum_{j,k\in\mathbb{Z}} s_{j+k} f_j \overline{(Jg)}_k = (f, Jg)_S.\end{aligned}$$

In the next step we use Theorem 3.5.6. We need only to write $j, k, \ldots \in \mathbb{Z}_0$ instead of $j, k, \ldots \in \mathbb{N}_0$. Let us sketch the proof of the theorem. As in Chap. 3, we will first assume that the expression (4.2.7) gives the scalar product in l_{fin}. That is, we assume that the given sequence $s = (s_n)$, $n \in \mathbb{Z}$ is not degenerate, i.e., if $(f, f)_S = 0$ for $f \in l_{\text{fin}}$, then $f = 0$.

Consider the operator J (4.2.5). It is Hermitian and defined in the space $\mathcal{H}_S$ that is a completion of l_{fin} with respect to the scalar product (4.2.7) and it has the domain $\text{Dom}(J) = l_{\text{fin}}$. Moreover, it is real in S with respect to the ordinary transformation from $f = (f_j)$, $j \in \mathbb{Z}$, to $\bar{f} = (\bar{f}_j)$, $j \in \mathbb{Z}$. Therefore it has equal defect numbers and has a self-adjoint extension in S.

Let us choose and fix some extension of A. We take and fixed some such extension A. We will apply to this operator A the general results of Sect. 4.1. But at first it is necessary to construct some equipment of the space $\mathcal{H}_S$.

Thus, we will consider the following rigging:

$$(l_2(p))_{-,S} \supset \mathcal{H}_S \supset l_2(p) \supset l_{\text{fin}}, \tag{4.2.8}$$

where $l_2(p)$ is the weighted l_2 type space (l_2 the space over $\mathbb{Z}$) with a weight $p = (p_n)$, $n \in \mathbb{Z}$, $p_n \geq 1$. The norm in $l_2(p)$ is given by $\|f\|^2_{l_2(p)} = \sum_{n\in\mathbb{Z}} |f_n|^2 p_n$;

$(l_2(p))_{-,S} = \mathcal{H}_-$ is negative space with respect to the positive space $l_2(p) = \mathcal{H}_+$ and $\mathcal{H}_S = \mathcal{H}$ is the zero space. The space $l_{\text{fin}} = \mathcal{D}$ is provided with the coordinate-wise uniformly finite convergence.

As in the proof of Theorem 3.5.6, we make sure that there is a sufficiently fast growing sequence p such that the imbedding $l_2(p) \hookrightarrow S$ is quasi-nuclear. In the next step we use the rigging (4.2.8) to construct a generalized eigenvectors. The inner structure of the space $(l_2(p))_{-,S}$ is complicated, because of the complicated structure of $\mathcal{H}_S$. This is a reason to introduce new auxiliary rigging

$$l = (l_{\text{fin}})' \supset (l_2(p^{-1})) \supset l_2 \supset l_2(p) \supset l_{\text{fin}}, \tag{4.2.9}$$

where $l_2(p^{-1})$, $p^{-1} = (p_n^{-1})$, $n \in \mathbb{Z}$, is the negative space with respect to the positive space $l_2(p)$ and the zero (initial) space l_2. Chains (4.2.8) and (4.2.9) have the same positive space $l_2(p)$. The general Lemma 2.7.5 in Chap. 2 establishes that the space $(l_2(p))_{-,S}$ is isometric to the space $l_2(p^{-1})$.

We go back to the operator A. It is some self-adjoint extension of J in the space $\mathcal{H}_S$. It is easy to understand that operator A is standardly connected with the rigging (4.2.8), but instead of riggings (4.2.8) we consider (4.2.9) and Lemma 2.7.5.

Let $\varphi(\lambda) \in (l_2(p))_{-,S}$ be a generalized eigenvector of the operator A in terms of the (4.2.8). Thus, in this case due to Theorem 4.1.2 and (4.1.4), we have

$$(\varphi(\lambda), Af)_S = \lambda(\varphi(\lambda), f)_S, \quad \lambda \in \mathbb{R}, \quad f \in l_{\text{fin}}. \tag{4.2.10}$$

Denote

$$P(\lambda) = U\varphi(\lambda) \in l_2(p^{-1}), \quad P(\lambda) = (P_n(\lambda)),\ n \in \mathbb{Z},\ \forall n \in \mathbb{Z},\ P_n(\lambda) \in \mathbb{R}$$

(here we apply Lemma 2.7.5 with $\mathcal{H}_- = (l_2(p))_{-,S}$ and $\mathcal{F}_- = l_2(p^{-1})$). By using (2.7.36), we can rewrite (4.2.10) in the form

$$(P(\lambda), Af)_{l_2} = \lambda(P(\lambda), f)_{l_2}, \quad \lambda \in \mathbb{R}, \quad f \in l_{\text{fin}}. \tag{4.2.11}$$

The correspondence Fourier transform (4.1.5) has the form

$$\mathcal{H}_S \supset l_{\text{fin}} \ni f \to (Ff)(\lambda) = \hat{f}(\lambda) = (f, P(\lambda))_{l_2} \in L_2(\mathbb{R}, d\sigma(\lambda)). \tag{4.2.12}$$

Let us calculate $P(\lambda)$. The operator A is a self-adjoint extension in $\mathcal{H}_S$ of the operator J with $\text{Dom}(J) = l_{\text{fin}}$, and therefore, obtained on l_{fin} by the formula (4.2.5), and hence (4.2.11) gives

$$\begin{aligned}\sum_{n\in\mathbb{Z}} \lambda P_n(\lambda)\bar{f}_n &= \lambda(P(\lambda), f)_{l_2} = (P(\lambda), Af)_{l_2} \\ &= (P(\lambda), Jf)_{l_2} = (J^+P(\lambda), f)_{l_2} = \sum_{n\in\mathbb{Z}} P_{n+1}(\lambda)\bar{f}_n, \quad \forall f \in l_{\text{fin}}.\end{aligned} \tag{4.2.13}$$

Hence, we have

$$\lambda P_n(\lambda) = P_{n+1}(\lambda), \quad n \in \mathbb{Z}. \tag{4.2.14}$$

Without loss of generality, we can take $P_0(\lambda) = 1, \ \lambda \in \mathbb{R}$. Then the equalities (4.2.14) yields

$$P_n(\lambda) = \lambda^n, \quad n \in \mathbb{Z}. \tag{4.2.15}$$

Thus, the Fourier transform (4.2.12) finally has the form

$$S \supset l_{\text{fin}} \ni f \to (Ff)(\lambda) = \hat{f}(\lambda) = \sum_{n \in \mathbb{Z}} f_n \lambda^n \in L_2(\mathbb{R}, d\sigma(\lambda)), \tag{4.2.16}$$

and the Parseval equality (4.1.5) has the form

$$(f, g)_S = \int_{\mathbb{R}} \hat{f}(\lambda)\overline{\hat{g}(\lambda)}\, d\sigma(\lambda), \quad f, g \in l_{\text{fin}}. \tag{4.2.17}$$

To construct the Fourier transform (4.2.12) and to verify formulas (4.2.13)–(4.2.17) it is necessary to note that for our operator A, the algebraically inverse operator A^{-1} on l_{fin}, exists and the vector $q = \delta_0 \in l_{\text{fin}}$ has the property $A^n q = J^n \delta_0 = \delta_n \in \mathcal{D}, n \in \mathbb{Z}$. The last set is total in l_{fin} and Theorem 4.1.2 is applicable.

The Parseval equality (4.2.17) immediately leads to the representation (4.2.1). According to (4.2.15) and (4.2.16) $\hat{\delta}_n = \lambda^n$ and $\hat{\delta}_0 = 1$; by (4.2.7), we obtain

$$s_n = (\delta_n, \delta_0)_S = (\hat{\delta}_n, \hat{\delta}_0)_{L_2(\mathbb{R}, d\sigma(\lambda))} = \int_{\mathbb{R}} \lambda^n \, d\sigma(\lambda), \quad n \in \mathbb{Z},$$

i.e., we get (4.2.1) with the measure $d\rho(\lambda) = d\sigma(\lambda)$.

If the operator J (4.2.5) is essentially self-adjoint in $\mathcal{H}_S$, then we can take the closure of J as A. In this case the measure $d\rho(\lambda)$ in the representation (4.2.1) is unique.

Thus, to complete the proof of the theorem for the case of the scalar product (4.2.7), it is sufficient to prove that the condition (4.2.3) guarantees essential self-adjointness of J in the space $\mathcal{H}_S$.

In what follows, we will use Theorem 2.6.8. Thus, it is necessary to prove that the operator J has a total set Q of quasi-analytic vectors in the space $\mathcal{H}_S$.

Denote $Q = \{\delta_p\}, \ p \in \mathbb{Z}$. This set is total in $\mathcal{H}_S$: its linear span is equal to l_{fin}. Check that each such vector δ_p is quasi-analytic. According to (4.2.5)–(4.2.7) and using Theorem 2.6.8, Lemma 2.6.7, and (2.6.21), we can write

$$\|J^n \delta_p\|_S^2 = \|\delta_{p+n}\|_S^2 = s_{2p+2n}, \quad n \in \mathbb{N}, \ p \in \mathbb{Z}.$$

Hence, we have

$$\sum_{n=1}^{\infty} \frac{1}{\sqrt[n]{\|J^n \delta_p\|}} = \sum_{n=1}^{\infty} \frac{1}{\sqrt[2n]{s_{2p+2n}}}. \tag{4.2.18}$$

But, since series $\sum\limits_{n=1}^{\infty} \frac{1}{\sqrt[2n]{s_{2p+2n}}}$ and $\sum\limits_{n=1}^{\infty} \frac{1}{\sqrt[2n]{s_{2n}}}$ are either convergent or divergent simultaneously, then the equality (4.2.18) (the condition (4.2.3)) give that vector δ_p is quasi-analytic.

Therefore, in the non-degenerate case, the theorem is proved.

Consider the case when the quadratic form is degenerate, that is, there exists a nonzero $f = (f_j)$, $j \in \mathbb{Z}$ such that

$$\sum_{j,k\in\mathbb{Z}} s_{j+k} f_j \bar{f}_k = 0. \tag{4.2.19}$$

In this case the expression (4.2.7) gives a quasi-scalar product and for the construction of the space $\mathcal{H}_S$ it is necessary to take factor space l_{fin} of all such f and after this to make the completion. The operator J is Hermitian with respect to the quasi-scalar product, therefore it is correctly defined on our $\mathcal{H}_S$ and it is Hermitian with respect to the introduced scalar product. After this, it is necessary to repeat given above scheme of the proof, applying what was said at the end of the proof of Theorem 3.5.6. □

Both for the proof of Theorem 3.5.6 and for the proof of the last theorem, one can make certain remarks about the support of the measure $d\rho(\lambda)$ in the degenerate case.

In the next section we develope the spectral theory of block Jacobi-Laurent matrices corresponding to the strong moment problem.

4.3 The Orthogonalization Procedure and the Construction of a Tri-Diagonal Block Matrix

We at first propose some orthogonalization procedure and the construction of a tri-diagonal block matrix of a self-adjoint operator related to the corresponding strong moment problem.

Instead of usual space l_2 of sequences $f = (f_n)$, $f_n \in \mathbb{C}$ throughout this section, in which the ordinary Jacobi matrix acts, we will use the "double" space l_2 defined as

$$\mathbf{l}_2 = \mathcal{H}_0 \oplus \mathcal{H}_1 \oplus \mathcal{H}_2 \oplus, \; \cdots, \qquad \mathcal{H}_0 = \mathbb{C}^1, \quad \mathcal{H}_1 = \mathcal{H}_2 = \cdots \mathbb{C}^2. \tag{4.3.1}$$

Ours tri-diagonal matrices act in this space (4.3.1). It is clear that this space actually coincides with the space $l_2 \times l_2$, but on $\mathbb{Z}$, that is, it is the space of sequences $f = (f_n)$, $f_n \in \mathbb{C}^1$. However, its representation in the form (4.3.1) is more convenient for us. In fact, because the coordinate f_0 is counted in $\mathbf{l}_2$ once, and not twice, as in $l_2 \times l_2$.

Let $d\rho(\lambda)$ be a Borel measure with *a bounded support* on the real axis $\mathbb{R}$ and $L_2 = L_2(\mathbb{R}, d\rho(\lambda))$ is the space of complex square integrable functions defined on $\mathbb{R}$. We suppose that the Borel measure $d\rho(\lambda)$ is such that all functions $\mathbb{R} \ni \lambda \longmapsto \lambda^m$, $m \in \mathbb{Z}$ belong to L_2, and all these functions are linearly independent.

In order to find an analog of usual Jacobi matrix J there is need to choose an order for the orthogonalization in L_2 applied to the family of linear independent functions

$$\mathbb{R} \ni \lambda \longrightarrow \lambda^m, \quad m \in \mathbb{Z}. \tag{4.3.2}$$

We choose the following order for the orthogonalization via the Gram-Schmidt procedure

$$\lambda^0; \quad \lambda^{-1}, \lambda^1; \quad \lambda^{-2}, \lambda^2; \quad \dots \quad ; \quad \lambda^{-n}, \lambda^n; \quad \dots \tag{4.3.3}$$

Applying the Gram-Schmidt orthogonalization procedure to (4.3.3) with real coefficients, we obtain an orthonormal polynomials system in the space L_2 (with respect to variables λ and λ^{-1}, so-called Laurent polynomials), numbered in such an order

$$P_{0;0}(\lambda); \quad P_{1;0}(\lambda), P_{1;1}(\lambda); \quad P_{2;0}(\lambda), P_{2;1}(\lambda); \quad \dots; \quad P_{n;0}(\lambda), P_{n;1}(\lambda); \quad \dots \tag{4.3.4}$$

where each polynomial has the form $P_{n;\alpha}(\lambda) = k_{n;\alpha}\lambda^{(-1)^{\alpha+1}n} + \cdots$, $n \in \mathbb{N}$, $\alpha = 0, 1$, $k_{n;\alpha} > 0$; here "$+\cdots$" denotes the next part of the corresponding polynomial; $P_0(\lambda) = P_{0;0}(\lambda) = 1$. In such a way $P_{n;\alpha}$ is some linear combination of

$$\begin{gathered} \{1;\ \lambda^{-1}, \lambda^1;\ \lambda^{-2}, \lambda^2;\ \dots;\ \lambda^{-(n-1)}, \lambda^{(n-1)};\ \dots; \lambda^{-n}\} \text{ for } \alpha = 0, \\ \{1;\ \lambda^{-1}, \lambda^1;\ \lambda^{-2}, \lambda^2;\ \dots;\ \lambda^{-(n-1)}, \lambda^{(n-1)}; \dots; \lambda^{-n}, \lambda^n\} \text{ for } \alpha = 1. \end{gathered} \tag{4.3.5}$$

Since *the family (4.3.2) is total in the space L_2, the sequence (4.3.4) is an orthonormal basis in this space.*

Denote by $\mathcal{P}_{n;\alpha}$ the real subspace spanned by elements $P_{n;\alpha}$, $n \in \mathbb{N}$, $\alpha = 0, 1$, from (4.3.5). It is clear that $\forall n \in \mathbb{N}$, we have

$$\begin{aligned} &\mathcal{P}_{0;0} \subset \mathcal{P}_{1;0} \subset \mathcal{P}_{1;1} \subset \mathcal{P}_{2;0} \subset \mathcal{P}_{2;1} \subset \cdots \subset \mathcal{P}_{n;0} \subset \mathcal{P}_{n;1} \subset \cdots, \\ &\mathcal{P}_{n;\alpha} = \{P_{0;0}(\lambda)\} \oplus \{P_{1;0}(\lambda)\} \\ &\qquad \oplus \{P_{1;1}(\lambda)\} \oplus \{P_{2;0}(\lambda)\} \oplus \{P_{2;1}(\lambda)\} \oplus \cdots \oplus \{P_{n;\alpha}(\lambda)\}, \end{aligned} \tag{4.3.6}$$

where $\{P_{m;\alpha}(\lambda)\}$, $m \in \mathbb{N}$, $\alpha = 0, 1$ denotes the one-dimensional real space generated by $P_{m;\alpha}(\lambda)$; $\mathcal{P}_{0;0} = \mathbb{R}$.

As was spoken above, in the sequel investigation, we need, instead of the space l_2, use the complex Hilbert space (4.3.1). Each vector $f \in \mathbf{l}_2$ has a form $f = (f_n)$, $f_n \in \mathcal{H}_n$, and hence,

$$\|f\|_{\mathbf{l}_2}^2 = \sum_{n=0}^{\infty} \|f_n\|_{\mathcal{H}_n}^2 < \infty, \quad (f, g)_{\mathbf{l}_2} = \sum_{n=0}^{\infty} (f_n, g_n)_{\mathcal{H}_n}, \quad \forall f, g \in \mathbf{l}_2.$$

For $n = 0$, the vector $f_0 \in \mathcal{H}_0$ has, in the standard orthonormal basis $\{e_{0;0}\}$ of the space $\mathbb{C}^1$, the representation $f_{0;0}$, hence, $f_0 = (f_{0;0})$. For $n \in \mathbb{N}$, coordinates of vector $f_n \in \mathcal{H}_n$, in the corresponding orthonormal basis $\{e_{n;0}, e_{n;1}\}$ of the space $\mathbb{C}^2$, are denoted by $(f_{n;0}, f_{n;1})$ and, hence, we have $f_n = (f_{n;0}, f_{n;1})$. By the way, it is clear, that the space $\mathbf{l}_2$ is almost isometric to the space $l_2 \times l_2$.

Using the orthonormal system (4.3.4) one can define a mapping of $\mathbf{l}_2$ into L_2. We put $\forall n \in \mathbb{N}_0$ and $\forall \lambda \in \mathbb{R}$, $P_n(\lambda) = (P_{n;0}, P_{n;1}(\lambda)) \in \mathcal{H}_n$, then

$$\mathbf{l}_2 \ni f = (f_n)_{n=0}^{\infty} \longmapsto (If)(\lambda) := \hat{f}(\lambda) = \sum_{n=0}^{\infty} (f_n, P_n(\lambda))_{\mathcal{H}_n} \in L_2. \tag{4.3.7}$$

Since for $n \in \mathbb{N}_0$ we get

$$(f_n, P_n(\lambda))_{\mathcal{H}_n} = f_{n;0} P_{n;0}(\lambda) + f_{n;1} P_{n;1}(\lambda)$$

and

$$\|f\|_{\mathbf{l}_2}^2 = \|(f_{0;0}, f_{1;0}, f_{1;1}, f_{2;0}, f_{2;1}, \ldots, f_{n;0}, f_{n;1}, \ldots)\|_{\mathbf{l}_2}^2,$$

we see that (4.3.7) is a mapping of the space $\mathbf{l}_2$ into L_2, and the use of the orthonormal system (4.3.4) shows that this mapping is isometric. The image of $\mathbf{l}_2$ under the mapping (4.3.7) coincides with the space L_2, because under our assumption the system (4.3.4) is an orthonormal basis in L_2 (the Laurent polynomials basis). Therefore, the mapping (4.3.7) is unitary transformation I that acts from $\mathbf{l}_2$ onto L_2.

Let us now dwell on some auxiliary constructions, which will be essentially used further. Namely, we prove technically important Theorems 4.3.5, 4.3.11, 4.3.12.

Let A be an arbitrary linear operator defined on $\mathrm{Dom}(A) = \mathbf{l}_{\text{fin}} \subset \mathbf{l}_2$, where $\mathbf{l}_{\text{fin}}$ denotes the set of finite vectors from $\mathbf{l}_2$. It is possible to construct the corresponding operator matrix $(a_{j,k})_{j,k=0}^{\infty}$, where for each $j, k \in \mathbb{N}_0$ the element $a_{j,k}$ is an operator acting from $\mathcal{H}_k$ into $\mathcal{H}_j$, so that $\forall f, g \in \mathrm{Dom}(A) = \mathbf{l}_{\text{fin}} \subset \mathbf{l}_2$, we have

$$(Af)_j = \sum_{k=0}^{\infty} a_{j,k} f_k, \quad j \in \mathbb{N}_0, \quad (Af, g)_{\mathbf{l}_2} = \sum_{j,k=0}^{\infty} (a_{j,k} f_k, g_j)_{\mathcal{H}_j}. \tag{4.3.8}$$

To proof (4.3.8) we only need to write the usual matrix of the operator A in the space $\mathbf{l}_2$ using the basis

$$(e_{0;0};\ e_{1;0}, e_{1;1};\ e_{2;0}, e_{2;1};\ \ldots;\ e_{n;0}, e_{n;1}, \ldots), \quad e_{0;0} = 1. \tag{4.3.9}$$

Then $a_{j,k}$, for each $j, k \in \mathbb{N}_0$, is the operator $\mathcal{H}_k \longrightarrow \mathcal{H}_j$ that has a matrix representation

$$a_{j,k;\alpha,\beta} = (Ae_{k;\beta}, e_{j;\alpha})_{\mathbf{l}_2}, \tag{4.3.10}$$

where $\alpha = 0, 1$ and $\beta = 0, 1$. We will write: $a_{j,k} = (a_{j,k;\alpha,\beta})_{\alpha,\beta=0}^{1,1}$, $j, k \in \mathbb{N}$ (taking into account cases: $a_{0,1} = (a_{0,1;\alpha,\beta})_{\alpha,\beta=0}^{0,1}$, $a_{1,0} = (a_{1,0;\alpha,\beta})_{\alpha,\beta=0}^{1,0}$ and $a_{0,0} = (a_{0,0;\alpha,\beta})_{\alpha,\beta=0}^{0,0} = a_{0,0;0,0}$).

We note that the first formula from (4.3.8) is fulfilled for $f \in \mathbf{l}_{\text{fin}}$; in the second formula $f \in \mathbf{l}_{\text{fin}}$, $g \in \mathbf{l}_2$.

Let us consider the image $\hat{A} = IAI^{-1} : L_2 \longrightarrow L_2$ of the above presented operator $A : \mathbf{l}_2 \longrightarrow \mathbf{l}_2$ by the mapping I (4.3.7). Its matrix in the basis (4.3.4), i.e.,

$$(P_{0;0}(\lambda);\ P_{1;0}(\lambda), P_{1;1}(\lambda);\ P_{2;0}(\lambda), P_{2,1}(\lambda);\ \ldots;\ P_{n;0}(\lambda), P_{n;1}(\lambda);\ \ldots),$$

is equal to the usual matrix of operator A regarded as the operator: $\mathbf{l}_2 \longrightarrow \mathbf{l}_2$ in the corresponding basis (4.3.9). By using (4.3.10) and the above mentioned procedure, we get the operator matrix $(a_{j,k})_{j,k=0}^{\infty}$ of $A : \mathbf{l}_2 \longrightarrow \mathbf{l}_2$. *By definition, this matrix is also the operator matrix of* $\hat{A} : L_2 \longrightarrow L_2$. It is clear that we can take an arbitrary essential self-adjoint operator in L_2 as the operator $\hat{A}$.

Return now to the objects connected with our measure $d\rho(\lambda)$ and sequences (4.3.3), (4.3.4).

Lemma 4.3.1 *For polynomials* $P_{n;\alpha}(\lambda)$ *in* (4.3.4) *and subspaces* $\mathcal{P}_{m,\beta}$ *with* (4.3.6), *relations*

$$\begin{aligned} \lambda P_{0;0}(\lambda) &= \lambda \in \mathcal{P}_{1;1}, \\ \lambda P_{n;0}(\lambda) &\in \mathcal{P}_{n;1}, \\ \lambda P_{n;1}(\lambda) &\in \mathcal{P}_{n+1;1}, \quad n \in \mathbb{N}. \end{aligned} \tag{4.3.11}$$

hold true.

Proof According to (4.3.4) the polynomial $P_{n;\alpha}(\lambda)$, $n \in \mathbb{N}$ is equal to some linear combination of

$$\{1; \lambda^{-1}, \lambda^{1}; \ldots; \lambda^{-(n-1)}, \lambda^{n-1}, \lambda^{(-1)^{\alpha+1}n}\}.$$

Hence, multiplying by λ, we obtain a linear combination

$$\{\lambda;\, 1, \lambda^2;\, \lambda^{-1}, \lambda^3;\, \ldots;\, \lambda^{-(n-2)}, \lambda^n, \lambda^{(-1)^{\alpha+1}n+1}\}$$

and such a linear combination belongs to $\mathcal{P}_{n;1}$ for $\alpha = 0$ and to $\mathcal{P}_{n+1;1}$ for $\alpha = 1$. The first inclusion in (4.3.11) is trivial. □

Lemma 4.3.2 *Let $\hat{A}$ be the (bounded and self-adjoint) multiplication by λ operator in the space L_2, namely,*

$$L_2 \ni \varphi(\lambda) \longmapsto (\hat{A}\varphi)(\lambda) = \lambda\varphi(\lambda) \in L_2.$$

The real operator matrix $(a_{j,k})_{j,k=0}^{\infty}$ of the operator $\hat{A}$ (i.e., $A = I^{-1}\hat{A}I$) has a tri-diagonal structure, i.e., $a_{j,k} = 0$ for $|j-k| > 1$.

Proof By using (4.3.10) for $e_{n;\gamma} = I^{-1}P_{n;\gamma}(\lambda)$, $n \in \mathbb{N}_0$, $\gamma = 0, 1$, we have

$$a_{j,k;\alpha,\beta} = (Ae_{k;\beta}, e_{j;\alpha})_{l_2} = \int\limits_{\mathbb{R}} \lambda P_{k;\beta}(\lambda) P_{j;\alpha}(\lambda)\, d\rho(\lambda), \quad \forall j, k \in \mathbb{N}_0, \tag{4.3.12}$$

where $\alpha, \beta = 0, 1$. From (4.3.11), we have $\lambda P_{k;\alpha} \in \mathcal{P}_{k+1;\alpha}$. According to (4.3.6), the integral in (4.3.12) is equal to zero for $j > k+1$ and for each $\alpha = 0, 1$.

On another hand side, in the integral (4.3.12), we can multiply by λ polynomial $P_{j;\alpha}(\lambda)$. Therefore, as above, we conclude that this integral is equal to zero for $k > j+1$ and for each $\beta = 0, 1$.

As result the integral in (4.3.12), i.e., elements $a_{j,k;\alpha,\beta}$, $j, k \in \mathbb{N}_0$, are equal to zero for $|j-k| > 1$, $\alpha, \beta = 0, 1$. (In previous considerations, it is necessary to take into account that $e_{0;0} = I^{-1}P_{0;0}(\lambda)$, $P_{0;0}(\lambda) = 1$). □

In such a way the matrix $(a_{j,k})_{j,k=0}^{\infty}$ of our multiplication operator $\hat{A}$ has a tri-diagonal block structure

$$\begin{bmatrix} a_{0,0} & a_{0,1} & 0 & 0 & 0 & \ldots \\ a_{1,0} & a_{1,1} & a_{1,2} & 0 & 0 & \ldots \\ 0 & a_{2,1} & a_{2,2} & a_{2,3} & 0 & \ldots \\ 0 & 0 & a_{3,2} & a_{3,3} & a_{3,4} & \ldots \\ \vdots & \vdots & \vdots & \vdots & \vdots & \ddots \end{bmatrix}. \tag{4.3.13}$$

From (4.3.12), we conclude that the symmetry

$$a_{j,k;\alpha,\beta} = a_{k,j;\beta,\alpha}, \quad j, k \in \mathbb{N}_0, \quad \alpha, \beta = 0, 1 \tag{4.3.14}$$

holds true.

A more detailed analysis of the expression (4.3.12) allows us to find out whether what elements of the $(a_{j,k;\alpha,\beta})_{\alpha,\beta=0}^{1,1}$ are zero and what are not zero in the general case for $|j-k| \leq 1$. We can also describe properties of the matrix with respect to the permutation of indexes j, k, and α, β.

Lemma 4.3.3 *Let $(a_{j,k})_{j,k=0}^{\infty}$ be the operator matrix* (4.3.13) *generates the multiplication by λ operator in L_2. Now, $a_{j,k} : \mathcal{H}_k \longrightarrow \mathcal{H}_j$, $a_{j,k} = (a_{j,k;\alpha,\beta})_{\alpha,\beta=0}^{1,1}$ are matrices of operators $a_{j,k}$ in the corresponding standard orthonormal basis. Then*

$$\begin{aligned} a_{j,j+1;0,0} &= a_{j,j+1;0,1} = 0, \\ a_{j+1,j;0,0} &= a_{j+1,j;1,0} = 0, \quad \forall j \in \mathbb{N}. \end{aligned} \tag{4.3.15}$$

If we choose inside of each pair $\{\lambda^{-m}, \lambda^{m}\}$ in (4.3.5) another order, then Lemma 4.3.3 is not true but it will be also possible to describe zero elements of matrices $(a_{j,k;\alpha,\beta})_{\alpha,\beta=0}^{1,1}$. Such matrices $(a_{j,k})_{j,k=0}^{\infty}$ have also tri-diagonal block structure and has zero elements but in another places.

Proof According to (4.3.12), we have

$$\begin{aligned} a_{j,j+1;0,0} &= \int\limits_{\mathbb{R}} \lambda P_{j+1;0}(\lambda) P_{j;0}(\lambda)\, d\rho(\lambda), \\ a_{j,j+1;0,1} &= \int\limits_{\mathbb{R}} \lambda P_{j+1;1}(\lambda) P_{j;0}(\lambda)\, d\rho(\lambda), \quad j \in \mathbb{N}. \end{aligned} \tag{4.3.16}$$

In the first integral in (4.3.16) according to (4.3.11), $\lambda P_{j;0}(\lambda) \in \mathcal{P}_{j;1}$ but $P_{j+1;0}(\lambda)$ is orthogonal to the last set (see (4.3.6)). Therefore, this integral is equal to zero. Similarly, the second integral in (4.3.16) also equal to zero: it is necessary to use the orthogonality of $P_{j+1;1}(\lambda)$ to $\mathcal{P}_{j;1}$.

Thus, the first two equalities in (4.3.15) are fulfilled. The second equalities are fulfilled according to (4.3.14). □

The mentioned above shows that the (2×2)-matrices in (4.3.13), $a_{j,j+1}$ and $a_{j+1,j}$, $j \in \mathbb{N}$ have the first rows and columns, respectively, zero. Taking into account (4.3.13), we can conclude that the self-adjoint matrix of the multiplication by λ operator is five-diagonal as usual scalar matrix in the usual basis of the space l_2.

Lemma 4.3.4 *Elements*

$$a_{0,1;0,1},\, a_{1,0;1,0}, \quad a_{j,j+1;1,1},\, a_{j+1,j;1,1}, \quad j \in \mathbb{N}, \tag{4.3.17}$$

of matrices $(a_{j,k})_{j,k=0}^{\infty}$ in (4.3.13), *are always positive.*

Proof The symmetry (4.3.14) shows that it is sufficient to show the positiveness of the second and the forth elements in (4.3.17). We start by $a_{1,0;1,0}$. Denote by $P'_{1;1}(\lambda)$ a non-normalized vector $P_{1;1}(\lambda)$ obtained from the Gram-Schmidt orthogonalization procedure. According to (4.3.3) and (4.3.4) we have

$$P'_{1;1}(\lambda) = \lambda - (\lambda, P_{1;0}(\lambda))_{L_2} P_{1;0}(\lambda) - (\lambda, 1)_{L_2}.$$

Therefore, using (4.3.12), we get

$$\begin{aligned} a_{1,0;1,0} &= \int_{\mathbb{R}} \lambda P_{1;1}(\lambda)\, d\rho(\lambda) = \|P'_{1;1}(\lambda)\|_{L_2}^{-1} \int_{\mathbb{R}} \lambda P'_{1;1}(\lambda)\, d\rho(\lambda) \\ &= \|P'_{1;1}(\lambda)\|_{L_2}^{-1} \int_{\mathbb{R}} \lambda(\lambda - (\lambda, P_{1;0}(\lambda))_{L_2} P_{1;0}(\lambda) - (\lambda, 1)_{L_2})\, d\rho(\lambda) \\ &= \|P'_{1;1}(\lambda)\|_{L_2}^{-1} (\|\lambda\|_{L_2}^2 - |(\lambda, P_{1;0}(\lambda))_{L_2}|^2 - |(\lambda, 1)_{L_2}|^2). \end{aligned} \tag{4.3.18}$$

The positiveness of the expression (4.3.18) follows from the Parseval equality for the decomposition of the function $\lambda \in L_2$ with respect to the orthonormal basis (4.3.4) in the space L_2. Namely

$$|(\lambda, 1)_{L_2}|^2 + |(\lambda, P_{1;0}(\lambda))_{L_2}|^2 + |(\lambda, P_{1;1}(\lambda))_{L_2}|^2 + \cdots = \|\lambda\|_{L_2}^2,$$

where $1 = P_{0;0}(\lambda)$.

Pass to the proof of the positiveness of $a_{j+1,j;1,1}$ where $j \in \mathbb{N}$. From (4.3.12), we have

$$a_{j+1,j;1,1} = \int_{\mathbb{R}} \lambda P_{j;1}(\lambda) P_{j+1;1}(\lambda)\, d\rho(\lambda). \tag{4.3.19}$$

According to (4.3.4) and (4.3.6), we have

$$P_{j;1}(\lambda) = k_{j;1}\lambda^j + R_{j;0}(\lambda), \tag{4.3.20}$$

where $R_{j;0}(\lambda)$ is some polynomial from $\mathcal{P}_{j;0}$, and coefficients $k_{j;1} > 0$. Multiply expression (4.3.20) by λ we get

$$\lambda P_{j;1}(\lambda) = k_{j;1}\lambda^{j+1} + \lambda R_{j;0}(\lambda), \quad \lambda R_{j;0}(\lambda) \in \mathcal{P}_{j;1} \tag{4.3.21}$$

(now, it is necessary to use the second inclusion from (4.3.11) and (4.3.6)).

Similarly to (4.3.20) we have

$$P_{j+1;1}(\lambda) = k_{j+1;1}\lambda^{j+1} + R_{j+1;0}(\lambda), \quad R_{j+1;0}(\lambda) \in \mathcal{P}_{j+1;0}, \tag{4.3.22}$$

where $k_{j+1;1} > 0$. Find λ^{j+1} from (4.3.22) and substitute it into (4.3.21). We get

$$\begin{aligned}\lambda P_{j;1}(\lambda) &= \frac{k_{j;1}}{k_{j+1;1}}(P_{j+1;1}(\lambda) - R_{j+1;0}(\lambda)) + \lambda R_{j;0}(\lambda) \\ &= \frac{k_{j;1}}{k_{j+1;1}} P_{j+1;1}(\lambda) - \frac{k_{j;1}}{k_{j+1;1}} R_{j+1;0}(\lambda) + \lambda R_{j;0}(\lambda).\end{aligned} \tag{4.3.23}$$

The second two terms in (4.3.23) belong to $\mathcal{P}_{j+1;0}$ and $\mathcal{P}_{j;1}$ respectively and are in any cases orthogonal to $P_{j+1;1}(\lambda)$. Therefore, the substitution of the expression (4.3.23) into (4.3.19) gives $a_{j,j+1;1,1} = \frac{k_{j;1}}{k_{j+1;1}} > 0$. □

In what follows we will use usual well known notations for elements $a_{j,k}$ of the Jacobi matrix (4.3.13):

$$\begin{aligned} a_n &= a_{n+1,n} : \mathcal{H}_n \longrightarrow \mathcal{H}_{n+1}, \\ b_n &= a_{n,n} \quad : \mathcal{H}_n \longrightarrow \mathcal{H}_n, \\ c_n &= a_{n,n+1} : \mathcal{H}_{n+1} \longrightarrow \mathcal{H}_n, \quad n \in \mathbb{N}_0. \end{aligned} \tag{4.3.24}$$

All previous investigation we summarize in the theorem.

Theorem 4.3.5 *The bounded self-adjoint multiplication by λ operator $\hat{A}$, (with a strong cyclic vector) in the space L_2, in the orthonormal basis* (4.3.4) *of polynomials, has the form of tri-diagonal block Jacobi type symmetric matrix $J = (a_{j,k})_{j,k=0}^{\infty}$ that acts in the space* (4.3.1), *namely,*

$$\mathbf{l}_2 = \mathcal{H}_0 \oplus \mathcal{H}_1 \oplus \mathcal{H}_2 \oplus, \quad \cdots, \quad \mathcal{H}_0 = \mathbb{C}^1, \quad \mathcal{H}_n = \mathbb{C}^2, \quad n \in \mathbb{N}. \tag{4.3.25}$$

In designations (4.3.24), *this matrix has the form*

$$J = \left[\begin{array}{c|cc|cc|cc|cc}
*b_0 & * & c_0 \; + & & & & & & \cdots \\
\hline
* & * & * & 0 & 0 & & & & \\
a_0 & & b_1 & & c_1 & & & 0 & \cdots \\
+ & * & * & * & + & & & & \\
\hline
& 0 & * & * & * & 0 & 0 & & \\
& & a_1 & & b_2 & & c_2 & & \cdots \\
& 0 & + & * & * & * & + & & \\
\hline
& & & 0 & * & * & * & 0 & 0 \\
& & 0 & & a_2 & & b_3 & & c_3 \;\; \cdots \\
& & & 0 & + & * & * & * & + \\
\hline
\vdots & \vdots & & \vdots & & \vdots & & \vdots & \ddots
\end{array}\right]. \tag{4.3.26}$$

In (4.3.26), $b_0 = b_{0;0,0}$ is a 1×1-matrix, i.e., scalar; the diagonal b_n consists 2×2-matrices: $b_n = (b_{n;\alpha,\beta})_{\alpha,\beta=0}^{1,1}$, $\forall n \in \mathbb{N}$; a_0 is a 1×2-matrix: $a_0 = (a_{0;\alpha,\beta})_{\alpha,\beta=0}^{1,0}$; a_n is a 2×2-matrix: $a_n = (a_{n;\alpha,\beta})_{\alpha,\beta=0}^{1,1}$ $\forall n \in \mathbb{N}$; c_0 is a 2×1-matrix: $c_0 = (c_{0;\alpha,\beta})_{\alpha,\beta=0}^{0,1}$; c_n is a 2×2-matrix: $c_n = (c_{n;\alpha,\beta})_{\alpha,\beta=0}^{1,1}$, $\forall n \in \mathbb{N}$. In the matrices a_n and c_n, elements

$$a_{n;0,0},\, a_{n;1,0}, \quad c_{n;0,0},\, c_{n;0,1}, \quad \forall n \in \mathbb{N} \tag{4.3.27}$$

are always equal to zero; elements

$$a_{0;1,0},\, c_{0;0,1},\, a_{n;1,1},\, c_{n;1,1}, \quad \forall n \in \mathbb{N} \tag{4.3.28}$$

are always positive.

Thus, it is possible to say, that $\forall n \in \mathbb{N}$, every left columns matrices a_n (starting from the $n = 1$) and every over rows of the matrices c_n (starting from the $n = 1$) contain zero elements. All positive elements in (4.3.26) *are denoted by "+".*

Thus, the matrix (4.3.26) *in the scalar form is five-diagonals. It is symmetric in the basis (4.3.4), i.e., $b_{n;\alpha,\beta} = b_{n;\beta,\alpha}$, $c_{n;\alpha,\beta} = a_{n;\beta,\alpha}$, $n \in \mathbb{N}_0$, $\alpha, \beta = 0, 1$.*

For the considered operator $A = I^{-1}\hat{A}I$, we have

$$\begin{gathered}(Af)_n = (Jf)_n = a_{n-1} f_{n-1} + b_n f_n + c_n f_{n+1}, \\ \forall f, g \in \mathrm{Dom}(A) = \mathbf{l}_{\mathrm{fin}} \subset \mathbf{l}_2, \quad f_{-1} := 0, \ n \in \mathbb{N}_0.\end{gathered} \tag{4.3.29}$$

We want to do some enough simple, but essential remarks concerning operator A generated in (4.3.25) by matrix J from (4.3.26) in the case when such operator is unbounded and has the inverse one A^{-1}.

Remark 4.3.6 Assume that the measure $d\rho(\lambda)$, in the beginning of Sect. 4.3, has an arbitrary support from $\mathbb{R}$ and all functions

$$\mathbb{R} \ni \lambda \longmapsto \lambda^m, \quad m \in \mathbb{Z} \tag{4.3.30}$$

belong to $L_2(\mathbb{R}, d\rho(\lambda)) = L_2$ and are linearly independent. In this case we can repeat constructions (4.3.3)–(4.3.29), but now the operator A, defined in the space $\mathbf{l}_2$ on the set $\mathbf{l}_{\mathrm{fin}}$, is only Hermitian with real elements (see (4.3.14)) and, therefore, has equal defect numbers. Moreover, *in what follows we will assume that the set of functions* (4.3.30) *is total in L_2.*

Remark 4.3.7 Consider the one-to-one mapping between $\mathbb{R} \setminus \{0\}$ and $\mathbb{R}$

$$\mathbb{R}_0 := \mathbb{R} \setminus \{0\} \ni \lambda \longmapsto \mu = \lambda^{-1} =: \varphi(\lambda) \in \mathbb{R}. \tag{4.3.31}$$

We will assume that our given measure $d\rho(\lambda)$ on $\mathbb{R}$ is such that the point "0" not belong to its support, i.e., $0 \notin \mathrm{supp}(d\rho(\lambda))$. In such a case the mapping (4.3.31) transfers this measure into the Borel measure $d\sigma(\mu)$ on $\mathbb{R}$ (i.e., $\forall\alpha \in \mathfrak{B}(\mathbb{R}), \sigma(\alpha) = \rho(\varphi^{-1}(\alpha))$). For an arbitrary function $\mathbb{R}_0 \ni \lambda \longmapsto F(\lambda) \in \mathbb{C}$ we have

$$\int\limits_{\mathbb{R}} F(\mu^{-1})\, d\sigma(\mu) = \int\limits_{\mathbb{R}} F(\lambda)\, d\rho(\lambda); \quad F(\mu^{-1}) =: (I_\varphi F)(\mu). \tag{4.3.32}$$

Consider the Hilbert space of complex-valued functions of a variable μ on the real axis $\mathbb{R}$, i.e., $L_2(\mathbb{R}, d\sigma(\mu)) = L_{2,\varphi}$. Then, form (4.3.32), we conclude that the operator $I_\varphi : L_2 \longmapsto L_{2,\varphi}$, acting by the rule (4.3.32), is isometric between spaces $L_2(\mathbb{R}, d\sigma(\mu)) = L_{2,\varphi}$ and $L_2(\mathbb{R}, d\rho(\lambda)) = L_2$.

Let the operator A, constructed in $\mathbf{l}_2$ according to (4.3.29) by the matrix J, is essentially self-adjoint and invertible. Then its L_2-image, i.e., the multiplication by λ operator $\hat{A}$, defined at first on linear combinations of functions (4.3.30), is also invertible and $0 \notin \operatorname{supp}(d\rho(\lambda))$. This inverse operator $\hat{A}^{-1}$ as an operator in $\mathbf{l}_2$ is generated by the algebraically inverse matrix J^{-1}.

The mapping (4.3.31) shows that this matrix, in the space constructed above *using the measure* $d\sigma(\mu)$, has also the form (4.3.26), but other polynomials of the type (4.3.4) correspond to it. Of course, it is easy to calculate this matrix in previous basis, connected with $d\rho(\lambda)$, i.e., the matrix J^{-1} which we denote now by K.

Note that the construction (4.3.31), (4.3.32) makes it possible to find interesting examples of measures $d\rho(\lambda)$ for which the set (4.3.30) is total in $L_2(\mathbb{R}, d\rho(\lambda))$.

It is very interesting and unexpectedly that this inverse matrix $J^{-1} = K$ to the tri-diagonal matrix J is also tri-diagonal. This result is a consequence of the way of construction of the basis (4.3.4): these polynomials are linear combinations of λ^m and λ^{-n}, $m, n \in \mathbb{N}_0$.

We will prove corresponding results since the form of matrix K is a little another as J. At first, instead of Lemma 4.3.3, we have

Lemma 4.3.8 *For polynomials $P_{n;\alpha}(\lambda)$ from* (4.3.4) *and subspaces $\mathcal{P}_{m,\beta}$, relations*

$$\begin{aligned} \lambda^{-1} P_{0;0}(\lambda) = \lambda^{-1} &\in \mathcal{P}_{1;0}, \\ \lambda^{-1} P_{n;0}(\lambda) &\in \mathcal{P}_{n+1;0}, \\ \lambda^{-1} P_{n;1}(\lambda) &\in \mathcal{P}_{n+1;0}, \quad n \in \mathbb{N}. \end{aligned} \tag{4.3.33}$$

hold true.

Proof This proof is the same as the proof of Lemma 4.3.3, and follows from (4.3.5), (4.3.6) and (4.3.7). □

Denote by $(p_{j,k})_{j,k=0}^{\infty}$, $(p_{j,k;\alpha,\beta})_{\alpha,\beta=0,0}^{1,1}$ the operator matrix K of the multiplication by λ^{-1} operator in the previous space $L_2(\mathbb{R}, d\rho(\lambda)) = L_2$; this matrix is constructed as before (see (4.3.8), (4.3.9) and (4.3.10)) using the basis (4.3.4).

We can repeat now Lemma 4.3.4 for K: using (4.3.33) instead of (4.3.11), we assert that the integral

$$p_{j,k;\alpha,\beta} = \int_{\mathbb{R}} \lambda^{-1} P_{k;\beta}(\lambda) P_{j;\alpha}(\lambda)\, d\rho(\lambda), \quad j,k \in \mathbb{N}_0, \quad \alpha,\beta = 0,1 \qquad (4.3.34)$$

is equal to zero if $|j-k| > 1$. Thus, our matrix K has a form (4.3.13). Of course, $K = J^{-1}$ is self-adjoint and symmetric.

Instead of equalities of Lemma 4.3.3, we have some other equalities.

Lemma 4.3.9 *For elements* $p_{j,k} = (p_{j,k;\alpha,\beta})_{\alpha,\beta=0,0}^{1,1}$ *of the matrix, we have equalities*

$$\begin{aligned} p_{0,1;0,1} &= p_{1,0;1,0} = 0, \\ p_{j,j+1;0,1} &= p_{j,j+1;1,1} = 0, \\ p_{j+1,j;1,0} &= p_{j+1,j;1,1} = 0, \quad j \in \mathbb{N}. \end{aligned} \qquad (4.3.35)$$

Proof It is also similar to the proof of Lemma 4.3.3 but with using (4.3.33) instead of (4.3.11). For example, we have according to (4.3.34),

$$p_{j,j+1;0,1} = \int_{\mathbb{R}} \lambda^{-1} P_{j+1,1}(\lambda) P_{j;0}(\lambda)\, d\rho(\lambda), \quad j \in \mathbb{N}_0. \qquad (4.3.36)$$

By using the second inclusion from (4.3.33), we get that $\lambda^{-1} P_{j;0}(\lambda) \in \mathcal{P}_{j+1;0}$, therefore, this function is orthogonal to $P_{j+1;1}(\lambda)$ and the integral (4.3.36) is equal to zero. Similarly, using the third inclusion from (4.3.33), we get that $p_{j,j+1;1,1} = 0$. The rest equalities in (4.3.35) are valid due to the symmetry of K. □

The analog of Lemma 4.3.4 is the following assertion.

Lemma 4.3.10 *Elements*

$$\begin{aligned} &p_{0,1;0,0}, \quad p_{1,0;0,0}, \\ &p_{j,j+1;0,0}, \quad p_{j+1,j;0,0}, \quad j \in \mathbb{N}, \end{aligned} \qquad (4.3.37)$$

of matrices $(p_{j,k})_{j,k=0}^{\infty}$, *are always positive.*

Proof As before, we start from $p_{1,0;0,0}$. Denote by $P'_{1;0}(\lambda)$ a non-normalized vector $P_{1;0}(\lambda)$ obtained from the orthogonalization procedure, i.e., $P'_{1;0}(\lambda) = \lambda^{-1} -$

$(\lambda^{-1}, 1)_{L_2}$. Therefore as in (4.3.18) due to the Parseval equality for λ^{-1}, we have

$$\begin{aligned} p_{1,0;0,0} &= \int_{\mathbb{R}} \lambda^{-1} P_{1;0}(\lambda)\, d\rho(\lambda) \\ &= \|P'_{1;0}(\lambda)\|_{L_2}^{-1} \int_{\mathbb{R}} \lambda^{-1}(\lambda^{-1} - (\lambda^{-1}, 1)_{L_2})\, d\rho(\lambda) \\ &= \|P'_{1;0}(\lambda)\|_{L_2}^{-1} (\|\lambda^{-1}\|_{L_2}^2 - |(\lambda^{-1}, 1)_{L_2}|^2) > 0. \end{aligned}$$

As in (4.3.19), consider the fourth element in (4.3.37). We have

$$p_{j+1,j;0,0} = \int_{\mathbb{R}} \lambda^{-1} P_{j;0}(\lambda) P_{j+1;0}(\lambda)\, d\rho(\lambda), \quad j \in \mathbb{N}. \tag{4.3.38}$$

According to (4.3.4) and (4.3.6), we have

$$P_{j;0}(\lambda) = k_{j;0}\lambda^{-j} + R_{j-1;1}(\lambda), \tag{4.3.39}$$

where $R_{j-1;1}(\lambda)$ is some polynomial from $\mathcal{P}_{j-1;1}$, $k_{j;0} > 0$. Multiplying (4.3.39) by λ^{-1}, we get

$$\lambda^{-1} P_{j;0}(\lambda) = k_{j;0}\lambda^{-(j+1)} + \lambda^{-1} R_{j-1;1}(\lambda), \tag{4.3.40}$$

where $\lambda^{-1} R_{j-1;1}(\lambda) \in \mathcal{P}_{j-1;0}$.

Similarly to (4.3.39), we have

$$P_{j+1;0}(\lambda) = k_{j+1;0}\lambda^{-(j+1)} + R_{j;1}(\lambda), \quad R_{j;1}(\lambda) \in \mathcal{P}_{j;1}; \quad k_{j+1;0} > 0. \tag{4.3.41}$$

Take $\lambda^{-(j+1)}$ from (4.3.41) and substitute it into (4.3.40), thus

$$\begin{aligned} \lambda^{-1} P_{j;0}(\lambda) &= \frac{k_{j;0}}{k_{j+1;0}} (P_{j+1;0}(\lambda) - R_{j;1}(\lambda)) + \lambda^{-1} R_{j-1;1}(\lambda) \\ &= \frac{k_{j;0}}{k_{j+1;0}} P_{j+1;0}(\lambda) - \frac{k_{j;0}}{k_{j+1;0}} R_{j;1}(\lambda) + \lambda^{-1} R_{j-1;1}(\lambda). \end{aligned} \tag{4.3.42}$$

In this expression the second two terms belong to $\mathcal{P}_{j;1}$ and $\mathcal{P}_{j-1;0}$ respectively and are orthogonal to $P_{j+1;0}(\lambda)$. After substitution of (4.3.42) into (4.3.38) we get: $p_{j+1,j;0,0} > 0$.

The positivity of the rest elements from (4.3.37) follows from the symmetry of K. □

The results of the last considerations we can formulate as a theorem.

Theorem 4.3.11 *Let the measure $d\rho(\lambda)$ be such that the multiplication by λ^{-1} operator $\hat{A}$ is self-adjoint and invertible in L_2. Then the bounded inverse to $\hat{A}$ operator is generated in the space* $\mathbf{l}_2$ (4.3.25) *by the tri-diagonal block Jacobi type symmetric matrix $J^{-1} = K$ of the form similar to* (4.3.26), *namely,*

$$K = (p_{j,k})_{j,k=0}^{\infty}, \quad p_n := p_{n+1,n}, \; q_n := p_{n,n}, \; r_n := p_{n,n+1}, \; n \in \mathbb{N}_0;$$

$$K = \begin{bmatrix} q_0\begin{pmatrix}*\end{pmatrix} & r_0\begin{pmatrix}+ & 0\end{pmatrix} & 0 & 0 & \cdots \\ p_0\begin{pmatrix}+\\0\end{pmatrix} & q_1\begin{pmatrix}* & *\\ * & *\end{pmatrix} & r_1\begin{pmatrix}+ & 0\\ * & 0\end{pmatrix} & 0 & \cdots \\ 0 & p_1\begin{pmatrix}+ & *\\ 0 & 0\end{pmatrix} & q_2\begin{pmatrix}* & *\\ * & *\end{pmatrix} & r_2\begin{pmatrix}+ & 0\\ * & 0\end{pmatrix} & \cdots \\ 0 & 0 & p_2\begin{pmatrix}+ & *\\ 0 & 0\end{pmatrix} & q_3\begin{pmatrix}* & *\\ * & *\end{pmatrix} & r_3\begin{pmatrix}+ & 0\\ * & 0\end{pmatrix} \; \cdots \\ \vdots & \vdots & \vdots & \vdots & \ddots \end{bmatrix}, \tag{4.3.43}$$

where

$$p_{0;1,0} = r_{0;0,1} = 0, \quad p_{n;1,0} = p_{n;1,1} = r_{n;0,1} = r_{n;1,1} = 0, \quad n \in \mathbb{N},$$
$$p_{n;0,0}, r_{n;0,0} > 0, \quad n \in \mathbb{N}_0. \tag{4.3.44}$$

The action of the operator is defined by the expression

$$(J^{-1}f)_n = (Kf)_n = p_{n-1}f_{n-1} + q_n f_n + r_n f_{n+1},$$
$$n \in \mathbb{N}_0, \quad f_{-1} := 0, \quad f \in \mathbf{l}_2. \tag{4.3.45}$$

We will consider now some simple but essential for us generalization of the last results. Namely, we will assume that the multiplication by λ operator $\hat{A}$ is self-adjoint in L_2, but the bounded $\hat{A}^{-1}$, possibly, not exist. Remind that all functions $\mathbb{R} \ni \lambda \longmapsto \lambda^m$, $m \in \mathbb{Z}_-$ belong to L_2 (according to the remarked above).

Therefore, the multiplication by λ^{-1} operator in L_2 exists and is defined on functions F from L_2, for which $\lambda^{-1}F(\lambda) \in L_2$. The set of such functions is linear and of course, dense in L_2. This operator is (algebraically) inverse to $\hat{A}$, we denote it also by $\hat{A}^{-1}$. Thus, we can formulate the following assertion: *consider the general situation: all functions* (4.3.30) *belong to L_2 and the corresponding multiplication by λ operator $\hat{A}$ is self-adjoint. Then, for $\hat{A}$, the algebraically inverse operator $\hat{A}^{-1}$*

exists in a dense domain $\mathrm{Dom}(\hat{A}^{-1}) \supset \mathrm{Ran}(\hat{A})$ in L_2.

$$(\hat{A}^{-1}\hat{A})F = \lambda^{-1}(\lambda F(\lambda)) = F(\lambda), \quad F \in \mathrm{Dom}(\hat{A}),$$

$$\mathrm{Dom}(\hat{A}^{-1}) = \{G \in L_2 \mid \lambda^{-1}G(\lambda) \in L_2\}. \tag{4.3.46}$$

Theorem 4.3.12 *Consider the Laurent polynomials basis* (4.3.4) *in* L_2 *and transfer this space into* $\mathbf{l}_2$*. The algebraically inverse operator* $\hat{A}^{-1}$ *can be rewritten as an operator in* $\mathbf{l}_2$ *generated by the matrix* K *of the form* (4.3.43) *on the set* $\mathbf{l}_{\mathrm{fin}}$*. Such a matrix has properties* (4.3.44) *and acts according to the rule* (4.3.45).

We will call it as the algebraically inverse matrix $J^{-1} = K$ *to* J*;* $J^{-1}Jf = f = JJ^{-1}f$, $f \in \mathbf{l}_{\mathrm{fin}}$.

Proof At first we note that every Laurent polynomial belongs to $\mathrm{Dom}(\hat{A}^{-1})$. Indeed such a polynomial is a linear combination of finite number of functions λ^m, $m \in \mathbb{Z}$. But according to the definition of $\mathrm{Dom}(\hat{A}^{-1})$ in (4.3.46), $\forall m \in \mathbb{Z}$, $\lambda^m \in \mathrm{Dom}(\hat{A}^{-1})$ since $\lambda^{-1}\lambda^m \in L_2$.

Therefore we can construct the matrix K of the type (4.3.10) of the operator $\hat{A}^{-1}$ using polynomials (4.3.4). This matrix has the structure (4.3.43) with properties (4.3.44) since we can repeat for this matrix Lemmas 4.3.24, 4.3.9, 4.3.10: all integrals of type (4.3.34) now exist. The last equality in the formulation of the theorem follows from (4.3.46) and, for example, from the remarks connected with (4.3.31). □

4.4 Direct and Inverse Spectral Problems Corresponding to Tri-Diagonal Block Jacobi-Laurent Matrices Generating Self-Adjoint Operators

In this section we will consider the operator $\mathbf{J}$ in the space $\mathbf{l}_2$ (4.3.1), (4.3.25), generated by the matrix J (4.3.26) with conditions on its elements, (4.3.27), (4.3.28) noted in Theorem 4.3.5. Moreover, we will demand that the algebraically inverse matrix J^{-1} exists and satisfies conditions of Theorem 4.3.11.

At first we remind some general facts concerning a rigging for the case of space $\mathbf{l}_2$ and the generalized eigenfunction expansion for self-adjoint operators acting on this space. In addition to the space $\mathbf{l}_2$ we consider its rigging

$$(\mathbf{l}_{\mathrm{fin}})' \supset \mathbf{l}_2(p^{-1}) \supset \mathbf{l}_2 \supset \mathbf{l}_2(p) \supset \mathbf{l}_{\mathrm{fin}}, \tag{4.4.1}$$

where $\mathbf{l}_2(p)$ is weighted $\mathbf{l}_2$ space with a weight $p = (p_n)_{n=0}^{\infty}$, $p_n \geq 1$, $(p^{-1} = (p_n^{-1})_{n=0}^{\infty})$. In our case $\mathbf{l}_2(p)$ is the Hilbert space of sequences $f = (f_n)$, $f_n \in \mathcal{H}_n$

for which we have

$$\|f\|^2_{\mathbf{l}_2(p)} = \sum_{n=0}^{\infty} \|f_n\|^2_{\mathcal{H}_n} p_n, \quad (f, g)_{\mathbf{l}_2(p)} = \sum_{n=0}^{\infty} (f_n, g_n)_{\mathcal{H}_n} p_n.$$

The space $\mathbf{l}_2(p^{-1})$ is defined similarly; recall that $\mathbf{l}_{\text{fin}}$ is the space of finite sequences and $(\mathbf{l}_{\text{fin}})'$ is the space conjugate to $\mathbf{l}_{\text{fin}}$ and equal to the space $\mathbf{l}$ of all sequences $f = (f_n)_{n=1}^{\infty}$, $f_n \in \mathcal{H}_n$. It is easy to show that the imbedding $\mathbf{l}_2(p) \hookrightarrow \mathbf{l}_2$ is quasi-nuclear if $\sum\limits_{n=0}^{\infty} p_n^{-1} < \infty$.

Let A be an arbitrary self-adjoint operator standardly connected with the chain (4.4.1). According to the projection spectral theorem (see Sect. 4.2), such an operator has a representation

$$Af = \int_{\mathbb{R}} \lambda \Phi(\lambda)\, d\sigma(\lambda) f, \quad f \in \mathbf{l}_2, \tag{4.4.2}$$

where $\Phi(\lambda) : \mathbf{l}_2(p) \longrightarrow \mathbf{l}_2(p^{-1})$ is an operator of the generalized projection and $d\sigma(\lambda)$ is spectral measure. For every $f \in \mathbf{l}_{\text{fin}}$ the projection $\Phi(\lambda)f \in \mathbf{l}_2(p^{-1})$ is a generalized eigenvector of the operator A with the corresponding eigenvalues λ. For all $f, g \in \mathbf{l}_{\text{fin}}$, we have the Parseval equality

$$(f, g)_{\mathbf{l}_2} = \int_{\mathbb{R}} (\Phi(\lambda) f, g)_{\mathbf{l}_2}\, d\sigma(\lambda); \tag{4.4.3}$$

after extending by continuity, the equality (4.4.3) holds true $\forall f, g \in \mathbf{l}_2$.

Let us denote by π_n the operator of orthogonal projection in $\mathbf{l}_2$ on $\mathcal{H}_n$, $n \in \mathbb{N}_0$. Hence, $\forall f = (f_n) \in \mathbf{l}_2$, we have $f_n = \pi_n f$. This operator acts similarly on the spaces $\mathbf{l}_2(p)$ and $\mathbf{l}_2(p^{-1})$ (but possibly with the norm which is not equal to one).

Let us consider the operator matrix $(\Phi_{j,k}(\lambda))_{j,k=0}^{\infty}$, where

$$\Phi_{j,k}(\lambda) = \pi_j \Phi(\lambda) \pi_k \ : \mathbf{l}_2 \longrightarrow \mathcal{H}_j \quad (\text{or} \quad \mathcal{H}_k \longrightarrow \mathcal{H}_j). \tag{4.4.4}$$

Now, we can rewrite the Parseval equality (4.4.3) in the form

$$\begin{aligned} (f, g)_{\mathbf{l}_2} &= \sum_{j,k=0}^{\infty} \int_{\mathbb{R}} (\Phi(\lambda)\pi_k f, \pi_j g)_{\mathbf{l}_2}\, d\sigma(\lambda) \\ &= \sum_{j,k=0}^{\infty} \int_{\mathbb{R}} (\pi_j \Phi(\lambda)\pi_k f, g)_{\mathbf{l}_2}\, d\sigma(\lambda) \\ &= \sum_{j,k=0}^{\infty} \int_{\mathbb{R}} (\Phi_{j,k}(\lambda) f_k, g_j)_{\mathbf{l}_2}\, d\sigma(\lambda), \quad \forall f, g \in \mathbf{l}_2. \end{aligned} \tag{4.4.5}$$

Let us now pass to the study of a more special self-adjoint operator A that acts on the space $\mathbf{l}_2$. Namely, let $A = \mathbf{J}$, where $\mathbf{J}$ is the closed operator generated on the space $\mathbf{l}_2$ by the matrix (4.3.26) with conditions (4.3.27), (4.3.28) by the rule

$$\mathbf{l}_2 \supset \mathbf{l}_{\text{fin}} \ni f \longmapsto \mathbf{J}f := Jf \in \mathbf{l}_2. \tag{4.4.6}$$

We will assume also that $\mathbf{J}$ is *self-adjoint*. From (4.4.6), (4.3.29), it is easy to conclude that our operator $\mathbf{J}$ is standardly connected with chain (4.4.1). Thus, above stated results of the type (4.4.2)—(4.4.5) we can apply to the operator $\mathbf{J}$.

Additionally, we will demand *that for matrix J, there exists its algebraically inverse matrix J^{-1} which satisfies the conditions of Theorem* 4.3.11: (4.3.44), (4.3.45). *Conditions for coefficients of the matrix J, under which it will have an inverse one are given in the next sections.*

The existence of such J^{-1} is very essential condition. *Such a matrix J we will be also called the (self-adjoint) Jacobi-Laurent matrix.*

Our first aim is to rewrite the Parseval equality (4.4.5) for our $A = \mathbf{J}$ in terms of generalized eigenvectors of $\mathbf{J}$. We prove two essential lemmas.

Lemma 4.4.1 *Let $\varphi(\lambda) = (\varphi_n(\lambda))_{n=0}^{\infty}$, $\varphi_n(\lambda) \in \mathcal{H}_n$, $\lambda \in \mathbb{R}$ be a generalized eigenvector from $(\mathbf{l}_{\text{fin}})'$ of the operator $\mathbf{J}$, constructed by self-adjoint Jacobi-Laurent matrix J.*

We assert that $\varphi(\lambda)$, $\forall\lambda \in \mathbb{R}$ is a solution in $(\mathbf{l}_{\text{fin}})'$ of the difference equation

$$(J\varphi(\lambda))_n = a_{n-1}\varphi_{n-1}(\lambda) + b_n\varphi_n(\lambda) + c_n\varphi_{n+1}(\lambda) = \lambda\varphi_n(\lambda),$$
$$n \in \mathbb{N}_0, \quad \varphi_{-1}(\lambda) = 0 \tag{4.4.7}$$

with the initial condition $\varphi_0(\lambda) = \varphi_0$ and has the representation

$$\varphi_n(\lambda) = S_n(\lambda)\varphi_0; \quad S_0(\lambda) = 1, \quad S_n(\lambda) = (S_{n;0}, S_{n;1}), \quad n \in \mathbb{N}. \tag{4.4.8}$$

Here $S_{n;\alpha}$, $\alpha = 0, 1$ are Laurent polynomials of variables λ, λ^{-1} and these polynomials have the form

$$S_{n;\alpha}(\lambda) = l_{n;\alpha}\lambda^{(-1)^{\alpha+1}n} + w_{n;\alpha}(\lambda), \quad n \in \mathbb{N}, \quad \alpha = 0, 1. \tag{4.4.9}$$

In (4.4.9) $l_{n;\alpha} > 0$ and $w_{n;\alpha}(\lambda)$ is some linear combination with real coefficients λ^j, $j \in \{0, -1, 1, -2, 2, \cdots, -(n-1), -\alpha n\}$, i.e., belongs to $\mathcal{P}_{n-1;1}$ if $\alpha = 0$ and to $\mathcal{P}_{n;0}$, if $\alpha = 1$.

Proof At first we remaind that by definition $\varphi(\lambda) \in (\mathbf{l}_{\text{fin}})' = \mathbf{l}$ is a generalized eigenvector with an eigenvalue λ for the operator $\mathbf{J}$ standardly connected with the rigging (4.4.1) if the equality

$$(\varphi(\lambda), Jf)_{\mathbf{l}_2} = (\varphi(\lambda), \mathbf{J}f)_{\mathbf{l}_2} = \lambda(\varphi(\lambda), f)_{\mathbf{l}_2}, \quad f \in \mathbf{l}_{\text{fin}}. \tag{4.4.10}$$

holds true.

By using (4.3.29) and the arbitrariness of f, we conclude, from (4.4.10), that

$$(J\varphi(\lambda))_n = a_{n-1}\varphi_{n-1}(\lambda) + b_n\varphi_n(\lambda) + c_n\varphi_{n+1}(\lambda) = \lambda\varphi_n(\lambda),$$
$$n \in \mathbb{N}_0, \quad \varphi_{-1}(\lambda) = 0. \tag{4.4.11}$$

For the matrix J^{-1} (4.3.43), we have also a similar equality. Namely, using Theorem 4.3.12 (the equality $JJ^{-1}f = f$, $f \in \mathbf{l}_{\text{fin}}$), we get

$$(\varphi(\lambda), f)_{\mathbf{l}_2} = (\varphi(\lambda), JJ^{-1}f)_{\mathbf{l}_2} = \lambda(\varphi(\lambda), J^{-1}f)_{\mathbf{l}_2}, \quad \text{i.e.,}$$
$$(\varphi(\lambda), J^{-1}f)_{\mathbf{l}_2} = \lambda^{-1}(\varphi(\lambda), f)_{\mathbf{l}_2}, \quad \forall f \in \mathbf{l}_{\text{fin}}. \tag{4.4.12}$$

Explain that the matrix J^{-1} is tri-diagonal, therefore, $J^{-1}f \in \mathbf{l}_{\text{fin}}$ and the second equality in (4.4.12) follows from (4.4.10).

Similarly to (4.4.11), the last equality in (4.4.12) and (4.3.43) give

$$(J^{-1}\varphi(\lambda))_n = p_{n-1}\varphi_{n-1}(\lambda) + q_n\varphi_n(\lambda) + r_n\varphi_{n+1}(\lambda) = \lambda^{-1}\varphi_n(\lambda),$$
$$n \in \mathbb{N}_0, \quad \varphi_{-1}(\lambda) = 0, \quad \forall\lambda \in \mathbb{R}\setminus\{0\}. \tag{4.4.13}$$

First, we will give some explanations for further transformations and calculations. In addition to the two equalities (4.4.11) and (4.4.13), we have

$$((J + J^{-1})\varphi(\lambda))_n = (\lambda + \lambda^{-1})\varphi_n(\lambda), \quad n \in \mathbb{N}_0, \quad \varphi_{-1}(\lambda) = 0. \tag{4.4.14}$$

The matrix $J + J^{-1}$ is also block tri-diagonal acting in the space $\mathbf{l}_2$. But from (4.3.26)–(4.3.28) and (4.3.43), (4.3.44), we see that its blocks on two side diagonals for $n \in \mathbb{N}$ are 2×2 invertible matrices. Such a form of the matrix $J + J^{-1}$ actually shows that similarly to the classical Jacobi matrices, we can, from (4.4.14) step by step, find a generalized eigenvector $\varphi(\lambda) = (\varphi_n(\lambda))_{n=0}^{\infty}$ and $\forall n \in \mathbb{N}$ $\varphi_n(\lambda)$ is a polynomial with respect to $\lambda + \frac{1}{\lambda}$, i.e., it is a Laurent polynomial. But it is essential for us to get for every $\varphi_n(\lambda)$ an exact representation (4.4.8), (4.4.9). Therefore we give below more precise calculation.

Consider equalities (4.4.11), (4.4.13) for $n = 1$. We have

$$b_0\varphi_0 + c_0\varphi_1(\lambda) = \lambda\varphi_0,$$
$$q_0\varphi_0 + r_0\varphi_1(\lambda) = \lambda^{-1}\varphi_0,$$

i.e.,

$$c_{0;0,0}\varphi_{1;0}(\lambda) + c_{0;0,1}\varphi_{1;1}(\lambda) = (\lambda - b_0)\varphi_0,$$
$$r_{0;0,0}\varphi_{1;0}(\lambda) + r_{0;0,1}\varphi_{1;1}(\lambda) = (\lambda^{-1} - q_0)\varphi_0.$$

The last two equalities we can regard as a linear system of equations with respect to unknowns $\varphi_{1;0}(\lambda)$, $\varphi_{1;1}(\lambda)$, $\varphi_0 \in \mathbb{C}$ are given. According to (4.3.26), (4.3.43), we have for the matrices of this system and its solutions

$$D_1 = \begin{bmatrix} c_{0;0,0} & c_{0;0,1} \\ r_{0;0,0} & r_{0;0,1} \end{bmatrix} = \begin{bmatrix} * & + \\ + & 0 \end{bmatrix},$$

$$\Delta_1 = \mathrm{Det} D_1 = |D_1| = -r_{0;0,0}c_{0;0,1} < 0,$$

$$\varphi_{1;0}(\lambda) = \Delta_1^{-1} \begin{vmatrix} (\lambda - b_0)\varphi_0 & + \\ (\lambda^{-1} - q_0)\varphi_0 & 0 \end{vmatrix},$$

$$\varphi_{1;1}(\lambda) = \Delta_1^{-1} \begin{vmatrix} * & (\lambda - b_0)\varphi_0 \\ + & (\lambda^{-1} - a_0)\varphi_0 \end{vmatrix}. \tag{4.4.15}$$

It is clear that these two functions have the required form (4.4.7).

Let $n \in \mathbb{N}$. Taking the equality (4.4.13) for the coordinate "0" and the equality (4.4.11) for the coordinate "1", we get

$$\begin{aligned} (p_{n-1}\varphi_{n-1}(\lambda))_0 + (q_n\varphi_n(\lambda))_0 + (r_n\varphi_{n+1}(\lambda))_0 &= \lambda^{-1}\varphi_{n;0}(\lambda), \\ (a_{n-1}\varphi_{n-1}(\lambda))_1 + (b_n\varphi_n(\lambda))_1 + (c_n\varphi_{n+1}(\lambda))_1 &= \lambda\varphi_{n;1}(\lambda). \end{aligned} \tag{4.4.16}$$

We can rewrite the equalities (4.4.16) in the form

$$\begin{aligned} \begin{bmatrix} r_{n;0,0} & r_{n;0,1} \\ c_{n;1,0} & c_{n;1,1} \end{bmatrix} \varphi_{n+1}(\lambda) &= \begin{bmatrix} + & 0 \\ * & + \end{bmatrix} \varphi_{n+1}(\lambda) \\ &= (\lambda^{-1}\varphi_{n;0}(\lambda) - (p_{n-1}\varphi_{n-1}(\lambda))_0 - (q_n\varphi_n(\lambda))_0, \\ &\quad \lambda\varphi_{n;1}(\lambda) - (a_{n-1}\varphi_{n-1}(\lambda))_1 - (b_{n-1}\varphi_{n-1}(\lambda))_1), \end{aligned} \tag{4.4.17}$$

i.e.,

$$\varphi_{n+1;0}(\lambda) = \frac{1}{r_{n;0,0}} \left(\lambda^{-1}\varphi_{n;0}(\lambda) - (p_{n-1}\varphi_{n-1}(\lambda))_0 - (q_n\varphi_n(\lambda))_0 \right),$$

$$\begin{aligned} \varphi_{n+1;1}(\lambda) = \frac{1}{c_{n;1,1}} \big(&\lambda\varphi_{n;1}(\lambda) - (a_{n-1}\varphi_{n-1}(\lambda))_1 \\ &- (b_{n-1}\varphi_{n-1}(\lambda))_1 - c_{n;1,0}\varphi_{n+1;0}(\lambda) \big), \quad n \in \mathbb{N}. \end{aligned}$$

Now we apply the induction: according to (4.4.15), $\varphi_1(\lambda)$ has the form (4.4.8), (4.4.9); φ_0 is the same. Let for $n \in \mathbb{N}$ $\varphi_{n-1}(\lambda)$, $\varphi_n(\lambda)$ have required form (4.4.8), (4.4.9). Then from the second equality in (4.4.17), it is easy to see, that $\varphi_{n+1;0}(\lambda)$ has the form (4.4.8), (4.4.9). The last equality in (4.4.17) shows that the same situation is also for $\varphi_{n+1;1}(\lambda)$. □

In what follows, it will be convenient to look at $S_n(\lambda)$, $\forall n \in \mathbb{N}_0$ with a fixed λ, as a linear operator that acts $\forall n \in \mathbb{N}$ from $\mathbb{R}^1$ into $\mathbb{R}^2$, i.e., $\mathbb{R}^1 \ni \varphi_0 \longmapsto S_n(\lambda)\varphi_0 \in \mathbb{R}^2$, and into $\mathbb{R}^1$, if $n = 0$. This operator is a standard extension on the corresponding complex space. As result we can write

$$\begin{gathered}\mathcal{H}_0 \ni \varphi_0 \longmapsto S_n(\lambda)\varphi_0 \in \mathcal{H}_n,\\ S_n^*(\lambda) = (S_n(\lambda))^* \,:\, \mathcal{H}_n \longmapsto \mathcal{H}_0, \quad n \in \mathbb{N}_0, \quad \lambda \in \mathbb{R}.\end{gathered} \tag{4.4.18}$$

We also regard on $Q_n(\lambda)$ as a vector-valued Laurent polynomial of $\lambda \in \mathbb{R}$ with real coefficients.

By using these polynomials $Q_n(\lambda)$, we construct the representation for $\Phi_{j,k}(\lambda)$, introduced by (4.4.4).

Lemma 4.4.2 *The operator $\Phi_{j,k}(\lambda)$, $\forall\lambda \in \mathbb{R}$ has the representation:*

$$\Phi_{j,k}(\lambda) = S_j(\lambda)\Phi_{0,0}(\lambda)S_k^*(\lambda) : \mathcal{H}_k \longrightarrow \mathcal{H}_j, \quad j, k \in \mathbb{N}_0, \tag{4.4.19}$$

where $\Phi_{0,0}(\lambda) \geq 0$ is a scalar.

Proof For a fixed $k \in \mathbb{N}_0$ and an arbitrary fixed $x \in \mathcal{H}_k \subset \mathbf{l}_2$ the vector $\varphi(\lambda) = (\varphi_j(\lambda))_{j=0}^\infty$, where

$$\varphi_j(\lambda) = \Phi_{j,k}(\lambda)x = \pi_j\Phi(\lambda)\pi_k x \in \mathcal{H}_j, \quad \lambda \in \mathbb{R}$$

is a generalized solution in the space $(\mathbf{l}_{\text{fin}})'$ of the equation $J\varphi(\lambda) = \lambda\varphi(\lambda)$, since $\Phi(\lambda)$ is a projection operator onto generalized eigenvectors of the operator A with a corresponding eigenvalue λ. Therefore, $\forall g \in \mathbf{l}_{\text{fin}}$, we have $(\varphi, Jg)_{\mathbf{l}_2} = \lambda(\varphi, g)_{\mathbf{l}_2}$. Hence, it follows that $\varphi = \varphi(\lambda) \in \mathbf{l}_2(p^{-1})$ exists as an usual solution of the equation $J\varphi(\lambda) = \lambda\varphi(\lambda)$ with initial conditions $\varphi_0(\lambda) = \pi_0\Phi(\lambda)\pi_k x \in \mathcal{H}_0$.

By using Lemma 4.4.2 and due to (4.4.8), we get

$$\Phi_{j,k}(\lambda)x = S_j(\lambda)(\Phi_{0,k}(\lambda)x), \text{ i.e., } \Phi_{j,k}(\lambda) = S_j(\lambda)\Phi_{0,k}(\lambda), \; j \in \mathbb{N}_0. \tag{4.4.20}$$

The operator $\Phi(\lambda) : \mathbf{l}_2(p) \longrightarrow \mathbf{l}_2(p^{-1})$ is formally self-adjoint on $\mathbf{l}_2$. Hence, according to (4.4.4), we get

$$(\Phi_{j,k}(\lambda))^* = (\pi_j\Phi(\lambda)\pi_k)^* = \pi_k\Phi(\lambda)\pi_j = \Phi_{k,j}(\lambda), \quad j, k \in \mathbb{N}_0. \tag{4.4.21}$$

For a fixed $j \in \mathbb{N}_0$ from (4.4.21) and previous discussion, it follows that the vector

$$\psi(\lambda) = (\psi_k(\lambda))_{k=0}^\infty, \quad \psi_k(\lambda) = \Phi_{k,j}(\lambda)y = (\Phi_{j,k}(\lambda))^*y, \quad y \in \mathcal{H}_j$$

is an usual solution of the equations $J\psi(\lambda) = \lambda\psi(\lambda)$ with initial conditions $\psi_0(\lambda) = \Phi_{0,j}(\lambda)y = (\Phi_{j,0}(\lambda))^*y$.

Again using the Lemma 4.4.2 and the arbitrariness of y, we obtain a representation of the form (4.4.20):

$$\Phi_{0,k}(\lambda) = S_k(\lambda)\Phi_{0,j}(\lambda), \quad k \in \mathbb{N}_0. \tag{4.4.22}$$

Taking into account (4.4.3) and (4.4.22), we get

$$\Phi_{0,k}(\lambda) = (\Phi_{k,0}(\lambda))^* = (S_k(\lambda)\Phi_{0,0}(\lambda))^* = \Phi_{0,0}(\lambda)(S_k(\lambda))^*, \quad k \in \mathbb{N}_0. \tag{4.4.23}$$

Here we used $\Phi_{0,0}(\lambda) \geq 0$, this inequality follows from (4.4.3) and (4.4.4)). Substituting (4.4.23) into (4.4.20), we get (4.4.19). □

Now we can write the Parseval equality (4.4.5) in concrete form. To this end, we substitute the expression (4.4.19) for $\Phi_{j,k}(\lambda)$ into (4.4.5) and get that

$$\begin{aligned}
(f, g)_{\mathbf{l}_2} &= \sum_{j,k=0}^{\infty} \int_{\mathbb{R}} (\Phi_{j,k}(\lambda) f_k, g_j)_{\mathbf{l}_2}\, d\sigma(\lambda) \\
&= \sum_{j,k=0}^{\infty} \int_{\mathbb{R}} (S_j(\lambda)\Phi_{0,0}(\lambda)S_k^*(\lambda) f_k, g_j)_{\mathbf{l}_2}\, d\sigma(\lambda) \\
&= \sum_{j,k=0}^{\infty} \int_{\mathbb{R}} (S_k^*(\lambda) f_k, S_j^*(\lambda) g_j)_{\mathbf{l}_2}\, d\rho(\lambda) \\
&= \int_{\mathbb{R}} \Big(\sum_{k=0}^{\infty} S_k^*(\lambda) f_k\Big)\overline{\Big(\sum_{j=0}^{\infty} S_j^*(\lambda) g_j\Big)}\, d\rho(\lambda), \quad \forall f, g \in \mathbf{l}_{\text{fin}},
\end{aligned} \tag{4.4.24}$$

$$d\rho(\lambda) = \Phi_{0,0}(\lambda)\, d\sigma(\lambda).$$

Introduce the Fourier transform "$\widehat{\ }$", induced by the self-adjoint operator $A = \mathbf{J}$ on the space $\mathbf{l}_2$ as some isometric map from $\mathbf{l}_2$ into $L_2(\mathbb{R}, d\rho(\lambda)) = L_2$. First, using (4.4.18), we define it on $\mathbf{l}_{\text{fin}}$

$$\mathbf{l}_2 \supset \mathbf{l}_{\text{fin}} \ni f = (f_n)_{n=0}^{\infty} \longmapsto \hat{f}(\lambda) = \sum_{n=0}^{\infty} S_n^*(\lambda) f_n \in L_2(\mathbb{R}, d\rho(\lambda)). \tag{4.4.25}$$

Hence, (4.4.24) gives the Parseval equality in a final form

$$(f, g)_{\mathbf{l}_2} = \int_{\mathbb{R}} \hat{f}(\lambda)\overline{\hat{g}(\lambda)}\, d\rho(\lambda), \quad f, g \in \mathbf{l}_{\text{fin}}. \tag{4.4.26}$$

The extension of (4.4.26) by continuity implies that (4.4.24) holds true for $\forall f, g \in \mathbf{l}_2$.

Polynomials $S_n^*(\lambda) = (S_{n;0}(\lambda), S_{n;1}(\lambda)) : \mathcal{H}_0 \longrightarrow \mathcal{H}_n$ are in some sense orthogonal. This orthogonality follows from (4.4.25) and (4.4.26). Namely, we take $f = (0, \ldots, 0, f_j, 0, \ldots)$, $f_j \in \mathcal{H}_j$, $g = (0, \ldots, 0, g_k, 0, \ldots)$, $g_k \in \mathcal{H}_k$ in (4.4.25) and (4.4.26). Then

$$\int\limits_{\mathbb{R}} (S_j^*(\lambda) f_j)\overline{(S_k^*(\lambda) g_k)}\, d\rho(\lambda) = \delta_{j,k}(f_j, g_j)_{\mathcal{H}_j},$$

$$f_j \in \mathcal{H}_j, \quad g_k \in \mathcal{H}_k, \quad j, k \in \mathbb{N}_0. \tag{4.4.27}$$

Let us rewrite the equality (4.4.27) in a simpler form. For this, we at first note that according to (4.4.18) and (4.4.8), we have for $n \in \mathbb{N}$ and $f_n = (f_{n,0}, f_{n,1}) \in \mathcal{H}_n$, $\lambda \in \mathbb{R}$, we obtain

$$S_n^*(\lambda) f_n = S_{n;0}(\lambda) f_{n;0} + S_{n;1}(\lambda) f_{n;1}, \quad S_0^*(\lambda) = 1. \tag{4.4.28}$$

Taking in (4.4.27) $f_j = e_\alpha$ and $g_k = e_\beta, \alpha, \beta = 0, 1$, we get from (4.4.27), (4.4.28) the final relation about the orthogonality

$$\int\limits_{\mathbb{R}} S_{j;\alpha}(\lambda) S_{k};\beta(\lambda)\, d\rho(\lambda) = \delta_{j,k}\delta_{\alpha,\beta},$$

$$j, k \in \mathbb{N}_0, \quad \alpha, \beta = 0, 1; \quad S_{0,0}(\lambda) = S_0(\lambda) = 1, \quad \lambda \in \mathbb{R}. \tag{4.4.29}$$

Let us remark that due to (4.4.28) the Fourier transform (4.4.25), $\forall f = (f_n) \in \mathbf{l}_2$, can be rewritten in the form

$$\hat{f}(\lambda) = f_{0;0} + \sum_{n=1}^{\infty} \sum_{\alpha=0}^{1} S_{n;\alpha}(\lambda) f_{n;\alpha}, \quad \lambda \in \mathbb{R}.$$

The stated above results of this section we can formulate as the spectral theorem for our operator $A = \mathbf{J}$.

Theorem 4.4.3 *Consider the space (4.3.25), namely,*

$$\mathbf{l}_2 = \mathcal{H}_0 \oplus \mathcal{H}_1 \oplus \mathcal{H}_2 \oplus \cdots, \quad \mathcal{H}_0 = \mathbb{C}^1, \quad \mathcal{H}_n = \mathbb{C}^2, \quad n \in \mathbb{N} \tag{4.4.30}$$

and the linear self-adjoint operator $A = \mathbf{J}$ which is defined by a block tri-diagonal Jacobi-Laurent matrix J on finite vectors $\mathbf{l}_{\text{fin}}$. So, we suppose that J of the form (4.3.26), (4.3.27), (4.3.28) *generates the self-adjoint operator $\mathbf{J}$ and for J the algebraically inverse matrix J^{-1} exists and has the form* (4.3.43), (4.3.44).

The generalized eigenfunction expansion of the operator $\mathbf{J}$ has the following form. According to Lemma 4.4.2, *we represent, starting from $\varphi_0 \in \mathbb{R}$, the solution*

$\varphi(\lambda) = (\varphi_n(\lambda))_{n=0}^{\infty}$, $\varphi_n(\lambda) \in \mathcal{H}_n$, of Eq. (4.4.7) *for $\lambda \in \mathbb{R}$*

$$\varphi_n(\lambda) = S_n(\lambda)\varphi_0 = (S_{n;0}(\lambda), S_{n;1}(\lambda))\varphi_0.$$

Here $S_0(\lambda) = 1$ and $S_{n;\alpha}(\lambda)$, $\alpha = 0, 1$, $n \in \mathbb{N}$ are real polynomials of variables λ and λ^{-1} of the form (4.4.8), (4.4.9). *The corresponding Fourier transform has the form*

$$\begin{aligned} \mathbf{l}_2 \supset \mathbf{l}_{\text{fin}} \ni f = (f_n)_{n=0}^{\infty} \longmapsto \hat{f}(\lambda) &= f_{0;0} + \sum_{n=1}^{\infty} Q_n^*(\lambda) f_n \\ &= S_{0;0}(\lambda) f_{0;0} + \sum_{n=1}^{\infty} \sum_{\alpha=0}^{1} S_{n;\alpha}(\lambda) f_{n;\alpha} \in L_2(\mathbb{R}, d\rho(\lambda)). \end{aligned} \tag{4.4.31}$$

Here $S_n^*(\lambda) : \mathcal{H}_n \longrightarrow \mathcal{H}_0$ is the operator adjoint $S_n(\lambda) : \mathcal{H}_0 \longrightarrow \mathcal{H}_n$, $d\rho(\lambda)$ is the Borel probability spectral measure of $\mathbf{J}$.

The Parseval equality has the form

$$(f, g)_{\mathbf{l}_2} = \int_{\mathbb{R}} \hat{f}(\lambda)\overline{\hat{g}(\lambda)}\, d\rho(\lambda), \quad (\mathbf{J}f, g)_{\mathbf{l}_2} = \int_{\mathbb{R}} \lambda \hat{f}(\lambda)\overline{\hat{g}(\lambda)}\, d\rho(\lambda), \quad \forall f, g \in \mathbf{l}_{\text{fin}}. \tag{4.4.32}$$

Formulas (4.4.31) *and* (4.4.32) *are extended by continuity for $\forall f, g \in \mathbf{l}_2$, hence, the operator* (4.4.31) *now is isometric, which maps $\mathbf{l}_2$ onto the whole $L_2(\mathbb{R}, d\rho(\lambda))$.*

Polynomials $S_{n;\alpha}(\lambda)$, $n \in \mathbb{N}$, $\alpha = 0, 1$, and $S_{0;0}(\lambda) = 1$ form an orthonormal system in $L_2(\mathbb{R}, d\rho(\lambda))$ in the sense of (4.4.29), and which is total in this space.

The last theorem solves the direct spectral problem for the self-adjoint operator $A = \mathbf{J}$, which is generated on the space $\mathbf{l}_2$ by the self-adjoint Jacobi-Laurent matrix J of the form (4.3.26) with algebraically inverse matrix J^{-1} of the form (4.3.43).

We remark that the self-adjoint operator in the space (4.4.30), generated by the matrix J with the structure (4.3.26), must have the spectral measure as a 2×2-matrix measure. It is very essential that in our case this measure is a scalar measure. This fact is a consequence of situation: our operator $\mathbf{J}$ has the algebraically inverse one and, therefore, we can prove Lemma 4.4.2.

Pass to the corresponding inverse spectral problem. Briefly speaking, the problem is as follows. We have the spectral measure $d\rho(\lambda)$ of the operator $\mathbf{J}$ in the space $\mathbf{l}_2$ generated by some self-adjoint Jacobi-Laurent matrix J; in what way we can find the matrix J?

The following theorem is the solution of such a problem.

Theorem 4.4.4 *The spectral measure $d\rho(\lambda)$ of the self-adjoint operator* **J** *generated on the space* $\mathbf{l}_2$ *by a self-adjoint Jacobi-Laurent matrix J has the following properties:*

(1) *the measure $d\rho(\lambda)$ is a Borel probability measure on $\mathbb{R}$;*
(2) *all functions*

$$\mathbb{R} \ni \lambda \longmapsto \lambda^m, \quad m \in \mathbb{Z} \tag{4.4.33}$$

belong to the space $L_2(\mathbb{R}, d\rho(\lambda))$ and are linearly independent in this space;
(3) *the set of functions* (4.4.33) *is total in the space $L_2(\mathbb{R}, d\rho(\lambda))$.*

Conversely, for each given measure $d\rho(\lambda)$ with properties (1)–(3), *it is possible to construct a self-adjoint Jacobi-Laurent matrix J for which this measure is spectral.*

For the construction of a matrix J it is necessary to repeat constructions of Sect. 4.3*: by the orthogonalization of* (4.3.3) *we find $P_{n;\alpha}(\lambda)$,* (4.3.4) *and then apply the formulas* (4.3.12) *and* (4.3.34). *If we start with the spectral measure $d\rho(\lambda)$ of the operator* **J**, *the last procedure gives the initial self-adjoint Jacobi-Laurent matrix J.*

Proof Let **J** be given self-adjoint operator constructed by J and $d\rho(\lambda)$ is its spectral measure. The condition (1) holds true. From Theorem 4.4.3 and the formula (4.4.9), it follows that each function (4.4.33) is some linear combination with real coefficients of the functions (4.4.9) and the main coefficient $l_{n;\alpha}$ by the function $\lambda^{(-1)^{\alpha+1}n}$ in $S_{n;\alpha}(\lambda)$, is positive. Functions $S_{n;\alpha}(\lambda)$, $n \in \mathbb{N}_0$, $\alpha = 0, 1$ belong to $L_2(\mathbb{R}, d\rho(\lambda)) = L_2$ and are orthogonal (see (4.4.29)), therefore all functions (4.4.33) belong to L_2 and are linear independent, i.e., the property (2) holds true.

Pass to the property (3). From the second equality in(4.4.32) and taking into account the isometry of the Fourier transform (4.4.31) between $\mathbf{l}_2$ and L_2, we conclude that our self-adjoint operator **J** is unitary equivalent to the closed multiplication on λ operator $\hat{A}$ in the space L_2 defined, at first, on the linear combination of functions $S_{n;\alpha}$, $n \in \mathbb{N}_0$, $\alpha = 0, 1$. These functions form a total set in L_2. By using the representation (4.4.9), we conclude that the set (4.4.33) is also total in L_2.

The other statement follows directly from results of Sects. 4.3 and 4.4. It is necessary only to note that the Laurent polynomials $P_{n;\alpha}(\lambda)$ and polynomials $Q_{n;\alpha}(\lambda)$ from Theorem 4.4.3 are the same: both classes of these polynomials are constructed from polynomials (4.3.3) or (4.4.9) and are orthonormal. □

In what follows, we denote the Laurent polynomials $S_{n;\alpha}(\lambda)$ for J in the standard way $P_{n;\alpha}(\lambda)$, $n \in \mathbb{N}_0$, $\alpha, \beta = 0, 1$.

4.5 Considerations of Hermitian Block Jacobi-Laurent Type Matrices

This section is a development of the last part of Section 4.3. Here we consider the matrices J and J^{-1} of the form (4.3.26) and (4.3.43) for which the conditions (4.3.27), (4.3.28) and (4.3.44) are fulfilled but the corresponding to J operator $\mathbf{J}$ defined in the space $\mathbf{l}_2$ (4.4.30) by relation (4.4.27) on $\mathbf{l}_{\text{fin}}$ is only Hermitian. In this case the set of functions (4.4.33) (i.e., (4.3.30)) is as earlier total in $L_2(\mathbb{R}, d\rho(\lambda))$. Previous constructions of course, we can carry out (including the construction of algebraically inverse matrix J^{-1}) but other results of Sect. 4.4, connected with self-adjointness $\mathbf{J}$, are not fulfilled.

So, we consider the matrices J, J^{-1} of the type (4.3.26), (4.3.43), with conditions (4.3.27), (4.3.28), (4.3.44). It is assumed that the matrix J^{-1} is algebraically inverse to J:

$$J^{-1}Jf = JJ^{-1}f = f, \quad f \in \mathbf{l}_{\text{fin}}.$$

Then the operator $\mathbf{J}$, constructed by (4.4.6) in $\mathbf{l}_2$ is only Hermitian. We will call such a matrix J as the Hermitian block Jacobi-Laurent type matrix.

At first we will formulate and prove some simple but essential theorem.

Theorem 4.5.1 *The defect indexes of* $\mathbf{J}$ *may be* $(0, 0)$ *or* $(1, 1)$. *Let* $P(z) = (P_n(z))_{n=0}^{\infty}$, $P_0(z) = 1 \in \mathbb{C}^1$, $P_n(z) = (P_{n;0}(z), P_{n;1}(z)) \in \mathbb{C}^2$, $z \in \mathbb{C} \setminus \mathbb{R}$, $n \in \mathbb{N}$ *be the solution of difference equation*

$$JP(z) = zP(z), \quad \textit{i.e.,}$$

$$a_{n-1}P_{n-1}(z) + b_nP_n(z) + c_nP_{n+1}(z) = zP_n(z), \quad n \in \mathbb{N}_0, \tag{4.5.1}$$

$P_{-1}(z) = 0$. *Consider the series*

$$\sum_{n=0}^{\infty} \|P_n(z)\|_{\mathcal{H}_n}^2. \tag{4.5.2}$$

If this series divergent for some $z \in \mathbb{C} \setminus \mathbb{R}$, *then it is divergent for every* $z \in \mathbb{C} \setminus \mathbb{R}$ *and the deficiency index of* $\mathbf{J}$ *is* $(0, 0)$. *If it is convergent for some* z *(and, therefore, for every* z*), then the deficiency index of* $\mathbf{J}$ *is* $(1, 1)$.

Proof First, we note that the proof of the existence of the solution $P(z)$ of the Eq. (4.5.1) is the same as in Lemma 4.4.1 with the change of λ to z. The proof of the lemma does not depend on the self-adjointness of $\mathbf{J}$, it is only necessary to use the fact that J^{-1} is algebraic inverse to J.

As above $P_n(z)$ must be of the form (4.4.8), (4.4.9) with change $S_{n;\alpha}(\lambda)$ to $P_{n;\alpha}(\lambda)$. Thus, we have:

$$P_0(z) = 1,$$
$$P_n(z) = (P_{n;0}(z), P_{n;1}(z)),$$
$$P_{n;\alpha}(z) = l_{n;\alpha} z^{(-1)^{\alpha+1}n} + w_{n;\alpha}(z), \quad n \in \mathbb{N}, \quad \alpha = 0, 1;$$

the structure of the polynomial $P_{n;\alpha}(z)$ is the same as $S_{n;\alpha}(z)$ in (4.4.9).

The operator $\mathbf{J}$ generated by J on $\mathbf{l}_{\text{fin}}$ is Hermitian, let $\mathbf{J}^*$ be its adjoint in $\mathbf{l}_2$. Let $z \in \mathbb{C} \setminus \mathbb{R}$, consider the corresponding to $\bar{z}$ the deficient subspace $N_{\bar{z}}$ of the operator $\mathbf{J}$, i.e., the subspace of $\mathbf{l}_2$ consisting of vectors $g \in \mathbf{l}_2$ for which,

$$((\mathbf{J} - \bar{z}\mathbb{I})f, g)_{\mathbf{l}_2} = ((J - \bar{z}\mathbb{I})f, g)_{\mathbf{l}_2} = 0, \quad f \in \mathbf{l}_{\text{fin}}. \tag{4.5.3}$$

Since f in (4.5.3) is finite, then we can transfer matrix $J - \bar{z}\mathbb{I}$ in (4.5.3) on g. As result we get

$$(f, (J - z\mathbb{I})g) = 0, \; \forall f \in \mathbf{l}_{\text{fin}}, \quad \text{i.e.,} \quad Jg = \bar{z}g$$
$$a_{n-1}g_{n-1} + b_n g_n + c_n g_{n+1} = z g_n, \quad n \in \mathbb{N}_0, \quad g_{-1} = 0 \tag{4.5.4}$$

(note that elements of J are real and the set $\mathbf{l}_{\text{fin}}$ is dense in $\mathbf{l}_2$). Thus, $g = (g_n)_{n=0}^{\infty}$ is a solution of the difference Eq. (4.5.4).

From the above explanation, it follows that this solution has the form

$$(g_n)_{n=0}^{\infty} = (g_0 P_n(z))_{n=0}^{\infty}. \tag{4.5.5}$$

In such a case, when the series (4.5.2) diverges, the sequence (4.5.5) belongs to $\mathbf{l}_2$ if and only if $g_0 = 0$, that is, if $g = 0$. Then the operator $\mathbf{J}$ is self-adjoint and its defect numbers are $(0, 0)$.

Let the series (4.5.2) be convergent. Then the sequence (4.5.5) belongs to $\mathbf{l}_2$ and the set of such sequences is one-dimensional, i.e. $\dim(N_{\bar{z}}) = 1$. The reality of elements of J implies $N_{\bar{z}} = N_z$ and the defect numbers are $(1, 1)$. □

Let $\mathbf{J}$ be an above introduced operator in the case of convergent series (4.5.2) $\forall z \in \mathbb{C} \setminus \mathbb{R}$, i.e. if this operator is only Hermitian. We can take its some self-adjoint extension $\tilde{\mathbf{J}}$ in the space $\mathbf{l}_2$. It is easy for this extension to repeat all constructions of Sect. 4.4 and to prove Theorems 4.4.3 and 4.4.4. But in this "indeterminate" case, the spectral measure $d\rho(\lambda)$ of J depends on an extension of $\tilde{\mathbf{J}}$. Therefore, the immediate goal is to give a description of all such measures corresponding to the strong moment problem, that is, to operators having (possibly densely defined) inverses.

Consider in (4.5.1) Laurent polynomials "of the first kind" as an analog of classical polynomials of the first kind in the usual theory of Jacobi matrices. In our Laurent case it is possible and usefully to introduce the analog of polynomials of the second kind.

Such Laurent polynomials $Q_n(z) = (Q_{n;0}(z), Q_{n;1}(z))$ of the second kind are introduced similarly to classical case by formula

$$Q_n(z) = \int\limits_{\mathbb{R}} \frac{P_n(\lambda) - P_n(z)}{\lambda - z} \, d\rho(\lambda), \quad z \in \mathbb{C} \setminus \mathbb{R}, \quad n \in \mathbb{N}_0. \tag{4.5.6}$$

Here $d\rho(\lambda)$ is the spectral measure of some fixed self-adjoint extension of operator $\mathbf{J}$ in the space $\mathbf{l}_2$ or of operator $\mathbf{J}$ if it is self-adjoint; we integrate in (4.5.6) vector-valued functions.

Lemma 4.5.2 *The sequence $Q(z) = (Q_n(z))_{n=0}^{\infty}$, $z \in \mathbb{C} \setminus \mathbb{R}$, where $Q_n(z)$ is given by formula (4.5.6), is a solution of the difference equation with indicated below initial data*

$$\begin{aligned} a_{n-1} Q_{n-1}(z) + b_n Q_n(z) + c_n Q_{n+1}(z) &= z Q_n(z), \\ p_{n-1} Q_{n-1}(z) + q_n Q_n(z) + r_n Q_{n+1}(z) &= z^{-1} Q_n(z), \quad n \in \mathbb{N}, \\ Q_0(z) = 0, \quad Q_1(z) &= (c_{0;0,1}^{-1} s_{-1} z^{-1}, c_{0;0,1}^{-1}(1 - c_{0;0,0} r_{0;0,0}^{-1} s_{-1} z^{-1})), \\ s_{-1} &= \int\limits_{\mathbb{R}} \lambda^{-1} \, d\rho(\lambda), \quad z \in \mathbb{C}. \end{aligned} \tag{4.5.7}$$

Let us first note a simple general fact, which will be used below.

Lemma 4.5.3 *Let $\varphi(z) = (\varphi_n(z))_{n=0}^{\infty}$, $\varphi_n(z) \in \mathcal{H}_n = \mathbb{C}^2$, $n \in \mathbb{N}$, $z \in \mathbb{C} \setminus \{0\}$ is some solution of the first equation in (4.5.7), then it is the solution of the second equation in (4.5.7) (a similar fact holds true for Eqs.* (4.4.11), (4.4.13)).

Proof Indeed, let us extend the sequence $\varphi(z) = (\varphi_n(z))_{n=1}^{\infty}$ on $\mathbb{N}_0$ by putting $\varphi_0(z) = 0$. Then, according to (4.5.7), such an extended sequence $\varphi'(z)$ is a solution of the equation $J\varphi'(z) = z\varphi'(z)$. The matrix J^{-1} is the algebraically inverse to J and also block tri-diagonal, therefore, we can write: $J^{-1}J = JJ^{-1} = 1$, i.e., $J^{-1}Jf = f$, $f \in \mathbf{l}$. Taking $f = \varphi'(z)$, we get

$$\varphi'(z) = J^{-1} J \varphi'(z) = z J^{-1} \varphi'(z),$$

i.e., the second equality in (4.5.7). □

Let us proceed to the proof of Lemma 4.5.2.

Proof Since $P_0(z) = 1$, we get from (4.5.6) that

$$Q_0(z) = 0, \quad z \in \mathbb{C}. \tag{4.5.8}$$

Let us calculate $Q_{1;0}(z)$. From (4.4.15) and (4.5.6), we conclude that

$$\begin{aligned} P_{1;0}(z) &= \frac{1}{r_{0;0,0}}(z^{-1} - q_0), \\ Q_{1;0}(z) &= \frac{1}{r_{0;0,0}} \int\limits_{\mathbb{R}} (\lambda^{-1} - z^{-1})(\lambda - z)^{-1}\, d\rho(\lambda) \\ &= \frac{1}{r_{0;0,0}} \int\limits_{\mathbb{R}} \lambda^{-1}\, d\rho(\lambda) z^{-1}, \quad \forall z \in \mathbb{C}. \end{aligned} \tag{4.5.9}$$

Let us calculate $Q_{1;1}(z)$. From (4.4.15) we get

$$P_{1;1}(z) = -\frac{c_{0;0,0}}{r_{0;0,0}c_{0;0,1}}(z^{-1} - q_0) + \frac{1}{c_{0;0,1}}(z - b_0), \quad z \in \mathbb{C}. \tag{4.5.10}$$

From (4.5.6) and (4.5.10), we conclude that after simple calculation

$$Q_{1;1}(z) = \frac{1}{c_{0;0,1}} - \frac{c_{0;0,0}}{r_{0;0,0}c_{0;0,1}} \int\limits_{\mathbb{R}} \lambda^{-1}\, d\rho(\lambda) z^{-1}, \quad z \in \mathbb{C}.$$

Formulas (4.5.8)–(4.5.10) give the initial data for (4.5.7).

For the proof of Lemma it is necessary to check that $Q_n(z)$, $n \in \mathbb{N}$, $Q_0(z) = 0$ satisfy the both Eq. (4.5.7). We start with the first equation. Since $Q_0(z) = 0$, then the left part of this equation is equal to $(JQ(z))_n$, $n \in \mathbb{N}$. But according to (4.5.6) and (4.5.1) we have

$$\begin{aligned} (JQ(z))_n &= \int\limits_{\mathbb{R}} \frac{(JP(\lambda))_n - (JP(z))_n}{\lambda - z}\, d\rho(\lambda) \\ &= \int\limits_{\mathbb{R}} \frac{\lambda P_n(\lambda) - zP_n(z)}{\lambda - z}\, d\rho(\lambda) \\ &= z \int\limits_{\mathbb{R}} \frac{P_n(\lambda) - P_n(z)}{\lambda - z}\, d\rho(\lambda) + \int\limits_{\mathbb{R}} P_n(\lambda)\, d\rho(\lambda) \\ &= zQ_n(z) + \int\limits_{\mathbb{R}} P_n(\lambda)\, d\rho(\lambda), \quad \forall z \in \mathbb{C} \setminus \mathbb{R},\ n \in \mathbb{N} \end{aligned} \tag{4.5.11}$$

According to (4.4.29) the last integral in (4.5.11) is equal to zero, therefore (4.5.11) gives $(JQ(z))_n = zQ_n(z)$, $n \in \mathbb{N}$. This means that the first equality in (4.5.7) is fulfilled.

The second equality in (4.5.7) is fulfilled due to Lemma 4.5.3. □

As a result, for Laurent polynomials $Q_n(z)$ of the second kind, we have a case similar to the first kind polynomial: their sequence is a solution for $n \in \mathbb{N}$ of the system (4.5.7) with the initial data $Q_0(z) = 0$, $Q_1(z)$ specified in (4.5.7). These polynomials (in the general case) are not orthonormal in the space $L_2(\mathbb{R}, d\rho(\lambda))$ with respect to the spectral measure $d\rho(\lambda)$ generated by the self-adjoint expansion $\tilde{\mathbf{J}}$ of the operator $\mathbf{J}$ in the space $\mathbf{l}_2$ (this case is analogous to the case of classical Jacobi matrices).

Polynomials of the second kind are needed to describe the set of all spectral measures generated by all self-adjoint expansions in $\mathbf{l}_2$ of the operator $\mathbf{J}$. It is possible successively to find step by step the solution of the system (4.5.7) by given $Q_0(z)$, $Q_1(z)$. For polynomials $Q_n(z)$, we can prove an analog of Lemma 4.4.2 about their precise structure, but such results are not necessary for us and we are restricted by following rough result.

Lemma 4.5.4 *All Laurent polynomials of the second kind have the form*

$$Q_0(z) = 0,$$
$$Q_n(z) = (Q_{n;0}(z), Q_{n;1}(z)) \in \mathbb{C}^2, \quad z \in \mathbb{C} \setminus \mathbb{R}, \quad n \in \mathbb{N}, \tag{4.5.12}$$

where $\forall \alpha = 0, 1$, $Q_{n;\alpha}(z)$ *is a linear combination of* $1, z^{-1}, z, \ldots, z^{-n}, z^n$ *with real coefficients.*

Proof Vectors $Q_n(z) \in \mathbb{C}^2$, $n \in \mathbb{N}$ are solutions of the system of difference Eq. (4.5.7). Here we assume that $a_0 = p_0 = 0$ since $Q_0(z) = 0$. In addition to the equality (4.5.7), we get

$$\begin{aligned}(a_{n-1} + p_{n-1})Q_{n-1}(z) + (b_n + q_n)Q_n(z) + (c_n + r_n)Q_{n+1}(z) \\ = (z + z^{-1})Q_n(z), \quad n \in \mathbb{N}, \quad a_0 + p_0 = 0.\end{aligned} \tag{4.5.13}$$

We can rewrite the last equation as some equation with tri-diagonal block (2×2) matrices acting in the space $\mathbb{C}^2 \oplus \mathbb{C}^2 \oplus \ldots$.

Elements on external diagonals of matrices $a_n + p_n$ and $c_n + r_n$ have inverses according to (4.3.26), (4.3.27), (4.3.28) and according to (4.3.43), (4.3.44), have positive elements on the main diagonals and zeros on positions $(1, 0)$ and $(0, 1)$, respectively.

Therefore, it is possible to find the solution (4.5.13) step by step, starting from $Q_1(z)$, which has the form $(C_1 z^{-1}, C_2 + C_3 z^{-1})$, $C_1, C_2, C_3 \in \mathbb{R}$. Each $n + 1$ step yields the multiplication of $Q_n(z)$ by $z + \frac{1}{z}$ plus some linear combination of $1, z^{-1}, z, \ldots, z^{-n}, z^n$. Which proves the lemma. □

Let us proceed to the description of the spectral measure $d\rho(\lambda)$ corresponding to the operator $\tilde{\mathbf{J}}$ acting in the space $\mathbf{l}_2$. Let $R_z : \mathbf{l}_2 \to \mathbf{l}_2$, $z \in \mathbb{C} \setminus \mathbb{R}$ be the resolvent of the operator $\tilde{\mathbf{J}}$.

If we use imbedding operators (4.4.1) and formulas (4.4.19), (4.4.31), then we can write the resolvent R_z of the operator A written in (4.3.8), (4.3.10). Hence,

$$(R_z f)_j = \sum_{k=0}^{\infty} R_{z;\,j,k} f_k,$$

$$\begin{aligned} R_{z;\,j,k;\alpha,\beta} &= (R_z e_{k,\beta}, e_{j,\alpha})_{\mathbf{l}_2} = \left(\left(\int_{\mathbb{R}} \frac{\Phi(\lambda)}{\lambda - z}\, d\sigma(\lambda)\right) e_{k;\beta}, e_{j,\alpha}\right)_{\mathbf{l}_2} \\ &= \int_{\mathbb{R}} \frac{P_{k;\beta}(\lambda) - P_{j;\alpha}(\lambda)}{\lambda - z}\, d\rho(\lambda), \\ & z \in \mathbb{C} \setminus \mathbb{R}, \quad j, k \in \mathbb{N}_0, \quad \alpha, \beta = 0, 1; \quad f \in \mathbf{l}_2. \end{aligned} \tag{4.5.14}$$

Introduce the *Weyl function* by an expression

$$m(z) = R_{z;0,0;0,0} = (R_z e_{0;0}, e_{0;0})_{\mathbf{l}_2} = \int_{\mathbb{R}} \frac{1}{\lambda - z}\, d\rho(\lambda), \quad z \in \mathbb{C} \setminus \mathbb{R}. \tag{4.5.15}$$

The classical fact states that the measure $d\rho(\lambda)$ is determined uniquely by the function $m(z)$, so instead of the measure $d\rho(\lambda)$ it suffices to obtain a description of all functions $m(z)$. To do this, we first write down some simple but important formulas.

Thus, using (4.5.14), (4.5.15) and the definition (4.5.6), we can write

$$\begin{aligned} (R_z e_{0;0})_{j;\alpha} = R_{z;\,j,0;\alpha,0} &= \int_{\mathbb{R}} \frac{P_{j;\alpha}(\lambda)}{\lambda - z}\, d\rho(\lambda) \\ &= \int_{\mathbb{R}} \frac{P_{j;\alpha}(\lambda) - P_{j;\alpha}(z)}{\lambda - z}\, d\rho(\lambda) + P_{j;\alpha}(z) \int_{\mathbb{R}} \frac{d\rho(\lambda)}{\lambda - z} \\ &= Q_{j;\alpha} z + m(z) P_{j;\alpha}(z), \quad \forall z \in \mathbb{C} \setminus \mathbb{R}, \\ & j \in \mathbb{N}_0, \alpha = 0, 1. \end{aligned}$$

Another form of this equality

$$(R_z e_{0;0})_j = Q_j(z) + m(z) P_j(z), \quad z \in \mathbb{C} \setminus \mathbb{R}, \quad j \in \mathbb{N}_0. \tag{4.5.16}$$

By using the Hilbert identity and (4.5.16), we get

$$\begin{aligned}\frac{\overline{m(\zeta)} - m(z)}{\bar{\zeta} - z} &= \left(\frac{R_{\bar{\zeta}} - R_z}{\bar{\zeta} - z} e_{0;0}, e_{0;0}\right)_{\mathbf{l}_2} = \left(R_{\bar{\zeta}} R_z e_{0;0}, e_{0;0}\right)_{\mathbf{l}_2} \\ &= \left(R_{\zeta}^* R_z e_{0;0}, e_{0;0}\right)_{\mathbf{l}_2} = (R_z e_{0;0}, R_{\zeta} e_{0;0})_{\mathbf{l}_2} \\ &= \sum_{j=0}^{\infty} (Q_j(z) + m(z)P_j(z), Q_j(\zeta)z + m(\zeta)P_j(\zeta))_{\mathcal{H}_j}, \\ &\forall z, \zeta \in \mathbb{C} \setminus \mathbb{R},\ z \neq \bar{\zeta}.\end{aligned} \tag{4.5.17}$$

Let $\zeta = z$, then (4.5.17) yields

$$\frac{\overline{m(z)} - m(z)}{\bar{z} - z} = \sum_{j=0}^{\infty} \|Q_j(z) + m(z)P_j(z)\|_{\mathcal{H}_j}^2, \quad z \in \mathbb{C} \setminus \mathbb{R}. \tag{4.5.18}$$

From this equality we can conclude some simple, but essential fact, which we formulate, as a theorem.

Theorem 4.5.5 *Let us consider two series*

$$\sum_{j=0}^{\infty} \|P_j(z)\|_{\mathcal{H}_j}^2, \quad \sum_{j=0}^{\infty} \|Q_j(z)\|_{\mathcal{H}_j}^2, \quad \forall z \in \mathbb{C} \setminus \mathbb{R}. \tag{4.5.19}$$

These series are simultaneously divergent or convergent. In the first case operator $\mathbf{J}$ *is self-adjoint in* $\mathbf{l}_2$*, in the second case it is not self-adjoint and the measure* $d\rho(\lambda)$*, appeared above, is the spectral measure of some its self-adjoint extension in the space* $\mathbf{l}_2$*.*

In the second case, both series (4.5.19) converge simultaneously on every bounded set $\mathbb{C}$ *that has a positive distance from the real axis.*

Proof For every $z \in \mathbb{C} \setminus \mathbb{R}$ and the function $m(z) \neq 0$, the series (4.5.18) is convergent. Therefore, if one of the series (4.5.19) diverges, then the second must also diverge.

Let both series are convergent in (4.5.19). It is necessary to prove that they converge uniformly in the domain G specified in the theorem.

Since the left part in equality (4.5.18) is a continuous function $z \in G$ and all terms in (4.5.18) are continuous positive functions, then by the Dini theorem these series converge uniformly on G.

In this case, the operator $\mathbf{J}$ is not self-adjoint in $\mathbf{l}_2$. The previously mentioned measure $d\rho(\lambda)$ and the function $m(z)$ are related to some fixed extension of $\mathbf{J}$ to the self-adjoint operator in $\mathbf{l}_2$. Let us take another its extension (such an extension exists because the deficiency index of $\mathbf{J}$ is $(1, 1)$). Let the measure $d\rho_1(\lambda)$ and the function $m_1(z)$ are corresponding to this extension the spectral measure and the Weyl function ; $m_1(z) \neq m(z)$, $z \in \mathbb{C} \setminus \mathbb{R}$. Here it is worth making the following general remark. *Laurent polynomials of the first and the second kind do not depend on the corresponding spectral measure $d\rho(\lambda)$: they are solutions of Eqs. (4.4.11) and (4.4.13) with λ replaced with z and corresponding initial data independent on $d\rho(\lambda)$*. Explain the last assertion. For $P_n(z)$ we have initial data $P_{-1}(z) = 0$, $P_0(z) = 1$, $n \in \mathbb{N}_0$. For $Q_n(z)$ the initial data are given in (4.5.7), $n \in \mathbb{N}_0$. These expressions contain s_{-1}, but using (4.3.34) we conclude that $s_{-1} = q_0$. Thus, the initial data for $Q_n(z)$ are defined only by elements of matrices J, J^{-1}.

Consider the equality (4.5.18) for $m_1(z)$. As above the corresponding to $m_1(z)$ series in (4.5.18) converges uniformly in G. We have

$$\sum_{j=0}^{n} \|(Q_j(z) + m(z)P_j(z)) - (Q_j(z) + m_1(z)P_j(z))\|^2_{\mathcal{H}_j}$$
$$= |m(z) - m_1(z)| \sum_{j=0}^{n} \|P_j(z)\|^2_{\mathcal{H}_j}, \quad z \in \mathbb{C} \setminus \mathbb{R}, \ \forall n \in \mathbb{N}_0. \tag{4.5.20}$$

The left part of (4.5.20) converges uniformly in G if $n \longrightarrow \infty$, since such a convergence exists in (4.5.18) for $m(z)$ and $m_1(z)$. Therefore, such a convergence should be for the right part in (4.5.20), that is, for the first row in (4.5.19). The uniform convergence for the second series in (4.5.19) follows from such a convergence for (4.5.18) and for the first series. □

Note, that it is possible to prove for the second case: the series (4.5.19) are uniformly convergent also in every bounded closed domain from $\mathbb{C} \setminus \{0\}$.

By means of simple calculations we can rewrite the equality (4.5.18) for $z \in \mathbb{C} \setminus \mathbb{R}$ in the form

$$\left(\sum_{j=0}^{\infty} \|P_j(z)\|^2_{\mathcal{H}_j}\right)|m(z)|^2 + \left(\frac{1}{z - \bar{z}} + \sum_{j=0}^{\infty}(Q_j(z), P_j(z))_{\mathcal{H}_j}\right)\overline{m(z)}$$
$$+ \overline{\left(\frac{1}{z - \bar{z}} + \sum_{j=0}^{\infty}(Q_j(z), P_j(z))_{\mathcal{H}_j}\right)}m(z) + \sum_{j=0}^{\infty} \|Q_j(z)\|^2_{\mathcal{H}_j} = 0. \tag{4.5.21}$$

The last equality in (4.5.21) means that the point $m(z)$ lies on some circle in the complex plane. We can now repeat all conclusions from Chap. 2, Sects. 2.9–2.10, but with some changes. These facts give the description of all spectral measures $d\rho(\lambda)$ for the classical Jacobi matrix in the indeterminate case, when it is only Hermitian, but not self-adjoint operator.

Such repeating gives in our case the new result in the form of a theorem.

Theorem 4.5.6 *Let the Jacobi-Laurent matrix J corresponds to the indeterminate case. Fix some point $z \in \mathbb{C} \setminus \mathbb{R}$ and consider the point $m(z) \in \mathbb{C}$ which is defined by the integral*

$$m(z) = \int_{\mathbb{R}} \frac{d\rho(\lambda)}{\lambda - z}, \tag{4.5.22}$$

where $d\rho(\lambda)$ is the spectral measure of some self-adjoint extension in the space $\mathbf{l}_2$ of the operator $\mathbf{J}$. Taking such sort of different spectral measures $d\rho(\lambda)$, we assert that the point (4.5.22) *$m(z)$ pass to some circle K_z (an analog of the Weyl-Hamburger circle) with the center $O(z)$ and the radius $R(z)$, namely,*

$$\begin{aligned} O(z) &= -\frac{1}{\sum\limits_{j=0}^{\infty} \|P_j(z)\|^2_{\mathcal{H}_j}} \left(\frac{1}{z - \bar{z}} + \sum_{j=0}^{\infty} (Q_j(z), P_j(z))_{\mathcal{H}_j} \right), \\ R(z) &= \left(|z - \bar{z}| \sum_{j=0}^{\infty} \|P_j(z)\|^2_{\mathcal{H}_j} \right)^{-1}. \end{aligned} \tag{4.5.23}$$

It is possible, as for classical Jacobi matrices, to give some procedure in finding the measure $d\rho(\lambda)$ by given point on the circle K_z described in (4.5.23).

We will not provide relevant considerations here. It is clear that the above developed approach to the spectral theory of Jacobi-Laurent matrices is similar to classical Jacobi matrices and, therefore, many results of last theory it is possible to transfer on our case. Note once again that in the previously developed approach, instead of the usual space l_2 on $\mathbb{N}_0$, the space $\mathbf{l}_2$ on $\mathbb{C}^2$ of vector-valued sequences $f = (f_n)$, but the first coordinate is one-dimensional: $f_0 \in \mathbb{C}^1$. This feature of the $\mathbf{l}_2$ space makes it possible to use the *scalar spectral measure $d\rho(\lambda)$*: in short, $\forall \alpha \in \mathfrak{B}(\mathbb{R})$, $P(\alpha) = (E(\alpha)e_{0,0}, e_{0,0})_{\mathbf{l}_2}$, where $e_{0,0} = (1, 0, 0, \ldots) \in \mathbf{l}_2$ and E is the expansion of the identity.

We will finish this section by an analog of the classical Carleman theorem (see Chap. 1, (2.6.26)) for Jacobi-Laurent matrices.

Theorem 4.5.7 *The operator $\mathbf{J}$ is self-adjoint in $\mathbf{l}_2$ if for the matrix J of the form* (4.3.26), *the condition*

$$\sum_{n=0}^{\infty} \frac{1}{\|a_n\|} = \infty \tag{4.5.24}$$

holds true.

Proof By simple calculations, it can be proved that the Green formula is fulfilled in this situation. Namely, for arbitrary sequences $f = (f_n)$, $g = (g_n)$, $f_n, g_n \in \mathcal{H}_n$, we have

$$\sum_{j=m}^{n} \Big(((Jf)_j, g_j)_{\mathcal{H}_j} - (f_j, (Jg)_j)_{\mathcal{H}_j} \Big)$$
$$= \big[(f_n, c_n g_{n+1})_{\mathcal{H}_n} - (c_n f_{n+1}, g_n)_{\mathcal{H}_n} \big] \tag{4.5.25}$$
$$- \big[(f_m, a_{m-1} g_{m-1})_{\mathcal{H}_m} - (a_{m-1} f_{m-1}, g_m)_{\mathcal{H}_m} \big],$$
$$m, n \in \mathbb{N}_0, \quad n > m,$$

where it is assumed that $f_{-1} = g_{-1} = 0$.

Let us explain that if $m > 0$, then all $\mathcal{H}_j$ in (4.5.25) are equal to $\mathbb{C}^2$ and calculations in (4.5.25) are simple. For case $m = 0$ we write at first the equality (4.5.25) for $m = 1$; then it is sufficient to add to each side a corresponding trivial equality

$$((Jf)_0, g_0)_{\mathbb{C}^1} - (f_0, (Jg)_0)_{\mathbb{C}^1} = (b_0 f_0 + c_0 f_1, g_0)_{\mathbb{C}^1} -$$
$$(f_0, b_0 g_0 + c_0 g_1)_{\mathbb{C}^1} = (c_0 f_1, g_0)_{\mathbb{C}^1} - (f_0, c_0 g_1)_{\mathbb{C}^1}$$
$$= (f_1, a_0 g_0)_{\mathbb{C}^2} - (a_0 f_0, g_1)_{\mathbb{C}^2}.$$

As result the last square brackets transfers into zero and we get (4.5.25) for the indicated m, n.

Let $z \in \mathbb{C} \setminus \mathbb{R}$ be fixed. Consider Laurent polynomials of the first kind $P_n(z)$, $n \in \mathbb{N}_0$. For the sequence

$$(f_n)_{n=0}^{\infty} = (g_n)_{n=0}^{\infty} = (P_n(z))_{n=0}^{\infty} = P(z),$$

the equality (4.5.25), with $m = 0$, $\forall n \in \mathbb{N}$, gives

$$(z - \bar{z}) \sum_{j=0}^{n} \|P_j(z)\|^2_{\mathcal{H}_j} = \sum_{j=0}^{n} \Big(((JP(z))_j, P_j(z))_{\mathcal{H}_j} - (P_j(z), (JP(z))_j)_{\mathcal{H}_j} \Big)$$
$$= \big(P_n(z), c_n P_{n+1}(z))_{\mathcal{H}_n} - (c_n P_{n+1}(z), P_n(z))_{\mathcal{H}_n} \big), \tag{4.5.26}$$

here is used (4.5.1)).

Since $P_0(z) = 1$, from (4.5.26), we get

$$|z - \bar{z}| \le |z - \bar{z}| \sum_{j=0}^{n} \|P_j(z)\|^2_{\mathcal{H}_j} \le 2\|c_n\| \|P_n(z)\|_{\mathbb{C}^2} \|P_{n+1}(z)\|_{\mathbb{C}^2},$$

therefore,

$$\|c_n\|^{-1} \leq 2\|P_n(z)\|_{\mathbb{C}^2}\|P_{n+1}(z)\|_{\mathbb{C}^2}|z-\bar{z}|^{-1}.$$

By using this inequality, (4.5.24), and the equality $\|c_n\| = \|a_n^*\| = \|a_n\|$, we get

$$\begin{aligned}
\infty = \sum_{n=1}^{\infty} \frac{1}{\|a_n\|} &\leq 2|z-\bar{z}|^{-1} \sum_{n=1}^{\infty} \|P_n(z)\|_{\mathbb{C}^2}\|P_{n+1}(z)\|_{\mathbb{C}^2} \\
&\leq 2|z-\bar{z}|^{-1}\Big(\sum_{n=1}^{\infty} \|P_n(z)\|_{\mathbb{C}^2}^2\Big)^{1/2}\Big(\sum_{n=1}^{\infty} \|P_{n+1}(z)\|_{\mathbb{C}^2}^2\Big)^{1/2} \\
&\leq 2|z-\bar{z}|^{-1} \sum_{n=0}^{\infty} \|P_n(z)\|_{\mathcal{H}_n}^2.
\end{aligned}$$

Therefore, the series (4.5.2) is divergent and our operator **J** is self-adjoint. □

4.6 The Connection Between the Strong Moment Problem and the Spectral Theory of Jacobi-Laurent Matrices

For the classical moment problem, the connection between such a problem and the spectral theory of Jacobi operators and matrices is well known. A more precise and consistent way of such connection with Jacobi matrices is given in Chap. 2 (Sect. 2.7). Here we will show this way in slightly different terms.

Let $s = (s_n)$, $n \in \mathbb{N}_0$, $s_n \in \mathbb{R}$ be some sequence of real numbers. As mentioned earlier, this is a moment sequence if there exists a Borel measure $d\rho(\lambda)$ on $\mathbb{R}$ for which

$$s_n = \int_{\mathbb{R}} \lambda^n \, d\rho(\lambda), \quad n \in \mathbb{N}_0. \tag{4.6.1}$$

The classical theorem about moments s_n asserts that the representation (4.6.1) exists if and only if the sequence s is positive defined, i.e., for an arbitrary finite sequence $f = (f_n)_{n=0}^{\infty}$, $f_n \in \mathbb{C}$, we have

$$\sum_{j,k=0}^{\infty} s_{j+k} f_j \bar{f}_k \geq 0.$$

Thus, let we have a moment sequence s with the representation (4.6.1). Consider the space $L_2(\mathbb{R}, d\rho(\lambda)) = L_2$ and the multiplication by λ operator $\hat{A}$ in this space, i.e., $(\hat{A}F)(\lambda) = \lambda F(\lambda)$, where $F \in L_2$ belongs to $\mathrm{Dom}(\hat{A})$ consisting all polynomials on λ with complex coefficients. All such polynomials $F(\lambda)$ belong to L_2 and $\hat{A}F$ is well defined, since all functions

$$\lambda^m, \quad m \in \mathbb{N}_0, \tag{4.6.2}$$

are summable according to (4.6.1). We assume that they are linearly independent in L_2 (i.e., we consider the non-degenerate moment problem (4.6.1)) and the set (4.6.2) is total in L_2. The operator $\hat{A}$ defined above is always Hermitian.

Consider the Gram-Schmidt orthogonalization procedure with real coefficients for functions $1, \lambda, \lambda^2, \ldots$. As a result, we get a sequence of polynomials

$$P_0(\lambda) = 1, P_1(\lambda), P_2(\lambda), \ldots$$

(polynomials of the first kind), which form an orthonormal basis in L_2. Now we can go from L_2 to the space l_2 of sequences $f = (f_n)$, $f_n \in \mathbb{C}^1$. Our operator $\hat{A}$ passes, as it is easy to understand, into some ordinary symmetric Jacobi matrix

$$\begin{bmatrix} b_0 & a_0 & 0 & 0 & 0 & \ldots \\ a_0 & b_1 & a_1 & 0 & 0 & \ldots \\ 0 & a_1 & b_2 & a_2 & 0 & \ldots \\ 0 & 0 & a_2 & b_3 & a_3 & \ldots \\ \vdots & \vdots & \vdots & \vdots & \vdots & \ddots \end{bmatrix}, \tag{4.6.3}$$

acting in the space l_2 at first on the vectors from $l_{\text{fin}} \subset l_2$. Its real elements a_n, b_n are calculated according to classical formulas. It is important that the *formulas for a_n, b_n can be rewritten through the given moments s_n, $n \in \mathbb{N}_0$: we orthogonalize the sequence λ^n, $n \in \mathbb{N}_0$. So, we can construct the Jacobi matrix J immediately by given moments s_n, $n \in \mathbb{N}_0$.*

In the case of essential self-adjointness of $\hat{A}$, the operator $\mathbf{J}$, generated in l_2 by the matrix J (4.6.3) on l_{fin}, is self-adjoint (after the extension by the continuity) and its spectral measure is equal to given above $d\rho(\lambda)$. Conversely, let us first have the matrix J and construct the corresponding self-adjoint operator $\mathbf{J}$ in l_2 with the spectral measure $d\rho(\lambda)$. Then, for the sequence of numbers s_n, constructed from $(a_n, b_n)_{n=0}^{\infty}$ by the reverse procedure to the indicated above one, we get the representation (4.6.1) (it is necessary to take $\mathbf{J}^n$; $\forall n \in \mathbb{N}_0$, $s_n = (\mathbf{J}^n e_0, e_0)_{l_2}$, $e_0 = (1, 0, 0, \ldots)$). If $\hat{A}$ is not self-adjoint in L_2, then a self-adjoint extension of the operator $\mathbf{J}$ in l_2 is necessary (possibly with an exit from the space). For all such expansions, we have the measure $d\rho(\lambda)$ in (4.6.1), but this measure is not unique for a given s. Each such a so-called "orthogonal" measure is defined by some self-adjoint extension of $\mathbf{J}$ in l_2. Thus, if we apply the description of all expansions in l_2 (as it exists in the classical theory of Jacobi matrices), then as a result we get a description of all "orthogonal" measures of $d\rho(\lambda)$, which give representations for the given moment sequences $s = (s_n)_{n=0}^{\infty}$.

This classical scheme can be repeated to consider the strong moment sequence $s = (s_n)_{n=-\infty}^{\infty}$ (4.2.1) and the Jacobi-Laurent matrix (4.3.26), (4.3.43). Assume for simplicity that the sequence s is non-degenerate. Let us formulate and prove the theorem.

Theorem 4.6.1 *Let $s = (s_n)$, $n \in \mathbb{Z}$ be a strong moment sequence and* (4.2.1) *is its representation by the integral with some Borel measure $d\rho(\lambda)$. We assume, that the measure in* (4.2.1) *is such that the set $\{\lambda^m\}$, $m \in \mathbb{Z}$ is total in $L_2(\mathbb{R}, d\rho(\lambda))$.*

Let us corresponds s to the Jacobi-Laurent matrix J according to the rule described above.

Now consider the operator $\mathbf{J}$ constructed in (4.4.6) *in the space $\mathbf{l}_2$* (4.4.30). *Suppose that this operator is essentially self-adjoint, then its spectral measure is equal to the measure $d\rho(\lambda)$ in the representation* (4.2.1).

If the operator $\mathbf{J}$ is only Hermitian, then all its self-adjoint extensions in $\mathbf{l}_2$, corresponding measures, are one of the measures $d\rho(\lambda)$ in the representation (4.2.1). *Taking all such extensions (see Theorem* 4.5.6), *we obtain all such measures in* (4.2.1) *for a fixed sequence s. The correspondence between measures and extensions is one-to-one.*

The rule for constructing matrices J for the sequence s is as follows. Let us introduce finite-dimensional matrices and denote their determinant

$$\mathcal{H}_k^{(l)} = \begin{bmatrix} s_l & s_{l+1} & \dots & s_{l+k} \\ s_{l+1} & s_{l+2} & \dots & s_{l+k+1} \\ \dots & \dots & \dots\dots \\ s_{l+k} & s_{l+k+1} & \dots & s_{l+2k} \end{bmatrix}, \qquad \begin{matrix} g_k := \mathrm{Det}\mathcal{H}_{2k}^{(-2k)} \geq 0, \\ h_k := \mathrm{Det}\mathcal{H}_{2k-1}^{(-2k)} \geq 0, \end{matrix} \quad k \in \mathbb{N}_0$$

and $\forall n \in \mathbb{N}$ Laurent polynomials $R_n(\lambda) = (R_{n;0}(\lambda), R_{n;0}(\lambda)) \in \mathbb{C}^2$, $\lambda \in \mathbb{R}$, where

$$R_{n;0}(\lambda) = -(g_n h_{n-1})^{1/2}\mathrm{Det}\begin{bmatrix} s_{-2n-1} & s_{-2n} & \dots & s_0 \\ s_{-2n} & s_{-2n+1} & \dots & s_1 \\ \dots & \dots & \dots\dots \\ s_{-1} & s_0 & \dots & s_{2n} \\ \lambda^{-n-1} & \lambda^{-n} & \dots & \lambda^n \end{bmatrix},$$

$$R_{n;1}(\lambda) = (g_n h_n)^{1/2}\mathrm{Det}\begin{bmatrix} s_{-2n} & s_{-2n+1} & \dots & s_0 \\ s_{-2n+1} & s_{-2n+2} & \dots & s_1 \\ \dots & \dots & \dots\dots \\ s_{-1} & s_0 & \dots & s_{2n-1} \\ \lambda^{-n} & \lambda^{-n+1} & \dots & \lambda^n \end{bmatrix}, \qquad (4.6.4)$$

$$R_{0;0}(\lambda) = 1.$$

On the other hand, on the linear set L of Laurent polynomials, we introduce the scalar product $(\cdot,\cdot)_S$, $\forall R, T \in L$, *by setting*

$$(R,T)_S = \sum_{j,k\in\mathbb{Z}} s_{j+k} r_j \bar{t}_k, \quad R(\lambda) = \sum_{j\in\mathbb{Z}} r_j \lambda^j, \quad T(\lambda) = \sum_{k\in\mathbb{Z}} t_k \lambda^k. \tag{4.6.5}$$

Elements $a_{j,k;\alpha,\beta}$ *of matrix J are introduced by the expression*

$$a_{j,k;\alpha,\beta} = (\lambda R_{k;\beta}(\lambda), R_{j;\alpha}(\lambda))_S, \quad j,k \in \mathbb{N}_0, \quad \alpha,\beta = 0,1. \tag{4.6.6}$$

Proof First, we note that the representation (4.2.1) shows that the scalar product (4.6.5) is equal to the scalar product in $L_2(\mathbb{R}, d\rho(\lambda))$ (the non-degeneracy of s gives that $(\cdot,\cdot)_S$ is a scalar, but not a quasi-scalar product). Therefore, formulas (4.6.6) are the same as (4.3.12).

The totality condition in $L_2(\mathbb{R}, d\rho(\lambda))$ of the set (4.3.30) for $m \in \mathbb{N}_0$ is assumed, therefore, it is possible again apply to this case the construction of Sects. 4.3 and 4.4. It is clear that our Laurent polynomials (4.6.4) are equal to polynomials of the first kind $P_{n;\alpha}(\lambda)$: formulas (4.6.4) are easily obtained by calculating orthogonal polynomials by the Gram-Schmidt procedure from the sequence (4.3.3), using notations (4.2.1) for the integral. As it follows from Theorem 4.4.3, every function $\mathbb{R} \ni \lambda \longmapsto \lambda^m$, $m \in \mathbb{Z}$ belongs to $L_2(\mathbb{R}, d\rho(\lambda))$ with respect to the spectral measure $d\rho(\lambda)$. Therefore, each self-adjoint extension of the operator $\mathbf{J}$ in $\mathbf{l}_2$ constructed by J of the type (4.6.6) gives some spectral measure $d\rho(\lambda)$ for which the representation (4.2.1) holds true. Since we can describe such extensions (see Theorem 4.5.6 and corresponding remarks in Sect. 4.4), then we can get the description of all measures for (4.2.1) with the given s and such that the set (4.3.30) for $m \in \mathbb{N}_0$ is total in $L_2(\mathbb{R}, d\rho(\lambda))$. □

4.7 Two Addition Facts

In this short section we will touch to two type of results: (1) in what manner the representation (4.2.1), i.e. the main result of Theorem 4.2.1 and corresponding results, relating to positive definite kernels outlined in [21] Chap. 8, have a connection; (2) how to explain the way to obtain results of the type of Sects. 4.2–4.6 for the matrices related to the strong moment problem.

(1) We will use the designations similar Chap. 2. Consider for the sequence of complex numbers $f = (f_n)$, $n \in \mathbb{Z}$, the difference expression of the order $r \in \mathbb{N}$

$$(Lf)_j = \sum_{\alpha=r_-}^{r_+} a_{j,\alpha} f_{j+\alpha}, \quad j \in \mathbb{Z}. \tag{4.7.1}$$

Here $a_{j,\alpha} \in \mathbb{C}$ and the expansion $r = r_- + r_+, r_-, r_+ \in \mathbb{N}_0$ is fixed. Let $K = (K_{j,k})$, $j, k \in \mathbb{Z}$, $K_{j,k} \in \mathbb{C}$ be a some positive defined kernel (matrix), i.e., the inequality

$$\sum_{j,k \in \mathbb{Z}} K_{j,k} f_j \bar{f}_k \geq 0, \tag{4.7.2}$$

holds true for an arbitrary finite sequence $f = (f_j)$, $j \in \mathbb{Z}$.

Such a kernel generates, on the set l_{fin} of finite sequences, (quasi)scalar product $(\cdot, \cdot)_K$ and a norm $\|\cdot\|_K$ for which $\|f\|_K^2$ is given by (4.7.2). We denote by H_K the corresponding Hilbert space.

Let kernel K and expression L are $*$-commuting, i.e., the equality

$$L_{(j)} K = \bar{L}_{(k)} K, \tag{4.7.3}$$

holds true on $\mathbb{Z} \times \mathbb{Z}$.

Here (j), (k) denote, that L of the form (4.7.1) acts on $K = (K_{j,k})$, $j, k \in \mathbb{Z}$ with respect to indexes (j), (k).

The main result of the theory of positive definite kernels K which are $*$-commuting with expression L is a representation of K by integral on the fundamental solution of difference equation

$$(Lf)_j = \lambda f_j, \quad j \in \mathbb{N}; \quad \lambda \in \mathbb{R}. \tag{4.7.4}$$

Remind that a system of r solutions on $j \in \mathbb{N}$ of (4.7.4) $\chi_{j;\alpha}(\lambda)$, $\alpha = 0, 1, \ldots, r-1$, is called fundamental if these solutions satisfy the following initial data with some fixed $p \in \mathbb{Z}$:

$$\chi_{j;\alpha}(\lambda) = \delta_{j,p-r_-+\alpha}, \quad j = p - r_-, \ldots, p + r_+ - 1, \quad \alpha = 0, 1, \ldots, r-1$$

in standard notation $\delta_{m,n}$ (fundamental system of solutions "with an initial data near a point p").

Conditions for coefficients $a_{j,\alpha}$ of the expression L in (4.7.1), which guarantee the existence of a fundamental solution, are obtained simply. It is assumed that such a solution exists for $p = 0$.

We will use the theorem (see [21], Chapter 8, Theorem 5.1).

Theorem 4.7.1 *The positive definite kernel* $K = (K_{j,k})$, $j, k \in \mathbb{Z}$ *due to the fundamental system* $\chi_{j;\alpha}(\lambda)$, $j \in \mathbb{Z}$, $\alpha = 0, 1, \ldots, r-1$, $\lambda \in \mathbb{R}$ *has the representation*

$$K_{j,k} = \int_{\mathbb{R}} \sum_{\alpha,\beta=0}^{r-1} \chi_{j;\alpha}(\lambda) \overline{\chi_{k;\beta}(\lambda)}\, d\rho_{\alpha,\beta}(\lambda), \quad j, k \in \mathbb{Z} \tag{4.7.5}$$

with non-negative $(r \times r)$*-matrix measure* $d\rho(\lambda) = (d\rho_{\alpha,\beta}(\lambda))_{\alpha,\beta=0}^{r-1}$ *on* $\mathfrak{B}(\mathbb{R})$ *if and only if* K *and* L *are connected by the equality* (4.7.3). *The measure in* (4.7.5) *is defined by* K *uniquely if and only if the closure of the operator in the space* $\mathcal{H}_K$, $l_{\text{fin}} \ni f \longmapsto Lf \in l_{\text{fin}}$ *is maximal.*

The proof of this theorem is similar to the proof of Theorem 4.1.1 and based on Theorem 2.11.5.

The representation (4.2.1) is a particular case of Theorem 4.7.1: it is necessary to take L from (4.7.1) of the form

$$(Lf)_j = f_{j-1}, \quad j \in \mathbb{Z},$$
$$(p = 0,\ r = 1,\ r_- = 1,\ r_+ = 0,\ a_{j,\alpha} = 1,\ j \in \mathbb{Z},\ \alpha = 0). \tag{4.7.6}$$

Now, $\forall j, k \in \mathbb{Z}$, $K_{j,k} = s_{j+k}$, conditions (4.7.3) for L (4.7.6) and for this K are fulfilled. Also, now $r = 1$, $\chi_{j;0}(\lambda) = \lambda^j$, $j \in \mathbb{Z}$, and the matrix measure $d\rho(\lambda) = d\rho_{0,0}(\lambda)$ is an ordinary probabilistic measure. The representation (4.7.5) is transferred into (4.2.1).

Note that this partial case was not considered in the literature, but similar situations in the usual classical moment problem were considered.

(2) Let us turn to the strong matrix moment problem. We fix some separable Hilbert space H and consider the operator-valued measure $d\rho(\lambda)$ on $\mathbb{R}$. More precisely, we have the mapping $\mathfrak{B}(\mathbb{R}) \ni \Delta \longmapsto \rho(\Delta)$, where $\rho(\Delta)$ is a non-negative operator in $\mathcal{H}$, $\rho(\mathbb{R}) = 1$ and $\forall u \in \mathcal{H}$, $(\rho(\Delta)u, u)_{\mathcal{H}}$ is an ordinary measure on $\mathfrak{B}(\mathbb{R})$.

It is easy to introduce the integral $\int_{\mathbb{R}} F(\lambda)\, d\rho(\lambda)$ for a scalar-valued function $\mathbb{R} \ni \lambda \longmapsto F(\lambda) \in \mathbb{C}$, the value of this integral is an operator in $\mathcal{H}$. This operator may or may not be bounded. In what follows, we denote by $L(\mathcal{H})$ the set of all bounded operators in $\mathcal{H}$.

The strong operator moment problem is formulated as follows. Let $s = (s_n)$, $n \in \mathbb{Z}$ be some sequence of bounded operators in the space $\mathcal{H}$, i.e., $s_n \in L(\mathcal{H})$, $n \in \mathbb{Z}$. These operators are called the moments if we have the representation of the type (4.2.1)

$$s_n = \int_{\mathbb{R}} \lambda^n\, d\rho(\lambda) \in L(\mathcal{H}), \quad n \in \mathbb{Z}, \tag{4.7.7}$$

but with a measure $d\rho(\lambda)$ which was introduced above as an operator-valued measure. If in (4.7.7) only $n \in \mathbb{N}_0 \subset \mathbb{Z}$, then we have classical operator moment problem. If $\dim(\mathcal{H}) < \infty$, then the corresponding problem are called the matrix problem.

Many works are devoted to the study of the operator moment problem, but the priority for authors are studying matrix and classical (not strong) problems.

This section only demonstrates the approach from Sects. 4.3–4.6. Similarly to the condition (4.2.2), we add that the sequence $s = (s_n)$, $n \in \mathbb{Z}$, $s_n \in L(\mathcal{H})$ is positive defined if for an arbitrary finite sequence $f = (f_n)$, $n \in \mathbb{Z}$, $f_n \in \mathcal{H}$ we have

$$\sum_{j,k \in \mathbb{Z}} (s_{j+k} f_k, f_j)_{\mathcal{H}} \geq 0 \tag{4.7.8}$$

(for $H = \mathbb{C}^1$ this condition is the same as (4.2.2)).

We formulate now the generalization of Theorem 4.2.1 in the simplest case: when $\mathcal{H}$ is finite dimensional, $\mathcal{H} = \mathbb{C}^d$, $d \in \mathbb{N}$, and in addition, each matrix s_n, $n \in \mathbb{Z}$, acting in the space $\mathbb{C}^d$, has real elements (in any case it is Hermitian, this follows from (4.7.8), but we require its reality). Hence, we have the theorem.

Theorem 4.7.2 *The matrix sequence $s = (s_n)$, $n \in \mathbb{Z}$, $s_n \in L(\mathbb{C}^d)$ with real matrix elements is a strong moment sequence, i.e., the representation* (4.7.7) *with some operator-valued measure $d\rho(\lambda)$ with real matrix valued $\rho(\Delta)$, $\Delta \in \mathfrak{B}(\mathbb{R})$, holds true if and only if it is a positive definite sequence, i.e., the condition* (4.7.8) *is fulfilled.*

Proof It is similar to the proof of Theorem 4.2.1 and we only show the necessary changes.

Introduce the linear set $l_{\text{fin}}(\mathcal{H}) = l_{\text{fin}}(\mathbb{C}^d)$ of finite sequences $f = (f_n)$, $n \in \mathbb{Z}$ of vectors $f_n \in \mathcal{H} = \mathbb{C}^d$. According to (4.7.8), we introduce in $l_{\text{fin}}(\mathcal{H})$ (quasi)scalar product

$$(f, g)_{S(\mathcal{H})} := \sum_{j,k \in \mathbb{Z}} (s_{j+k} f_k, g_j)_{\mathcal{H}}, \quad f, g \in l_{\text{fin}}(\mathcal{H}).$$

Let $S(\mathcal{H})$ be the corresponding Hilbert space (with an appropriate factorization of the set $l_{\text{fin}}(\mathcal{H})$ if necessary). In our case this space is "real": $\forall f \in S(\mathcal{H})$ the mapping $f \longmapsto \bar{f} \in S(\mathcal{H})$ exists, that $\forall f, g \in S(\mathcal{H})$, $(\bar{f}, \bar{g})_{S(\mathcal{H})} = \overline{(f, g)_{S(\mathcal{H})}}$, $\bar{\bar{f}} = f$. This fact follows from the reality of s_n, $n \in \mathbb{Z}$.

As in Sect. 3.2, we define on $f \in l_{\text{fin}}(\mathcal{H})$ the operator of the type (4.2.5):

$$(Jf)_j = f_{j-1}, \quad j \in \mathbb{Z}; \quad \text{Dom}(J) = l_{\text{fin}}(\mathcal{H}).$$

As in (4.2.7) we check that this operator J is Hermitian in the space $S(\mathcal{H})$.

Riggings of type (4.2.8) and (4.2.9) are constructed similarly as in the proof of Theorem 4.2.1: since $\mathcal{H} = \mathbb{C}^d$ is finite-dimensional, then the imbedding of type $l_2(p) \hookrightarrow l_2$ can be do quasi-nuclear obtained by taking the sequence (p_n), $n \in \mathbb{Z}$, which grows sufficiently strongly on ∞.

The other part of the proof of our theorem is similar to the proof of Theorem 4.2.1. At first we take some self-adjoint extension A of operator J in the space $S(\mathcal{H})$. It is possible, since the defect numbers of J are mutually equal. Then it is necessary to use some generalization of Theorem 4.1.2: instead of one vector q it is necessary to take d vectors of the type $(1, 0, \dots, 0)$, $(0, 1, \dots, 0)$, $\dots$, $(0, 0, \dots, 1) \in \mathbb{C}^d$. The spectral measure $d\sigma(\lambda)$ must be now matrix-valued. It is possible also to apply directly the Theorem 4.1.1 instead of a variant of Theorem 4.1.2. □

If the matrices s_n, $n \in \mathbb{Z}$ are only Hermitian, it is necessary to add a condition to Theorem 4.7.2 about the equality of the defect numbers of the operator J.

Remark 4.7.3 3. The situation with $\dim(\mathcal{H}) = \infty$ is more complicated: for the construction of a quasi-nuclear rigging of the type $l_2(p) \hookrightarrow l_2$ (the spaces of sequences (f_n), $n \in \mathbb{Z}$) increases with weight $p = (p_n)$, $n \in \mathbb{Z}$ on ∞, that is not satisfactory. It is necessary to take for spaces $l_2(p)$ and (l_2) the sequences of vectors from $\mathcal{H}_1$, $(\mathcal{H}_2)$ and the imbedding $\mathcal{H}_1 \hookrightarrow \mathcal{H}_2$ must be quasi-nuclear. Such a situation requires changes in the formulation of Theorem of the type 4.7.2.

An analog of Theorem 4.7.2, of course, is true for classical matrix moment problem. In this case, the corresponding spectral theory for block Jacobi matrices is partially constructed in Chap. 2. Now the situation is a little more complicated: the defect numbers can be different.

4.8 The Inner Structure of the Jacobi-Laurent Matrix

Not every block matrix of the form (4.3.26) described in previous sections can have an inverse one of the form (4.3.43). For this, (to have an inverse one) it must satisfy additional conditions on coefficients. This section specifies the necessary conditions. Let us redefine elements of the matrix J for the next consideration, namely,

$$J = \begin{pmatrix} a_1 & b_1 & c_1 & & & & & & \\ b_1 & a_2 & b_2 & & & 0 & & & \\ c_1 & b_2 & a_3 & b_3 & c_3 & & & & \\ & & b_3 & a_4 & b_4 & & & & \\ & & c_3 & b_4 & a_5 & b_5 & c_5 & & \\ & 0 & & & b_5 & a_6 & b_6 & & \\ & & & & c_5 & b_6 & a_7 & b_7 & c_7 \\ & & & & & & \ddots & \ddots & \ddots \end{pmatrix}, \tag{4.8.1}$$

where $a_k, b_k \in \mathbb{R}$, $c_{2k-1} > 0$, $k \in \mathbb{N}$.

We consider an algebraically inverse matrix J^{-1} satisfying the equality

$$JJ^{-1} = E \tag{4.8.2}$$

on finite vectors, where E is the identity matrix. Together with (4.8.2) we do not consider $J^{-1}J = E$, since J and J^{-1} generate bounded symmetric operators.

Suppose the matrix J has the algebraically inverse one J^{-1}. We redefine elements (4.3.43) in the form

$$J^{-1} = \begin{pmatrix} \alpha_1 & \beta_1 & & & & & & \\ \beta_1 & \alpha_2 & \beta_2 & \gamma_2 & & 0 & & \\ & \beta_2 & \alpha_3 & \beta_3 & & & & \\ & \gamma_2 & \beta_3 & \alpha_4 & \beta_4 & \gamma_4 & & \\ & & & \beta_4 & \alpha_5 & \beta_5 & & \\ & 0 & & \gamma_4 & \beta_5 & \alpha_6 & \beta_6 & \gamma_6 \\ & & & & & \ddots & \ddots & \ddots \end{pmatrix}, \quad (4.8.3)$$

where $\alpha_k, \beta_k \in \mathbb{R}$, $\gamma_{2k} > 0$, $k \in \mathbb{N}$.

Let us investigate the following, in a certain sense, inverse problem. Let us take a matrix of the form (4.8.1) with arbitrary coefficients. Can we expect that there exists J^{-1} of the form (4.8.3) such that its J is its algebraic inverse, that is, (4.8.2) holds true? This question is not trivial, since the operator, generated by matrix (4.8.1) with elements (for example) $a_k = 0$, $c_{2k-1} = 1$, $k \in \mathbb{N}$, $b_1 = 1$, $b_k = -1$, $k > 1$ has zero in the point spectrum and hence does not inverse every where defined bounded operator in l_2. It is not difficult to construct an example with some elements $a_k = 2$, $b_k = 1$, $c_{2k-1} = 1$, $k \in \mathbb{N}$ such that matrix J has inverse one, bat it is not of the form (4.8.3).

The necessary and sufficient conditions that give an answer for discussed question, are presented in the theorem.

Theorem 4.8.1 *The matrix J of the form* (4.8.1) *has the algebraically inverse one of the form* (4.8.3) *if and only if*

$$\det \begin{vmatrix} b_{n-1} & c_{n-1} \\ a_n & b_n \end{vmatrix} = 0, \quad (4.8.4)$$

$$\det \begin{vmatrix} a_1 & b_1 \\ b_1 & a_2 \end{vmatrix} \neq 0, \ \det \begin{vmatrix} a_n & b_n & 0 \\ b_n & a_{n+1} & b_{n+1} \\ 0 & b_{n+1} & a_{n+2} \end{vmatrix} \neq 0, \ \ n = 2k, \ k \in \mathbb{N}. \quad (4.8.5)$$

Remark 4.8.2 Conditions (4.8.5) are natural in the sense that if we have the matrix (4.8.1) with some determinants equal zero, than we obtain linear dependent corresponding rows and columns that we can every time exclude from the matrix. In such a case we have linear dependent corresponding polynomials $P_{k;\alpha}(\lambda)$, $k \in \mathbb{N}$, $\alpha = 0, 1$ and, hence, linear dependent functions $\lambda^m \in L_2(\mathbb{R}, d\rho(\lambda))$, $m \in \mathbb{Z}$. On the other hand, the condition (4.8.5) guarantees that zero is not a point of the point spectrum of the operator corresponding to J.

Proof We consider the equality (4.8.2) and solve the problem: for a given elements of the matrix (4.8.1) $a_k, b_k, c_{2k-1}, k \in \mathbb{N}$ to find (and calculate) uniquely elements of the matrix (4.8.3) $\alpha_k, \beta_k, \gamma_{2k}, k \in \mathbb{N}$.

Denote the positions of elements of the matrix JJ^{-1} as (x, y), where x is the row and y the column of the matrix JJ^{-1}, also by (x, y) we denote corresponding scalar equations generated by (4.8.2).

First, consider the pairs of elements of the JJ^{-1} matrix, which are at the positions

$$(n-1, n+1),\ (n, n+1);\quad ((n-1, n+2)\ (n, n+2)),\qquad n = 2k,\ k \in \mathbb{N},$$

Elements of these couple must equal zero due to (4.8.2): for $n = 2k,\ k \in \mathbb{N}$

$$\begin{cases} b_{n-1}\beta_n + c_{n-1}\alpha_{n+1} = 0 \\ a_n\beta_n + b_n\alpha_{n+1} = 0, \end{cases},\quad \left(\begin{cases} b_{n-1}\gamma_n + c_{n-1}\beta_{n+1} = 0 \\ a_n\gamma_n + b_n\beta_{n+1} = 0, \end{cases}\right). \tag{4.8.6}$$

Non-zero elements β_n, α_{n+1} and γ_n, β_{n+1}, $n = 2k,\ k \in \mathbb{N}$ exist as the solution of corresponding homogeneous system (4.8.6) if and only if the corresponding determinants are equal zero, i.e., the condition (4.8.4) is fulfilled.

Similar conclusions can be drawn from the elements of the matrix JJ^{-1}, namely,

$$(1, 2)\ (1, 3);\ (n+2, n)\ (n+3, n);\quad ((n+2, n+1)\ (n+3, n+1)),\qquad n = 2k,\ k \in \mathbb{N}.$$

In such simple way we had proved the necessity of the condition (4.8.4). Indeed, this condition is also sufficient in the theorem, taking into account (4.8.5). But this fact requires consideration of all elements from (4.8.2). We apply the method of the mathematical induction. To begin the induction, we consider a non-standard zero step.

Step 0 In this step we consider such a part of the equality (4.8.2):

$$\begin{pmatrix} a_1\alpha_1 + b_1\beta_1 & a_1\beta_1 + b_1\alpha_2 + c_1\beta_2 & * \\ b_1\alpha_1 + a_2\beta_1 & b_1\beta_1 + a_2\alpha_2 + b_2\beta_2 & * \\ c_1\alpha_1 + b_2\beta_1 & * & * \\ 0 & * & * \end{pmatrix} = \begin{pmatrix} 1 & 0 & * \\ 0 & 1 & * \\ 0 & * & * \end{pmatrix}, \tag{4.8.7}$$

where "$*$" denotes a non-considering in this step element. We consider only the elements that form the equation at positions (1,1), (2,1), (3,1), (1,2), (2,2). Since the equations

$$\begin{cases} b_1\alpha_1 + a_2\beta_1 = 0 \\ c_1\alpha_1 + b_2\beta_1 = 0 \end{cases}$$

were previously considered the conditions (4.8.5) were taken into account, we can consider the system generated by the elements at positions (1,1) and (2,1):

$$\begin{cases} a_1\alpha_1 + b_1\beta_1 = 1 \\ b_1\alpha_1 + a_2\beta_1 = 0. \end{cases}$$

The mean determinant of this non-homogeneous system, due to (4.8.5), is not equal zero and hence this system has unique solution α_1, β_1.

$$\alpha_1 = \frac{a_2}{a_1a_2 - b_1^2}, \quad \beta_1 = \frac{-b_1}{a_1a_2 - b_1^2}. \tag{4.8.8}$$

Let us show that elements (1,2) and (2,2) of JJ^{-1} form the system of equations in (4.8.2) with linear dependent coefficients taking in to account that α_1 and β_1 obtained in (4.8.8). Indeed, for the coefficients of the system

$$\begin{cases} a_1\beta_1 + b_1\alpha_2 + c_1\beta_2 = 0 \\ b_1\beta_1 + a_2\alpha_2 + b_2\beta_2 = 1 \end{cases},$$

using condition (4.8.5) and (4.8.8), we get

$$\frac{b_1}{a_2} = \frac{c_1}{b_2} = \frac{-a_1\beta_1}{1 - b_1\beta_1} = \frac{a_1\frac{b_1}{a_1a_2-b_1^2}}{1 + b_1\frac{b_1}{a_1a_2-b_1^2}} = \frac{a_1b_1}{a_1a_2 - b_1^2 + b_1^1} = \frac{a_1b_1}{a_1a_2} = \frac{b_1}{a_2}.$$

Hence in the next step we can consider only the position (2,2) instead of considering both (1,2) and (2,2) together.

Step 1 In this step we consider the following positions of elements of the matrix JJ^{-1}, namely,

$$(2,2), (3,2), (4,2), (5,2), (1,3), (2,3), (3,3), (4,3), (5,3), (1,4),$$
$$(2,4), (3,4), (4,4),$$

i.e., the corresponding part of (4.8.2) has a form

$$\begin{pmatrix} ** & b_1\beta_2 + c_1\alpha_3 \\ *\, b_1\beta_1 + a_2\alpha_2 + b_2\beta_2 & a_2\beta_2 + b_2\alpha_3 \\ *\, c_1\beta_1 + b_2\alpha_2 + a_3\beta_2 + b_3\gamma_2 & b_2\beta_2 + a_3\alpha_3 + b_3\beta_3 \\ *\, b_3\beta_2 + a_4\gamma_2 & b_3\alpha_3 + a_4\beta_3 \\ *\, c_3\beta_2 + b_4\gamma_2 & c_3\alpha_3 + b_4\beta_3 \\ *\, 0 & 0 \end{pmatrix}$$

$$
\begin{pmatrix}
b_1\gamma_2 + c_1\beta_3 & * \\
a_2\gamma_2 + b_2\beta_3 & * \\
b_2\gamma_2 + a_3\beta_3 + b_3\alpha_4 + c_3\beta_4 & * \\
b_3\beta_3 + a_4\alpha_4 + b_4\beta_4 & * \\
* & * \\
* & *
\end{pmatrix} = \begin{pmatrix}
* & * & 0 & 0 & * \\
* & 1 & 0 & 0 & * \\
* & 0 & 1 & 0 & * \\
* & 0 & 0 & 1 & * \\
* & 0 & 0 & * & * \\
* & 0 & 0 & * & *
\end{pmatrix}, \tag{4.8.9}
$$

where, as usual, "$*$" denotes a non-considering element.

Since elements (4,2) and (5,2) form linear dependent system (condition (4.8.5)), then let us consider elements (2,2), (3,2) and (4,2) of JJ^{-1} and show that corresponding equations of (4.8.2) form linear independent system that gives unique solution $\alpha_2, \beta_2, \gamma_2$, taking in to account that β_1 obtained in (4.8.8), namely,

$$
\begin{cases}
b_1\beta_1 + a_2\alpha_2 + b_2\beta_2 = 1 \\
c_1\beta_1 + b_2\alpha_2 + a_3\beta_2 + b_3\gamma_2 = 0 \\
b_3\beta_2 + a_4\gamma_2 = 0
\end{cases}
$$

and

$$
\begin{cases}
a_2\alpha_2 + b_2\beta_2 = 1 - b_1\beta_1 \\
b_2\alpha_2 + a_3\beta_2 + b_3\gamma_2 = -c_1\beta_1 \\
b_3\beta_2 + a_4\gamma_2 = 0.
\end{cases} \tag{4.8.10}
$$

Here we use the condition (4.8.5)

$$
\det \begin{vmatrix} a_2 & b_2 & 0 \\ b_2 & a_3 & b_3 \\ 0 & b_3 & a_4 \end{vmatrix} \neq 0,
$$

that yields the uniqueness of the solution $\alpha_2, \beta_2, \gamma_2$. Substituting γ_2 from the third equation in to the second one in (4.8.10), we get

$$
\begin{cases}
a_2\alpha_2 + b_2\beta_2 = 1 - b_1\beta_1 \\
b_2\alpha_2 + (a_3 - \frac{b_3^2}{a_4})\beta_2 = -c_1\beta_1.
\end{cases}
$$

From the last system we have

$$
\alpha_2 = \frac{(1 - b_1\beta_1)(a_3 - \frac{b_3^2}{a_4}) + c_1\beta_1 b_2}{a_2(a_3 - \frac{b_3^2}{a_4}) - b_2^2}
$$

$$
\beta_2 = \frac{-a_2c_1\beta_1 - b_2(1 - b_1\beta_1)}{a_2(a_3 - \frac{b_3^2}{a_4}) - b_2^2} = \frac{(-a_2c_1 + b_1b_2)\beta_1 - b_2}{a_2(a_3 - \frac{b_3^2}{a_4}) - b_2^2}.
$$

Taking in to account the condition (4.8.4) in the form $a_2c_1 - b_1b_2 = 0$, we get

$$\beta_2 = \frac{-b_2}{a_2(a_3 - \frac{b_3^2}{a_4}) - b_2^2}. \tag{4.8.11}$$

From the position (4.2) we get

$$\gamma_2 = -\frac{b_3}{a_4}\beta_2. \tag{4.8.12}$$

By using positions, for example, (1,3) and (1,4) (also we can consider position (2,3), (2,4)), we get

$$\alpha_3 = -\frac{b_1}{c_1}\beta_2, \tag{4.8.13}$$

$$\beta_3 = -\frac{b_1}{c_1}\gamma_2. \tag{4.8.14}$$

Let us show that β_3 obtained in (4.8.14) is the same as in the position (4,3) (also we can consider (5.3)). Indeed, the substitution γ_2 from (4.8.12) into (4.8.14),

$$\beta_3 = \frac{b_1}{c_1}\frac{b_3}{a_4}\beta_2 \tag{4.8.15}$$

is the same as β_3 given from the position (4,3) and we substitute α_3 from (4.8.13) and get

$$\beta_3 = -\frac{b_3}{a_4}\alpha_3 = \frac{b_3}{a_4}\frac{b_1}{c_1}\beta_2.$$

In the next local step we show that the position (3,3) is fulfilled with obtained above α_3 and β_3 in (4.8.13) and (4.8.15), respectively, i.e., we must to verify the equality

$$b_2\beta_2 + a_3\alpha_3 + b_3\beta_3 = 1.$$

Substitutions α_3 and β_3 from (4.8.13) and (4.8.15) give

$$b_2\beta_2 - a_3\frac{b_1}{c_1}\beta_2 + b_3\frac{b_1}{c_1}\frac{b_3}{a_4}\beta_2 = 1, \quad \text{i.e.,} \quad \beta_2(b_2 - a_3\frac{b_1}{c_1} + b_3\frac{b_1}{c_1}\frac{b_3}{a_4}) = 1,$$

$$\beta_2(b_2 - \frac{b_1}{c_1}(a_3 - \frac{b_3^2}{a_4})) = 1, \quad \text{i.e.,} \quad \beta_2 = \frac{1}{b_2 - \frac{b_1}{c_1}(a_3 - \frac{b_3^2}{a_4})}.$$

We use the condition (4.8.4) in the form $b_1 b_2 = c_1 a_2$ and get

$$\beta_2 = \frac{1}{b_2 - \frac{a_2}{b_2}(a_3 - \frac{b_3^2}{a_4})} = \frac{b_2}{b_2^2 - a_2(a_3 - \frac{b_3^2}{a_4})},$$

that is equal β_2 obtained in (4.8.11).

Let us consider position (3,4) and (4,4) and show that elements on positions (3,4) and (4,4) of JJ^{-1} form the system with linear dependent coefficients in (4.8.2) taking in to account that γ_2 and β_3 obtained in (4.8.12) and (4.8.15). Indeed, for the coefficients of the system

$$\begin{cases} b_2\gamma_2 + a_3\beta_3 + b_3\alpha_4 + c_3\beta_4 = 0 \\ \qquad b_3\beta_3 + a_4\alpha_4 + b_4\beta_4 = 1, \end{cases}$$

on the form

$$\begin{cases} b_3\alpha_4 + c_3\beta_4 = -b_2\gamma_2 - a_3\beta_3 \\ a_4\alpha_4 + b_4\beta_4 = 1 - b_3\beta_3, \end{cases}$$

using condition (4.8.4), (4.8.11), (4.8.12) and (4.8.15), we have

$$\frac{b_3}{a_4} = \frac{c_3}{b_4} = \frac{-b_2\gamma_2 - a_3\beta_3}{1 - b_3\beta_3} = \frac{b_2\frac{b_3}{a_4}\beta_2 - a_3\frac{b_1}{c_1}\frac{b_3}{a_4}\beta_2}{1 - b_3\frac{b_1}{c_1}\frac{b_3}{a_4}\beta_2}$$

$$= \frac{\frac{b_3}{a_4}\beta_2(b_2 - a_3\frac{b_1}{c_1})}{1 - \frac{b_1}{c_1}\frac{b_3^2}{a_4}\beta_2} = \frac{b_3}{a_4}\frac{(b_2 - a_3\frac{b_1}{c_1})\frac{-b_2}{a_2(a_3 - \frac{b_3^2}{a_4}) - b_2^2}}{1 + \frac{b_1}{c_1}\frac{b_3^2}{a_4}\frac{b_2}{a_2(a_3 - \frac{b_3^2}{a_4}) - b_2^2}}$$

$$= \frac{b_3}{a_4}\frac{-b_2(b_2 - a_3\frac{b_1}{c_1})c_1 a_4}{-c_1 a_4 b_2^2 + c_1 a_4 a_2(a_3 - \frac{b_3^2}{a_4}) + b_1 b_3^2 b_2}$$

$$= \frac{b_3}{a_4}\frac{-b_2^2 c_1 a_4 + b_1 b_2 a_3 a_4}{-c_1 a_4 b_2^2 + c_1 a_2 a_3 a_4 - c_1 a_2 b_3^2 + b_1 b_2 b_3^2} =$$

$$= \frac{b_3}{a_4}\frac{-b_2^2 c_1 a_4 + b_1 b_2 a_3 a_4}{-b_2^2 c_1 a_4 + b_1 b_2 a_3 a_4 + b_3^2(b_1 b_2 - c_1 a_2)} = \frac{b_3}{a_4},$$

where the condition (4.8.4) is used twice in the form $(b_1 b_2 - c_1 a_2) = 0$.

Hence, in the next step we can consider only the position (4,4) instead of the consideration both (3,4) and (4,4) together.

From the position (4,3), we have

$$b_3\alpha_3 + a_4\beta_3 = 0, \qquad \text{i.e.,} \quad \beta_3 = -\frac{b_3}{a_4}\alpha_3.$$

Since (1,3) gives $\alpha_3 = -\frac{b_1}{c_1}\beta_2$, then we have

$$\beta_3 = \frac{b_3}{a_4}\frac{b_1}{c_1}\beta_2.$$

Also from position (1,4), we have

$$\beta_3 = -\frac{b_1}{c_1}\gamma_2.$$

Since (4,2) gives

$$\gamma_2 = -\frac{b_3}{a_4}\beta_2,$$

we have the same for β_3, namely,

$$\beta_3 = \frac{b_1}{c_1}\frac{b_3}{a_4}\beta_2.$$

This completes the first inductive step.

Step 2 We consider the n-th inductive step. In this step, we use indexes $n = 2k$, $k \in \mathbb{N}$.

At this step, we consider the following elements of the matrix JJ^{-1} at positions (n, n), $(n+1, n)$, $(n+2, n)$, $(n+3, n)$, $(n-1, n+1)$, $(n, n+1)$, $(n+1, n+1)$, $(n+2, n+1)$, $(n+3, n+1)$, $(n-1, n+2)$, $(n, n+2)$, $(n+1, n+2)$, $(n+2, n+2)$, i.e., the corresponding part of (4.8.2) has a form

$$\begin{pmatrix}
* & b_{n-1}\beta_n + c_{n-1}\alpha_{n+1} \\
b_{n-1}\beta_{n-1} + a_n\alpha_n + b_n\beta_n & a_n\beta_n + b_n\alpha_{n+1} \\
c_{n-1}\beta_{n-1} + b_n\alpha_n + a_{n+1}\beta_n + b_{n+1}\gamma_n & b_n\beta_n + a_{n+1}\alpha_{n+1} + b_{n+1}\beta_{n+1} \\
b_{n+1}\beta_n + a_{n+2}\gamma_n & b_{n+1}\alpha_{n+1} + a_{n+2}\beta_{n+1} \\
c_{n+1}\beta_n + b_{n+2}\gamma_n & c_{n+1}\alpha_{n+1} + b_{n+2}\beta_{n+1} \\
0 & 0
\end{pmatrix}$$

$$\left.\begin{matrix} b_{n-1}\gamma_n + c_{n-1}\beta_{n+1} \\ a_n\gamma_n + b_n\beta_{n+1} \\ b_n\gamma_n + a_{n+1}\beta_{n+1} + b_{n+1}\alpha_{n+2} + c_{n+1}\beta_{n+2} \\ b_{n+1}\beta_{n+1} + a_{n+2}\alpha_{n+2} + b_{n+2}\beta_{n+2} \\ * \\ * \end{matrix}\right) = \begin{pmatrix} * & 0 & 0 \\ 1 & 0 & 0 \\ 0 & 1 & 0 \\ 0 & 0 & 1 \\ 0 & 0 & * \\ 0 & 0 & * \end{pmatrix}, \quad n \in \mathbb{N}, \tag{4.8.16}$$

where, as usual, "$*$" denotes a non-considering element.

Since elements $(n+2, n)$ and $(n+3, n)$ form linear dependent system (condition (4.8.5)), then let us consider elements (n, n), $(n+1, n)$ and $(n+2, n)$ of JJ^{-1} and show that these elements form in (4.8.2) linear independent system that gives unique solution $\alpha_n, \beta_n, \gamma_n$, taking in to account that β_{n-1} obtained above by induction assumption

$$\begin{cases} b_{n-1}\beta_{n-1} + a_n\alpha_n + b_n\beta_n = 1 \\ c_{n-1}\beta_{n-1} + b_n\alpha_n + a_{n+1}\beta_n + b_{n+1}\gamma_n = 0 \\ b_{n+1}\beta_n + a_{n+2}\gamma_n = 0 \end{cases}$$

and

$$\begin{cases} a_n\alpha_n + b_n\beta_n = 1 - b_{n-1}\beta_{n-1} \\ b_n\alpha_n + a_{n+1}\beta_n + b_{n+1}\gamma_n = -c_{n-1}\beta_{n-1} \\ b_{n+1}\beta_n + a_{n+2}\gamma_n = 0 \end{cases}. \tag{4.8.17}$$

Here we use the condition (4.8.5), namely,

$$\det \begin{vmatrix} a_n & b_n & 0 \\ b_n & a_{n+1} & b_{n+1} \\ 0 & b_{n+1} & a_{n+2} \end{vmatrix} \neq 0,$$

that gives the uniqueness of the solution $\alpha_n, \beta_n, \gamma_n$. Substitute γ_n from the third equation in to the second in (4.8.17) and we get

$$\begin{cases} a_n\alpha_n + b_n\beta_n = 1 - b_{n-1}\beta_{n-1} \\ b_n\alpha_n + (a_{n+1} - \frac{b_{n+1}^2}{a_{n+2}})\beta_n = -c_{n-1}\beta_{n-1} \end{cases}$$

From the last system we have

$$\alpha_n = \frac{(1 - b_{n-1}\beta_{n-1})(a_{n+1} - \frac{b_{n+1}^2}{a_{n+2}}) + c_{n-1}\beta_{n-1}b_n}{a_n(a_{n+1} - \frac{b_{n+1}^2}{a_{n+2}}) - b_n^2}$$

$$\beta_n = \frac{-a_n c_{n-1}\beta_{n-1} - b_n(1 - b_{n-1}\beta_{n-1})}{a_n(a_{n+1} - \frac{b_{n+1}^2}{a_{n+2}}) - b_n^2} = \frac{(-a_n c_{n-1} + b_{n-1}b_n)\beta_{n-1} - b_n}{a_n(a_{n+1} - \frac{b_{n+1}^2}{a_{n+2}}) - b_n^2}.$$

Taking in to account the condition (4.8.4) in the form

$$a_n c_{n-1} - b_{n-1} b_{n-1} = 0,$$

we get

$$\beta_n = \frac{-b_n}{a_n(a_{n+1} - \frac{b_{n+1}^2}{a_{n+2}}) - b_n^2}. \tag{4.8.18}$$

From the position $(n+2, n)$, we get

$$\gamma_n = -\frac{b_{n+1}}{a_{n+2}} \beta_n. \tag{4.8.19}$$

By using positions, for example, $(n-1, n+1)$ and $(n-1, n+2)$ (also we can consider position $(n, n+1)$, $(n, n+2)$), we get

$$\alpha_{n+1} = -\frac{b_{n-1}}{c_{n-1}} \beta_n, \tag{4.8.20}$$

$$\beta_{n+1} = -\frac{b_{n-1}}{c_{n-1}} \gamma_n. \tag{4.8.21}$$

Let us show that β_{n+1}, obtained in (4.8.21), is the same as in the position $(n+2, n+1)$ (also we can consider $(n+3, n+1)$). Indeed, the substitution γ_n from (4.8.19) into (4.8.21) gives

$$\beta_{n+1} = \frac{b_{n-1}}{c_{n-1}} \frac{b_{n+1}}{a_{n+2}} \beta_n, \tag{4.8.22}$$

and it is the same as β_{n+1} given from the position $(n+2, n+1)$, and we substitute α_{n+1} from (4.8.20) and get

$$\beta_{n+1} = -\frac{b_{n+1}}{a_{n+2}} \alpha_{n+1} = \frac{b_{n+1}}{a_{n+2}} \frac{b_{n-1}}{c_{n-1}} \beta_n.$$

In the next local step we show that the position $(n+1, n+1)$ is fulfilled with obtained α_{n+1} and β_{n+1} in (4.8.20) and (4.8.22), respectively, i.e., we must to verify the equality

$$b_n \beta_n + a_{n+1} \alpha_{n+1} + b_{n+1} \beta_{n+1} = 1.$$

Substitutions α_{n+1} and β_{n+1} from (4.8.20) and (4.8.22) give

$$b_n\beta_n - a_{n+1}\frac{b_{n-1}}{c_{n-1}}\beta_n + b_{n+1}\frac{b_{n-1}}{c_{n-1}}\frac{b_{n+1}}{a_{n+2}}\beta_n = 1,$$

$$\beta_n(b_n - a_{n+1}\frac{b_{n-1}}{c_{n-1}} + b_{n+1}\frac{b_{n-1}}{c_{n-1}}\frac{b_{n+1}}{a_{n+2}}) = 1,$$

$$\beta_n(b_n - \frac{b_{n-1}}{c_{n-1}}(a_{n+1} - \frac{b_{n+1}^2}{a_{n+2}})) = 1, \quad \text{i.e.,} \quad \beta_n = \frac{1}{b_n - \frac{b_{n-1}}{c_{n-1}}(a_{n+1} - \frac{b_{n+1}^2}{a_{n+2}})}.$$

By the use of the condition (4.8.4) in the form

$$\frac{b_{n-1}}{c_{n-1}} = \frac{a_n}{b_n},$$

we get

$$\beta_n = \frac{1}{b_n - \frac{a_n}{b_n}(a_{n+1} - \frac{b_{n+1}^2}{a_{n+2}})} = \frac{b_n}{b_n^2 - a_n(a_{n+1} - \frac{b_{n+1}^2}{a_{n+2}})},$$

that is equal β_n obtained in (4.8.18).

Let us consider position $(n+1, n+2)$ and $(n+2, n+2)$ and show that elements on positions $(n+1, n+2)$ and $(n+2, n+2)$ of JJ^{-1} form the system with linear dependent coefficients in (4.8.2) taking in to account that γ_n and β_{n+1} obtained in (4.8.19) and (4.8.22). Indeed, for the coefficients of the system

$$\begin{cases} b_n\gamma_n + a_{n+1}\beta_{n+1} + b_{n+1}\alpha_{n+2} + c_{n+1}\beta_{n+2} = 0 \\ b_{n+1}\beta_{n+1} + a_{n+2}\alpha_{n+2} + b_{n+2}\beta_{n+2} = 1, \end{cases}$$

on the form

$$\begin{cases} b_{n+1}\alpha_{n+2} + c_{n+1}\beta_{n+2} = -b_n\gamma_n - a_{n+1}\beta_{n+1} \\ a_{n+2}\alpha_{n+2} + b_{n+2}\beta_{n+2} = 1 - b_{n+1}\beta_{n+1}, \end{cases}$$

using condition (4.8.4), (4.8.18), (4.8.19) and (4.8.22), we have

$$\frac{b_{n+1}}{a_{n+2}} = \frac{c_{n+1}}{b_{n+2}} = \frac{-b_n\gamma_n - a_{n+1}\beta_{n+1}}{1 - b_{n+1}\beta_{n+1}} = \frac{b_n\frac{b_{n+1}}{a_{n+2}}\beta_n - a_{n+1}\frac{b_{n-1}}{c_{n-1}}\frac{b_{n+1}}{a_{n+2}}\beta_n}{1 - b_{n+1}\frac{b_{n-1}}{c_{n-1}}\frac{b_{n+1}}{a_{n+2}}\beta_n}$$

$$= \frac{\frac{b_{n+1}}{a_{n+2}}\beta_n(b_n - a_{n+1}\frac{b_{n-1}}{c_{n-1}})}{1 - \frac{b_{n-1}}{c_{n-1}}\frac{b_{n+1}^2}{a_{n+2}}\beta_n} = (\frac{b_{n+1}}{a_{n+2}})\frac{(b_n - a_{n+1}\frac{b_{n-1}}{c_{n-1}})\frac{-b_n}{a_n(a_{n+1}-\frac{b_{n+1}^2}{a_{n+2}})-b_n^2}}{1 + \frac{b_{n-1}}{c_{n-1}}\frac{b_{n+1}^2}{a_{n+2}}\frac{b_n}{a_n(a_{n+1}-\frac{b_{n+1}^2}{a_{n+2}})-b_n^2}}$$

$$= (\frac{b_{n+1}}{a_{n+2}})\frac{-b_n(b_n - a_{n+1}\frac{b_{n-1}}{c_{n-1}})c_{n-1}a_{n+2}}{-c_{n-1}a_{n+2}b_n^2 + c_{n-1}a_{n+2}a_n(a_{n+1} - \frac{b_{n+1}^2}{a_{n+2}}) + b_{n-1}b_{n+1}^2 b_n}$$

$$= (\frac{b_{n+1}}{a_{n+2}})\frac{-b_n^2 c_{n-1}a_{n+2} + b_{n-1}b_n a_{n+1}a_{n+2}}{-c_{n-1}a_{n+2}b_n^2 + c_{n-1}a_n a_{n+1}a_{n+2} - c_{n-1}a_n b_{n+1}^2 + b_{n-1}b_n b_{n+1}^2}$$

$$= (\frac{b_{n+1}}{a_{n+2}})\frac{-b_n^2 c_{n-1}a_{n+2} + b_{n-1}b_n a_{n+1}a_{n+2}}{-b_n^2 c_{n-1}a_{n+2} + b_{n-1}b_n a_{n+1}a_{n+2} + b_{n+1}^2(b_{n-1}b_n - c_{n-1}a_n)} = \frac{b_{n+1}}{a_{n+2}},$$

where the condition (4.8.4) is used twice in the form $(b_{n-1}b_n - c_{n-1}a_n) = 0$.

Hence in the next local step we can consider only the position $(n+2, n+2)$ instead of the consideration of both $(n+1, n+2)$ and $(n+2, n+2)$ together (if it is need).

From the position $(n+2, n+1)$, we have

$$b_{n+1}\alpha_{n+1} + a_{n+2}\beta_{n+1} = 0,$$

i.e.,

$$\beta_{n+1} = -\frac{b_{n+1}}{a_{n+2}}\alpha_{n+1}.$$

Since $(n-1, n+1)$ yields

$$\alpha_{n+1} = -\frac{b_{n-1}}{c_{n-1}}\beta_n,$$

then we have

$$\beta_{n+1} = \frac{b_{n+1}}{a_{n+2}}\frac{b_{n-1}}{c_{n-1}}\beta_n.$$

Also, from the position $(n-1, n+2)$, we have

$$\beta_{n+1} = -\frac{b_{n-1}}{c_{n-1}}\gamma_n.$$

Since $(n+2, n)$ gives

$$\gamma_n = -\frac{b_{n+1}}{a_{n+2}}\beta_n,$$

we have the same for β_{n+1}, i.e.,

$$\beta_{n+1} = \frac{b_{n-1}}{c_{n-1}}\frac{b_{n+1}}{a_{n+2}}\beta_n.$$

This completes the n-th step of the induction and the proof of the theorem as a whole.

□

The next corollary from Theorem 4.8.1 gives a recommendations how to construct the Jacobi type matrix corresponding to the strong Hamburger moment problem.

Corollary 4.8.3 *If the sequences of real bounded numbers* $\{a_k\}$, $\{b_k\}$, *and* $\{c_k\}$, $k \in \mathbb{N}$ *satisfy conditions*

$$\det\begin{vmatrix} b_{n-1} & c_{n-1} \\ a_n & b_n \end{vmatrix} = 0,$$

$$b_1 \det\begin{vmatrix} a_1 & b_1 \\ b_1 & a_2 \end{vmatrix} < 0, \quad b_n b_{n+1} \det\begin{vmatrix} a_n & b_n & 0 \\ b_n & a_{n+1} & b_{n+1} \\ 0 & b_{n+1} & a_{n+2} \end{vmatrix} > 0, \qquad n = 2k,\ k \in \mathbb{N},$$

then we construct by (4.8.1) the Jacobi type matrix corresponds to the strong Hamburger moment problem.

Inequalities in the corollary guarantee that γ_n are positive in the matrix J^{-1}.
For Theorem 4.8.1 we have (in some sense) an inverse theorem.

Theorem 4.8.4 *The matrix* J^{-1} *on the form (4.8.3) has the algebraically inverse one* J *on the form (4.8.1) if and only if*

$$\det\begin{vmatrix} \beta_n & \gamma_n \\ \alpha_{n+1} & \beta_{n+1} \end{vmatrix} = 0, \ n = 2k, \ k \in \mathbb{N}, \tag{4.8.23}$$

$$\det\begin{vmatrix} \alpha_{n-1} & \beta_{n-1} & 0 \\ \beta_{n-1} & \alpha_n & \beta_n \\ 0 & \beta_n & \alpha_{n+1} \end{vmatrix} \neq 0, \ n = 2k, \ k \in \mathbb{N}. \tag{4.8.24}$$

Proof The proof of this theorem is similar to the proof of Theorem 4.8.1. □

Bibliographical Notes

The strong moment problem is preceded by works of Akhiezer N. Y., [2, 3], and directly Kats I. S., Nudel'man A. A. [153]. Also, the general provisions are contained in works of Schmüdgen K. [255], Szafraniec F. H. [275]. The strong, in particular Stiltjes and matrix, moment problem were considered in the works of Derevyagin M. S., Derkach V. O., Malamud M. M., Simonov K. K., Hendriksen E., Nijhuis C., Jones W. B., Njåstad O., Thron W. J., Waadeland H. [79, 80, 137, 142–144, 171, 228, 261, 267] To prove the integral representation, works of Berezansky Y. M. are used significantly. [17, 18]. The general theory of orthogonal polynomials that arise by considering the strong moment problem, is in books of Geronimus Ya. L. [117], Szegö, G. [278].

The main material of the chapter is taken from works of Berezansky Yu. M., Dudkin M. E. [40, 86]. The works of Carleman T. [64], Nussbaum A. E. [231, 232], Nelson E. [225] are used to prove the uniqueness of the representation in the case of an determinate moment problem.

In general, the strong moment problem has many generalization, see for example, Devinatz A. [81]. General information about the Weyl function can be found in the work of Gesztesy F., Simon B. [120].

Chapter 5
Block Jacobi Type Matrices in the Complex Moment Problem

This chapter considers the generalization of the connection between the classical moment problem and the spectral theory of self-adjoint Jacobi matrices. Namely, an analogue of the Jacobi matrix is proposed, which is related to the complex moment problem and also to the system of polynomials orthogonal with respect to some measure with a compact support on the complex plane. Such a matrix has a block tri-diagonal structure and is a normal operator acting in the space of a type l_2. The existence of a one-to-one correspondence between such measures on the complex plane and block tri-diagonal Jacobi-type normal matrices is proved. For the sake of simplicity, only the bounded normal operator is considered. From the point of view of the complex moment problem, this restriction means that the measure associated with orthogonal polynomials has a compact support. The chapter ends with some solution of the complex moment problem.

5.1 Construction of the Tri-Diagonal Block Jacobi Type Matrix of a Bounded Normal Operator

Let us first recall the construction of the corresponding operator J based on the classical Jacobi matrix J (see Chap. 3). This operator acts in the space l_2 of sequences $f = (f_n)_{n=0}^{\infty}$, $f_n \in \mathbb{C}$ and is constructed as an extension of the mapping $l_2 \supset l_{\mathrm{fin}} \ni f \mapsto Jf \in l_2$ in the space l_2. This operator will be Hermitian and self-adjoint for a certain behavior on "∞" of elements a_n, b_n. If a_n, b_n are uniformly bounded, then J is a bounded self-adjoint operator.

The generalized eigenfunction expansion $(P_n(\lambda))_{n=0}^{\infty}$ of the operator J $(P_n(\lambda) \in \mathbb{R}$, $\lambda \in \mathbb{R}$ has the following form. We considered the difference equation

$$(Jf)_n = a_{n-1} f_{n-1} + b_n f_n + a_n f_{n+1} = \lambda f_n, \quad n \in \mathbb{N}_0, \ f_{-1} = 0, \tag{5.1.1}$$

Y. M. Berezansky, M. E. Dudkin, *Jacobi Matrices and the Moment Problem*,
Operator Theory: Advances and Applications 294,
https://doi.org/10.1007/978-3-031-46387-7_5

where $\lambda \in \mathbb{R}$ is a spectral parameter. Due to conditions $a_n > 0$, $n \in \mathbb{N}_0$, the solution $f_n = P_n(\lambda)$ of the Eq. (5.1.1) exists, where $P_n(\lambda)$ is a polynomial of the n-th degree of the variable λ. A sequence $P(\lambda) = (P_n(\lambda))_{n=0}^{\infty}$ is in a certain sense (see Chap. 3) a generalized eigenvector of the operator J with a corresponding eigenvalue λ : $JP(\lambda) = \lambda P(\lambda)$.

The corresponding Fourier transform has the form

$$l_2 \supset l_{\text{fin}} \ni f = (f_n)_{n=0}^{\infty} \mapsto \hat{f}(\lambda) = \sum_{n=0}^{\infty} f_n P_n(\lambda) \in L_2, \tag{5.1.2}$$

where $L_2 = L_2(\mathbb{R}, d\rho(\lambda))$ is the space of summable with the square functions of the spectral measure $d\rho(\lambda)$ of the operator J. After the extension by continuity, the mapping (5.1.2) will be an isometry operator transferring the space l_2 into the space L_2. Due to the Fourier transform (5.1.2), the operator J transfers into the multiplication by λ operator in L_2.

This construction is closely related to the theory of orthogonal polynomials of a variable λ on the real axis related to the spectral measure. Namely, it follows from (5.1.2) that $(P_n(\lambda))_{n=0}^{\infty}$ is an orthogonal polynomial basis in L_2. Considering the spectral measure as given, this basis can be constructed directly using the Gram-Schmidt orthogonalization procedure in L_2 space, applying it to a sequence of linearly independent functions

$$1, \lambda, \lambda^2, \dots, \lambda^n, \dots, \quad \lambda \in \mathbb{R} \tag{5.1.3}$$

assuming that the system (5.1.3) is linearly independent and complete in L_2. As a result of orthogonalization, we obtain the sequence $P(\lambda) = (P_n(\lambda))_{n=0}^{\infty}$ that is the generalized eigenfunction of the operator J, whose spectral measure will be given by the measure $d\rho(\lambda)$?

For elements a_n, b_n of matrix J, we now have simple formulas (see Chap. 3). This is the simplest inverse problem of spectral analysis: how to recover the operator J for a given spectral measure $d\rho(\lambda)$.

After this reminder, let us move on to the main goal of Chap. 5: to find out the form of the block Jacobi type matrix J, which corresponds to the complex moment problem when the corresponding operator J will not be self-adjoint, but a bounded normal operator acting not in space l_2, but in the Hilbert space adapted to the complex moment problem. First of all, we note that the spectrum of such an operator is a bounded set in the complex plane $\mathbb{C}$, symmetric with respect to the real axis, so the system (5.1.3) will need to be replaced with a more complicated one.

Let $d\rho(z)$ be a Borel (probability) measure on $\mathbb{C}$ with a compact support and $L_2 = L_2(\mathbb{C}, d\rho(z))$ the space of square integrable complex valued functions defined on $\mathbb{C}$. Assume that the support of the measure is not finite, but a compact set containing an open subset and, hence, functions $\mathbb{C} \ni z \longmapsto z^m \bar{z}^n$, $m, n \in \mathbb{N}_0$ are linearly independent in L_2.

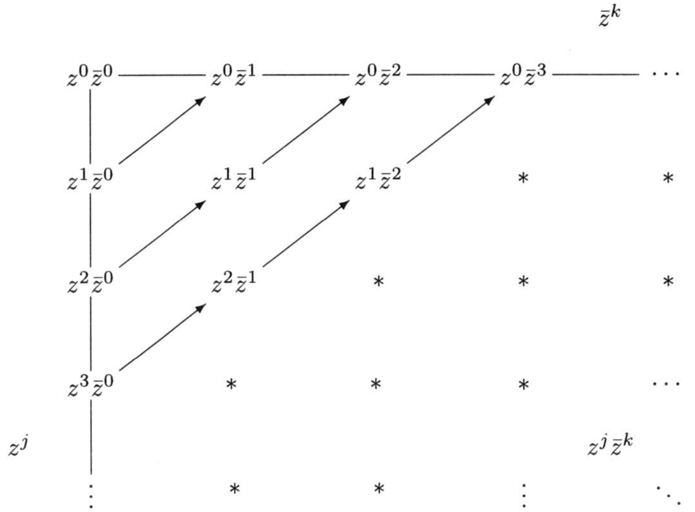

Fig. 5.1 The orthogonalization order

In order that to find an analog of Jacobi matrix J, there is need to choose an order of the orthogonalization in L_2 for the family of functions

$$\{z^j\bar{z}^k\}, \quad j, k \in \mathbb{N}_0. \tag{5.1.4}$$

Our assumption about the compactness of $\mathrm{supp}(d\rho(z))$ implies that this family is total in L_2. The following order of orthogonalization according to the Gram-Schmidt procedure is proposed (see Fig. 5.1).

According to Fig. 5.1 we have:

$$z^0\bar{z}^0; \quad z^1\bar{z}^0, z^0\bar{z}^1; \quad z^2\bar{z}^0, z^1\bar{z}^1, z^0\bar{z}^2; \quad \ldots; \quad z^n\bar{z}^0, z^{n-1}\bar{z}^1, \ldots, z^0\bar{z}^n; \quad \ldots \tag{5.1.5}$$

Considering the sequence of functions (5.1.5) we start the orthogonalization according to the Gram-Schmidt procedure. As a result we obtain an orthonormal system of polynomials of variables $z^m\bar{z}^n$, $m, n \in \mathbb{N}_0$:

$$\begin{array}{llllll} P_{0;0}(z); & P_{1;0}(z), & P_{2;0}(z), & \ldots; & P_{n;0}(z), & \ldots \\ & P_{1;1}(z); & P_{2;1}(z), & & P_{n;1}(z), & \\ & & P_{2;2}(z); & & P_{n;2}(z), & \\ & & & & \ldots & \\ & & & & P_{n;n}(z); & \end{array} \tag{5.1.6}$$

where each polynomial has a form

$$P_{n;\alpha}(z) = k_{n;\alpha} z^{n-\alpha} \bar{z}^{\alpha} + \cdots, \quad n \in \mathbb{N}_0, \quad \alpha = 0, 1, \dots, n, k_{n;\alpha} > 0;$$

here "$+\cdots$" denotes the next part of the corresponding polynomial; $P_{0;0}(z) = 1$. In such a way $P_{n;\alpha}$ is some linear combination of

$$\{1;\ z^1\bar{z}^0, z^0\bar{z}^1;\ \dots;\ z^n\bar{z}^0, z^{n-1}\bar{z}^1, \dots, z^{n-\alpha}\bar{z}^{\alpha}\}. \tag{5.1.7}$$

According to the totality of the family (5.1.4) in the space L_2, the sequence (5.1.6) is an orthonormal basis in this space.

Let $\mathcal{P}_{n;\alpha}$, $\forall n \in \mathbb{N}$ be denote the subspace spanned by (5.1.7). It is clear that $\forall n \in \mathbb{N}$, we get

$$\begin{aligned} &\mathcal{P}_{0;0} \subset \mathcal{P}_{1;0} \subset \mathcal{P}_{1;1} \subset \mathcal{P}_{2;0} \subset \mathcal{P}_{2;1} \subset \mathcal{P}_{2;2} \subset \cdots \\ &\quad \subset \mathcal{P}_{n;0} \subset \mathcal{P}_{n;1} \subset \cdots \subset \mathcal{P}_{n;n} \subset \cdots, \\ &\mathcal{P}_{n;\alpha} = \{P_{0;0}(z)\} \oplus \{P_{1;0}(z)\} \oplus \{P_{1;1}(z)\} \oplus \{P_{2;0}(z)\} \oplus \{P_{2;1}(z)\} \oplus \\ &\quad \{P_{2;2}(z)\} \oplus \cdots \oplus \{P_{n;0}(z)\} \oplus \{P_{n;1}(z)\} \oplus \cdots \oplus \{P_{n;\alpha}(z)\}, \end{aligned} \tag{5.1.8}$$

where $\{P_{n;\alpha}(z)\}$, $n \in \mathbb{N}$, $\alpha = 0, 1, \dots n$ denotes one dimensional space spanned by $P_{n;\alpha}(z)$; $\mathcal{P}_{0;0} = \mathbb{C}$.

For further constructions, we propose to consider a new space

$$\mathbf{l}_2 = \mathcal{H}_0 \oplus \mathcal{H}_1 \oplus \mathcal{H}_2 \oplus \cdots, \quad \mathcal{H}_n = \mathbb{C}^{n+1}, \quad n \in \mathbb{N}_0, \tag{5.1.9}$$

instead of usual traditional space l_2 used in the classical Hamburger (one-dimensional) moment problem.

Each vector $f \in \mathbf{l}_2$ has a form $f = (f_n)_{n=0}^{\infty}$, $f_n \in \mathcal{H}_n$, hence,

$$\|f\|_{\mathbf{l}_2}^2 = \sum_{n=0}^{\infty} \|f_n\|_{\mathcal{H}_n}^2 < \infty, \quad (f, g)_{\mathbf{l}_2} = \sum_{n=0}^{\infty} (f_n, g_n)_{\mathcal{H}_n}, \quad \forall f, g \in \mathbf{l}_2.$$

For $n \in \mathbb{N}_0$, coordinates of vector $f_n \in \mathcal{H}_n$ in the some orthonormal basis

$$\{e_{n;0}, e_{n;1}, e_{n;2}, \dots, e_{n;n}\}$$

in the space $\mathbb{C}^{n+1}$, we denote by $(f_{n;0}, f_{n;1}, f_{n;2}, \dots, f_{n;n})$ and, hence, we have $f_n = (f_{n;0}, f_{n;1}, f_{n;2}, \dots, f_{n;n})$. It is clear that the space $\mathbf{l}_2$ is almost isometric to $l_2 \times l_2$ (the notation "almost isometric" denotes that the space $l_2 \times l_2$ contains a vector $e_0 \times e_0$, and $\mathbf{l}_2$ has only $e_{0,0}$).

By using the orthonormal system (5.1.6), one can define a mapping of $\mathbf{l}_2$ into L_2. Denote $\forall n \in \mathbb{N}_0$ and $\forall z \in \mathbb{C}$,

$$P_n(z) = (P_{n;0}, P_{n;1}(z), P_{n;2}(z), \ldots, P_{n;n}(z)) \in \mathcal{H}_n,$$

then instead of (5.1.2) we have:

$$\mathbf{l}_2 \ni f = (f_n)_{n=0}^{\infty} \longmapsto \hat{f}(z) = \sum_{n=0}^{\infty} (f_n, P_n(z))_{\mathcal{H}_n} \in L_2. \tag{5.1.10}$$

Since for $n \in \mathbb{N}_0$

$$(f_n, P_n(z))_{\mathcal{H}_n} = f_{n;0}\overline{P_{n;0}(z)} + f_{n;1}\overline{P_{n;1}(z)} + f_{n;2}\overline{P_{n;2}(z)} + \cdots + f_{n;n}\overline{P_{n;n}(z)}$$

and

$$\|f\|_{\mathbf{l}_2}^2 = \|(f_{0;0}, f_{1;0}, f_{1;1}, f_{2;0}, f_{2;1}, f_{2;2}, \ldots, f_{n;0}, f_{n;1}, \ldots, f_{n;n}, \ldots)\|_{l_2}^2,$$

then we can see that (5.1.10) can be regarded as a mapping of $\mathbf{l}_2$ into L_2, and using the orthonormal system (5.1.6) shows that this imbedding is an almost isometry. The image of $\mathbf{l}_2$ by the mapping (5.1.10) is L_2, because under our assumption the system (5.1.6) is an orthonormal basis in L_2. Therefore, the mapping (5.1.10) is an isometric transform (denoted by I) acting from $\mathbf{l}_2$ into L_2.

Let A be a bounded linear operator defined on the space $\mathbf{l}_2$. Let us construct the operator matrix $(a_{j,k})_{j,k=0}^{\infty}$, where for each $j, k \in \mathbb{N}_0$, the element $a_{j,k}$ is an operator acting from $\mathcal{H}_k$ to $\mathcal{H}_j$, so that we have

$$(Af)_j = \sum_{k=0}^{\infty} a_{j,k} f_k, \ j \in \mathbb{N}_0, \quad (Af, g)_{\mathbf{l}_2} = \sum_{j,k=0}^{\infty} (a_{j,k} f_k, g_j)_{\mathcal{H}_j}, \quad \forall f, g \in \mathbf{l}_2. \tag{5.1.11}$$

For the proof of (5.1.11), we need only to write the usual matrix of the operator A in the space $l_2 \times l_2$ using the basis

$$(e_{0;0};\ e_{1;0}, e_{1;1};\ e_{2;0}, e_{2;1}, e_{2;2};\ \ldots;\ e_{n;0}, e_{n;1}, \ldots, e_{n;n};\ \ldots), \quad e_{0;0} = 1. \tag{5.1.12}$$

Then $a_{j,k}$ for each $j, k \in \mathbb{N}_0$ is the operator $\mathcal{H}_k \longrightarrow \mathcal{H}_j$ that has a matrix representation

$$a_{j,k;\alpha,\beta} = (Ae_{k;\beta}, e_{j;\alpha})_{\mathbf{l}_2}, \tag{5.1.13}$$

where $\alpha = 0, 1, \ldots, j$ and $\beta = 0, 1, \ldots, k$. We will write $a_{j,k} = (a_{j,k;\alpha,\beta})_{\alpha,\beta=0}^{j,k}$ (including cases:

$$a_{0,k} = (a_{0,k;\alpha,\beta})_{\alpha,\beta=0}^{0,k},\ a_{j,0} = (a_{j,0;\alpha,\beta})_{\alpha,\beta=0}^{j,0},\ a_{0,0} = (a_{0,0;\alpha,\beta})_{\alpha,\beta=0}^{0,0} = a_{0,0;0,0}).$$

Note, that the same representation (5.1.11) is also valid for the general operator A on the space $\mathbf{l}_2$ with the domain $\mathfrak{D}(A) = \mathbf{l}_{\text{fin}} \subset \mathbf{l}_2$, where $\mathbf{l}_{\text{fin}}$ denotes the set of finite vectors from $\mathbf{l}_2$. In this case the first formula in (5.1.11) is fulfilled for $f \in \mathbf{l}_{\text{fin}}$ and the second formula is valid for $f \in \mathbf{l}_{\text{fin}}$, $g \in \mathbf{l}_2$.

Let us consider the image $\hat{A} = IAI^{-1} : L_2 \longrightarrow L_2$ the mentioned above operator $A : \mathbf{l}_2 \longrightarrow \mathbf{l}_2$ that is defined by the imbedding (5.1.10). Its matrix in the basis (5.1.6)

$$(P_{0;0}(z);\ P_{1;0}(z), P_{1;1}(z);\ P_{2;0}(z), P_{2,1}(z), P_{2,2}(z);\ \ldots;$$
$$P_{n;0}(z), P_{n;1}(z), \ldots, P_{n;n}(z); \ldots)$$

is equal to the usual matrix of the operator A as the operator: $\mathbf{l}_2 \longrightarrow \mathbf{l}_2$ in the corresponding basis (5.1.12). By using (5.1.13) and the previous construction, we obtain the operator matrix $(a_{j,k})_{j,k=0}^{\infty}$ of the operator $A : \mathbf{l}_2 \longrightarrow \mathbf{l}_2$. *By the definition this matrix is also the operator matrix of* $\hat{A} : L_2 \longrightarrow L_2$.

It is clear that $\hat{A}$ can be an arbitrary linear bounded operator in L_2.

Lemma 5.1.1 *For polynomials $P_{n;\alpha}(z)$, and subspaces $\mathcal{P}_{m,\beta}$, $n, m \in \mathbb{N}_0$, $\alpha = 0, 1, \ldots, n$, $\beta = 0, 1, \ldots, m$, relations*

$$zP_{n;\alpha}(z) \in \mathcal{P}_{n+1;\alpha}, \quad \bar{z}P_{n;\alpha}(z) \in \mathcal{P}_{n+1;\alpha+1}, \tag{5.1.14}$$

holds true.

Proof According to (5.1.6) the polynomial $P_{n;\alpha}(z)$, $n \in \mathbb{N}_0$, is equal to some linear combination of

$$\{1; z^1\bar{z}^0, z^0\bar{z}^1; \ldots; z^n\bar{z}^0, z^{n-1}\bar{z}^1, \ldots, z^{n-\alpha}\bar{z}^{\alpha}\}.$$

Hence, using the multiplication by z we get a linear combination of

$$\{z; z^2\bar{z}^0, z^1\bar{z}^1; \ldots; z^{n+1}\bar{z}^0, z^n\bar{z}^1, \ldots, z^{n+1-\alpha}\bar{z}^{\alpha}\},$$

and such a linear combination belongs to $\mathcal{P}_{n+1;\alpha}$. Similarly, the multiplication by $\bar{z}$ gives a linear combination of

$$\{\bar{z}^1; z^1\bar{z}^1, z^0\bar{z}^2; \ldots; z^n\bar{z}^1, z^{n-1}\bar{z}^2, \ldots, z^{n-\alpha}\bar{z}^{\alpha+1}\},$$

and such a linear combination belongs to $\mathcal{P}_{n+1;\alpha+1}$, since $z^{n-\alpha}\bar{z}^{\alpha+1} \in \mathcal{P}_{n+1,\alpha+1}$.

□

Lemma 5.1.2 *Let $\hat{A}$ be the normal bounded operator of multiplication by z in the space L_2, namely,*

$$L_2 \ni \varphi(z) \longmapsto (\hat{A}\varphi)(z) = z\varphi(z) \in L_2.$$

The operator matrix $(a_{j,k})_{j,k=0}^{\infty}$ of the operator $\hat{A}$ (i.e., $A = I^{-1}\hat{A}I$) has a tri-diagonal structure: $a_{j,k} = 0$ for $|j-k| > 1$.

Proof By using (5.1.13) for $e_{n;\gamma} = I^{-1}P_{n;\gamma}(z)$, $n \in \mathbb{N}_0$, $\gamma = 0, 1, \ldots, n$, we have

$$a_{j,k;\alpha,\beta} = (Ae_{k;\beta}, e_{j;\alpha})_{l_2} = \int\limits_{\mathbb{C}} zP_{k;\beta}(z)\overline{P_{j;\alpha}(z)}\, d\rho(z), \quad \forall j, k \in \mathbb{N}_0, \tag{5.1.15}$$

where $\alpha = 0, 1, \ldots, j$, $\beta = 0, 1, \ldots, k$. From (5.1.14), we have $zP_{k;\alpha}(z) \in \mathcal{P}_{k+1;\alpha}$. According to (5.1.8) the integral in (5.1.15) is equal to zero for $j > k+1$ and for each $\alpha = 0, 1, \ldots, j$.

On the other hand, the integral in (5.1.15) has a form

$$a_{j,k;\alpha,\beta} = \overline{\int\limits_{\mathbb{C}} \bar{z}P_{j;\alpha}(z)\overline{P_{k;\beta}(z)}\, d\rho(z)}. \tag{5.1.16}$$

From (5.1.14), we have now that $\bar{z}P_{j;\alpha}(z) \in \mathcal{P}_{j+1;\alpha+1}$. According to (5.1.8), the last integral is equal to zero for $k > j+1$ and for each $\beta = 0, 1, \ldots, k$.

As result, the integral in (5.1.16), i.e., coefficients $a_{j,k;\alpha,\beta}$, $j, k \in \mathbb{N}_0$, are equal to zero for $|j-k| > 1$; $\alpha = 0, 1, \ldots, j$, $\beta = 0, 1, \ldots, k$. (Here, it is necessary to take into account, that $e_{0;0} = I^{-1}P_{0;0}(z) = 1$). □

In such a way we found out that the matrix $(a_{j,k})_{j,k=0}^{\infty}$ of the operator $\hat{A}$ has a tri-diagonal block structure

$$\begin{pmatrix} a_{0,0} & a_{0,1} & 0 & 0 & 0 & \ldots \\ a_{1,0} & a_{1,1} & a_{1,2} & 0 & 0 & \ldots \\ 0 & a_{2,1} & a_{2,2} & a_{2,3} & 0 & \ldots \\ 0 & 0 & a_{3,2} & a_{3,3} & a_{3,4} & \ldots \\ \vdots & \vdots & \vdots & \vdots & \vdots & \ddots \end{pmatrix}. \tag{5.1.17}$$

A detailed analysis of the expression (5.1.15) allows us to find out which elements of the matrix $(a_{j,k;\alpha,\beta})_{\alpha,\beta=0}^{j,k}$ are zero and which are not, in each case for $|j-k| \le 1$. By the describing properties of the matrix, the permutation property of indices j, k and α, β is used.

Let us denote by $((a^*)_{j,k})_{j,k=0}^{\infty}$ the operator matrix of the operator $(\hat{A})^*$, which is adjoint to $\hat{A}$. We remark that $(\hat{A})^*$ is the multiplication by $\bar{z}$ operator. Taking into account the expression (5.1.15) for $j, k \in \mathbb{N}_0$, we have

$$(a^*)_{j,k;\alpha,\beta} = \int_{\mathbb{C}} \bar{z} P_{k;\beta}(z)\overline{P_{j;\alpha}(z)}\, d\rho(z) = \overline{\int_{\mathbb{C}} z P_{j;\alpha}(z)\overline{P_{k;\beta}(z)}\, d\rho(z)} = \overline{a_{k,j;\beta,\alpha}}, \tag{5.1.18}$$

where $\alpha = 0, 1, \dots, j$ and $\beta = 0, 1, \dots, k$.

Lemma 5.1.3 *Let $(a_{j,k})_{j,k=0}^{\infty}$ be an multiplication by z operator matrix in L_2, where $a_{j,k} : \mathcal{H}_k \longrightarrow \mathcal{H}_j$; $a_{j,k} = (a_{j,k;\alpha,\beta})_{\alpha,\beta=0}^{j,k}$ are matrices of operators $a_{j,k}$ in the corresponding orthonormal basis. Then, $\forall j \in \mathbb{N}_0$, we have*

$$\begin{aligned} &\forall\alpha = 0, 1, \dots j-1, \; a_{j,j+1;\alpha,\alpha+2} = a_{j,j+1;\alpha,\alpha+3} = \cdots = a_{j,j+1;\alpha,j+1} = 0;\\ &\forall\beta = 0, 1, \dots j, \quad\; a_{j+1,j;\beta+1,\beta} = a_{j+1,j;\beta+2,\beta} = \cdots = a_{j+1,j;j+1,\beta} = 0. \end{aligned} \tag{5.1.19}$$

If we choose inside of each diagonal $\{z^n\bar{z}^0, z^{n-1}\bar{z}^1, z^{n-2}\bar{z}^2, \dots, z^0\bar{z}^n\}$ (in Fig. 5.1) another order (preserving the order of diagonals), then Lemma 5.1.3 does not hold true, but it is also possible to describe zero elements of matrices $(a_{j,k;\alpha,\beta})_{\alpha,\beta=0}^{j,k}$. Such matrices $(a_{j,k})_{j,k=0}^{\infty}$ have also tri-diagonal block structure and have zeros but in other places.

Proof According to (5.1.15) and (5.1.14), for $j \in \mathbb{N}_0$, we have $\forall\alpha = 0, 1, \dots, j$ and $\forall\beta = 0, 1, \dots, j+1$,

$$a_{j,j+1;\alpha,\beta} = \int_{\mathbb{C}} z P_{j+1,\beta}(z)\overline{P_{j;\alpha}(z)}\, d\rho(z) = \overline{\int_{\mathbb{C}} \bar{z} P_{j,\alpha}(z)\overline{P_{j+1;\beta}(z)}\, d\rho(z)},$$

where $\bar{z}P_{j;\alpha}(z) \in \mathcal{P}_{j+1;\alpha+1}$. But according to (5.1.8), $P_{j+1;\beta}(z)$ is orthogonal to $\mathcal{P}_{j+1;\alpha+1}$ for $\beta > \alpha+1$ and, hence, the last integral is equal to zero. This gives the first equalities in (5.1.19).

Similarly, from (5.1.15) and (5.1.14), for $j \in \mathbb{N}_0$, we get $\forall\alpha = 0, 1, \dots, j+1$ and $\forall\beta = 0, 1, \dots, j$,

$$a_{j+1,j;\alpha,\beta} = \int_{\mathbb{C}} z P_{j,\beta}(z)\overline{P_{j+1;\alpha}(z)}\, d\rho(z),$$

where $zP_{j;\beta}(z) \in \mathcal{P}_{j+1;\beta}$. But according to (5.1.8), $P_{j+1;\alpha}(z)$ is orthogonal to $\mathcal{P}_{j+1;\beta}$, if $\alpha > \beta$ and, hence, the last integral is equal to zero. This gives the second equality in (5.1.19). □

Thus, after these investigations, we conclude that in (5.1.17) for $\forall j \in \mathbb{N}$, the over right corner of every $((j+1)\times(j+2))$-matrix $a_{j,j+1}$ (starting from the third diagonal) and the lower left corner of each $((j+2)\times(j+1))$-matrix $a_{j+1,j}$ (starting from the second diagonal) contains of zero elements. Taking into account (5.1.17), we can conclude that the normal matrix of the multiplication by z operator is a multi-diagonal usual scalar matrix, i.e., in the usual basis of the space $\mathbf{l}_2$.

Lemma 5.1.4 *Elements*

$$a_{0,1;0,1},\, a_{1,0;0,0};\quad a_{j,j+1;\alpha,\alpha+1},\, a_{j+1,j;\alpha,\alpha},\quad j\in\mathbb{N},\quad \alpha=0,1,\ldots,j, \tag{5.1.20}$$

of matrices $(a_{j,k})_{j,k=0}^{\infty}$ *in Lemma* 5.1.3 *are always positive.*

Proof We start from the investigation of $a_{1,0;0,0}$. By using (5.1.15) and a non-normalized vector $P'_{1;0}(z) = z-(z,1)_{L_2}$ obtained from the Gram-Schmidt orthogonalization procedure $P_{1;0}(z)$, we get

$$\begin{aligned} a_{1,0;0,0} &= \int_{\mathbb{C}} z\overline{P_{1;0}(z)}\,d\rho(z) = \|P'_{1;0}(z)\|_{L_2}^{-1}\int_{\mathbb{C}} z\overline{(z-(z,1)_{L_2})}\,d\rho(z) \\ &= \|P'_{1;0}(z)\|_{L_2}^{-1}(\|\bar z\|_{L_2}^2 - |(\bar z,1)_{L_2}|^2). \end{aligned} \tag{5.1.21}$$

The last difference is positive (see below, (5.1.23)), therefore, $a_{1,0;0,0}>0$.

Consider $a_{0,1;0,1}$. Denote $P'_{1;1}(z)$ a non-normalized vector $P_{1;1}(z)$ obtained from the Gram-Schmidt orthogonalization procedure. According to (5.1.5) and (5.1.6), we get

$$P'_{1;1}(z) = \bar z - (\bar z, P_{1;0}(z))_{L_2}P_{1;0}(z) - (\bar z,1)_{L_2}.$$

Therefore, using (5.1.15), we get

$$\begin{aligned} a_{0,1;0,1} &= \int_{\mathbb{C}} zP_{1;1}(z)\,d\rho(z) = \|P'_{1;1}(z)\|_{L_2}^{-1}\int_{\mathbb{C}} zP'_{1;1}(z)\,d\rho(z) \\ &= \|P'_{1;1}(z)\|_{L_2}^{-1}\int_{\mathbb{C}} z(\bar z-(\bar z,P_{1;0}(z))_{L_2}P_{1;0}(z)-(\bar z,1)_{L_2})\,d\rho(z) \\ &= \|P'_{1;1}(z)\|_{L_2}^{-1}(\|\bar z\|_{L_2}^2 - |(\bar z,P_{1;0}(z))_{L_2}|^2 - |(\bar z,1)_{L_2}|^2). \end{aligned} \tag{5.1.22}$$

Also using (5.1.23), we conclude that the last expression is positive and, therefore, $a_{0,1;0,1}>0$.

The positiveness in (5.1.21) and (5.1.22) follows from the Parseval equality for the decomposition of the function $\bar z\in L_2$ respect to the orthonormal basis (5.1.6) in the space L_2, namely,

$$|(\bar{z}, 1)_{L_2}|^2 + |(\bar{z}, P_{1;0}(z))_{L_2}|^2 + |(\bar{z}, P_{1;1}(z))_{L_2}|^2 + \cdots = \|\bar{z}\|^2_{L_2}, \tag{5.1.23}$$

where $P_{0;0}(z) = 1$.

Let us now pass to show the positivity of $a_{j+1,j;\alpha,\alpha}$, where $j \in \mathbb{N}$, $\alpha = 0, 1, \ldots, j$. From (5.1.15) we have

$$a_{j+1,j;\alpha,\alpha} = \int_{\mathbb{C}} z P_{j;\alpha}(z) \overline{P_{j+1;\alpha}(z)}\, d\rho(z). \tag{5.1.24}$$

According to (5.1.6) and (5.1.8),

$$P_{j;\alpha}(z) = k_{j;\alpha} z^{j-\alpha} \bar{z}^{\alpha} + R_{j;\alpha}(z), \tag{5.1.25}$$

where $R_{j;\alpha}(z)$ is some polynomial from $\mathcal{P}_{j;\alpha-1}$ if $\alpha > 0$, or from $\mathcal{P}_{j-1;j-1}$ if $\alpha = 0$. Therefore, $zR_{j;\alpha}(z)$ is some polynomial from $\mathcal{P}_{j+1;\alpha-1}$, or from $\mathcal{P}_{j;j-1}$ (see (5.1.14) and (5.1.8)). Multiplying (5.1.25) by z, we conclude

$$\begin{aligned} zP_{j;\alpha}(z) &= k_{j;\alpha} z^{j+1-\alpha} \bar{z}^{\alpha} + zR_{j;\alpha}(z); \\ &\qquad zR_{j;\alpha}(z) \in \mathcal{P}_{j+1;\alpha-1} \quad \text{or} \quad \mathcal{P}_{j;j-1} \subset \mathcal{P}_{j;j}. \end{aligned} \tag{5.1.26}$$

On the other hand, the equality (5.1.25) for $P_{j+1;\alpha}(z)$ gives

$$\begin{aligned} P_{j+1;\alpha}(z) &= k_{j+1;\alpha} z^{j+1-\alpha} \bar{z}^{\alpha} + R_{j+1;\alpha}(z); \\ &\qquad R_{j+1;\alpha}(z) \in \mathcal{P}_{j+1;\alpha-1} \quad \text{or} \quad \mathcal{P}_{j;j}. \end{aligned} \tag{5.1.27}$$

Take $z^{j+1-\alpha}\bar{z}^{\alpha}$ from (5.1.27) and substitute it into (5.1.26). Hence, we get

$$\begin{aligned} zP_{j;\alpha}(z) &= \frac{k_{j;\alpha}}{k_{j+1;\alpha}} (P_{j+1;\alpha}(z) - R_{j+1;\alpha}(z)) + zR_{j;\alpha}(z) \\ &= \frac{k_{j;\alpha}}{k_{j+1;\alpha}} P_{j+1;\alpha}(z) - \frac{k_{j;\alpha}}{k_{j+1;\alpha}} R_{j+1;\alpha}(z) + zR_{j;\alpha(z)}(z), \end{aligned} \tag{5.1.28}$$

where the second term belongs to $\mathcal{P}_{j+1;\alpha-1}$ or to $\mathcal{P}_{j;j}$ and in any case are orthogonal to $P_{j+1;\alpha}(z)$.

Therefore, after the substitution of the expression (5.1.28) into (5.1.24), we get that

$$a_{j+1,j;\alpha,\alpha} = \frac{k_{j;\alpha}}{k_{j+1;\alpha}} > 0.$$

Consider, at last, the elements $a_{j,j+1;\alpha,\alpha+1}$, where $j \in \mathbb{N}$, $\alpha = 0, 1, \ldots, j$. From (5.1.15), we get

$$a_{j,j+1;\alpha,\alpha+1} = \int\limits_{\mathbb{C}} z P_{j+1,\alpha+1}(z)\overline{P_{j;\alpha}(z)}\, d\rho(z) = \overline{\int\limits_{\mathbb{C}} \bar{z} P_{j,\alpha}(z)\overline{P_{j+1;\alpha+1}(z)}\, d\rho(z)}. \tag{5.1.29}$$

For $P_{j;\alpha}(z)$, we have the expression (5.1.25). Multiplying it by $\bar{z}$, we get an expression similar to (5.1.26).

$$\bar{z} P_{j;\alpha}(z) = k_{j;\alpha} z^{j-\alpha}\bar{z}^{\alpha+1} + \bar{z} R_{j;\alpha}(z), \quad \bar{z} R_{j;\alpha}(z) \in \mathcal{P}_{j+1;\alpha} \quad \text{or} \quad \mathcal{P}_{j;j}, \tag{5.1.30}$$

(but this time it is necessary to use the second inclusion (5.1.14) and (5.1.8)).

Now, the equality (5.1.25) gives

$$P_{j+1;\alpha+1}(z) = k_{j+1;\alpha+1} z^{j-\alpha}\bar{z}^{\alpha+1} + R_{j+1;\alpha+1}(z), \quad R_{j+1;\alpha+1}(z) \in \mathcal{P}_{j+1;\alpha}. \tag{5.1.31}$$

By taking $z^{j-\alpha}\bar{z}^{\alpha+1}$ from (5.1.31) and substituting it into (5.1.30), we get

$$\begin{aligned} \bar{z} P_{j;\alpha}(z) &= \frac{k_{j;\alpha}}{k_{j+1;\alpha+1}} (P_{j+1;\alpha+1}(z) - R_{j+1;\alpha+1}(z)) + \bar{z} R_{j;\alpha}(z) \\ &= \frac{k_{j;\alpha}}{k_{j+1;\alpha+1}} P_{j+1;\alpha+1}(z) - \frac{k_{j;\alpha}}{k_{j+1;\alpha+1}} R_{j+1;\alpha+1}(z) + \bar{z} R_{j;\alpha}(z). \end{aligned} \tag{5.1.32}$$

As above, the second term in (5.1.32) belongs to $\mathcal{P}_{j+1;\alpha}$, or to $\mathcal{P}_{j;j}$ and in any case is orthogonal to $P_{j+1;\alpha+1}(z)$. Therefore, the substitution of the expression (5.1.32) into (5.1.29) gives

$$a_{j,j+1;\alpha,\alpha+1} = \frac{k_{j;\alpha}}{k_{j+1;\alpha+1}} > 0.$$

□

To denote elements $a_{j,k}$ of the Jacobi matrix, we use the standard notation:

$$\begin{aligned} a_n &= a_{n+1,n} &&: \ \mathcal{H}_n \longrightarrow \mathcal{H}_{n+1}, \\ b_n &= a_{n,n} &&: \ \mathcal{H}_n \longrightarrow \mathcal{H}_n, \\ c_n &= a_{n,n+1} &&: \ \mathcal{H}_{n+1} \longrightarrow \mathcal{H}_n, \quad n \in \mathbb{N}_0. \end{aligned} \tag{5.1.33}$$

All previous investigations we summarize in a theorem.

Theorem 5.1.5 *The bounded normal multiplication by the independent variable z operator $\hat{A}$, (with a strong cyclic vector) in the space L_2, in the orthonormal basis* (5.1.6) *of polynomials, has the form of a block trt-diagonal Jacobi type normal matrix $J = (a_{j,k})_{j,k=0}^{\infty}$ acting in the space* (5.1.9)*, namely,*

$$\mathbf{l}_2 = \mathcal{H}_0 \oplus \mathcal{H}_1 \oplus \mathcal{H}_2 \oplus \cdots, \quad \mathcal{H}_n = \mathbb{C}^{n+1}, \quad n \in \mathbb{N}_0. \tag{5.1.34}$$

Norms of all operators $a_{j,k} : \mathcal{H}_k \longrightarrow \mathcal{H}_j$ are uniformly bounded with respect to $j, k \in \mathbb{N}_0$. In desihnations of (5.1.33)*, this matrix has the form*

$$J = \left(\begin{array}{c|c|c|c|c|c} b_0 & c_0 & 0 & 0 & 0 & \dots \\ \hline a_0 & b_1 & c_1 & 0 & 0 & \dots \\ \hline 0 & a_1 & b_2 & c_2 & 0 & \dots \\ \hline 0 & 0 & a_2 & b_3 & c_3 & \dots \\ \vdots & \vdots & \vdots & \vdots & \vdots & \ddots \end{array}\right) \tag{5.1.35}$$

$$J = \left(\begin{array}{c|cc|ccc|cccc|ccccc} * & * & + & & & & & & & & & & & & \\ \hline + & * & * & * & + & 0 & & & & & & & & & \\ 0 & * & * & * & * & + & & & & & & 0 & & & \\ \hline & + & * & * & * & * & * & + & 0 & 0 & & & & & \\ & 0 & + & * & * & * & * & * & + & 0 & & & & & \\ & 0 & 0 & * & * & * & * & * & * & + & & & & & \\ \hline & & & + & * & * & * & * & * & * & * & + & 0 & 0 & 0 \\ & & & 0 & + & * & * & * & * & * & * & * & + & 0 & 0 \\ & 0 & & 0 & 0 & + & * & * & * & * & * & * & * & + & 0 \\ & & & 0 & 0 & 0 & * & * & * & * & * & * & * & * & + \\ \hline & & & & \ddots & & & & \ddots & & & & \ddots & & \end{array}\right). \tag{5.1.36}$$

Namely, in (5.1.36), elements are as follows: $\forall n \in \mathbb{N}_0$, b_n is a $((n+1) \times (n+1))$-matrix, $b_n = (b_{n;\alpha,\beta})_{\alpha,\beta=0}^{n,n}$ ($b_0 = b_{0;0,0}$ is a scalar); a_n is a $((n+2) \times (n+1))$-matrix, $a_n = (a_{n;\alpha,\beta})_{\alpha,\beta=0}^{n+1,n}$; c_n is a $((n+1) \times (n+2))$-matrix, $c_n = (c_{n;\alpha,\beta})_{\alpha,\beta=0}^{n,n+1}$. Some elements in matrices a_n and c_n are always equal to zero $\forall n \in \mathbb{N}$, namely,

$$\begin{aligned} &a_{0;1,0} = 0; \\ &a_{n;\beta+1,\beta} = a_{n;\beta+2,\beta} = \cdots = a_{n;n+1,\beta} = 0, \quad \beta = 0, 1, \dots, n; \\ &c_{n;\alpha,\alpha+2} = c_{n;\alpha,\alpha+3} = \cdots = c_{n;\alpha,n+1} = 0, \quad \alpha = 0, 1, \dots, n-1. \end{aligned} \tag{5.1.37}$$

Some elements in matrices a_n and c_n are always positive, namaely,

$$a_{n;\alpha,\alpha};\ c_{n;\alpha,\alpha+1} > 0, \quad \alpha = 0, 1, \dots, n, \quad \forall n \in \mathbb{N}_0. \tag{5.1.38}$$

Thus, it is possible to say, that $\forall n \in \mathbb{N}_0$ *each lower left corner of the matrix* a_n *(starting from the second diagonal from above) and each upper right corner of the matrix* c_n *(starting from the third diagonal from below) contain zero elements. All positive elements in* (5.1.36) *are denoted by "+". Possibly zero and non-zero elements are indicated by "*".*

The matrix (5.1.36) *in scalar form has a multi-diagonal form.*

The adjoint operator $(\hat{A})^*$ *in the basis (5.1.6) has a similar form, that is, the block tri-diagonal Jacobi matrix* J^+.

These matrices J, J^+ *act by the rule*

$$\begin{aligned}(Jf)_n &= a_{n-1}f_{n-1} + b_n f_n + c_n f_{n+1},\\ (J^+f)_n &= c_{n-1}^* f_{n-1} + b_n^* f_n + a_n^* f_{n+1},\\ n \in \mathbb{N}_0, &\quad f_{-1} = 0, \quad \forall f = (f_n)_{n=0}^{\infty} \in \mathbf{l}_2,\end{aligned} \tag{5.1.39}$$

here "" denotes the usual adjoint matrix.*

The form of the coefficients J^+ follows from (5.1.18) and (5.1.33), the Eq. (5.1.39) generalizes (5.1.1).

5.2 Direct and Inverse Spectral Problems for the Tri-Diagonal Block Jacobi Type Matrices of Bounded Normal Operators

As it was mentioned above the main result of the previous section is really the solution of an inverse problem for a corresponding direct one appearing in the title of this section.

We consider operators in the space $\mathbf{l}_2$ of the form (5.1.9). Additionally to the space $\mathbf{l}_2$ we consider its equipment, i.e., we have a rigging

$$(\mathbf{l}_{\mathrm{fin}})' \supset \mathbf{l}_2(p^{-1}) \supset \mathbf{l}_2 \supset \mathbf{l}_2(p) \supset \mathbf{l}_{\mathrm{fin}}, \tag{5.2.1}$$

where $\mathbf{l}_2(p)$ is a weighted space $\mathbf{l}_2$ with a weight $p = (p_n)_{n=0}^{\infty}$, $p_n \geq 1$, ($p^{-1} = (p_n^{-1})_{n=0}^{\infty}$). In our case $\mathbf{l}_2(p)$ is the Hilbert space of sequences $f = (f_n)_{n=0}^{\infty}$, $f_n \in \mathcal{H}_n$, for which we have

$$\|f\|_{\mathbf{l}_2(p)}^2 = \sum_{n=0}^{\infty} \|f_n\|_{\mathcal{H}_n}^2 p_n, \quad (f, g)_{\mathbf{l}_2(p)} = \sum_{n=0}^{\infty} (f_n, g_n)_{\mathcal{H}_n} p_n. \tag{5.2.2}$$

The space $\mathbf{l}_2(p^{-1})$ is defined similarly; remind that $\mathbf{l}_{\text{fin}}$ is the space of finite sequences and $(\mathbf{l}_{\text{fin}})'$ is the space adjoint to $\mathbf{l}_{\text{fin}}$. It is easy to show that the imbedding $\mathbf{l}_2(p) \hookrightarrow \mathbf{l}_2$ is quasi-nuclear if $\sum_{n=0}^{\infty} np_n^{-1} < \infty$ (see, for example, Chap. 2).

Let A be a normal operator standardly connected with the chain (5.2.1). (About the standard connection see the beginning of the last section of this chapter.) According to the projection spectral theorem such an operator has a representation

$$Af = \int_{\mathbb{C}} z\Phi(z)\, d\sigma(z) f, \quad f \in \mathbf{l}_2, \tag{5.2.3}$$

where $\Phi(z) : \mathbf{l}_2(p) \longrightarrow \mathbf{l}_2(p^{-1})$ is the operator of generalized projection and $d\sigma(z)$ is a spectral measure. The adjoint to A operator A^*, has the same representation (5.2.3), where $z\Phi(z)$ is replaced with $\bar{z}\Phi(z)$. For every $f \in \mathbf{l}_{\text{fin}}$ the projection $\Phi(z)f \in \mathbf{l}_2(p^{-1})$ is a generalized eigenvector of the operators A and A^* with corresponding eigenvalues z and $\bar{z}$, respectively. For all $f, g \in \mathbf{l}_{\text{fin}}$ we have the Parseval equality

$$(f, g)_{\mathbf{l}_2} = \int_{\mathbb{C}} (\Phi(z)f, g)_{\mathbf{l}_2}\, d\sigma(z). \tag{5.2.4}$$

After the extension by continuity, the equality (5.2.4) holds true $\forall f, g \in \mathbf{l}_2$.

Let us denote by π_n the operator of orthogonal projection in $\mathbf{l}_2$ on $\mathcal{H}_n$, $n \in \mathbb{N}_0$. Hence, $\forall f = (f_n)_{n=0}^{\infty} \in \mathbf{l}_2$ we have $f_n = \pi_n f$. This operator acts similarly on the spaces $\mathbf{l}_2(p)$ and $\mathbf{l}_2(p^{-1})$, but possibly with the norm which is not equal to one.

Let us consider the operator matrix $(\Phi_{j,k}(z))_{j,k=0}^{\infty}$, where

$$\Phi_{j,k}(z) = \pi_j \Phi(z) \pi_k \; : \; \mathbf{l}_2 \longrightarrow \mathcal{H}_j, \quad (\text{or} \quad \mathcal{H}_k \longrightarrow \mathcal{H}_j). \tag{5.2.5}$$

Then the Parseval equality (5.2.4) is written in the form

$$\begin{aligned} (f, g)_{\mathbf{l}_2} &= \sum_{j,k=0}^{\infty} \int_{\mathbb{C}} (\Phi(z)\pi_k f, \pi_j g)_{\mathbf{l}_2}\, d\sigma(z) \\ &= \sum_{j,k=0}^{\infty} \int_{\mathbb{C}} (\pi_j \Phi(z)\pi_k f, g)_{\mathbf{l}_2}\, d\sigma(z) \\ &= \sum_{j,k=0}^{\infty} \int_{\mathbb{C}} (\Phi_{j,k}(z) f_k, g_j)_{\mathbf{l}_2}\, d\sigma(z), \quad \forall f, g \in \mathbf{l}_2. \end{aligned} \tag{5.2.6}$$

Let us now consider a more special bounded operator A acting in space $\mathbf{l}_2$. More precisely, let it be given by a block tri-diagonal Jacobi type matrix J of the form (5.1.36). So, this operator A is defined by the first expression in (5.1.39); the adjoint operator defined similarly by the second expression in (5.1.39). Remind that the norm of all elements a_n, b_n and c_n are uniformly bounded with respect to $n \in \mathbb{N}_0$.

For further research, let us assume that conditions in (5.1.37) and (5.1.38) are fulfilled, and, additionally, the operator A given in (5.1.36) is bounded normal on $\mathbf{l}_2$. *Conditions of a boundedness and the normality will be investigated in the next section.*

At the next step we will rewrite the Parseval equality (5.2.6) in terms of generalized eigenvectors of the operator A. At first we prove the lemma.

Lemma 5.2.1 *Let $\varphi(z) = (\varphi_n(z))_{n=0}^{\infty}$, $\varphi_n(z) \in \mathcal{H}_n$, $z \in \mathbb{C}$ be a generalized eigenvector from $(\mathbf{l}_{\text{fin}})'$ of the operator A with an eigenvalue z and, hence, as we have reminded above, it is also a generalized eigenvector of A^* with the corresponding eigenvalue $\bar{z}$; $\varphi_0(z) = \varphi_0$ is independent of z. Thus, $\varphi(z)$ is a solution in $(\mathbf{l}_{\text{fin}})'$ of the system containing two difference equations (see (5.1.39)), namely,*

$$\begin{gathered}(J\varphi(z))_n = a_{n-1}\varphi_{n-1}(z) + b_n\varphi_n(z) + c_n\varphi_{n+1}(z) = z\varphi_n(z),\\ (J^+\varphi(z))_n = c_{n-1}^*\varphi_{n-1}(z) + b_n^*\varphi_n(z) + a_n^*\varphi_{n+1}(z) = \bar{z}\varphi_n(z),\\ n \in \mathbb{N}_0, \quad \varphi_{-1}(z) = 0,\end{gathered} \tag{5.2.7}$$

with an initial condition $\varphi_0 \in \mathbb{C}$.

This solution has a form

$$\varphi_n(z) = Q_n(z)\varphi_0 = (Q_{n;0}, Q_{n;1}, \ldots, Q_{n;n},)\varphi_0, \quad \forall n \in \mathbb{N}, \tag{5.2.8}$$

where $Q_{n;\alpha}$, $\alpha = 0, 1, \ldots, n$ are polynomials of variables z and $\bar{z}$, and these polynomials have the form

$$Q_{n;\alpha}(z) = l_{n;\alpha}\bar{z}^{n-\alpha}z^{\alpha} + q_{n;\alpha}(z, \bar{z}), \quad \alpha = 1, \ldots, n, \tag{5.2.9}$$

where $l_{n;\alpha} > 0$ and $q_{n;\alpha}(z)$ is some linear combinations of $\bar{z}^j z^k$, $0 \le j + k \le n - 1$ and $\bar{z}^{n-(\alpha-1)}z^{\alpha-1}$ for $\alpha = 1, \ldots n$.

Proof For $n = 0$ the system (5.2.7) has the form

$$\begin{aligned}b_0\varphi_0 + c_0\varphi_1 &= z\varphi_0, & \bar{a}_{0;0,0}\varphi_{1;0} + \bar{a}_{0;0,1}\varphi_{1;1} &= (\bar{z} - \bar{b}_{0;0,0})\varphi_0,\\ & & \text{or} & \\ b_0^*\varphi_0 + a_0^*\varphi_1 &= \bar{z}\varphi_0, & c_{0;0,0}\varphi_{1;0} + c_{0;0,1}\varphi_{1;1} &= (z - b_{0;0,0})\varphi_0.\end{aligned} \tag{5.2.10}$$

Here and in what follows we denote

$$\varphi_n(z) = (\varphi_{n;0}(z), \varphi_{n;1}(z), \ldots, \varphi_{n;n}(z)) \in \mathcal{H}_n, \quad \forall n \in \mathbb{N}, \ \varphi_0 = \varphi_{0;0}.$$

By using the assumption (5.1.37) and (5.1.38), we rewrite the last two equalities of (5.2.10) in the form

$$\Delta_0\varphi_1(z) = ((\bar{z} - \bar{b}_{0;0,0})\varphi_0, (z - b_{0;0,0})\varphi_0);$$
$$\Delta_0 = \begin{pmatrix} a_{0;0,0} & 0 \\ c_{0;0,0} & c_{0;0,1} \end{pmatrix}, \quad a_{0;0,0} > 0, \quad c_{0;0,1} > 0. \tag{5.2.11}$$

Therefore,

$$\varphi_{1;0}(z) = \frac{1}{a_{0;0,0}}(\bar{z} - \bar{b}_{0;0,0})\varphi_0 = Q_{1;0}(z)\varphi_0,$$
$$\varphi_{1;1}(z) = (r_1(\bar{z} - \bar{b}_{0;0,0}) + r_2(z - b_{0;0,0}) + r_3)\varphi_0 = Q_{1;1}(z)\varphi_0, \tag{5.2.12}$$

where $r_1 > 0$, r_2 and r_3 are some constants. In other words, the solution of $\varphi_n(z)$ of the Eq. (5.2.7), for $n = 1$ has a form (5.2.8) and (5.2.9).

Suppose, using induction, that for $n \in \mathbb{N}$ coordinates $\varphi_{n-1}(z)$ and $\varphi(z)$ of our generalized eigenvector $\varphi(z) = (\varphi_n(z))_{n=0}^{\infty}$ have the form (5.2.8) and (5.2.9) and prove that $\varphi_{n+1}(z)$ also have the form (5.2.8) and (5.2.9).

Our eigenvector $\varphi(z)$ satisfies the system (5.2.7) of two equations. But this system is overdetermined; it consists of $2(n + 1)$ scalar equations from which it is necessary to find only $n + 2$ unknowns $\varphi_{n+1;0}, \varphi_{n+1;1}, \ldots, \varphi_{n+1;n+1}$ using as an initial data the previous $n+1$ values $\varphi_{n;0}, \varphi_{n;1}, \ldots, \varphi_{n;n}$ of coordinates of the vector $\varphi_n(z)$.

We preside by the following manner. According to Theorem 5.1.5, especially to (5.1.37) and (5.1.38), it follows that $((n + 1) \times (n + 2))$-matrices a_n^* and c_n act on $\psi_{n+1} \in \mathcal{H}_n$ by the rule

$$a_n^*\psi_{n+1}(z) = \begin{pmatrix} a_{n;0,0} & 0 & 0 & \ldots 0 & 0 \\ \bar{a}_{n;1,0} & a_{n;1,1} & 0 & \ldots 0 & 0 \\ \bar{a}_{n;2,0} & \bar{a}_{n;2,1} & a_{n;2,2} & \ldots 0 & 0 \\ \vdots & \vdots & \vdots & \ddots \vdots & \vdots \\ \bar{a}_{n;n-1,0} & \bar{a}_{n;n-1,1} & \bar{a}_{n;n-1,2} & \ldots 0 & 0 \\ \bar{a}_{n;n,0} & \bar{a}_{n;n,1} & \bar{a}_{n;n,2} & \ldots a_{n;n,n} & 0 \end{pmatrix} \psi_{n+1}(z), \tag{5.2.13}$$

$$c_n\psi_{n+1}(z) = \begin{pmatrix} c_{n;0,0} & c_{n;0,1} & 0 & \ldots 0 & 0 \\ c_{n;1,0} & c_{n;1,1} & c_{n;1,2} & \ldots 0 & 0 \\ c_{n;2,0} & c_{n;2,1} & c_{n;2,2} & \ldots 0 & 0 \\ \vdots & \vdots & \vdots & \ddots \vdots & \vdots \\ c_{n;n-1,0} & c_{n;n-1,1} & c_{n;n-1,2} & \ldots c_{n;n-1,n} & 0 \\ c_{n;n,0} & c_{n;n,1} & c_{n;n,2} & \ldots c_{n;n,n} & c_{n;n,n+1} \end{pmatrix} \psi_{n+1}(z), \tag{5.2.14}$$

where $\psi_{n+1}(z) = (\psi_{n+1;0}(z), \psi_{n+1;1}(z), \ldots, \psi_{n+1;n+1}(z))$.

Similarly to (5.2.11), we construct such a combination of matrix elements (5.2.13) and (5.2.14), which is the $((n+2)\times(n+2))$-matrix of the form

$$\Delta_n \psi_{n+1}(z) = \begin{pmatrix} a_{n;0,0} & 0 & 0 & \ldots 0 & 0 \\ c_{n;0,0} & c_{n;0,1} & 0 & \ldots 0 & 0 \\ c_{n;1,0} & c_{n;1,1} & c_{n;1,2} & \ldots 0 & 0 \\ \vdots & \vdots & \vdots & \ddots \vdots & \vdots \\ c_{n;n-1,0} & c_{n;n-1,1} & c_{n;n-1,2} & \ldots c_{n;n-1,n} & 0 \\ c_{n;n,0} & c_{n;n,1} & c_{n;n,2} & \ldots c_{n;n,n} & c_{n;n,n+1} \end{pmatrix} \psi_{n+1}(z), \tag{5.2.15}$$

where $\psi_{n+1}(z) = (\psi_{n+1;0}(z), \psi_{n+1;1}(z), \ldots, \psi_{n+1;n+1}(z))$.

The matrix (5.2.15) is invertible because its elements on the main diagonal are positive (see (5.1.38)). Rewrite the equalities (5.2.7) in the form

$$\begin{aligned} a_n^*\varphi_{n+1}(z) &= \bar{z}\varphi_n(z) - c_{n-1}^*\varphi_{n-1}(z) - b_n^*\varphi_n(z), \\ c_n\varphi_{n+1}(z) &= z\varphi_n(z) - a_{n-1}\varphi_{n-1}(z) - b_n\varphi_n(z), \quad n\in\mathbb{N}. \end{aligned} \tag{5.2.16}$$

We see that the first $n+2$ scalar equations (of $2(n+1)$ scalar equations (5.2.16)) have the form

$$\begin{aligned} \Delta_n\varphi_{n+1}(z) = \Big(&\bar{z}Q_{n;0}(z) - (c_{n-1}^*Q_{n-1}(z) - (b^*Q_n(z))_{n;0}, \\ &zQ_{n;0}(z) - (a_{n-1}Q_{n-1}(z))_{n;0} - (b_nQ_n(z))_{n;0}, \ldots, \\ &zQ_{n;n}(z) - (a_{n-1}Q_{n-1}(z))_{n;n} - (b_nQ_n(z))_{n;n}\Big)\varphi_0. \end{aligned} \tag{5.2.17}$$

The construction of the matrix Δ_n and the vector form on the right-hand side in (5.2.17) and (5.2.8), (5.2.9) yield

$$\begin{aligned} \varphi_{n+1;0}(z) &= Q_{n+1;0}(z)\varphi_0 \\ &= \frac{1}{a_{n;0,0}}(\bar{z}Q_{n;0}(z) - (c_{n-1}^*Q_{n-1}(z))_{n;0}) - (b_n^*Q_n(z))_{n;0})\varphi_0 \\ &= \frac{1}{a_{n;0,0}}(\bar{z}(l_{n;0}\bar{z}^n + q_{n;0}(z)) - (c_{n-1}^*Q_{n-1}(z))_{n;0} \\ &- (b_n^*Q_n(z))_{n;0})\varphi_0, \end{aligned} \tag{5.2.18}$$

that is, the main term on the right-hand side (5.2.18) is equal to $\frac{l_{n;0}}{a_{n;0,0}}\bar{z}_{n+1}z^0$, hence, it has the form (5.2.9).

Similarly we get $\varphi_{n+1;1}(z), \ldots, \varphi_{n+1;n+1}(z)$. It is necessary to take into account that the diagonal of the matrix Δ_n has also a positive elements $c_{n;0,1}, c_{n;1,2}, \ldots, c_{n;n,n+1}$ due to (5.1.38). This completes the induction and finishes the proof. □

Remark 5.2.2 Note that the previous lemma does not assert the existence of a solution of the overdetermined system (5.2.7) for an arbitrary initial data $\varphi_0 \in \mathbb{C}$; it is only proved that the generalized eigenvector from $(\mathbf{l}_{\mathrm{fin}})'$ of the operator A is a solution of (5.2.7) and has the form (5.2.8) and (5.2.9).

In what follows, it will be convenient to look at $Q_n(z)$ with fixed z as a linear operator that acts from $\mathcal{H}_0$ into $\mathcal{H}_n$, i.e., $\mathcal{H}_0 \ni \varphi_0 \longmapsto Q_n(z)\varphi_0 \in \mathcal{H}_n$. We regard also $Q_n(z)$ as an operator valued polynomial of variables $z, \bar{z} \in \mathbb{C}$; hence, for the adjoint operator we have $Q_n^*(z) = (Q_n(z))^* : \mathcal{H}_n \longrightarrow \mathcal{H}_0$. By using these polynomials $Q_n(z)$, we construct the representation for $\Phi_{j,k}(z)$.

Lemma 5.2.3 *The operator $\Phi_{j,k}(z)$, $\forall z \in \mathbb{C}$ has the representation:*

$$\Phi_{j,k}(z) = Q_j(z)\Phi_{0,0}(z)Q_k^*(z) : \mathcal{H}_k \longrightarrow \mathcal{H}_j, \quad j, k \in \mathbb{N}_0, \tag{5.2.19}$$

where $\Phi_{0,0}(z) \geq 0$ is a scalar.

Proof For a fixed $k \in \mathbb{N}_0$, the vector $\varphi = \varphi(z) = (\varphi_j(z))_{j=0}^{\infty}$, where

$$\varphi_j(z) = \Phi_{j,k}(z) = \pi_j \Phi(z)\pi_k \in \mathcal{H}_j, \quad z \in \mathbb{C} \tag{5.2.20}$$

is a solution and simultaneously, a generalized eigenvector in $(\mathbf{l}_{\mathrm{fin}})'$ of the equation $J\varphi(z) = z\varphi(z)$, since $\Phi(z)$ is a projector on to generalized eigenvectors of the operator A with correspondence generalized eigenvalues z. Therefore, $\forall g \in \mathbf{l}_{\mathrm{fin}}$, we have $(\varphi, J^+ g)_{\mathbf{l}_2} = z(\varphi, g)_{\mathbf{l}_2}$. Applying the first finite-difference equation J^+ to the vector φ, we obtain $(J\varphi, g)_{\mathbf{l}_2} = z(\varphi, g)_{\mathbf{l}_2}$. Hence, $\varphi = \varphi(z) \in \mathbf{l}_2(p^{-1})$ exists as an usual solution of the equation $J\varphi = z\varphi$ with an initial condition $\varphi_0 = \pi_0 \Phi(z)\pi_k \in \mathcal{H}_0$.

Since $\forall f \in \mathbf{l}_{\mathrm{fin}}$, the vector $\Phi(z) f \in \mathbf{l}_2(p^{-1})$ is also a generalized eigenvector of the operator A^* with the corresponding eigenvalue $\bar{z}$ (because A is normal operator), the same vector $\varphi = \varphi(z)$ in (5.2.20) is also the solution of the equation $J^+\varphi = \bar{z}\varphi$ with the same initial condition $\varphi_0 = \pi_0 \Phi(z)\pi_k$.

By using Lemma 5.2.1, in particular (5.2.8), we get

$$\Phi_{j,k}(z) = Q_j(z)(\Phi_{0,k}(z)), \quad j \in \mathbb{N}_0. \tag{5.2.21}$$

The operator $\Phi(z) : \mathbf{l}_2(p) \longrightarrow \mathbf{l}_2(p^{-1})$ is formally normal on $\mathbf{l}_2$, The operator as a derivative of the resolution of the identity of the operator A in $\mathbf{l}_2$ with respect to the spectral measure. Hence, according to (5.2.19) we get

$$(\Phi_{j,k}(z))^* = (\pi_j \Phi(z)\pi_k)^* = \pi_k \Phi(z)\pi_j = \Phi_{k,j}(z), \quad j, k \in \mathbb{N}_0. \tag{5.2.22}$$

For a fixed $j \in \mathbb{N}_0$ from (5.2.22) and previous conversation, it follows that the vector

$$\psi = \psi(z) = (\psi_k(z))_{k=0}^{\infty}, \quad \psi_k(z) = \Phi_{k,j}(z) = (\Phi_{j,k}(z))^*$$

is an usual solution of equations $J\psi = z\psi$ and $J^+\psi = \bar{z}\psi$ with an initial condition $\psi_0 = \Phi_{0,j}(z) = (\Phi_{j,0}(z))^*$.

By using Lemma 5.2.1, we also obtain a representation of the type (5.2.21),

$$\Phi_{k,j}(z) = Q_k(z)(\Phi_{0,j}(z)), \quad k \in \mathbb{N}_0. \tag{5.2.23}$$

Taking into account (5.2.22) and (5.2.23), we get

$$\Phi_{0,k}(z) = (\Phi_{k,0}(z))^* = (Q_k(z)\Phi_{0,0}(z))^* = \Phi_{0,0}(z)(Q_k(z))^*, \quad k \in \mathbb{N}_0 \tag{5.2.24}$$

(here we used $\Phi_{0,0}(z) \geq 0$; this inequality follows from (5.2.4) and (5.2.5)). Substituting (5.2.24) into (5.2.21), we obtain (5.2.19). □

Now, we can rewrite the Parseval equality (5.2.6) in concrete form. To do this, we substitute the expression (5.2.19) for $\Phi_{j,k}(z)$ in (5.2.6), therefore, we get

$$\begin{aligned}
(f, g)_{\mathbf{l}_2} &= \sum_{j,k=0}^{\infty} \int_{\mathbb{C}} (\Phi_{j,k}(z) f_k, g_j)_{\mathbf{l}_2}\, d\sigma(z) \\
&= \sum_{j,k=0}^{\infty} \int_{\mathbb{C}} (Q_j(z)\Phi_{0,0}(z) Q_k^*(z) f_k, g_j)_{\mathbf{l}_2}\, d\sigma(z) \\
&= \sum_{j,k=0}^{\infty} \int_{\mathbb{C}} (Q_k^*(z) f_k, Q_j^*(z) g_j)_{\mathbf{l}_2}\, d\rho(z) \\
&= \int_{\mathbb{C}} \Big(\sum_{k=0}^{\infty} Q_k^*(z) f_k\Big) \overline{\Big(\sum_{j=0}^{\infty} Q_j^*(z) g_j\Big)}\, d\rho(z),, \quad \forall f, g \in \mathbf{l}_{\text{fin}},
\end{aligned} \tag{5.2.25}$$

$$d\rho(z) = \Phi_{0,0}(z)\, d\sigma(z).$$

We denote the Fourier transform "$\widehat{\ }$" generated by the normal operator A in the space $\mathbf{l}_2$. Hence,

$$\mathbf{l}_2 \supset \mathbf{l}_{\text{fin}} \ni f = (f_n)_{n=0}^{\infty} \longmapsto \hat{f}(z) = \sum_{n=0}^{\infty} Q_n^*(z) f_n \in L_2(\mathbb{C}, d\rho(z)) \quad \forall f \in \mathbf{l}_{\text{fin}}. \tag{5.2.26}$$

Thus, (5.2.25) gives the Parseval equality in a final

$$(f, g)_{\mathbf{l}_2} = \int_{\mathbb{C}} \hat{f}(z)\overline{\hat{g}(z)}\, d\rho(z), \quad \forall f, g \in \mathbf{l}_{\mathrm{fin}}. \tag{5.2.27}$$

Extending (5.2.27) by continuity, we get that the last expression is valid $\forall f, g \in \mathbf{l}_2$.

The orthogonality of polynomials $Q_n^*(z)$ follows from (5.2.26) and (5.2.27). Namely, it is sufficient only to take

$$f = (0, \ldots, 0, f_k, 0, \ldots), \; f_k \in \mathcal{H}_k, \; g = (0, \ldots, 0, g_j, 0, \ldots), \; g_j \in \mathcal{H}_j$$

in (5.2.26) and (5.2.27). Then $\forall k, j \in \mathbb{N}_0$

$$\int_{\mathbb{C}} (Q_k^*(z) f_k)\overline{(Q_j^*(z) g_j)}\, d\rho(z) = \delta_{j,k}(f_j, g_j)_{\mathcal{H}_j}. \tag{5.2.28}$$

By using the representation (5.2.8) for these polynomials we rewrite the equality (5.2.28) in a more classical scalar form. To do this, we remark that $Q_0^*(z) = \bar{Q}_0(z)$ and for $n \in \mathbb{N}$, according to (5.2.8), we have

$$Q_n(z) = (Q_{n;0}(z), Q_{n;1}(z), \ldots, Q_{n;n}(z)) : \mathcal{H}_0 \longrightarrow \mathcal{H}_n.$$

Hence, for the adjoint operator $Q_n^*(z) : \mathcal{H}_n \longrightarrow \mathcal{H}_0$, $\forall x \in \mathcal{H}_0$, $y = (y_0, y_1, \ldots, y_n) \in \mathcal{H}_n$, we have

$$\begin{aligned}
(Q_n(z)x, y)_{\mathcal{H}_n} &= ((Q_{n;0}(z)x, Q_{n;1}(z)x, \ldots, Q_{n;n}(z)x), (y_0, y_1, \ldots, y_n))_{\mathcal{H}_n} \\
&= Q_{n;0}(z)x\bar{y}_0 + Q_{n;1}(z)x\bar{y}_1 + \cdots + Q_{n;n}(z)x\bar{y}_n \\
&= x\overline{(\overline{Q_{n;0}(z)}y_0 + \overline{Q_{n;1}(z)}y_1 + \cdots + \overline{Q_{n;n}(z)}y_n)} \\
&= (x, Q_n^*(z)y)_{\mathcal{H}_0},
\end{aligned}$$

i.e., $Q_n^*(z)y = \overline{Q_{n;0}(z)}y_0 + \overline{Q_{n;1}(z)}y_1 + \cdots + \overline{Q_{n;n}(z)}y_n$.

Due to the last equality for $n \in \mathbb{N}$ and $f_n = (f_{n,0}, f_{n,1}, \ldots, f_{n,n}) \in \mathcal{H}_n$, $z \in \mathbb{C}$, we get

$$Q_n^*(z) f_n = \overline{Q_{n;0}(z)} f_{n;0} + \overline{Q_{n;1}(z)} f_{n;1} + \cdots + \overline{Q_{n;n}(z)} f_{n;n}, \quad Q_0^*(z) = 1. \tag{5.2.29}$$

Therefore, (5.2.28) has the form

$$\int_{\mathbb{C}} \left(\sum_{\alpha=0}^{k} \overline{Q_{k;\alpha}(z)} f_{k;\alpha} \right) \overline{\left(\sum_{\beta=0}^{j} \overline{Q_{j;\beta}(z)} g_{j;\beta} \right)} \, d\rho(z) = \delta_{j,k} \sum_{\alpha=0}^{j} f_{j;\alpha} \bar{g}_{j;\alpha},$$

$$\forall f_{k;0}, f_{k;1}, \ldots, f_{k;k}, \ g_{j;0}, g_{j;1}, \ldots, g_{j;j} \in \mathbb{C}, \quad j, k \in \mathbb{N}_0.$$

This equality is equivalent to the following orthogonality relation in the usual classical form

$$\int_{\mathbb{C}} \overline{Q_{k;\beta}(z)} Q_{j;\alpha} \, d\rho(z) = \delta_{j,k} \delta_{\alpha,\beta}, \quad Q_{0;0} = Q_0(z), \tag{5.2.30}$$

$$\forall j, k \in \mathbb{N}_0, \quad \forall \alpha = 0, 1, \ldots, j, \ \beta = 0, 1, \ldots, k.$$

Let us remark that due to (5.2.29) the Fourier transform (5.2.26) can be rewritten

$$\hat{f}(z) = \sum_{n=0}^{\infty} \sum_{\alpha=0}^{n} \overline{Q_{n;\alpha}(z)} f_{n;\alpha}, \quad z \in \mathbb{C}, \quad \forall f = (f_n)_{n=0}^{\infty} \in \mathbf{l}_2. \tag{5.2.31}$$

By using the previous considerations, it is possible to formulate a version of the spectral theorem for the bounded normal operator A.

Theorem 5.2.4 *Consider the space (5.1.9), namely,*

$$\mathbf{l}_2 = \mathcal{H}_0 \oplus \mathcal{H}_1 \oplus \mathcal{H}_2 \oplus, \cdots, \quad \mathcal{H}_n = \mathbb{C}^{n+1}, \quad n \in \mathbb{N}_0 \tag{5.2.32}$$

and the linear operator A which is defined on finite vectors $\mathbf{l}_{\mathrm{fin}}$ by a block tri-diagonal Jacobi type matrix J of the form (5.1.36) *using the first expression in* (5.1.39). *We suppose that all its coefficients a_n, b_n and c_n, $n \in \mathbb{N}_0$ are uniformly bounded, some elements of these matrices are always equal to zero or positive according to* (5.1.37), (5.1.38) *and the extension of A by continuity is bounded normal operator on this space.*

The generalized eigenvectors expansion of the operator A has the following form. According to Lemma 5.2.1 *we represent, using $\varphi_0 \in \mathbb{C}$, the solution $\varphi(z) = (\varphi_n(z))_{n=0}^{\infty}$, $\varphi_n(z) \in \mathcal{H}_n$ of equations* (5.2.7) *(which exists due to the projection spectral theorem) for $z \in \mathbb{C}$,*

$$\varphi_n(z) = Q_n(z)\varphi_0 = (Q_{n;0}(z), Q_{n;1}(z), \cdots, Q_{n;n}(z))\varphi_0,$$

where $Q_{n;\alpha}(z)$*,* $\alpha = 0, 1, \dots, n$ *are polynomials of variables* z *and* $\bar{z}$*. Then the Fourier transform has the form*

$$\mathbf{l}_2 \supset \mathbf{l}_{\text{fin}} \ni= (f_n)_{n=0}^{\infty} \longmapsto$$

$$\hat{f}(z) = \sum_{n=0}^{\infty} Q_n^*(z) f_n = \sum_{n=0}^{\infty}\sum_{\alpha=0}^{n} \overline{Q_{n;\alpha}(z)} f_{n;\alpha} \in L_2(\mathbb{C}, d\rho(z)). \qquad (5.2.33)$$

Here $Q_n^*(z) : \mathcal{H}_n \longrightarrow \mathcal{H}_0$ *is the operator adjoint to* $Q_n(z) : \mathcal{H}_0 \longrightarrow \mathcal{H}_n$*,* $d\rho(z)$ *is the probability spectral measure of* A*.*

The Parseval equality has the form

$$(f, g)_{\mathbf{l}_2} = \int_{\mathbb{C}} \hat{f}(z)\overline{\hat{g}(z)}\, d\rho(z); \quad (Jf, g)_{\mathbf{l}_2} = \int_{\mathbb{C}} z\hat{f}(z)\overline{\hat{g}(z)}\, d\rho(z), \quad \forall f, g \in \mathbf{l}_{\text{fin}}. \qquad (5.2.34)$$

Identities (5.2.33) *and* (5.2.34) *are extended by continuity* $\forall f, g \in \mathbf{l}_2$ *making the operator* (5.2.33) *isometric which maps whole* $\mathbf{l}_2$ *onto the whole* $L_2(\mathbb{C}, d\rho(z))$*.*

Polynomials $\overline{Q_{n;\alpha}(z)}$*,* $n \in \mathbb{N}$*,* $\alpha = 0, 1, \dots, n$*, and* $Q_{0;0}(z) = 1$ *form an orthonormal system in* $L_2(\mathbb{C}, d\rho(z))$ *in the sense of* (5.2.30) *and such a set is total in this space.*

Proof It is only necessary to show that the orthogonal polynomials $\overline{Q_{n;\alpha}(z)}$, $n \in \mathbb{N}$, $\alpha = 0, 1, \dots, n$, and $Q_{0;0}(z) = 1$ form a total set in the space $L_2(\mathbb{C}, d\rho(z))$. For this reason we remark at first that due to the compactness of the support of the measure $d\rho(z)$ in $\mathbb{C}$ elements $z^j\bar{z}^k$, $j, k \in \mathbb{N}_0$ form a total set in $L_2(\mathbb{C}, d\rho(z))$.

Assume the contrary, that is, our system of polynomials is not total. Then there exist non-zero function $h(z) \in L_2(\mathbb{C}, d\rho(z))$ that is orthogonal to all these polynomials and hence, according to (5.2.9) to all $z^j\bar{z}^k$, $j, k \in \mathbb{N}_0$. Hence, $h(z) = 0$ that is a contradiction. □

The last theorem solves the direct problem for the bounded normal operator A, which is generated by the matrix J by the matrix (5.1.36) on the space $\mathbf{l}_2$.

The inverse problem consists in constructing according to a given measure $d\rho(z)$ in $\mathbb{C}$ with a compact support of a bounded normal matrix J of the form (5.1.36), whose spectral measure coincides with $d\rho(z)$. *This construction is carried out according to Theorem* 5.1.5 *using the Gram-Schmidt orthogonalization of the system* (5.1.5)*. For the matrix* J *of the form* (5.1.36)*, constructed by* $d\rho(z)$*, the spectral measure of the corresponding bounded normal operator* A *coincides with the initial measure.*

Proof This fact is true, since the system of orthogonal polynomials, connected with A, $\overline{Q_{n,\alpha}(z)}$, $\alpha = 0, 1, \dots, n$, $n \in \mathbb{N}_0$, is orthonormal in $L_2(\mathbb{C}, d\rho(z))$ and, according to Lemma 5.2.1, is constructed by $\bar{z}^j z^k$, $z \in \mathbb{C}$ as well as the system

(5.1.6) is constructed by $z^j \bar{z}^k$, $j, k \in \mathbb{N}_0$. Hence, $\forall n \in \mathbb{N}$

$$Q_0(z) = 1 = P_0(z), \quad \overline{Q_{n,\alpha}(z)} = P_{n;\alpha}(z), \quad \alpha = 0, 1, \ldots, n, \tag{5.2.35}$$

Since both systems of polynomials form total sets in the space $L_2(\mathbb{C}, d\rho(z))$, then (5.2.35) shows that the spectral measure constructed by the operator coincides with the given one above. □

Let us remark that the expression (5.1.15) (as it is known in the classical theory of Jacobi matrices) restores the original (initial) matrix (5.1.36) by the spectral measure $d\rho(z)$ of the operator generated by J in $\mathbf{l}_2$.

5.3 Normality Conditions of Block Jacobi Type Matrices

First, we will find the conditions that would guarantee the formal normality of the matrix J of the type (5.1.35). Formally, the adjoint matrix J^+ has the form

$$J^+ = \begin{pmatrix} b_0^* & a_0^* & 0 & 0 & \cdots \\ c_0^* & b_1^* & a_1^* & 0 & \cdots \\ 0 & c_1^* & b_2^* & a_2^* & \cdots \\ \vdots & \vdots & \vdots & \vdots & \ddots \end{pmatrix}, \quad \begin{aligned} c_n^* &: \mathcal{H}_n \longrightarrow \mathcal{H}_{n+1}, \\ b_n^* &: \mathcal{H}_n \longrightarrow \mathcal{H}_n, \\ a_n^* &: \mathcal{H}_{n+1} \longrightarrow \mathcal{H}_n, \quad n \in \mathbb{N}_0. \end{aligned} \tag{5.3.1}$$

Multiplying matrices (5.1.35) and (5.3.1), we get

$$JJ^+ = \begin{pmatrix} b_0b_0^* + c_0c_0^* & b_0a_0^* + c_0b_1^* & c_0a_1^* & 0 & 0 & \cdots \\ a_0b_0^* + b_1c_0^* & a_0a_0^* + b_1b_1^* + c_1c_1^* & b_1a_1^* + c_1b_2^* & c_1a_2^* & 0 & \cdots \\ a_1c_0^* & a_1b_1^* + b_2c_1^* & a_1a_1^* + b_2b_2^* + c_2c_2^* & b_2a_2^* + c_2b_3^* & c_2a_3^* & \cdots \\ \vdots & \vdots & \vdots & \vdots & \vdots & \ddots \end{pmatrix}. \tag{5.3.2}$$

The expression for J^+J is similar to (5.3.2) if a_n, b_n and c_n are replaced with c_n^*, b_n^* and a_n^*, respectively and vice versa.

Comparing these expressions for JJ^+ and J^+J we conclude that the equality $JJ^+ = J^+J$ is equivalent to fulfillment the system of equalities

$$\begin{aligned} &c_0 c_0^* = a_0^* a_0; \\ &c_n a_{n+1}^* = a_n^* c_{n+1}, \\ &b_n a_n^* + c_n b_{n+1}^* = b_n^* c_n + a_n^* b_{n+1}, \\ &a_n a_n^* + b_{n+1} b_{n+1}^* + c_{n+1} c_{n+1}^* = c_n^* c_n + b_{n+1}^* b_{n+1} + a_{n+1}^* a_{n+1}, \quad n \in \mathbb{N}_0, \end{aligned} \tag{5.3.3}$$

where we take into account that b_0 is a scalar, $b_0^* = \bar{b}_0$. Note that the necessity of equalities

$$a_n b_n^* + b_{n+1} c_n^* = c_n^* b_n + b_{n+1}^* a_n, \quad a_{n+1} c_n^* = c_{n+1}^* a_n, \quad n \in \mathbb{N}_0,$$

follows from the third and the second equalities in (5.3.3) taking into account the adjointness "$*$".

So, conditions (5.3.3) are necessary and sufficient for matrix equality $JJ^+ = J^+J$. If norms of operators generated by a_n, b_n and c_n are uniformly bounded with respect to $n \in \mathbb{N}_0$, then $\tilde{J}$ as an operator in $\mathbf{l}_2$ is bounded and the condition (5.3.3) gives the normality of the operator.

We take as initial matrices a_0, b_0, c_0 and find a_1, b_1, c_1; a_2, b_2, c_2; ... from (5.3.3) step by step and similarly for further indices (but non-unique manner) the normal matrix J is constructed. But for such a matrix, Theorem 5.2.4, in general, is not valid, because it is necessary to find these matrices in such way that a_n and c_n must be of the form (5.1.35). Only in this case according to Lemma 5.2.1, functions (5.1.4) are linearly independent and Theorem 5.2.4 is applicable.

The finding of matrices a_n, b_n and c_n, $n \in \mathbb{N}_0$, which are a solution of equations (5.3.3) and a_n, c_n have the form (5.1.35), is an enough complicated problem and we investigate here only some special cases.

That is, it is assumed, first of all, that all matrices b_n are self-adjoint, i.e., $b_n^* = b_n$, $n \in \mathbb{N}_0$. Then conditions (5.3.3) obtain a slightly simplified form

$$\begin{aligned} c_0 c_0^* &= a_0^* a_0; \\ c_n a_{n+1}^* &= a_n^* c_{n+1}, \\ b_n (a_n^* - c_n) &= (a_n^* - c_n) b_{n+1}, \\ a_n a_n^* + c_{n+1} c_{n+1}^* &= c_n^* c_n + a_{n+1}^* a_{n+1}, \quad n \in \mathbb{N}_0. \end{aligned} \tag{5.3.4}$$

Here, we will consider that all matrices a_n, c_n, $n \in \mathbb{N}_0$ have a form as in (5.1.35) and (5.1.36), where $a_{n;0,0}, a_{n;1,1}, \ldots, a_{n;n,n}$ and $c_{n;0,1}, c_{n;1,2}, \ldots, c_{n;n,n+1}$, $\forall n \in \mathbb{N}_0$ are positive and all other elements are equal to zero. So, the matrices, $\forall n \in \mathbb{N}_0$, have the form

$$a_n = \underbrace{\begin{pmatrix} a_{n;0,0} & 0 & \ldots & 0 \\ 0 & a_{n;1,1} & \ldots & 0 \\ \vdots & \vdots & \ddots & \vdots \\ 0 & 0 & \ldots & a_{n;n,n} \\ 0 & 0 & \ldots & 0 \end{pmatrix}}_{n+1} \left.\vphantom{\begin{pmatrix} a \\ a \\ a \\ a \\ a \end{pmatrix}}\right\} n+2, \tag{5.3.5}$$

$$a_n^* = \underbrace{\begin{pmatrix} a_{n;0,0} & 0 & \ldots & 0 & 0 \\ 0 & a_{n;1,1} & \ldots & 0 & 0 \\ \vdots & \vdots & \ddots & \vdots & \vdots \\ 0 & 0 & \ldots & a_{n;n,n} & 0 \end{pmatrix}}_{n+2} \left.\vphantom{\begin{pmatrix} a \\ a \\ a \\ a \end{pmatrix}}\right\} n+1, \tag{5.3.6}$$

$$c_n = \underbrace{\begin{pmatrix} 0 & c_{n;0,1} & 0 & \ldots & 0 \\ 0 & 0 & c_{n;1,2} & \ldots & 0 \\ \vdots & \vdots & \vdots & \ddots & \vdots \\ 0 & 0 & 0 & \ldots & c_{n;n,n+1} \end{pmatrix}}_{n+2} \left.\vphantom{\begin{pmatrix} a \\ a \\ a \\ a \end{pmatrix}}\right\} n+1, \tag{5.3.7}$$

$$c_n^* = \underbrace{\begin{pmatrix} 0 & 0 & \ldots & 0 \\ c_{n;0,1} & 0 & \ldots & 0 \\ 0 & c_{n;1,2} & \ldots & 0 \\ \vdots & \vdots & \ddots & \vdots \\ 0 & 0 & \ldots & c_{n;n,n+1} \end{pmatrix}}_{n+1} \left.\vphantom{\begin{pmatrix} a \\ a \\ a \\ a \\ a \end{pmatrix}}\right\} n+2, \tag{5.3.8}$$

Multiplying matrices of the type (5.3.5)–(5.3.8), we can rewrite the first, the second and the fourth equality of (5.3.4) in the form of corresponding equalities for elements of these matrices: $c_{0;0,1} = a_{0;0,0}$ and $\forall n \in \mathbb{N}_0$,

$$\begin{aligned}
c_{n;0,1}a_{n+1;1,1} &= a_{n;0,0}c_{n+1;0,1},\\
c_{n;1,2}a_{n+1;2,2} &= a_{n;1,1}c_{n+1;1,2},\\
c_{n;2,3}a_{n+1;3,3} &= a_{n;2,2}c_{n+1;2,3},\\
&\dots\\
c_{n;n-1,n}a_{n+1;n,n} &= a_{n;n-1,n-1}c_{n+1;n-1,n},\\
c_{n;n,n+1}a_{n+1;n+1,n+1} &= a_{n;n,n}c_{n+1;n,n+1};
\end{aligned}
\tag{5.3.9}$$

$$\begin{aligned}
a^2_{n;0,0} + c^2_{n+1;0,1} &= a^2_{n+1;0,0},\\
a^2_{n;1,1} + c^2_{n+1;1,2} &= c^2_{n;0,1} + a^2_{n+1;1,1},\\
a^2_{n;2,2} + c^2_{n+1;2,3} &= c^2_{n;1,2} + a^2_{n+1;2,2},\\
&\dots\\
a^2_{n;n,n} + c^2_{n+1;n,n+1} &= c^2_{n;n-1,n} + a^2_{n+1;n,n},\\
c^2_{n+1;n+1,n+2} &= c^2_{n;n,n+1} + a^2_{n+1;n+1,n+1}.
\end{aligned}$$

The system of equalities (5.3.9) is equivalent to the system (5.3.4) for the case $b_n = 0$, $n \in \mathbb{N}_0$. At first we find step by step the solution $c_{n;\alpha,\beta}$, $a_{n;\gamma,\gamma}$.

Let $n \in \mathbb{N}_0$ be fixed. Denote elements of (5.3.9):

$$\begin{array}{llll}
c^2_{n;0,1} = x_0, & c^2_{n;1,2} = x_1, & \dots, & c^2_{n;n,n+1} = x_n;\\
a^2_{n;n,n} = \xi_0, & a^2_{n;n-1,n-1} = \xi_1, & \dots, & a^2_{n;0,0} = \xi_n;\\
c^2_{n+1;0,1} = y_0, & c^2_{n+1;1,2} = y_1, & \dots, & c^2_{n+1;n+1,n+2} = y_{n+1};\\
a^2_{n+1;n+1,n+1} = \eta_0, & a^2_{n+1;n,n} = \eta_1, & \dots, & a^2_{n+1;0,0} = \eta_{n+1}.
\end{array}
\tag{5.3.10}$$

Taking the square of the equalities from (5.3.9), we can rewrite (5.3.9) in the form

$$
\begin{aligned}
x_0\eta_n &= \xi_n y_0, & \xi_n + y_0 &= \eta_{n+1},\\
x_1\eta_{n-1} &= \xi_{n-1} y_1, & \xi_{n-1} + y_1 &= x_0 + \eta_n,\\
x_2\eta_{n-2} &= \xi_{n-2} y_2, & \xi_{n-2} + y_2 &= x_1 + \eta_{n-1},\\
&\dots\\
x_{n-1}\eta_1 &= \xi_1 y_{n-1}, & \xi_0 + y_n &= x_{n-1} + \eta_1,\\
x_n\eta_0 &= \xi_0 y_n; & y_{n+1} &= x_n + \eta_0.
\end{aligned}
\tag{5.3.11}
$$

We regard conditions (5.3.11) as a system of $2n+3$ linear equations of $2(n+2)$ positive unknowns $y_0, y_1, \dots, y_{n+1}; \eta_0, \eta_1, \dots, \eta_{n+1}$ with coefficients depending on $x_0, x_1, \dots, x_n; \xi_0, \xi_1, \dots, \xi_n$.

We start from initial data $c_{0;0,1} = a_{0;0,0} = \delta_0^{1/2}$, where $\delta_0 > 0$ is a fixed number, i.e., for $n = 0$ from $x_0 = \xi_0 = \delta_0$. Then (5.3.11) gives

$$
x_0\eta_0 = \xi_0 y_0, \quad \xi_0 + y_0 = \eta_1, \quad y_1 = x_0 + \eta_0.
$$

The general solution of this system has a form

$$
y_0 = \eta_0 = \delta_1, \quad y_1 = \eta_1 = \delta_0 + \delta_1, \tag{5.3.12}
$$

where $\delta_1 > 0$ is an arbitrary given number.

Consider the system (5.3.11) for $n = 1$ and take the coefficients $(x_0, x_1; \xi_0, \xi_1)$, equal to the solution $(y_0, y_1; \eta_0, \eta_1)$ (5.3.12) of the previous system. Thus, we get

$$
\begin{aligned}
x_0\eta_1 &= \xi_1 y_0, & & \delta_1\eta_1 = (\delta_0 + \delta_1) y_0,\\
x_1\eta_0 &= \xi_0 y_1, & & (\delta_0 + \delta_1)\eta_0 = \delta_1 y_1,\\
\xi_1 + y_0 &= \eta_2, & \text{i.e.,}\quad & (\delta_0 + \delta_1) + y_0 = \eta_2,\\
\xi_0 + y_1 &= x_0 + \eta_1, & & \delta_1 + y_1 = \delta_1 + \eta_1,\\
y_2 &= x_1 + \eta_0, & & y_2 = (\delta_0 + \delta_1) + \eta_0.
\end{aligned}
\tag{5.3.13}
$$

The general solution of the system (5.3.13) has a form

$$
y_0 = \eta_0 = \delta_2, \quad y_1 = \eta_1 = (\delta_0 + \delta_1)\delta_2\delta_1^{-1}, \quad y_2 = \eta_2 = \delta_0 + \delta_1 + \delta_2, \tag{5.3.14}
$$

where $\delta_2 > 0$ is an arbitrary number. It is obvious that the solution is symmetric, i.e., $y_k = \eta_k$, $k = 0, 1, 2$, and, therefore, the system (5.3.11) has more simple form for $n = 2$. In this system $x_k = \xi_k$, $k = 0, 1, 2$.

In what follow we assert such a symmetry for an arbitrary $n = 2, 3, \dots$.

Lemma 5.3.1 *Let coefficients of the system* (5.3.11) *be given in the form* $x_k = \xi_k > 0$, $k = 0, 1, \ldots, n$, $n = 2, 3, \ldots$. *Then its solution*

$$(y_0, y_1, \ldots, y_{n+1}; \eta_0, \eta_1, \ldots, \eta_{n+1})$$

with initial data $y_0 = \eta_0 = \delta_{n+1} > 0$ *is symmetric, i.e.,* $y_k = \eta_k$, $k = 0, 1, \ldots, n+1$.

Proof From the first and the last equations of the first column in (5.3.11), we get

$$\eta_n = \xi_n y_0 x_0^{-1} = x_n \eta_0 \xi_0^{-1} = y_n.$$

This equality and the first and the last equations from the second column in (5.3.11) give

$$\eta_{n+1} = \xi_n + y_0 = x_n + \eta_0 = y_{n+1}.$$

The second and the fourth equation of the second column in (5.3.11) and $\eta_n = y_n$ give

$$\xi_{n-1} + y_1 = x_0 + \eta_n = x_0 + y_n = \xi_0 + y_n = x_{n-1} + \eta_1 = \xi_{n-1} + \eta_1,$$

from which we conclude: $\eta_1 = y_1$.

This equality and the second and the fourth equation of the first column in (5.3.11) give

$$\eta_{n-1} = \xi_{n-1} y_1 x_1^{-1} = x_{n-1} \eta_1 \xi_1^{-1} = y_{n-1}.$$

This equality, the third equation of the second column in (5.3.11) and non-written equation $\xi_1 + y_{n-1} = x_{n-2} + \eta_2$ of the second column give:

$$\xi_{n-2} + y_2 = x_1 + \eta_{n-1} = x_1 + y_{n-1} = \xi_1 + y_{n-1} = x_{n-2} + \eta_2 = \xi_{n-2} + \eta_2,$$

and, hence, $\eta_2 = y_2$.

This equality and the third equation of the first column in (5.3.11) and non-written equation $x_{n-2}\eta_2 = \xi_2 y_{n-2}$ of this column give

$$\eta_{n-2} = \xi_{n-2} y_2 x_2^{-1} = x_{n-2} \eta_2 \xi_2^{-1} = y_{n-2}.$$

Repeating the last two steps, we get equalities $\eta_3 = y_3$, $\eta_{n-3} = y_{n-3}$; $\eta_4 = y_4, \ldots$. As result, we prove that $\eta_k = y_k$ for all $k = 0, 1, \ldots, n+1$. □

By using this lemma, we can instead of the system (5.3.11) consider the more simple system $\forall n \in \mathbb{N}$, we have

$$
\begin{aligned}
x_0 y_n &= x_n y_0, \\
x_1 y_{n-1} &= x_{n-1} y_1, \\
x_2 y_{n-2} &= x_{n-2} y_2, \\
&\dots \\
x_{[n/2]} y_{[n/2]+1} &= x_{[n/2]+1} y_{[n/2]}, && n \text{ is odd}, \\
x_{[n/2]-1} y_{[n/2]+1} &= x_{[n/2]+1} y_{[n/2]-1}, && n \text{ is even}.
\end{aligned}
\tag{5.3.15}
$$

$$
\begin{aligned}
x_n + y_0 &= y_{n+1}, \\
x_{n-1} + y_1 &= x_0 + y_n, \\
x_{n-2} + y_2 &= x_1 + y_{n-1}, \\
&\dots \\
x_{[n/2]+1} + y_{[n/2]} &= x_{[n/2]-1} + y_{[n/2]+2}, && n \text{ is odd}, \\
x_{[n/2]} + y_{[n/2]} &= x_{[n/2]-1} + y_{[n/2]+1} && n \text{ is even}.
\end{aligned}
$$

(Here, the square brackets in (5.3.15) denote the integer part of the number.) So, we have the following procedure for finding matrices of the form (5.3.5)–(5.3.8) that satisfy (5.3.4), where all $b_n = 0$, $n \in \mathbb{N}_0$.

Consider the system (5.3.15) of $n + 1$ equations, where $(x_0, x_1, \dots, x_n)$ are positive coefficients and $(y_0, y_1, \dots, y_{n+1})$ are real unknowns.

For $n = 1$ we put $x_0 = \delta_1$, $x_1 = \delta_0 + \delta_1$ ($\delta_0, \delta_1 > 0$—are arbitrary initial data) and find its solution (5.3.14) with an arbitrary $y_0 = \delta_2 > 0$.

Consider the system (5.3.15) for $n = 2$ with coefficients (x_0, x_1, x_2) equal to the previous solution (y_0, y_1, y_2) and find the solution (y_0, y_1, y_2, y_3), where $y_0 = \delta_3 > 0$ is an arbitrary number.

Assume that $y_k > 0$, $k = 0, 1, 2, 3$, and consider (5.3.15) for $n = 3$ with coefficients (x_0, x_1, x_2, x_3) equal to (y_0, y_1, y_2, y_3) and initial data $y_0 = \delta_4 > 0$ for a new solution. Continue such procedure, assuming on each step that the solution is positive (such positiveness depends on choice of $\delta_0, \delta_1, \dots, > 0$). As result, we find $(y_0, y_1, \dots, y_{n+1})$ for every $n \in \mathbb{N}_0$.

By means of formulas (5.3.10) (where $\xi_k = x_k$ and $\eta_k = y_k$), we calculate elements of matrices (5.3.5)–(5.3.8). For these matrices and $b_n = 0$, $n \in \mathbb{N}_0$ the matrix J of the type (5.1.36) is formally normal. If all y_k in the found solution $(y_0, y_1, \dots, y_{n+1})$ of (5.3.15) are uniformly bounded with respect to $n \in \mathbb{N}_0$, then the operator A generated by J is bounded and normal.

A similar situation occurs in the more general case, that is, when b_n are not zero. Now it is necessary to find b_n, $n \in \mathbb{N}_0$, step by step, starting by $b_0 \in \mathbb{R}$ and then we fine $b_{n+1} = b_{n+1}^*$ with $b_n = b_n^*$ using the third equality of (5.3.4).

Example 5.3.2 We will use the procedure described above to find the solution of the system (5.3.15) in the case

$$\delta_n = q^n, \quad q > 0, \quad n \in \mathbb{N}_0. \tag{5.3.16}$$

It is easy to calculate that this solution at the n-th step looks like

$$\begin{aligned} y_0 &= q^{n+1}, \\ y_1 &= q^n + q^{n+1}, \\ y_2 &= q^{n-1} + q^n + q^{n+1}, \ldots, \\ y_{n+1} &= 1 + q + q^2 + \ldots + q^{n+1}. \end{aligned} \tag{5.3.17}$$

So, for the case $0 < q < 1$ the corresponding to the matrix J operator A is bounded. In this case A is a normal operator. If $q \geq 1$, then a generated by J operator is unbounded and formally normal.

In the case of (5.3.16), (5.3.17) it is not difficult to find self-adjoint matrices b_n, $n \in \mathbb{N}_0$, for which the third equation in (5.3.4) is satisfied . Of course, it is possible to obtain matrices corresponding to bounded normal operators of the type (5.1.35) (i.e., (5.1.36)) in more general situations as well.

Example 5.3.3 Let us give an example of the Jacobi type block matrix, corresponding to some complex moment problem, in which $b_n \not\equiv 0$. It is not difficult to verify that the matrix

$$J = \left(\begin{array}{c|cc|ccc|cccc|ccccc}
0 & 0 & \sqrt{3} & & & & & & & & & & & & \\
\sqrt{3} & 0 & 0 & 0 & \sqrt{2} & 0 & & & & & & & & & \\
0 & 2 & 0 & 0 & 0 & 1 & & & & & & 0 & & & \\
 & 1 & 0 & 0 & 0 & 0 & 0 & \sqrt{2} & 0 & 0 & & & & & \\
 & 0 & \sqrt{2} & \sqrt{2} & 0 & 0 & 0 & 0 & 1 & 0 & & & & & \\
 & 0 & 0 & 0 & \sqrt{2} & 0 & 0 & 0 & 0 & 1 & & & & & \\
 & & & 1 & 0 & 0 & 0 & 0 & 0 & 0 & 0 & \sqrt{2} & 0 & 0 & 0 \\
 & & & 0 & 1 & 0 & \sqrt{2} & 0 & 0 & 0 & 0 & 0 & 1 & 0 & 0 \\
 & 0 & & 0 & 0 & \sqrt{2} & 0 & 1 & 0 & 0 & 0 & 0 & 0 & 1 & 0 \\
 & & & 0 & 0 & 0 & 0 & 0 & \sqrt{2} & 0 & 0 & 0 & 0 & 0 & 1 \\
 & & & & \ddots & & & & \ddots & & & & & \ddots &
\end{array}\right)$$

corresponds to a bounded and normal operator, that is, it satisfies the identity $JJ^* = J^*J$ on finite vectors.

Additionally, we note that such a matrix is related to the Chebyshev polynomials on the Steiner domain on the complex plane. Recall that the Steiner domain is a closed set on the complex plane bounded by a curve of the hypocycloid, namely,

$$z^2\bar{z}^2 - 4z^3 - 4\bar{z}^3 + 18z\bar{z} - 27 = 0.$$

And Chebyshev polynomials of the first kind on the Steiner domain have the form

$$T_{0;0}(z) = 1, \quad T_{1;0}(z) = z, \ T_{0;1}(z) = \bar{z},$$

$$T_{2;0}(z) = z^2 - 2\bar{z}, \ T_{1;1}(z) = z\bar{z} - 3, \ T_{0;2}(z) = \bar{z}^2 - 2z, \ldots,$$

which are orthogonal on the Steiner domain with the weight function

$$h(z) = \frac{1}{\sqrt{-(z^2\bar{z}^2 - 4z^3 - 4\bar{z}^3 + 18z\bar{z} - 27)}}.$$

The Chebyshev polynomials are related to the polynomials discussed above by relations

$$P_{n;\alpha}(z) = \begin{cases} \frac{1}{\sqrt{6}} T_{n-\alpha,\alpha}(z), \ n \geq 2, \ \alpha = 1, 2, \ldots, n-1, \\ \frac{1}{\sqrt{3}} T_{n-\alpha,\alpha}(z), \ n \geq 2, \ \alpha = 0, n; \end{cases}$$

and in particular $P_{0;0}(z) = 1, \ P_{1;0}(z) = \frac{1}{\sqrt{3}} z, \ P_{1;1}(z) = \frac{1}{\sqrt{3}} \bar{z}$.

Remark 5.3.4 The study of the detailed internal structure of the Jacobi matrix of the block type of a normal operator, as well as for a unitary operator (see the next chapter), and objects similar to the Verblunsky coefficients, Schur parameters, corresponding to the spectral measure of a normal operator given by a normal matrix, remain an important and unsolved problem until today.

5.4 The Solution of the Complex Moment Problem

The section is devoted to the solution of the complex moment problem in the power form. By the complex moment problem in power form we will understand the problem in finding conditions for the given sequence $(s_{m,n})$, $m, n \in \mathbb{N}_0 = \{0, 1, 2, \ldots\}$ of complex numbers, which corresponds to the Borel measure $d\rho(z)$ on the complex plane $\mathbb{C}$, so that

$$s_{m,n} = \int_{\mathbb{C}} z^m \bar{z}^n \, d\rho(z), \ m, n \in \mathbb{N}_0. \tag{5.4.1}$$

The problem is solved under the condition that the set $\{z^m \bar{z}^n\}$, $m, n \in \mathbb{N}_0$ is dense in $L_2(\mathbb{C}, d\rho(z))$. There are many known works devoted to the representation of (5.4.1) in various versions, i.e., $(m, n) \in \mathbb{Z} \times \mathbb{Z}$, $(m, n) \in \mathbb{N}_0 \times \mathbb{Z}$, $(m+n) \in \mathbb{N}_0$, and others, and a measure $d\rho(z)$ is concentrated on the whole plane $\mathbb{C}$, on some half-plane in $\mathbb{C}$, on a parallelogram, on the circle of radius R in $\mathbb{C}$ and many others.

The section considers the case $m, n \in \mathbb{N}_0$ and $d\rho(z)$ is defined without restrictions on the plane $\mathbb{C}$. The solution approach is also based on the generalization of the method derived from the works of Krein M. G. By using the given sequence $(s_{m,n})$, $m, n \in \mathbb{N}_0$, we will consider the (quasi)scalar product

$$(f, g)_S = \sum_{j,k,m,n=0}^{\infty} f_{j,k} \bar{g}_{m,n} s_{j+n,k+m}, \tag{5.4.2}$$

on finite sequences $f = (f_{j,k})_{j,k=0}^{\infty}$ and $g = (g_{j,k})_{j,k=0}^{\infty}$ under the condition $(f, f)_S \geq 0$ for an arbitrary finite sequence $f = (f_{j,k})_{j,k=0}^{\infty}$. In the Hilbert space generated by the scalar product $(\cdot, \cdot)_S$, we consider the operator N acting on finite vectors as follows:

$$(Nf)_{j,k} = f_{j,k-1}, \quad (\bar{N} f)_{j,k} = f_{j-1,k}, \quad j, k \in \mathbb{N}_0, \tag{5.4.3}$$

where $\bar{N}$ is an operator formally adjoint to N, and we put $f_{j,-1} = f_{-1,k} = 0$.

It is not difficult to verify that the operator N is formally normal. Now, the theory of the generalized eigenvectors expansion is applied to this operator (more precisely, to its normal expansion). In this way, the generalized eigenvectors $P(\lambda)$ corresponding according to (5.4.3) to the generalized eigenvalues of $\lambda \in \mathbb{C}$ have the form of linear combinations $(\lambda^m \bar{\lambda}^n)$, $m, n \in \mathbb{N}_0$.

Therefore, the corresponding Parseval equality, using the scalar product (5.4.2) for arbitrary f and g in terms of "Fourier coefficients", leads directly to the representation (5.4.1).

The quasi-analytic criterion of self-adjointness of an operators plays the main role by solving the question about the uniqueness of the measure $d\rho(z)$, which gives the representation (5.4.1).

The main result of the section is contained in the theorem.

Theorem 5.4.1 *If a given two-index sequence of complex numbers* $(s_{m,n})$, $m, n \in \mathbb{N}_0$ *has the representation* (5.4.1), *then*

$$\sum_{j,k,m,n=0}^{\infty} f_{j,k} \bar{f}_{m,n} s_{j+n,k+m} \geq 0, \tag{5.4.4}$$

for an arbitrary finite sequence of complex numbers $(f_{j,k})_{j,k=0}^{\infty}$, $f_{j,k} \in \mathbb{C}$.

The representation (5.4.1) *exists and is unique if this sequence* $(s_{m,n})$, $m, n \in \mathbb{N}_0$ *is positive definite, that is, it satisfies the condition (5.4.4), and*

$$\sum_{p=1}^{\infty} \frac{1}{\sqrt[2p]{s_{2p,2p}}} = \infty. \tag{5.4.5}$$

Note that the condition (5.4.4) is obviously necessary for the representation (5.4.1). The conditions (5.4.4) and (5.4.5) lead to the representation (5.4.1) in the theorem with a unique measure (we will explain further that the condition (5.4.5) guarantees the uniqueness).

Let us first recall some constructions (see Chap. 2, Sect. 2.7 and 2.1). Let the normal operator N defined on $\mathfrak{D}(N)$ be given in the separable Hilbert space $\mathcal{H}$; N^* is its adjoint one, $\mathfrak{D}(N^*) = \mathfrak{D}(N)$. Consider the equipment of the space $\mathcal{H}$:

$$\mathcal{H}_- \supset \mathcal{H} \supset \mathcal{H}_+ \supset \mathcal{D}, \tag{5.4.6}$$

where $\mathcal{H}_+$ is a positive Hilbert space topologically and quasi-nuclear imbedded into $\mathcal{H}$ (topologically means densely and continuously; quasi-nuclear means that the imbedding operator is of Hilbert-Schmidt type); $\mathcal{H}_-$ is the dual to $\mathcal{H}_+$ with respect to space $\mathcal{H}$; $\mathcal{D}$ is a linear, topological space, topologically embedded into $\mathcal{H}_+$.

The operator N is called standardly connected with the chain (5.4.6) if $\mathcal{D} \subset \mathfrak{D}(N)$ and restrictions $N \restriction \mathcal{D}$, $N^* \restriction \mathcal{D}$ act continuously from $\mathcal{D}$ into $\mathcal{H}_+$. This case is under consideration.

Let us recall that a vector $\Omega \in \mathcal{D}$ is called a strong cyclic vector of operators N and N^* if for any $p, q \in \mathbb{N}$, we have

$$\Omega \in \mathfrak{D}(N^p) \cap \mathfrak{D}((N^*)^q), \quad N^p (N^*)^q \Omega \in \mathcal{D}$$

and the set of all these vectors $N^p (N^*)^q \Omega$, as $p, q = 0, 1, 2, \ldots$, is total in the space $\mathcal{H}_+$ (and, hence, also in $\mathcal{H}$).

Assuming that the strong cyclic vector exists, we formulate some short variant of the projection spectral theorem (see Chap. 2, Sect. 2.11).

Theorem 5.4.2 *For the normal operator N with a strong cyclic vector in the separable Hilbert space $\mathcal{H}$, there exists a non-negative finite Borel measure $d\rho(z)$ with support on the complex plane such that for ρ-almost every $\lambda \in \mathbb{C}$, there exists a generalized joint eigenvector $\xi_\lambda \in \mathcal{H}_-$, i.e.,*

$$(\xi_\lambda, N^* f)_{\mathcal{H}} = \lambda (\xi_\lambda, f)_{\mathcal{H}}, \quad (\xi_\lambda, N f)_{\mathcal{H}} = \bar{\lambda} (\xi_\lambda, f)_{\mathcal{H}}, \quad \lambda \in \mathbb{C}, \ f \in \mathcal{D}, \ \xi_\lambda \neq 0. \tag{5.4.7}$$

The corresponding Fourier transform F is given by the rule

$$\mathcal{H} \supset \mathcal{H}_+ \ni f \mapsto (Ff)(\lambda) = \hat{f}(\lambda) = (f, \xi_\lambda)_{\mathcal{H}} \in L_2(\mathbb{C}, d\rho(\lambda)), \tag{5.4.8}$$

is an isometry operator (after the extension by continuity) acting from the space $\mathcal{H}$ into $L_2(\mathbb{C}, d\rho(\lambda))$. The image of the operator N (N^*), by the transformation F, is the multiplication by λ ($\bar{\lambda}$) operator in $L_2(\mathbb{C}, d\rho(\lambda))$.

We remark that the notation $(\cdot,\cdot)_{\mathcal{H}}$ denotes the same as $\langle\cdot,\cdot\rangle_{\mathcal{H}}$ (see Chap. 2, Sect. 2.7). We also recall (see Chap. 2, Sect. 2.6) that for the operator A defined on $\mathfrak{D}(A)$ in $\mathcal{H}$, the vector $f \in \bigcap_{n=0}^{\infty} \mathfrak{D}(A^n)$ is called quasi-analytic if the class $C\{m_n\}$, where in this case $m_n = \|A^n f\|_{\mathcal{H}}$, is quasi-analytic. The quasi-analyticity of the vector f for the Hermitian operator is equivalent to the condition

$$\sum_{n=1}^{\infty} \frac{1}{\sqrt[n]{\|A^n f\|_{\mathcal{H}}}} = \infty. \tag{5.4.9}$$

The quasi-analyticity is used in the criterion of self-adjointness and commutativity (see Chap. 2, Sect. 2.6).

Proof of Theorem 5.4.2 The necessity of the condition (5.4.4) is obvious. Indeed, if the sequence $(s_{m,n})$, $m, n \in \mathbb{N}_0$ has the representation (5.4.1), then for an arbitrary finite sequence $f = (f_{m,n})_{m,n=0}^{\infty}$ of complex numbers $f_{m,n} \in \mathbb{C}$, we have

$$\sum_{j,k,m,n=0}^{\infty} f_{j,k}\bar{f}_{m,n}s_{j+n,k+m} = \int_{\mathbb{C}} \left| \sum_{m,n=0}^{\infty} f_{m,n}z^m\bar{z}^n \right|^2 d\rho(z) \geq 0.$$

Let us denote by l the linear space of sequences $f = (f_{m,n})_{m,n=0}^{\infty}$, $f_{m,n} \in \mathbb{C}$ and by l_{fin} its linear subspace consisting of finite sequences $f = (f_{m,n})_{m,n=0}^{\infty}$, i.e., the sequences such that $f_{m,n} \neq 0$ for only finite numbers n and m. Let $\delta_{m,n}$, $m, n \in \mathbb{N}_0$ be the δ-sequence such that each $f \in l_{\text{fin}}$ has representation $f = \sum_{n,m=0}^{\infty} f_{m,n}\delta_{m,n}$.

Let us consider a linear operators on l_{fin}, namely,

$$J : (Jf)_{j,k} = f_{j,k-1}, \quad J^+ : (J^+f)_{j,k} = f_{j-1,k}, \quad j, k \in \mathbb{N}_0, \tag{5.4.10}$$

where always $f_{j,-1} = f_{-1,k} \equiv 0$. The operators J and J^+ are the "creation" type. For the δ-sequence we get

$$J\delta_{j,k} = \delta_{j,k+1}, \quad J^+\delta_{j,k} = \delta_{j+1,k}. \tag{5.4.11}$$

The operator J is formally adjoint to J^+ with respect to the (quasi)scalar product

$$(f, g)_S = \sum_{j,k,m,n=0}^{\infty} f_{j,k}\bar{g}_{m,n}s_{j+n,k+m}, \quad f, g \in l_{\text{fin}}. \tag{5.4.12}$$

Indeed,

$$\begin{aligned}(Jf, g)_S &= \sum_{j,k,m,n=0}^{\infty} (Jf)_{j,k}\bar{g}_{m,n}s_{j+n,k+m}\\ &= \sum_{j,k,m,n=0}^{\infty} f_{j,k-1}\bar{g}_{m,n}s_{j+n,k+m}\\ &= \sum_{j,k,m,n=0}^{\infty} f_{j,k}\bar{g}_{m,n}s_{j+n,k+m+1}\\ &= \sum_{j,k,m,n=0}^{\infty} f_{j,k}\bar{g}_{m-1,n}s_{j+n,k+m}\\ &= \sum_{j,k,m,n=0}^{\infty} f_{j,k}\overline{(J^+g)}_{m,n}s_{j+n,k+m}\\ &= (f, J^+g)_S.\end{aligned}$$

The operator J commute with J^+ on l_{fin}, namely,

$$(J^+Jf)_{j,k} = f_{j-1,k-1} = (JJ^+f)_{j,k}.$$

Hence the operator J is a formally normal and its adjoint is J^+.

Let S be the Hilbert space obtained as the completion of the factor space

$$\dot{l}_{\text{fin}} := l_{\text{fin}}/\{h \in l_{\text{fin}} \mid (h, h)_S = 0\}.$$

The element f of the space S is a representative of the class $\dot{f}$ of equivalent elements from $\dot{l}_{\text{fin}}$. Hence operators $\dot{J}$ and $\dot{J}^+$ are correctly defined in S. These facts in the case of self-adjoint operators are described in detail in [21], Chapter 8, §1, SubSect. 4, and [23], Chapter 5, §5, SubSect. 2. Therefore,

$$\dot{J}\dot{f} = (Jf)^{\cdot},\ f \in \mathfrak{D}(\dot{J}) = \dot{l}_{\text{fin}};\quad \dot{J}^+\dot{f} = (J^+f)^{\cdot},\ f \in \mathfrak{D}(\dot{J}^*) = \dot{l}_{\text{fin}}.$$

Let us denote, for the next considerations, by N and N^+ the closure "$\sim$" of $\dot{J}$ and $\dot{J}^+$ in S.

In the next step we use Theorem 5.4.2. For simplicity, we suppose that the given sequence $(s_{m,n})$ is non-degenerate, i.e., if $(f, f)_S = 0$ for $f \in l_{\text{fin}}$, then $f = 0$ and, hence, $\dot{f} = f$ and $\tilde{\dot{J}} = N$. Studies in the general case are more complicated, see [21], Chapter 8, §1, SubSect. 4, and [23], Chapter 5, §5, Subsections 1–3.

We assume also for a moment that the operator N is normal one. Here, it will be proved that N is normal under the condition (5.4.5). In general, the conditions for the existence of normal extensions of a formally normal operator, especially in a connection with the complex moment problem, were considered in [272] (see also [83, 93]).

Let us construct the rigging of spaces

$$(l_2(p))_{-,S} \supset S \supset l_2(p) \supset l_{\text{fin}}, \tag{5.4.13}$$

where $l_2(p)$ is weighted l_2-space with a weight $p = (p_{m,n})_{m,n=0}^{\infty}$, $p_{m,n} \geq 1$. The norm in $l_2(p)$ is given by the expression

$$\|f\|^2_{l_2(p)} = \sum_{m,n=0}^{\infty} |f_{m,n}|^2 p_{m,n};$$

where $(l_2(p))_{-,S} = \mathcal{H}_-$ is negative space with respect to the positive space $l_2(p) = \mathcal{H}_+$ and the zero space $S = \mathcal{H}$.

Lemma 5.4.3 *If the sequence $(p_{m,n})$, $m, n \in \mathbb{N}_0$ decreases enough fast, then the imbedding $l_2(p) \hookrightarrow S$ is quasi-nuclear.*

Proof The inequality (5.4.4) means that a multimatrix $(K_{j,k;m,n})_{j,k,m,n=0}^{\infty}$, where $K_{j,k;m,n} = s_{j+n,k+m}$, is non-negative definite and, therefore, for $j, k, m, n \in \mathbb{N}_0$, we have

$$|s_{j+n,k+m}|^2 = |K_{j,k;m,n}|^2 \leq K_{j,k;j,k} K_{m,n;m,n} = s_{j+k,j+k} s_{m+n,m+n}. \tag{5.4.14}$$

Let the weight $q = (q_{j,k})_{j,k=0}^{\infty}$, $q_{j,k} \geq 1$, be such that $\sum_{j,k=0}^{\infty} s_{j+k,j+k} q_{j,k}^{-1} < \infty$. Then from (5.4.12) and (5.4.14) it follows that

$$\|f\|^2_S = \sum_{j,k,m,n=0}^{\infty} f_{j,k} \bar{f}_{m,n} s_{j+n,k+m} \leq \left(\sum_{j,k=0}^{\infty} \frac{s_{j+k,j+k}}{q_{j,k}} \right) \|f\|^2_{l_2(q)}, \quad f \in l_{\text{fin}}.$$

Therefore, the imbedding $l_2(q) \hookrightarrow S$ is topological. And if $\sum_{j,k=0}^{\infty} q_{j,k} p_{j,k}^{-1} < \infty$, then $l_2(p) \hookrightarrow l_2(q)$ is quasi-nuclear. The composition $l_2(p) \hookrightarrow S$ of the quasi-nuclear and topological imbedding is also quasi-nuclear one. □

In the next step we use the rigging (5.4.13) to construct generalized eigenvectors. The inner structure of the space $(l_2(p))_{-,S}$ is complicated, because of the complicated structure of S. This is a reason to introduce a new auxiliary rigging

$$l = (l_{\text{fin}})' \supset (l_2(p^{-1})) \supset l_2 \supset l_2(p) \supset l_{\text{fin}}, \tag{5.4.15}$$

where $l_2(p^{-1})$, $p^{-1} = (p_{m,n}^{-1})_{m,n=0}^{\infty}$ is the negative space with respect to the positive space $l_2(p)$ and the zero space l_2. Chains (5.4.13) and (5.4.15) have the same positive space $l_2(p)$.

We use the Lemma 2.7.5 that establishes an isomorphism between spaces $(l_2(p))_{-,S}$ and $l_2(p^{-1})$.

Instead of riggings (2.7.35), we consider (5.4.13) and (5.4.15). Let $\xi_\lambda \in (l_2(p))_{-,S}$ be a generalized eigenvector of the operator N in terms of the chain (5.4.13). Thus, in this case due to Theorem 5.4.2, we have

$$(\xi_\lambda, N^* f)_S = \lambda(\xi_\lambda, f)_S, \quad (\xi_\lambda, Nf)_S = \bar{\lambda}(\xi_\lambda, f)_S, \ \lambda \in \mathbb{C}, \ f \in l_{\text{fin}}. \qquad (5.4.16)$$

Denote

$$P(\lambda) = U\xi_\lambda \in l_2(p^{-1}) \subset l, \ P(\lambda) = (P_{m,n}(\lambda))_{m,n=0}^{\infty}, \ P_{m,n}(\lambda) \in \mathbb{C}.$$

By using (2.7.36) we can rewrite (5.4.16) in the form

$$\begin{aligned} &(P(\lambda), N^* f)_{l_2} = \lambda(P(\lambda), f)_{l_2}, \\ &(P(\lambda), Nf)_{l_2} = \bar{\lambda}(P(\lambda), f)_{l_2}, \ \lambda \in \mathbb{C}^1, \ f \in l_{\text{fin}}. \end{aligned} \qquad (5.4.17)$$

The correspondence Fourier transform has the form

$$S \supset l_{\text{fin}} \ni f \to (Ff)(\lambda) = \hat{f}(\lambda) = (f, P(\lambda))_{l_2} \in L_2(\mathbb{C}, d\rho(\lambda)). \qquad (5.4.18)$$

Let us calculate $P(\lambda)$. The operator N^* is defined by the second expression in (5.4.10) and, hence, (5.4.17) gives

$$\begin{aligned} \sum_{m,n=0}^{\infty} \lambda P_{m,n}(\lambda) \bar{f}_{m,n} &= \lambda(P(\lambda), f)_{l_2} = (P(\lambda), N^* f)_{l_2} \\ &= (P(\lambda), J^+ f)_{l_2} = \sum_{m,n=0}^{\infty} P_{m+1,n}(\lambda) \bar{f}_{m,n}, \quad \forall f \in l_{\text{fin}}. \end{aligned} \qquad (5.4.19)$$

Similarly, using (5.4.10) and (5.4.17), we have

$$\begin{aligned} \sum_{m,n=0}^{\infty} \bar{\lambda} P_{m,n}(\lambda) \bar{f}_{m,n} &= \bar{\lambda}(P(\lambda), f)_{l_2} = (P(\lambda), Nf)_{l_2} \\ &= (P(\lambda), Jf)_{l_2} = \sum_{m,n=0}^{\infty} P_{m,n+1}(\lambda) \bar{f}_{m,n}, \quad \forall f \in l_{\text{fin}}. \end{aligned} \qquad (5.4.20)$$

Hence, we have

$$\lambda P_{m,n}(\lambda) = P_{m+1,n}(\lambda), \ \bar{\lambda} P_{m,n}(\lambda) = P_{m,n+1}(\lambda), \ m, n \in \mathbb{N}_0.$$

Without loss of generality, we can take $P_{0,0}(\lambda) = 1$, $\lambda \in \mathbb{C}$. Hence, last two equalities give

$$P_{m,n}(\lambda) = \lambda^m \bar{\lambda}^n, \quad m, n \in \mathbb{N}_0. \tag{5.4.21}$$

Thus, the Fourier transform (5.4.18) has the form

$$S \supset l_{\text{fin}} \ni f \to (Ff)(\lambda) = \hat{f}(\lambda) = \sum_{m,n=0}^{\infty} f_{m,n} \lambda^m \bar{\lambda}^n \in L_2(\mathbb{C}, d\rho(\lambda)) \tag{5.4.22}$$

and

$$(f, g)_S = \int_{\mathbb{C}} \hat{f}(\lambda) \overline{\hat{g}(\lambda)} \, d\rho(\lambda), \quad f, g \in l_{\text{fin}}. \tag{5.4.23}$$

To construct the Fourier transform (5.4.18) and verify formulas (5.4.19)–(5.4.23) it is necessary to check that, for our operators N and N^*, the vector $\Omega = \delta_{0,0} \in l_{\text{fin}}$ is strong cyclic in the sense of the chain (5.4.13). But it is evidently true, since by (5.4.11) we have

$$N^p (N^*)^q \Omega = J^p (J^+)^q \delta_{0,0} = \delta_{q,p}.$$

The Parseval equality (5.4.23) immediately leads to the representation (5.4.1): according to (5.4.21), (5.4.22) $\hat{\delta}_{m,n} = \lambda^m \bar{\lambda}^n$ and $\hat{\delta}_{0,0} = 1$; by (5.4.12), we get

$$\begin{aligned} s_{m,n} = (\delta_{m,n}, \delta_{0,0})_S &= (\hat{\delta}_{m,n}, \hat{\delta}_{0,0})_{L_2(\mathbb{C}, d\rho(\lambda))} \\ &= \int_{\mathbb{C}} \lambda^m \bar{\lambda}^n \, d\rho(\lambda), \quad m, n \in \mathbb{N}_0, \ \lambda \in \mathbb{C}. \end{aligned}$$

The uniqueness of representation (5.4.1) follows from the fact that the operator N is normal (not only just a formally normal). Thus, for the finish of the proof of Theorem 5.4.1, it is necessary only to check that the condition (5.4.5) provides the normality of N. For this reason we introduce two closed Hermitian operators defined on, invariant under their action, linear set

$$\mathcal{D} = l_{\text{fin}} = \text{span}\{\delta_{m,n} \mid m, n \in \mathbb{N}_0\}$$

by expressions

$$A_1 = \frac{1}{2}(N + N^*), \quad A_2 = \frac{1}{2i}(N - N^*).$$

For the normality of N it is sufficiently to show that operators A_1 and A_2 are self-adjoint both and commutative in the strong resolvent sense. But for this (see Chap. 2, Theorems 2.6.8 and 2.6.11) it must to show that the operator $A_1^2 + A_2^2$ is essentially self-adjoint, i.e., for example, has a total set $\mathcal{D}$ of quasi-analytic vectors.

Due to (5.4.10), the operator $\mathcal{A} := A_1^2 + A_2^2$ acts on $\delta_{m,n} \in \mathcal{D}$ as follows

$$\mathcal{A}\delta_{m,n} = (A_1^2 + A_2^2)\delta_{m,n} \equiv NN^+\delta_{m,n} = \delta_{m+1,n+1},$$

and for $p \geq 1$, we have

$$\mathcal{A}^p\delta_{m,n} = \delta_{m+p,n+p}.$$

According to (5.4.12), the space S has the norm $\|f\|_S = \sqrt{(f, f)_S}$. Hence, $\forall\delta_{m,n} \in \mathcal{D}$ we get

$$\|\mathcal{A}^p\delta_{m,n}\|_S^2 = \|\delta_{m+p,n+p}\|_S^2 = s_{m+n+2p,m+n+2p}.$$

Since

$$\sum_{p=1}^{\infty} \frac{1}{\sqrt[p]{\|\mathcal{A}^p\delta_{m,n}\|}} = \sum_{p=1}^{\infty} \frac{1}{\sqrt[2p]{s_{m+n+2p,m+n+2p}}},$$

then we conclude that the quasi-analyticity of the class $C\{\|\mathcal{A}^p\delta_{m,n}\|\}$ is equivalent to the quasi-analyticity of the class $C\{\sqrt{s_{m+n+2p,m+n+2p}}\}$ and due to the quasi-analyticity properties it is equivalent to the quasi-analyticity of the class $C\{\sqrt{s_{2p,2p}}\}$.

In any case, the operator $\mathcal{A}$ must be positive regardless of whether it is essentially self-adjoint. Therefore, the theorem 5.4.1 is proved. □

Remark 5.4.4 If considered as a partial case: the real sequence $c_n := s_{n,n}$, $n \in \mathbb{N}_0$ instead of the complex $(s_{m,n})$ (that is, in which $s_{m,n} = 0$ for $m \neq n, m, n \in \mathbb{N}_0$), and, therefore, the corresponding one-dimensional real moment problem, then the well-known result can be obtained (see Chap. 3). The sequence c_n has the representation

$$c_n = \int_0^{+\infty} t^{2n}\, d\rho(t)$$

if and only if it is the Stiltjes sequence and it is unique if

$$\sum_{n=1}^{\infty} \frac{1}{\sqrt[2n]{c_{2n}}} = \infty,$$

provided the measure is supported on $[0, \infty)$. Here Theorem 2.6.8, Lemma 2.6.7, and (2.6.21) are used.

Remark 5.4.5 The condition (5.4.4) is only necessary in Theorem 5.4.1, i.e., for the representation (5.4.1). For the clearness we propose the simple contrexample.

Let A_1 and A_2 be two self-adjoint operators commuting on the linear set $\mathcal{D}$ dense in the Hilbert space $\mathcal{H}$, where $\mathcal{D}$ is invariant under the action of A_1 and A_2. Suppose A_1 and A_2 are essentially self-adjoint on $\mathcal{D}$, i.e., $A_1 = (A_1|\mathcal{D})^{\sim}$, $A_2 = (A_2|\mathcal{D})^{\sim}$, but A_1 and A_2 are not commute in the strong resolvent sense. The existence of such sort operators guarantees the Nelson example [225] (see also [48], Chapter 13, §9).

For the next consideration, we put $\dot{N} = (A_1+iA_2) \restriction \mathcal{D}$ and $\bar{N} = (A_1-iA_2) \restriction \mathcal{D}$. In this case, $\dot{N}$ is a formally normal operator that does not have a normal extension even in an arbitrarily wide Hilbert space [66, 253].

Hence, the sequence

$$s_{m,n} := (\dot{N}^m f_0, \dot{N}^n f_0)_{\mathcal{H}}, \quad f_0 \in \mathcal{D}$$

obviously satisfies the condition (5.4.4), but we do not have a representation (5.4.1). Indeed, for a finite sequence $f = (f_{j,k})_{j,k=0}^{\infty}$, we have

$$\begin{aligned}
&\sum_{j,k,m,n}^{\infty} f_{j,k}\bar{f}_{m,n}s_{j+n,k+m} \\
&\quad = \sum_{j,k,m,n}^{\infty} f_{j,k}\bar{f}_{m,n}(N^{j+n}f_0, N^{k+m}f_0)_{\mathcal{H}} \\
&\quad = \left(\sum_{j,k}^{\infty} f_{j,k}N^j(N^*)^k f_0, \sum_{m,n}^{\infty} f_{m,n}N^m(N^*)^n f_0\right)_{\mathcal{H}} \\
&\quad = \|\sum_{m,n}^{\infty} f_{m,n}N^m(N^*)^n f_0\|_{\mathcal{H}}^2 \geq 0.
\end{aligned}$$

Finally, we recall that the classical problem of moments is closely related to the theory of ordinary Jacobi matrices: see Chap. 3, Sect. 3.4, Theorem 3.4.3 about the Weyl-Hamburger circle. In general, such a connection is also possible for a complex moment problem, but the corresponding theory of polynomials of the first kine $P_j(z)$ and the second kind Q_j is not developed here because the internal description of block Jacobi matrices related to such a problem is not investigated.

5.5 The Weyl Function and Polynomials of the Second Kind in the Complex Moment Problem

A Weyl-type function can be introduced for the case of a complex moment problem in power form, similarly to the way such a function is introduced in the classical moment problem (see Chap. 3). Let us set for the resolvent $R_z = (N - z\mathbb{I})^{-1}$ of a bounded normal operator N, the functino

$$M(z) := (R_z\delta_0, R_z\delta_0)_{\mathbf{l}_2} = (R_z^* R_z\delta_0, \delta_0)_{\mathbf{l}_2}, \quad z \in \mathbb{C} \setminus S, \tag{5.5.1}$$

where S is a spectrum of the operator N, δ_0 is corresponding cyclic vector. The Weyl type function introduced in this way preserves its basic traditional property, which is described by the theorem.

Theorem 5.5.1 *Let J be a fixed block Jacobi type matrix that generates a bounded normal operator N in $\mathbf{l}_2$. Let R_z and S be its resolvent and spectrum. Then the function $M(z) := (R_z\delta_0, R_z\delta_0)_{\mathbf{l}_2}$ for $z \in \mathbb{C} \setminus S$ determines uniquely the spectral measure $d\rho(z)$, $z \in S$ of this operator.*

Proof According to the spectral theorem, the function $M(z)$ has a representation

$$M(z) := (R_z\delta_0, R_z\delta_0)_{\mathbf{l}_2} = (R_z^* R_z\delta_0, \delta_0)_{\mathbf{l}_2} = \int_S \frac{d\rho(\xi)}{(\xi - z)(\bar{\xi} - \bar{z})}, \quad z \in \mathbb{C} \setminus S, \tag{5.5.2}$$

Since the spectrum is located in the circle $|z| \leq \|N\|$, then outside $|z| \geq \|N\|$ function (5.5.2) can be written in the form

$$\begin{aligned} M(z) &= \int_S \frac{d\rho(\xi)}{(\xi - z)(\bar{\xi} - \bar{z})} \\ &= \frac{1}{|z|^2} \int_S \frac{d\rho(\xi)}{(1 - \xi z^{-1})(1 - \bar{\xi}\bar{z}^{-1})} \\ &= \frac{1}{|z|^2} \int_S \sum_{m=0}^{\infty} \left(\frac{\xi}{z}\right)^m \sum_{n=0}^{\infty} \left(\frac{\bar{\xi}}{\bar{z}}\right)^n d\rho(\xi) \\ &= \frac{1}{|z|^2} \int_S \sum_{m,n=0}^{\infty} \frac{1}{z^m \bar{z}^n} \xi^m \bar{\xi}^n \, d\rho(\xi) \\ &= \frac{1}{|z|^2} \sum_{m,n=0}^{\infty} \frac{s_{m,n}}{z^m \bar{z}^n}, \end{aligned} \tag{5.5.3}$$

where $(s_{m,n})$, $m, n \in \mathbb{N}_0$ are moments of the measure $d\rho(z)$ (5.4.1). Form the representation (5.5.3) of the function $M(z)$ in the form of a Taylor series of variables $z^{-m}\bar{z}^{-n}$ under the condition $|z| \geq \|N\|$, it follows that coefficients $s_{m,n}$ of this series are restored uniquely by $M(z)$. But these coefficients are (complex) moments of the measure $d\rho(z)$, which allow the estimation of $|s_{m,n}| \leq \|N\|^{m+n}$, $m, n \in \mathbb{N}_0$. On the other hand, the measure $d\rho(z)$ is uniquely restored by moments $s_{m,n}$, $m, n \in \mathbb{N}_0$, and therefore, it is uniquely restored by to the function $M(z)$. □

Polynomials of the second kind corresponding to the complex moment problem can also be introduced similarly to the way as it is done in the classical moment problem (see Chap. 3). Take

$$Q_n(z) := \int_{\mathbb{C}} \frac{P_n(\xi) - P_n(z)}{\xi - z} \, d\rho(\xi), \tag{5.5.4}$$

where $P_n(z)$ are polynomials of the first kind, $z, \xi \in \mathbb{C} \setminus S$, $d\rho(\xi)$ is a measure with a compact support on the complex plane corresponding to bounded normal operator and its adjoint one generated by block Jacobi type matrices (see Sects. 5.1–5.3, and, in particular 5.4). For the polynomials introduced in (5.5.4), their main property are described by the theorem.

Theorem 5.5.2 *The sequence $Q(z) = (Q_n(z))_{n=0}^{\infty}$, $z \in \mathbb{C} \setminus S$ with $Q_n(z)$ given by the formula* (5.5.4), *is a solution of the system* (5.2.7), *namely,*

$$\begin{aligned} (JQ(z))_n &= a_{n-1}Q_{n-1}(z) + b_n Q_n(z) + c_n Q_{n+1}(z) = zQ_n(z), \\ (J^+Q(z))_n &= c_{n-1}^* Q_{n-1}(z) + b_n^* Q_n(z) + a_n^* Q_{n+1}(z) = \bar{z}Q_n(z), \\ & n \in \mathbb{N}_0, \quad Q_{-1}(z) := -1, \quad a_{-1} := 1. \end{aligned} \tag{5.5.5}$$

In particular, $Q_0(z) = 0$.

Proof Since $P_0 = P_{0;0} = 1$, then from (5.5.4), it follows immediately $Q_0 = Q_{0;0} = 0$.

Let us check that $Q_n(z) = (Q_{n;0}, Q_{n;1}, \ldots, Q_{n;n})$, $n \in \mathbb{N}$ satisfy also (5.5.5). Indeed,

$$\begin{aligned} (JQ(z))_n &= \int_{\mathbb{C}} \frac{(JP(\xi))_n - (JP(z))_n}{\xi - z} \, d\rho(\xi) \\ &= \int_{\mathbb{C}} \frac{\xi P_n(\xi) - zP_n(z)}{\xi - z} \, d\rho(\xi) \\ &= z \int_{\mathbb{C}} \frac{P_n(\xi) - P_n(z)}{\xi - z} \, d\rho(\xi) + \int_{\mathbb{C}} P_n(z) \, d\rho(\xi). \end{aligned} \tag{5.5.6}$$

The last integral in (5.5.6) is zero, so the first equality in (5.5.5) is proved. The second equality in (5.5.4) is proved similarly. □

Bibliographical Notes

The complex moment problem is a generalization of the (real one-dimensional) classical moment problem, which is described in works of Akhiezer N. I. [2, 3]. To obtain an integral representation, the method of generalized eigenvectors expansions for a normal operator (for a pairs of commutative self-adjoint operators) developed by Berezansky Yu. M. is used [17, 18], and the spectral theory of sets of self-adjoint commutative operators Samoilenko Yu. S. [250] is also used. Historically, the first information about the complex moment problem is by Kilpi Y. [157–160]. The main content of the chapter is presented in papers of Berezansky Y. M., Dudkin M. E. [34, 37–39] and also by Ivasiuk I. Ya. [141]. Papers of Stochel J. Szafraniec F. H. [269–272, 275] are directly related to investigations of the complex moment problem and some of its generalizations. The indeterminate case of the complex moment problem is related to the problem of describing the commutative of self-adjoint extensions of operators, which is in works of Bokhonov Yu. E. [57], Vaĭnerman L. Ĭ. [287], Višik M. I. [290], also Jørgensen P. E. T. [145]. During research, orthogonal polynomials on the complex plane arise. General information about orthogonal polynomials can be found in works of Geronimus Ya. L. [117], Szegö G. [278], Suetin P. K. [273, 274]. Constructed block matrices can generate operators with non-trivial deficiency indices, see Dyukarev Yu. M. [97]. Scales of Hilbert spaces are constructed for normal operators, as for self-adjoint operators in the book of Koshmanenko V. D. and Dudkin M. E. [171]. The Weyl function for the case of a normal operator is obtained as a generalization of the paper of Brasche J. F., Malamud M. M., Neidhardt H. [59], where the Weyl function in the self-adjoint case was considered; and also by Gesztesy F., Simon B. in [120]. About normal extensions of a formally normal operator are discussed by Biriuk G., Coddington E. [53, 65, 66], Dudkin M. E., Nizhnik L. P. [83, 93], Szafraniec F. H. [276], Schmüdgen K. [252, 254] and in particular other related discussions [255]. Papers of Bisgaard T. M., Devinatz A. [55, 56, 81] considered generalizations of the complex moment problem for the case of two-sided, and other cases. The question of the uniqueness of the representation is solved with the help of papers Carleman T. [64], Mandelbrojt S. [206], Nelson E. [225], Nussbaum A. E. [231, 232].

Chapter 6
Unitary Block Jacobi Type Matrices and the Trigonometric Moment Problem

This chapter considers yet another, in comparing with Chap. 5, generalization of the connection between the classical moment problem and the spectral theory of self-adjoint Jacobi matrices. Namely, an analogue of the Jacobi matrix is proposed for consideration, which related to the trigonometric moment problem, as well as the system of polynomials orthogonal with respect to some probability measure with a support on the unit circle. Such a matrix also has a block tri-diagonal structure and generates a unitary operator that acts in the l_2 type space. By using this connection, the existence of a one-to-one correspondence between probability measures on the unit circle and the block tri-diagonal Jacobi type unitary matrices is established.

Thus, the main question of the section is the following: in what way is it possible to generalize the construction given in the second and the fourth chapters, in particular, in studies of orthogonal polynomials to the case of the unit circle $\mathbb{T} \subset \mathbb{C}$? As Chaps. 3 and 5, this one ends with a statement of the corresponding (trigonometric) moment problem.

6.1 Introduction to the Unitary Block Jacobi Type Matrices

Let us dwell on the case of the trigonometric problem of moments. In other words, the construction of Chaps. 3, 4 and 5 applies to the case when the matrix J of the form (5.1.35), (5.1.36) is a unitary operator. The difference with Chap. 5 is that the spectrum of the corresponding operator generated by a matrix J is supported on the unit circle $\mathbb{T} \subset \mathbb{C}$ and, hence, the set of functions (5.1.4) is linearly dependent, because in such a case

$$z^{j+n}\bar{z}^{k+n} = z^j\bar{z}^k, \quad j, k \in \mathbb{N}_0, \quad \forall n \in \mathbb{N}_0, \quad z \in \mathbb{T}.$$

Y. M. Berezansky, M. E. Dudkin, *Jacobi Matrices and the Moment Problem*,
Operator Theory: Advances and Applications 294,
https://doi.org/10.1007/978-3-031-46387-7_6

Therefore, only those functions from (5.1.4) are selected for which $j \cdot k = 0$, i.e.,

$$\begin{aligned} &z^0\bar{z}^0 = 1; \\ &z^1\bar{z}^0 = z^1, \quad z^0\bar{z}^1 = \bar{z}^1; \\ &z^2\bar{z}^0 = z^2, \quad z^0\bar{z}^2 = \bar{z}^2; \qquad \dots \\ &z^n\bar{z}^0 = z^n, \quad z^0\bar{z}^n = \bar{z}^n; \qquad \dots, \quad z \in \mathbb{T}. \end{aligned} \tag{6.1.1}$$

In the case of classical Jacobi matrices $\bar{z} = z$ for $z \in \mathbb{R}$. We consider only infinite Jacobi-type matrices, therefore the support $d\rho(z)$ on $\mathbb{T}$ is not contained in a finite number of points, therefore, functions (6.1.1) are linearly independent in $L_2(\mathbb{T}, d\rho(z))$.

The global linear order of the sequence (6.1.1) is the same as in (5.1.5) Fig. 5.1. But now, each "diagonal" $n = 1, 2, \dots$ has only 2 non-zero elements (instead of $n + 1$ as before). In this case, the operator $\tilde{J}$ is considered in some subspace $\mathbf{l}_{2,u}$ of the space $\mathbf{l}_2$ given by the expression (5.1.9):

$$\mathbf{l}_{2,u} = \mathcal{H}_0 \oplus \mathcal{H}_1 \oplus \mathcal{H}_2 \oplus \cdots, \quad \text{where} \quad \mathcal{H}_0 = \mathbb{C}, \mathcal{H}_1 = \mathcal{H}_2 = \cdots = \mathbb{C}^2; \tag{6.1.2}$$

$$\mathbf{l}_{2,u} \subset \mathbf{l}_2.$$

The corresponding block matrix acts as an operator in the space (6.1.2) and its elements have the form

$$\begin{aligned} a_0 :& \quad \mathbb{C} \longrightarrow \mathbb{C}^2, \\ b_0 :& \quad \mathbb{C} \longrightarrow \mathbb{C}, \\ c_0 :& \quad \mathbb{C}^2 \longrightarrow \mathbb{C}, \\ a_n, b_n, c_n :& \quad \mathbb{C}^2 \longrightarrow \mathbb{C}^2, \qquad n \in \mathbb{N}, \end{aligned} \tag{6.1.3}$$

with conditions

$$\begin{gathered} a_0 = \underbrace{\begin{pmatrix} a_{0;0,0} \\ 0 \end{pmatrix}}_{1} \Big\} 2, \quad b_0 = \underbrace{\big(b_{0;0,0}\big)}_{1} \} 1, \quad c_0 = \underbrace{\big(c_{0;0,0}\ c_{0;0,1}\big)}_{2} \} 1; \\ a_n = \underbrace{\begin{pmatrix} a_{n;0,0} & a_{n;0,1} \\ 0 & 0 \end{pmatrix}}_{2} \Big\} 2, \quad c_n = \underbrace{\begin{pmatrix} 0 & 0 \\ c_{n;1,0} & c_{n;1,1} \end{pmatrix}}_{2} \Big\} 2; \\ a_{0;0,0}, c_{0;0,1}, a_{n;0,0}, c_{n;1,1} > 0, \quad n \in \mathbb{N}. \end{gathered} \tag{6.1.4}$$

Therefore, the general results concerning normal operators also hold true for the unitary operator $\tilde{J}$, acting in the space $\mathbf{l}_{2,u}$ (6.1.2) and obtained as a block Jacobi type matrix with elements from (6.1.3) and with condition (6.1.4). That is, the direct spectral problem leads to a Fourier transform of the type (5.2.33) acting from the space $\mathbf{l}_{2,u}$ into $L_2(\mathbb{T}, d\rho(z))$. The inverse spectral problem has a solution similar to that described in Chap. 5. At the same time, we used the orthogonalization of the sequence (6.1.1). This problem is, of course, closely related to the trigonometric moment problem.

6.2 Construction of the Tri-Diagonal Block Jacobi Type Matrices of a Unitary Operator

This section contains, discussed above, the direct and inverse problems for the block tri-diagonal Jacobi type matrix of the unitary operator associated with the trigonometric moment problem on the unit circle. The solution of the inverse problem is presented in Theorem 6.2.5. Namely, the unitary operator, given by its spectral measure on the unit circle, is represented in the form of a block tri-diagonal type Jacobi matrix acting in the orthonormal basis of polynomials on the unit circle. The direct problem is presented in Theorem 6.3.3. Namely, for a given tri-diagonal Jacobi type matrix, the spectral measure (with support on the unit circle) is restored in the sense of Fourier coefficients that obtained from the generalized eigenvectors expansion into series, respectively to corresponding unitary operator.

The direct and inverse problems together obviously establish a one-to-one correspondence between unitary operators and infinite (five-diagonal) block tri-diagonal Jacobi matrices.

Let us denote $\mathbb{T} = \{z \in \mathbb{C} \mid |z| = 1\} = \{e^{i\theta} \mid \theta \in [0, 2\pi)\}$ a unit circle on the complex plane $\mathbb{C}$. Let $d\rho(z) = d\rho(\theta)$ be a Borel probability measure on $\mathbb{T}$ and $L_2 = L_2(\mathbb{T}, d\rho(\theta))$ be the space of square integrable complex valued functions defined on $\mathbb{T}$. Suppose that the measure is not concentrated in a finite number of points and, therefore, all functions $[0, 2\pi) \ni \theta \longmapsto e^{li\theta}$, $l \in \mathbb{Z}$ are linear independent in L_2.

Let us consider the sequence of functions

$$1; \quad e^{i\theta}, e^{-i\theta}; \quad e^{2i\theta}, e^{-2i\theta}; \quad \ldots; \quad e^{ni\theta}, e^{-ni\theta}; \quad \ldots, \tag{6.2.1}$$

and start the orthogonalization according to the Gram-Schmidt procedure. As a result we get the orthonormal system of polynomials

$$
\begin{aligned}
P_0(\theta) &= 1;\\
P_{1;1}(\theta) &= k_{1;1}e^{i\theta} + \cdots,\\
P_{1;2}(\theta) &= k_{1;2}e^{-i\theta} + \cdots;\\
P_{2;1}(\theta) &= k_{2;1}e^{2i\theta} + \cdots,\\
P_{2;2}(\theta) &= k_{2;2}e^{-2i\theta} + \cdots;\\
&\cdots\\
P_{n;1}(\theta) &= k_{1;1}e^{ni\theta} + \cdots,\\
P_{n;2}(\theta) &= k_{1;2}e^{-ni\theta} + \cdots;\\
&\cdots
\end{aligned}
\tag{6.2.2}
$$

of variables $e^{i\theta}$ and $e^{-i\theta}$, where $k_{n;1} > 0$, $k_{n;2} > 0$ and "$+\cdots$" denotes terms after the first in the corresponding polynomial. Thus, $P_{n;1}$ is a linear combination of

$$
\{1, e^{i\theta}, e^{-i\theta}, \ldots, e^{(n-1)i\theta}, e^{-(n-1)i\theta}, e^{ni\theta}\}, \tag{6.2.3}
$$

and $P_{n;2}$ is a linear combination of

$$
\{1, e^{i\theta}, e^{-i\theta}, \ldots, e^{(n-1)i\theta}, e^{-(n-1)i\theta}, e^{ni\theta}, e^{-ni\theta}\}. \tag{6.2.4}
$$

For all $n \in \mathbb{N}$, we denote $\mathcal{P}_{n;1}$ and $\mathcal{P}_{n;2}$ spaces spanned by (6.2.3) and (6.2.4), respectively. Hence, $P_{n;1}(\theta)$ ($P_{n;2}(\theta)$) are orthogonal to $\mathcal{P}_{n-1;2}$ ($\mathcal{P}_{n;1}$). It is clear that $\forall n \in \mathbb{N}$, we have

$$
\begin{aligned}
&\mathcal{P}_0 \subset \mathcal{P}_{1;1} \subset \mathcal{P}_{1;2} \subset \cdots \subset \mathcal{P}_{n;1} \subset \mathcal{P}_{n;2} \subset \cdots,\\
&\mathcal{P}_{n;1} = \{P_0(\theta)\} \oplus \{P_{1;1}(\theta)\} \oplus \cdots \oplus \{P_{n-1;1}(\theta)\} \oplus \{P_{n-1;2}(\theta)\} \oplus \{P_{n;1}(\theta)\},\\
&\mathcal{P}_{n;2} = \mathcal{P}_{n;1} \oplus \{P_{n;2}(\theta)\} = \{P_0(\theta)\} \oplus \{P_{1;1}(\theta)\} \oplus \cdots \oplus \{P_{n;1}(\theta)\} \oplus \{P_{n;2}(\theta)\},
\end{aligned}
\tag{6.2.5}
$$

where $\{P_{n;\alpha}(\theta)\}$, $n \in \mathbb{N}, \alpha = 1, 2$ denotes a one-dimensional space generated by $P_{n;\alpha}(\theta)$; $\mathcal{P}_0 = \mathbb{C}$.

In the sequel investigation we need the special Hilbert space

$$
\mathbf{l}_{2,u} = \mathcal{H}_0 \oplus \mathcal{H}_1 \oplus \mathcal{H}_2 \oplus \cdots, \quad \mathcal{H}_0 = \mathbb{C}, \quad \mathcal{H}_1 = \mathcal{H}_2 = \cdots = \mathbb{C}^2, \tag{6.2.6}
$$

i.e., (6.1.2).

Each vector $f \in \mathbf{l}_{2,u}$ has a form $f = (f_n)_{n=0}^{\infty}$, $f_n \in \mathcal{H}_n$, and hence,

$$\|f\|_{\mathbf{l}_{2,u}}^2 = \sum_{n=0}^{\infty} \|f_n\|_{\mathcal{H}_n}^2 < \infty, \quad (f, g)_{\mathbf{l}_{2,u}} = \sum_{n=0}^{\infty} (f_n, g_n)_{\mathcal{H}_n}, \quad \forall f, g \in \mathbf{l}_{2,u}.$$

For $n \in \mathbb{N}$ we denote $(f_{n;1}, f_{n;2})$ coordinates of a vector $f_n \in \mathcal{H}$ in the space $\mathbb{C}^2$ with a usual orthonormal basis $\{e_{n;1}, e_{n;2}\}$, hence, for a vector we have a representation $f_n = (f_{n;1}, f_{n;2})$.

By using the orthonormal system (6.2.2), we define a mapping $\mathbf{l}_{2,u}$ into L_2. We put

$$\forall \theta \in [0, 2\pi), \ P_n(\theta) = (P_{n;1}(\theta), P_{n;2}(\theta)) \in \mathcal{H}_n,$$

then

$$\mathbf{l}_{2,u} \ni f = (f_n)_{n=0}^{\infty} \longmapsto \hat{f}(\theta) = \sum_{n=0}^{\infty} (f_n, P_n(\theta))_{\mathcal{H}_n} \in L_2. \tag{6.2.7}$$

Since for $n \in \mathbb{N}$

$$(f_n, P_n(\theta))_{\mathcal{H}_n} = f_{n;1}\overline{P_{n;1}(\theta)} + f_{n;2}\overline{P_{n;2}(\theta)}$$

and

$$\|f\|_{\mathbf{l}_{2,u}}^2 = \|(f_0, f_{1;1}, f_{1;2}, f_{2,1}, \ldots)\|_{l_2}^2,$$

hence, it is obvious, that (6.2.7) is the mapping of the usual space l_2 into L_2 defined due to the orthonormal system (6.2.2), thus this mapping is an isometry. The image of $\mathbf{l}_{2,u}$ by the transformation (6.2.7) coincides with the space L_2. It is so, because the linear combination (6.2.1) approximates uniformly any continuous function on $\mathbb{T}$ and such a set of functions is dense in L_2; hence, the system (6.2.2) is total in L_2. Therefore the mapping (6.2.7) is an isometric operator (denotes by I) which imbeds $\mathbf{l}_{2,u}$ into L_2.

Let A be a bounded linear operator defined on the space $\mathbf{l}_{2,u}$. We will construct an operator-valued matrix $(a_{j,k})_{j,k=0}^{\infty}$, where for all $j, k \in \mathbb{N}_0$ each element $a_{j,k}$ is an operator from $\mathcal{H}_k$ into $\mathcal{H}_j$, i.e., $\forall f, g \in \mathbf{l}_{2,u}$, we have

$$(Af)_j = \sum_{k=0}^{\infty} a_{j,k} f_k, \ j \in \mathbb{N}_0, \quad (Af, g)_{\mathbf{l}_{2,u}} = \sum_{j,k=0}^{\infty} (a_{j,k} f_k, g_j)_{\mathcal{H}_j}. \tag{6.2.8}$$

For the proof of (6.2.8) we need only to write the usual matrix of the operator A using the basis

$$e_0 = 1, e_{1;1}, e_{1;2}, e_{2;1}, \ldots$$

in the space l_2. Then for each $j, k \in \mathbb{N}$, $a_{j,k}$ is an operator $\mathcal{H}_k \longrightarrow \mathcal{H}_j$ that has a matrix representation $(a_{j,k;\alpha,\beta})_{\alpha,\beta=1}^2$, so that

$$a_{j,k;\alpha,\beta} = (Ae_{k;\beta}, e_{j;\alpha})_{l_2}. \tag{6.2.9}$$

If $j = 0, k = 1, 2, \ldots$, then $(a_{0,k;\beta})_{\beta=1}^2$ is a (1×2)-matrix operator $\mathcal{H}_k \longrightarrow \mathcal{H}_0$, where $a_{0,k;\beta} = (Ae_{k;\beta}, e_0)_{l_2}$; if $k = 0, j = 1, 2, \ldots$, then $(a_{j,0;\alpha})_{\alpha=1}^2$ is a (2×1)-matrix operator $\mathcal{H}_0 \longrightarrow \mathcal{H}_j$, where $a_{j,0;\alpha} = (Ae_0, e_{j;\alpha})_{l_2}$. And $a_{0,0}$ is a (1×1)-matrix operator (the scalar) $\mathcal{H}_0 \longrightarrow \mathcal{H}_0$, where $a_{0,0} = (Ae_0, e_0)_{l_2}$.

Let us consider the image $\hat{A} = IAI^{-1} : L_2 \longrightarrow L_2$ of the above operator A by the mapping (6.2.7). Its matrix, in the basis (6.2.2), namely,

$$P_0(\theta),\ P_{1;1}(\theta),\ P_{1;2}(\theta),\ P_{2;1}(\theta), \ldots$$

is equal to the usual matrix of operator A: $l_2 \longrightarrow l_2$ in the corresponding basis (e_0, $e_{1;1}, e_{1;2}, e_{2;1}, \ldots$). By using (6.2.9) and the above mentioned procedure, we get the operator matrix $(a_{j,k})_{j,k=0}^{\infty}$ for $A : l_2 \longrightarrow l_2$. *By definition this matrix is also the operator matrix of* $\hat{A} : L_2 \longrightarrow L_2$.

It is clear that $\hat{A}$ can be an arbitrary linear bounded operator in L_2.

Lemma 6.2.1 *Let $\hat{A}$ be a unitary operator of the multiplication by $e^{i\theta}$ in the space L_2, namely,*

$$L_2 \ni \varphi(\theta) \longmapsto (\hat{A}\varphi)(\theta) = e^{i\theta}\varphi(\theta) \in L_2.$$

The operator matrix $(a_{j,k})_{j,k=0}^{\infty}$ of the operator $\hat{A}$ (i.e., $A = I^{-1}\hat{A}I$) has a tri-diagonal structure, i.e., $a_{j,k} = 0$ for $|j - k| > 1$.

Proof By using (6.2.9) for $e_{n;\gamma} = I^{-1}P_{n;\gamma}(\theta)$, $n \in \mathbb{N}$, $\gamma = 1, 2$, we have for $j, k \in \mathbb{N}$

$$a_{j,k;\alpha,\beta} = (Ae_{k;\beta}, e_{j;\alpha})_{l_2} = \int_0^{2\pi} e^{i\theta} P_{k;\beta}(\theta)\overline{P_{j;\alpha}(\theta)}\, d\rho(\theta), \quad \alpha, \beta = 1, 2. \tag{6.2.10}$$

From (6.2.2) we have $e^{i\theta}P_{k;1}(\theta) \in \mathcal{P}_{k+1;1}$ and $e^{i\theta}P_{k;2}(\theta) \in \mathcal{P}_{k+1;1}$. According to (6.2.5), the integral in (6.2.10) is equal to zero for $j > k + 1$ and for each $\alpha = 1, 2$.

On the other hand, the integral in (6.2.10) can be written

$$a_{j,k;\alpha,\beta} = \overline{\int_0^{2\pi} e^{-i\theta} P_{j;\alpha}(\theta)\overline{P_{k;\beta}(\theta)}\, d\rho(\theta)}.$$

From (6.2.2), we have now that $e^{-i\theta}P_{j;1}(\theta) \in \mathcal{P}_{j;2}$ and $e^{-i\theta}P_{j;2}(\theta) \in \mathcal{P}_{j+1;2}$. According to (6.2.5), the last integral is equal to zero for $k > j + 1$ and for each $\beta = 1, 2$.

As result, we have integrals in (6.2.10), i.e., coefficients $a_{j,k;\alpha,\beta}$, $j, k \in \mathbb{N}$, which are equal zero for $|j - k| > 1$; $\alpha, \beta = 1, 2$. The cases $j = 1, 2, \ldots, k = 0$ and $j = 0, k = 1, 2, \ldots$ are considered similarly (it is necessary to take into account that $e_0 = I^{-1}P_0(\theta)$, $P_0(\theta) = 1$). □

Therefore, the matrix $(a_{j,k})_{j,k=0}^{\infty}$ of the operator $\hat{A}$ has a three diagonal block structure

$$\begin{pmatrix} a_{0,0} & a_{0,1} & 0 & 0 & 0 & \ldots \\ a_{1,0} & a_{1,1} & a_{1,2} & 0 & 0 & \ldots \\ 0 & a_{2,1} & a_{2,2} & a_{2,3} & 0 & \ldots \\ 0 & 0 & a_{3,2} & a_{3,3} & a_{3,4} & \ldots \\ \vdots & \vdots & \vdots & \vdots & \vdots & \ddots \end{pmatrix}. \tag{6.2.11}$$

A detailed analysis of the expression (6.2.10) and a similar expression for $j, k \in \mathbb{N}_0$ allows us to find out the zero elements of the matrices $(a_{j,k;\alpha,\beta})_{\alpha,\beta=1}^{2}$ in the case of $|j - k| \leq 1$. Let us describe the properties of the matrices using permutations of indices j, k and α, β. Note that (6.2.10) implies the uniform boundedness of the norms $\|a_{j,k}\|$ with respect to $j, k \in \mathbb{N}_0$.

Lemma 6.2.2 *For polynomials $P_{n;\alpha}(\theta)$ and subspaces $\mathcal{P}_{m,\beta}$, $n, m \in \mathbb{N}_0$, $\alpha, \beta = 1, 2$, relations*

$$\begin{aligned} e^{i\theta}P_{n;1}(\theta) &\in \mathcal{P}_{n+1;1}, & e^{-i\theta}P_{n;1}(\theta) &\in \mathcal{P}_{n;2}; \\ e^{i\theta}P_{n;2}(\theta) &\in \mathcal{P}_{n+1;1}, & e^{-i\theta}P_{n;2}(\theta) &\in \mathcal{P}_{n+1;2}; \\ e^{i\theta}P_0(\theta) &\in \mathcal{P}_{1;1}, & e^{-i\theta}P_0(\theta) &\in \mathcal{P}_{1;2}. \end{aligned} \tag{6.2.12}$$

hold true.

Proof According to (6.2.2), the polynomial $P_{n;1}(\theta)$, $n \in \mathbb{N}$ is some linear combination of

$$\{1, e^{i\theta}, e^{-i\theta}, \ldots, e^{(n-1)i\theta}, e^{-(n-1)i\theta}, e^{ni\theta}\}.$$

Thus, the multiplication by $e^{i\theta}$ gives a linear combination

$$\{e^{i\theta}, e^{2i\theta}, 1, \ldots, e^{ni\theta}, e^{-(n-2)i\theta} e^{(n+1)i\theta}\},$$

and such a linear combination belongs to $\mathcal{P}_{n+1;1}$. Similarly, the multiplication by $e^{-i\theta}$ gives a linear combination

$$\{e^{-i\theta}, 1, e^{-2i\theta}, \ldots, e^{(n-2)i\theta}, e^{-ni\theta}, e^{(n-1)i\theta}\},$$

which belongs to $\mathcal{P}_{n;2}$.

According to (6.2.2), the polynomial $P_{n;2}(\theta)$, $n \in \mathbb{N}$ is some linear combination of

$$\{1, e^{i\theta}, e^{-i\theta}, \ldots, e^{(n-1)i\theta}, e^{-(n-1)i\theta}, e^{ni\theta}, e^{-ni\theta}\}.$$

Hence, the multiplication by $e^{i\theta}$ gives a linear combination

$$\{e^{i\theta}, e^{2i\theta}, 1, \ldots, e^{ni\theta}, e^{-(n-2)i\theta}, e^{(n+1)i\theta}, e^{-(n-1)i\theta}\}$$

and it belongs to $\mathcal{P}_{n+1;1}$. Similarly the multiplication by $e^{-i\theta}$ gives a linear combination

$$\{e^{-i\theta}, 1, e^{-2i\theta}, \ldots, e^{(n-2)i\theta}, e^{-ni\theta}, e^{(n-1)i\theta}, e^{-(n+1)i\theta}\}$$

and such a combination belongs to $\mathcal{P}_{n+1;2}$.

The case $n = 0$ is obvious, hence all expressions in (6.2.12) are proved. □

Let us denote $((a^*)_{j,k})_{j,k=0}^{\infty}$ the operator matrix of the operator $(\hat{A})^*$, adjoint to $\hat{A}$. We remark that $(\hat{A})^* = (\hat{A})^{-1}$ is the multiplication by $e^{-i\theta}$ operator. Taking into account the expression (6.2.10) for $j, k \in \mathbb{N}$, $\alpha, \beta = 1, 2$, we have

$$(a^*)_{j,k;\alpha,\beta} = \int_0^{2\pi} e^{-i\theta} P_{k;\beta}(\theta)\overline{P_{j;\alpha}(\theta)}\, d\rho(\theta) = \overline{\int_0^{2\pi} e^{i\theta} P_{j;\alpha}(\theta)\overline{P_{k;\beta}(\theta)}\, d\rho(\theta)} = \overline{a_{k,j;\beta,\alpha}}. \tag{6.2.13}$$

In cases $j = 0$, $k \in \mathbb{N}$, $k = 0$, $j \in \mathbb{N}$, $j = 0$, $k = 0$ instead of (6.2.10), we have

$$\begin{aligned}
a_{0,k;\beta} &= \int_0^{2\pi} e^{i\theta} P_{k;\beta}(\theta)\, d\rho(\theta), \quad k \in \mathbb{N}, \quad \beta = 1, 2;\\
a_{j,0;\alpha} &= \int_0^{2\pi} e^{i\theta} \overline{P_{j;\alpha}(\theta)}\, d\rho(\theta), \quad j \in \mathbb{N}, \quad \alpha = 1, 2;\\
a_{0,0} &= \int_0^{2\pi} e^{i\theta}\, d\rho(\theta).
\end{aligned} \tag{6.2.14}$$

In these cases the equality (6.2.13) has a form

$$(a^*)_{0,k;\beta} = \bar{a}_{k,0;\beta},\ (a^*)_{j,0;\alpha} = \bar{a}_{0,j;\alpha},\ a_{0,0} = \bar{a}_{0,0},\ j,k = \mathbb{N},\ \alpha,\beta = 1,2. \tag{6.2.15}$$

Lemma 6.2.3 *Let $(a_{j,k})_{j,k=0}^{\infty}$ be the multiplication by $e^{i\theta}$ operator matrix in L_2, where $a_{j,k} : \mathcal{H}_k \longrightarrow \mathcal{H}_j$; $a_{0,0}$, $a_{0,k} = (a_{0,k;\beta})_{\beta=1}^{2}$, $a_{j,0} = (a_{j,0;\alpha})_{\alpha=1}^{2}$, $a_{j,k} = (a_{j,k;\alpha,\beta})_{\alpha,\beta=1}^{2}$ are matrices of operators $a_{j,k}$ in the standard (usual) basis. Then*

$$a_{j,0;2} = a_{j,j+1;1,1} = a_{j,j+1;1,2} = a_{j+1,j;2,1} = a_{j+1,j;2,2} = 0, \quad j \in \mathbb{N}. \tag{6.2.16}$$

Proof According to (6.2.14), for $j \in \mathbb{N}$, we have

$$a_{j,0;2} = \int_0^{2\pi} e^{i\theta}\overline{P_{j,2}(\theta)}\,d\rho(\theta). \tag{6.2.17}$$

Due to (6.2.12), $e^{i\theta} = e^{i\theta}P_0(\theta) \in \mathcal{P}_{1;1}$, but $P_{j;2}(\theta)$ is orthogonal to $\mathcal{P}_{1;1}$ for $j \in \mathbb{N}$ (see (6.2.5)). Hence, in such a case the integral is equal to zero in (6.2.17).

According to (6.2.10) and (6.2.12), for $j \in \mathbb{N}$, we have

$$a_{j,j+1;1,1} = \int_0^{2\pi} e^{i\theta}P_{j+1,1}(\theta)\overline{P_{j;1}(\theta)}\,d\rho(\theta) = \overline{\int_0^{2\pi} e^{-i\theta}P_{j,1}(\theta)\overline{P_{j+1;1}(\theta)}\,d\rho(\theta)},$$

where $e^{-i\theta}P_{j;1}(\theta) \in \mathcal{P}_{j;2}$. But, according to (6.2.5), $P_{j+1;1}(\theta)$ is orthogonal to $\mathcal{P}_{j;2}$ and, hence, the last integral is equal to zero.

Similarly with (6.2.10), (6.2.12) and (6.2.5) for $j \in \mathbb{N}$, we have

$$a_{j,j+1;1,2} = \int_0^{2\pi} e^{i\theta}P_{j+1,2}(\theta)\overline{P_{j;1}(\theta)}\,d\rho(\theta) = \overline{\int_0^{2\pi} e^{-i\theta}P_{j,1}(\theta)\overline{P_{j+1;2}(\theta)}\,d\rho(\theta)},$$

where $e^{-i\theta}P_{j;1}(\theta) \in \mathcal{P}_{j;2}$. But according to (6.2.5), $P_{j+1;2}(\theta)$ is orthogonal to $\mathcal{P}_{j;2}$ and, hence, the last integral is equal to zero.

Also, from (6.2.10), (6.2.12) and (6.2.5) for $j \in \mathbb{N}$, we have

$$a_{j+1,j;2,1} = \int_0^{2\pi} e^{i\theta}P_{j;1}(\theta)\overline{P_{j+1;2}(\theta)}\,d\rho(\theta),$$

$$a_{j+1,j;2,2} = \int_0^{2\pi} e^{i\theta} P_{j;2}(\theta)\overline{P_{j+1;2}(\theta)}\, d\rho(\theta),$$

where $e^{i\theta} P_{j;1}(\theta), e^{i\theta} P_{j;2}(\theta) \in \mathcal{P}_{j+1;1}$. But, according to (6.2.5), $P_{j+1;2}(\theta)$ is orthogonal to $\mathcal{P}_{j+1;1}$ and, hence, both last integrals are also equal to zero. □

Thus, the last line of the (2×1)-matrix $a_{1,0}$ and the top and bottom lines of the (2×2)-matrix $a_{j,j+1}$ $(a_{j+1,j})$ contain always zero elements. Taking into account (6.2.11), we assert that the unitary matrix of the multiplication by $e^{i\theta}$ operator is five-diagonal in the usual sense, as a scalar matrix.

Lemma 6.2.4 *Elements*

$$a_{0,1;2}, a_{1,0;1}; \quad a_{j,j+1;2,2}, a_{j+1,j;1,1}, \quad j \in \mathbb{N} \tag{6.2.18}$$

of the matrix $(a_{j,k})_{j,k=0}^{\infty}$ *in Lemma 6.2.3 are positive.*

Proof Let us denote by $P'_{1;1}(\theta)$ a non-normalized vector $P_{1;1}(\theta)$ obtained from the Gram-Schmidt orthogonalization procedure of the sequence (6.2.1), $P'_{1;1}(\theta) = e^{i\theta} - (e^{i\theta}, 1)_{L_2}$. Then, according to (6.2.14), we get

$$\begin{aligned} a_{1,0;1} &= \int_0^{2\pi} e^{i\theta}\overline{P_{1;1}(\theta)}\, d\rho(\theta) = \|P'_{1;1}(\theta)\|_{L_2}^{-1} \int_0^{2\pi} e^{i\theta}\overline{P'_{1;1}(\theta)}\, d\rho(\theta) \\ &= \|P'_{1;1}(\theta)\|_{L_2}^{-1} \int_0^{2\pi} e^{i\theta}\overline{(e^{i\theta} - (e^{i\theta}, 1)_{L_2})}\, d\rho(\theta) \\ &= \|P'_{1;1}(\theta)\|_{L_2}^{-1} \left(1 - |(e^{-i\theta}, 1)_{L_2}|^2\right). \end{aligned} \tag{6.2.19}$$

Later we will show that

$$|(e^{-i\theta}, 1)_{L_2}|^2 + |(e^{-i\theta}, P_{1;1}(\theta))_{L_2}|^2 < 1. \tag{6.2.20}$$

Hence, from (6.2.19), we get $a_{1,0;1} > 0$.

Let us consider $a_{0,1;2}$. As earlier we denote $P'_{1;2}(\theta)$ a non-normalized vector $P_{1;2}(\theta)$, obtained from the Gram-Schmidt orthogonalization procedure of the sequence (6.2.1),

$$P'_{1;2}(\theta) = e^{-i\theta} - (e^{-i\theta}, P_{1;1}(\theta))_{L_2} P_{1;1}(\theta) - (e^{-i\theta}, 1)_{L_2}.$$

Then, according to (6.2.14) and (6.2.20), we get

$$a_{0,1;2} = \int_0^{2\pi} e^{i\theta} P_{1;2}(\theta)\, d\rho(\theta)$$

$$= \|P'_{1;2}(\theta)\|_{L_2}^{-1} \int_0^{2\pi} e^{i\theta} \left(e^{-i\theta} - (e^{-i\theta}, P_{1;1}(\theta))_{L_2} P_{1;1}(\theta) - (e^{-i\theta}, 1)_{L_2}\right) d\rho(\theta)$$

$$= \|P'_{1;2}(\theta)\|_{L_2}^{-1} \left(1 - |(e^{-i\theta}, P_{1;1}(\theta))_{L_2}|^2 - |(e^{-i\theta}, 1)_{L_2}|^2\right) > 0.$$

In the next step we consider $a_{j,j+1;2,2}$ and $a_{j+1,j;1,1}$, $j \in \mathbb{N}$. According to (6.2.10), we have

$$a_{j,j+1;2,2} = \int_0^{2\pi} e^{i\theta} P_{j+1;2}(\theta)\overline{P_{j;2}(\theta)}\, d\rho(\theta) = \overline{\int_0^{2\pi} e^{-i\theta} P_{j;2}(\theta)\overline{P_{j+1;2}(\theta)}\, d\rho(\theta)}, \tag{6.2.21}$$

$$a_{j+1,j;1,1} = \int_0^{2\pi} e^{i\theta} P_{j;1}(\theta)\overline{P_{j+1;1}(\theta)}\, d\rho(\theta). \tag{6.2.22}$$

In the case of the expression (6.2.21) for $a_{j,j+1;2,2}$, we have $e^{-i\theta} P_{j;2}(\theta) \in \mathcal{P}_{j+1;2}$ (see (6.2.12)) and the coefficient of this function by $P_{j+1;2}(\theta)$ is positive in the decomposition (6.2.5). This fact follows from (6.2.1) and (6.2.2), since

$$e^{-i\theta} P_{j;2}(\theta) = k_{j;2} e^{-(j+1)i\theta} + e^{-i\theta}(\cdots) = \frac{k_{j;2}}{k_{j+1;2}} P_{j+1;2}(\theta) + Q_j(\theta), \tag{6.2.23}$$

where dots indicate the following part of the corresponding decomposition $P_{j;2}(\theta)$ in (6.2.2). In (6.2.23), $Q_j(\theta)$ is a linear combination of $e^{li\theta}$, $l = -j, -(j-1), \ldots, (j+1)$, that are orthogonal to $P_{j+1;2}(\theta)$. Substituting (6.2.23) into (6.2.21), we get the equality $a_{j,j+1;2,2} = k_{j;2}(k_{j+1,2})^{-1} > 0$.

The positiveness of $a_{j+1;1,1}$ follows from (6.2.22) in a similar way; here we used the fact that according to (6.2.12), $e^{i\theta} P_{j;1}(\theta) \in \mathcal{P}_{j+1;1}$ and a coefficient of the function $P_{j+1;1}(\theta)$ is positive, as it follows from equality of the type (6.2.23).

Let us finally prove the inequality (6.2.20). Indeed, it follows from the Parseval equality by the decomposition of the function $e^{-i\theta}$ with respect to the orthonormal basis (6.2.2), namely,

$$|(e^{-i\theta}, P_0(\theta))_{L_2}|^2 + |(e^{-i\theta}, P_{1;1}(\theta))_{L_2}|^2 + |(e^{-i\theta}, P_{1;2}(\theta))_{L_2}|^2 + \cdots$$
$$= \|e^{-i\theta}\|_{L_2}^2 = 1. \tag{6.2.24}$$

Hence, the lemma is proved. □

In what follows we will use usual well known notations for elements $a_{j,k}$ of the matrix:

$$
\begin{aligned}
a_n &= a_{n+1,n} &&: \quad \mathcal{H}_n \longrightarrow \mathcal{H}_{n+1},\\
b_n &= a_{n,n} &&: \quad \mathcal{H}_n \longrightarrow \mathcal{H}_n,\\
c_n &= a_{n,n+1} &&: \quad \mathcal{H}_{n+1} \longrightarrow \mathcal{H}_n, \quad n \in \mathbb{N}_0.
\end{aligned}
\tag{6.2.25}
$$

All previous investigations we summarize in the theorem.

Theorem 6.2.5 *The unitary multiplication by $e^{i\theta}$ operator $\hat{\mathrm{A}}$, (with a strong cyclic vector) in the orthonormal basis* (6.2.2) *of polynomials in the space L_2, has the form of tri-diagonal block Jacobi type unitary matrix $J = (a_{j,k})_{j,k=0}^{\infty}$ acting as the operator in the space* (6.2.6)*, namely,*

$$
\mathbf{l}_{2,u} = \mathcal{H}_0 \oplus \mathcal{H}_1 \oplus \mathcal{H}_2 \oplus \cdots, \quad \mathcal{H}_0 = \mathbb{C}, \quad \mathcal{H}_n = \mathbb{C}^2, \quad n \in \mathbb{N}. \tag{6.2.26}
$$

The norms of all operators $a_{j,k} : \mathcal{H}_k \longrightarrow \mathcal{H}_j$ are uniformly bounded with respect to $j, k \in \mathbb{N}_0$. In designations (6.2.25)*, this matrix has the form*

$$
J = \begin{pmatrix}
b_0 & c_0 & 0 & 0 & 0 & \ldots\\
a_0 & b_1 & c_1 & 0 & 0 & \ldots\\
0 & a_1 & b_2 & c_2 & 0 & \ldots\\
0 & 0 & a_2 & b_3 & c_3 & \ldots\\
\vdots & \vdots & \vdots & \vdots & \vdots & \ddots
\end{pmatrix},
\quad
\begin{aligned}
a_n &: \mathcal{H}_n \longrightarrow \mathcal{H}_{n+1},\\
b_n &: \mathcal{H}_n \longrightarrow \mathcal{H}_n,\\
c_n &: \mathcal{H}_{n+1} \longrightarrow \mathcal{H}_n, \quad n \in \mathbb{N}_0.
\end{aligned}
\tag{6.2.27}
$$

and in more detail

$$
J = \left(\begin{array}{c|cc|cc|cc|cc|c}
*b_0 & * & c_0\, + & & & & & & & \cdots\\
\hline
+ & * & * & 0 & 0 & & & & & \\
a_0 & \multicolumn{2}{c|}{b_1} & \multicolumn{2}{c|}{c_1} & & 0 & & & \cdots\\
0 & * & * & * & + & & & & & \\
\hline
 & + & * & * & * & 0 & 0 & & & \\
 & \multicolumn{2}{c|}{a_1} & \multicolumn{2}{c|}{b_2} & \multicolumn{2}{c|}{c_2} & & & \cdots\\
 & 0 & 0 & * & * & * & + & & & \\
\hline
 & & & + & * & * & * & 0 & 0 & \\
 & 0 & & \multicolumn{2}{c|}{a_2} & \multicolumn{2}{c|}{b_3} & \multicolumn{2}{c|}{c_3} & \cdots\\
 & & & 0 & 0 & * & * & * & + & \\
\hline
\vdots & \vdots & & \vdots & & \vdots & & \vdots & & \ddots
\end{array}\right). \tag{6.2.28}
$$

In (6.2.28)*, b_0 is a (1×1)-matrix (i.e., a scalar), a_0 is a (2×1)-matrix: $a_0 = (a_{0;\alpha})_{\alpha=1}^{2}$, c_0 is a (1×2)-matrix: $c_0 = (c_{0;\beta})_{\beta=1}^{2}$; for $j \in \mathbb{N}$, elements $a_j = (a_{j;\alpha,\beta})_{\alpha,\beta=1}^{2}$, $b_j = (b_{j;\alpha,\beta})_{\alpha,\beta=1}^{2}$, $c_j = (c_{j;\alpha,\beta})_{\alpha,\beta=1}^{2}$ are (2×2)-matrices.*

Matrices a_j, b_j, and c_j have always zero elements on some positions, namely,

$$\begin{aligned} &a_{0;1} > 0,\ a_{0;2} = 0; && c_{0;2} > 0; \\ &a_{n;2,1} = a_{n;2,2} = 0,\ a_{n;1,1} > 0; && c_{n;1,1} = c_{n;1,2} = 0,\ c_{n;2,2} > 0;\ n \in \mathbb{N}. \end{aligned} \tag{6.2.29}$$

Possibly zero and non-zero elements of the matrix are indicated by "$*$".

The matrix (6.2.28) *has five diagonals in the scalar form.*

The adjoint operator $(\hat{A})^*$ has a similar block tri-diagonal Jacobi type form J^+ in the basis (6.2.2).

Matrices J, J^+ act as operators act by the rule

$$\begin{gathered} (Jf)_n = a_{n-1} f_{n-1} + b_n f_n + c_n f_{n+1}, \\ (J^+ f)_n = c^*_{n-1} f_{n-1} + b^*_n f_n + a^*_n f_{n+1}, \\ n \in \mathbb{N}_0, \quad f_{-1} = 0, \quad \forall f = (f_n)_{n=0}^{\infty} \in \mathbf{l}_{2,u}, \end{gathered} \tag{6.2.30}$$

here "$*$" denotes the usual adjoint matrix.

The form of coefficients J^+ follows from (6.2.13), (6.2.15), and (6.2.25).

6.3 Direct and Inverse Spectral Problems for the Tri-Diagonal Block Jacobi Type Matrices of Unitary Operators

As mentioned at the beginning of the chapter, one of the main results is the solution of the inverse spectral problem for the corresponding direct one.

Let us consider the space $\mathbf{l}_{2,u}$ of the form (6.2.6). In addition to the space $\mathbf{l}_{2,u}$ consider its equipment

$$(\mathbf{l}_{\text{fin}})' \supset \mathbf{l}_{2,u}(p^{-1}) \supset \mathbf{l}_{2,u} \supset \mathbf{l}_{2,u}(p) \supset \mathbf{l}_{\text{fin}}, \tag{6.3.1}$$

where $\mathbf{l}_{2,u}(p)$ is the weighted $\mathbf{l}_{2,u}$ space with a weight $p = (p_n)_{n=0}^{\infty}$, $p_n \geq 1$, $(p^{-1} = (p_n^{-1})_{n=0}^{\infty})$. In such a case $\mathbf{l}_{2,u}(p)$ is a the Hilbert space of sequences $f = (f_n)_{n=0}^{\infty}$, $f_n \in \mathcal{H}_n$, for which

$$\|f\|^2_{\mathbf{l}_{2,u}(p)} = \sum_{n=0}^{\infty} \|f_n\|^2_{\mathcal{H}_n} p_n, \quad (f, g)_{\mathbf{l}_{2,u}(p)} = \sum_{n=0}^{\infty} (f_n, g_n)_{\mathcal{H}_n} p_n. \tag{6.3.2}$$

The space $\mathbf{l}_{2,u}(p^{-1})$ is defined similarly; $\mathbf{l}_{\text{fin}}$ is the space of finite sequences and $(\mathbf{l}_{\text{fin}})'$ is the space adjoint to $\mathbf{l}_{\text{fin}}$. It is easy to show that the imbedding $\mathbf{l}_{2,u}(p) \hookrightarrow \mathbf{l}_{2,u}$ is quasi-nuclear if $\sum\limits_{n=0}^{\infty} p_n^{-1} < \infty$.

Let A be a unitary operator standardly connected with the chain (6.3.1). (About the standard connection see the beginning of the last section of this chapter.) According to the projection spectral theorem, such an operator has a representation

$$Af = \int\limits_{\mathbb{T}} z\Phi(z)\, d\sigma(z) f, \quad f \in \mathbf{l}_{2,u}, \quad z = e^{i\theta} \in \mathbb{T}, \quad \theta \in [0, 2\pi), \tag{6.3.3}$$

where $\Phi(z) : \mathbf{l}_{2,u}(p) \longrightarrow \mathbf{l}_{2,u}(p^{-1})$ is the operator of generalized projection and $d\sigma(z)$ is a spectral measure. The adjoint to A operator $A^* = A^{-1}$ has a similar representation (6.2.12), where $z\Phi(z)$ is replaced with $\bar{z}\Phi(z)$. For all $f \in \mathbf{l}_{\text{fin}}$, the projection $\Phi(z)f \in \mathbf{l}_{2,u}(p^{-1})$ is a generalized eigenvector of the operators A and A^* with corresponding eigenvalues z and $\bar{z}$. For all $f, g \in \mathbf{l}_{\text{fin}}$, we have the Parseval equality

$$(f, g)_{\mathbf{l}_{2,u}} = \int\limits_{\mathbb{T}} (\Phi(z) f, g)_{\mathbf{l}_{2,u}}\, d\sigma(z), \quad z = e^{i\theta} \in \mathbb{T}, \tag{6.3.4}$$

that after the extension by continuity (6.3.4) is defined for all $f, g \in \mathbf{l}_{2,u}$.

We denote by π_n the orthogonal projection operator from $\mathbf{l}_{2,u}$ in $\mathcal{H}_n$, $n \in \mathbb{N}_0$. Hence, $\forall f = (f_n)_{n=0}^{\infty} \in \mathbf{l}_{2,u}$, we have $f_n = \pi_n f$. This operator acts similarly in the spaces $\mathbf{l}_{2,u}(p)$ and $\mathbf{l}_{2,u}(p^{-1})$, but perhaps its norm does not equal one.

Consider the operator matrix $(\Phi_{j,k}(z))_{j,k=0}^{\infty}$, where

$$\Phi_{j,k}(z) = \pi_j \Phi(z) \pi_k \; : \; \mathbf{l}_{2,u} \longrightarrow \mathcal{H}_j, \; (\text{or } \mathcal{H}_k \longrightarrow \mathcal{H}_j). \tag{6.3.5}$$

The Parseval equality (6.3.4) has the form

$$\begin{aligned} (f, g)_{\mathbf{l}_{2,u}} &= \sum_{j,k=0}^{\infty} \int\limits_{\mathbb{T}} (\Phi(z)\pi_k f, \pi_j g)_{\mathbf{l}_{2,u}}\, d\sigma(z) \\ &= \sum_{j,k=0}^{\infty} \int\limits_{\mathbb{T}} (\pi_j \Phi(z)\pi_k f, g)_{\mathbf{l}_{2,u}}\, d\sigma(z) \\ &= \sum_{j,k=0}^{\infty} \int\limits_{\mathbb{T}} (\Phi_{j,k}(z) f_k, g_j)_{\mathbf{l}_{2,u}}\, d\sigma(z), \\ & z = e^{i\theta} \in \mathbb{T}, \quad \theta \in [0, 2\pi), \quad \forall f, g \in \mathbf{l}_{2,u}. \end{aligned} \tag{6.3.6}$$

Consider some special bounded operator A acting in the space $\mathbf{l}_{2,u}$. More precisely, let this operator be given by the matrix J, which has a tri-diagonal block structure of the form (6.2.28), where all norms of the coefficients a_n, b_n and c_n are uniformly bounded with respect to $n \in \mathbb{N}_0$. Therefore, the action of the operator A is determined by the first expression in (6.2.30) and the action of its adjoint operator is determined, similarly, by the second expression in (6.2.30).

In what follows, *it is assumed that conditions* (6.2.29) *hold true, and, additionally, the operator* A, *given in* (6.2.28), *is unitary in the space* $\mathbf{l}_{2,u}$. Without analyzing the conditions for the coefficients a_n, b_n and c_n under which the operator A is unitary, we note that such conditions follow from the equality $AA^* = A^*A = 1$. We make this analysis in the next section.

In the next step, we will rewrite the Parseval equality (6.3.6) in terms of the generalized eigenvectors of the operator A. But first, consider a lemma.

Lemma 6.3.1 *Let* $\varphi(z) = (\varphi_n(z))_{n=0}^{\infty}$, $\varphi_n(z) \in \mathcal{H}_n$, $z \in \mathbb{C}$ *be a solution in* $(\mathbf{l}_{\text{fin}})'$ *of the system*

$$\begin{aligned}(J\varphi(z))_n &= a_{n-1}\varphi_{n-1}(z) + b_n\varphi_n(z) + c_n\varphi_{n+1}(z) = z\varphi_n(z),\\ (J^+\varphi(z))_n &= c_{n-1}^*\varphi_{n-1}(z) + b_n^*\varphi_n(z) + a_n^*\varphi_{n+1}(z) = \bar{z}\varphi_n(z),\\ &n \in \mathbb{N}_0, \quad \varphi_{-1}(z) = 0, \quad z = e^{i\theta} \in \mathbb{T}, \quad \theta \in [0, 2\pi),\end{aligned} \tag{6.3.7}$$

with an initial condition $\varphi_0(z) = \varphi_0 \in \mathbb{C}$. *Then, this solution exists* $\forall \varphi_0$ *and has a form*

$$\varphi_n(z) = Q_n(z)\varphi_0 = (Q_{n;1}, Q_{n;2})\varphi_0, \quad \forall n \in \mathbb{N}, \tag{6.3.8}$$

where $Q_{n;1}$ *and* $Q_{n;2}$ *are polynomials of variables* $z = e^{i\theta}$ *and* $\bar{z} = e^{-i\theta}$ *and such polynomials have a form*

$$Q_{n;1}(z) = l_{n;1}\bar{z}^n + q_{n;1}(z), \quad Q_{n;2}(z) = l_{n;2}z^n + q_{n;2}(z), \quad \forall n \in \mathbb{N}. \tag{6.3.9}$$

Here $l_{n;1} > 0$, $l_{n;2} > 0$ *and* $q_{n;1}(z)$, $q_{n;2}(z)$ *are linear combinations of* $z^j\bar{z}^k$ *for* $0 \le j + k \le n - 1$, $\forall n \in \mathbb{N}$, $Q_0(z) = 1$, $z = e^{i\theta} \in \mathbb{T} \subset \mathbb{C}$.

Proof For $n = 0$, the system (6.3.7) has a form

$$\begin{aligned} b_0\varphi_0 + c_0\varphi_1 &= z\varphi_0, \\ b_0^*\varphi_0 + a_0^*\varphi_1 &= \bar{z}\varphi_0, \end{aligned} \quad \text{or} \quad \begin{aligned} c_{0;1}\varphi_{1;1} + c_{0;2}\varphi_{1;2} &= (z - b_0)\varphi_0, \\ a_{0;1}^*\varphi_{1;1} + a_{0;2}^*\varphi_{1;2} &= (\bar{z} - \bar{b}_0)\varphi_0. \end{aligned} \tag{6.3.10}$$

Here and in what follows, we denote $\varphi_n(z) = (\varphi_{n;1}(z), \varphi_{n;2}(z)) \in \mathcal{H}_n$, $n \in \mathbb{N}$. By using the assumption (6.2.29) for the matrix (6.2.28), we rewrite the last two equalities from (6.3.10) in the form

$$d_0\varphi_1(z) = ((z - b_0)\varphi_0, (\bar{z} - \bar{b}_0)\varphi_0); \; d_0 = \begin{pmatrix} c_{0;1} & c_{0;2} \\ a_{0;1} & 0 \end{pmatrix}, \; c_{0;2} > 0, \; a_{0;1} > 0.$$

Hence, there exist an inverse matrix d_0^{-1}, thus

$$(\varphi_{1;1}(z), \varphi_{1;2}(z)) = \varphi_1(z) = d_0^{-1}((z - b_0)\varphi_0, (\bar{z} - \bar{b}_0)\varphi_0).$$

Therefore,

$$\begin{aligned} \varphi_{1;1}(z) &= \frac{1}{a_{0;1}}(\bar{z} - \bar{b}_0)\varphi_0 = Q_{1;1}(z)\varphi_0, \\ \varphi_{1;2}(z) &= (r_1(z - b_0) + r_2(\bar{z} - \bar{b}_0) + r_3)\varphi_0 = Q_{1;2}(z)\varphi_0, \end{aligned} \tag{6.3.11}$$

where $r_1 > 0$, r_2, and r_3 are some constants. In other words, the solution $\varphi_n(z)$ of the system (6.3.7), for $n = 1$, has the form (6.3.8) and (6.3.9).

Assume by induction that $\varphi_n(z)$ exists for $n \in \mathbb{N}$ and has the form (6.3.8) and (6.3.9), and show that $\varphi_{n+1}(z)$ exists and has the form (6.3.8) and (6.3.9).

From (6.3.7), we get

$$\begin{aligned} c_n\varphi_{n+1}(z) &= (z\mathbb{1} - b_n)\varphi_n(z) - a_{n-1}\varphi_{n-1}(z), \\ a_n^*\varphi_{n+1}(z) &= (\bar{z}\mathbb{1} - b_n^*)\varphi_n(z) - c_{n-1}^*\varphi_{n-1}(z). \end{aligned}$$

Adding the last two equalities and taking into account (6.2.29), (6.3.8) and (6.3.9), we obtain

$$\begin{aligned} d_n\varphi_{n+1}(z) &= ((z + \bar{z})\mathbb{1} - (b_n + b_n^*))\varphi_n(z) - (a_{n-1} + c_{n-1}^*)\varphi_{n-1}(z) \\ &= ((z + \bar{z})\mathbb{1} - (b_n + b_n^*))(l_{n;1}\bar{z}^n + q_{n;1}(z), l_{n;2}z^n + q_{n;2}(z))\varphi_0 \\ &\quad - (a_{n-1} + c_{n-1}^*)(l_{n-1;1}\bar{z}^{n-1} + q_{n-1;1}(z), l_{n-1;2}z^{n-1} + q_{n-1;2}(z))\varphi_0 \\ &= (s_{n;1}(z), s_{n;2}(z))\varphi_0; \\ d_n &= \begin{pmatrix} a_{n;1,1} & 0 \\ c_{n;2,1} + a_{n;1,2} & c_{n;2,2} \end{pmatrix}, \quad a_{n;1,1} > 0, \quad c_{n;2,2} > 0. \end{aligned} \tag{6.3.12}$$

It follows from (6.3.12) that the matrix d_n^{-1} exists. Thus, it is possible to write the expression for $\varphi_{n+1;1}(z)$ and $\varphi_{n+1;2}(z)$ using the right-hand side (6.3.12), which is denoted $(s_{n;1}(z), s_{n;2}(z))\varphi_0$, i.e.,

$$\begin{aligned} (\varphi_{n+1;1}(z), \varphi_{n+1;2}(z)) &= d_n^{-1}(s_{n;1}(z), s_{n;2}(z))\varphi_0, \\ \varphi_{n+1;1}(z) &= \frac{1}{a_{n;1,1}}s_{n;1}\varphi_0, \\ \varphi_{n+1;2}(z) &= (d_n^{-1})_{2,1}s_{n;1}(z) + (d_n^{-1})_{2,2}s_{n;2}(z). \end{aligned} \tag{6.3.13}$$

It can be seen from expressions (6.3.12) and (6.3.13) that an increase by "1" of the power of z and $\bar{z}$ on the right-hand side in these formulas leads to multiplication by $(z-\bar{z})$. This shows that $s_{n;1}(z)$ and $s_{n;2}(z)$ have the form $l_{n+1;1}\bar{z}^{n+1}+q_{n+1;1}(z)$ and $l_{n+1;2}z^{n+1}+q_{n+1;2}(z)$, respectively.

It is not difficult to see that new higher coefficients $l_{n+1;1}$ and $l_{n+1;2}$ are also positive. For $l_{n+1,1}$ this fact follows from the second expression in (6.3.13) and for $l_{n+1,2}$ it follows from the last expression in (6.3.13), that is, from the positiveness of the diagonal elements of the matrix d_n and the location of the coefficient $l_{n;2}z^n$ at the second coordinate of the vector $(l_{n;1}\bar{z}^n+q_{n;1}(z), l_{n;2}z^n+q_{n;2}(z))$. This completes the induction and proof. □

Consider $Q_n(z)$ with fixed $z=e^{i\theta}$ as a linear operator acting from $\mathcal{H}_0$ to $\mathcal{H}_n$, i.e., $\mathcal{H}_0 \ni \varphi_0 \longmapsto Q_n(z)\varphi_0 \in \mathcal{H}_n$. We also consider $Q_n(z)$ as an operator-valued polynomial of variables $z-e^{i\theta}, \bar{z}=e^{-i\theta} \in \mathbb{C}$. For the adjoint operator we have $Q_n^*(z)=(Q_n(z))^* : \mathcal{H}_n \longrightarrow \mathcal{H}_0$. By using polynomials $Q_n(z)$, we construct a representation $\Phi_{j,k}(z)$.

Lemma 6.3.2 *The operator $\Phi_{j,k}(z)$, $\forall z=e^{i\theta}\in\mathbb{T}$ has a representation*

$$\Phi_{j,k}(z)=Q_j(z)\Phi_{0,0}(z)Q_k^*(z) : \mathcal{H}_k \longrightarrow \mathcal{H}_j, \quad j,k\in\mathbb{N}_0, \tag{6.3.14}$$

where $\Phi_{0,0}(z)\geq 0$ is a scalar.

Proof For a fixed $k\in\mathbb{N}_0$, the vector $\varphi=\varphi(z)=(\varphi_j(z))_{j=0}^{\infty}$, where

$$\varphi_j(z)=\Phi_{j,k}(z)=\pi_j\Phi(z)\pi_k\in\mathcal{H}_j, \quad z=e^{i\theta}\in\mathbb{T}, \quad \theta\in[0,2\pi), \tag{6.3.15}$$

is a generalized solution in $(\mathbf{l}_{\text{fin}})'$, of the equation $J\varphi(z)=z\varphi(z)$, since $\Phi(z)$ is a projector on generalized eigenvectors corresponding to generalized eigenvalues z of the operator A. Therefore, $\forall g\in\mathbf{l}_{\text{fin}}$ we have an equality $(\varphi, J^+g)_{\mathbf{l}_{2,u}}=z(\varphi,g)_{\mathbf{l}_{2,u}}$. Substituting φ into the finite-difference equation for J^+, we get $(J\varphi,g)_{\mathbf{l}_{2,u}}=z(\varphi,g)_{\mathbf{l}_{2,u}}$. Hence, $\varphi=\varphi(z)\in\mathbf{l}_{2,u}(p^{-1})$ is a solution of the equation $J\varphi=z\varphi$ with initial conditions $\varphi_0=\pi_0\Phi(z)\pi_k\in\mathcal{H}_0$.

Since $\forall f\in\mathbf{l}_{\text{fin}}$, the vector $\Phi(z)f\in\mathbf{l}_2(p^{-1})$ is also the generalized eigenvector of the operator A^* with corresponding eigenvalue $\bar{z}$ (because A is unitary operator). Similarly, $\varphi=\varphi(z)$ in (6.3.15) is also the solution of the equation $J^+\varphi=\bar{z}\varphi$ with the same initial conditions $\varphi_0=\pi_0\Phi(z)\pi_k$.

By using the Lemma 6.3.1 and (6.3.8), we obtain

$$\Phi_{j,k}(z)=Q_j(z)(\Phi_{0,k}(z)), \quad j\in\mathbb{N}_0. \tag{6.3.16}$$

The operator $\Phi(z):\mathbf{l}_{2,u}(p)\longrightarrow\mathbf{l}_{2,u}(p^{-1})$ is formally self-adjoint in $\mathbf{l}_{2,u}$, and it is a derivation with respect to the corresponding resolution of identity in $\mathbf{l}_{2,u}$ on the spectral measure of the operator A. Hence, according to (6.3.14), we get

$$(\Phi_{j,k}(z))^*=(\pi_j\Phi(z)\pi_k)^*=\pi_k\Phi(z)\pi_j=\Phi_{k,j}(z), \quad j,k\in\mathbb{N}_0. \tag{6.3.17}$$

For a fixed $j \in \mathbb{N}_0$ from (6.3.17) and previous conversation it follows that the vector

$$\psi = \psi(z) = (\psi_k(z))_{k=0}^{\infty}, \quad \psi_k(z) = \Phi_{k,j}(z) = (\Phi_{j,k}(z))^*$$

is an usual solution of equations $J\psi = z\psi$ and $J^+\psi = \bar{z}\psi$ with initial conditions $\psi_0 = \Phi_{0,j}(z) = (\Phi_{j,0}(z))^*$.

Using again Lemma 6.3.1, we obtain the representation of the type (6.3.16),

$$\Phi_{k,j}(z) = Q_k(z)(\Phi_{0,j}(z)), \quad k \in \mathbb{N}_0. \tag{6.3.18}$$

Taking into account (6.3.17) and (6.3.18), we get

$$\Phi_{0,k}(z) = (\Phi_{k,0}(z))^* = (Q_k(z)\Phi_{0,0}(z))^* = \Phi_{0,0}(z)(Q_k(z))^*, \quad k \in \mathbb{N}_0 \tag{6.3.19}$$

(here we used $\Phi_{0,0}(z) \geq 0$ following from (6.3.4) and (6.3.5)). Substituting (6.3.19) into (6.3.16), we obtain (6.3.14). □

Now, we obtain the possibility to rewrite the Parseval equality (6.3.6) in more concrete form. For this end we substitute the expression (6.3.14) for $\Phi_{j,k}(z)$ into (6.3.6) and get

$$\begin{aligned}
(f, g)_{\mathbf{l}_{2,u}} &= \sum_{j,k=0}^{\infty} \int_{\mathbb{T}} (\Phi_{j,k}(z) f_k, g_j)_{\mathbf{l}_{2,u}} \, d\sigma(z) \\
&= \sum_{j,k=0}^{\infty} \int_{\mathbb{T}} (Q_j(z)\Phi_{0,0}(z) Q_k^*(z) f_k, g_j)_{\mathbf{l}_{2,u}} \, d\sigma(z) \\
&= \sum_{j,k=0}^{\infty} \int_{\mathbb{T}} (Q_k^*(z) f_k, Q_j^*(z) g_j)_{\mathbf{l}_{2,u}} \, d\rho(z) \\
&= \int_{\mathbb{T}} \Big(\sum_{k=0}^{\infty} Q_k^*(z) f_k \Big) \overline{\Big(\sum_{j=0}^{\infty} Q_j^*(z) g_j \Big)} \, d\rho(z), \ \forall f, g \in \mathbf{l}_{\text{fin}}
\end{aligned} \tag{6.3.20}$$

$$d\rho(z) = \Phi_{0,0}(z) \, d\sigma(z), \quad z = e^{i\theta} \in \mathbb{T}, \quad \theta \in [0, 2\pi).$$

Consider the Fourier transform "$\widehat{\ }$" generated by the unitary operator A in the space $\mathbf{l}_{2,u}$, namely,

$$\mathbf{l}_{2,u} \supset \mathbf{l}_{\text{fin}} \ni f = (f_n)_{n=0}^{\infty} \longmapsto \hat{f}(z) = \sum_{n=0}^{\infty} Q_n^*(z) f_n \in L_2(\mathbb{T}, d\rho(z)). \tag{6.3.21}$$

Hence, (6.3.20) means the Parseval equality in the final form

$$(f, g)_{\mathbf{l}_{2,u}} = \int_{\mathbb{T}} \hat{f}(z)\overline{\hat{g}(z)}\, d\rho(z), \quad \forall f, g \in \mathbf{l}_{\text{fin}}. \tag{6.3.22}$$

After an extension by continuity, (6.3.22) is valid $\forall f, g \in \mathbf{l}_{2,u}$.

The orthogonality of polynomials $Q_n^*(z)$ follows from (6.3.21) and (6.3.22). Namely, there it is sufficient only to take, in (6.3.21) and (6.3.22), vectors

$$f = (0, \dots, 0, f_k, 0, \dots), \ f_k \in \mathcal{H}_k, \ g = (0, \dots, 0, g_j, 0, \dots), \ g_j \in \mathcal{H}_j.$$

Then,

$$\int_{\mathbb{T}} (Q_k^*(z) f_k)\overline{(Q_j^*(z) g_j)}\, d\rho(z) = \delta_{j,k}(f_j, g_j)_{\mathcal{H}_j}, \quad \forall k, j \in \mathbb{N}_0. \tag{6.3.23}$$

By using the representation (6.3.8) for these polynomials, we can rewrite the equality (6.3.23) in the usual (classical) scalar form. For this reason we remark that $Q_0^*(z) = \bar{Q}_0(z)$, for $n \in \mathbb{N}$ according to (6.3.8), we have

$$Q_n(z) = (Q_{n;1}(z), Q_{n;2}(z)) : \mathcal{H}_0 \longrightarrow \mathcal{H}_n.$$

Hence, for the adjoint operator $Q_n^*(z) : \mathcal{H}_n \longrightarrow \mathcal{H}_0$ we have $\forall x \in \mathcal{H}_0$, $y = (y_1, y_2) \in \mathcal{H}_n$

$$\begin{aligned}(Q_n(z)x, y)_{\mathcal{H}_n} &= ((Q_{n;1}(z)x, Q_{n;2}(z)x), (y_1, y_2))_{\mathcal{H}_n} \\ &= Q_{n;1}(z)x\bar{y}_1 + Q_{n;2}(z)x\bar{y}_2 \\ &= x\overline{(\overline{Q_{n;1}(z)}y_1 + \overline{Q_{n;2}(z)}y_2)} = (x, Q_n^*(z)y)_{\mathcal{H}_0},\end{aligned}$$

i.e.,

$$Q_n^*(z)y = \overline{Q_{n;1}(z)}y_1 + \overline{Q_{n;2}(z)}y_2.$$

Due to the last equality for $n \in \mathbb{N}$ and $f_n = (f_{n,1}, f_{n_2}) \in \mathcal{H}_n$, $z \in \mathbb{T}$, we get

$$Q_n^*(z) f_n = \overline{Q_{n;1}(z)} f_{n;1} + \overline{Q_{n;2}(z)} f_{n;2}, \quad Q_0^*(z) = 1. \tag{6.3.24}$$

Therefore, (6.3.23) has the form

$$\begin{aligned}&\int_{\mathbb{T}} (\overline{Q_{k;1}(z)} f_{k;1} + \overline{Q_{k;2}(z)} f_{k;2})\overline{(\overline{Q_{j;1}(z)} f_{j;1} + \overline{Q_{j;2}(z)} f_{j;2})}\, d\rho(z) \\ &= \delta_{j,k}(f_{j;1}\bar{g}_{j;1} + f_{j;2}\bar{g}_{j;2}), \quad \forall f_0, g_0, f_{k;1}, f_{k;2}, g_{j;1}, g_{j;2} \in \mathbb{C}, \ j, k \in \mathbb{N}.\end{aligned}$$

This equality is equivalent to the following relation of the orthogonality in the usual classical form

$$\int_{\mathbb{T}} \overline{Q^*_{k;\beta}(z) Q_{j;\alpha}}\, d\rho(z) = \delta_{j,k}\delta_{\alpha,\beta}, \quad Q_{0;\alpha} = Q_0(z), \quad \forall j, k \in \mathbb{N}_0, \forall \alpha, \beta = 1, 2. \tag{6.3.25}$$

Let us remark that due to (6.3.24), the Fourier transform (6.3.21) can be rewritten in the form $\forall f = (f_n)_{n=0}^{\infty} \in \mathbf{l}_{2,u}$,

$$\hat{f}(z) = f_0 + \sum_{n=1}^{\infty} (\overline{Q_{n;1}(z)} f_{n;1} + \overline{Q_{n;2}(z)} f_{n;2}), \quad z \in \mathbb{T}. \tag{6.3.26}$$

By using the above stated results of this section, we can formulate the spectral theorem for our unitary operator A.

Theorem 6.3.3 *Consider the space (6.2.6), i.e.,*

$$\mathbf{l}_{2,u} = \mathcal{H}_0 \oplus \mathcal{H}_1 \oplus \mathcal{H}_2 \oplus, \cdots, \quad \mathcal{H}_0 = \mathbb{C}, \quad \mathcal{H}_1 = \mathcal{H}_2 = \cdots = \mathbb{C}^2, \tag{6.3.27}$$

and the linear operator A which is defined on finite vectors $\mathbf{l}_{\mathrm{fin}}$ by a block tri-diagonal Jacobi type matrix J of the form (6.2.28) *using the first expression in* (6.2.30)*. Suppose, all its coefficients a_n, b_n and c_n are uniformly bounded, some elements of these matrices are always equal to zero and some elements are always positive according to* (6.2.29)*, and the extension of A by continuity is unitary operator in the space* (6.3.27)*.*

The generalized eigenfunction expansion of the operator A has the following form. According to Lemma 6.3.1, we construct, by $\varphi_0 \in \mathbb{C}$, the solution $\varphi(z) = (\varphi_n(z))_{n=0}^{\infty}$, $\varphi_n(z) \in \mathcal{H}_n$ of equations (6.3.7) *for $z = e^{i\theta} \in \mathbb{T}$, $\theta \in [0, 2\pi)$ in the form*

$$\varphi_n(z) = Q_n(z)\varphi_0 = (Q_{n;1}(z), Q_{n;2}(z))\varphi_0,$$

where $Q_{n;1}(z)$ and $Q_{n;2}(z)$ are polynomials of variables $z = e^{i\theta}$ and $\bar{z} = e^{-i\theta}$. Then the Fourier transform has a form

$$\mathbf{l}_{2,u} \supset \mathbf{l}_{\mathrm{fin}} \ni f = (f_n)_{n=0}^{\infty} \longmapsto \hat{f}(z) = \sum_{n=0}^{\infty} Q^*_n(z) f_n$$

$$= f_0 + \sum_{n=1}^{\infty} (\overline{Q_{n;1}(z)} f_{n;1} + \overline{Q_{n;2}(z)} f_{n;2}) \in L_2(\mathbb{T}, d\rho(z)). \tag{6.3.28}$$

Here $Q^*_n(z) : \mathcal{H}_n \longrightarrow \mathcal{H}_0$ is the operator adjoint to $Q_n(z) : \mathcal{H}_0 \longrightarrow \mathcal{H}_n$, $d\rho(z)$ is the probability spectral measure of A.

The Parseval equality has a form

$$(f, g)_{\mathbf{l}_{2,u}} = \int_{\mathbb{T}} \hat{f}(z)\overline{\hat{g}(z)}\, d\rho(z),$$
$$(Jf, g)_{\mathbf{l}_{2,u}} = \int_{\mathbb{T}} z\hat{f}(z)\overline{\hat{g}(z)}\, d\rho(z), \quad \forall f, g \in \mathbf{l}_{\text{fin}}. \tag{6.3.29}$$

Equalities (6.3.28) *and* (6.3.29) *have extensions by continuity for all* $f, g \in \mathbf{l}_{2,u}$. *Thus, the operator* (6.3.28) *becomes an isometry mapping whole* $\mathbf{l}_{2,u}$ *into whole* $L_2(\mathbb{T}, d\rho(z))$.

Polynomials $\overline{Q_{n;\alpha}(z)}$, $n \in \mathbb{N}$, $\alpha = 1, 2$, *and* $Q_{0,\alpha}(z) = 1$ *form the orthonormal system in* $L_2(\mathbb{T}, d\rho(z))$ *in the sense of (6.3.25) and it is total in this space.*

Proof It remains only to show that orthogonal polynomials of the form $\overline{Q_{n;\alpha}(z)}$, $n \in \mathbb{N}$, $\alpha = 1.2$, and $Q_{0,\alpha}(z) = 1$ form a total set in the space $L_2(\mathbb{T}, d\rho(z))$. To do this, note that due to the finiteness of the measure $d\rho(z)$ on $\mathbb{T}$, elements $z^j\bar{z}^k$, $j, k \in \mathbb{N}_0$, form a total set in $L_2(\mathbb{T}, d\rho(z))$.

Assume the opposite, that is, the system of polynomials is not total. Then there exists a non-zero function $h(z) \in L_2(\mathbb{T}, d\rho(z))$ which is orthogonal to these polynomials, and therefore, according to (6.3.9), for of all $z^j\bar{z}^k$, $j, k \in \mathbb{N}_0$. Thus, $h(z) = 0$. □

The last theorem solves the direct problem for the unitary operator A defined in the space $\mathbf{l}_{2,u}$ by the matrix J of the form (6.2.28).

The inverse problem consists in construction by the given measure $d\rho(z)$ on $\mathbb{T}$ the unitary matrix J of the form (6.2.28) that has its spectral measure equal to $d\rho(z)$. *This construction is carried out according to Theorem* 6.2.5 *due to the Gram-Schmidt orthogonalization procedure for the system* (6.2.1). *For the matrix J of the form* (6.2.28) *constructed by $d\rho(z)$, the spectral measure of corresponding unitary operator A coincides with start (given) measure.*

Proof This fact is true since the system of orthogonal polynomials, connected with A, $\overline{Q_{n,1}(z)}$ and $\overline{Q_{n,2}(z)}$, $n \in \mathbb{N}$, and $Q_0(z)$, is orthonormal in $L_2(\mathbb{T}, d\rho(z))$ and according to Lemma 6.3.1 it is constructed by $\bar{z}^j$, z^k, $z \in \mathbb{T}$ similarly as the system (6.2.2) constructed by $e^{ji\theta}$, $e^{-ki\theta}$, $j, k \in \mathbb{N}_0$. Therefore,

$$Q_0(z) = 1 = P_0(z), \quad \overline{Q_{n,1}(z)} = P_{n;1}(\theta), \quad \overline{Q_{n,2}(z)} = P_{n;2}(\theta),$$
$$\forall n \in \mathbb{N}, \quad z = e^{i\theta}, \quad \theta \in [0, 2\pi). \tag{6.3.30}$$

Since both system of polynomials form a total set in $L_2(\mathbb{T}, d\rho(z))$, then (6.3.30) shows that the spectral measure, constructed by the operator, coincides with the given one. □

Let us remark that expressions (6.2.10) and (6.2.14) (as it was known in the classical theory of Jacobi matrices) restore the initial matrix (6.2.28) by the spectral measure $d\rho(z) = d\rho(\theta)$ of the operator constructed by J in $\mathbf{l}_{2,u}$.

6.4 The Detailed Internal Structure of the Block Matrix of the Unitary Operator

The section solves the open question of Cantero M. J., Moral L., Velázquez L. about the representation of an arbitrary unitary operator by CMV matrices. This issue is considered from the point of view of a one-to-one correspondence between unitary operators and (five-diagonal) Jacobi-type block tri-diagonal matrices in the form of the corresponding direct and inverse spectral problems for the trigonometric moment problem.

Let us find the conditions for the matrix J of the form (6.2.28) so that it is a unitary operator. In the space $\mathbf{l}_{2,u}$, the formally adjoint matrix J^+ has the form

$$J^+ = \begin{pmatrix} b_0^* & a_0^* & 0 & 0 & \cdots \\ c_0^* & b_1^* & a_1^* & 0 & \cdots \\ 0 & c_1^* & b_2^* & a_2^* & \cdots \\ \vdots & \vdots & \vdots & \vdots & \ddots \end{pmatrix}, \quad \begin{aligned} c_n^* &: \mathcal{H}_n \longrightarrow \mathcal{H}_{n+1}, \\ b_n^* &: \mathcal{H}_n \longrightarrow \mathcal{H}_n, \\ a_n^* &: \mathcal{H}_{n+1} \longrightarrow \mathcal{H}_n, \quad n \in \mathbb{N}_0. \end{aligned} \tag{6.4.1}$$

Multiplying the matrices (6.2.27) and (6.4.1), we get

$$JJ^+ = \begin{pmatrix} b_0b_0^* + c_0c_0^* & b_0a_0^* + c_0b_1^* & c_0a_1^* & 0 & 0 & \cdots \\ a_0b_0^* + b_1c_0^* & a_0a_0^* + b_1b_1^* + c_1c_1^* & b_1a_1^* + c_1b_2^* & c_1a_2^* & 0 & \cdots \\ a_1c_0^* & a_1b_1^* + b_2c_1^* & a_1a_1^* + b_2b_2^* + c_2c_2^* & b_2a_2^* + c_2b_3^* & c_2a_3^* & \cdots \\ \vdots & \vdots & \vdots & \vdots & \vdots & \ddots \end{pmatrix}. \tag{6.4.2}$$

The expression for J^+J is similar to (6.4.2) if a_n, b_n and c_n are replaced with c_n^*, b_n^* and a_n^*, respectively and vice versa.

Comparing expressions JJ^+ and J^+J with the single operator I, we conclude that equalities $JJ^+ = J^+J = I$ are equivalent to the system of equations

$$\begin{aligned}
&b_0 b_0^* + c_0 c_0^* = b_0^* b_0 + a_0^* a_0 = 1,\\
&c_n a_{n+1}^* = a_n^* c_{n+1} = 0,\\
&b_n a_n^* + c_n b_{n+1}^* = b_n^* c_n + a_n^* b_{n+1} = 0,\\
&a_n a_n^* + b_{n+1} b_{n+1}^* + c_{n+1} c_{n+1}^* = c_n^* c_n + b_{n+1}^* b_{n+1} + a_{n+1}^* a_{n+1} = I_n,\\
&n \in \mathbb{N}_0,
\end{aligned} \tag{6.4.3}$$

where I_n is the identity matrix (operator) in a appropriate space.

Note that the necessity of equalities

$$a_n b_n^* + b_{n+1} c_n^* = c_n^* b_n + b_{n+1}^* a_n = 0, \quad a_{n+1} c_n^* = c_{n+1}^* a_n = 0, \quad n \in \mathbb{N}_0,$$

follows from the third and the second equalities in (6.4.3), if we take into account the the adjointness "*".

Taking initial matrices a_0, b_0, c_0, and finding, from (6.4.3), elements a_1, b_1, c_1; a_2, b_2, c_2; ..., etc., (in a non-unique manner) we construct some unitary matrix J. But for such a matrix, Theorem 6.3.3 in general is not valid, because it is necessary to find these matrices in such a way that a_n and c_n must be of the form (6.2.28). Only in this case all functions (6.2.1) are linearly independent and Theorem 6.3.3 is applicable.

For an analysis we write (6.4.3) in the matrix form

$$b_0 b_0^* + \begin{pmatrix} c_{0;2,1} & c_{0;2,2} \end{pmatrix} \begin{pmatrix} \bar{c}_{0;2,1} \\ \bar{c}_{0;2,2} \end{pmatrix} = b_0^* b_0 + \begin{pmatrix} a_{0;1,1} & 0 \end{pmatrix} \begin{pmatrix} \bar{a}_{0;1,1} \\ 0 \end{pmatrix}, \tag{6.4.4}$$

and for $n \in \mathbb{N}$, we have

$$\begin{aligned}
&\begin{pmatrix} 0 & 0 \\ c_{n;2,1} & c_{n;2,2} \end{pmatrix} \begin{pmatrix} \bar{a}_{n+1;1,1} & 0 \\ \bar{a}_{n+1;2,1} & 0 \end{pmatrix}\\
&\qquad = \begin{pmatrix} \bar{a}_{n;1,1} & 0 \\ \bar{a}_{n;2,1} & 0 \end{pmatrix} \begin{pmatrix} 0 & 0 \\ c_{n+1;2,1} & c_{n+1;2,2} \end{pmatrix},
\end{aligned} \tag{6.4.5}$$

$$\begin{aligned}
&\begin{pmatrix} b_{n;1,1} & b_{n;1,2} \\ b_{n;2,1} & b_{n;2,2} \end{pmatrix} \begin{pmatrix} \bar{a}_{n;1,1} & 0 \\ \bar{a}_{n;1,2} & 0 \end{pmatrix}\\
&\qquad + \begin{pmatrix} 0 & 0 \\ c_{n;2,1} & c_{n;2,2} \end{pmatrix} \begin{pmatrix} \bar{b}_{n+1;1,1} & \bar{b}_{n+1;2,1} \\ \bar{b}_{n+1;1,2} & \bar{b}_{n+1;2,2} \end{pmatrix} = 0,
\end{aligned} \tag{6.4.6}$$

$$\begin{pmatrix} \bar{b}_{n;1,1} & \bar{b}_{n;2,1} \\ \bar{b}_{n;1,2} & \bar{b}_{n;2,2} \end{pmatrix} \begin{pmatrix} 0 & 0 \\ c_{n;2,1} & c_{n;2,2} \end{pmatrix} + \begin{pmatrix} \bar{a}_{n;1,1} & 0 \\ \bar{a}_{n;1,2} & 0 \end{pmatrix} \begin{pmatrix} b_{n+1;1,1} & b_{n+1;1,2} \\ b_{n+1;2,1} & b_{n+1;2,2} \end{pmatrix} = 0, \tag{6.4.7}$$

$$\begin{pmatrix} a_{n;1,1} & a_{n;1,2} \\ 0 & 0 \end{pmatrix} \begin{pmatrix} \bar{a}_{n;1,1} & 0 \\ \bar{a}_{n;1,2} & 0 \end{pmatrix} + \begin{pmatrix} b_{n+1;1,1} & b_{n+1;1,2} \\ b_{n+1;2,1} & b_{n+1;2,2} \end{pmatrix} \begin{pmatrix} \bar{b}_{n+1;1,1} & \bar{b}_{n+1;2,1} \\ \bar{b}_{n+1;1,2} & \bar{b}_{n+1;2,2} \end{pmatrix} + \\ + \begin{pmatrix} 0 & 0 \\ c_{n+1;2,1} & c_{n+1;2,2} \end{pmatrix} \begin{pmatrix} 0 & \bar{c}_{n+1;2,1} \\ 0 & \bar{c}_{n+1;2,2} \end{pmatrix} = \begin{pmatrix} 1 & 0 \\ 0 & 1 \end{pmatrix}, \tag{6.4.8}$$

$$\begin{pmatrix} 0 & \bar{c}_{n;2,1} \\ 0 & \bar{c}_{n;2,2} \end{pmatrix} \begin{pmatrix} 0 & 0 \\ c_{n;2,1} & c_{n;2,2} \end{pmatrix} + \begin{pmatrix} \bar{b}_{n+1;1,1} & \bar{b}_{n+1;2,1} \\ \bar{b}_{n+1;1,2} & \bar{b}_{n+1;2,2} \end{pmatrix} \begin{pmatrix} b_{n+1;1,1} & b_{n+1;1,2} \\ b_{n+1;2,1} & b_{n+1;2,2} \end{pmatrix} + \\ + \begin{pmatrix} \bar{a}_{n+1;1,1} & 0 \\ \bar{a}_{n+1;1,2} & 0 \end{pmatrix} \begin{pmatrix} a_{n+1;1,1} & a_{n+1;1,2} \\ 0 & 0 \end{pmatrix} = \begin{pmatrix} 1 & 0 \\ 0 & 1 \end{pmatrix}. \tag{6.4.9}$$

If we suppose for a convenience that b_0, c_0, and a_0 are of the form

$$b_0 = \begin{pmatrix} 0 & 0 \\ 0 & b_{0;2,2} \end{pmatrix} \quad a_0 = \begin{pmatrix} 0 & a_{0;1,2} \\ 0 & 0 \end{pmatrix} \quad b_0 = \begin{pmatrix} 0 & 0 \\ c_{0;2,1} & c_{0;2,2} \end{pmatrix}, \tag{6.4.10}$$

then we can say that equalities (6.4.5)–(6.4.9) hold true for $n \in \mathbb{N}_0$.

From (6.4.4), we obtain

$$|c_{0;2,1}|^2 + |c_{0;2,2}|^2 + |b_{0;2,2}|^2 = |a_{0;1,1}|^2 + |b_{0;2,2}|^2 = 1. \tag{6.4.11}$$

Taking into account (6.4.10) for equalities (6.4.5)–(6.4.9), we find from (6.4.5):

$$c_{n;2,1}\bar{a}_{n+1;1,1} + c_{n;2,2}\bar{a}_{n+1;1,2} = 0; \tag{6.4.12}$$

from (6.4.6), we get

$$\begin{aligned} b_{n;1,1}\bar{a}_{n;1,1} + b_{n;1,2}\bar{a}_{n;1,2} &= 0, \\ b_{n;2,1}\bar{a}_{n;1,1} + b_{n;2,2}\bar{a}_{n;1,2} + c_{n;2,1}\bar{b}_{n+1;1,1} + c_{n;2,2}\bar{b}_{n+1;1,2} &= 0, \\ c_{n;2,1}\bar{b}_{n+1;2,1} + c_{n;2,2}\bar{b}_{n+1;2,2} &= 0; \end{aligned} \tag{6.4.13}$$

from (6.4.7), we get

$$\begin{aligned}\bar{b}_{n;2,1}c_{n;2,1} + \bar{a}_{n;1,1}b_{n+1;1,1} &= 0,\\ \bar{b}_{n;2,1}c_{n;2,2} + \bar{a}_{n;1,1}b_{n+1;1,2} &= 0,\\ \bar{b}_{n;2,2}c_{n;2,1} + \bar{a}_{n;1,2}b_{n+1;1,1} &= 0,\\ \bar{b}_{n;2,2}c_{n;2,2} + \bar{a}_{n;1,2}b_{n+1;1,2} &= 0;\end{aligned} \tag{6.4.14}$$

from (6.4.8), we get

$$\begin{aligned}|a_{n;1,1}|^2 + |a_{n;1,2}|^2 + |b_{n+1;1,1}|^2 + |b_{n+1;1,2}|^2 &= 1,\\ b_{n+1;1,1}\bar{b}_{n+1;2,1} + b_{n+1;1,2}\bar{b}_{n+1;2,2} &= 0,\\ b_{n+1;2,1}\bar{b}_{n+1;1,1} + b_{n+1;2,2}\bar{b}_{n+1;1,2} &= 0,\\ |b_{n+1;2,1}|^2 + |b_{n+1;2,2}|^2 + |c_{n+1;2,1}|^2 + |c_{n+1;2,2}|^2 &= 1;\end{aligned} \tag{6.4.15}$$

from (6.4.9), we get

$$\begin{aligned}|c_{n;2,1}|^2 + |b_{n+1;1,1}|^2 + |b_{n+1;2,1}|^2 + |a_{n+1;1,1}|^2 &= 1,\\ \bar{c}_{n;2,1}c_{n;2,2} + \bar{b}_{n+1;1,1}b_{n+1;1,2} + \bar{b}_{n+1;2,1}b_{n+1;2,2} + \bar{a}_{n+1;1,1}a_{n+1;1,2} &= 0,\\ \bar{c}_{n;2,2}c_{n;2,1} + \bar{b}_{n+1;1,2}b_{n+1;1,1} + \bar{b}_{n+1;2,2}b_{n+1;2,1} + \bar{a}_{n+1;1,2}a_{n+1;1,1} &= 0,\\ |c_{n;2,2}|^2 + |b_{n+1;1,2}|^2 + |b_{n+1;2,2}|^2 + |a_{n+1;1,2}|^2 &= 1.\end{aligned} \tag{6.4.16}$$

We remark that the second equalities in (6.4.15) and in (6.4.16) are adjoint to the third equalities correspondingly.

Denote h_k and v_k, $k = 0, 1, 2, \ldots$ rows and columns of the matrix (6.2.28). We regard each h_k and v_k as a finite vector in l_2. Such a vector contains no more than fore non-zero elements. Put $(h_i, \bar{h}_j) = (h_i, \bar{h}_j)_{l_2}$, $(v_i, \bar{v}_j) = (v_i, \bar{v}_j)_{l_2}$ as the usual scalar product in l_2; $(\bar{h}_j)$, $(\bar{v}_j)$ denote vectors with conjugate elements).

Lemma 6.4.1 *Systems of Eqs.* (6.4.11)–(6.4.16) *are equivalent to equalities*

$$(v_m, \bar{v}_n) = 0,\ m \neq n,\ (v_m, \bar{v}_m) = \|v_m\|^2 = 1,$$

$$(h_m, \bar{h}_n) = 0,\ m \neq n,\ (h_m, \bar{h}_m) = \|h_m\|^2 = 1,\ m, n \in \mathbb{N}_0.$$

Proof follows immediately from the direct observation of (6.4.11)–(6.4.16), using (6.2.28). Namely, the equalities (6.4.11)–(6.4.16) are equivalent (6.4.17)–(6.4.22), correspondingly

$$\|v_0\| = \|h_0\| = 1; \tag{6.4.17}$$

$$(h_{2n+1}, \bar{h}_{2(n+2)}) = 0; \tag{6.4.18}$$

$$(h_{2n}, \bar{h}_{2(n+1)}) = 0, \; (h_{2n+1}, \bar{h}_{2(n+1)}) = 0, \; (h_{2n+1}, \bar{h}_{2(n+1)+1}) = 0; \tag{6.4.19}$$

$$\begin{gathered}(\bar{v}_{2n}, v_{2(n+1)}) = 0, \; (\bar{v}_{2n}, v_{2(n+1)+1}) = 0, \\ (\bar{v}_{2n+1}, v_{2(n+1)}) = 0, \; (\bar{v}_{2n+1}, v_{2(n+1)+1}) = 0;\end{gathered} \tag{6.4.20}$$

$$\begin{gathered}\|h_{2(n+1)}\| = 1, \; (h_{2(n+1)}, \bar{h}_{2(n+1)+1}) = 0, \\ (\bar{h}_{2(n+1)+1}, h_{2(n+1)}) = 0, \; \|h_{2(n+1)+1}\| = 1;\end{gathered} \tag{6.4.21}$$

$$\begin{gathered}\|v_{2(n+1)}\| = 1, \; (\bar{v}_{2(n+1)}, v_{2(n+1)+1}) = 0, \\ (v_{2(n+1)+1}, \bar{v}_{2(n+1)}) = 0, \; \|v_{2(n+1)+1}\| = 1.\end{gathered} \tag{6.4.22}$$

□

Now, using Lemma 6.4.1, we will obtain the inner structure of J. Let us start with $b_{0;2,2} := \bar{\alpha}_0 \in \mathbb{C}$. Since $(v_0, \bar{v}_0) = \|v_0\|^2 = 1$, i.e., (6.4.17), we get $|\alpha_0| \leq 1$ and $a_{0;1,2} := \rho_0 = \sqrt{1 - |\alpha|^2}$. At the "zero" step we have

$$J = \left(\begin{array}{c|ccc|cc|cc} \bar{\alpha}_0 & * & c_0 & + & & & & \cdots \\ \rho_0 & * & & * & 0 & 0 & & \\ & & b_1 & & & c_1 & 0 & \cdots \\ 0 & * & & * & * & + & & \\ \hline \vdots & \vdots & & & \vdots & & \vdots & \ddots \end{array} \right). \tag{6.4.23}$$

Then we use $(v_0, \bar{v}_1) = 0$, $(v_0, \bar{v}_2) = 0$, and Lemma 6.4.1, i.e., the first and the second equalities in (6.4.20) with $n = 0$. Put

$$c_{0;2,1} := \bar{\alpha}_1 \rho_0, \; c_{0;2,2} := \rho_1 \rho_0, \; b_{1;1,1} := -\bar{\alpha}_1 \alpha_0, \; b_{1;1,2} := -\rho_1 \alpha_0,$$

where $\alpha_1 \in \mathbb{C}$ and $\rho_1 \in \mathbb{R}$ are some coefficients. Since $(h_0, \bar{h}_0) = 1$, i.e., (6.4.17), then

$$|\alpha_0|^2 + |\bar{\alpha}_1 \rho_0|^2 + |\rho_1 \rho_0|^2 = 1,$$

and taking into account $|\alpha_0|^2 + |\rho_0|^2 = 1$, we get $|\alpha_1|^2 + |\rho_1|^2 = 1$.

After the "zero" and the "first" steps, we have

$$J = \left(\begin{array}{c|cc|cc|cc} \bar{\alpha}_0 & \bar{\alpha}_1 \rho_0 & \rho_1 \rho_0 & & & & \cdots \\ \rho_0 & -\bar{\alpha}_1 \alpha_0 & -\rho_1 \alpha_0 & 0 & 0 & & \\ & & & & c_1 & 0 & \cdots \\ 0 & * & * & * & + & & \\ \hline \vdots & \vdots & \vdots & \vdots & & \vdots & \ddots \end{array} \right). \tag{6.4.24}$$

In particular, by using $(h_1, \bar{h}_1) = 1$, $(h_0, \bar{h}_1) = 0$, and Lemma 6.4.1, we have

$$|\rho_0|^2 + |\bar{\alpha}_1\alpha_0|^2 + |\rho_1\alpha_0|^2 = 1,$$

and

$$(\bar{\alpha}_0)(\rho_0) + (\bar{\alpha}_1\rho_0)(-\alpha_1\rho_0) + (\rho_1\rho_0)(-\rho_1\bar{\alpha}_0) = 0.$$

At the "second" step $(h_0, \bar{h}_2) = 0$ and Lemma 6.4.1 are used, i.e., the third equality (6.4.19) for $n = 0$ and $(h_0, \bar{h}_3) = 0$ and Lemma 6.4.1, i.e., (6.4.18) for $n = 0$. (Obviously, we can use $(h_1, \bar{h}_2) = 0$ and $(h_1, \bar{h}_3) = 0$, i.e., the second equality of (6.4.21) and the first equality of(6.4.19)). Put

$$b_{1;2,1} := \bar{\alpha}_2\rho_1,\ b_{1;2,2} := -\bar{\alpha}_2\alpha_1,\ a_{1;1,1} := \rho_2\rho_1,\ a_{1;1,2} := -\rho_2\alpha_0,$$

where $\alpha_2 \in \mathbb{C}$ and $\rho_2 \in \mathbb{R}$ some new proportion coefficients.

By using $(v_1, \bar{v}_1) = 1$, Lemma 6.4.1, and taking into account $|\alpha_0|^2 + |\rho_0|^2 = 1$, $|\alpha_1|^2 + |\rho_1|^2 = 1$, we get $|\alpha_2|^2 + |\rho_2|^2 = 1$.

In particular, using $(v_1, \bar{v}_2) = 0$ and Lemma 6.4.1, we get the third equality of (6.4.22).

Hence, after the "zero" (6.4.23), "first" (6.4.24) and the "second" steps, we have

$$J = \left(\begin{array}{c|cc|cc|cc|c}
\bar{\alpha}_0 & \bar{\alpha}_1\rho_0 & \rho_1\rho_0 & & & & & \cdots \\
\rho_0 & -\bar{\alpha}_1\alpha_0 & -\rho_1\alpha_0 & 0 & 0 & & & \\
 & & & & c_1 & & 0 & \cdots \\
0 & \bar{\alpha}_2\rho_1 & -\bar{\alpha}_2\alpha_1 & * & + & & & \\
\hline
 & \rho_2\rho_1 & -\rho_2\alpha_1 & * & * & 0 & 0 & \\
 & & & & b_2 & & c_2 & \cdots \\
 & 0 & 0 & * & * & * & + & \\
\vdots & \vdots & \vdots & \vdots & \vdots & \vdots & \vdots & \ddots
\end{array}\right).$$

Now, obviously using the above steps, we can prove such a theorem by the induction.

Theorem 6.4.2 *The matrix J in (6.2.28) has a form*

$$J = \left(\begin{array}{c|cc|cc|cc|cc|c}
\bar{\alpha}_0 & \bar{\alpha}_1\rho_0 & \rho_1\rho_0 & & & & & & & \cdots \\
\rho_0 & -\bar{\alpha}_1\alpha_0 & -\rho_1\alpha_0 & 0 & 0 & & & 0 & & \cdots \\
0 & \bar{\alpha}_2\rho_1 & -\bar{\alpha}_2\alpha_1 & \bar{\alpha}_3\rho_2 & \rho_3\rho_2 & & & & & \cdots \\
\hline
 & \rho_2\rho_1 & -\rho_2\alpha_1 & -\bar{\alpha}_3\alpha_2 & -\rho_3\alpha_2 & 0 & 0 & & & \cdots \\
 & 0 & 0 & \bar{\alpha}_4\rho_3 & -\bar{\alpha}_4\alpha_3 & \bar{\alpha}_5\rho_4 & \rho_5\rho_4 & & & \cdots \\
\hline
 & & & \rho_4\rho_3 & -\rho_4\alpha_3 & -\bar{\alpha}_5\alpha_4 & -\rho_5\alpha_4 & 0 & 0 & \cdots \\
 & 0 & & 0 & 0 & \bar{\alpha}_6\rho_5 & -\bar{\alpha}_6\alpha_5 & \bar{\alpha}_7\rho_6 & \rho_7\rho_6 & \cdots \\
\vdots & \vdots & \vdots & \vdots & \vdots & \vdots & \vdots & \vdots & \vdots & \ddots
\end{array}\right), \tag{6.4.25}$$

where $\alpha_k \in \mathbb{C}$, $k \in \mathbb{N}_0$ any coefficients such that $|\alpha_k| < 1$ and $\rho_k = \sqrt{1 - |\alpha_k|^2}$. Conversely, each matrix of the form (6.4.25) *is a unitary operator in l_2 as a five-diagonal one or in $\mathbf{l}_{2,u}$ as a block tree-diagonal one.*

6.5 The Solution of the Trigonometric Moment Problem

The trigonometric moment problem consists in finding conditions for the sequence of complex numbers (s_n), $n \in \mathbb{Z}$, $s_n \in \mathbb{C}$, which are necessary and sufficient for the existence of the representation

$$s_n = \int_{\mathbb{T}} e^{i\lambda n}\, d\rho(\lambda), \quad n \in \mathbb{Z}, \tag{6.5.1}$$

where $d\rho(\lambda)$ is a Borel measure on the σ-algebra of Borel sets on the unit circle $\mathbb{T} = \{z \in \mathbb{C} \mid |z| = 1\}$ in a usual topology. It covers the questions of existence, uniqueness of such a measure, and others.

There are several approaches to researching such questions. Here, in our opinion, the spectral approach first proposed by Krein M. G. for the classical moment problem, is more natural. It uses a spectral representation for the unitary operator and enables broad generalizations, particularly important for mathematical physics.

This approach is closely related to the theory of block tri-diagonal matrices of the Jacobi type, presented at the beginning of this chapter.

For the solution of the trigonometric moment problem, we need some auxiliary aspects of the corresponding spectral theory. Let a unitary operator V be given in a separable Hilbert space $\mathcal{H}$, i.e., $\mathfrak{D}(V) = \mathcal{H}$. We consider an equipment of the space $\mathcal{H}$:

$$\mathcal{H}_- \supset \mathcal{H} \supset \mathcal{H}_+ \supset \mathcal{D}, \tag{6.5.2}$$

where $\mathcal{H}_+$ is a Hilbert space topologically and quasi-nuclear imbedded into $\mathcal{H}$. We denote, as usual, $\mathcal{H}_-$ the space dual to $\mathcal{H}_+$ with respect to the space $\mathcal{H}$; the set $\mathcal{D}$ is a linear topological space, topologically imbedded into $\mathcal{H}_+$.

The operator V is called standardly connected with the chain (6.5.2), if the restriction $V \upharpoonright \mathcal{D}$ acts from $\mathcal{D}$ into $\mathcal{H}_+$ continuously. In what follows, we consider just such a case.

Recall that the vector $\Omega \in \mathcal{D}$ is called strong cyclic for the operator V if for all $n \in \mathbb{Z}$ $\Omega \in \mathfrak{D}(V^n) \in \mathcal{D}$ and the set of all vectors $A^n\Omega$ for $n = \mathbb{Z}$ is total in the space $\mathcal{H}_+$ (and therefore in $\mathcal{H}$).

Let us assume that a strong cyclic vector exists and formulate some simplified variant of the projective spectral theorem (see [23], Chapter 3, Theorem 2.7, or [21], Chapter 5, [48], Chapter 15).

Theorem 6.5.1 *For the unitary operator V with a strong cyclic vector in the separable Hilbert space $\mathcal{H}$ there exists a non-negative finite Borel measure $d\rho(\lambda)$ on the unit circle $\mathbb{T}$ such that for ρ-almost every $e^{i\lambda} \in \mathbb{T}$ there exists a generalized eigenvector $\xi_\lambda \in \mathcal{H}_-$, i.e., for all $f \in \mathcal{D}$*

$$(\xi_\lambda, Vf)_{\mathcal{H}} = e^{-i\lambda}(\xi_\lambda, f)_{\mathcal{H}}, \quad \xi_\lambda \neq 0. \tag{6.5.3}$$

The corresponding Fourier transform F, given by the rule

$$\mathcal{H} \supset \mathcal{H}_+ \ni f \mapsto (Ff)(\lambda) = \hat{f}(\lambda) = (f, \xi_\lambda)_{\mathcal{H}} \in L_2(\mathbb{T}, d\rho(\lambda)), \tag{6.5.4}$$

is an isometry operator (after the extension by continuity) acting from the space $\mathcal{H}$ in $L_2(\mathbb{T}, d\rho(\lambda))$. The image of the operator V by the transformation F is the multiplication by $e^{i\lambda}$ operator in $L_2(\mathbb{T}, d\rho(\lambda))$, i.e., $(FA)f(\lambda) = e^{i\lambda} f(\lambda)$.

We remark that the notation $(\cdot, \cdot)_{\mathcal{H}}$ denotes the same as $\langle \cdot, \cdot \rangle_{\mathcal{H}}$ (see Chap. 2, Sect. 2.7).

The main result of this section is contained in the theorem.

Theorem 6.5.2 *The sequence of complex numbers (s_n), $n \in \mathbb{Z}$ has a representation* (6.5.1) *if and only if it is positive definite, i.e.,*

$$\sum_{m,n \in \mathbb{Z}} s_{n-m} f_m \bar{f}_n \geq 0 \tag{6.5.5}$$

on finite sequences of complex numbers $(f_n)_{n\in\mathbb{Z}}$, $f_n \in \mathbb{C}$.

If the representation (6.5.1) *exists, then the measure $d\rho(\lambda)$ is unique.*

Note that the condition (6.5.5) is necessary and sufficient for the existence of the representation (6.5.1), and the representation (the measure) is unique.

Proof Let us prove the necessity of the condition (6.5.5). If the sequence (s_n), $n \in \mathbb{Z}$ has a representation (6.5.1), then for an arbitrary finite sequence $f = (f_n)_{n\in\mathbb{Z}}$, $f_n \in \mathbb{C}$, we have

$$\sum_{m,n \in \mathbb{Z}} s_{n-m} f_m \bar{f}_n = \int_{\mathbb{T}} \left| \sum_{n \in \mathbb{Z}} e^{i\lambda n} f_n \right|^2 d\rho(\lambda) \geq 0,$$

which proves the necessity.

Let us proceed to the proof of sufficiency. We will use the scheme described in Theorem 3.5.2. Thus, the role of the first of the chains (3.5.13) will be played an equipment for not the usual space l_2, but two-sided sequences, which we denote by $l_{u,2} = G_0$, namely,

$$f = (f_j)_{j\in\mathbb{Z}}, \ f_j \in \mathbb{C}, \quad \|f\|_{l_2}^2 = \sum_{j\in\mathbb{Z}} |f_j|^2 < \infty.$$

The role of spaces G_+ and G_- will play the spaces $l_{u,2}$ with a corresponding weight, namely, we denote by $l_{u,2}(q)$ the space $l_{u,2}$ with the weight $q = (q_j)_{j\in\mathbb{Z}}$, $q_j > 0$; it consists of sequences $f = (f_j)_{j\in\mathbb{Z}}$, $f_j \in \mathbb{C}$, for which

$$\|f\|^2_{l_2(q)} = \sum_{j\in\mathbb{Z}} |f_j|^2 q_j < \infty, \quad (f, g)_{l_2(q)} = \sum_{j\in\mathbb{Z}} f_j \bar{g}_j q_j. \tag{6.5.6}$$

Let us fix $q = (q_j)_{j\in\mathbb{Z}}$, $q_j \ge 1$. Then the firs of chains (3.5.13) (i.e., (3.5.1)) has a form

$$l_{u,2}(q^{-1}) \supset l_{u,2} \supset l_{u,2}(q) \supset l_{u,\mathrm{fin}}, \quad q^{-1} = (q_j^{-1})_{j\in\mathbb{Z}}, \; q = (q_j)_{j\in\mathbb{Z}}, \; q_j \ge 1, \tag{6.5.7}$$

where $l_{u,\mathrm{fin}}$ denotes the space of two-sided finite sequences. The involution "$*$" is now a usual complex conjugations $(f_j)_{j\in\mathbb{Z}} = f \mapsto f^* = (\bar{f}_j)_{j\in\mathbb{Z}}$. Accordingly, it is convenient to change the notation of vectors in Theorem 3.5.2 from φ to f.

Let us specify the choice of weight q. The condition (6.5.5) in Theorem say that the matrix $K = (K_{jk})_{j,k\in\mathbb{Z}}$, where $K_{jk} = s_{j-k}$, is positive defined. Then

$$|K_{jk}|^2 \le K_{jj} K_{kk}, \text{ i.e., } |s_{j,k}|^2 \le s_{2j} s_{2k}, \; j, k \in \mathbb{Z}. \tag{6.5.8}$$

Let $q = (q_n)_{n\in\mathbb{Z}}$, $q_n \ge 1$ be a sequence such that

$$C_u(s, q) := \sum_{n\in\mathbb{Z}} s_{2n} q_n^{-1} < \infty. \tag{6.5.9}$$

By using (6.5.8) and (6.5.9), we get

$$\begin{aligned}
0 &\le \sum_{j,k\in\mathbb{Z}} s_{j-k} f_k \bar{f}_j \le \sum_{j,k\in\mathbb{Z}}^{\infty} |s_{j-k}||f_k||f_j| \\
&\le \sum_{j,k\in\mathbb{Z}} \sqrt{s_{2j}s_{2k}}|f_k||f_j| \le \sum_{j,k\in\mathbb{Z}} \sqrt{s_{2j}s_{2k}q_j^{-1}q_k^{-1}}|f_k q_k^{1/2}||f_j q_j^{1/2}| \\
&\le \left(\sum_{j,k\in\mathbb{Z}} s_{2j}s_{2k}q_j^{-1}q_k^{-1}\right)^{1/2} \left(\sum_{j,k\in\mathbb{Z}} |f_k|^2 q_k |f_j|^2 q_j\right)^{1/2} \\
&= \left(\sum_{j\in\mathbb{Z}} s_{2j} q_j^{-1}\right)\left(\sum_{j\in\mathbb{Z}} |f_j|^2 q_j\right) = C_u(s, q)\|f\|^2_{l_2(q)}, \quad f \in l_{u,\mathrm{fin}}.
\end{aligned} \tag{6.5.10}$$

Passing here to the limit in two-sided finite sequences $l_{u,\text{fin}}$ to $l_{u,2}(q)$, we obtain the chain

$$\begin{gathered} l_{u,2}(q^{-1}) \supset l_{u,2} \supset l_{u,2}(q) \supset l_{u,\text{fin}}, \ q = (q_n)_{n\in\mathbb{Z}}, \ q_n \geq 1, \\ C_u(s, q) = \sum_{n\in\mathbb{Z}} \frac{s_{2n}}{q_n} < \infty. \end{gathered} \tag{6.5.11}$$

We fix the weight q for which the condition (6.5.11) is fulfilled (the sequence (s_n), $n \in \mathbb{Z}$ is considered given). Then the chain of spaces (6.5.11) plays the role of the first of the chains (3.5.13). As already said, the matrix $K = (K_{jk})_{j,k\in\mathbb{Z}}$, where $K_{jk} = s_{j-k}$ is the kernel K of the discrete variable $j \in \mathbb{Z}$; we will denote $S_u = K$. It will be positive definite, since the condition (3.5.3) is fulfilled: it coincides with (6.5.5).

The next step in the scheme of Theorem 3.5.2 is the construction of a chain (3.5.17). Here we require that the imbedding $H_+ \hookrightarrow G_+$ will be Hilbert-Schmidt type. In the space $G_+ = l_{u,2}(q)$, where q satisfies the condition (6.5.11), the space $l_{u,2}(p)$ embedded quasi-nucleary, where $p = (p_n)_{n\in\mathbb{Z}}$, if

$$\sum_{n\in\mathbb{Z}} p_n^2 q_n^2 < \infty. \tag{6.5.12}$$

Thus, as a chain (3.5.17) can be written

$$(l_{u,\text{fin}})' \supset l_{u,2}(p^{-1}) \supset l_{u,2}(q^{-1}) = l_{u,2} \supset l_{u,2}(q) \supset l_{u,2}(p) \supset l_{u,\text{fin}}, \tag{6.5.13}$$

where the weight q is chosen such that the condition (6.5.11) is fulfilled, and the weight p is so that the condition (6.5.12) is fulfilled. Fix the weight $p = (p_n)_{n\in\mathbb{Z}}$.

According to Theorem 3.5.2, the necessary chain (3.5.18) is constructed, where we set $K = S_u$ and, hence, get

$$H_{-,S_u} \supset H_{S_u} \supset H_{+,S_u} \supset l_{u,\text{fin}}. \tag{6.5.14}$$

Let us first prove the theorem for the case of the non-degenerate kernel $K = S_u$, i.e., when the inequality (6.5.5) contains the sign "$>$" for an arbitrary finite sequence $(f_n)_{n\in\mathbb{Z}} \in l_{u,\text{fin}}$, which is not equal to zero (that is, at least for one n, $f_n \neq 0$).

The proof is completely based on Theorem 3.5.2.

In this case, the expression

$$(f, g)_{S_u} := \sum_{j\in\mathbb{Z}} s_{j-k} f_k \bar{g}_j, \quad f, g \in l_{u,\text{fin}} \tag{6.5.15}$$

is a scalar (rather than a quasi-scalar) product in the space $l_{u,\text{fin}}$. We denote the completion of this space relative to (6.5.15) by H_{S_u}.

The linear space $\mathbb{C}^\infty$ of two-sided finite sequences $f = (f_n)_{n\in\mathbb{Z}}$, $f_n \in \mathbb{C}$, we denote by l_u. Let $\delta_n = (\ldots, 0, 1, 0, \ldots)$, $n \in \mathbb{Z}$ be the δ-sequence ("1" is situated on the n-th position), then for all $f \in l_{u,\text{fin}}$ we have $f = \sum_{n\in\mathbb{Z}} f_n\delta_n$. Thus, we have a chain of linear spaces

$$l_u \supset H_{S_u} \supset l_{u,\text{fin}}, \quad \forall f \in l_{u,\text{fin}},\ f = (f_n)_{n\in\mathbb{Z}} = \sum_{n\in\mathbb{Z}} f_n\delta_n. \tag{6.5.16}$$

In the case of a non-degenerate kernel S_u, chains (6.5.14) and (6.5.13) give two chains of Hilbert spaces

$$(l_{u,2}(p))_{-,S_u} \supset H_{S_u} \supset l_{u,2}(p) \supset l_{u,\text{fin}}, \tag{6.5.17}$$

$$l_u = (l_{u,2})' \supset l_{u,2}(p^{-1}) \supset l_{u,2} \supset l_{u,2}(p) \supset l_{u,\text{fin}}, \tag{6.5.18}$$

where the weight p satisfies requirements (6.5.12) and (6.5.9).

Consider a linear expression and its algebraic inverse in the space l_u, namely,

$$J : (Jf)_k = f_{k-1}, \quad J^{-1} : (J^{-1}f)_k = f_{k+1}, \quad k \in \mathbb{Z},\ f \in l_u. \tag{6.5.19}$$

For the δ-sequence, we have relations

$$J\delta_n = \delta_{n+1}, \quad J^{-1}\delta_n = \delta_{n-1}, \quad n \in \mathbb{Z}. \tag{6.5.20}$$

It will be a unitary operator in H_S, which is given by the scalar product (6.5.15). Indeed, for $f, g \in l_{u,\text{fin}}$, we have

$$\begin{aligned}(Jf, g)_{S_u} &= \sum_{j,k\in\mathbb{Z}} s_{j-k}(Jf)_k\bar{g}_j = \sum_{j,k\in\mathbb{Z}} s_{j-k}f_{k-1}\bar{g}_j \\ &= \sum_{j,k\in\mathbb{Z}} s_{j-k+1}f_k\bar{g}_j = \sum_{j,k\in\mathbb{Z}} s_{j-k}f_k\overline{(Jg)_j} = (f, J^{-1}g)_{S_u},\end{aligned} \tag{6.5.21}$$

(the definition (6.5.19) and $J^* = J^{-1}$ are taken into account here).

Thus, the operator $l_{u,\text{fin}} \ni f \mapsto Jf \in l_{u,\text{fin}}$ will be unitary in the space H_{S_u}. Hence, we have

$$Vf = Jf, \quad f \in l_{u,\text{fin}} \subset \mathfrak{D}(V) = H_{S_u}. \tag{6.5.22}$$

Let us now apply Lemma 2.7.5. As chains (2.7.35) now we consider (6.5.17) and (6.5.18). Let $\xi_\lambda \in (l_{u,2}(p))_{-,S_u}$ be generalized eigenvector of the operator V with respect to the rigging (6.5.17). In such a case, due to Theorem 6.5.1,

$$(\xi_\lambda, Vf)_{S_u} = e^{-i\lambda}(\xi_\lambda, f)_{S_u}, \quad \lambda \in \mathbb{R},\ f \in l_{u,\text{fin}}. \tag{6.5.23}$$

Denote

$$P(\lambda) = U\xi_\lambda \in l_{u,2}(p^{-1}) \subset l_u, \ \ P(\lambda) = (P_n(\lambda))_{n\in\mathbb{Z}}.$$

By using (2.7.36), the expression (6.5.23) obtain a form

$$(P(\lambda), Vf)_{l_{2,u}} = e^{-i\lambda}(P(\lambda), f)_{l_2}, \ \ \lambda \in \mathbb{R}, \ f \in l_{u,\mathrm{fin}}. \tag{6.5.24}$$

The corresponding Fourier transform has a form

$$H_{S_u} \supset l_{\mathrm{fin}} \ni f \to (Ff)(\lambda) = \hat{f}(\lambda) = (f, P(\lambda))_{l_{u,2}} \in L_2(\mathbb{T}, d\rho(\lambda)). \tag{6.5.25}$$

Let us calculate $P(\lambda)$. The operator V is defined in (6.5.22) due to the expression (6.5.19), hence, (6.5.24) gives $\forall f \in l_{u,\mathrm{fin}}$

$$\begin{aligned}\sum_{n\in\mathbb{Z}} e^{i\lambda} P_n(\lambda)\bar{f}_n &= e^{i\lambda}(P(\lambda), f)_{l_{u,2}} = (P(\lambda), V^{-1}f)_{l_{u,2}} \\ &= (P(\lambda), J^{-1}f)_{l_{u,2}} = (JP(\lambda), f)_{l_{u,2}} = \sum_{n\in\mathbb{Z}} P_{n+1}(\lambda)\bar{f}_n.\end{aligned} \tag{6.5.26}$$

Hence

$$e^{i\lambda}P_n(\lambda) = P_{n+1}(\lambda), \quad n \in \mathbb{Z}.$$

Without loss of generality, we assume $P_0(\lambda) = 1$. Now the last two formulas give

$$P_n(\lambda) = e^{i\lambda n}, \quad n \in \mathbb{Z}. \tag{6.5.27}$$

Therefore, the Fourier transform (6.5.25) has the form

$$H_{S_u} \supset l_{u,\mathrm{fin}} \ni f \to (Ff)(\lambda) = \hat{f}(\lambda) = \sum_{n\in\mathbb{Z}} f_n e^{i\lambda n} \in L_2(\mathbb{T}, d\rho(\lambda)), \tag{6.5.28}$$

and we get the Parseval equality

$$(f, g)_{S_u} = \int_{\mathbb{T}} \hat{f}(\lambda)\overline{\hat{g}(\lambda)}\, d\rho(\lambda), \ \ f, g \in l_{u,\mathrm{fin}}. \tag{6.5.29}$$

To construct the Fourier transform (6.5.25) and check formulas (6.5.26)–(6.5.29) it is necessary to find out the existence of a strong cyclic vector for the operator V $\Omega = \delta_0 \in l_{u,\mathrm{fin}}$ in the sense of the rigging (6.5.2).

The last is really so, since from (6.5.20), we have

$$V^p\Omega = J^p\delta_0 = \delta_p, \quad p \in \mathbb{Z}.$$

The Parseval equality (6.5.29) directly leads to the representation (6.5.1): according to (6.5.27), (6.5.28) $\hat{\delta}_n = e^{i\lambda n}$ and in particular $\hat{\delta}_0 = 1$; the scalar product (6.5.15) gives

$$s_n = (\delta_n, \delta_0)_{S_u} = (\hat{\delta}_n, \hat{\delta}_0)_{L_2(\mathbb{T}, d\rho(\lambda))} = \int\limits_{\mathbb{T}} e^{i\lambda n}\, d\rho(\lambda), \ n \in \mathbb{Z}.$$

Therefore, the first part of the theorem in the case of the non-degenerate kernel $K = S_u$ is proved. To prove its second part, it is necessary to make sure that the condition (3.5.26) leads to the unitary operator V in the space H_{S_u}.

Therefore, Theorem 6.5.2 is proven in the case considered above.

Consider the case of a degenerate kernel $S_u = K$, that is, when the left expression in the inequality (6.5.5) is equal to zero for some non-zero sequence $f \in l_{u,\text{fin}}$. Consider the (linear) set of all such sequences. Since the inequality (6.5.8) holds true, then such a set will be closed in the space $l_{u,2}(q)$ constructed according to (6.5.11); that is, in the space G_+ from the chain (3.5.17).

Thus, we get the general situation of the quasi-scalar product considered in Theorem 3.5.2. The expression $(f, g)_{S_u}$ in (6.5.15) will now be a quasi-scalar product, but the operator J according to the equality (6.5.21) will be unitary in the quasi-scalar product (3.5.4). According to Lemma 3.5.4, this operator transfers a class into a class and the proof our theorems starting with the definition (6.5.22) can be repeated.

Remark 6.5.3 Let conditions of Theorem 6.5.2 be fulfilled and we have a representation of the moment sequence (6.5.1) by some measure $d\rho(\lambda)$ with a support the circle $\mathbb{T}$. Suppose that the support of this measure is wider than the set consisting of a finite number of points on the circle. Then the inequality (6.5.5) should contain the sign ">", that is, the corresponding product (6.5.15) will not be quasi-scalar, but scalar.

In fact, for the support α of the measure $d\rho(\lambda)$, we have the relation which follows from (6.5.1) and (6.5.5) for $\alpha \subset \mathbb{T}$, namely,

$$\begin{aligned} \int\limits_{\alpha} |F_n(\lambda)|^2\, d\rho(\lambda) &= \int\limits_{\mathbb{T}} |F_n(\lambda)|^2\, d\rho(\lambda) = \int\limits_{\alpha} |\sum_{j=0}^{\pm n} f_j e^{i\lambda j}|^2\, d\rho(\lambda) \\ &= \sum_{j,k=0}^{\pm n} s_{j-k} f_j \bar{f}_k \geq 0, \text{ where } F_n(\lambda) = \sum_{j=0}^{\pm n} f_j e^{i\lambda j}. \end{aligned} \tag{6.5.30}$$

Suppose that the product $(\,\cdot\,,\,\cdot\,)_{S_u}$ is quasi-scalar, then there will be $n \in \mathbb{Z}$, $f_{-n}, \dots, f_0, \dots, f_n$, which are not all equal to zero, so that in (6.5.30), the sign "$\geq$" will be replaced with the sign "$=$" and for the corresponding polynomial $F_n(\lambda)$, due to (6.5.30), we get $F_n(\lambda) = 0$, $e^{i\lambda} \in \alpha$. But the polynomial $F_n(\lambda) = 0$ has a maximum of $2n + 1$ different zeros, and α under the condition has more than $2n + 1$ points. The obtained contradiction completes the proof. □

Bibliographical Notes

The results of the chapter are a generalization of Akhiezer N. I. classical works [2, 3]. To obtain these results, the works of Berezansky Yu. M. were used. [17, 18]. When considering the trigonometric moment problem, orthogonal polynomials arise, studied in works of Golinskii L. B. [126], Golinskii L., Totik V. [127], Simon B. [262–266], Atzmon A.[13]. The general theory of orthogonal polynomials is found in Geronimus Ya. L. [117], Szegö G. [278]. Koshmanenko V. D. and Dudkin M. E. [171] have the basic information about equipped spaces. The main content of the chapter is contained in works of Berezansky Yu. M., Dudkin M. E. [36, 84]. Five-diagonal (block tri-diagonal) matrices were first constructed in the works of Cantero M. J., Moral L., Velázquez L. [62, 63]. Works of Killip R., Nenciu I. [156], Arlinskiĭ Yu., Golinskiĭ L., Tsekanovskiĭ E. [11, 12] contain a lot of information about the trigonometric moment problem and orthogonal polynomials on the unit circle. The book of Schmüdgen K. [255] also has a relation to the trigonometric moment problem.

Chapter 7
Block Jacobi Type Matrices and the Complex Moment Problem in the Exponential Form

This chapter considers the generalization of the information given in Chaps. 3 and 5. Namely, an analogue of the Jacobi type block matrix related to the complex moment problem in the exponential form is proposed and corresponding polynomials, orthogonal with respect to some probability measure on the complex plane are investigated. In this case, two commutative block matrices are obtained, which both have a tri-diagonal block structure, one of them generates a unitary and the second one generates a self-adjoint operator in the l_2-type space. A one-to-one correspondence is also established between probability measures on a compact set of the complex plane and such couple of matrices.

7.1 Construction of the Tri-Diagonal Block Matrix Corresponding to the Complex Moment Problem in the Exponential Form

Let $d\rho(z)$ be a Borel measure with a compact support on $\mathbb{C}$ and $L_2 = L_2(\mathbb{C}, d\rho(z))$ be the space of square integrable complex-valued functions defined on $\mathbb{C}$. We suppose that functions $\mathbb{C} \ni z \longmapsto r^m e^{in\theta}$, $m \in \mathbb{N}_0, n \in \mathbb{Z}$ are linear independent and forms total set in L_2, here $z = re^{i\theta}, r \geq 0, \theta \in [0, 2\pi)$.

In order to find an analog of Jacobi matrix J of the form (3.1.3) or (5.1.36), there is need to choose an order of the orthogonalization in L_2 applied to the set of functions

$$\{r^m e^{in\theta}\}, \quad m \in \mathbb{N}_0,\ n \in \mathbb{Z}. \tag{7.1.1}$$

We use the linear order for the orthogonalization according to the Gram-Schmidt procedure (see Fig. 7.1).

Y. M. Berezansky, M. E. Dudkin, *Jacobi Matrices and the Moment Problem*,
Operator Theory: Advances and Applications 294,
https://doi.org/10.1007/978-3-031-46387-7_7

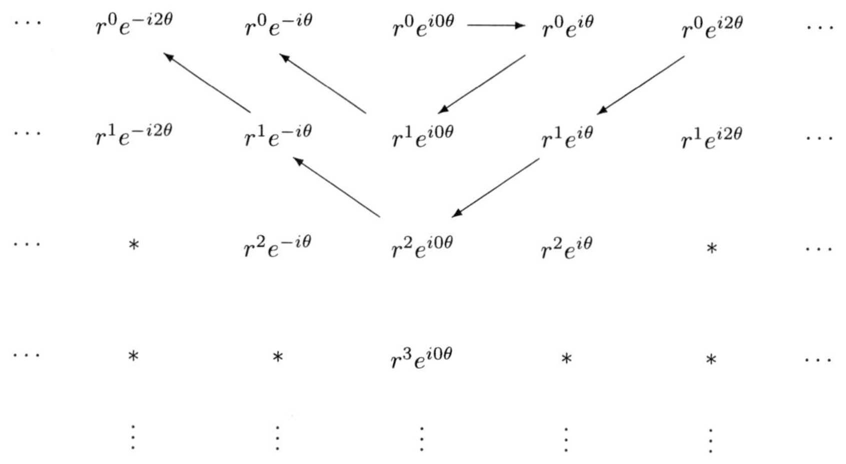

Fig. 7.1 The orthogonalization order

According to Fig. 7.1, we get

$$
\begin{aligned}
&r^0e^{i0\theta} = 1;\\
&r^0e^{i\theta}, r^1e^{i0\theta}, r^0e^{-i\theta};\\
&r^0e^{i2\theta}, r^1e^{i\theta}, r^2e^{i0\theta}, r^1e^{-i\theta}, r^0e^{-i2\theta}; \quad \dots;\\
&r^0e^{in\theta}, r^1e^{i(n-1)\theta}, \dots, r^{n-1}e^{i\theta}, r^ne^{i0\theta}, r^{n-1}e^{-i\theta} \dots, r^1e^{-i(n-1)\theta}, r^0e^{-in\theta};\\
&\dots
\end{aligned}
\tag{7.1.2}
$$

That is, as some sets on "lines" starting at $r^0e^{in\theta}$ and ending at $r^0e^{-in\theta}$ and "with an angle" at $r^ne^{i0\theta}$.

Considering the sequence of functions (7.1.2), we carry out orthogonalization according to the Gram-Schmidt procedure. As a result, we get a system of orthogonal polynomials (each polynomial consists of $(r^me^{in\theta})$, $m \in \mathbb{N}_0$, $n \in \mathbb{Z}$), which are placed in groups

$$
\begin{array}{lllll}
P_{0;0}(z); & P_{1;0}(z), & P_{2;0}(z), & \dots; & P_{n;0}(z), \quad \dots\\
& P_{1;1}(z), & P_{2;1}(z), & & P_{n;1}(z),\\
& P_{1;2}(z); & P_{2;2}(z), & & P_{n;2}(z),\\
& & P_{2;3}(z), & & P_{n;3}(z),\\
& & P_{2;4}(z); & & P_{n;4}(z),\\
& & & & \dots\\
& & & & P_{n;2n}(z);
\end{array}
\tag{7.1.3}
$$

where each polynomial has a form

$$P_{n;\alpha}(z) = k_{n;\alpha} r^{n-|\alpha-n|} e^{i(n-\alpha)\theta} + R_{n;\alpha},$$
$$\alpha = 0, 1, \dots, 2n,\ k_{n;\alpha} > 0,\ n \in \mathbb{N}_0, \qquad (7.1.4)$$

here $R_{n;\alpha}$ denote the next part of the corresponding polynomial; $P_{0;0}(z) = 1$. In such a way $P_{n;\alpha}$ is some linear combination of

$$\{1 = r^0 e^{i0\theta};\ r^0 e^{i\theta}, r^1 e^{i0\theta}, r^0 e^{-i\theta};\ \dots;$$
$$r^0 e^{in\theta}, r^1 e^{i(n-1)\theta}, \dots, r^{n-|\alpha-n|} e^{i(n-\alpha)\theta}\}. \qquad (7.1.5)$$

Since the family (7.1.1) is total in the space L_2, the sequence (7.1.3) form an orthonormal basis in this space.

Let $\mathcal{P}_{n;\alpha}$, $\forall n \in \mathbb{N}$ denotes a subspace spanned by (7.1.5). Hence, $\forall n \in \mathbb{N}$, we have

$$\begin{aligned}
&\mathcal{P}_{0;0} \subset \mathcal{P}_{1;0} \subset \mathcal{P}_{1;1} \subset \mathcal{P}_{1;2} \subset \mathcal{P}_{2;0} \subset \mathcal{P}_{2;1} \subset \mathcal{P}_{2;2} \subset \mathcal{P}_{2;3} \subset \mathcal{P}_{2;4} \subset \cdots\\
&\quad \subset \mathcal{P}_{n;0} \subset \mathcal{P}_{n;1} \subset \cdots \subset \mathcal{P}_{n;2n} \subset \cdots,\\
&\mathcal{P}_{n;\alpha} = \{P_{0;0}(z)\} \oplus \{P_{1;0}(z)\} \oplus \{P_{1;1}(z)\} \oplus \{P_{1;2}(z)\} \oplus\\
&\quad \oplus \{P_{2;0}(z)\} \oplus \{P_{2;1}(z)\} \oplus \{P_{2;2}(z)\} \oplus \{P_{2;3}(z)\} \oplus \{P_{2;4}(z)\} \oplus \cdots\\
&\quad \oplus \{P_{n;0}(z)\} \oplus \{P_{n;1}(z)\} \oplus \cdots \oplus \{P_{n;\alpha}(z)\},
\end{aligned} \qquad (7.1.6)$$

where $\{P_{n;\alpha}(z)\}$, $n \in \mathbb{N}$, $\alpha = 0, 1, \dots 2n$ denotes a one-dimensional space spanned by $P_{n;\alpha}(z)$; $\mathcal{P}_{0;0} = \mathbb{C}$.

For further research, it is necessary to consider the Hilbert space

$$\mathbf{l}_2 = \mathcal{H}_0 \oplus \mathcal{H}_1 \oplus \mathcal{H}_2 \oplus \cdots, \quad \mathcal{H}_n = \mathbb{C}^{2n+1}, \quad n \in \mathbb{N}_0, \qquad (7.1.7)$$

instead of usual traditional space l_2 used in the classical Hamburger (one-dimensional) moment problem.

Each vector $f \in \mathbf{l}_2$ has a form $f = (f_n)_{n=0}^{\infty}$, $f_n \in \mathcal{H}_n$, and, hence,

$$\|f\|^2_{\mathbf{l}_2} = \sum_{n=0}^{\infty} \|f_n\|^2_{\mathcal{H}_n} < \infty, \quad (f, g)_{\mathbf{l}_2} = \sum_{n=0}^{\infty} (f_n, g_n)_{\mathcal{H}_n}, \quad \forall f, g \in \mathbf{l}_2.$$

Coordinates of a vector $f_n \in \mathcal{H}_n$, $n \in \mathbb{N}_0$ in some orthonormal basis $\{e_{n;0}, e_{n;1}, e_{n;2}, \dots, e_{n;2n}\}$ of the space $\mathbb{C}^{n+1}$, are denoted by $(f_{n;0}, f_{n;1}, f_{n;2}, \dots, f_{n;2n})$ and, hence, we have $f_n = (f_{n;0}, f_{n;1}, f_{n;2}, \dots, f_{n;2n})$. It is clear, that the space $\mathbf{l}_2$ is almost isometric $l_2 \times l_2$ (the vector $e_{0;0}$ is taken only one time in $\mathbf{l}_2$).

By using the orthonormal system (7.1.3), one can define a mapping of $\mathbf{l}_2$ into L_2. Denote,

$$P_n(z) = (P_{n;0}, P_{n;1}(z), P_{n;2}(z), \dots, P_{n;2n}) \in \mathcal{H}_n, \ \forall n \in \mathbb{N}_0, \ \forall z = re^{i\theta} \in \mathbb{C},$$

then

$$\mathbf{l}_2 \ni f = (f_n)_{n=0}^{\infty} \longmapsto \hat{f}(z) = \sum_{n=0}^{\infty} (f_n, P_n(z))_{\mathcal{H}_n} \in L_2. \tag{7.1.8}$$

Since, for $n \in \mathbb{N}_0$, we have

$$(f_n, P_n(z))_{\mathcal{H}_n} = f_{n;0}\overline{P_{n;0}(z)} + f_{n;1}\overline{P_{n;1}(z)} + f_{n;2}\overline{P_{n;2}(z)} + \dots + f_{n;2n}\overline{P_{n;2n}(z)}$$

and

$$\|f\|_{\mathbf{l}_2}^2 = \|(f_{0;0}, f_{1;0}, f_{1;1}, f_{1;2}, \dots, f_{n;0}, f_{n;1}, \dots, f_{n;2n}, \dots)\|_{l_2}^2,$$

then (7.1.8) is the mapping of the space $\mathbf{l}_2$ into L_2, using the orthonormal system (7.1.3) we conclude that this mapping is isometric. The image $\mathbf{l}_2$ by the mapping (7.1.8) coincides with the space L_2, because under our assumption, the system (7.1.3) is an orthonormal basis in L_2. Therefore, the mapping (7.1.8) is an isometric transformation (denoted by I), acting from whole $\mathbf{l}_2$ into whole L_2.

Let A be a bounded linear operator defined on the space $\mathbf{l}_2$. It is possible to construct the operator matrix $(a_{j,k})_{j,k=0}^{\infty}$, where for each $j, k \in \mathbb{N}_0$, the element $a_{j,k}$ is an operator from $\mathcal{H}_k$ into $\mathcal{H}_j$, so that we have

$$(Af)_j = \sum_{k=0}^{\infty} a_{j,k} f_k, \ j \in \mathbb{N}_0, \quad (Af, g)_{\mathbf{l}_2} = \sum_{j,k=0}^{\infty} (a_{j,k} f_k, g_j)_{\mathcal{H}_j}, \quad \forall f, g \in \mathbf{l}_2. \tag{7.1.9}$$

For the proof of (7.1.9) we need only to write the usual matrix of the operator A in the space $\mathbf{l}_2$ using the basis

$$(e_{0;0}; \ e_{1;0}, e_{1;1}, e_{1;2}; \ \dots; \ e_{n;0}, e_{n;1}, \dots, e_{n;2n}; \ \dots), \quad e_{0;0} = 1. \tag{7.1.10}$$

Then, for each $j, k \in \mathbb{N}_0$, $a_{j,k}$ is the operator $\mathcal{H}_k \longrightarrow \mathcal{H}_j$ that has a matrix representation

$$a_{j,k;\alpha,\beta} = (Ae_{k;\beta}, e_{j;\alpha})_{\mathbf{l}_2}, \tag{7.1.11}$$

where $\alpha = 0, 1, \ldots, 2j$, $\beta = 0, 1, \ldots, 2k$. We will write $a_{j,k} = (a_{j,k;\alpha,\beta})_{\alpha,\beta=0}^{j,k}$, including the cases

$$a_{0,0} = (a_{0,0;\alpha,\beta})_{\alpha,\beta=0}^{0,0} = a_{0,0;0,0},\ a_{0,1} = (a_{0,1;\alpha,\beta})_{\alpha,\beta=0}^{0,2},\ a_{1,0} = (a_{1,0;\alpha,\beta})_{\alpha,\beta=0}^{2,0}.$$

Note, that the same representation (7.1.9) is also valid for the general operator A on the space $\mathbf{l}_2$ defined on $\mathbf{l}_{\text{fin}} \subset \mathbf{l}_2$, where $\mathbf{l}_{\text{fin}}$ denotes the set of finite vectors of $\mathbf{l}_2$. In this case the first formula in (7.1.9) holds true for $f \in \mathbf{l}_{\text{fin}}$; the second formula is valid for $f \in \mathbf{l}_{\text{fin}}$, $g \in \mathbf{l}_2$.

Let us consider the image $\hat{A} = IAI^{-1} : L_2 \longrightarrow L_2$ of the above bounded operator $A : \mathbf{l}_2 \longrightarrow \mathbf{l}_2$ given by the mapping (7.1.8). Its matrix in the basis (7.1.3),

$$(P_{0;0}(z);\ P_{1;0}(z), P_{1;1}(z), P_{1;2}(z);\ \ldots;\ P_{n;0}(z), P_{n;1}(z), \ldots, P_{n;2n}(z); \ldots),$$

coincides with a usual matrix of the operator A regarded as an operator: $\mathbf{l}_2 \longrightarrow \mathbf{l}_2$ in the corresponding basis (7.1.10). By using (7.1.11) and the above mentioned procedure, we get the operator matrix $(a_{j,k})_{j,k=0}^{\infty}$ of the operator $A : \mathbf{l}_2 \longrightarrow \mathbf{l}_2$. *By the definition this matrix is also the operator matrix of* $\hat{A} : L_2 \longrightarrow L_2$.

It is clear that $\hat{A}$ can be an arbitrary linear bounded operator in L_2.

Lemma 7.1.1 *For polynomials $P_{n;\alpha}(z)$ and subspaces $\mathcal{P}_{m,\beta}$, $n, m \in \mathbb{N}_0$, $\alpha = 0, 1, \ldots, 2n$, $\beta = 0, 1, \ldots, 2m$, relations*

$$r P_{n;\alpha}(z) \in \mathcal{P}_{n+1;\alpha+1}, \quad \alpha = 0, 1, ..., 2n; \tag{7.1.12}$$

$$e^{i\theta} P_{n;\alpha}(z) \in \mathcal{P}_{n+1;\alpha}, \quad \alpha = 0, 1, ..., n; \tag{7.1.13}$$

$$e^{i\theta} P_{n;\alpha}(z) \in \mathcal{P}_{n+1;n}, \quad \alpha = n+1, n+2, ..., 2n; \tag{7.1.14}$$

$$e^{-i\theta} P_{n;\alpha}(z) \in \mathcal{P}_{n+1;\alpha+2}, \quad \alpha = n, n+1, ..., 2n; \tag{7.1.15}$$

$$e^{-i\theta} P_{n;\alpha}(z) \in \mathcal{P}_{n;2n}, \quad \alpha = 0, 1, ..., n-1,\ n \in \mathbb{N}_0. \tag{7.1.16}$$

hold true.

Proof According to (7.1.3), the polynomial $P_{n;\alpha}(z)$, $n \in \mathbb{N}_0$ is a linear combination of (7.1.5), i.e.,

$$\{1 = r^0 e^{i0\theta};\ r^0 e^{i\theta}, r^1 e^{i0\theta}, r^0 e^{-i\theta};\ \ldots;$$
$$r^0 e^{in\theta}, r^1 e^{i(n-1)\theta}, \ldots, r^{n-|\alpha-n|} e^{i(n-\alpha)\theta}\}.$$

Hence, multiplying by r each element in (7.1.5), we obtain the the set

$$\{r^1 e^{i0\theta};\ r^1 e^{i\theta}, r^2 e^{i0\theta}, r^1 e^{-i\theta};\ \ldots;$$
$$r^1 e^{in\theta}, r^2 e^{i(n-1)\theta}, \ldots, r^{n-|\alpha-n|+1} e^{i(n-\alpha)\theta}\},$$

and a linear combination of such elements belongs to $\mathcal{P}_{n+1;\alpha+1}$ for $\alpha = 0, 1, ..., 2n$, since for the last element (according to the order (7.1.2)), we have

$$r^{n-|\alpha-n|+1}e^{i(n-\alpha)\theta} = r^{(n+1)-|(\alpha+1)-(n+1)|}e^{i((n+1)-(\alpha+1))\theta}.$$

Hence, we proved (7.1.12).

The multiplication by $e^{i\theta}$ of each element in (7.1.5) gives the set

$$\begin{aligned}\{r^0e^{i\theta};\ r^0e^{i2\theta}, r^1e^{i\theta}, r^0e^{i0\theta};\ \dots;\\ r^0e^{i(n+1)\theta}, r^1e^{i(n)\theta}, \dots, r^{n-|\alpha-n|}e^{i(n-\alpha+1)\theta}\}\end{aligned} \tag{7.1.17}$$

that a linear combination belongs to $\mathcal{P}_{n+1;\alpha}$, if $\alpha = 0, 1, ..., n$, since for the last element (according to the order (7.1.2)) gives

$$\begin{aligned}r^{n-|\alpha-n|}e^{i(n-\alpha+1)\theta} &= r^{(n+1)-|n-\alpha|-1}e^{i((n+1)-\alpha)\theta}\\ &= r^{(n+1)-|\alpha-(n+1)|}e^{i((n+1)-\alpha)\theta}.\end{aligned}$$

The last equality is true since $\alpha \le n$. Hence, we proved (7.1.13). And the linear combinations of elements (7.1.17) belongs to $\mathcal{P}_{n+1;n}$, if $\alpha = n + 1, ..., 2n$, since for the last element (according to the order (7.1.2)), we have

$$r^{n-|\alpha-n|}e^{i(n-\alpha+1)\theta} = r^{(n+1)-|\alpha-(n+1)|}e^{i((n+1)-\alpha)\theta}.$$

Since for $\alpha = n + 1, n + 2, ..., 2n$, $P_{n;\alpha}(z)$ contains the element $r^n e^{i0\theta}$, then $r^n e^{i\theta}$ is contained in $e^{i\theta} P_{n;\alpha}(z)$ and $r^n e^{i\theta} \in \mathcal{P}_{n+1;n}$. Hence, we proved (7.1.14).

Similarly, multiplying by $e^{-i\theta}$ each ellement in (7.1.5), we get

$$\begin{aligned}\{r^0e^{-i\theta};\ r^0e^{0\theta}, r^1e^{-i\theta}, r^0e^{-i2\theta};\ \dots;\\ r^0e^{i(n-1)\theta}, r^1e^{i(n-2)\theta}, \dots, r^{n-|\alpha-n|}e^{i(n-\alpha-1)\theta}\}.\end{aligned} \tag{7.1.18}$$

The linear combination of elements of (7.1.18) belongs to the subset $\mathcal{P}_{n+1;\alpha+2}$, if $\alpha = n, n + 1, ..., 2n$, since for the element (according to the order (7.1.2)), we have

$$\begin{aligned}r^{n-|\alpha-n|}e^{i(n-\alpha-1)\theta} &= r^{(n+1)-|\alpha-n|+1}e^{i(n-\alpha-1)\theta}\\ &= r^{(n+1)-|(\alpha+1)-(n+1)|+1}e^{i((n+1)-(\alpha+2))\theta}\\ &= r^{(n+1)-|(\alpha+2)-(n+1)|}e^{i((n+1)-(\alpha+2))\theta}\\ &= r^{(n+1)-|(n+1)-(\alpha+2)|}e^{i((n+1)-(\alpha+2))\theta}.\end{aligned}$$

Hence, we proved (7.1.15).

The linear combination of elements in (7.1.18) belongs in any case to $\mathcal{P}_{n;2n}$, if $\alpha = 0, 1, ..., n - 1$, since $r^0e^{-i(n-1)\theta}$ is contained in $P_{n;\alpha}(z)$, $\forall\alpha = 0, 1, ..., n - 1$.

Hence, for $\alpha = 0, 1, ..., n-1$, $e^{-i\theta} P_{n;\alpha}(z)$ contains $r^0 e^{-in\theta}$. Hence, we proved (7.1.16). □

Lemma 7.1.2 *Let $\hat{A}$ be the self-adjoint multiplication by r operator on the space L_2, namely,*

$$L_2 \ni \varphi(z) \longmapsto (\hat{A}\varphi)(z) = r\varphi(z) \in L_2.$$

The operator matrix $(a_{j,k})_{j,k=0}^{\infty}$ of $\hat{A}$ (i.e., $A = I^{-1}\hat{A}I$) has a tri-diagonal structure, i.e., $a_{j,k} = 0$ for $|j-k| > 1$.

Proof By using (7.1.11), for $e_{n;\gamma} = I^{-1} P_{n;\gamma}(z)$, $n \in \mathbb{N}_0$, $\gamma = 0, 1, \dots, 2n$, we have

$$a_{j,k;\alpha,\beta} = (Ae_{k;\beta}, e_{j;\alpha})_{l_2} = \int_{\mathbb{C}} r P_{k;\beta}(z)\overline{P_{j;\alpha}(z)}\, d\rho(z), \quad \forall j, k \in \mathbb{N}_0, \tag{7.1.19}$$

where $\alpha = 0, 1, \dots, 2j$, $\beta = 0, 1, \dots, 2k$. From (7.1.12), we have $r P_{k;\alpha}(z) \in \mathcal{P}_{k+1;\alpha+1}$ for $\alpha = 0, 1, ..., 2j$. According to (7.1.6), the integral in (7.1.19) is equal to zero for $j > k+1$.

On the other hand, the integral in (7.1.19) has the form

$$a_{j,k;\alpha,\beta} = \overline{\int_{\mathbb{C}} \bar{r} P_{j;\alpha}(z)\overline{P_{k;\beta}(z)}\, d\rho(z)}. \tag{7.1.20}$$

From (7.1.12), we have $\bar{r} P_{j;\alpha}(z) = r P_{j;\alpha}(z) \in \mathcal{P}_{j+1;\alpha+1}$. According to (7.1.6), the last integral is equal to zero for $k > j+1$ and for each $\alpha = 0, 1, \dots, 2j$, $\beta = 0, 1, \dots, 2k$.

As a result the integral in (7.1.20), i.e., coefficients $a_{j,k;\alpha,\beta}$, $j, k \in \mathbb{N}_0$, are equal to zero for $|j-k| > 1$, $\alpha = 0, 1, \dots, 2j$, $\beta = 0, 1, \dots, 2k$. In the previous considerations it is necessary to take into account $e_{0;0} = I^{-1} P_{0;0}(z)$, $P_{0;0}(z) = 1$. □

In such a way the matrix $(a_{j,k})_{j,k=0}^{\infty}$ of the operator $\hat{A}$ has the tri-diagonal block structure

$$\begin{bmatrix} a_{0,0} & a_{0,1} & 0 & 0 & 0 & \dots \\ a_{1,0} & a_{1,1} & a_{1,2} & 0 & 0 & \dots \\ 0 & a_{2,1} & a_{2,2} & a_{2,3} & 0 & \dots \\ 0 & 0 & a_{3,2} & a_{3,3} & a_{3,4} & \dots \\ \vdots & \vdots & \vdots & \vdots & \vdots & \ddots \end{bmatrix}. \tag{7.1.21}$$

A more detailed analysis of the expression (7.1.19) makes it possible to know about zero and non-zero matrix elements $(a_{j,k;\alpha,\beta})_{\alpha,\beta=0}^{2j,2k}$ for $|j-k| \leq 1$. We use actively also the permutation properties of indexes of matrix elements. j, k and α, β.

Let us denote by $((a^*)_{j,k})_{j,k=0}^{\infty}$ the operator matrix of the operator $(\hat{A})^*$, which is adjoint to $\hat{A}$. We remark that $(\hat{A})^* = \hat{A}$ is also the multiplication by $r = \bar{r}$ operator. Taking into account the expression (7.1.19) for $j, k \in \mathbb{N}_0$, we have

$$\begin{aligned}(a^*)_{j,k;\alpha,\beta} &= \int_{\mathbb{C}} \bar{r} P_{k;\beta}(z)\overline{P_{j;\alpha}(z)}\, d\rho(z) \\ &= \overline{\int_{\mathbb{C}} r P_{j;\alpha}(z)\overline{P_{k;\beta}(z)}\, d\rho(z)} = \bar{a}_{k,j;\beta,\alpha},\end{aligned} \tag{7.1.22}$$

where $\alpha = 0, 1, \ldots, 2j$, $\beta = 0, 1, \ldots, 2k$. Since $\bar{r} = r$, then the matrix (7.1.21) is Hermitian $(a_{j,k;\alpha,\beta} = \bar{a}_{k,j;\beta,\alpha})$.

Lemma 7.1.3 *Let* $(a_{j,k})_{j,k=0}^{\infty}$ *be an operator matrix of the multiplication by* r *operator in* L_2*, where* $a_{j,k} : \mathcal{H}_k \longrightarrow \mathcal{H}_j$*;* $a_{j,k} = (a_{j,k;\alpha,\beta})_{\alpha,\beta=0}^{2j,2k}$ *are matrices of operators* $a_{j,k}$ *in the corresponding orthonormal basis. Then, for* $j \in \mathbb{N}_0$*, we have*

$$\begin{aligned} a_{j,j+1;\alpha,\beta} &= 0,\ \alpha = 0, 1, \ldots 2j, \quad \beta = \alpha + 2, \alpha + 3, \ldots, 2j + 2; \\ a_{j+1,j;\alpha,\beta} &= 0,\ \beta = 0, 1, \ldots 2j, \quad \alpha = \beta + 2, \beta + 3, \ldots, 2j + 2, \end{aligned} \tag{7.1.23}$$

If we choose another order inside of each direction “in the line with a corner”

$$\{r^0 e^{in\theta}, r^1 e^{i(n-1)\theta}, r^2 e^{i(n-2)\theta}, \ldots, r^1 e^{-i(n-1)\theta}, r^0 e^{-in\theta}\},$$

(see Fig. 7.1 and comments after (7.1.2))) preserving the order of lines, then Lemma 7.1.3 is valid and it will be also possible to describe zero elements of matrices $(a_{j,k;\alpha,\beta})_{\alpha,\beta=0}^{2j,2k}$. Such a matrix $(a_{j,k})_{j,k=0}^{\infty}$ have also a tri-diagonal block structure but has zeros at the same places.

Proof of Lemma 7.1.3 According to (7.1.19) and (7.1.12), for $\forall\alpha = 0, 1, \ldots, 2j$, $\forall\beta = \alpha + 2, \alpha + 3, \ldots, 2j + 2$, and $j \in \mathbb{N}_0$, we have

$$a_{j,j+1;\alpha,\beta} = \int_{\mathbb{C}} r P_{j+1,\beta}(z)\overline{P_{j;\alpha}(z)}\, d\rho(z) = \overline{\int_{\mathbb{C}} r P_{j,\alpha}(z)\overline{P_{j+1;\beta}(z)}\, d\rho(z)},$$

where $r P_{j;\alpha}(z) \in \mathcal{P}_{j+1;\alpha+1}$. But according to (7.1.6) $P_{j+1;\beta}(z)$ is orthogonal to $\mathcal{P}_{j+1;\alpha+1}$ for $\beta > \alpha + 1$ and, hence, the last integral is equal to zero. This gives the first equalities in (7.1.23).

Similarly to (7.1.19) and (7.1.12), $\forall\beta = 0, 1, \ldots, 2j$, $\forall\alpha = \beta + 2, \beta + 3, \ldots, 2j + 2$, and $j \in \mathbb{N}_0$, we have

$$a_{j+1,j;\alpha,\beta} = \int\limits_{\mathbb{C}} r P_{j,\beta}(z)\overline{P_{j+1;\alpha}(z)}\, d\rho(z),$$

where $r P_{j;\beta}(z) \in \mathcal{P}_{j+1;\beta+1}$. But according to (7.1.6), $P_{j+1;\alpha}(z)$ is orthogonal to $\mathcal{P}_{j+1;\beta+1}$ if $\alpha > \beta + 1$ and, hence, the last integral is also equal to zero. This gives the second equalities in (7.1.23). □

Thus, after these studies, we claim that, in (7.1.21) for $\forall j \in \mathbb{N}$, the upper right corner of each $((2j + 1) \times (2j + 3))$-matrices $a_{j,j+1}$, $j \in \mathbb{N}_0$ (starting from the third diagonal) and the lower left corner of each $((2j+3) \times (2j+1))$-matrix $a_{j+1,j}$ (starting from the third diagonal) contains zero elements. Considering (7.1.21), it can be asserted that the Hermitian matrix as a multiplication by r operator is multi-diagonal as usual, that is, in the usual basis of the space $\mathbf{l}_2$.

Lemma 7.1.4 *Elements*

$$a_{j,j+1;\alpha,\alpha+1},\ a_{j+1,j;\alpha+1,\alpha};\quad \alpha = 0, 1, \ldots 2j,\quad j \in \mathbb{N}, \tag{7.1.24}$$

of matrices $(a_{j,k})_{j,k=0}^{\infty}$ *in Lemma 7.1.3, are positive.*

Proof We start with the study of $a_{0,1;0,1}$. Denote by $P'_{1;1}(z)$ a non-normalized vector $P_{1;1}(z)$, obtained due to the Gram-Schmidt orthogonalization procedure taking the vector $re^{i0\theta}$. According to (7.1.2) and (7.1.3), we have:

$$P'_{1;1}(z) = r - (r, P_{1;0}(z))_{L_2} P_{1;0}(z) - (r, 1)_{L_2},$$

where $1 = P_{0;0}(z)$. Therefore, using (7.1.19), we get

$$\begin{aligned} a_{0,1;0,1} &= \int\limits_{\mathbb{C}} r P_{1;1}(z)\, d\rho(z) = \|P'_{1;1}(z)\|_{L_2}^{-1} \int\limits_{\mathbb{C}} r P'_{1;1}(z)\, d\rho(z) \\ &= \|P'_{1;1}(z)\|_{L_2}^{-1} \int\limits_{\mathbb{C}} r(r - (r, P_{1;0}(z))_{L_2} P_{1;0}(z) - (r, 1)_{L_2})\, d\rho(z) \\ &= \|P'_{1;1}(z)\|_{L_2}^{-1} (\|r\|_{L_2}^2 - |(r, P_{1;0}(z))_{L_2}|^2 - |(r, 1)_{L_2}|^2). \end{aligned} \tag{7.1.25}$$

By using also (7.1.26), we conclude that the last expression is positive and, therefore, $a_{0,1;0,1} > 0$, since (7.1.1) is the total set of linear independent vectors in L_2. The element $a_{1,0;1,0}$ is also positive since the matrix is Hermitian.

The positiveness (7.1.25) follows from the Parseval equality for the decomposition of the function $r \in L_2$ with respect to the orthonormal basis L_2:

$$|(r, 1)_{L_2}|^2 + |(r, P_{1;0}(z))_{L_2}|^2 + |(r, P_{1;1}(z))_{L_2}|^2 + \cdots = \|r\|_{L_2}^2. \tag{7.1.26}$$

Let us pass to the proof of the positiveness of $a_{j,j+1;\alpha,\alpha+1}$, where $\alpha = 0, 1, \ldots, 2j$, $j \in \mathbb{N}$. From (7.1.19), we have:

$$a_{j,j+1;\alpha,\alpha+1} = \int_{\mathbb{C}} r P_{j+1;\alpha+1}(z)\overline{P_{j;\alpha}(z)}\, d\rho(z). \tag{7.1.27}$$

According to (7.1.4) and (7.1.6), we have

$$P_{j;\alpha}(z) = k_{j;\alpha} r^{j-|\alpha-j|} e^{i(j-\alpha)\theta} + R_{j;\alpha}(z), \tag{7.1.28}$$

where $R_{j;\alpha}(z)$ is some polynomial from $\mathcal{P}_{j;\alpha-1}$ if $\alpha > 0$, or from $\mathcal{P}_{j-1;2j-2}$ if $\alpha = 0$. Therefore, $rR_{j;\alpha}(z)$ is some polynomial from $\mathcal{P}_{j+1;\alpha}$ (if $\alpha = 0$, then $rR_{j;\alpha}(z) \in \mathcal{P}_{j;2j-1}$ (see (7.1.12). Multiplying (7.1.28) by r, we get

$$rP_{j;\alpha}(z) = k_{j;\alpha} r^{(j+1)-|\alpha-j|} e^{i(j-\alpha)\theta} + rR_{j;\alpha}(z), \tag{7.1.29}$$

where $rR_{j;\alpha}(z) \in \mathcal{P}_{j+1;\alpha}$ or to $\mathcal{P}_{j;2j-2}$.

On the other hand, the equality (7.1.28) in the case of $P_{j+1;\alpha}(z)$ gives

$$\begin{aligned} P_{j+1;\alpha+1}(z) &= k_{j+1;\alpha+1} r^{(j+1)-|(\alpha+1)-(j+1)|} e^{i((j+1)-(\alpha+1))\theta} + R_{j+1;\alpha+1}(z) \\ &= k_{j+1;\alpha+1} r^{(j+1)-|(\alpha-j)|} e^{i(j-\alpha)\theta} + R_{j+1;\alpha+1}(z), \end{aligned} \tag{7.1.30}$$

where $R_{j+1;\alpha+1}(z) \in \mathcal{P}_{j+1;\alpha+1}$ since $\alpha + 1 > 0$.

Take

$$r^{(j+1)-|(\alpha-j)|} e^{i(j-\alpha)\theta} = r^{(j+1)-|((\alpha+1)-(j+1))|} e^{i((j+1)-(\alpha+1))\theta}$$

from (7.1.30) and substitute it into (7.1.29). Thus, we get:

$$\begin{aligned} rP_{j;\alpha}(z) &= \frac{k_{j;\alpha}}{k_{j+1;\alpha+1}}(P_{j+1;\alpha+1}(z) - R_{j+1;\alpha+1}(z)) + rR_{j;\alpha}(z) \\ &= \frac{k_{j;\alpha}}{k_{j+1;\alpha+1}} P_{j+1;\alpha+1}(z) - \frac{k_{j;\alpha}}{k_{j+1;\alpha+1}} R_{j+1;\alpha+1}(z) + rR_{j;\alpha}(z), \end{aligned} \tag{7.1.31}$$

where the last two terms belong to $\mathcal{P}_{j+1;\alpha}$ and to $\mathcal{P}_{j+1;\alpha}$ or to $\mathcal{P}_{j;2j-1}$, correspondingly and are in any case orthogonal to $P_{j+1;\alpha+1}(z)$.

Hence, the substituting of the expression (7.1.31) into (7.1.27) gives $a_{j+1,j;\alpha,\alpha} = k_{j;\alpha}(k_{j+1;\alpha+1})^{-1} > 0$. □

In the sequel investigation, we use well-known convenient notations for the $a_{j,k}$ elements of the Jacobi matrix

$$\begin{aligned} a_n &= a_{n+1,n} : & \mathcal{H}_n &\longrightarrow \mathcal{H}_{n+1}, \\ b_n &= a_{n,n} \quad : & \mathcal{H}_n &\longrightarrow \mathcal{H}_n, \\ c_n &= a_{n,n+1} : & \mathcal{H}_{n+1} &\longrightarrow \mathcal{H}_n, \quad n \in \mathbb{N}_0. \end{aligned} \tag{7.1.32}$$

Let U be a bounded linear operator defined on the space $\mathbf{l}_2$. Construct the operator matrix $(u_{j,k})_{j,k=0}^{\infty}$, where for each $j, k \in \mathbb{N}_0$ the element $u_{j,k}$ is an operator from $\mathcal{H}_k$ into $\mathcal{H}_j$, so that like (7.1.9), $\forall f, g \in \mathbf{l}_2$, we have

$$(Uf)_j = \sum_{k=0}^{\infty} u_{j,k} f_k, \ j \in \mathbb{N}_0, \quad (Uf, g)_{\mathbf{l}_2} = \sum_{j,k=0}^{\infty} (u_{j,k} f_k, g_j)_{\mathcal{H}_j}. \tag{7.1.33}$$

To prove (7.1.33), it is only necessary to rewrite the usual matrix of the operator U in the space $\mathbf{l}_2$, using a basis

$$(e_{0;0};\ e_{1;0}, e_{1;1}, e_{1;2};\ \ldots;\ e_{n;0}, e_{n;1}, \ldots, e_{n;2n};\ \ldots), \quad e_{0;0} = 1. \tag{7.1.34}$$

Then $u_{j,k}$ for each $j, k \in \mathbb{N}_0$ is the operator $\mathcal{H}_k \longrightarrow \mathcal{H}_j$ that has a matrix representation

$$u_{j,k;\alpha,\beta} = (Ue_{k;\beta}, e_{j;\alpha})_{\mathbf{l}_2}, \tag{7.1.35}$$

where $\alpha = 0, 1, \ldots, 2j$, $\beta = 0, 1, \ldots, 2k$.

Let us write $u_{j,k} = (u_{j,k;\alpha,\beta})_{\alpha,\beta=0}^{2j,2k}$ including cases

$$u_{0,0} = (u_{0,0;\alpha,\beta})_{\alpha,\beta=0}^{0,0} = u_{0,0;0,0}, \ u_{0,1} = (u_{0,1;\alpha,\beta})_{\alpha,\beta=0}^{0,2}, \ u_{1,0} = (u_{1,0;\alpha,\beta})_{\alpha,\beta=0}^{2,0}.$$

Note, that the same representation (7.1.33) is also valid for the general operator U on the space $\mathbf{l}_2$ defined on $\mathbf{l}_{\mathrm{in}} \subset \mathbf{l}_2$, where $\mathbf{l}_{\mathrm{fin}}$ denotes, as usually, the set of finite vectors from $\mathbf{l}_2$. In this case the first formula in (7.1.33) is valid for $f \in \mathbf{l}_{\mathrm{fin}}$; in the second formula $f \in \mathbf{l}_{\mathrm{fin}}$, $g \in \mathbf{l}_2$.

Let us consider the image $\hat{U} = IUI^{-1} : L_2 \longrightarrow L_2$ of the bounded operator U: $\mathbf{l}_2 \longrightarrow \mathbf{l}_2$ by the mapping (7.1.8). Its matrix in the basis (7.1.3)

$$(P_{0;0}(z);\ P_{1;0}(z), P_{1;1}(z), P_{1;2}(z);\ \ldots;\ P_{n;0}(z), P_{n;1}(z), \ldots, P_{n;2n}(z); \ldots)$$

is equal to the usual matrix of operator U regared as the operator acting $\mathbf{l}_2 \longrightarrow \mathbf{l}_2$ in the corresponding basis (7.1.34). By using (7.1.35) and the above mentioned procedure, we get the operator matrix $(u_{j,k})_{j,k=0}^{\infty}$ of $U : \mathbf{l}_2 \longrightarrow \mathbf{l}_2$. *By the definition this matrix is also the operator matrix of* $\hat{U} : L_2 \longrightarrow L_2$.

It is clear that $\hat{U}$ can be an arbitrary linear bounded operator in L_2.

Lemma 7.1.5 *Let $\hat{U}$ be the unitary multiplication by $e^{i\theta}$ operator on the space L_2, namely,*

$$L_2 \ni \varphi(z) \longmapsto (\hat{U}\varphi)(z) = e^{i\theta}\varphi(z) \in L_2.$$

The operator matrix $(u_{j,k})_{j,k=0}^{\infty}$ of the operator $\hat{U}$ (i.e., $U = I^{-1}\hat{U}I$) has a tri-diagonal structure, i.e., $u_{j,k} = 0$ for $|j-k| > 1$.

Proof By using (7.1.35) for $e_{n;\gamma} = I^{-1}P_{n;\gamma}(z)$, $n \in \mathbb{N}_0$, $\gamma = 0, 1, \ldots, 2n$, $\forall j, k \in \mathbb{N}_0$, we have

$$u_{j,k;\alpha,\beta} = (Ue_{k;\beta}, e_{j;\alpha})_{l_2} = \int\limits_{\mathbb{C}} e^{i\theta} P_{k;\beta}(z)\overline{P_{j;\alpha}(z)}\, d\rho(z), \tag{7.1.36}$$

where $\alpha = 0, 1, \ldots, 2j$, $\beta = 0, 1, \ldots, 2k$. From (7.1.13) and (7.1.14), we have $e^{i\theta}P_{k;\beta}(z) \in \mathcal{P}_{k+1;\beta}$ if $\beta = 0, 1, \ldots, k$ and $e^{i\theta}P_{k;\beta}(z) \in \mathcal{P}_{k+1;k}$ if $\beta = k+1, k+2, \ldots, 2k$. According to (7.1.6), the integral in (7.1.36) is equal to zero for $j > k+1$ and for each $\beta = 0, 1, \ldots, 2k$.

On the other hand, the integral in (7.1.36) has a form

$$u_{j,k;\alpha,\beta} = \overline{\int\limits_{\mathbb{C}} e^{-i\theta} P_{j;\alpha}(z)\overline{P_{k;\beta}(z)}\, d\rho(z)}. \tag{7.1.37}$$

From (7.1.15), (7.1.16) we have now that $e^{-i\theta}P_{j;\alpha}(z) \in \mathcal{P}_{j+1;\alpha+2}$ if $\alpha = j, j+1, \ldots, 2j$ and $e^{-i\theta}P_{j;\alpha}(z) \in \mathcal{P}_{j;2j}$ if $\alpha = 0, 1, \ldots, j-1$. According to (7.1.6), the last integral is equal to zero for $k > j+1$ and for each $\alpha = 0, 1, \ldots, 2j$.

As result, the integral in(7.1.37), i.e., coefficients $u_{j,k;\alpha,\beta}$, $j, k \in \mathbb{N}_0$, are equal to zero for $|j-k| > 1$, $\alpha = 0, 1, \ldots, 2j$, $\beta = 0, 1, \ldots, 2k$.

In previous considerations it is necessary to take into account that $e_{0;0} = I^{-1}P_{0;0}(z)$, $P_{0;0}(z) = 1$. □

In such a way the matrix $(u_{j,k})_{j,k=0}^{\infty}$ of the operator $\hat{U}$ has a tri-diagonal block structure

$$\begin{bmatrix} u_{0,0} & u_{0,1} & 0 & 0 & 0 & \cdots \\ u_{1,0} & u_{1,1} & u_{1,2} & 0 & 0 & \cdots \\ 0 & u_{2,1} & u_{2,2} & u_{2,3} & 0 & \cdots \\ 0 & 0 & u_{3,2} & u_{3,3} & u_{3,4} & \cdots \\ \vdots & \vdots & \vdots & \vdots & \vdots & \ddots \end{bmatrix}. \tag{7.1.38}$$

A more detailed analysis of the expression (7.1.36) makes it possible to know about the zero and non-zero elements of the matrices $(u_{j,k;\alpha,\beta})_{\alpha,\beta=0}^{2j,2k}$ in each case for $|j-k| \leq 1$. Permutational properties of matrix indices j, k and α, β are also used.

Let us denote by $((u^*)_{j,k})_{j,k=0}^{\infty}$ the operator matrix of the operator $(\hat{U})^*$ which is adjoint to $\hat{U}$. We remark that $(\hat{U})^*$ is the multiplication by $e^{-i\theta}$ operator. Taking into account the expression (7.1.36) for $j, k \in \mathbb{N}_0$, we have

$$
\begin{aligned}
(u^*)_{j,k;\alpha,\beta} &= \int_{\mathbb{C}} e^{-i\theta} P_{k;\beta}(z) \overline{P_{j;\alpha}(z)}\, d\rho(z) \\
&= \overline{\int_{\mathbb{C}} e^{i\theta} P_{j;\alpha}(z) \overline{P_{k;\beta}(z)}\, d\rho(z)} = \bar{u}_{k,j:\beta,\alpha},
\end{aligned}
\tag{7.1.39}
$$

where $\alpha = 0, 1, \ldots, 2j$ and $\beta = 0, 1, \ldots, 2k$.

Lemma 7.1.6 *Let $(u_{j,k})_{j,k=0}^{\infty}$ be an operator matrix of the multiplication by $e^{i\theta}$ operator in L_2, where $u_{j,k} : \mathcal{H}_k \longrightarrow \mathcal{H}_j$;are matrices of operators $u_{j,k}$ in the corresponding orthonormal basis. Then $\forall j \in \mathbb{N}$, we have*

$$u_{j,j+1;\alpha,\beta} = 0, \quad \alpha = 0, 1, \ldots, j-1, \qquad \beta = 0, 1, \ldots, 2j+2; \tag{7.1.40}$$

$$u_{j,j+1;\alpha,\beta} = 0, \quad \alpha = j, j+1, \ldots, 2j-1, \quad \beta = \alpha+3, \alpha+4, \ldots, 2j+2; \tag{7.1.41}$$

$$u_{j+1,j;\alpha,\beta} = 0, \quad \beta = 0, 1, \ldots, j, \qquad \alpha = \beta+1, \beta+2, \ldots, 2j+2; \tag{7.1.42}$$

$$u_{j+1,j;\alpha,\beta} = 0, \quad \beta = j+1, j+1 \ldots, 2j, \quad \alpha = j+1, j+2, \ldots, 2j+2; \tag{7.1.43}$$

and $u_{1,0;1,0} = u_{1,0;2,0} = 0.$

If we choose inside of each direction

$$\{r^0 e^{in\theta}, r^1 e^{i(n-1)\theta}, r^2 e^{i(n-2)\theta}, \ldots, r^1 e^{-i(n-1)\theta}, r^0 e^{-in\theta}\},$$

(see Fig. 7.1) another order (preserving the order of "lines with the corner" (see the comments after (7.1.2)), then Lemma 7.1.6 is not valid, but it will still be possible to describe the zeros of matrices $(u_{j,k;\alpha,\beta})_{\alpha,\beta=0}^{2j,2k}$. Such matrices $(u_{j,k})_{j,k=0}^{\infty}$ also have a tri-diagonal block structure and zero elements are possibly but in another places.

Proof of Lemma 7.1.6 According to (7.1.36), (7.1.15), and (7.1.16), for $j \in \mathbb{N}$, $\alpha = 0, 1, \ldots, 2j$, $\beta = 0, 1, \ldots, 2k$, we have

$$u_{j,j+1;\alpha,\beta} = \int_{\mathbb{C}} e^{i\theta} P_{j+1,\beta}(z) \overline{P_{j;\alpha}(z)}\, d\rho(z) = \overline{\int_{\mathbb{C}} e^{-i\theta} P_{j,\alpha}(z) \overline{P_{j+1;\beta}(z)}\, d\rho(z)},$$

where $e^{-i\theta} P_{j;\alpha}(z) \in \mathcal{P}_{j+1;\alpha+2}$ for $\alpha = j, j+1, \ldots, 2j$. But according to (7.1.6), $\mathcal{P}_{j+1;\alpha+2}$ is orthogonal to $\mathcal{P}_{j+1;\beta}$ for $\beta > \alpha+2$ and, hence, the last integral is equal to zero. This gives the equalities in (7.1.41).

And $e^{-i\theta} P_{j;\alpha}(z) \in \mathcal{P}_{j;2j}$ for $\alpha = 0, 1, \ldots, j-1$. But, according to (7.1.6), $\mathcal{P}_{j;2j}$ is orthogonal to $\mathcal{P}_{j+1;\beta}$ for $\beta = 0, 1, \ldots, 2j+2$ and, hence, the last integral is also equal to zero. This gives the equalities in (7.1.40).

Similarly, from (7.1.36), (7.1.13), and (7.1.14) for $j \in \mathbb{N}_0$, $\alpha = 0, 1, \ldots, 2j+2$, $\beta = 0, 1, \ldots, 2j$, we have

$$u_{j+1,j;\alpha,\beta} = \int\limits_{\mathbb{C}} e^{i\theta} P_{j,\beta}(z) \overline{P_{j+1;\alpha}(z)}\, d\rho(z),$$

where $e^{i\theta} P_{j;\beta}(z) \in \mathcal{P}_{j+1;\beta}$ for $\beta = 0, 1, \ldots, j$. But according to (7.1.6), $\mathcal{P}_{j+1;\beta}$ is orthogonal to $\mathcal{P}_{j+1;\alpha}$ for $\beta < \alpha$ and, hence, the last integral is equal to zero for $\alpha = \beta+1, \beta+2, \ldots, 2j+2$. This gives the equalities in (7.1.42).

And $e^{i\theta} P_{j;\alpha}(z) \in \mathcal{P}_{j+1;j}$ for $\beta = j+1, j+2, \ldots, 2j$. But according to (7.1.6), $\mathcal{P}_{j+1;j}$ is orthogonal to $\mathcal{P}_{j+1;\alpha}$ for $\alpha > j$ and, hence, the last integral is also equal to zero for $\alpha = j+1, j+2, \ldots, 2j+2$. This gives the equalities in (7.1.43) and $u_{1,0;1,0} = u_{1,0;2,0} = 0$.

Note that we do not have additional information about $u_{0,1}$. □

Thus, after last investigations we conclude that in (7.1.38) for $\forall j \in \mathbb{N}$, each lower left corner of the matrix $u_{j+1,j}$ (starting from the second diagonal, which located in the upper left corner) and $j+2$ of the last rows and each upper right corner of the matrix $u_{j,j+1}$ (starting from the second diagonal located from the lower right corner) and n of the first rows contain always zero elements. Considering (7.1.38), it can be stated that the unitary matrix corresponding to the multiplication by $e^{i\theta}$ operator, is multi-diagonal in the usual sense, that is, in the usual basis of the space $\mathbf{l}_2$.

Lemma 7.1.7 *Elements*

$$\begin{aligned} &u_{0,1;0,2}, \quad u_{1,0;0,0}; \\ &u_{j,j+1;\alpha,\alpha+2}, \qquad \alpha = j+1, j+2, \ldots, 2j; \\ &u_{j+1,j;\alpha,\alpha}, \qquad \alpha = 0, 1, \ldots j; \quad j \in \mathbb{N}, \end{aligned} \tag{7.1.44}$$

of matrices $(u_{j,k})_{j,k=0}^{\infty}$, *in Lemma 7.1.6, are positive.*

Proof We start with the study of $u_{1,0;0,0}$. By using (7.1.36), we denote $P'_{1;0}(z) = e^{i\theta} - (e^{i\theta}, 1)_{L_2}$ $(1 = P_{0;0}(z))$ the non-normalized vector $P_{1;0}(z)$ obtained due to the Gram-Schmidt orthogonalization procedure taking the vector $r^0 e^{i\theta}$. Therefore, we have

$$\begin{aligned} u_{1,0;0,0} = \int\limits_{\mathbb{C}} e^{i\theta} \overline{P_{1;0}(z)}\, d\rho(z) &= \|P'_{1;0}(z)\|_{L_2}^{-1} \int\limits_{\mathbb{C}} e^{i\theta} \overline{(e^{i\theta} - (e^{i\theta}, 1)_{L_2})}\, d\rho(z) \\ &= \|P'_{1;0}(z)\|_{L_2}^{-1} (\|e^{i\theta}\|_{L_2}^2 - |(e^{-i\theta}, 1)_{L_2}|^2). \end{aligned} \tag{7.1.45}$$

The last difference is positive (see below, (7.1.47)), hence, $a_{1,0;0,0} > 0$.

Let us consider $a_{0,1;0,2}$. Denote, as before, $P'_{1;2}(z)$ the non-normalized vector $P_{1;2}(z)$, obtained due to the Gram-Schmidt orthogonalization procedure taking the vector $r^0 e^{-i\theta}$. According to (7.1.2) and (7.1.3), we have

$$P'_{1;2}(z) = e^{-i\theta} - (e^{-i\theta}, P_{1;1}(z))_{L_2} P_{1;1}(z) - (e^{-i\theta}, P_{1;0}(z))_{L_2} P_{1;0}(z) - (e^{-i\theta}, 1)_{L_2}.$$

Therefore, using (7.1.36), we get

$$\begin{aligned} u_{0,1;0,2} &= \int_{\mathbb{C}} e^{-i\theta} P_{1;2}(z)\, d\rho(z) = \|P'_{1;2}(z)\|_{L_2}^{-1} \int_{\mathbb{C}} e^{i\theta} P'_{1;2}(z)\, d\rho(z) \\ &= \|P'_{1;2}(z)\|_{L_2}^{-1} \int_{\mathbb{C}} e^{i\theta} (e^{-i\theta} - (e^{-i\theta}, P_{1;1}(z))_{L_2} P_{1;1}(z) \\ &\quad - (e^{-i\theta}, P_{1;0}(z))_{L_2} P_{1;0}(z) - (e^{-i\theta}, 1)_{L_2})\, d\rho(z) \\ &= \|P'_{1;2}(z)\|_{L_2}^{-1} (\|e^{-i\theta}\|_{L_2}^2 - |(e^{-i\theta}, P_{1;1}(z))_{L_2}|^2 \\ &\quad - |(e^{-i\theta}, P_{1;0}(z))_{L_2}|^2 - |(e^{-i\theta}, 1)_{L_2}|^2). \end{aligned} \tag{7.1.46}$$

Also using (7.1.47), we conclude that the last expression is positive and therefore $u_{0,1;0,2} > 0$.

The positiveness in (7.1.45) and (7.1.46) follows from the Parseval equality applied to the decomposition of the function $e^{-i\theta} \in L_2$ with respect to the orthonormal basis (7.1.3) of the space L_2, namely,

$$|(e^{-i\theta}, 1)_{L_2}|^2 + |(e^{-i\theta}, P_{1;0}(z))_{L_2}|^2 + |(e^{-i\theta}, P_{1;1}(z))_{L_2}|^2 + \cdots = \|e^{-i\theta}\|_{L_2}^2. \tag{7.1.47}$$

Let us pass to the proof of the positiveness of $u_{j+1,j;\alpha,\alpha}$, where $j \in \mathbb{N}$, $\alpha = 0, 1, \ldots, j$. From (7.1.36) we have:

$$u_{j+1,j;\alpha,\alpha} = \int_{\mathbb{C}} e^{i\theta} P_{j;\alpha}(z) \overline{P_{j+1;\alpha}(z)}\, d\rho(z). \tag{7.1.48}$$

According to (7.1.4) and (7.1.6),

$$P_{j;\alpha}(z) = k_{j;\alpha} r^{j-|\alpha-j|} e^{i(j-\alpha)\theta} + R_{j;\alpha}(z), \tag{7.1.49}$$

where $R_{j;\alpha}(z)$ is some polynomial from $\mathcal{P}_{j;\alpha-1}$ if $\alpha > 0$ or from $\mathcal{P}_{j-1;2(j-1)}$ if $\alpha = 0$. Therefore, $e^{i\theta} R_{j;\alpha}(z)$ is some polynomial from $\mathcal{P}_{j+1;\alpha-1}$ for $\alpha = 1, 2, \ldots, j$ or from $\mathcal{P}_{j+1;j}$ for $\alpha = j+1, j+2, \ldots, 2j$, or from $\mathcal{P}_{j;2(j-1)}$ for $\alpha = 0$ (see (7.1.13), (7.1.14), and (7.1.6)). Multiplying (7.1.49) by $e^{i\theta}$, we get

$$\begin{aligned} e^{i\theta} P_{j;\alpha}(z) &= k_{j;\alpha} r^{j-|\alpha-j)|} e^{i(j-\alpha+1)\theta} + e^{i\theta} R_{j;\alpha}(z); \\ &= k_{j;\alpha} r^{j+1-|\alpha-(j+1))|} e^{i(j+1-\alpha)\theta} + e^{i\theta} R_{j;\alpha}(z); \\ e^{i\theta} R_{j;\alpha}(z) &\in \begin{cases} \mathcal{P}_{j;2(j-1)}, \ \alpha = 0; \\ \mathcal{P}_{j+1;\alpha-1}, \ \alpha = 1, 2, \ldots, j; \\ \mathcal{P}_{j+1;j}, \quad \alpha = j+1, j+2, \ldots, 2j. \end{cases} \end{aligned} \tag{7.1.50}$$

On the other hand, equality (7.1.49), for $P_{j+1;\alpha}(z)$, gives

$$\begin{aligned} P_{j+1;\alpha}(z) &= k_{j+1;\alpha} r^{j+1-|\alpha-(j+1)|} e^{i(j+1-\alpha)\theta} + R_{j+1;\alpha}(z); \\ R_{j+1;\alpha}(z) &\in \begin{cases} \mathcal{P}_{j;2j}, \qquad \alpha = 0; \\ \mathcal{P}_{j+1;\alpha-1}, \ \alpha = 1, 2, \ldots, 2j. \end{cases} \end{aligned} \tag{7.1.51}$$

Let us find $r^{j+1-|\alpha-(j+1)|} e^{i(j+1-\alpha)\theta}$ from (7.1.51) and substitute it in (7.1.50). Thus, we have:

$$\begin{aligned} e^{i\theta} P_{j;\alpha}(z) &= \frac{k_{j;\alpha}}{k_{j+1;\alpha}} (P_{j+1;\alpha}(z) - R_{j+1;\alpha}(z)) + e^{i\theta} R_{j;\alpha}(z) \\ &= \frac{k_{j;\alpha}}{k_{j+1;\alpha}} P_{j+1;\alpha}(z) - \frac{k_{j;\alpha}}{k_{j+1;\alpha}} R_{j+1;\alpha}(z) + e^{i\theta} R_{j;\alpha}(z), \end{aligned} \tag{7.1.52}$$

where the second term belong to $\mathcal{P}_{j+1;\alpha-1}$ ($\alpha \neq 0$) or $\mathcal{P}_{j;2j}$ ($\alpha = 0$) and the third term belong to $\mathcal{P}_{j;2(j-1)}$ if $\alpha = 0$, to $\mathcal{P}_{j+1;\alpha-1}$ if $\alpha = 1, 2, \ldots, j$, to $\mathcal{P}_{j+1;j}$ if $\alpha = j+1, j+2, \ldots, 2j$ and is in any case orthogonal to $P_{j+1;\alpha}(z)$ if $\alpha \leq j$. Therefore, after the substitution the expression (7.1.52) into (7.1.48) we get that $u_{j+1,j;\alpha,\alpha} = k_{j;\alpha}(k_{j+1;\alpha})^{-1} > 0$.

Consider at last the elements $u_{j,j+1;\alpha,\alpha+2}$ where $j \in \mathbb{N}$ and $\alpha = j+1, j+2, \ldots, 2j$. From(7.1.36), we get

$$\begin{aligned} u_{j,j+1;\alpha,\alpha+2} &= \int_{\mathbb{C}} e^{i\theta} P_{j+1,\alpha+2}(z) \overline{P_{j;\alpha}(z)}\, d\rho(z) \\ &= \overline{\int_{\mathbb{C}} e^{-i\theta} P_{j,\alpha}(z) \overline{P_{j+1;\alpha+2}(z)}\, d\rho(z)}. \end{aligned} \tag{7.1.53}$$

For $P_{j;\alpha}(z)$ we have expression (7.1.49). Multiply it on $e^{-i\theta}$, similar to (7.1.50), we get

$$\begin{aligned} e^{-i\theta}P_{j;\alpha}(z) &= k_{j;\alpha}r^{(j-|(\alpha-1)|}e^{i(j-\alpha-1)\theta}+e^{-i\theta}R_{j;\alpha}(z), \\ &= k_{j;\alpha}r^{(j+1)-|(\alpha+2)-(j+1)|}e^{i(j+1-(\alpha+2))\theta}+e^{-i\theta}R_{j;\alpha}(z), \end{aligned}$$

$$e^{-i\theta}R_{j;\alpha}(z) \in \begin{cases} \mathcal{P}_{j-1;2(j-1)}, & \alpha = 0; \\ \mathcal{P}_{j;2j}, & \alpha = 1, 2, \ldots, j-1; \\ \mathcal{P}_{j+1;\alpha+2}, & \alpha = j, j+1, \ldots, 2j. \end{cases} \tag{7.1.54}$$

Now the equality (7.1.49), for $\mathcal{P}_{j+1;\alpha+2}$, has a form

$$P_{j+1;\alpha+2}(z) = k_{j+1;\alpha+2}r^{j+1-|(\alpha+2)-(j+1)|}e^{i((j+1)-(\alpha+2))\theta}+R_{j+1;\alpha+2}(z);$$

$$R_{j+1;\alpha+2}(z) \in \begin{cases} \mathcal{P}_{j+1;\alpha+1}, & \alpha = 1, 2, \ldots, 2j; \\ \mathcal{P}_{j;2j}, & \alpha = 0. \end{cases} \tag{7.1.55}$$

We take $r^{j+1-|(\alpha+2)-(j+1)|}e^{i((j+1)-(\alpha+2))\theta}$ and (7.1.55) and substitute it into (7.1.54). Thus, we get

$$\begin{aligned} e^{-i\theta}P_{j;\alpha}(z) &= \frac{k_{j;\alpha}}{k_{j+1;\alpha+2}}(P_{j+1;\alpha+2}(z)-R_{j+1;\alpha+2}(z))+e^{-i\theta}R_{j;\alpha}(z) \\ &= \frac{k_{j;\alpha}}{k_{j+1;\alpha+2}}P_{j+1;\alpha+2}(z)-\frac{k_{j;\alpha}}{k_{j+1;\alpha+2}}R_{j+1;\alpha+2}(z)+e^{-i\theta}R_{j;\alpha}(z). \end{aligned} \tag{7.1.56}$$

As before the second term in (7.1.56) belongs to $\mathcal{P}_{j+1;\alpha+1}$ for $\alpha = 1, 2, \ldots, 2j$, or to $\mathcal{P}_{j;2j}$ and for $\alpha = 0$, and the third term belongs to $\mathcal{P}_{j+1;\alpha+1}$ for $\alpha = j+1, j+2, \ldots, 2j$, or to $\mathcal{P}_{j;2j}$ for $\alpha = 1, 2, \ldots, j-1$ and to $\mathcal{P}_{j-1;2(j-1)}$ if $\alpha = 0$, and it is in any case orthogonal to $P_{j+1;\alpha+2}(z)$ for $\alpha = j+1, j+2, \ldots, 2j$.

Therefore, the substituting of the expression (7.1.56) into (7.1.53) gives: $u_{j,j+1;\alpha,\alpha+1} = k_{j;\alpha}(k_{j+1;\alpha+2})^{-1} > 0$. □

In what follows we will use usual convenient notations for elements $u_{j,k}$ of the Jacobi matrix, namely,

$$\begin{aligned} u_n &= u_{n+1,n} : & \mathcal{H}_n &\longrightarrow \mathcal{H}_{n+1}, \\ w_n &= u_{n,n} : & \mathcal{H}_n &\longrightarrow \mathcal{H}_n, \\ v_n &= u_{n,n+1} : & \mathcal{H}_{n+1} &\longrightarrow \mathcal{H}_n, \qquad n \in \mathbb{N}_0. \end{aligned} \tag{7.1.57}$$

All previous investigation are summarized in the theorem.

Theorem 7.1.8 *The bounded Hermitian multiplication by r operator $\hat{A}$ commuting with the unitary multiplication by $e^{i\theta}$ operator $\hat{U}$, (with a strong cyclic vector) on the space L_2 in the orthonormal basis* (7.1.3) *of polynomials, have the form of a tri-diagonal block Jacobi type symmetric matrix $J_A = (a_{j,k})_{j,k=0}^{\infty}$ and a unitary matrix $J_U = (u_{j,k})_{j,k=0}^{\infty}$, that act on the space* (7.1.7), *namely,*

$$\mathbf{l}_2 = \mathcal{H}_0 \oplus \mathcal{H}_1 \oplus \mathcal{H}_2 \oplus \cdots, \quad \mathcal{H}_n = \mathbb{C}^{n+1}, \quad n \in \mathbb{N}_0. \tag{7.1.58}$$

Norms of all operators $a_{j,k} : \mathcal{H}_k \longrightarrow \mathcal{H}_j$ and $u_{j,k} : \mathcal{H}_k \longrightarrow \mathcal{H}_j$ are uniformly bounded with respect to $j, k \in \mathbb{N}_0$.

In designations (7.1.32), *the symmetric matrix has the form*

$$J_A = \left[\begin{array}{cccccc} b_0 & c_0 & 0 & 0 & 0 & \dots \\ a_0 & b_1 & c_1 & 0 & 0 & \dots \\ 0 & a_1 & b_2 & c_2 & 0 & \dots \\ 0 & 0 & a_2 & b_3 & c_3 & \dots \\ \vdots & \vdots & \vdots & \vdots & \vdots & \ddots \end{array}\right]$$

$$= \left[\begin{array}{c|ccc|ccccc|ccccccc} * & * & + & 0 & & & & & & & & & & & & \\ \hline * & * & * & * & * & + & 0 & 0 & 0 & & & & & & & \\ + & * & * & * & * & * & + & 0 & 0 & & & & 0 & & & \\ 0 & * & * & * & * & * & * & + & 0 & & & & & & & \\ \hline & * & * & * & * & * & * & * & * & * & + & 0 & 0 & 0 & 0 & 0 \\ & + & * & * & * & * & * & * & * & * & * & + & 0 & 0 & 0 & 0 \\ & 0 & + & * & * & * & * & * & * & * & * & * & + & 0 & 0 & 0 \\ & 0 & 0 & + & * & * & * & * & * & * & * & * & * & + & 0 & 0 \\ & 0 & 0 & 0 & * & * & * & * & * & * & * & * & * & * & + & 0 \\ \hline & & & & * & * & * & * & * & * & * & * & * & * & * & * \\ & & & & + & * & * & * & * & * & * & * & * & * & * & * \\ & & & & 0 & + & * & * & * & * & * & * & * & * & * & * \\ & & 0 & & 0 & 0 & + & * & * & * & * & * & * & * & * & * \\ & & & & 0 & 0 & 0 & + & * & * & * & * & * & * & * & * \\ & & & & 0 & 0 & 0 & 0 & + & * & * & * & * & * & * & * \\ & & & & 0 & 0 & 0 & 0 & 0 & * & * & * & * & * & * & * \\ & & \ddots & & & & \ddots & & & & & & & \ddots & & \end{array}\right]. \tag{7.1.59}$$

In (7.1.59), *$\forall n \in \mathbb{N}_0$, b_n is a $((2n+1) \times (2n+1))$-matrix: $b_n = (b_{n;\alpha,\beta})_{\alpha,\beta=0}^{2n,2n}$, ($b_0 = b_{0;0,0}$ is a scalar); a_n is a $((2n+3) \times (2n+1))$-matrix: $a_n = (a_{n;\alpha,\beta})_{\alpha,\beta=0}^{2n+2,2n}$; c_n is a $((2n+1) \times (2n+3))$-natrix: $c_n = (c_{n;\alpha,\beta})_{\alpha,\beta=0}^{2n,2n+2}$. Matrices a_n and c_n have* *some elements always equal to zero, namely,* $\forall n \in \mathbb{N}_0$, *we have*

$$\begin{aligned} c_{n;\alpha,\beta} &= 0, \quad \alpha = 0, 1, \dots, 2n, \ \beta = \alpha + 2, \alpha + 3, \dots, 2n + 2; \\ a_{n;\alpha,\beta} &= 0, \quad \beta = 0, 1, \dots, 2n, \ \alpha = \beta + 2, \beta + 3, \dots, 2n + 2. \end{aligned} \tag{7.1.60}$$

Some other their elements are always positive, namely, $\forall n \in \mathbb{N}_0$, *we have*

$$c_{n;\alpha,\alpha+1} > 0, \ a_{n;\alpha+1,\alpha} > 0, \quad \alpha = 0, 1, \dots, 2n. \tag{7.1.61}$$

Therefore, we can say that $\forall n \in \mathbb{N}_0$ *each lower left corner of the matrix* a_n *(starting from the third diagonal) and each upper right corner of the matrix* c_n *(starting from the third diagonal) contain always zeros elements. All positive elements in* (7.1.59) *are marked "+". Possible zero and non-zero elements are indicated by "∗".*

Thus, the matrix (7.1.59) *is multi-diagonal in the scalar form.*

The adjoint operator $(\hat{A})^* = \hat{A}$ *in the basis* (7.1.3) *has the same form similar to the tri-diagonal block Jacobi type matrix* J_A.

In designations from (7.1.57), *the unitary matrix has the form*

$$J_U = \begin{bmatrix} w_0 & v_0 & 0 & 0 & 0 & \dots \\ u_0 & w_1 & v_1 & 0 & 0 & \dots \\ 0 & u_1 & w_2 & v_2 & 0 & \dots \\ 0 & 0 & u_2 & w_3 & v_3 & \dots \\ \vdots & \vdots & \vdots & \vdots & \vdots & \ddots \end{bmatrix}$$

$$= \left[\begin{array}{c|ccc|ccccc|ccccccc} * & * & * & + & & & & & & & & & & & & \\ + & * & * & * & 0 & 0 & 0 & 0 & 0 & & & & & & & \\ 0 & * & * & * & * & * & * & + & 0 & & & & 0 & & & \\ 0 & * & * & * & * & * & * & * & + & & & & & & & \\ & + & * & * & * & * & * & * & * & 0 & 0 & 0 & 0 & 0 & 0 & 0 \\ & 0 & + & * & * & * & * & * & * & 0 & 0 & 0 & 0 & 0 & 0 & 0 \\ & 0 & 0 & 0 & * & * & * & * & * & * & * & * & * & + & 0 & 0 \\ & 0 & 0 & 0 & * & * & * & * & * & * & * & * & * & * & + & 0 \\ & 0 & 0 & 0 & * & * & * & * & * & * & * & * & * & * & * & + \\ & & & & + & * & * & * & * & * & * & * & * & * & * & * \\ & & & & 0 & + & * & * & * & * & * & * & * & * & * & * \\ & & & & 0 & 0 & + & * & * & * & * & * & * & * & * & * \\ & & 0 & & 0 & 0 & 0 & 0 & 0 & * & * & * & * & * & * & * \\ & & & & 0 & 0 & 0 & 0 & 0 & * & * & * & * & * & * & * \\ & & & & 0 & 0 & 0 & 0 & 0 & * & * & * & * & * & * & * \\ & & & & 0 & 0 & 0 & 0 & 0 & * & * & * & * & * & * & * \\ & & \ddots & & & & \ddots & & & & & & \ddots & & & \end{array}\right]. \tag{7.1.62}$$

In (7.1.62), $\forall n \in \mathbb{N}_0$, w_n *is a* $((2n+1)\times(2n+1))$*-matrix:* $w_n = (b_{n;\alpha,\beta})_{\alpha,\beta=0}^{2n,2n}$, ($w_0 = w_{0;0,0}$ *is a scalar*)*;* u_n *is a* $((2n+3)\times(2n+1))$*-matrix:* $u_n = (u_{n;\alpha,\beta})_{\alpha,\beta=0}^{2n+2,2n}$*;* v_n *is a* $((2n+1)\times(2n+3))$*-matrix:* $v_n = (c_{n;\alpha,\beta})_{\alpha,\beta=0}^{2n,2n+2}$. *Matrices* u_n *and* v_n *have some elements that are always equal to zero, namely,* $\forall n \in \mathbb{N}$*, we have*

$$\begin{aligned}
v_{n;\alpha,\beta} &= 0,\ \alpha = 0, 1, \dots, n-1, && \beta = 0, 1, \dots, 2n+2;\\
v_{n;\alpha,\beta} &= 0,\ \alpha = n, n+1, \dots, 2n-1, && \beta = \alpha+3, \alpha+4, \dots 2n+2;\\
u_{n;\alpha,\beta} &= 0,\ \beta = 0, 1, \dots, n, && \alpha = \beta+1, \beta+2, \dots, 2n+2;\\
u_{n;\alpha,\beta} &= 0,\ \beta = n+1, n+2 \dots, 2n, && \alpha = n+1, n+2, \dots, 2n+2
\end{aligned} \tag{7.1.63}$$

and $u_{0;1,0} = u_{0;2,0} = 0$.

Some other their elements are always positive, namely, $\forall n \in \mathbb{N}_0$*, we have*

$$\begin{aligned}
&v_{0;0,2},\, u_{0;0,0} > 0;\\
&v_{n;\alpha,\alpha+2} > 0, \quad \alpha = n+1, n+2, \dots, 2n;\\
&u_{n;\alpha,\alpha} > 0, \quad \alpha = 0, 1, \dots, n; \quad n \in \mathbb{N}.
\end{aligned} \tag{7.1.64}$$

Therefore, we can say that $\forall n \in \mathbb{N}_0$*, each lower left corner of the matrix* u_n *(starting from the second diagonal, which is located in the upper left corner) and* $n+2$ *the last rows, and each upper right corner of the matrix* v_n *(starting from the second diagonal, which is located in the lower right corner) and* n *of the first rows contain always zero elements. All positive elements in* (7.1.62) *are denoted by "+". Possible zero and non-zero elements are indicated by "∗".*

Thus, the matrix (7.1.62) *is multi-diagonal in the scalar form.*

The adjoint operator $(\hat{U})^*$ *in the basis* (7.1.3) *has the form similar to tri-diagonal block Jacobi type matrix* J_{U^*}.

These matrices J_A*,* J_U*, and* J_{U^*} *act by the rule*

$$\begin{aligned}
(J_A f)_n &= a_{n-1} f_{n-1} + b_n f_n + c_n f_{n+1},\\
(J_U f)_n &= u_{n-1} f_{n-1} + w_n f_n + v_n f_{n+1},\\
(J_{U^*} f)_n &= v_{n-1}^* f_{n-1} + w_n^* f_n + u_n^* f_{n+1},\\
&n \in \mathbb{N}_0, \quad f_{-1} = 0, \quad \forall f = (f_n)_{n=0}^{\infty} \in \mathbf{l}_2
\end{aligned} \tag{7.1.65}$$

(here "∗" denotes the usual adjoint matrix).

Let us explain that the form of the coefficients in the expression for J_{U^*} follows from (7.1.39) and (7.1.57).

7.2 Direct and Inverse Spectral Problems for Tri-Diagonal Block Jacobi Type Matrices Corresponding to the Complex Moment Problem in the Exponential Form

As it was mentioned above, the main result of the previous section is actually the solution of the inverse problem corresponding to the direct one presented in the title of the section.

We consider operators on the space $\mathbf{l}_2$ of the form (7.1.7). In addition to the space $\mathbf{l}_2$, consider its equipment

$$(\mathbf{l}_{\text{fin}})' \supset \mathbf{l}_2(p^{-1}) \supset \mathbf{l}_2 \supset \mathbf{l}_2(p) \supset \mathbf{l}_{\text{fin}}, \tag{7.2.1}$$

where $\mathbf{l}_2(p)$ is the weighted $\mathbf{l}_2$ space with a weight $p = (p_n)_{n=0}^{\infty}$, where $p_n \geq 1$, and $p^{-1} = (p_n^{-1})_{n=0}^{\infty}$. In our case $\mathbf{l}_2(p)$ is the Hilbert space of sequences $f = (f_n)_{n=0}^{\infty}$, $f_n \in \mathcal{H}_n$ for which we have

$$\|f\|^2_{\mathbf{l}_2(p)} = \sum_{n=0}^{\infty} \|f_n\|^2_{\mathcal{H}_n} p_n, \quad (f, g)_{\mathbf{l}_2(p)} = \sum_{n=0}^{\infty} (f_n, g_n)_{\mathcal{H}_n} p_n. \tag{7.2.2}$$

The space $\mathbf{l}_2(p^{-1})$ is defined similarly; recall that $\mathbf{l}_{\text{fin}}$ is the space of finite sequences and $(\mathbf{l}_{\text{fin}})'$ is the space conjugate to $\mathbf{l}_{\text{fin}}$. It is easy to show that the imbedding $\mathbf{l}_2(p) \hookrightarrow \mathbf{l}_2$ is quasi-nuclear if $\sum_{n=0}^{\infty} n p_n^{-1} < \infty$.

Let A be a bounded self-adjoint operator strongly commuting with a unitary operators U. Suppose A and U standardly connected with the chain (7.2.1). (About the standard connection see the beginning of the last section of this chapter.) According to the projection spectral theorem, such operators have representations

$$Af = \int_{\mathbb{C}} r\Phi(z)\, d\sigma(z) f, \quad Uf = \int_{\mathbb{C}} e^{i\theta}\Phi(z)\, d\sigma(z) f, \quad f \in \mathbf{l}_2, \tag{7.2.3}$$

where $\Phi(z) : \mathbf{l}_2(p) \longrightarrow \mathbf{l}_2(p^{-1})$ is the generalized projection operator and $d\sigma(z)$ is a spectral measure. The adjoint to U operator U^* has the same representation as in (7.2.3), where $e^{i\theta}\Phi(z)$ is replaced with $e^{-i\theta}\Phi(z)$. For every $f \in \mathbf{l}_{\text{fin}}$, the projection $\Phi(z) f \in \mathbf{l}_2(p^{-1})$ is a generalized eigenvector of the operators U and U^* and A with corresponding eigenvalues $e^{i\theta}$, $e^{-i\theta}$, and r. For all $f, g \in \mathbf{l}_{\text{fin}}$, we have the Parseval equality

$$(f, g)_{\mathbf{l}_2} = \int_{\mathbb{C}} (\Phi(z) f, g)_{\mathbf{l}_2}\, d\sigma(z); \tag{7.2.4}$$

after the extension by continuity the equality (7.2.4) is valid for all $f, g \in \mathbf{l}_2$.

Let us denote by π_n the operator of orthogonal projection in $\mathbf{l}_2$ on $\mathcal{H}_n$, $n \in \mathbb{N}_0$. Hence, $\forall f = (f_n)_{n=0}^{\infty} \in \mathbf{l}_2$, we have $f_n = \pi_n f$. This operator acts similarly on the spaces $\mathbf{l}_2(p)$ and $\mathbf{l}_2(p^{-1})$ but possibly with the norm which is not equal to one.

Let us consider the operator matrix $(\Phi_{j,k}(z))_{j,k=0}^{\infty}$, where

$$\Phi_{j,k}(z) = \pi_j \Phi(z) \pi_k \; : \; \mathbf{l}_2 \longrightarrow \mathcal{H}_j, \; (\text{or } \mathcal{H}_k \longrightarrow \mathcal{H}_j). \tag{7.2.5}$$

The Parseval equality (7.2.4) can be written in the form

$$\begin{aligned}(f, g)_{\mathbf{l}_2} &= \sum_{j,k=0}^{\infty} \int_{\mathbb{C}} (\Phi(z)\pi_k f, \pi_j g)_{\mathbf{l}_2} \, d\sigma(z) \\ &= \sum_{j,k=0}^{\infty} \int_{\mathbb{C}} (\pi_j \Phi(z)\pi_k f, g)_{\mathbf{l}_2} \, d\sigma(z) \\ &= \sum_{j,k=0}^{\infty} \int_{\mathbb{C}} (\Phi_{j,k}(z) f_k, g_j)_{\mathbf{l}_2} d\,\sigma(z), \quad \forall f, g \in \mathbf{l}_2.\end{aligned} \tag{7.2.6}$$

Let us now pass to a study of a more special bounded operators A and U that act on the space $\mathbf{l}_2$. Namely, let them be given by matrices J_A and J_U having a tri-diagonal block structure of the form (7.1.59) and (7.1.62), respectively. Thus, these operators A and U are defined by the first and the second expressions (7.1.65), the adjoint operator U^* is similarly defined by the third expression in (7.1.65). Recall that norms of elements a_n, b_n, and c_n, as well as u_n, w_n, and v_n are uniformly bounded for all $n \in \mathbb{N}_0$.

For further research, *assume that the conditions* (7.1.60), (7.1.61) *and* (7.1.63), (7.1.64) *are satisfied and, additionally, operators A and U specified in* (7.1.59) *and* (7.1.62) *are self-adjoint and unitary, respectively, and they commute on* $\mathbf{l}_2$. Conditions under which the operators A and U are self-adjoint, unitary and commute, are investigated in the next section.

Here, we write the Parseval equality (7.2.6) in terms of generalized eigenvectors of the operator A. First, let us prove the lemma.

Lemma 7.2.1 *Let* $\varphi(z) = (\varphi_n(z))_{n=0}^{\infty}$, $\varphi_n(z) \in \mathcal{H}_n$, $z \in \mathbb{C}$ *be a generalized eigenvector from* $(\mathbf{l}_{\text{fin}})'$ *of commuting a self-adjoint operator A and a unitary operator U* (U^*) *with eigenvalues r and* $e^{i\theta}$, $(e^{-i\theta})$, *respectively. Thus,* $\varphi(z)$ *is a solution in* $(\mathbf{l}_{\text{fin}})'$ *of the system of three difference equations (see* (7.1.65)*), namely,*

$$\begin{aligned}(J_A\varphi(z))_n &= a_{n-1}\varphi_{n-1}(z) + b_n\varphi_n(z) + c_n\varphi_{n+1}(z) = r\varphi_n(z), \\ (J_U\varphi(z))_n &= u_{n-1}\varphi_{n-1}(z) + w_n\varphi_n(z) + v_n\varphi_{n+1}(z) = e^{i\theta}\varphi_n(z), \\ (J_{U^*}\varphi(z))_n &= v_{n-1}^*\varphi_{n-1}(z) + w_n^*\varphi_n(z) + u_n^*\varphi_{n+1}(z) = e^{-i\theta}\varphi_n(z), \\ &\qquad n \in \mathbb{N}_0, \quad \varphi_{-1}(z) = 0\end{aligned} \tag{7.2.7}$$

with an initial condition $\varphi_0 \in \mathbb{C}$.

It is claimed that this solution has the form

$$\varphi_n(z) = Q_n(z)\varphi_0 = (Q_{n;0}, Q_{n;1}, \ldots, Q_{n;2n},)\varphi_0, \quad \forall n \in \mathbb{N}. \tag{7.2.8}$$

Here $Q_{n;\alpha}$, $\alpha = 0, 1, \ldots, 2n$ *are polynomials of variables* r, $e^{i\theta}$, *and* $e^{-i\theta}$, *and these polynomials have the form*

$$Q_{n;\alpha}(z) = k_{n;\alpha} r^{n-|\alpha-n|} e^{-i(n-\alpha)\theta} + R_{n;\alpha}(z), \quad \alpha = 0, 1, \ldots, 2n, \; n \in \mathbb{N}_0; \tag{7.2.9}$$

where $k_{n;\alpha} > 0$ *and* $R_{n;\alpha}(z)$ *is some linear combinations of* r, $e^{i\theta}$, *and* $e^{-i\theta}$ *(of lesser order with respect to (7.1.2) and Fig. 7.1).*

Proof For $n = 0$, the system (7.2.7) (in a convenient order of equations) has the form

$$\begin{aligned} J_{U^*}\varphi_0 &= e^{-i\theta}\varphi_0, \\ J_A\varphi_0 &= r\varphi_0, \\ J_U\varphi_0 &= e^{i\theta}\varphi_0, \end{aligned} \tag{7.2.10}$$

i.e.,

$$\begin{aligned} &\bar{w}_{0;0,0}\varphi_{0;0} + \bar{u}_{0;0,0}\varphi_{1;0} &&= e^{-i\theta}\varphi_{0;0}, \\ &b_{0;0,0}\varphi_{0;0} + c_{0;0,0}\varphi_{1;0} + c_{0;0,1}\varphi_{1;1} &&= r\varphi_{0;0} \\ &w_{0;0,0}\varphi_{0;0} + v_{0;0,0}\varphi_{1;0} + v_{0;0,1}\varphi_{1;1} + v_{0;0,2}\varphi_{1;2} &&= e^{i\theta}\varphi_{0;0} \end{aligned}$$

or

$$\begin{aligned} &\bar{u}_{0;0,0}\varphi_{1;0} &&= (e^{-i\theta} - \bar{w}_{0;0,0})\varphi_{0;0}, \\ &c_{0;0,0}\varphi_{1;0} + c_{0;0,1}\varphi_{1;1} &&= (r - b_{0;0,0})\varphi_{0;0} \\ &v_{0;0,0}\varphi_{1;0} + v_{0;0,1}\varphi_{1;1} + v_{0;0,2}\varphi_{1;2} &&= (e^{i\theta} - w_{0;0,0})\varphi_{0;0}. \end{aligned} \tag{7.2.11}$$

Here and in what follows we denote

$$\varphi_0 = \varphi_{0;0} := Q_{0;0}; \quad \varphi_n(z) = (\varphi_{n;0}(z), \varphi_{n;1}(z), \ldots, \varphi_{n;2n}(z)) \in \mathcal{H}_n, \quad \forall n \in \mathbb{N}.$$

We rewrite equalities of (7.2.10) in the form

$$\begin{aligned} \Delta_0\varphi_1(z) &= \Big((e^{-i\theta} - \bar{w}_{0;0,0})\varphi_0), (r - b_{0;0,0})\varphi_0, (e^{i\theta} - w_{0;0,0})\varphi_0\Big); \\ \Delta_0 &= \begin{pmatrix} \bar{u}_{0;0,0} & 0 & 0 \\ c_{0;0,0} & c_{0;0,1} & 0 \\ v_{0;0,0} & c_{0;0,1} & c_{0;0,2} \end{pmatrix}, \end{aligned} \tag{7.2.12}$$

where due to conditions (7.1.60), (7.1.61) and (7.1.63), (7.1.64), we have always $\bar{u}_{0;0,0} > 0$, $c_{0;0,1} > 0$, $u_{0;0,2} > 0$. Hence, $\Delta_0 > 0$. Therefore, from (7.2.11) we obtain consequently

$$\varphi_{1;0}(z) = -\frac{\bar{w}_{0;0,0} - e^{-i\theta}}{\bar{u}_{0;0,0}}\varphi_0 =: Q_{1;0}(z)\varphi_0,$$

$$\varphi_{1;1}(z) = -\left(\frac{b_{0;0,0} - r}{c_{0;0,1}} + \frac{c_{0;0,0}}{c_{0;0,1}}\frac{e^{-i\theta} - \bar{w}_{0;0,0}}{\bar{u}_{0;0,0}}\right)\varphi_0 =: Q_{1;1}(z)\varphi_0,$$

$$\varphi_{1;2}(z) = -\left(\frac{w_{0;0,0} - e^{i\theta}}{v_{0;0,2}} - \frac{v_{0;0,0}}{v_{0;0,2}}[\frac{\bar{w}_{0;0,0} - e^{-i\theta}}{\bar{u}_{0;0,0}}]\right.$$
$$\left. - \frac{v_{0;0,1}}{v_{0;0,2}}[\frac{b_{0;0,0} - r}{c_{0;0,1}} + \frac{c_{0;0,0}}{c_{0;0,1}}\frac{e^{-i\theta} - \bar{w}_{0;0,0}}{\bar{u}_{0;0,0}}]\right)\varphi_0 =: Q_{1;2}(z)\varphi_0. \tag{7.2.13}$$

In other words, a solution $\varphi_n(z)$ by $n = 1$ from (7.2.7) has a form (7.2.8) with (7.2.9).

Suppose, by induction, that for $n \in \mathbb{N}$ the coordinates $\varphi_{n-1}(z)$ and $\varphi_n(z)$ of our generalized eigenvector $\varphi(z) = (\varphi_n(z))_{n=0}^{\infty}$ have the form (7.2.8) with (7.2.9), and we will show that $\varphi_{n+1}(z)$ also has the form (7.2.8) with (7.2.9).

Our eigenvector $\varphi_{n+1}(z)$ satisfies the system of equations (7.2.7). But this system is overdetermined: it consists of $3(2n+3)$ scalar equations from which it is necessary to find only $2n+3$ unknowns $\varphi_{n+1;0}, \varphi_{n+1;1}, \ldots, \varphi_{n+1;2(n+1)}$ using as an initial data the previous $2n+1$ values $\varphi_{n;0}, \varphi_{n;1}, \ldots, \varphi_{n;2n}$ of coordinates of the vector $\varphi_n(z)$. Similarly to (7.2.10), for $j = 0, 1, \ldots, 2n$, we have

$$J_{U^*}\varphi_{n;j} = e^{-i\theta}\varphi_{n;j},$$
$$J_A\varphi_{n;j} = r\varphi_{n;j},$$
$$J_U\varphi_{n;j} = e^{i\theta}\varphi_{n;j},$$

We act in this way. According to Theorem 7.1.8, namely, (7.1.60), (7.1.61) and (7.1.63), (7.1.64), a $((2n+3)\times(2n+1))$-matrices $\bar{u}_n, c_n, v_n$ and their application to $\varphi_{n+1} \in \mathcal{H}_n$ have the form

$$u_n^*\varphi_{n+1}(z) = \begin{bmatrix} \bar{u}_{n;0,0} & 0 & \ldots 0 & 0\ldots 0 \\ \bar{u}_{n;0,1} & \bar{u}_{n;1,1} & \ldots 0 & 0\ldots 0 \\ \vdots & \vdots & \ddots\vdots & \vdots\ddots\vdots \\ \bar{u}_{n;0,n} & \bar{u}_{n;1,n} & \ldots \bar{u}_{n;n,n} & 0\ldots 0 \\ \vdots & \vdots & \ddots\vdots & \vdots\ddots\vdots \\ \bar{u}_{n;0,2n-1} & \bar{u}_{n;1,2n-1} & \ldots \bar{u}_{n;n,2n-1} & 0\ldots 0 \\ \bar{u}_{n;0,2n} & \bar{u}_{n;1,2n} & \ldots \bar{u}_{n;n,2n} & 0\ldots 0 \end{bmatrix}\varphi_{n+1}(z), \tag{7.2.14}$$

$$c_n\varphi_{n+1}(z) = \begin{bmatrix} c_{n;0,0} & c_{n;0,1} & 0 & \dots 0 & 0 \\ c_{n;1,0} & c_{n;1,1} & c_{n;1,2} & \dots 0 & 0 \\ c_{n;2,0} & c_{n;2,1} & c_{n;2,2} & \dots 0 & 0 \\ \vdots & \vdots & \vdots & \ddots \vdots & \vdots \\ c_{n;2n-1,0} & c_{n;2n-1,1} & c_{n;2n-1,2} & \dots 0 & 0 \\ c_{n;2n,0} & c_{n;2n,1} & c_{n;2n,2} & \dots c_{n;2n,2n+1} & 0 \end{bmatrix} \varphi_{n+1}(z), \tag{7.2.15}$$

$$v_n\varphi_{n+1}(z) = \begin{bmatrix} 0 & \dots 0 & \dots 0 & 0 \\ \vdots & \ddots \vdots & \ddots \vdots & \vdots \\ 0 & \dots 0 & \dots 0 & 0 \\ v_{n;n,0} & \dots v_{n;n,n+2} & \dots 0 & 0 \\ \vdots & \ddots \vdots & \ddots \vdots & \vdots \\ v_{n;2n-1,0} & \dots v_{n;2n-1,n+2} & \dots v_{n;2n-1,2n+1} & 0 \\ v_{n;2n,0} & \dots v_{n;2n,n+2} & \dots v_{n;2n,2n+1} & v_{n;2n,2n+2} \end{bmatrix} \varphi_{n+1}(z), \tag{7.2.16}$$

where $\varphi_{n+1}(z) = (\varphi_{n+1;0}(z), \varphi_{n+1;1}(z), \dots, \varphi_{n+1;2n+2}(z))$.

Let us construct the combination similarly to (7.2.12), using matrices from (7.2.14)–(7.2.16), we get a $((2n+2) \times (2n+2))$-matrix

$$\Delta_n\varphi_{n+1}(z) = \begin{bmatrix} \bar{u}_{n;0,0} & 0 & 0 & \dots 0 & 0 \\ c_{n;0,0} & c_{n;0,1} & 0 & \dots 0 & 0 \\ c_{n;1,0} & c_{n;1,1} & c_{n;1,2} & \dots 0 & 0 \\ \vdots & \vdots & \vdots & \ddots \vdots & \vdots \\ c_{n;2n,0} & c_{n;2n,1} & c_{n;2n,2} & \dots c_{n;2n,2n+1} & 0 \\ v_{n;2n,0} & v_{n;2n,1} & v_{n;2n,2} & \dots v_{n;2n,2n+1} & v_{n;2n,2n+2} \end{bmatrix} \varphi_{n+1}(z), \tag{7.2.17}$$

where $\varphi_{n+1}(z) = (\varphi_{n+1;0}(z), \varphi_{n+1;1}(z), \dots, \varphi_{n+1;2n+2}(z))$.

The matrix (7.2.17) has an inverse one, since its elements are positive on the main diagonal (see (7.1.60), (7.1.61) and (7.1.63), (7.1.64)). Rewrite the equalities (7.2.7) in the form

$$u_n^*\varphi_{n+1}(z) = e^{-i\theta}\varphi_n(z) - v_{n-1}^*\varphi_{n-1}(z) - w_n^*\varphi_n(z),$$
$$c_n\varphi_{n+1}(z) = r\varphi_n(z) - a_{n-1}\varphi_{n-1}(z) - b_n\varphi_n(z),$$
$$v_n\varphi_{n+1}(z) = e^{i\theta}\varphi_n(z) - u_{n-1}\varphi_{n-1}(z) - w_n\varphi_n(z), \quad n \in \mathbb{N},$$

i.e.,

$$\begin{aligned} u_n^* \varphi_{n+1}(z) &= (e^{-i\theta} - w_n^*)\varphi_n(z) - v_{n-1}^* \varphi_{n-1}(z), \\ c_n \varphi_{n+1}(z) &= (r - b_n)\varphi_n(z) - a_{n-1}\varphi_{n-1}(z), \\ v_n \varphi_{n+1}(z) &= (e^{i\theta} - w_n)\varphi_n(z) - u_{n-1}\varphi_{n-1}(z), \quad n \in \mathbb{N}, \end{aligned} \tag{7.2.18}$$

We see, that scalar equations (7.2.18)) have the form

$$\bar{u}_{n;0,0}\varphi_{n+1,0}(z) = \Big\{(e^{-i\theta}1 - w_n^*)Q_n(z) - v_{n-1}^* Q_{n-1}(z)\Big\}_{n;j},$$

where "1" is the identity $(2n+2) \times (2n+2)$-matrix. Thus, for $j = 0$ we obtain

$$Q_{n+1,0}(z) := \varphi_{n+1,0}(z) = \frac{1}{\bar{u}_{n;0,0}} \Big\{(e^{-i\theta}1 - w_n^*)Q_n(z) - v_{n-1}^* Q_{n-1}(z)\Big\}_{n;0}.$$

Hence, $Q_{n+1,0}(z) = k_{n+1,0}e^{-i(n+1)\theta} + \dots$.

For the case $j = 1$, we have

$$c_{n;0,0}\varphi_{n+1,0}(z) + c_{n;0,1}\varphi_{n+1,1}(z) = \{(r1 - b_n)Q_n(z) - a_{n-1}Q_{n-1}(z)\}_{n;1},$$

$$\begin{aligned} Q_{n+1,1}(z) &:= \varphi_{n+1,1}(z) \\ &= \frac{1}{\bar{c}_{n;0,1}} \{-c_{n;0,0}Q_{n+1,0}(z) + (r1 - b_n)Q_n(z) - a_{n-1}Q_{n-1}(z)\}_{n;1}. \end{aligned}$$

Hence, $Q_{n+1,1}(z) = k_{n+1,1}re^{-in\theta} + \dots$.

For $j = 2, 3, \dots, 2n+1$, we have

$$\begin{aligned} c_{n;j,0}\varphi_{n+1,0}(z) + c_{n;j,1}\varphi_{n+1,1}(z) + \dots, + c_{n;j,j+1}\varphi_{n+1,j}(z) \\ = \{(r1 - b_n)Q_n(z) - a_{n-1}Q_{n-1}(z)\}_{n;j}, \end{aligned}$$

$$\begin{aligned} Q_{n+1,j}(z) := \varphi_{n+1,j}(z) = \frac{1}{c_{n;j,j+1}} \{&-c_{n;j,0}\varphi_{n+1,0}(z) - c_{n;j,1}\varphi_{n+1,1}(z) - \dots \\ &- c_{n;j,j}\varphi_{n+1,j-1}(z) + (r1 - b_n)Q_n(z) - a_{n-1}Q_{n-1}(z)\}_{n;j}. \end{aligned}$$

Hence, $Q_{n+1,j}(z) = k_{n+1,j}r^{n+1-|j-(n+1)|}e^{i(n+1-j)\theta} + \dots$.

And for the last equation from (7.2.18) (i.e., $j = 2n+2$), we have

$$\begin{aligned} v_{n;2n,0}\varphi_{n+1,0}(z) + v_{n;2n,1}\varphi_{n+1,1}(z) + \dots + v_{n;2n,2n+2}\varphi_{n+1,2n+2}(z) \\ = \Big\{(e^{i\theta}1 - w_n)Q_n(z) - u_{n-1}Q_{n-1}(z)\Big\}_{n;2n+2}, \end{aligned}$$

$$
\begin{aligned}
Q_{n+1,2n+2}(z) :=& \varphi_{n+1,2n+2}(z) = \frac{1}{v_{n;2n,2n+2}} \big\{ -v_{n;j,0}\varphi_{n+1,0}(z) \\
& - v_{n;j,1}\varphi_{n+1,1}(z) - \ldots - v_{n;2n,2n+1}\varphi_{n+1,2n+1}(z) \\
& + (e^{i\theta}1 - w_n)Q_n(z) - u_{n-1}Q_{n-1}(z) \big\}_{n;2n+2}.
\end{aligned}
$$

Hence, $Q_{n+1,2n+2}(z) = k_{n+1,2n+2}r^{n+1-|(2n+2)-(n+1)|}e^{i(n+1)-(2n+2))\theta} + \ldots$.

It is necessary to take into account that diagonal elements $\bar{u}_{n;0,0}$, $c_{n;0,1}$, $c_{n;1,2}$, ..., $c_{n;2n,2n+1}$, $v_{n;2n,2n+2}$ of the matrix Δ_n are positive due to (7.1.61) and (7.1.64). This completes the induction and finishes the proof. □

Remark 7.2.2 Note, that we did not assert, that the solution of the overdetermined system (7.2.7) exist for an arbitrary initial data $\varphi_0 \in \mathbb{C}$: we prove only, that the generalized eigenvector from $(\mathbf{l}_{\text{fin}})'$ of operators A and U is a solution of (7.2.7) and has the form (7.2.8) and (7.2.9).

In what follows, it will be convenient to look at $Q_n(z)$ with fixed z as a linear operator that acts from $\mathcal{H}_0$ into $\mathcal{H}_n$, i.e., $\mathcal{H}_0 \ni \varphi_0 \longmapsto Q_n(z)\varphi_0 \in \mathcal{H}_n$. We also regard $Q_n(z)$ as an operator valued polynomial of variables $z = re^{i\theta} \in \mathbb{C}$, i.e., r, $e^{i\theta}$, and $e^{-i\theta}$; hence, for the adjoint operator we have $Q_n^*(z) = (Q_n(z))^* : \mathcal{H}_n \longrightarrow \mathcal{H}_0$. By using these polynomials $Q_n(z)$ we construct the representation for $\Phi_{j,k}(z)$.

Lemma 7.2.3 *The operator $\Phi_{j,k}(z)$, $\forall z \in \mathbb{C}$ has the representation*

$$
\Phi_{j,k}(z) = Q_j(z)\Phi_{0,0}(z)Q_k^*(z) : \mathcal{H}_k \longrightarrow \mathcal{H}_j, \quad j, k \in \mathbb{N}_0, \tag{7.2.19}
$$

where $\Phi_{0,0}(z) \geq 0$ is a scalar.

Proof For a fixed $k \in \mathbb{N}_0$ the vector $\varphi = \varphi(z) = (\varphi_j(z))_{j=0}^{\infty}$, where

$$
\varphi_j(z) = \Phi_{j,k}(z) = \pi_j\Phi(z)\pi_k \in \mathcal{H}_j, \quad z \in \mathbb{C} \tag{7.2.20}
$$

is a generalized solution in $(\mathbf{l}_{\text{fin}})'$ of the system of equations

$$
\begin{aligned}
J_{U^*}\varphi &= e^{-i\theta}\varphi, \\
J_A\varphi &= r\varphi, \\
J_U\varphi &= e^{i\theta}\varphi,
\end{aligned} \tag{7.2.21}
$$

since $\Phi(z)$ is a projector on to generalized eigenvectors of operators A and U (U^*) with corresponding generalized eigenvalues r, $e^{i\theta}$, and $(e^{-i\theta})$. Therefore, $\forall g \in \mathbf{l}_{\text{fin}}$, we have

$$
\begin{aligned}
(\varphi, J_A g)_{\mathbf{l}_2} &= r(\varphi, g)_{\mathbf{l}_2}, \\
(\varphi, J_{U^*} g)_{\mathbf{l}_2} &= e^{i\theta}(\varphi, g)_{\mathbf{l}_2}, \\
(\varphi, J_U g)_{\mathbf{l}_2} &= e^{-i\theta}(\varphi, g)_{\mathbf{l}_2}.
\end{aligned}
$$

Hence, it follows that $\varphi = \varphi(z) \in \mathbf{l}_2(p^{-1})$ exists as an usual solution of equations (7.2.21) with initial condition $\varphi_0 = \pi_0\Phi(z)\pi_k \in \mathcal{H}_0$.

By using Lemma 7.2.1 and due to (7.2.8), we obtain

$$\Phi_{j,k}(z) = Q_j(z)(\Phi_{0,k}(z)), \quad j \in \mathbb{N}_0. \tag{7.2.22}$$

The operator $\Phi(z) : \mathbf{l}_2(p) \longrightarrow \mathbf{l}_2(p^{-1})$ is formally self-adjoint on $\mathbf{l}_2$ as derivative of the resolution of the identity of operators A and U (U^*) on the space $\mathbf{l}_2$ with respect to the spectral measure. Hence, according to (7.2.19) we get

$$(\Phi_{j,k}(z))^* = (\pi_j \Phi(z)\pi_k)^* = \pi_k \Phi(z)\pi_j = \Phi_{k,j}(z), \quad j,k \in \mathbb{N}_0. \tag{7.2.23}$$

For a fixed $j \in \mathbb{N}_0$ from (7.2.23) and previous considerations it follows that the vector

$$\psi = \psi(z) = (\psi_k(z))_{k=0}^{\infty}, \quad \psi_k(z) = \Phi_{k,j}(z) = (\Phi_{j,k}(z))^*$$

is an usual solution of the equations (7.2.21) with the initial condition $\psi_0 = \Phi_{0,j}(z) = (\Phi_{j,0}(z))^*$.

Again using Lemma 7.2.1, we obtain the representation for (7.2.22), namely,

$$\Phi_{k,j}(z) = Q_k(z)(\Phi_{0,j}(z)), \quad k \in \mathbb{N}_0. \tag{7.2.24}$$

Taking into account (7.2.23) and (7.2.24), we get

$$\Phi_{0,k}(z) = (\Phi_{k,0}(z))^* = (Q_k(z)\Phi_{0,0}(z))^* = \Phi_{0,0}(z)(Q_k(z))^*, \quad k \in \mathbb{N}_0 \tag{7.2.25}$$

(here we used $\Phi_{0,0}(z) \geq 0$, this inequality follows from (7.2.4) and (7.2.5)). Substituting (7.2.25) into (7.2.22) we get (7.2.19). □

Now we can rewrite the Parseval equality (7.2.6) in a refined form. To do this, we substitute the expression (7.2.19) for $\Phi_{j,k}(z)$ into (7.2.6) and get

$$\begin{aligned} (f,g)_{\mathbf{l}_2} &= \sum_{j,k=0}^{\infty} \int_{\mathbb{C}} (\Phi_{j,k}(z) f_k, g_j)_{\mathbf{l}_2}\, d\sigma(z) \\ &= \sum_{j,k=0}^{\infty} \int_{\mathbb{C}} (Q_j(z)\Phi_{0,0}(z)Q_k^*(z) f_k, g_j)_{\mathbf{l}_2}\, d\sigma(z) \\ &= \sum_{j,k=0}^{\infty} \int_{\mathbb{C}} (Q_k^*(z) f_k, Q_j^*(z) g_j)_{\mathbf{l}_2}\, d\rho(z) \\ &= \int_{\mathbb{C}} \left(\sum_{k=0}^{\infty} Q_k^*(z) f_k \right) \overline{\left(\sum_{j=0}^{\infty} Q_j^*(z) g_j \right)} d\rho(z), \quad \forall f, g \in \mathbf{l}_{\mathrm{fin}}, \end{aligned} \tag{7.2.26}$$

$$d\rho(z) = \Phi_{0,0}(z)\, d\sigma(z).$$

Let us write the Fourier transform "$\widehat{\ }$" generated by the bounded self-adjoint operator A, which commutes with the unitary operator U in the space $\mathbf{l}_2$, namely,

$$\mathbf{l}_2 \supset \mathbf{l}_{\text{fin}} \ni f = (f_n)_{n=0}^{\infty} \longmapsto \hat{f}(z) = \sum_{n=0}^{\infty} Q_n^*(z) f_n \in L_2(\mathbb{C}, d\rho(z)). \tag{7.2.27}$$

Hens, (7.2.26) gives the Parseval equality in a final form

$$(f, g)_{\mathbf{l}_2} = \int_{\mathbb{C}} \hat{f}(z)\overline{\hat{g}(z)}\, d\rho(z) \quad \forall f, g \in \mathbf{l}_{\text{fin}}. \tag{7.2.28}$$

Extending (7.2.28) by continuity, it becomes valid $\forall f, g \in \mathbf{l}_2$.

The orthogonality of polynomials $Q_n^*(z)$ follows from (7.2.27) and (7.2.28). Namely, it is sufficient only to take

$$f = (0, \ldots, 0, f_k, 0, \ldots),\ f_k \in \mathcal{H}_k, \quad g = (0, \ldots, 0, g_j, 0, \ldots),\ g_j \in \mathcal{H}_j$$

in (7.2.27) and (7.2.28). Then, we have

$$\int_{\mathbb{C}} (Q_k^*(z) f_k)\overline{(Q_j^*(z) g_j)}\, d\rho(z) = \delta_{j,k}(f_j, g_j)_{\mathcal{H}_j}, \quad \forall k, j \in \mathbb{N}_0. \tag{7.2.29}$$

By using the representation (7.2.8) for these polynomials we can rewrite the equality (7.2.29) in a (standard) classical scalar form. To do this, we remark that $Q_0^*(z) = \bar{Q}_0(z)$ and according to (7.2.8), we get

$$Q_n(z) = (Q_{n;0}(z), Q_{n;1}(z), \ldots, Q_{n;2n}(z)) : \mathcal{H}_0 \longrightarrow \mathcal{H}_n \quad n \in \mathbb{N}.$$

Hence, for the adjoint operator $Q_n^*(z) : \mathcal{H}_n \longrightarrow \mathcal{H}_0$, we have

$$\begin{aligned}
(Q_n(z)x, y)_{\mathcal{H}_n} &= ((Q_{n;0}(z)x, Q_{n;1}(z)x, \ldots, Q_{n;2n}(z)x), (y_0, y_1, \ldots, y_{2n}))_{\mathcal{H}_n} \\
&= Q_{n;0}(z)x\bar{y}_0 + Q_{n;1}(z)x\bar{y}_1 + \cdots + Q_{n;2n}(z)x\bar{y}_{2n} \\
&= x\overline{(\overline{Q_{n;0}(z)}y_0 + \overline{Q_{n;1}(z)}y_1 + \cdots + \overline{Q_{n;2n}(z)}y_{2n})} \\
&= (x, Q_n^*(z)y)_{\mathcal{H}_0}, \quad \forall x \in \mathcal{H}_0,\ y = (y_0, y_1, \ldots, y_{2n}) \in \mathcal{H}_n.
\end{aligned}$$

that is, $Q_n^*(z)y = \overline{Q_{n;0}(z)}y_0 + \overline{Q_{n;1}(z)}y_1 + \cdots + \overline{Q_{n;2n}(z)}y_{2n}$.

Due to the last equality for $n \in \mathbb{N}$ and $f_n = (f_{n,0}, f_{n,1}, \ldots, f_{n,2n}) \in \mathcal{H}_n$, $z \in \mathbb{C}$, we obtain

$$Q_n^*(z) f_n = \overline{Q_{n;0}(z)} f_{n;0} + \overline{Q_{n;1}(z)} f_{n;1} + \cdots + \overline{Q_{n;2n}(z)} f_{n;2n}, \quad Q_0^*(z) = 1. \tag{7.2.30}$$

Therefore, (7.2.29) has the form

$$\int_{\mathbb{C}} \left(\sum_{\alpha=0}^{2k} \overline{Q_{k;\alpha}(z)} f_{k;\alpha} \right) \overline{\left(\sum_{\beta=0}^{2j} \overline{Q_{j;\beta}(z)} f_{j;\beta} \right)} d\rho(z) = \delta_{j,k} \sum_{\alpha=0}^{2j} f_{j;\alpha} \bar{g}_{j;\alpha},$$

$$\forall f_{k;0}, f_{k;1}, \ldots, f_{k;2k}, g_{j;0}, g_{j;1}, \ldots, g_{j;2j} \in \mathbb{C}, \; j, k \in \mathbb{N}_0, \; (j < k).$$

This equality is equivalent to the orthogonality relation in the usual classical form

$$\int_{\mathbb{C}} \overline{Q^*_{k;\beta}(z)} Q_{j;\alpha} \, d\rho(z) = \delta_{j,k} \delta_{\alpha,\beta}, \quad Q_{0;0} = Q_0(z), \tag{7.2.31}$$

$$\forall j, k \in \mathbb{N}_0, \; \forall \alpha = 0, 1, \ldots, 2j, \; \beta = 0, 1, \ldots, 2k.$$

We note that due to (7.2.30) the Fourier transform (7.2.27) can be rewritten

$$\hat{f}(z) = \sum_{n=0}^{\infty} \sum_{\alpha=0}^{2n} \overline{Q_{n;\alpha}(z)} f_{n;\alpha}, \quad z \in \mathbb{C}, \quad \forall f = (f_n)_{n=0}^{\infty} \in \mathbf{l}_2. \tag{7.2.32}$$

By using the stated above results of this section, we can formulate the spectral theorem for our commuting bounded self-adjoint operator A and the unitary operator U.

Theorem 7.2.4 *Consider the space (7.1.7), namely,*

$$\mathbf{l}_2 = \mathcal{H}_0 \oplus \mathcal{H}_1 \oplus \mathcal{H}_2 \oplus, \cdots, \quad \mathcal{H}_n = \mathbb{C}^{n+1}, \quad n \in \mathbb{N}_0, \tag{7.2.33}$$

and commuting linear operators A and U defined on finite vectors $\mathbf{l}_{\text{fin}}$ *by a block tri-diagonal Jacobi type matrix* J_A *of the form* (7.1.59) *and* J_U *of the form* (7.1.62) *due to expressions in* (7.1.65). *Suppose that all its elements* a_n, b_n, c_n, *and* u_n, w_n, c_n, $n \in \mathbb{N}_0$ *are uniformly bounded, some elements of these matrices are always equal to zero or positive according to* (7.1.60), (7.1.61) *and* (7.1.63), (7.1.64) *and the extension A by the continuity is a bounded self-adjoint operator and the extension U by the continuity is a unitary operator.*

The generalized eigenfunction expansion of operators A and U has the following form. According to Lemma 7.2.1 *using* $\varphi_0 \in \mathbb{C}$, *we represent the solution* $\varphi(z) = (\varphi_n(z))_{n=0}^{\infty}$, $\varphi_n(z) \in \mathcal{H}_n$ *of the system* (7.2.7) *(which exists due to the projection spectral theorem) with* $z = re^{i\theta} \in \mathbb{C}$,

$$\varphi_n(z) = Q_n(z)\varphi_0 = (Q_{n;0}(z), Q_{n;1}(z), \cdots, Q_{n;2n}(z))\varphi_0,$$

where $Q_{n;\alpha}(z)$, $\alpha = 0, 1, \ldots, 2n$ *are polynomials of variables* r, $e^{i\theta}$, *and* $e^{i\theta}$. *Then the Fourier transform has the form*

$$\begin{aligned} \mathbf{l}_2 \supset \mathbf{l}_{\text{fin}} \ni f = (f_n)_{n=0}^{\infty} \longmapsto \hat{f}(z) &= \sum_{n=0}^{\infty} Q_n^*(z) f_n \\ &= \sum_{n=0}^{\infty} \sum_{\alpha=0}^{2n} \overline{Q_{n;\alpha}(z)} f_{n;\alpha} \in L_2(\mathbb{C}, d\rho(z)). \end{aligned} \tag{7.2.34}$$

Here $Q_n^*(z) : \mathcal{H}_n \longrightarrow \mathcal{H}_0$ *is adjoint to the operator* $Q_n(z) : \mathcal{H}_0 \longrightarrow \mathcal{H}_n$, $d\rho(z)$ *is the probability spectral measure of* A *and* U.

The Parseval equality, for all $f, g \in \mathbf{l}_{\text{fin}}$, *has the form*

$$\begin{aligned} (f, g)_{\mathbf{l}_2} &= \int_{\mathbb{C}} \hat{f}(z) \overline{\hat{g}(z)}\, d\rho(z); \\ (J_A f, g)_{\mathbf{l}_2} &= \int_{\mathbb{C}} r \hat{f}(z) \overline{\hat{g}(z)}\, d\rho(z); \\ (J_U f, g)_{\mathbf{l}_2} &= \int_{\mathbb{C}} e^{i\theta} \hat{f}(z) \overline{\hat{g}(z)}\, d\rho(z); \\ (J_{U^*} f, g)_{\mathbf{l}_2} &= \int_{\mathbb{C}} e^{-i\theta} \hat{f}(z) \overline{\hat{g}(z)}\, d\rho(z). \end{aligned} \tag{7.2.35}$$

The transform (7.2.34) *and equalities* (7.2.35) *are extended by continuity* $\forall f, g \in \mathbf{l}_2$ *and the operator* (7.2.34) *becomes isometry, which maps whole* $\mathbf{l}_2$ *into whole* $L_2(\mathbb{C}, d\rho(z))$.

Polynomials $\overline{Q_{n;\alpha}(z)}$, $n \in \mathbb{N}$, $\alpha = 0, 1, \ldots, 2n$ *and* $Q_{0;0}(z) = 1$ *form an orthonormal system in* $L_2(\mathbb{C}, d\rho(z))$ *in the sense of* (7.2.31) *and it is total on this space.*

Proof It is only necessary here to show that orthogonal polynomials $\overline{Q_{n;\alpha}(z)}$, $n \in \mathbb{N}$, $\alpha = 0, 1, \ldots, 2n$, and $Q_{0;0}(z) = 1$ form a total set in the space $L_2(\mathbb{C}, d\rho(z))$. For this reason we remark at first that due to the compactness of the support of the measure $d\rho(z)$ on $\mathbb{C}$, elements $r^t e^{ij\theta}$, $t \in \mathbb{N}_0$, $j \in \mathbb{Z}$ form a total set in $L_2(\mathbb{C}, d\rho(z))$.

Suppose the contrary, i.e., that our system of polynomials is not total. Then there exist a non-zero function $h(z) \in L_2(\mathbb{C}, d\rho(z))$, that is orthogonal to all these polynomials and hence, according to (7.2.9) orthogonal to all $r^t e^{ij\theta}$, $t \in \mathbb{N}_0$, $j \in \mathbb{Z}$. Then $h(z) = 0$. Which is a contradiction. □

The last theorem solves the direct problem for the bounded self-adjoint operator A commuting with the unitary operator U that are generated on the space $\mathbf{l}_2$ by the matrix J_A of the form (7.1.59) and J_U of the form (7.1.62).

The inverse problem consists in the construction, according to a given measure $d\rho(z)$ on $\mathbb{C}$ with a compact support, of bounded matrices: J_A of the form (7.1.59), which commutes with a unitary matrix J_U of the form (7.1.62), and such that their spectral measure coincides with $d\rho(z)$. *This construction is carried out according to Theorem* 7.1.8, *using the Gram-Schmidt orthogonalization procedure of the system* (7.1.2). *For matrices J_A of the form* (7.1.59) *and J_U of the form* (7.1.62), *constructed by $d\rho(z)$, the spectral measure of the corresponding bounded self-adjoint operator A and the unitary operator U coincides with the initial (starting) measure.*

Proof This holds true, since the system of orthogonal polynomials associated with A, $\overline{Q_{n,\alpha}(z)}$ $\alpha = 0, 1, \ldots, 2n$, $n \in \mathbb{N}_0$, is orthonormal in $L_2(\mathbb{C}, d\rho(z))$ and, according to Lemma 7.2.1, is constructed by $r^t e^{ij\theta}$, $t \in \mathbb{N}_0$, $j \in \mathbb{Z}$ in the same way as the system (7.1.3) is constructed by $r^t e^{ij\theta}$, $t \in \mathbb{N}_0$, $j \in \mathbb{Z}$. Hence,

$$Q_0(z) = 1 = P_0(z), \quad \overline{Q_{n,\alpha}(z)} = P_{n;\alpha}(z), \quad \alpha = 0, 1, \ldots, 2n, \quad \forall n \in \mathbb{N}. \tag{7.2.36}$$

Since both systems of polynomials form a total set in $L_2(\mathbb{C}, d\rho(z))$, then (7.2.36) shows that the spectral measures of the constructed operators coincides with the given one. □

Note that expressions (7.1.19) and (7.1.36) (as is known in the classical theory of Jacobi matrices) restore the original matrices (7.1.59) and (7.1.62) by the spectral measure $d\rho(z)$ operators formed by J_A and J_U in $\mathbf{l}_2$.

7.3 Conditions of Unitarity and Commutativity of Matrices Corresponding to the Complex Moment Problem in the Exponential Form

We will find the condition which guarantee that the matrix J_U of the type (7.1.62) is a unitary operator and commute with the matrix J_A of type (7.1.59) which is a bounded Hermitian operator. The formally adjoint matrix J_{U^*} has the form

$$J_{U^*} = \begin{bmatrix} w_0^* & u_0^* & 0 & 0 & \cdots \\ v_0^* & w_1^* & u_1^* & 0 & \cdots \\ 0 & v_1^* & w_2^* & u_2^* & \cdots \\ \vdots & \vdots & \vdots & \vdots & \ddots \end{bmatrix}, \quad \begin{aligned} v_n^* &: \mathcal{H}_n \longrightarrow \mathcal{H}_{n+1}, \\ w_n^* &: \mathcal{H}_n \longrightarrow \mathcal{H}_n, \\ u_n^* &: \mathcal{H}_{n+1} \longrightarrow \mathcal{H}_n, \ n \in \mathbb{N}_0. \end{aligned} \tag{7.3.1}$$

Multiplying matrices (7.1.62) and (7.3.1) we get

$$J_U J_{U^*} = \begin{bmatrix} w_0 w_0^* + v_0 v_0^* & w_0 u_0^* + v_0 w_1^* & v_0 u_1^* & 0 & \cdots \\ u_0 w_0^* + w_1 v_0^* & u_0 u_0^* + w_1 w_1^* + v_1 v_1^* & w_1 u_1^* + v_1 w_2^* & v_1 u_2^* & \cdots \\ u_1 v_0^* & u_1 w_1^* + w_2 v_1^* & u_1 u_1^* + w_2 w_2^* + v_2 v_2^* & w_2 u_2^* + v_2 w_3^* & \cdots \\ \vdots & \vdots & \vdots & \vdots & \ddots \end{bmatrix}. \quad (7.3.2)$$

The expression for $J_{U^*} J_U$ is the similar but with the change in (7.3.2) u_n, w_n, v_n onto v_n^*, w_n^* u_n^* and vice versa.

Comparing $J_U J_{U^*}$ and $J_{U^*} J_U$, we get that the equality $J_U J_{U^*} = J_{U^*} J_U$ is equivalent to the system of equations

$$\begin{aligned} & v_0 v_0^* = u_0^* u_0; \\ & v_n u_{n+1}^* = u_n^* v_{n+1}, \\ & w_n u_n^* + v_n w_{n+1}^* = w_n^* v_n + u_n^* w_{n+1} \equiv 0, \\ & u_n u_n^* + w_{n+1} w_{n+1}^* + v_{n+1} v_{n+1}^* = v_n^* v_n + w_{n+1}^* w_{n+1} + u_{n+1}^* u_{n+1}, \qquad n \in \mathbb{N}_0. \end{aligned} \quad (7.3.3)$$

Here we take into account that w_0 is a scalar, $w_0^* = \bar{w}_0$. Note that the necessity of equalities

$$u_n w_n^* + w_{n+1} v_n^* = v_n^* w_n + w_{n+1}^* u_n, \quad u_{n+1} v_n^* = v_{n+1}^* u_n, \quad n \in \mathbb{N}_0$$

follows from the third and second equalities from (7.3.3), by passing to adjoint matrices "$*$". Thus, conditions (7.3.3) are necessary and sufficient for the matrix equality $J_U J_{U^*} = J_{U^*} J_U$. Norms of operators u_n, w_n, and v_n are uniformly bounded for all $n \in \mathbb{N}_0$, therefore, the operator J_U is bounded in $\mathbf{l}_2$. The validity of (7.3.3) gives the unitarity of this operator.

Taking initial matrices u_0, w_0, v_0 and finding from (7.3.3) step by step u_1, w_1, v_1; u_2, w_2, v_2; ... etc. (in non-uniquely manner) we construct some unitary matrix J_U.

Multiplying matrices (7.1.59) and (7.1.62) we get

$$J_A J_U = \begin{bmatrix} b_0 w_0 + c_0 u_0 & b_0 v_0 + c_0 w_1 & c_0 v_1 & 0 & \cdots \\ a_0 w_0 + b_1 u_0 & a_0 v_0 + b_1 w_1 + c_1 u_1 & b_1 v_1 + c_1 w_2 & c_1 v_2 & \cdots \\ a_1 u_0 & a_1 w_1 + b_2 u_1 & a_1 v_1 + b_2 w_2 + c_2 u_2 & b_2 v_2 + c_2 w_3 & \cdots \\ \vdots & \vdots & \vdots & \vdots & \ddots \end{bmatrix}. \quad (7.3.4)$$

The expression for $J_U J_A$ is the similar to (7.3.4) with the change of u_n, w_n, v_n onto a_n, b_n, c_n.

$$J_A J_U =$$
$$\begin{bmatrix} w_0b_0 + v_0a_0 & w_0c_0 + v_0b_1 & v_0c_1 & 0 & \cdots \\ u_0b_0 + w_1a_0 & u_0c_0 + w_1b_1 + v_1a_1 & w_1c_1 + v_1b_2 & v_1c_2 & \cdots \\ u_1a_0 & u_1b_1 + w_2a_1 & u_1c_1 + w_2b_2 + v_2a_2 & w_2c_2 + v_2b_3 & \cdots \\ \vdots & \vdots & \vdots & \vdots & \ddots \end{bmatrix}. \tag{7.3.5}$$

Comparing $J_A J_U$ and $J_U J_A$ we get that the matrix equality $J_U J_A = J_A J_U$ is equivalent to the system of scalar equalities

$$\begin{aligned} &c_0u_0 = v_0a_0;\\ &c_nv_{n+1} = v_nc_{n+1}\\ &a_{n+1}u_n = u_{n+1}a_n\\ &b_nv_n + c_nw_{n+1} = w_nc_n + v_1b_{n+1},\\ &a_nw_n + b_{n+1}u_n = u_nb_n + w_{n+1}a_n,\\ &a_nv_n + b_{n+1}w_{n+1} + c_{n+1}u_{n+1} = u_nc_n + w_{n+1}b_{n+1} + v_{n+1}a_{n+1}, \quad n \in \mathbb{N}_0. \end{aligned} \tag{7.3.6}$$

Here we take into account that w_0 and b_0 are scalars and $b_0 = \bar{b}_0$.

Let us note about the necessity of equalities. Thus, conditions (7.3.6) are necessary and sufficient for the matrix equality $J_U J_A = J_A J_U$. Norms of operators a_n, b_n and c_n are uniformly bounded for all $n \in \mathbb{N}_0$ and the operator J_A is bounded on $\mathbf{l}_2$. Conditions (7.3.6) gives the mutual commutativity of operators A and U.

Taking a_0, b_0, c_0 as initial matrices and the matrix J_U, using (7.3.6), we get step by step a_1, b_1, c_1; a_2, b_2, c_2; ... etc. (in non-uniquely manner). In such a way we can construct some matrix J_A which commutes with J_U.

Remark 7.3.1 Finding matrices u_n, w_n and v_n, $n \in \mathbb{N}_0$, which are solutions of equations (7.3.3), (7.3.6) and u_n, v_n, a_n, c_n, have the form (7.1.59), (7.1.62), respectively, is a rather difficult problem. Therefore, only a partial case is considered below.

Suppose that matrices w_n are Hermitian: $w_n^* = w_n$, $n \in \mathbb{N}_0$. Then conditions (7.3.3) become the form

$$\begin{aligned} &v_0v_0^* = u_0^*u_0;\\ &v_nu_{n+1}^* = 0,\\ &w_n(u_n^* - v_n) = (u_n^* - v_n)w_{n+1},\\ &u_nu_n^* + v_{n+1}v_{n+1}^* = v_n^*v_n + u_{n+1}^*u_{n+1}, \quad n \in \mathbb{N}_0. \end{aligned} \tag{7.3.7}$$

In what follows, suppose that all matrices a_n, c_n, $n \in \mathbb{N}_0$ have a form, where

$$a_{n;0,0}, a_{n;1,1}, \dots, a_{n;n,n}, \quad c_{n;0,1}, c_{n;1,2}, \dots, c_{n;n,n+1}, \quad \forall n \in \mathbb{N}_0,$$

are positive and all other elements are equal to zero. Thus, our matrices have a form

$$a_n = \underbrace{\left.\begin{bmatrix} 0 & 0 & 0 \dots 0 \\ a_{n;1,0} & 0 & 0 \dots 0 \\ 0 & a_{n;2,1} & 0 \dots 0 \\ \vdots & \vdots & \vdots \ddots \vdots \\ 0 & 0 & 0 \dots a_{n;2n+1,2n} \\ 0 & 0 & 0 \dots 0 \end{bmatrix}\right\}}_{2n+1} 2n+3, \tag{7.3.8}$$

$$c_n = \underbrace{\left.\begin{bmatrix} 0 & c_{n;0,1} & 0 & \dots 0 & 0 & 0 \\ 0 & 0 & c_{n;1,2} & \dots 0 & 0 & 0 \\ \vdots & \vdots & \vdots & \ddots \vdots & \vdots & \vdots \\ 0 & 0 & 0 & \dots c_{n;2n-1,2n} & 0 & 0 \\ 0 & 0 & 0 & \dots 0 & c_{n;2n,2n+1} & 0 \end{bmatrix}\right\}}_{2n+3} 2n+1, \tag{7.3.9}$$

$$\begin{aligned} &a_{n;1,0}, a_{n;2,1}, \dots, a_{n;2n+1,2n} > 0, \\ &c_{n;0,1}, c_{n;1,2}, \dots, c_{n;2n,2n+1} > 0, \\ &a_{n;i,j} = c_{n;j,i}, \quad n, i, j \in \mathbb{N}_0. \end{aligned} \tag{7.3.10}$$

Suppose also that all matrices u_n, $n \in \mathbb{N}_0$ have a form where $u_{n;0,0}, u_{n;1,1}, \dots,$ $u_{n;n,n}$, $\forall n \in \mathbb{N}_0$ are positive and other elements are equal to zero. Thus, our matrices have a form

$$u_n = \begin{bmatrix} u_{n;0,0} & 0 & 0 & \dots 0 & 0 & 0 \dots 0 \\ 0 & u_{n;1,1} & 0 & \dots 0 & 0 & 0 \dots 0 \\ 0 & 0 & u_{n;2,2} & \dots 0 & 0 & 0 \dots 0 \\ \vdots & \vdots & \vdots & \ddots \vdots & \vdots & \vdots \ddots \vdots \\ 0 & 0 & 0 & \dots u_{n;n-1,n-1} & 0 & 0 \dots 0 \\ 0 & 0 & 0 & \dots 0 & u_{n;n,n} & 0 \dots 0 \\ 0 & 0 & 0 & \dots 0 & 0 & 0 \dots 0 \\ \vdots & \vdots & \vdots & \ddots \vdots & \vdots & \vdots \ddots \vdots \\ 0 & 0 & 0 & \dots 0 & 0 & 0 \dots 0 \end{bmatrix} \begin{matrix} \left.\vphantom{\begin{matrix}0\\0\\0\\0\\0\\0\end{matrix}}\right\} n+1 \\ \left.\vphantom{\begin{matrix}0\\0\\0\end{matrix}}\right\} n+2 \end{matrix} \tag{7.3.11}$$

(column groups: the first $n+1$ columns, the last n columns)

$$v_n = \begin{bmatrix} 0\ldots 0 & 0 & \ldots 0 & 0 & 0 \\ \vdots\ddots\vdots & \vdots & \ddots\vdots & \vdots & \vdots \\ 0\ldots 0 & 0 & \ldots 0 & 0 & 0 \\ 0\ldots 0 & 0 & \ldots 0 & 0 & 0 \\ 0\ldots 0 & v_{n;n,n+2} & \ldots 0 & 0 & 0 \\ \vdots\ddots\vdots & \vdots & \ddots\vdots & \vdots & \vdots \\ 0\ldots 0 & 0 & \ldots v_{n;2n-2,2n} & 0 & 0 \\ 0\ldots 0 & 0 & \ldots 0 & v_{n;2n-1,2n+1} & 0 \\ 0\ldots 0 & 0 & \ldots 0 & 0 & v_{n;2n,2n+2} \end{bmatrix} \begin{matrix} \left.\right\} n \\ \left.\right\} n+1 \end{matrix} \quad \underbrace{}_{n+2} \ \underbrace{}_{n+1} \tag{7.3.12}$$

By multiplying matrices of the type (7.3.8)–(7.3.10), (7.3.11) and (7.3.12), it is possible to rewrite the first, second and fourth equalities from (7.3.7) in the form of the corresponding equalities for matrix elements, namely, $c_{0;0,1} = a_{0;0,0}$ and $\forall n \in \mathbb{N}_0$,

$$\begin{aligned} &u^2_{n;0,0} = u^2_{n+1;0,0}, && v^2_{n;n,n+2} = v^2_{n+1;n+2,n+4}, \\ &u^2_{n;1,1} = u^2_{n+1;1,1}, && v^2_{n;n+1,n+3} = v^2_{n+1;n+3,n+5}, \\ &\ldots && \ldots \\ &u^2_{n;n-1,n-1} = u^2_{n+1;n-1,n-1}, && v^2_{2n;2n-1,2n+1} = v^2_{n+1;2n+1,2n+3}, \\ &u^2_{n;n,n} = u^2_{n+1;n,n}; && v^2_{n;2n,2n+2} = v^2_{n+1;2n+2,2n+4}, \\ & && v^2_{n+1;n+1,n+3} = u^2_{n+1;n+1,n+1}. \end{aligned} \tag{7.3.13}$$

The system of equalities (7.3.13) is equivalent to the system (7.3.3) for the case when $w_n = 0$, $n \in \mathbb{N}_0$. The solution of this system step by step gives

$$u_{n;0,0}, u_{n;1,1}, \ldots, u_{n;n,n} = 1, \quad v_{n;n,n+2}, v_{n;n+1,n+2}, \ldots, v_{n;2n,2n+2} = 1.$$

In this case, namely $w_n = 0$, $n \in \mathbb{N}_0$, a_n and c_n, $n \in \mathbb{N}_0$ in the form (7.3.8)–(7.3.10), the system (7.3.6) has a form

$$\begin{aligned} &c_0 u_0 = v_0 a_0; \\ &c_n v_{n+1} = v_n c_{n+1} \\ &a_{n+1} u_n = u_{n+1} a_n \\ &a_n v_n + c_{n+1} u_{n+1} = u_n c_n + v_{n+1} a_{n+1}, \quad n \in \mathbb{N}_0. \end{aligned} \tag{7.3.14}$$

The first equality in (7.3.14) obviously holds true. The second equality in (7.3.14) gives $c_{n+1;j+2,j+3} = c_{n;j,j+1}$, $j = n, n+1, \ldots, 2n$. The third equality in (7.3.14) gives $a_{n+1;j+1,j} = a_{n;j+1,j}$, $j = 0, 1, \dots, n$. The fourth equality in (7.3.14) follows from the second and third ones. By using the last inclusion, we redefine

$$c_{n;j,j+1} = \alpha_{n-|n-j|}, \quad j = 0, 1, \dots, 2n, \ n \in \mathbb{N}_0. \tag{7.3.15}$$

Hence, the matrix c_n has a form

$$c_n = \begin{bmatrix} 0 & \alpha_0 & 0 & 0 & \dots & 0 & \dots & 0 & 0 & 0 & 0 \\ 0 & 0 & \alpha_1 & 0 & \dots & 0 & \dots & 0 & 0 & 0 & 0 \\ 0 & 0 & 0 & \alpha_2 & \dots & 0 & \dots & 0 & 0 & 0 & 0 \\ \vdots & \vdots & \vdots & \vdots & \ddots & \vdots & \ddots & \vdots & \vdots & \vdots & \vdots \\ 0 & 0 & 0 & 0 & \dots & \alpha_n & \dots & 0 & 0 & 0 & 0 \\ \vdots & \vdots & \vdots & \vdots & \ddots & \vdots & \ddots & \vdots & \vdots & \vdots & \vdots \\ 0 & 0 & 0 & 0 & \dots & 0 & \dots & \alpha_2 & 0 & 0 & 0 \\ 0 & 0 & 0 & 0 & \dots & 0 & \dots & 0 & \alpha_1 & 0 & 0 \\ 0 & 0 & 0 & 0 & \dots & 0 & \dots & 0 & 0 & \alpha_0 & 0 \end{bmatrix} \tag{7.3.16}$$

Matrices a_n have a symmetric form with respect to c_n.

Example Take as a sequence $\{\alpha_n\}_{n=0}^{\infty}$ for (7.3.16) such that $\alpha_n < c$ for some fixed number $c > 0$. In this case, all matrices a_n and c_n, as operators (see (7.1.32)), are uniformly bounded and the matrix J_A defines uniquely a bounded self-adjoint operator . If the sequence $\{\alpha_n\}_{n=0}^{\infty}$ is not bounded, then we obtain formally commutative matrices J_A and J_U, but the matrix J_A may not uniquely define a self-adjoint operator.

7.4 The Solution of the Complex Moment Problem in the Exponential Form

Let $\mathcal{H}$ be a separable Hilbert space and let A be a self-adjoint operator defined on $\mathfrak{D}(A)$ in $\mathcal{H}$, U be a unitary commuting in the strong resolvent sense with A operator. Consider an equipmment of $\mathcal{H}$:

$$\mathcal{H}_- \supset \mathcal{H} \supset \mathcal{H}_+ \supset \mathcal{D} \tag{7.4.1}$$

such that $\mathcal{H}_+$ is a Hilbert space topologically and quasi-nuclear embedded into $\mathcal{H}$ (topologically means densely and continuously; quasi-nuclear means that the embedded operator is of Hilbert-Schmidt type); $\mathcal{H}_-$ is the space dual to $\mathcal{H}_+$ with respect to space $\mathcal{H}$; $\mathcal{D}$ is a linear, topological space, topologically embedded into $\mathcal{H}_+$.

Operators A and U are called standard connected with the chain (7.4.1), if $\mathcal{D} \subset \mathfrak{D}(A)$ and restrictions $A \upharpoonright \mathcal{D}$, $U \upharpoonright \mathcal{D}$ act from $\mathcal{D}$ into $\mathcal{H}_+$ continuously.

Let us recall that a vector $\omega_0 \in \mathcal{D}$ is called a strong cyclic vector of operators A and U if for $p \in \mathbb{N}_0$, $q \in \mathbb{Z}$, we have $\omega_0 \in \mathfrak{D}(U^q A^p)$, $U^q A^p \omega_0 \in \mathcal{D}$ and the set of all these vectors with ω_0 is total in the space $\mathcal{H}_+$ (and, hence, also in $\mathcal{H}$).

Assuming that the strong cyclic vector exists we formulate the short variant of the projection spectral theorem.

Theorem 7.4.1 *For the self-adjoint operator A commuting in the strong resolvent sense with the unitary operator U with a strong cyclic vector in the separable Hilbert space $\mathcal{H}$, there exists a non-negative finite Borel measure $d\rho(z)$ such that for ρ-almost every $z \in \mathbb{C}$ there exists a generalized joint eigenvector $\xi_z \in \mathcal{H}_-$, i.e.,*

$$(\xi_z, Af)_{\mathcal{H}} = r(\xi_z, f)_{\mathcal{H}}, \quad (\xi_z, Uf)_{\mathcal{H}} = e^{-i\theta}(\xi_z, f)_{\mathcal{H}}, \quad f \in \mathcal{D},\ \xi_z \neq 0,$$

where $z = re^{i\theta}$ is a two-parameter (r and θ) eigenvalue.

The corresponding Fourier transform F is given by the rule:

$$\mathcal{H} \supset \mathcal{H}_+ \ni f \mapsto (Ff)(z) = \hat{f}(z) = (f, \xi_z)_{\mathcal{H}} \in L_2(\mathbb{C}, d\rho(z)),$$

is an isometric operator (after an extension by continuity) acting from $\mathcal{H}$ into $L_2(\mathbb{C}, d\rho(z))$. Images of operators A and U under the transformation F are multiplication by r and $e^{i\theta}$, respectively, operators in the space $L_2(\mathbb{C}, d\rho(z))$.

We remark that the notation $(\cdot, \cdot)_{\mathcal{H}}$ denotes the same as $\langle \cdot, \cdot \rangle_{\mathcal{H}}$ (see Chap. 2, Sect. 2.7).

Theorem 7.4.2 *The sequence of complex numbers $(c_{t,j})$, $t \in \mathbb{N}_0$, $j \in \mathbb{Z}$ has the representation*

$$c_{t,j} = \int_0^\infty \int_0^{2\pi} r^t e^{ij\theta}\, d\rho(r, \theta), \quad t \in \mathbb{N}_0,\ j \in \mathbb{Z} \tag{7.4.2}$$

if and only if it is positive defined, i.e., satisfies two inequalities

$$\sum_{\substack{t,q \in \mathbb{N}_0 \\ j,k \in \mathbb{Z}}} f_{t,j} \bar{f}_{q,k} c_{t+q,j-k} \geq 0 \tag{7.4.3}$$

and

$$\sum_{\substack{t,q \in \mathbb{N}_0 \\ j,k \in \mathbb{Z}}} f_{t,j} \bar{f}_{q,k} c_{t+q+1,j-k} \geq 0 \tag{7.4.4}$$

for all finite sequences of complex numbers $(f_{t,j})$, $t \in \mathbb{N}_0$, $j \in \mathbb{Z}$, $(f_{t,j} \in \mathbb{C})$.

And additionally, to (7.4.3) *and* (7.4.4), *the representation* (7.4.2) *is unique if*

$$\sum_{p=1}^{\infty} \frac{1}{\sqrt[2p]{|c_{2p,0}|}} = \infty. \tag{7.4.5}$$

Proof The necessity of the condition (7.4.3) is obvious. Indeed, if the sequence $(c_{t,j})$, $t \in \mathbb{N}_0$, $j \in \mathbb{Z}$ has the representation (7.4.2), then for an arbitrary finite sequence $f = (f_{t,j})$, $f_{t,j} \in \mathbb{C}$ we have

$$\sum_{\substack{t,q\in\mathbb{N}_0\\ j,k\in\mathbb{Z}}} f_{t,j}\bar{f}_{q,k}c_{t+q,j-k} = \sum_{\substack{t,q\in\mathbb{N}_0\\ j,k\in\mathbb{Z}}} f_{t,j}\bar{f}_{q,k} \int_0^{\infty}\int_0^{2\pi} r^{t+q}e^{i(j-k)\theta}\, d\rho(r,\theta)$$

$$= \int_0^{\infty}\int_0^{2\pi} \left(\sum_{\substack{t\in\mathbb{N}_0\\ j\in\mathbb{Z}}} f_{t,j}r^t e^{ij\theta}\right)\overline{\left(\sum_{\substack{q\in\mathbb{N}_0\\ k\in\mathbb{Z}}} f_{q,k}r^q e^{ik\theta}\right)}\, d\rho(r,\theta) \geq 0.$$

Hence, we get (7.4.3).

Let us denote by l the linear space $\mathbb{C}^{\infty}$ of sequences $(f_{t,j})$, $t \in \mathbb{N}_0$, $j \in \mathbb{Z}$, $(f_{t,j} \in \mathbb{C})$ and by l_{fin} its linear subset consisting finite sequences f, i.e., sequences such that $f_{t,j} \neq 0$ for only finite numbers t and j. Let $\delta_{t,j}$, $t \in \mathbb{N}_0$, $j \in \mathbb{Z}$ be a δ-sequence such that each $f \in l_{\text{fin}}$ has the representation $f = \sum_{t\in\mathbb{N}_0,\, j\in\mathbb{Z}} f_{t,j}\delta_{t,j}$.

Let us consider linear operators defined on l_{fin} by expressions

$$(Af)_{t,j} = f_{t-1,j}, \quad (Uf)_{t,j} = f_{t,j-1}, \quad t \in \mathbb{N}_0, \ j \in \mathbb{Z}, \tag{7.4.6}$$

where we put always $f_{-1,j} \equiv 0$. Operators A and U are the "creation" type. For δ-sequence we get

$$A\delta_{t,j} = \delta_{t+1,j}, \quad U\delta_{t,j} = \delta_{t,j+1}, \quad t \in \mathbb{N}_0, \ j \in \mathbb{Z}. \tag{7.4.7}$$

The operator A is positive Hermitian and U is isometric with respect to the (quasi)scalar product

$$(f,g)_C = \sum_{\substack{t,q\in\mathbb{N}_0\\ j,k\in\mathbb{Z}}} f_{t,j}\bar{g}_{q,k}c_{t+q,j-k}, \quad f,g \in l_{\text{fin}}. \tag{7.4.8}$$

Indeed, $\forall f, g \in l_{\mathrm{fin}}$, we have

$$\begin{aligned}(Af, g)_C &= \sum_{\substack{t,q\in\mathbb{N}_0\\ j,k\in\mathbb{Z}}} (Af)_{t,j}\bar{g}_{q,k}c_{t+q,j-k}\\ &= \sum_{\substack{t,q\in\mathbb{N}_0\\ j,k\in\mathbb{Z}}} f_{t-1,j}\bar{g}_{q,k}c_{t+q,j-k}\\ &= \sum_{\substack{t,q\in\mathbb{N}_0\\ j,k\in\mathbb{Z}}} f_{t,j}\bar{g}_{q,k}c_{t+q+1,j-k}\end{aligned}$$

and

$$\begin{aligned}(f, Ag)_C &= \sum_{\substack{t,q\in\mathbb{N}_0\\ j,k\in\mathbb{Z}}} f_{t,j}\overline{(Ag)}_{q,k}c_{t+q,j-k}\\ &= \sum_{\substack{t,q\in\mathbb{N}_0\\ j,k\in\mathbb{Z}}} f_{t,j}\bar{g}_{q-1,k}c_{t+q,j-k}\\ &= \sum_{\substack{t,q\in\mathbb{N}_0\\ j,k\in\mathbb{Z}}} f_{t,j}\bar{g}_{q,k}c_{t+q+1,j-k}.\end{aligned}$$

Hence, $(Af, g) = (f, Ag)$, $\forall f, g \in l_{\mathrm{fin}}$. We prove the positivity of A taking into account the condition (7.4.4), namely,

$$\begin{aligned}(Af, f)_C &= \sum_{\substack{t,q\in\mathbb{N}_0\\ j,k\in\mathbb{Z}}} (Af)_{t,j}\bar{f}_{q,k}c_{t+q,j-k}\\ &= \sum_{\substack{t,q\in\mathbb{N}_0\\ j,k\in\mathbb{Z}}} f_{t-1,j}\bar{f}_{q,k}c_{t+q,j-k}\\ &= \sum_{\substack{t,q\in\mathbb{N}_0\\ j,k\in\mathbb{Z}}} f_{t,j}\bar{f}_{q,k}c_{t+q+1,j-k} \geq 0.\end{aligned}$$

For the operator U, we have

$$(Uf, Ug)_C = \sum_{\substack{t,q\in\mathbb{N}_0 \\ j,k\in\mathbb{Z}}} (Uf)_{t,j}\overline{(Uf)}_{q,k} c_{t+q,j-k}$$

$$= \sum_{\substack{t,q\in\mathbb{N}_0 \\ j,k\in\mathbb{Z}}} f_{t,j-1}\bar{f}_{q,k-1} c_{t+q,j-k} = (f, g)_C.$$

Let C be a Hilbert space obtained as the completion of the factor space

$$\dot{l}_{\text{fin}} := l_{\text{fin}}/\{h \in l_{\text{fin}} \mid (h, h)_C = 0\}.$$

Each element f of C is a representative of the class $\dot{f}$ of equivalent elements of $\dot{l}_{\text{fin}}$. This fact in case of self-adjoint and unitary operator is described in Sect. 2.12. Hence, operators $\dot{A}$ and $\dot{U}$ are correctly defined in C by the rule

$$\dot{A}\dot{f} = (Af)^{\cdot}, \quad f \in \mathfrak{D}(\dot{A}) = \dot{l}_{\text{fin}}, \quad \dot{U}\dot{f} = (Uf)^{\cdot}, \quad f \in \dot{l}_{\text{fin}}.$$

Let us denote also by A and U the closure $\dot{A}$ and $\dot{U}$, respectively, in C.

It is obvious that the operator A commutes with U on $G := l_{\text{fin}}$, namely $AUg = UAg$, $g \in G$. Suppose in general that the operator A is not self-adjoint but Hermitian with non-trivial deficiency indices. Let A_F be its Friedrichs extension and $\tilde{U}$ is the closure of U in C with respect to the scalar product (7.4.8). The following lemma proves that A_F commutes with $\tilde{U}$ in a strong resolvent sense. □

Lemma 7.4.3 *If the semi-bounded Hermitian operator A commutes with an isometric operator U on a subset G which is dense in a Hilbert space C, i.e.,*

$$AUg = UAg, \quad g \in G, \quad \tilde{G} = C, \tag{7.4.9}$$

then the Friedrichs extension A_F of A commutes with the closure $\tilde{U}$ of the operator U in C in the strong resolvent sense.

Proof Without loss of generality we assume that A is non-negative operator. Otherwise, $A + dI$, $d > 0$ is considered instead of A. First of all, we will show that

$$A_F\tilde{U}f = \tilde{U}A_F f, \quad \forall f \in \{f \in \mathfrak{D}(A_F) \mid \tilde{U}f \in \mathfrak{D}(A_F)\}. \tag{7.4.10}$$

Consider the operator $B = A + 1$, $\mathfrak{D}(B) = G$ and introduce the space C_+ as the completion of G with respect to the norm $(\varphi, \psi)_{C_+} = (B\varphi, \psi)_C$, $\forall\varphi, \psi \in G$. Now, we have

$$(g, g)_{C_+} = (Bg, g)_C \geq (g, g)_C, \quad g \in G$$

and there is an imbedding $C_+ \subset C$. The bilinear form

$$b(f,h) = (f,h)_{C_+}, \quad f,h \in \mathfrak{D}(b) := C_+$$

is closed and has the representation

$$b(f,h) = (B_F f,h)_C, \quad f \in \mathfrak{D}(B_F) \subset \mathfrak{D}(b) = C_+, \quad h \in \mathfrak{D}(b) = C_+.$$

The operator B_F in the space C is self-adjoint, $(B_F f, f) \geq \|f\|^2$, $f \in \mathfrak{D}(B_F)$ and it is the Friedrichs extension of B.

Let us show that

$$B_F \tilde{U} f = \tilde{U} B_F f, \quad f \in \{f \in \mathfrak{D}(B_F) \mid \tilde{U} f \in \mathfrak{D}(B_F)\} \tag{7.4.11}$$

Indeed, $\forall g \in G \subset \mathfrak{D}(B_F)$, using the unitarity property of $\tilde{U}$ in C and considering (7.4.9), we can write

$$\begin{aligned}(B_F \tilde{U} f, Ug)_C =&(\tilde{U} f, B_F U g)_C = (\tilde{U} f, BUg)_C \\ =&(\tilde{U} f, (A+1)Ug)_C = (\tilde{U} f, U(A+1)g)_C \\ =&(\tilde{U} f, UBg)_C = (\tilde{U} f, \tilde{U} Bg)_C = (f, Bg)_C.\end{aligned} \tag{7.4.12}$$

From (7.4.11),for $g \in G \subset \mathfrak{D}(B_F)$, we have

$$(\tilde{U} B_F f, Ug)_C = (\tilde{U} B_F f, \tilde{U} g)_C = (B_F f, g)_C = (f, B_F g)_C = (f, Bg)_C. \tag{7.4.13}$$

Comparing expressions (7.4.12) and (7.4.13), we get

$$(B_F \tilde{U} f, Ug)_C = (\tilde{U} B_F f, \tilde{U} g)_C, \quad \forall g \in G. \tag{7.4.14}$$

Since G is dense in C, then from (7.4.14), we get (7.4.11). Taking into account the definition of the Friedrichs extension, the equality (7.4.14), and $A_F = B_F - 1$, we get (7.4.10).

Expressions (7.4.10) and (7.4.11) mean that

$$(A_F + 1)\tilde{U} f = \tilde{U}(A_F + 1) f, \quad f \in \{f \in \mathfrak{D}(A_F) \mid \tilde{U} f \in \mathfrak{D}(A_F)\}. \tag{7.4.15}$$

Since the operator $A_F + 1$ has inverse one defined everywhere in C, then from (7.4.15) we get

$$(A_F + 1)^{-1} \tilde{U}^{-1} f = \tilde{U}^{-1} (A_F + 1)^{-1} f, \quad f \in C.$$

□

Note that in general the existence of conditions for a self-adjoint expansions that commute with a given unitary operator is closely related to the problem of extension of a formally normal operator to normal one and is related to the complex moment problem in the power form considered in Chap. 5.

In the next step we use Theorem 2.6.8. For the simplicity of the next consideration, we suppose that the complex moment sequence $(c_{t,j})$ is non-degenerate, i.e., if $(f, f)_C = 0$ $f \in l_{\text{fin}}$, then $f = 0$ and, hence, $\dot{f} = f$, $\dot{A} = A$ and $\dot{U} = U$. The investigation in general case is more complicated.

Let us consider one of the self-adjoint extensions of the operator A, which commutes with the unitary operator U (as it was shown in Lemma 7.4.3, the Friedrichs extension has such a property). Later, we will prove that A is essentially self-adjoint under the condition (7.4.5).

Let us construct the rigging of spaces

$$(l_2(p))_{-,C} \supset C \supset l_2(p) \supset l_{\text{fin}}, \tag{7.4.16}$$

where $l_2(p)$ is a weighted l_2 type space with a weight $p = (p_{t,j})$, $t \in \mathbb{N}_0$, $j \in \mathbb{Z}$, $p_{t,j} \geq 1$. The norm in $l_2(p)$ is given by $\|f\|_{l_2}^2 = \sum_{t\in\mathbb{N}_0,\ j\in\mathbb{Z}} |f_{t,j}|^2 p_{t,j}$; $(l_2(p))_{-,C} = \mathcal{H}_-$ is negative space with respect to the positive $l_2(p) = \mathcal{H}_+$ and the zero space $C = \mathcal{H}$. The space $l_{\text{fin}} = \mathcal{D}$ is finite with the coordinate-wise convergence.

Lemma 7.4.4 *If the sequence $(p_{t,j})$, is enough fast increasing, then the imbedding $l_2(p) \hookrightarrow C$ is quasi-nuclear.*

Proof The equality (7.4.3) means that the multimatrix $(K_{t,j;q,k})$, $t, q \in \mathbb{N}_0$, $j, k \in \mathbb{Z}$, where $K_{t,j;q,k} = c_{t+q,j-k}$ is non-negative definite and, therefore,

$$|c_{t+q,j-k}|^2 = |K_{t,j;q,k}|^2 \leq K_{t,j;t,j} K_{q,k;q,k} = c_{2t,0}c_{2q,0}, \quad t, j \in \mathbb{N}_0, \quad q, k \in \mathbb{Z}. \tag{7.4.17}$$

Let $(q_{t,j})$, $t \in \mathbb{N}_0$, $j \in \mathbb{Z}$, $q_{t,j} \geq 1$ be such that $\sum_{t,j\in\mathbb{N}_0,\ q,k\in\mathbb{Z}} c_{2t,0}q_{t,j}^{-1} < \infty$. Then from (7.4.8) and (7.4.17), it follows that

$$\|f\|_C^2 = \sum_{\substack{t,q\in\mathbb{N}_0 \\ j,k\in\mathbb{Z}}} f_{t,j}\bar{f}_{q,k}c_{t+q,j-k} \leq \|f\|_{l_2(q)}^2 \sum_{\substack{t\in\mathbb{N}_0 \\ j\in\mathbb{Z}}} \frac{c_{2t,0}}{q_{t,j}}, \qquad f \in l_{\text{fin}}.$$

Therefore, the imbedding $l_2(q) \hookrightarrow C$ is topological. And if $\sum_{t\in\mathbb{N}_0,\ j\in\mathbb{Z}} q_{t,j}p_{t,j}^{-1} < \infty$, then $l_2(p) \hookrightarrow l_2(q)$ is quasi-nuclear. The composition $l_2(p) \hookrightarrow C$ of the quasi-nuclear and topological imbedding operators is also quasi-nuclear one. □

In the next step we use the rigging (7.4.16) to construct generalized eigenvectors. The inner structure of the space $(l_2(p))_{-,C}$ is complicated, because of the complicated structure of C, This is a reason to introduce new auxiliary rigging

$$l = (l_{\text{fin}})' \supset (l_2(p^{-1})) \supset l_2 \supset l_2(p) \supset l_{\text{fin}}, \qquad (7.4.18)$$

where $l_2(p^{-1})$, $p^{-1} = (p_{t,j}^{-1})$, $t \in \mathbb{N}_0$, $j \in \mathbb{Z}$ is the negative space with respect to the positive $l_2(p)$ and the zero space l_2. Chains (7.4.16) and (7.4.18) have the same positive space $l_2(p)$. The general Lemma 2.7.5 establishes that the spaces $(l_2(p))_{-,C}$ is isometric to the space $l_2(p^{-1})$.

We consider (7.4.16) and (7.4.18) instead of riggings (2.7.35). Let $\xi_z \in (l_2(p))_{-,C}$ be a generalized eigenvector of operators A and U in terms of the chain (7.4.16). In such a case due to Theorem 7.4.1 with $z = re^{i\theta} \in \mathbb{C}$, we have

$$(\xi_z, Af)_C = r(\xi_z, f)_C, \quad (\xi_z, Uf)_C = e^{-i\theta}(\xi_z, f)_C, \quad f \in l_{\text{fin}}. \qquad (7.4.19)$$

Denote

$$P(z) = V\xi_z \in l_2(p^{-1}) \subset l, \;\; P(z) = (P_{t,j}(z)), \; t \in \mathbb{N}_0, j \in \mathbb{Z}, \;\; P_{t,j}(z) \in \mathbb{C}.$$

By using (2.7.36), we can rewrite (7.4.19) in the form

$$\begin{aligned}(P(z), Af)_{l_2} &= r(P(z), f)_{l_2}, \quad (P(z), Uf)_{l_2} = e^{-i\theta}(P(z), f)_{l_2},\\ z &= re^{i\theta}, \quad r \in \mathbb{R}^1, \quad \theta \in [0, 2\pi), \quad f \in l_{\text{fin}}.\end{aligned} \qquad (7.4.20)$$

The corresponding Fourier transform has the form

$$S \supset l_{\text{fin}} \ni f \to (Ff)(z) = \hat{f}(z) = (f, P(z))_{l_2} \in L_2(\mathbb{C}, d\rho(z)). \qquad (7.4.21)$$

Let us calculate $P(z)$. The operator A is generated by the rule (7.4.6) and, hence, (7.4.20) gives

$$\begin{aligned}\sum_{\substack{t\in\mathbb{N}_0\\ j\in\mathbb{Z}}} rP_{t,j}(z)\bar{f}_{t,j} &= r(P(z), f)_{l_2} = (P(z), Af)_{l_2}\\ &= (AP(z), f)_{l_2} = \sum_{\substack{t\in\mathbb{N}_0\\ j\in\mathbb{Z}}} P_{t+1,j}(z)\bar{f}_{t,j}, \quad \forall f \in l_{\text{fin}}.\end{aligned} \qquad (7.4.22)$$

Similarly, using (7.4.20), we have

$$\sum_{\substack{t\in\mathbb{N}_0\\ j\in\mathbb{Z}}} zP_{t,j}(z)\bar{f}_{t,j} = e^{i\theta}(P(z), f)_{l_2} = (P(z), Uf)_{l_2}$$
$$= (U^{-1}P(z), f)_{l_2} = \sum_{\substack{t\in\mathbb{N}_0\\ j\in\mathbb{Z}}} P_{t,j+1}(z)\bar{f}_{t,j} \quad \forall f \in l_{\text{fin}}. \tag{7.4.23}$$

Hence, we have

$$rP_{t,j}(z) = P_{t+1,j}(z),\ e^{i\theta}P_{t,j}(z) = P_{t,j+1}(z),\ t \in \mathbb{N}_0,\ j \in \mathbb{Z}.$$

Without loss of generality, we can put $P_{0,0}(z) = 1,\ z \in \mathbb{C}$. Then (7.4.22) and (7.4.23) gives

$$P_{t,j}(z) = r^t e^{ij\theta}, \quad t \in \mathbb{N}_0, \quad j \in \mathbb{Z}. \tag{7.4.24}$$

Thus the Fourier transform (7.4.21) has the form

$$S \supset l_{\text{fin}} \ni f \to (Ff)(z) = \hat{f}(z) = \sum_{\substack{t\in\mathbb{N}_0\\ j\in\mathbb{Z}}} f_{t,j}r^t e^{ij\theta} \in L_2(\mathbb{C}, d\rho(r,\theta)), \tag{7.4.25}$$

and we get the corresponding Parseval equality

$$(f, g)_C = \int_{\mathbb{C}^1} \hat{f}(z)\overline{\hat{g}(z)}\, d\rho(z), \quad f, g \in l_{\text{fin}}. \tag{7.4.26}$$

To construct the Fourier transform (7.4.21) and verify the formulas (7.4.22)–(7.4.26) there is still necessary to check that, for our operator A and U, the vector $\omega_0 = \delta_{0,0} \in l_{\text{fin}}$ is strong cyclic vector in the sense of the chain (7.4.16). But it is evidently true, since by (7.4.6) we have $A^t U^j \omega_0 = A^t U^j \delta_{0,0} = \delta_{t,j}$.

The Parseval equality (7.4.26) immediately leads to the representation (7.4.2); according to (7.4.24), (7.4.25) $\hat{\delta}_{t,j} = r^t e^{ij\theta}$ and by (7.4.8) we get

$$c_{t,j} = (\delta_{t,j}, \delta_{0,0})_C = (\hat{\delta}_{t,j}, \hat{\delta}_{0,0})_{L_2(\mathbb{C}, d\rho(z))} = \int_{\mathbb{C}} r^t e^{ij\theta}\, d\rho(r,\theta),$$

$$t \in \mathbb{N}_0, \quad j \in \mathbb{Z}, \quad z = re^{i\theta}, \quad r \in \mathbb{R}^1, \quad \theta \in [0, 2\pi).$$

According to well known facts, Theorem 2.6.8, Lemma 2.6.7 and (2.6.21), the measure in the representation (7.4.26) is unique if and only if the operator $l_{\text{fin}} \ni f \mapsto Af \in l_{\text{fin}}$ is essentially self-adjoint in the space C. Due to the quasi-analytic criterion of self-adjointness, i.e., Theorem 2.6.8, in our case it is enough to check that for each vector $\delta_{t,j}$, $t \in \mathbb{N}_0$, $j \in \mathbb{Z}$, the class $\mathrm{C}\{\|A^p\delta_{t,j}\|_C\}$ is quasi-analytic, since the set of vectors $\delta_{t,j}$ is total in C. According to (7.4.6), we have

$$\|A^p\delta_{t,j}\|_C^2 = \|\delta_{t+p,j}\|_C^2 = |c_{2t+2p,0}|.$$

Due to the quasi-analyticity properties, the quasi-analyticity of the class $\mathrm{C}\{m_{l+n}\}$ with a fixed $l \in \mathbb{N}$ is equivalent to the quasi-analyticity of the class $\mathrm{C}\{m_n\}$. This completes the proof of the Theorem 7.4.2.

Bibliographical Notes

The results of the chapter are also some generalization of the classical results of works of Akhiezer N. I. [2, 3]. The integral representation was obtained due to works of Berezansky Yu. M. [17, 18]. The theory of equipped spaces is used form the book of Koshmanenko V. D. and Dudkin M. E. [171]. The general theory of orthogonal polynomials see books of Geronimus Ya. L. [117], Szegö G. [278]. In research, there arose questions about the commutativity of symmetric and unitary operators which was discussed in works of Kočubeĭ A. N. [164], Schmüdgen K. [254, 255]. The main content of the chapter is in works of Berezansky Yu. M., Dudkin M. E. [35], [85]. Papers of Bisgaard T. M. [55, 56], Devinatz A. [81] Stochel J. Szafraniec F.H. [269–272, 275] are close to the situation considered in the chapter. By considerations the uniqueness of the integral representation, we use also actively works of Carleman T. [64], Nelson E. [225], Nussbaum A. E. [231, 232].

Chapter 8
Block Jacobi Type Matrices and the Two Dimensional Real Moment Problem

The chapter proposes an analogue of Jacobi type bock matrices, which are related to the real power two-dimensional moment problem and the corresponding polynomials orthogonal with respect to some probability measure on the real plane. In this case, two commutative block matrices are also obtained, which have a tri-diagonal block structure and are self-adjoint operators in the l_2 type space. The existence of a one-to-one correspondence between the probability measure on the compact set of the real plane and such matrices is also proved.

8.1 Construction of the Tri-Diagonal Block Jacobi Type Matrices Corresponding to the Two Dimensional Real Moment Problem

The inverse spectral problem for block Jacobi type matrices corresponding to the two-dimensional moment problem consists in constructing block tri-diagonal block matrices corresponding either to the two-dimensional problem of moments or, one might say, corresponding to the Borel measure on the real plane with the condition that the measure has all moments. The section gives detailed proof this inverse problem. We would like to note that historically, such matrices appeared for the first time in the work of Gekhtman M. I., Kalyuzhnyĭ A. A.

Let $d\rho(x, y)$ be a Borel probability measure with compact support on the real plane $\mathbb{R}^2$ and $L_2 = L_2(\mathbb{R}^2, d\rho(x, y))$ be the space of square integrable functions with respect to a Borel probability measure $d\rho(x, y)$ with a compact support on the

Y. M. Berezansky, M. E. Dudkin, *Jacobi Matrices and the Moment Problem*,
Operator Theory: Advances and Applications 294,
https://doi.org/10.1007/978-3-031-46387-7_8

real plane $\mathbb{R}^2$. We consider that the support of this measure is not only a compact set, but also one that contains a non-empty open subset. Hence functions $\mathbb{R}^2 \ni (x, y) \longmapsto x^m y^n,\ m, n \in \mathbb{N}_0$ can be linear independent and form total set in L_2.

Let us consider the multiplication by x and y operators, namely,

$$\hat{A} f(x, y) = x f(x, y), \quad \hat{B} f(x, y) = y f(x, y),$$

on the space L_2. It is obvious, that these operators are bounded and self-adjoint. To find the Jacobi type matrices of operators $\hat{A}$ and $\hat{B}$, some order of orthogonalization in L_2 is chosen for the family of functions:

$$\{x^m y^n\}, \quad m, n \in \mathbb{N}_0. \tag{8.1.1}$$

We use an order for orthogonalization according to the Gram-Schmidt procedure (see Fig. 8.1):

$$x^0y^0;\ x^0y^1, x^1y^0;\ x^0y^2, x^1y^1, x^2y^0;\ \dots;\quad x^0y^n, x^1y^{n-1}, \dots, x^ny^0;\ \dots. \tag{8.1.2}$$

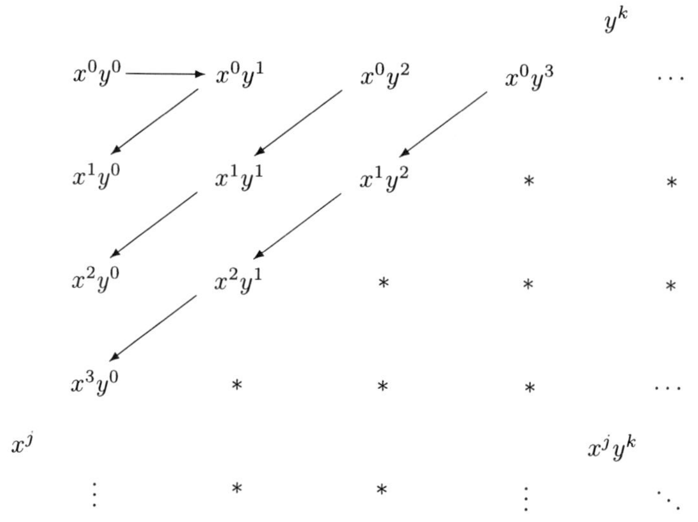

Fig. 8.1 The orthogonalization order

As a result, we get an orthonormal system of polynomials (each polynomial of variables $x^m y^n$, $m, n \in \mathbb{N}_0$), which we denote and arrange in groups

$$\begin{array}{lllll} P_{0;0}(x,y); & P_{1;0}(x,y), & P_{2;0}(x,y), & \ldots; & P_{n;0}(x,y), & \ldots, \\ & P_{1;1}(x,y); & P_{2;1}(x,y), & & P_{n;1}(x,y), & \\ & & P_{2;2}(x,y); & & P_{n;2}(x,y), & \\ & & & & \ldots & \\ & & & & P_{n;n}(x,y); & \end{array} \tag{8.1.3}$$

where each polynomial has a form $P_{n;\alpha}(x,y) = k_{n;\alpha}x^\alpha y^{n-\alpha} + \cdots$, $n \in \mathbb{N}_0$, $\alpha = 0, 1, \ldots, n$, $k_{n;\alpha} > 0$; here "$+\cdots$" denotes the next part of the corresponding polynomial; for convenience we put $P_{0;0}(x,y) = 1$. Therefore, $P_{n;\alpha}$ is some linear combination of elements

$$\{1;\ x^0y^1, x^1y^0;\ \ldots;\ x^0y^n, x^1y^{n-1}, \ldots, x^\alpha y^{n-\alpha}\}. \tag{8.1.4}$$

Since the set (8.1.1) is dense in the space L_2, the sequence (8.1.3) is an orthonormal basis in this space. Let the subspace $\mathcal{P}_{n;\alpha}$, $\forall n \in \mathbb{N}$ be spanned by (8.1.4). It is obvious that

$$\begin{aligned} &\mathcal{P}_{0;0} \subset \mathcal{P}_{1;0} \subset \mathcal{P}_{1;1} \subset \mathcal{P}_{2;0} \subset \mathcal{P}_{2;1} \subset \mathcal{P}_{2;2} \subset \cdots \\ &\qquad \subset \mathcal{P}_{n;0} \subset \mathcal{P}_{n;1} \subset \cdots \subset \mathcal{P}_{n;n} \subset \cdots, \\ &\mathcal{P}_{n;\alpha} = \{P_{0;0}(x,y)\} \oplus \{P_{1;0}(x,y)\} \oplus \{P_{1;1}(x,y)\} \oplus \\ &\qquad \oplus \{P_{2;0}(x,y)\} \oplus \{P_{2;1}(x,y)\} \oplus \{P_{2;2}(x,y)\} \oplus \cdots \oplus \\ &\qquad \oplus \{P_{n;0}(x,y)\} \oplus \{P_{n;1}(x,y)\} \oplus \cdots \oplus \{P_{n;\alpha}(x,y)\}, \quad \forall n \in \mathbb{N}, \end{aligned} \tag{8.1.5}$$

where $\{P_{n;\alpha}(x,y)\}$, $n \in \mathbb{N}$, $\alpha = 0, 1, \ldots n$ denote one dimensional space spanned by $P_{n;\alpha}(x,y)$; $\mathcal{P}_{0;0} = \mathbb{R}$.

In the sequel investigation we need the Hilbert space

$$\mathbf{l}_2 = \mathcal{H}_0 \oplus \mathcal{H}_1 \oplus \mathcal{H}_2 \oplus \cdots, \quad \mathcal{H}_n = \mathbb{C}^{n+1}, \quad n \in \mathbb{N}_0 \tag{8.1.6}$$

instead of usual traditional space l_2 used in the classical Hamburger (one-dimensional) moment problem.

Each vector $f \in \mathbf{l}_2$ has a form $f = (f_n)_{n=0}^\infty$, $f_n \in \mathcal{H}_n$, hence,

$$\|f\|_{\mathbf{l}_2}^2 = \sum_{n=0}^\infty \|f_n\|_{\mathcal{H}_n}^2 < \infty, \quad (f,g)_{\mathbf{l}_2} = \sum_{n=0}^\infty (f_n, g_n)_{\mathcal{H}_n}, \quad \forall f, g \in \mathbf{l}_2.$$

Coordinates of a vector $f_n \in \mathcal{H}_n$, $n \in \mathbb{N}_0$, in the some orthonormal basis $\{e_{n;0}, e_{n;1}, e_{n;2}, \ldots, e_{n;n}\}$ in the space $\mathbb{C}^{n+1}$, we denote by $(f_{n;0}, f_{n;1}, f_{n;2}, \ldots, f_{n;n})$ hence, we have $f_n = (f_{n;0}, f_{n;1}, f_{n;2}, \ldots, f_{n;n})$.

By using the orthonormal system (8.1.3), one can define a mapping of $\mathbf{l}_2$ into L_2. Denote

$$P_n(x, y) = (P_{n;0}, P_{n;1}(x, y), P_{n;2}(x, y), \dots, P_{n;n}) \in \mathcal{H}_n, \ \forall (x, y) \in \mathbb{R}^2, \ \forall n \in \mathbb{N}_0.$$

Then

$$\mathbf{l}_2 \ni f = (f_n)_{n=0}^{\infty} \longmapsto (If)(x, y) := \hat{f}(x, y) = \sum_{n=0}^{\infty} (f_n, P_n(x, y))_{\mathcal{H}_n} \in L_2. \tag{8.1.7}$$

Thus, for $n \in \mathbb{N}_0$ we get

$$(f_n, P_n(x, y))_{\mathcal{H}_n} = f_{n;0}\overline{P_{n;0}(x, y)} + f_{n;1}\overline{P_{n;1}(x, y)} + f_{n;2}\overline{P_{n;2}(x, y)} + \dots + f_{n;n}\overline{P_{n;n}(x, y)},$$

$$\|f\|^2_{\mathbf{l}_2} = \|(f_{0;0}, f_{1;0}, f_{1;1}, f_{2;0}, f_{2;1}, f_{2;2}, \dots, f_{n;0}, , f_{n;1}, \dots, f_{n;n}, \dots)\|^2_{l_2}.$$

Then, (8.1.7) is the mapping of the space $\mathbf{l}_2$ into L_2 and, taking into account the density of the orthonormal system (8.1.3), we get that this mapping is isometric. The mapping (8.1.7) maps whole space l_2 into whole space L_2, because the system (8.1.3) is an orthonormal basis in L_2. Therefore, the mapping(8.1.7) is an isometric transformation (denoted by I) acting from $\mathbf{l}_2$ into L_2.

Let T be a bounded linear operator defined on the space $\mathbf{l}_2$. Then there exists a unique sequence $(\tau_{j,k})_{j,k=0}^{\infty}$, where for each $j, k \in \mathbb{N}_0$, the element $\tau_{j,k}$ is an operator from $\mathcal{H}_k$ into $\mathcal{H}_j$, so that

$$(Tf)_j = \sum_{k=0}^{\infty} \tau_{j,k} f_k, \ j \in \mathbb{N}_0, \quad (Tf, g)_{\mathbf{l}_2} = \sum_{j,k=0}^{\infty} (\tau_{j,k} f_k, g_j)_{\mathcal{H}_j}, \quad f, g \in \mathbf{l}_2. \tag{8.1.8}$$

For the proof of (8.1.8) we need only to write the usual matrix of the operator T in $\mathbf{l}_2$ using the basis

$$(e_{0;0};\ e_{1;0}, e_{1;1};\ e_{2;0}, e_{2;1}, e_{2;2};\ \dots;\ e_{n;0}, e_{n;1}, \dots, e_{n;n};\ \dots), \quad e_{0;0} = 1. \tag{8.1.9}$$

Then $\tau_{j,k}$ is the operator $\mathcal{H}_k \longrightarrow \mathcal{H}_j$ for each $j, k \in \mathbb{N}_0$. This operator has a matrix representation

$$\tau_{j,k;\alpha,\beta} = (Te_{k;\beta}, e_{j;\alpha})_{\mathbf{l}_2} \tag{8.1.10}$$

where $\alpha = 0, 1, \ldots, j$ and $\beta = 0, 1, \ldots, k$. We rewrite $\tau_{j,k} = (\tau_{j,k;\alpha,\beta})_{\alpha,\beta=0}^{j,k}$, including cases: $\tau_{0,k} = (\tau_{0,k;\alpha,\beta})_{\alpha,\beta=0}^{0,k}$, $\tau_{j,0} = (\tau_{j,0;\alpha,\beta})_{\alpha,\beta=0}^{j,0}$ and $\tau_{0,0} = (\tau_{0,0;\alpha,\beta})_{\alpha,\beta=0}^{0,0} = \tau_{0,0;0,0}$.

It should be noted that the image (8.1.8) is also correct for the general operator T acting in the space $\mathbf{l}_2$ with the domain $\mathrm{Dom}(T) = \mathbf{l}_{\mathrm{fin}} \subset \mathbf{l}_2$, where $\mathbf{l}_{\mathrm{fin}}$ denotes the subspace of finite vectors of $\mathbf{l}_2$. In this case, the first formula in (8.1.8) is valid for $f \in \mathbf{l}_{\mathrm{fin}}$; the second formula is valid for $f \in \mathbf{l}_{\mathrm{fin}}$, $g \in \mathbf{l}_2$.

Let us consider an image $\hat{T} = ITI^{-1} : L_2 \longrightarrow L_2$ of the above mentioned bounded operator $T : \mathbf{l}_2 \longrightarrow \mathbf{l}_2$ by the mapping (8.1.7). This matrix in the basis (8.1.3)

$$P_{0;0}(x,y);\ P_{1;0}(x,y), P_{1;1}(x,y);\ P_{2;0}(x,y), P_{2,1}(x,y), P_{2,2}(x,y);\ \ldots;$$
$$P_{n;0}(x,y), P_{n;1}(x,y), \ldots, P_{n;n}(x,y); \ldots$$

is equal to the usual matrix of the operator T that we regard as the operator: $\mathbf{l}_2 \longrightarrow \mathbf{l}_2$ in the corresponding basis (8.1.9). By using (8.1.10) and the above mentioned procedure, we get the operator matrix $(\tau_{j,k})_{j,k=0}^{\infty}$ $T : \mathbf{l}_2 \longrightarrow \mathbf{l}_2$. *By the definition this matrix is also the operator matrix of* $\hat{T} : L_2 \longrightarrow L_2$. It is obvious that $\hat{T}$ can be an arbitrary linear bounded operator in L_2. In the next text we consider T instead of $\hat{T}$, and the role of T will play A and B, corresponding to our matrices J_A and J_B.

Lemma 8.1.1 *For polynomials $P_{n;\alpha}(x,y)$ and subspaces $\mathcal{P}_{m,\beta}$, $n, m \in \mathbb{N}_0$, $\alpha = 0, 1, \ldots, n$, $\beta = 0, 1, \ldots, m$, relations*

$$xP_{n;\alpha}(x,y) \in \mathcal{P}_{n+1;\alpha+1}, \qquad yP_{n;\alpha}(x,y) \in \mathcal{P}_{n+1;\alpha}, \tag{8.1.11}$$

hold true.

Proof According to (8.1.3) the polynomial $P_{n;\alpha}(x,y)$, $n \in \mathbb{N}_0$, is some linear combination

$$\{1; x^0y^1, x^1y^0; \ldots; x^0y^n, x^1y^{n-1}, \ldots, x^\alpha y^{n-\alpha}\}.$$

Hence, the multiplication by x of the last set gives a linear combination

$$\{x; x^1y^1, x^2y^0; \ldots; x^1y^n, x^2y^{n-1}, \ldots, x^{\alpha+1}y^{n-\alpha}\},$$

and such a linear combination belongs to $\mathcal{P}_{n+1;\alpha+1}$. The similar multiplication by y gives a linear combination

$$\{y^1; x^0y^2, x^1y^1; \ldots; x^0y^{n+1}, x^1y^n, \ldots, x^\alpha y^{n-\alpha+1}\},$$

and such a linear combination belongs to $\mathcal{P}_{n+1;\alpha}$. $\square$

Lemma 8.1.2 *Let $\hat{A}$ be an operator of the multiplication by x in the space L_2, namely,*

$$L_2 \ni \varphi(x, y) \longmapsto (\hat{A}\varphi)(x, y) = x\varphi(x, y) \in L_2.$$

(It is obvious that $\hat{A}$ is self-adjoint and bounded.) Then, the matrix $\hat{A} = (a_{j,k})_{j,k=0}^{\infty}$, in basis (8.1.3) (i.e., $A = I^{-1}\hat{A}I$), has a tri-diagonal structure, i.e., $a_{j,k} = 0$ for $|j-k| > 1$.

Proof By using (8.1.10) for $e_{n;\gamma} = I^{-1}P_{n;\gamma}(x, y)$, $n \in \mathbb{N}_0$, $\gamma = 0, 1, \ldots, n$, we get

$$a_{j,k;\alpha,\beta} = (Ae_{k;\beta}, e_{j;\alpha})_{l_2} = \int_{\mathbb{R}^2} x P_{k;\beta}(x, y)\overline{P_{j;\alpha}(x, y)}\, d\rho(x, y), \quad \forall j, k \in \mathbb{N}_0, \tag{8.1.12}$$

where $\alpha = 0, 1, \ldots, j$, $\beta = 0, 1, \ldots, k$. From (8.1.11), we get $xP_{k;\beta}(x, y) \in \mathcal{P}_{k+1;\beta+1}$. According to expressions from (8.1.5), the integral in (8.1.12) is equal to zero for $j > k+1$ and for each $\beta = 0, 1, \ldots, j$.

On the other hand, the integral in (8.1.12) has the form

$$\begin{aligned}(a^*)_{j,k;\alpha,\beta} &= \int_{\mathbb{R}^2} x P_{k;\beta}(x, y)\overline{P_{j;\alpha}(x, y)}\, d\rho(x, y) \\ &= \overline{\int_{\mathbb{R}^2} x P_{j;\alpha}(x, y)\overline{P_{k;\beta}(x, y)}\, d\rho(x, y)} = \overline{a_{k,j;\beta,\alpha}},\end{aligned} \tag{8.1.13}$$

where $\alpha = 0, 1, \ldots, j$ and $\beta = 0, 1, \ldots, k$. Since the operator $\hat{A}$ is symmetric, from (8.1.11), we have $xP_{j;\alpha}(x, y) \in \mathcal{P}_{j+1;\alpha+1}$. According to (8.1.5), the last integral is equal to zero for $k > j+1$ and for each $\alpha = 0, 1, \ldots, k$, $\beta = 0, 1, \ldots, k$. As result the integral in (8.1.12), i.e., coefficients $a_{j,k;\alpha,\beta}$, $j, k \in \mathbb{N}_0$ are equal to zero for $|j-k| > 1$; $\alpha = 0, 1, \ldots, j$, $\beta = 0, 1, \ldots, k$. (In the previous considerations it is necessary to take into account that $P_{0;0}(x, y) = 1$, $e_{0;0} = I^{-1}P_{0;0}(x, y)$). □

Therefore, the matrix $(a_{j,k})_{j,k=0}^{\infty}$ of the operator $\hat{A}$ has a tri-diagonal block structure

$$\begin{bmatrix} a_{0,0} & a_{0,1} & 0 & 0 & 0 & \ldots \\ a_{1,0} & a_{1,1} & a_{1,2} & 0 & 0 & \ldots \\ 0 & a_{2,1} & a_{2,2} & a_{2,3} & 0 & \ldots \\ \vdots & \vdots & \vdots & \vdots & \vdots & \ddots \end{bmatrix}. \tag{8.1.14}$$

Further detailed analysis of expressions (8.1.12) makes it possible to identify zero and non-zero elements of matrices $(a_{j,k;\alpha,\beta})_{\alpha,\beta=0}^{j,k}$ in each case for $|j-k| \le 1$. We also use properties of permutations of matrix indices j, k and α, β.

Lemma 8.1.3 *Let $(a_{j,k})_{j,k=0}^{\infty}$ be an operator matrix of the multiplication by x operator in L_2, where $a_{j,k} : \mathcal{H}_k \longrightarrow \mathcal{H}_j$; $a_{j,k} = (a_{j,k;\alpha,\beta})_{\alpha,\beta=0}^{j,k}$ are matrices of operators $a_{j,k}$ in the corresponding orthonormal basis. Then, $\forall j \in \mathbb{N}_0$, we get*

$$\begin{aligned}
&\forall\alpha = 0, 1, \ldots j-1,\ a_{j,j+1;\alpha,\alpha+2} = a_{j,j+1;\alpha,\alpha+3} = \cdots = a_{j,j+1;\alpha,j+1} = 0;\\
&\forall\beta = 0, 1, \ldots j-1,\ a_{j+1,j;\beta+2,\beta} = a_{j+1,j;\beta+3,\beta} = \cdots = a_{j+1,j;j+1,\beta} = 0.
\end{aligned} \tag{8.1.15}$$

If we choose inside of each diagonal

$$\{x^0y^n, x^ny^{n-1}, x^0y^{n-2}, \ldots, x^ny^0\}$$

another order (preserving the order of diagonals), then Lemma 8.1.3 is not valid but it is also possible to describe zeros of matrices $(a_{j,k;\alpha,\beta})_{\alpha,\beta=0}^{j,k}$. Such matrices $(a_{j,k})_{j,k=0}^{\infty}$ have also tri-diagonal block structure and have another elements always equal zero.

Proof According to (8.1.12) and (8.1.11), for $j \in \mathbb{N}_0$, $\forall\alpha = 0, 1, \ldots, j$ and $\forall\beta = 0, 1, \ldots, j+1$, we have

$$\begin{aligned}
a_{j,j+1;\alpha,\beta} &= \int\limits_{\mathbb{R}^2} xP_{j+1,\beta}(x,y)\overline{P_{j;\alpha}(x,y)}\,d\rho(x,y)\\
&= \overline{\int\limits_{\mathbb{R}^2} xP_{j,\alpha}(x,y)\overline{P_{j+1;\beta}(x,y)}\,d\rho(x,y)},
\end{aligned}$$

where $xP_{j;\alpha}(x,y) \in \mathcal{P}_{j+1;\alpha+1}$. But according to (8.1.5), $P_{j+1;\beta}(x,y)$ is orthogonal to $\mathcal{P}_{j+1;\alpha+1}$ for $\beta > \alpha+1$ and, hence, the last integral is equal to zero. This gives the first equalities in (8.1.15).

Similarly, from (8.1.12) and (8.1.11), for $j \in \mathbb{N}_0$ $\forall\alpha = 0, 1, \ldots, j+1$ and $\forall\beta = 0, 1, \ldots, j$, we have

$$a_{j+1,j;\alpha,\beta} = \int\limits_{\mathbb{R}^2} xP_{j,\beta}(x,y)\overline{P_{j+1;\alpha}(x,y)}\,d\rho(x,y),$$

where $xP_{j;\beta}(x,y) \in \mathcal{P}_{j+1;\beta+1}$. But according to (8.1.5), $P_{j+1;\alpha}(x,y)$ is orthogonal to $\mathcal{P}_{j+1;\beta+1}$ for $\alpha > \beta+1$ and, hence, the last integral is equal to zero. This gives the second equalities in (8.1.15). □

Therefore, after last investigations we conclude that, in (8.1.14) for $\forall j \in \mathbb{N}$, the right corner for each $((j+1) \times (j+2))$-matrix $a_{j,j+1}$ (starting from the third diagonal) and the left corner for each $((j+2) \times (j+1))$-matrix $a_{j+1,j}$ (starting from the third diagonal) consist always zero elements. Taking into account (8.1.14), we can conclude that the matrix of the multiplication by x operator is multi-diagonal as a usual scalar matrix, i.e., in the usual basis of the space $\mathbf{l}_2$.

Lemma 8.1.4 *Elements*

$$a_{0,1;0,1},\, a_{1,0;1,0}; \quad a_{j,j+1;\alpha,\alpha+1},\, a_{j+1,j;\alpha+1,\alpha}; \quad j \in \mathbb{N}, \alpha = 0, 1, \ldots j, \tag{8.1.16}$$

of matrices $(a_{j,k})_{j,k=0}^{\infty}$ *in Lemma* 8.1.7, *are always positive.*

Proof We start from the investigation of $a_{0,1;0,1}$. Denote $P'_{1;1}(x,y)$ a non-normalized vector $P_{1;1}(x,y)$ obtained from the Gram-Schmidt orthogonalization procedure. According to (8.1.2) and (8.1.3), we have

$$P'_{1;1}(x,y) = x - (x, P_{1;0}(x,y))_{L_2} P_{1;0}(x,y) - (x,1)_{L_2}.$$

Therefore, using (8.1.30), we get

$$\begin{aligned} a_{0,1;0,1} = & \int_{\mathbb{R}^2} x P_{0;0} P_{1;1}(x,y)\, d\rho(x,y) = \|P'_{1;1}(x,y)\|_{L_2}^{-1} \int_{\mathbb{R}^2} x P'_{1;1}(x,y)\, d\rho(x,y) \\ = & \|P'_{1;1}(x,y)\|_{L_2}^{-1} \int_{\mathbb{R}^2} x(x - (x, P_{1;0}(x,y))_{L_2} P_{1;0}(x,y) - (x,1)_{L_2})\, d\rho(x,y) \\ = & \|P'_{1;1}(x,y)\|_{L_2}^{-1} (\|x\|_{L_2}^2 - |(x, P_{1;0}(x,y))_{L_2}|^2 - |(x,1)_{L_2}|^2), \end{aligned} \tag{8.1.17}$$

where $(1 = P_{0;0}(x,y))$.

Due to (8.1.36) also, we conclude that the last expression is positive and, hence, $a_{0,1;0,1} > 0$. Since the operator A is symmetric, than $a_{0,1;0,1} = a_{1,0;1,0} > 0$. The positiveness in (8.1.17) follows from the Parseval equality applied by the decomposition of the function $x \in L_2$ with respect to the orthonormal basis (8.1.3) in the space L_2, namely,

$$|(x,1)_{L_2}|^2 + |(x, P_{1;0}(x,y))_{L_2}|^2 + |(x, P_{1;1}(x,y))_{L_2}|^2 + \cdots = \|x\|_{L_2}^2. \tag{8.1.18}$$

Consider elements $a_{j,j+1;\alpha,\alpha+1}$, for $j \in \mathbb{N}$, $\alpha = 0, 1, \ldots, j$. From (8.1.30), we get

$$a_{j,j+1;\alpha,\alpha+1} = \int_{\mathbb{R}^2} x P_{j+1,\alpha+1}(x, y) \overline{P_{j;\alpha}(x, y)}\, d\rho(x, y) \\ = \overline{\int_{\mathbb{R}^2} x P_{j,\alpha}(x, y) \overline{P_{j+1;\alpha+1}(x, y)}\, d\rho(x, y)}. \tag{8.1.19}$$

For $P_{j;\alpha}(x, y)$, according to (8.1.3) and (8.1.5), we have

$$P_{j;\alpha}(x, y) = k_{j;\alpha} x^{\alpha} y^{j-\alpha} + R_{j;\alpha}(x, y), \tag{8.1.20}$$

where $R_{j;\alpha}(x, y)$ is some polynomial from $\mathcal{P}_{j;\alpha-1}$ if $\alpha > 0$, or from $\mathcal{P}_{j-1;j-1}$ if $\alpha = 0$. Thus, $x R_{j;\alpha}(x, y)$ is some polynomial from $\mathcal{P}_{j+1;\alpha}$ or from $\mathcal{P}_{j;j}$ (see (8.1.11) and (8.1.5)). Multiplying it on x, we get

$$x P_{j;\alpha}(x, y) = k_{j;\alpha} x^{\alpha+1} y^{j-\alpha} + x R_{j;\alpha}(x, y), \tag{8.1.21}$$

where $x R_{j;\alpha}(x, y) \in \mathcal{P}_{j+1;\alpha}$ or belongs to $\mathcal{P}_{j;j}$. On the other hand, equality (8.1.20) gives:

$$P_{j+1;\alpha}(x, y) = k_{j+1;\alpha} x^{\alpha+1} y^{j-\alpha} + R_{j+1;\alpha}(x, y), \tag{8.1.22}$$

where $R_{j+1;\alpha}(x, y) \in \mathcal{P}_{j+1;\alpha}$ if $\alpha > 0$ or belongs to $\mathcal{P}_{j;j}$ if $\alpha = 0$.

We take $x^{\alpha+1} y^{j-\alpha}$ from (8.1.22) and substitute it into (8.1.21). Thus, we get

$$x P_{j;\alpha}(x, y) = \frac{k_{j;\alpha}}{k_{j+1;\alpha}} (P_{j+1;\alpha}(x, y) - R_{j+1;\alpha}(x, y)) + x R_{j;\alpha}(x, y) \\ = \frac{k_{j;\alpha}}{k_{j+1;\alpha}} P_{j+1;\alpha}(x, y) - \frac{k_{j;\alpha}}{k_{j+1;\alpha}} R_{j+1;\alpha}(x, y) + x R_{j;\alpha}(x, y). \tag{8.1.23}$$

The second two terms in (8.1.23) belong to $\mathcal{P}_{j+1;\alpha}$ or belong to $\mathcal{P}_{j;j}$ and in any case orthogonal to $P_{j+1;\alpha+1}(x, y)$.

Thus, the substitution of the expression (8.1.23) into (8.1.19) gives $a_{j,j+1;\alpha,\alpha+1} = k_{j;\alpha}(k_{j+1;\alpha+1})^{-1} > 0$.

Since the matrix (8.1.32) is symmetric, then elements $a_{j,j+1;\alpha,\alpha+1} = a_{j,j+1;\alpha,\alpha+1}$ are also positive $j \in \mathbb{N}$, $\alpha = 0, 1, \ldots, j$. □

In what follows we will use convenient notations for elements $a_{j,k}$ of the Jacobi matrix (8.1.14), namely,

$$\begin{aligned} a_n &= a_{n+1,n} : \quad \mathcal{H}_n \longrightarrow \mathcal{H}_{n+1}, \\ b_n &= a_{n,n} \quad : \quad \mathcal{H}_n \longrightarrow \mathcal{H}_n, \\ c_n &= a_{n,n+1} : \quad \mathcal{H}_{n+1} \longrightarrow \mathcal{H}_n, \qquad n \in \mathbb{N}_0. \end{aligned} \tag{8.1.24}$$

All previous investigations are summarized in the theorem.

Theorem 8.1.5 *The bounded self-adjoint multiplication by x operator $\hat{A}$, (with a strong cyclic vector) in the space L_2 in the orthonormal basis* (8.1.3) *of polynomials has the form of tri-diagonal block Jacobi type symmetric matrix $J_A = (a_{j,k})_{j,k=0}^{\infty}$, which acts in the space* (8.1.6), *namely,*

$$\mathbf{l}_2 = \mathcal{H}_0 \oplus \mathcal{H}_1 \oplus \mathcal{H}_2 \oplus \cdots, \quad \mathcal{H}_n = \mathbb{C}^{n+1}, \quad n \in \mathbb{N}_0. \tag{8.1.25}$$

Norms of all operators $a_{j,k} : \mathcal{H}_k \longrightarrow \mathcal{H}_j$ are uniformly bounded with respect to $j, k \in \mathbb{N}_0$. In designations (8.1.24), *this matrix has the form*

$$J_A = \begin{bmatrix} b_0 & c_0 & 0 & 0 & 0 & \dots \\ a_0 & b_1 & c_1 & 0 & 0 & \dots \\ 0 & a_1 & b_2 & c_2 & 0 & \dots \\ 0 & 0 & a_2 & b_3 & c_3 & \dots \\ \vdots & \vdots & \vdots & \vdots & \vdots & \ddots \end{bmatrix},$$

i.e.,

$$J_A = \left[\begin{array}{c|cc|ccc|cccc|ccccc} * & * & + & & & & & & & & & & & & \\ \hline * & * & * & * & + & 0 & & & & & & & 0 & & \\ + & * & * & * & * & + & & & & & & & & & \\ \hline & * & * & * & * & * & * & + & 0 & 0 & & & & & \\ & + & * & * & * & * & * & * & + & 0 & & & & & \\ & 0 & + & * & * & * & * & * & * & + & & & & & \\ \hline & & & * & * & * & * & * & * & * & * & + & 0 & 0 & 0 \\ & & & + & * & * & * & * & * & * & * & * & + & 0 & 0 \\ & 0 & & 0 & + & * & * & * & * & * & * & * & * & + & 0 \\ & & & 0 & 0 & + & * & * & * & * & * & * & * & * & + \\ & & & & \ddots & & & & \ddots & & & & \ddots & & \end{array}\right]. \tag{8.1.26}$$

In (8.1.26), *$\forall n \in \mathbb{N}_0$ b_n is a $((n+1) \times (n+1))$-matrix: $b_n = (b_{n;\alpha,\beta})_{\alpha,\beta=0}^{n,n}$, ($b_0 = b_{0;0,0}$ is a scalar); a_n is a $((n+2) \times (n+1))$-matrix: $a_n = (a_{n;\alpha,\beta})_{\alpha,\beta=0}^{n+1,n}$; c_n is a $((n+1) \times (n+2))$-matrix: $c_n = (c_{n;\alpha,\beta})_{\alpha,\beta=0}^{n,n+1}$. Matrices a_n and c_n have some* *elements always equal to zero* $\forall n \in \mathbb{N}$, *namely,*

$$\begin{aligned} a_{n;\beta+2,\beta} &= a_{n;\beta+3,\beta} = \cdots = a_{n;n+1,\beta} = 0, \quad \beta = 0, 1, \ldots, n-1; \\ c_{n;\alpha,\alpha+2} &= c_{n;\alpha,\alpha+3} = \cdots = c_{n;\alpha,n+1} = 0, \quad \alpha = 0, 1, \ldots, n-1. \end{aligned} \tag{8.1.27}$$

Some another their elements are always positive, namely

$$a_{n;\alpha+1,\alpha};\, c_{n;\alpha,\alpha+1} > 0, \quad \alpha = 0, 1, \ldots, n, \quad \forall n \in \mathbb{N}_0. \tag{8.1.28}$$

Thus, it is possible to say, that $\forall n \in \mathbb{N}_0$, *each upper left corner of the matrix* a_n *(starting from the third diagonal) and each right corner of the matrix* c_n *(starting from the third diagonal) consist always of zero elements. All positive elements of* (8.1.26) *are denoted by "+". Therefore, the matrix* (8.1.26) *in scalar form is multi-diagonal in the indicated structure. Possibly zero and non-zero elements are marked with "∗". The symmetry of the operator* $(\hat{A})^* = \hat{A}$ *gives*

$$a_{n;\alpha,\beta} = c_{n;\beta,\alpha}, \quad \beta = 0, 1, 2, \ldots, n, \ \alpha = 0, 1, \ldots, \beta, \beta+1, \ n \in \mathbb{N}_0.$$

The matrix J_A *acts by the rule*

$$(J_A f)_n = a_{n-1} f_{n-1} + b_n f_n + c_n f_{n+1}, \quad n \in \mathbb{N}_0, \quad \forall f = (f_n)_{n=0}^{\infty} \in \mathbf{l}_2, \tag{8.1.29}$$

were, we put, usually, $a_{-1} = 0$, $f_{-1} = 0$.

Let us now pass to the study of the matrix J_B associated with the variable y. For this we need a lemma similar to Lemma 8.1.1.

Lemma 8.1.6 *Let* $\hat{B}$ *be the multiplication by* y *operator in the space* L_2, *namely,*

$$L_2 \ni \varphi(x, y) \longmapsto (\hat{B}\varphi)(x, y) = y\varphi(x, y) \in L_2.$$

(Obviously, $\hat{B}$ *is bounded and self-adjoint). The matrix* $(b_{j,k})_{j,k=0}^{\infty}$ *of the operator* $\hat{B}$ *in the basis* (8.1.3) *(i.e.,* $B = I^{-1}\hat{B}I$ *) has a tri-diagonal structure:* $b_{j,k} = 0$ *for* $|j-k| > 1$.

Proof By using (8.1.10) for $e_{n;\gamma} = I^{-1}P_{n;\gamma}(x, y)$, $n \in \mathbb{N}_0$, $\gamma = 0, 1, \ldots, n$, we get

$$b_{j,k;\alpha,\beta} = (Be_{k;\beta}, e_{j;\alpha})_{l_2} = \int_{\mathbb{R}^2} y P_{k;\beta}(x, y)\overline{P_{j;\alpha}(x, y)}\, d\rho(x, y), \tag{8.1.30}$$

where $\alpha = 0, 1, \ldots, j$, $\beta = 0, 1, \ldots, k$, $\forall j, k \in \mathbb{N}_0$. From (8.1.11), we have $yP_{k;\alpha}(x, y) \in \mathcal{P}_{k+1;\alpha}$. According to (8.1.5), the integral in (8.1.30) is equal to zero for $j > k+1$ and for each $\alpha = 0, 1, \ldots, j$.

On the other hand, the integral in (8.1.30) has the form

$$
\begin{aligned}
(b^*)_{j,k;\alpha,\beta} &= \int_{\mathbb{R}^2} y P_{k;\beta}(x,y)\overline{P_{j;\alpha}(x,y)}\, d\rho(x,y) \\
&= \overline{\int_{\mathbb{R}^2} y P_{j;\alpha}(x,y)\overline{P_{k;\beta}(x,y)}\, d\rho(x,y)} = \overline{b_{k,j;\beta,\alpha}},
\end{aligned}
\tag{8.1.31}
$$

where $\alpha = 0, 1, \ldots, j$ and $\beta = 0, 1, \ldots, k$. From (8.1.11), we have $yP_{j;\alpha}(x,y) \subset \mathcal{P}_{j+1,\alpha}$. According to (8.1.5) the integral in (8.1.31) is equal to zero for $k > j+1$ for each $\alpha = 0, 1, \ldots, j$, $\beta = 0, 1, \ldots, k$.

As result, the integral in (8.1.30), i.e., coefficients $b_{j,k;\alpha,\beta}$, $j, k \in \mathbb{N}_0$ are equal to zero for $|j-k| > 1$, $\alpha = 0, 1, \ldots, j$, $\beta = 0, 1, \ldots, k$. (In the previous considerations it is necessary to take into account that $e_{0;0} = I^{-1}P_{0;0}(x,y)$, $P_{0;0}(x,y) = 1$). □

Thus, the matrix $(b_{j,k})_{j,k=0}^{\infty}$ of the operator $\hat{B}$ has a tri-diagonal block structure

$$
\begin{bmatrix}
b_{0,0} & b_{0,1} & 0 & 0 & 0 & \ldots \\
b_{1,0} & b_{1,1} & b_{1,2} & 0 & 0 & \ldots \\
0 & b_{2,1} & b_{2,2} & b_{2,3} & 0 & \ldots \\
\vdots & \vdots & \vdots & \vdots & \vdots & \ddots
\end{bmatrix}.
\tag{8.1.32}
$$

Further detailed analysis of the expressions (8.1.30) makes it possible know about the zero and non-zero elements of the matrix $(b_{j,k;\alpha,\beta})_{\alpha,\beta=0}^{j,k}$ in each case for $|j-k| \le 1$. We also use permutations properties of matrix indices j, k and α, β.

Lemma 8.1.7 *Let $(b_{j,k})_{j,k=0}^{\infty}$ be an operator matrix of the multiplication by y operator in L_2, $b_{j,k} : \mathcal{H}_k \longrightarrow \mathcal{H}_j$; $b_{j,k} = (b_{j,k;\alpha,\beta})_{\alpha,\beta=0}^{j,k}$ are matrices of operators $b_{j,k}$ in the corresponding orthonormal basis. Then, $\forall j \in \mathbb{N}_0$, we get*

$$
\begin{aligned}
&\forall\alpha = 0, 1, \ldots j,\ b_{j,j+1;\alpha,\alpha+1} = b_{j,j+1;\alpha,\alpha+2} = \cdots = b_{j,j+1;\alpha,j+1} = 0;\\
&\forall\beta = 0, 1, \ldots j,\ b_{j+1,j;\beta+1,\beta} = b_{j+1,j;\beta+2,\beta} = \cdots = b_{j+1,j;j+1,\beta} = 0.
\end{aligned}
\tag{8.1.33}
$$

If we choose inside of each diagonal

$$\{x^0y^n, x^1y^{n-1}, x^2y^{n-2}, \ldots, x^ny^0\},$$

(see Fig. 8.1) another order (preserving the order of diagonals), then Lemma 8.1.7 is not valid but it is also possible to describe zero elements of matrices $(b_{j,k;\alpha,\beta})_{\alpha,\beta=0}^{j,k}$. Such matrices $(b_{j,k})_{j,k=0}^{\infty}$ have also a tri-diagonal block structure and has zero elements but possibly in another places.

Proof According to (8.1.30) and (8.1.11) for $j \in \mathbb{N}_0$, $\forall\alpha = 0, 1, \dots, j$ and $\forall\beta = 0, 1, \dots, j$, we have

$$b_{j,j+1;\alpha,\beta} = \int\limits_{\mathbb{R}^2} y P_{j+1,\beta}(x, y)\overline{P_{j;\alpha}(x, y)}\, d\rho(x, y)$$
$$= \overline{\int\limits_{\mathbb{R}^2} y P_{j,\alpha}(x, y)\overline{P_{j+1;\beta}(x, y)}\, d\rho(x, y)} = \bar{b}_{j+1,j;\beta,\alpha},$$

where $yP_{j;\alpha}(x, y) \in \mathcal{P}_{j+1;\alpha}$. According to (8.1.5), $P_{j+1;\beta}(x, y)$ is orthogonal to $\mathcal{P}_{j+1;\alpha}$ for $\beta > \alpha$ and, hence, the last integral is equal to zero. This gives the first equalities in (8.1.33).

Similarly to (8.1.30) and (8.1.11), for $j \in \mathbb{N}_0$, $\forall\alpha = 0, 1, \dots, j + 1$ and $\forall\beta = 0, 1, \dots, j$, we get

$$b_{j+1,j;\alpha,\beta} = \int\limits_{\mathbb{R}^2} y P_{j,\beta}(x, y)\overline{P_{j+1;\alpha}(x, y)}\, d\rho(x, y),$$

where $yP_{j;\beta}(x, y) \in \mathcal{P}_{j+1;\beta}$. But according to (8.1.5) $P_{j+1;\alpha}(x, y)$ is orthogonal to $\mathcal{P}_{j+1;\beta}$ if $\alpha > \beta$ and, hence, the last integral is equal to zero. This gives the second equalities in (8.1.33). □

Thus, after these investigations, we can conclude that in (8.1.32) for $\forall j \in \mathbb{N}$, the right corner of each $((j + 1) \times (j + 2))$-matrix $b_{j,j+1}$ (starting from the second diagonal) and in the left corner of each $((j + 2) \times (j + 1))$-matrix $b_{j+1,j}$ (starting from the second diagonal) consist always zero elements. Taking into account (8.1.14), we can conclude that the symmetric matrix of the multiplication by y operator is a multi-diagonal usual scalar matrix, that is, in the usual (standard) basis of the $\mathbf{l}_2$ space.

Lemma 8.1.8 *Elements*

$$b_{0,1;0,0}, b_{1,0;0,0}; \quad b_{j,j+1;\alpha,\alpha}, b_{j+1,j;\alpha,\alpha}; \quad j \in \mathbb{N}, \alpha = 0, 1, \dots j \tag{8.1.34}$$

of matrices $(b_{j,k})_{j,k=0}^{\infty}$ *from Lemma* 8.1.7, *are always positive.*

Proof We start from the investigation of $b_{1,0;0,0}$. By using (8.1.12) we denote $P'_{1;0}(x, y) = y - (y, 1)_{L_2}$ a non-normalized vector $P_{1;0}(x, y)$ obtained due to the Gram-Schmidt orthogonalization procedure. Hence, we get

$$\begin{aligned} b_{1,0;0,0} &= \int\limits_{\mathbb{R}^2} y P_{0;0}(x, y)\overline{P_{1;0}(x, y)}\, d\rho(x, y) \\ &= \|P'_{1;0}(x, y)\|_{L_2}^{-1} \int\limits_{\mathbb{R}^2} y\overline{(y - (y, 1)_{L_2})}\, d\rho(x, y) \\ &= \|P'_{1;0}(x, y)\|_{L_2}^{-1} (\|y\|_{L_2}^2 - |(y, 1)_{L_2}|^2), \end{aligned} \tag{8.1.35}$$

where $P_{0;0}(x, y) = 1$. The last difference is positive (see below (8.1.36)), therefore $b_{1,0;0,0} > 0$. The element $b_{0,1;0,0}$ is also positive since the matrix B is symmetric, i.e., $b_{1,0;0,0} = b_{0,1;0,0}$.

The positiveness (8.1.35) follows from the Parseval equality when the function $y \in L_2$ is expanded by an orthonormal basis (8.1.3) in the space L_2, namely,

$$|(y, 1)_{L_2}|^2 + |(y, P_{1;0}(x, y))_{L_2}|^2 + |(y, P_{1;1}(x, y))_{L_2}|^2 + \cdots = \|y\|^2_{L_2}. \quad (8.1.36)$$

Pass to the proof of the positiveness $b_{j+1,j;\alpha,\alpha}$, where $j \in \mathbb{N}$, $\alpha = 0, 1, \ldots, j$. From (8.1.12) we have:

$$b_{j+1,j;\alpha,\alpha} = \int_{\mathbb{R}^2} y P_{j;\alpha}(x, y) \overline{P_{j+1;\alpha}(x, y)}\, d\rho(x, y). \quad (8.1.37)$$

According to (8.1.3) and (8.1.5), we get

$$P_{j;\alpha}(x, y) = k_{j;\alpha} x^\alpha y^{j-\alpha} + R_{j;\alpha}(x, y), \quad (8.1.38)$$

where $R_{j;\alpha}(x, y)$ is some polynomial from $\mathcal{P}_{j;\alpha-1}$ if $\alpha > 0$, or from $\mathcal{P}_{j-1;j-1}$ if $\alpha = 0$. Therefore, $yR_{j;\alpha}(x, y)$ is some polynomial from $\mathcal{P}_{j+1;\alpha}$ or from $\mathcal{P}_{j;j-1}$ (see (8.1.11) and (8.1.5)). Multiplying (8.1.38) by x, we conclude

$$y P_{j;\alpha}(x, y) = k_{j;\alpha} x^\alpha y^{j-\alpha+1} + x R_{j;\alpha}(x, y), \quad (8.1.39)$$

where $yR_{j;\alpha}(x, y) \in \mathcal{P}_{j+1;\alpha-1}$ or $\mathcal{P}_{j;j-1} \subset \mathcal{P}_{j;j}$.

On the other hand, the equality (8.1.38) for $P_{j+1;\alpha}(x, y)$ gives

$$P_{j+1;\alpha}(x, y) = k_{j+1;\alpha} x^\alpha y^{j-\alpha+1} + R_{j+1;\alpha}(x, y), \quad (8.1.40)$$

where $R_{j+1;\alpha}(x, y) \in \mathcal{P}_{j+1;\alpha-1}$ or $\mathcal{P}_{j;j}$.

Find $x^\alpha y^{j-\alpha+1}$ from (8.1.40) and substitute it into (8.1.39). Thus, we get

$$\begin{aligned} y P_{j;\alpha}(x, y) &= \frac{k_{j;\alpha}}{k_{j+1;\alpha}} (P_{j+1;\alpha}(x, y) - R_{j+1;\alpha}(x, y)) + y R_{j;\alpha}(x, y) \\ &= \frac{k_{j;\alpha}}{k_{j+1;\alpha}} P_{j+1;\alpha}(x, y) - \frac{k_{j;\alpha}}{k_{j+1;\alpha}} R_{j+1;\alpha}(x, y) + y R_{j;\alpha(x,y)}(x, y), \end{aligned} \quad (8.1.41)$$

where both last terms belong to $\mathcal{P}_{j+1;\alpha-1}$ or to $\mathcal{P}_{j;j}$ and are always orthogonal to $P_{j+1;\alpha}(x, y)$. Therefore, after the substitution of the expression (8.1.41) into (8.1.37) we get that $b_{j+1,j;\alpha,\alpha} = k_{j;\alpha}(k_{j+1;\alpha})^{-1} > 0$.

Since the matrix (8.1.14) is symmetric, then elements $b_{j,j+1;\alpha,\alpha+1} = b_{j+1,j;\alpha+1,\alpha}$ are also positive where $j \in \mathbb{N}$, $\alpha = 0, 1, \ldots, j$. □

In what follows we will use convenient notations for elements $b_{j,k}$ of the matrix

$$\begin{aligned} u_n &= b_{n+1,n} : \quad \mathcal{H}_n \longrightarrow \mathcal{H}_{n+1}, \\ w_n &= b_{n,n} \quad : \quad \mathcal{H}_n \longrightarrow \mathcal{H}_n, \\ v_n &= b_{n,n+1} : \quad \mathcal{H}_{n+1} \longrightarrow \mathcal{H}_n, \quad n \in \mathbb{N}_0. \end{aligned} \tag{8.1.42}$$

All last previous investigation we summarize also in the theorem.

Theorem 8.1.9 *The bounded self-adjoint multiplication by y operator $\hat{B}$, (with a strong cyclic vector) in the space L_2 in the orthonormal basis* (8.1.3) *of polynomials, has the form of tri-diagonal block Jacobi type symmetric matrix $J_B = (b_{j,k})_{j,k=0}^{\infty}$ which acts in the space* (8.1.6), *namely,*

$$\mathbf{l}_2 = \mathcal{H}_0 \oplus \mathcal{H}_1 \oplus \mathcal{H}_2 \oplus \cdots, \quad \mathcal{H}_n = \mathbb{C}^{n+1}, \quad n \in \mathbb{N}_0. \tag{8.1.43}$$

Norms of all operators $b_{j,k} : \mathcal{H}_k \longrightarrow \mathcal{H}_j$ are uniformly bounded with respect to $j, k \in \mathbb{N}_0$. In designations (8.1.24), *this matrix has the form*

$$J_B = \begin{bmatrix} w_0 & v_0 & 0 & 0 & 0 & \dots \\ u_0 & w_1 & v_1 & 0 & 0 & \dots \\ 0 & u_1 & w_2 & v_2 & 0 & \dots \\ 0 & 0 & u_2 & w_3 & v_3 & \dots \\ \vdots & \vdots & \vdots & \vdots & \vdots & \ddots \end{bmatrix},$$

i.e.,

$$J_B = \left[\begin{array}{c|cc|ccc|cccc|ccccc} * & + & 0 & & & & & & & & & & & & \\ \hline + & * & * & + & 0 & 0 & & & & & & & & & \\ 0 & * & * & * & + & 0 & & & & & & 0 & & & \\ \hline & + & * & * & * & * & + & 0 & 0 & 0 & & & & & \\ & 0 & + & * & * & * & * & + & 0 & 0 & & & & & \\ & 0 & 0 & * & * & * & * & * & + & 0 & & & & & \\ \hline & & & + & * & * & * & * & * & * & + & 0 & 0 & 0 & 0 \\ & & & 0 & + & * & * & * & * & * & * & + & 0 & 0 & 0 \\ & 0 & & 0 & 0 & + & * & * & * & * & * & * & + & 0 & 0 \\ & & & 0 & 0 & 0 & * & * & * & * & * & * & * & + & 0 \\ \hline & & & & \ddots & & & \ddots & & & & \ddots & & & \end{array}\right]. \tag{8.1.44}$$

In (8.1.44), $\forall n \in \mathbb{N}_0$, *$w_n$ is a $((n+1) \times (n+1))$-matrix: $w_n = (w_{n;\alpha,\beta})_{\alpha,\beta=0}^{n,n}$, ($w_0 = w_{0;0,0}$ a scalar); u_n is a $((n+2) \times (n+1))$-matrix: $u_n = (a_{n;\alpha,\beta})_{\alpha,\beta=0}^{n+1,n}$;*

v_n is a $((n+1)\times(n+2))$-matrix: $v_n = (c_{n;\alpha,\beta})_{\alpha,\beta=0}^{n,n+1}$. In these matrices u_n and v_n some elements are always equal to zero, namely $\forall n \in \mathbb{N}_0$, we have

$$\begin{aligned} u_{n;\beta+1,\beta} &= u_{n;\beta+2,\beta} = \cdots = u_{n;n+1,\beta} = 0, \quad \beta = 0,1,\ldots,n; \\ v_{n;\alpha,\alpha+1} &= v_{n;\alpha,\alpha+2} = \cdots = v_{n;\alpha,n+1} = 0, \quad \alpha = 0,1,\ldots,n. \end{aligned} \tag{8.1.45}$$

Some another their elements are always positive, namely

$$u_{n;\alpha,\alpha};\, v_{n;\alpha,\alpha+1} > 0, \quad \alpha = 0,1,\ldots,n, \quad \forall n \in \mathbb{N}_0. \tag{8.1.46}$$

Thus, we can assume that $\forall n \in \mathbb{N}_0$ each element in the left corner of the matrix u_n (starting from the second diagonal) and each element in the right corner of the matrix v_n (starting from the second diagonal) is the zero element. All positive elements of (8.1.27) *are denoted by "+". Therefore, the matrix* (8.1.27) *in a scalar form is multi-diagonal in the indicated structure. Both zero and non-zero elements are marked with "$*$".*

The symmetry of the operator $(\hat{B})^* = \hat{B}$ gives

$$u_{n;\alpha,\beta} = v_{n;\beta,\alpha}, \quad \alpha = 0,1,2,\ldots,n, \ \beta = \alpha,\ldots,n, \ n \in \mathbb{N}.$$

The matrix J_B acts by the rule

$$(J_B f)_n = u_{n-1} f_{n-1} + w_n f_n + v_n f_{n+1}, \quad n \in \mathbb{N}_0, \quad \forall f = (f_n)_{n=0}^{\infty} \in \mathbf{l}_2, \tag{8.1.47}$$

where we put, usually, $u_{-1} = 0$, $f_{-1} = 0$.

8.2 Direct and Inverse Spectral Problems for the Tri-Diagonal Block Jacobi Type Matrices Corresponding to the Two-Dimensional Real Moment Problem

As it was mentioned above the main result of the previous section is really the solution of an inverse problem for a corresponding direct one appearing in the title of this section. Indeed, one of the main results of the section is the solution of the direct spectral problem, i.e., the restoration of the measure by the given block matrices. By the restoration we understand usually the writing of the corresponding Parseval equalities in terms of generalized eigenvectors.

We consider operators in the space $\mathbf{l}_2$ of the form (8.1.6). Additionally to the space $\mathbf{l}_2$, we consider its equipment, i.e., we have a rigging

$$(\mathbf{l}_{\text{fin}})' \supset \mathbf{l}_2(p^{-1}) \supset \mathbf{l}_2 \supset \mathbf{l}_2(p) \supset \mathbf{l}_{\text{fin}}, \tag{8.2.1}$$

where $\mathbf{l}_2(p)$ is a weighted space $\mathbf{l}_2$ with a weight $p = (p_n)_{n=0}^{\infty}$, $p_n \geq 1$, $(p^{-1} = (p_n^{-1})_{n=0}^{\infty})$. In our case $\mathbf{l}_2(p)$ is the Hilbert space of sequences $f = (f_n)_{n=0}^{\infty}$, $f_n \in \mathcal{H}_n$ with a norm and a scalar product:

$$\|f\|_{\mathbf{l}_2(p)}^2 = \sum_{n=0}^{\infty} \|f_n\|_{\mathcal{H}_n}^2 p_n, \quad (f, g)_{\mathbf{l}_2(p)} = \sum_{n=0}^{\infty} (f_n, g_n)_{\mathcal{H}_n} p_n. \tag{8.2.2}$$

The space $\mathbf{l}_2(p^{-1})$ is defined similarly; remind that $\mathbf{l}_{\text{fin}}$ is the space of finite sequences and $(\mathbf{l}_{\text{fin}})'$ is the space adjoint to $\mathbf{l}_{\text{fin}}$. It is easy to show that the imbedding $\mathbf{l}_2(p) \hookrightarrow \mathbf{l}_2$ is quasi-nuclear if $\sum\limits_{n=0}^{\infty} n p_n^{-1} < \infty$.

Let A and B be pare of commuting bounded self-adjoint operators standardly connected with the chain (8.2.1). (About the standard connection see the beginning of Sect. 8.4 of this chapter.) According to the projection spectral theorem, such operators have representations:

$$Af = \int\limits_{\mathbb{R}^2} x\Phi(x, y)\, d\sigma(x, y) f, \quad Bf = \int\limits_{\mathbb{R}^2} y\Phi(x, y)\, d\sigma(x, y) f, \quad f \in \mathbf{l}_2, \tag{8.2.3}$$

where $\Phi(x, y) : \mathbf{l}_2(p) \longrightarrow \mathbf{l}_2(p^{-1})$ is the operator of generalized projection and $d\sigma(x, y)$ is a spectral measure. For every $f \in \mathbf{l}_{\text{fin}}$, the projection $\Phi(x, y) f \in \mathbf{l}_2(p^{-1})$ is a generalized eigenvector of operators A and B with corresponding eigenvalues x and y. For all $f, g \in \mathbf{l}_{\text{fin}}$, we have Parseval equality

$$(f, g)_{\mathbf{l}_2} = \int\limits_{\mathbb{R}^2} (\Phi(x, y) f, g)_{\mathbf{l}_2}\, d\sigma(x, y); \tag{8.2.4}$$

after the extension by continuity the equality (8.2.4) is valid for all $f, g \in \mathbf{l}_2$.

Let us denote by π_n the operator of orthogonal projection in $\mathbf{l}_2$ on $\mathcal{H}_n$, $n \in \mathbb{N}_0$. Hence, $\forall f = (f_n)_{n=0}^{\infty} \in \mathbf{l}_2$ we have $f_n = \pi_n f$. This operator acts similarly on spaces $\mathbf{l}_2(p)$ and $\mathbf{l}_2(p^{-1})$.

After closure, the operator corresponding to the matrix $(\Phi_{j,k}(x, y))_{j,k=0}^{\infty}$, has a representation

$$\Phi_{j,k}(x, y) = \pi_j \Phi(x, y) \pi_k \; : \; \mathbf{l}_2 \longrightarrow \mathcal{H}_j, \quad (\mathcal{H}_k \longrightarrow \mathcal{H}_j). \tag{8.2.5}$$

The Parseval equality (8.2.4), we can rewrite in the form

$$\begin{aligned}(f,g)_{\mathbf{l}_2} &= \sum_{j,k=0}^{\infty}\int_{\mathbb{R}^2}(\Phi(x,y)\pi_k f,\pi_j g)_{\mathbf{l}_2}\,d\sigma(x,y)\\ &= \sum_{j,k=0}^{\infty}\int_{\mathbb{R}^2}(\pi_j\Phi(x,y)\pi_k f,g)_{\mathbf{l}_2}\,d\sigma(x,y)\\ &= \sum_{j,k=0}^{\infty}\int_{\mathbb{R}^2}(\Phi_{j,k}(x,y)f_k,g_j)_{\mathbf{l}_2}\,d\sigma(x,y),\quad \forall f,g\in\mathbf{l}_2.\end{aligned} \tag{8.2.6}$$

Let us now pass to a study of a more special bounded operator A and B that act on the space $\mathbf{l}_2$. Namely, let their be given by matrices J_A and J_B which have a tri-diagonal block structure of forms (8.1.26) and (8.1.44). So, these operators A and B are defined by expressions (8.1.29) and (8.1.47). Remind that the norm of all elements a_n, b_n, c_n and u_n, w_n, v_n are uniformly bounded $n\in\mathbb{N}_0$.

For further research, *we assume that conditions* (8.1.27), (8.1.28) *and* (8.1.45), (8.1.46) *are fulfilled and, additionally, operators A and B generated by matrices* (8.1.26) *and* (8.1.44) *are bounded, commutative, self-adjoint on* $\mathbf{l}_2$, (8.1.6).

At the next step we will rewrite the Parseval equality (8.2.6) in terms of generalized eigenvectors of commuting self-adjoint operators A and B. At first we prove the lemma.

Lemma 8.2.1 *Let* $\varphi(x,y) = (\varphi_n(x,y))_{n=0}^{\infty}$, $\varphi_n(x,y)\in\mathcal{H}_n$, $(x,y)\in\mathbb{R}^2$*—be a generalized eigenvector from* $(\mathbf{l}_{\text{fin}})'$ *of the operator A with the eigenvalue x and also a generalized eigenvector of B with the eigenvalue y. Then* $\varphi(x,y)$ *is a solution in* $(\mathbf{l}_{\text{fin}})'$ *of two difference Eqs.* (8.1.29) *and* (8.1.47), *namely,*

$$\begin{aligned}(J_A\varphi(x,y))_n &= a_{n-1}\varphi_{n-1}(x,y)+b_n\varphi_n(x,y)+c_n\varphi_{n+1}(x,y)=x\varphi_n(x,y),\\ (J_B\varphi(x,y))_n &= u_{n-1}\varphi_{n-1}(x,y)+w_n\varphi_n(x,y)+v_n\varphi_{n+1}(x,y)=y\varphi_n(x,y),\end{aligned} \tag{8.2.7}$$

with an initial condition $\varphi_0\in\mathbb{R}$, *for* $n\in\mathbb{N}_0$, $\varphi_{-1}(x,y)=:0$.

We assert that this solution has a form

$$\varphi_n(x,y)=Q_n(x,y)\varphi_0=(Q_{n;0},Q_{n;1},\dots,Q_{n;n},)\varphi_0,\quad \forall n\in\mathbb{N}. \tag{8.2.8}$$

Here $Q_{n;\alpha}$, $\alpha = 0,1,\dots,n$ *are polynomials of variables x and y and these polynomials have the form*

$$Q_{n;\alpha}(x,y)=l_{n;\alpha}y^{n-\alpha}x^{\alpha}+q_{n;\alpha}(x,y),\quad \alpha=1,\dots,n, \tag{8.2.9}$$

where $l_{n;\alpha} > 0$ *and* $q_{n;\alpha}(x, y)$ *is a previous part of the polynomial due to* (8.1.2) *and it is some linear combination of* $y^j x^k$, $0 \le j + k \le n - 1$, $y^{n-(\alpha-1)}x^{\alpha-1}$. *The last expressions are described in the case* $\alpha = 1, \dots n$ *and* $y^{n-1}x^{n-1}$ *if* $\alpha = 0$.

Proof For $n = 0$, the system (8.2.7) has the form

$$\begin{aligned} w_0\varphi_0 + v_0\varphi_1 &= y\varphi_0, \\ b_0\varphi_0 + c_0\varphi_1 &= x\varphi_0, \end{aligned} \quad \text{i.e.,} \quad \begin{aligned} v_{0;0,0}\varphi_{1;0} &= (y - w_{0;0,0})\varphi_0, \\ c_{0;0,0}\varphi_{1;0} + c_{0;0,1}\varphi_{1;1} &= (x - b_{0;0,0})\varphi_0. \end{aligned} \tag{8.2.10}$$

Here and in what follows we denote

$$\varphi_n(x, y) = (\varphi_{n;0}(x, y), \varphi_{n;1}(x, y), \dots, \varphi_{n;n}(x, y)) \in \mathcal{H}_n, \ \forall n \in \mathbb{N}; \quad \varphi_0 = \varphi_{0;0}.$$

By using assumptions (8.1.27), (8.1.28) and (8.1.45), (8.1.46), rewrite the last two equalities (8.2.10) in the form

$$\begin{gathered} \Delta_0\varphi_1(x, y) = ((y - w_{0;0,0})\varphi_0, (x - b_{0;0,0})\varphi_0); \\ \Delta_0 = \begin{pmatrix} v_{0;0,0} & 0 \\ c_{0;0,0} & c_{0;0,1} \end{pmatrix}, \quad v_{0;0,0} > 0, \quad c_{0;0,1} > 0. \end{gathered} \tag{8.2.11}$$

Therefore,

$$\begin{aligned} \varphi_{1;0}(x, y) &= \frac{1}{v_{0;0,0}}(y - w_{0;0,0})\varphi_0 = Q_{1;0}(x, y)\varphi_0, \\ \varphi_{1;1}(x, y) &= \left(\frac{(x - b_{0;0,0})}{c_{0;0,1}} - \frac{c_{0;0,0}(y - w_{0;0,0})}{c_{0;0,1}v_{0;0,0}}\right)\varphi_0 = Q_{1;1}(x, y)\varphi_0. \end{aligned} \tag{8.2.12}$$

In another words the solution $\varphi_n(x, y)$ of (8.2.7) for $n = 0$ has a form (8.2.8) and (8.2.9).

Suppose, using the induction, that for $n \in \mathbb{N}$, elements $\varphi_{n-1}(x, y)$ and $\varphi(x, y)$ of our generalized eigenvector $\varphi(x, y) = (\varphi_n(x, y))_{n=0}^{\infty}$, have the form (8.2.8) and (8.2.9) and we will prove that $\varphi_{n+1}(x, y)$ can also be written in the form (8.2.8) and (8.2.9).

Our eigenvector $\varphi(x, y)$ satisfies the system (8.2.7) of two equations. But this system is overdetermined: it consists of $2(n + 1)$ scalar equations from which it is necessary to find only $n + 2$ unknowns $\varphi_{n+1;0}, \varphi_{n+1;1}, \dots, \varphi_{n+1;n+1}$, using as an initial data the previous $n + 1$ values $\varphi_{n;0}, \varphi_{n;1}, \dots, \varphi_{n;n}$ which are coordinates of the vector $\varphi_n(x, y)$.

According to Theorems 8.1.9 and 8.1.5, especially to (8.1.27), (8.1.28) and (8.1.45), (8.1.46), $((n+1)\times(n+2))$-matrices c_n, v_n, their application to $\varphi_{n+1}\in\mathcal{H}_n$ have a form

$$v_n\varphi_{n+1}(x,y)=\begin{bmatrix} v_{n;0,0} & 0 & 0 & \ldots 0 & 0\\ v_{n;1,0} & v_{n;1,1} & 0 & \ldots 0 & 0\\ v_{n;2,0} & v_{n;2,1} & v_{n;2,2} & \ldots 0 & 0\\ \vdots & \vdots & \vdots & \ddots\vdots & \vdots\\ v_{n;n-1,0} & v_{n;n-1,1} & v_{n;n-1,2} & \ldots 0 & 0\\ v_{n;n,0} & v_{n;n,1} & v_{n;n,2} & \ldots v_{n;n,n} & 0\end{bmatrix}\varphi_{n+1}(x,y),$$

$$c_n\varphi_{n+1}(x,y)=\begin{bmatrix} c_{n;0,0} & c_{n;0,1} & 0 & \ldots 0 & 0\\ c_{n;1,0} & c_{n;1,1} & c_{n;1,2} & \ldots 0 & 0\\ c_{n;2,0} & c_{n;2,1} & c_{n;2,2} & \ldots 0 & 0\\ \vdots & \vdots & \vdots & \ddots\vdots & \vdots\\ c_{n;n-1,0} & c_{n;n-1,1} & c_{n;n-1,2} & \ldots c_{n;n-1,n} & 0\\ c_{n;n,0} & c_{n;n,1} & c_{n;n,2} & \ldots c_{n;n,n} & c_{n;n,n+1}\end{bmatrix}\varphi_{n+1}(x,y), \tag{8.2.13}$$

where $\varphi_{n+1}(x,y)=(\varphi_{n+1;0}(x,y),\varphi_{n+1;1}(x,y),\ldots,\varphi_{n+1;n+1}(x,y))$.

Construct, similarly to (8.2.11), the combination of matrices (8.2.13), i.e., the $((n+2)\times(n+2))$-matrix

$$\Delta_n\varphi_{n+1}(x,y)=\begin{bmatrix} v_{n;0,0} & 0 & 0 & \ldots 0 & 0\\ c_{n;0,0} & c_{n;0,1} & 0 & \ldots 0 & 0\\ c_{n;1,0} & c_{n;1,1} & c_{n;1,2} & \ldots 0 & 0\\ \vdots & \vdots & \vdots & \ddots\vdots & \vdots\\ c_{n;n-1,0} & c_{n;n-1,1} & c_{n;n-1,2} & \ldots c_{n;n-1,n} & 0\\ c_{n;n,0} & c_{n;n,1} & c_{n;n,2} & \ldots c_{n;n,n} & c_{n;n,n+1}\end{bmatrix}\varphi_{n+1}(x,y), \tag{8.2.14}$$

where $\varphi_{n+1}(x,y)=(\varphi_{n+1;0}(x,y),\varphi_{n+1;1}(x,y),\ldots,\varphi_{n+1;n+1}(x,y))$.

The matrix (8.2.14) is invertible because its elements on the main diagonal are positive (see (8.1.28) and (8.1.46)). Rewrite equalities (8.2.7) in the form

$$\begin{aligned} c_n\varphi_{n+1}(x,y)&=x\varphi_n(x,y)-a_{n-1}\varphi_{n-1}(x,y)-b_n\varphi_n(x,y),\\ v_n\varphi_{n+1}(x,y)&=y\varphi_n(x,y)-u_{n-1}\varphi_{n-1}(x,y)-v_n\varphi_n(x,y), \qquad n\in\mathbb{N}. \end{aligned} \tag{8.2.15}$$

The first $n+2$ scalar equations (from $2(n+1)$ scalar Eq. (8.2.15)) have a form

$$\Delta_n \varphi_{n+1}(x, y) = (x Q_{n;0}(x, y) - (u_{n-1} Q_{n-1}(x, y) - (w_n Q_n(x, y))_{n;0}, \\ y Q_{n;0}(x, y) - (a_{n-1} Q_{n-1}(x, y))_{n;0} - (b_n Q_n(x, y))_{n;0}, \dots, \\ y Q_{n;n}(x, y) - (a_{n-1} Q_{n-1}(x, y))_{n;n} - (b_n Q_n(x, y))_{n;n})\varphi_0. \tag{8.2.16}$$

The construction of matrix Δ_n and the form of vector on the right-hand side (8.2.16) and (8.2.8), (8.2.9) give

$$\varphi_{n+1;0}(x, y) = Q_{n+1;0}(x, y)\varphi_0 \\ = \frac{1}{a_{n;0,0}}(x Q_{n;0}(x, y) - (u_{n-1} Q_{n-1}(x, y))_{n;0}) - (w_n Q_n(x, y))_{n;0})\varphi_0 \\ = \frac{1}{a_{n;0,0}}(x(l_{n;0}x^n + q_{n;0}(x, y)) - (u_{n-1} Q_{n-1}(x, y))_{n;0} - \\ (w_n Q_n(x, y))_{n;0})\varphi_0, \tag{8.2.17}$$

that is, the first term on the right-hand side in (8.2.17) is equal to $\frac{l_{n;0}}{a_{n;0,0}} x^{n+1} y^0$, so it has the form (8.2.9).

Similar calculations give the same effect for

$$\varphi_{n+1;1}(x, y), \dots, \varphi_{n;n}(x, y), \varphi_{n+1;n+1}(x, y).$$

It is necessary to take into account that the diagonal elements

$$v_{n;0,0}, c_{n;0,1}, c_{n;1,2}, \dots, c_{n;n,n+1}$$

of the matrix Δ_n are positive due to (8.1.46) and (8.1.28). This completes the induction and finishes the proof. □

By the formulation and by the proof of Lemma8.2.1, it is not asserted that the solution of the overdefined system (8.2.7) exists for any initial data $\varphi_0 \in \mathbb{R}$: it is only proved that the generalized eigenvector from $(\mathbf{l}_{\text{fin}})'$ of operators A and B is a solution (8.2.7) and has the form (8.2.8) and (8.2.9).

Consider $Q_n(x, y)$ with fixed x and y as a linear operator that acts from $\mathcal{H}_0$ to $\mathcal{H}_n$, that is, $\mathcal{H}_0 \ni \varphi_0 \longmapsto Q_n(x, y)\varphi_0 \in \mathcal{H}_n$. We also regard $Q_n(x, y)$ as an operator-valued polynomial of variables $x, y \in \mathbb{R}$; therefore, the adjoint operator has the form

$$Q_n^*(x, y) = (Q_n(x, y))^* : \mathcal{H}_n \longrightarrow \mathcal{H}_0.$$

By using these polynomials $Q_n(x, y)$, we construct the representation for $\Phi_{j,k}(x, y)$.

Lemma 8.2.2 *The operator $\Phi_{j,k}(x, y)$, $\forall(x, y) \in \mathbb{R}^2$ has the representation*

$$\Phi_{j,k}(x, y) = Q_j(x, y)\Phi_{0,0}(x, y)Q_k^*(x, y) : \mathcal{H}_k \longrightarrow \mathcal{H}_j, \quad j, k \in \mathbb{N}_0, \tag{8.2.18}$$

where $\Phi_{0,0}(x, y) \geq 0$ is a scalar.

Proof For a fixed $k \in \mathbb{N}_0$, the vector $\varphi = \varphi(x, y) = (\varphi_j(x, y))_{j=0}^{\infty}$, where

$$\varphi_j(x, y) = \Phi_{j,k}(x, y) = \pi_j\Phi(x, y)\pi_k \in \mathcal{H}_j, \quad (x, y) \in \mathbb{R}^2 \tag{8.2.19}$$

is a generalized solution, in $(\mathbf{l}_{\text{fin}})'$ of equations

$$J_A\varphi(x, y) = x\varphi(x, y), \quad J_B\varphi(x, y) = y\varphi(x, y)$$

since $\Phi(x, y)$ is a projector on to generalized eigenvectors of operators A and B with correspondence generalized eigenvalues (x, y). Hence, $\varphi = \varphi(x, y) \in \mathbf{l}_2(p^{-1})$, exists as an usual solution of equations $J_A\varphi = x\varphi$, $J_B\varphi = y\varphi$ with an initial condition $\varphi_0 = \pi_0\Phi(x, y)\pi_k \in \mathcal{H}_0$.

By using Lemma 8.2.1 and according to (8.2.8), we get

$$\Phi_{j,k}(x, y) = Q_j(x, y)(\Phi_{0,k}(x, y)), \quad j \in \mathbb{N}_0. \tag{8.2.20}$$

The operator $\Phi(x, y) : \mathbf{l}_2(p) \longrightarrow \mathbf{l}_2(p^{-1})$ is formally self-adjoint on $\mathbf{l}_2$ and it is the derivative of the resolution of identity of the operator A (B) on $\mathbf{l}_2$ with respect to the spectral measure. Hence, according to (8.2.18), we get

$$(\Phi_{j,k}(x, y))^* = (\pi_j\Phi(x, y)\pi_k)^* = \pi_k\Phi(x, y)\pi_j = \Phi_{k,j}(x, y), \quad j, k \in \mathbb{N}_0. \tag{8.2.21}$$

For a fixed $j \in \mathbb{N}_0$, from (8.2.21) and previous conversation, it follows that the vector

$$\varphi = \varphi(x, y) = (\varphi_k(x, y))_{k=0}^{\infty}, \quad \varphi_k(x, y) = \Phi_{k,j}(x, y) = (\Phi_{j,k}(x, y))^*$$

is an usual solution of equations $J_A\varphi = x\varphi$ and $J_B\varphi = y\varphi$ with an initial condition $\varphi_0 = \Phi_{0,j}(x, y) = (\Phi_{j,0}(x, y))^*$.

Again using Lemma 8.2.1, we obtain the representation of the (8.2.20), namely,

$$\Phi_{k,j}(x, y) = Q_k(x, y)(\Phi_{0,j}(x, y)), \quad k \in \mathbb{N}_0. \tag{8.2.22}$$

Taking into account (8.2.21) and (8.2.22), we get

$$\begin{aligned}\Phi_{0,k}(x,y) = (\Phi_{k,0}(x,y))^* &= (Q_k(x,y)\Phi_{0,0}(x,y))^* \\ &= \Phi_{0,0}(x,y)(Q_k(x,y))^*, \quad k \in \mathbb{N}_0\end{aligned} \tag{8.2.23}$$

(here we used $\Phi_{0,0}(x,y) \geq 0$, this inequality follows from (8.2.4) and (8.2.5)). By substituting (8.2.23) into (8.2.20), we obtain (8.2.18). □

Now it is possible to rewrite the Parseval equality (8.2.6) in a more concrete form. To this end, we substitute the expression (8.2.18) for $\Phi_{j,k}(x,y)$ into (8.2.6) and get

$$\begin{aligned}(f,g)_{\mathbf{l}_2} &= \sum_{j,k=0}^{\infty} \int_{\mathbb{R}^2} (\Phi_{j,k}(x,y) f_k, g_j)_{\mathbf{l}_2}\, d\sigma(x,y) = \\ &= \sum_{j,k=0}^{\infty} \int_{\mathbb{R}^2} (Q_j(x,y)\Phi_{0,0}(x,y)Q_k^*(x,y) f_k, g_j)_{\mathbf{l}_2}\, d\sigma(x,y) \\ &= \sum_{j,k=0}^{\infty} \int_{\mathbb{R}^2} (Q_k^*(x,y) f_k, Q_j^*(x,y) g_j)_{\mathbf{l}_2}\, d\rho(x,y) = \\ &= \int_{\mathbb{R}^2} \Big(\sum_{k=0}^{\infty} Q_k^*(x,y) f_k\Big)\overline{\Big(\sum_{j=0}^{\infty} Q_j^*(x,y) g_j\Big)}\, d\rho(x,y),\end{aligned} \tag{8.2.24}$$

$$d\rho(x,y) = \Phi_{0,0}(x,y)\, d\sigma(x,y), \quad \forall f,g \in \mathbf{l}_{\text{fin}}.$$

Denote the Fourier transform "$\widehat{\ }$" induced by commuting self-adjoint operators A and B on the space $\mathbf{l}_2$

$$\mathbf{l}_2 \supset \mathbf{l}_{\text{fin}} \ni f = (f_n)_{n=0}^{\infty} \longmapsto \hat{f}(x,y) = \sum_{n=0}^{\infty} Q_n^*(x,y) f_n \in L_2(\mathbb{R}^2, d\rho(x,y)). \tag{8.2.25}$$

Hence, (8.2.24) gives the Parseval equality in a final form

$$(f,g)_{\mathbf{l}_2} = \int_{\mathbb{R}^2} \hat{f}(x,y)\overline{\hat{g}(x,y)}\, d\rho(x,y), \quad \forall f,g \in \mathbf{l}_{\text{fin}}. \tag{8.2.26}$$

We extend (8.2.26) by continuity and it becomes valid $\forall f,g \in \mathbf{l}_2$. The orthogonality of $Q_n^*(x,y)$ follows from (8.2.25) and (8.2.26). Namely, it is sufficient only to take

$$f = (0,\dots,0,f_k,0,\dots),\ f_k \in \mathcal{H}_k, \quad g = (0,\dots,0,g_j,0,\dots),\ g_j \in \mathcal{H}_j$$

in (8.2.25) and (8.2.26). Then

$$\int_{\mathbb{R}^2} (Q_k^*(x,y)f_k)\overline{(Q_j^*(x,y)g_j)}\,d\rho(x,y) = \delta_{j,k}(f_j,g_j)_{\mathcal{H}_j}, \quad \forall k,j \in \mathbb{N}_0. \tag{8.2.27}$$

By using the representation (8.2.8) for these polynomials we can rewrite the equality (8.2.27) in a usual classical scalar form. To do this, we remark that in general $Q_0^*(x,y) = \bar{Q}_0(x,y)$ and, for $n \in \mathbb{N}$ according to (8.2.8),

$$Q_n(x,y) = (Q_{n;0}(x,y),\ Q_{n;1}(x,y),\dots,Q_{n;n}(x,y)) : \mathcal{H}_0 \longrightarrow \mathcal{H}_n.$$

Hence, for the adjoint operator $Q_n^*(x,y) : \mathcal{H}_n \longrightarrow \mathcal{H}_0$ we have

$$\begin{aligned}
(Q_n(x,y)q,p)_{\mathcal{H}_n} &= \\
&= ((Q_{n;0}(x,y)q, Q_{n;1}(x,y)q,\dots,Q_{n;n}(x,y)q),(p_0,p_1,\dots,p_n))_{\mathcal{H}_n}\\
&= Q_{n;0}(x,y)q\bar{p}_0 + Q_{n;1}(x,y)q\bar{p}_1 + \dots + Q_{n;n}(x,y)q\bar{p}_n\\
&= q\overline{(\overline{Q_{n;0}(x,y)}p_0 + \overline{Q_{n;1}(x,y)}p_1 + \dots + \overline{Q_{n;n}(x,y)}p_n)}\\
&= (q, Q_n^*(x,y)p)_{\mathcal{H}_0},
\end{aligned}$$

where

$$Q_n^*(x,y)p = \overline{Q_{n;0}(x,y)}p_0 + \overline{Q_{n;1}(x,y)}p_1 + \dots + \overline{Q_{n;n}(x,y)}p_n,\ \forall q \in \mathcal{H}_0,$$

and $p = (p_0,p_1,\dots,p_n) \in \mathcal{H}_n$.

Due to the last equality for $n \in \mathbb{N}$ and $f_n = (f_{n,0}, f_{n,1},\dots,f_{n,n}) \in \mathcal{H}_n$ we get

$$\begin{aligned}
Q_n^*(x,y)f_n &= \overline{Q_{n;0}(x,y)}f_{n;0} + \overline{Q_{n;1}(x,y)}f_{n;1} + \dots + \overline{Q_{n;n}(x,y)}f_{n;n},\\
Q_0^*(x,y) &= 1.
\end{aligned} \tag{8.2.28}$$

Therefore (8.2.27) has the form

$$\int_{\mathbb{R}^2} \left(\sum_{\alpha=0}^{k} \overline{Q_{k;\alpha}(x,y)}f_{k;\alpha}\right)\overline{\left(\sum_{\beta=0}^{j} \overline{Q_{j;\beta}(x,y)}f_{j;\beta}\right)}\,d\rho(x,y) = \delta_{j,k}\sum_{\alpha=0}^{j} f_{j;\alpha}\bar{g}_{j;\alpha},$$

$\forall f_{k;0}, f_{k;1}, \ldots, f_{k;k}, \ g_{j;0}, g_{j;1}, \ldots, g_{j;j} \ \in \ \mathbb{C}, \ j, k \ \in \ \mathbb{N}_0$. This equality is equivalent to the orthogonality relation in the usual classical form

$$\int_{\mathbb{R}^2} \overline{Q^*_{k;\beta}(x,y)} Q_{j;\alpha}\, d\rho(x,y) = \delta_{j,k}\delta_{\alpha,\beta} \quad (Q_{0;0} = Q_0(x,y)), \tag{8.2.29}$$

$\forall j, k \in \mathbb{N}_0, \forall \alpha = 0, 1, \ldots, j, \beta = 0, 1, \ldots, k.$

Let us remark that due to (8.2.28), the Fourier transform (8.2.25) can be rewritten in the form

$$\hat{f}(x,y) = \sum_{n=0}^{\infty}\sum_{\alpha=0}^{n} \overline{Q_{n;\alpha}(x,y)} f_{n;\alpha}, \quad \forall f = (f_n)_{n=0}^{\infty} \in \mathbf{l}_2, \quad (x,y) \in \mathbb{R}^2. \tag{8.2.30}$$

By using the stated above results of this section, we can formulate the spectral theorem for our operators A and B.

Theorem 8.2.3 *Let us consider the space*

$$\mathbf{l}_2 = \mathcal{H}_0 \oplus \mathcal{H}_1 \oplus \mathcal{H}_2 \oplus, \cdots, \quad \mathcal{H}_n = \mathbb{C}^{n+1}, \quad n \in \mathbb{N}_0 \tag{8.2.31}$$

and the linear operators A and B which are defined on finite vectors $\mathbf{l}_{\text{fin}}$ by block tri-diagonal Jacobi type matrices J_A and J_B of forms (8.1.26) *and* (8.1.44) *due to expressions* (8.1.29) *and* (8.1.47), *respectively. We suppose that all coefficients a_n, b_n, c_n, and u_n, v_n, w_n, $n \in \mathbb{N}_0$, are uniformly bounded, some elements of these matrices are always equal to zero and some elements are always positive according to* (8.1.27), (8.1.28) *and* (8.1.45), (8.1.46), *and extensions of A and B by continuity are bounded commuting self-adjoint operators on this space.*

The generalized eigenvectors expansion of operators A and B has the following form. According to Lemma 8.2.1 *we get, using $\varphi_0 \in \mathbb{R}$, the solution $\varphi(x,y) = (\varphi_n(x,y))_{n=0}^{\infty}$, $\varphi_n(x,y) \in \mathcal{H}_n$ of Eq.* (8.2.7) *(which exists due to the projection spectral theorem)*

$$\begin{aligned} \varphi_n(x,y) &= Q_n(x,y)\varphi_0 \\ &= (Q_{n;0}(x,y), Q_{n;1}(x,y), \cdots, Q_{n;n}(x,y))\varphi_0, \quad (x,y) \in \mathbb{R}^2, \end{aligned}$$

where $Q_{n;\alpha}(x,y)$, $\alpha = 0, 1, \ldots, n$, are polynomials of variables x and y. Then, the Fourier transform has the form

$$\begin{aligned} \mathbf{l}_2 \supset \mathbf{l}_{\text{fin}} \ni f = (f_n)_{n=0}^{\infty} \longmapsto \hat{f}(x,y) &= \sum_{n=0}^{\infty} Q_n^*(x,y) f_n \\ &= \sum_{n=0}^{\infty}\sum_{\alpha=0}^{n} \overline{Q_{n;\alpha}(x,y)} f_{n;\alpha} \in L_2(\mathbb{R}^2, d\rho(x,y)). \end{aligned} \tag{8.2.32}$$

Here $Q_n^*(x,y) : \mathcal{H}_n \longrightarrow \mathcal{H}_0$ *is the operator adjoint to* $Q_n(x,y) : \mathcal{H}_0 \longrightarrow \mathcal{H}_n$, $d\rho(x,y)$ *is the probability spectral measure of operators A and B.*

The Parseval equality, $\forall f, g \in \mathbf{l}_{\text{fin}}$, *has the form*

$$\begin{aligned}(f,g)_{\mathbf{l}_2} &= \int\limits_{\mathbb{R}^2} \hat{f}(x,y)\overline{\hat{g}(x,y)}\,d\rho(x,y),\\ (J_A f,g)_{\mathbf{l}_2} &= \int\limits_{\mathbb{R}^2} x\hat{f}(x,y)\overline{\hat{g}(x,y)}\,d\rho(x,y),\\ (J_B f,g)_{\mathbf{l}_2} &= \int\limits_{\mathbb{R}^2} y\hat{f}(x,y)\overline{\hat{g}(x,y)}\,d\rho(x,y).\end{aligned} \tag{8.2.33}$$

Transformations (8.2.32) *and identities* (8.2.33) *are extended by the continuity* $\forall f, g \in \mathbf{l}_2$, *so that the operator* (8.2.32) *is isometric and maps the whole* $\mathbf{l}_2$ *into the whole* $L_2(\mathbb{R}^2, d\rho(x,y))$.

Polynomials $\overline{Q_{n;\alpha}(x,y)}$, $n \in \mathbb{N}$, $\alpha = 0, 1, \ldots, n$, *and* $Q_{0;0}(x,y) = 1$ *form an orthonormal system in* $L_2(\mathbb{R}^2, d\rho(x,y))$ *in the sense of* (8.2.29) *and it is total in this space.*

Matrixes

$$J_A = (\tau_{j,k})_{j,k=0}^{\infty},\ \tau_{j,k} = (\tau_{j,k;\alpha,\beta})_{\alpha,\beta=0}^{j,k},$$

$$J_B = (\theta_{j,k})_{j,k=0}^{\infty},\ \theta_{j,k} = (\theta_{j,k;\alpha,\beta})_{\alpha,\beta=0}^{j,k},$$

are restored according to formulas

$$\begin{aligned}\tau_{j,k;\alpha,\beta} &= (J_A\delta_{k,\beta}, \delta_{j,\alpha})_{\mathbf{l}_2} = \iint\limits_{\mathbb{R}^2} x\overline{P_{k;\beta}(x,y)}P_{j;\alpha}(x,y)d\rho(x,y),\\ \theta_{j,k;\alpha,\beta} &= (J_B\delta_{k,\beta}, \delta_{j,\alpha})_{\mathbf{l}_2} = \iint\limits_{\mathbb{R}^2} y\overline{P_{k;\beta}(x,y)}P_{j;\alpha}(x,y)d\rho(x,y).\end{aligned} \tag{8.2.34}$$

Here, for convenience, it is re-designated $b_j = \tau_{j,j}$, $c_j = \tau_{j,j+1}$, $a_j = \tau_{j+1,j}$, *and* $w_j = \theta_{j,j}$, $v_j = \theta_{j,j+1}$, $u_j = \theta_{j+1,j}$, $j \in \mathbb{N}_0$.

Proof It is only necessary to show that orthogonal polynomials $\overline{Q_{n;\alpha}(x,y)}$, $n \in \mathbb{N}$, $\alpha = 0, 1, \ldots, n$ and $Q_{0;0}(x,y) = 1$ form a total set in the space $L_2(\mathbb{R}^2, d\rho(x,y))$. For this reason, we remark at first that due to the compactness of the support of the measure $d\rho(x,y)$ on $\mathbb{R}^2$, elements $x^j y^k$, $j,k \in \mathbb{N}_0$ form a total set in $L_2(\mathbb{R}^2, d\rho(x,y))$. Indeed, suppose the contrary, i.e., that our system of polynomials is not total. Then there exist non-zero function $h(x,y) \in L_2(\mathbb{R}^2, d\rho(x,y))$ that is orthogonal to all these polynomials and hence, according to (8.2.9) it is also orthogonal to all $x^j y^k$, $j, k \in \mathbb{N}_0$. Hence, $h(x, y) = 0$. The obtained contradiction completes the proof. □

The last theorem solves the direct problem for the bounded symmetric commuting operators A and B that are generated on the space $\mathbf{l}_2$ by the matrices J_A and J_B of the form (8.1.44) and (8.1.26).

The inverse problem consists in construction by a given measure $d\rho(x, y)$ with compact support on $\mathbb{R}^2$, the bounded symmetric commuting matrices J_A and J_B of the form (8.1.44) and (8.1.26), respectively, that have as operators their spectral measure equal to $d\rho(x, y)$. *This construction is conducted according to Theorem* 8.1.5 *and* 8.1.9 *with the use of the Gram-Schmidt orthogonalization procedure for the system (8.1.2). For matrices J_A and J_B of the form* (8.1.26) *and* (8.1.44), *that are constructed by $d\rho(x, y)$, the spectral measure of corresponding bounded symmetric commuting operators A and B coincides with the start measure.*

Proof This fact is true, since the system of orthogonal polynomials, connected with A and B, $\overline{Q_{n,\alpha}(x, y)}$, $\alpha = 0, 1, \ldots, n$, $n \in \mathbb{N}_0$, are orthonormal in $L_2(\mathbb{R}^2, d\rho(x, y))$. And, polynomials $Q_{n,\alpha}(x, y)$ of variables $y^j x^k$, $(x, y) \in \mathbb{R}$ are constructed according to Lemma 8.2.1 in the same way as the system (8.1.3) is constructed by $x^j y^k$, $j, k \in \mathbb{N}_0$. Hence,

$$
\begin{aligned}
Q_0(x, y) &= P_0(x, y) = 1, \\
\overline{Q_{n,\alpha}(x, y)} &= P_{n;\alpha}(x, y), \quad \alpha = 0, 1, \ldots, n, \ \forall n \in \mathbb{N}.
\end{aligned}
\tag{8.2.35}
$$

Since both system of polynomials form a total set in $L_2(\mathbb{R}^2, d\rho(x, y))$, then (8.2.35) shows that the spectral measure constructed by the operator coincides with the given one. □

Let us remark that expressions (8.1.12) and (8.1.30) (as it was known in the classical theory of Jacobi matrices) restore initial matrices (8.1.44) and (8.1.26) from the spectral measure $d\rho(x, y)$ of operators generated by J_A and J_B on $\mathbf{l}_2$.

8.3 An Analogue of the Weyl Function Corresponding to the Two-Dimensional Real Moment Problem

Similarly to the classical case, we introduced in Sect. 5.5 the Weyl function for the complex moment problem. We also introduce an analogue of the Weyl function for the two-dimensional real moment problem:

$$
M(z_1, z_2) = (R_{z_1}(A)\delta_0, R_{\bar{z}_2}(B)\delta_0)_{\mathbf{l}_2}, \tag{8.3.1}
$$

$z_1, z_2 \in \mathbb{C} \setminus \mathbb{R}$. Similarly, as in the (classical) Hamburger moment problem, the introduced function (8.3.1) is characterized by the theorem.

Theorem 8.3.1 *Let J_A and J_B be Jacobi type block matrices of the form* (8.1.26) *and* (8.1.44)) *with the corresponding properties according to which they generate bounded commutative self-adjoint in* $\mathbf{l}_2$ *operators J_A and J_B. Denote by $R_{z_1}(A)$ and $R_{z_2}(B)$ their resolvents* ($\text{Im} z_i \neq 0,\ i = 1, 2$)*. Then, the function $M(z_1, z_2)$ in* (8.3.1) *determines uniquely the spectral measure of these operators.*

Proof Due to the spectral theorem for a self adjoint operators, we have

$$M(z_1, z_2) = (R_{z_1}(A)\delta_0, R_{\bar{z}_2}(B)\delta_0)_{\mathbf{l}_2} = \int\limits_{\mathbb{R}^2} \frac{1}{(\lambda - z_1)(\mu - z_2)}\, d\rho(\lambda, \mu).$$

Since corresponding to matrices J_A and J_B operators are considered bounded, then their specter and, hence, for $z_1, z_2 \in \mathbb{C} \setminus \mathbb{R}$, we can write

$$\begin{aligned}
M(z_1, z_2) &= \int\limits_{\mathbb{R}^2} \frac{1}{(\lambda - z_1)(\mu - z_2)}\, d\rho(\lambda, \mu) \\
&= \frac{1}{z_1 z_2} \int\limits_{\mathbb{R}^2} \frac{1}{(1 - \frac{\lambda}{z_1})(1 - \frac{\mu}{z_2})}\, d\rho(\lambda, \mu) \\
&= \frac{1}{z_1 z_2} \int\limits_{\mathbb{R}^2} \left(\sum_{m=0}^{\infty} \left(\frac{\lambda}{z_1} \right)^m \right) \left(\sum_{n=0}^{\infty} \left(\frac{\lambda}{z_2} \right)^n \right) d\rho(\lambda, \mu) \\
&= \frac{1}{z_1 z_2} \int\limits_{\mathbb{R}^2} \left(\sum_{m,n=0}^{\infty} \frac{1}{z_1^m z_2^n} \lambda^m \mu^n \right) d\rho(\lambda, \mu) \\
&= \frac{1}{z_1 z_2} \sum_{n,m=0}^{\infty} \frac{1}{z_1^m z_2^n} \int\limits_{\mathbb{R}^2} \lambda^m \mu^n\, d\rho(\lambda, \mu) = \sum_{m,n=0}^{\infty} \frac{s_{m,n}}{z_1^{m+1} z_2^{n+1}},
\end{aligned}$$

where (see 8.4.4)

$$s_{m,n} = \int\limits_{\mathbb{R}^2} \lambda^m \mu^n\, d\rho(\lambda, \mu), \quad m, n \in \mathbb{N}_0.$$

It follows from the previous representation of the function $M(z_1, z_2)$ as a series $z_1^m z_2^n$ that coefficients $s_{m,n}$ of this series can be uniquely recovered from $M(z_1, z_2)$. But these coefficients are moments of the measure $d\rho(\lambda, \mu)$ and satisfy the estimate $|s_{n,m}| \leq \|J_A\|^m \|J_B\|^n$. Since the measure $d\rho(\lambda, \mu)$ is uniquely recovered by moments $s_{m,n}$, then the function $M(z_1, z_2)$ is also uniquely recovered. $\square$

8.4 The Solution of the Two-Dimensional Real Moment Problem

In this section, for completeness of research, a detailed solution of the two-dimensional power moment problem is given. The solution is written using Parseval equality, using the corresponding generalized eigenvectors expansion.

Let us introduce the basic concepts and notations necessary for formulating results of the section. Let $\mathcal{H}$ be a separable Hilbert space and let A and B be a commuting self-adjoint operators defined on $\mathrm{Dom}(A)$ and $\mathrm{Dom}(B)$ in $\mathcal{H}$. Consider an equipment of $\mathcal{H}$:

$$\mathcal{H}_- \supset \mathcal{H} \supset \mathcal{H}_+ \supset \mathcal{D}, \tag{8.4.1}$$

where $\mathcal{H}_+$ is a Hilbert space topologically and quasi-nuclear imbedded into $\mathcal{H}$ (we recall that as usual, topologically means densely and continuously; quasi-nuclear means that the imbedding operator is of the Hilbert-Schmidt type); $\mathcal{H}_-$ is the dual one to $\mathcal{H}_+$ with respect to the space $\mathcal{H}$; $\mathcal{D}$ is a linear set, i.e., topological space topologically embedded into $\mathcal{H}_+$.

Operators A and B are called standardly connected with the chain (8.4.1), if $\mathcal{D} \subset \mathrm{Dom}(A)$, $\mathcal{D} \subset \mathrm{Dom}(B)$ and restrictions $A \upharpoonright \mathcal{D}$, $B \upharpoonright \mathcal{D}$ act from $\mathcal{D}$ into $\mathcal{H}_+$ continuously.

Let us recall also that a vector $\Omega \in \mathcal{D}$ is called a strong cyclic vector of operators A and B if for any $p, q \in \mathbb{N}$, we have $\Omega \in \mathrm{Dom}(A^p) \cap \mathrm{Dom}(B^q)$, $A^p B^q \Omega \in \mathcal{D}$ and the set of all these vectors and Ω, as $p, q = \mathbb{N}_0$, is total in the space $\mathcal{H}_+$ (and, hence, also in $\mathcal{H}$).

Assuming that the strong cyclic vector exists we formulate the short variant of the projection spectral theorem. The fool version and corresponding prove one can fine in the bibliographical notes to this chapter.

Theorem 8.4.1 *For commuting self-adjoint operators A and B with a strong cyclic vector in the separable Hilbert space $\mathcal{H}$, there exists a non-negative finite Borel measure $d\rho(x, y)$ such that for ρ-almost every points $(x, y) \in \mathbb{R}^2$ there exists a generalized joint eigenvector $\xi_{x,y} \in \mathcal{H}_-$, i.e.,*

$$(\xi_{x,y}, Af)_{\mathcal{H}} = x(\xi_{x,y}, f)_{\mathcal{H}}, \quad (\xi_{x,y}, Bf)_{\mathcal{H}} = y(\xi_{x,y}, f)_{\mathcal{H}}, \quad f \in \mathcal{D}; \ \xi_{x,y} \neq 0. \tag{8.4.2}$$

The corresponding Fourier transform F is given by the rule

$$\mathcal{H} \supset \mathcal{H}_+ \ni f \mapsto (Ff)(x, y) = \hat{f}(x, y) = (f, \xi_{x,y})_{\mathcal{H}} \in L_2(\mathbb{R}^2, d\rho(x, y)), \tag{8.4.3}$$

is the isometry operator (after its extension by continuity) acting from $\mathcal{H}$ into L_2. The image of the operators A and B by the actin of F are multiplication by x and y, respectively, operators in L_2.

We remark that the notation $(\cdot,\cdot)_{\mathcal{H}}$ denotes the same as $\langle\cdot,\cdot\rangle_{\mathcal{H}}$ (see Chap. 2, Sect. 2.7).

The two-dimensional real (power) moment problem consists in finding condition on the two-indexed sequence $(s_{m,n})$, $m, n \in \mathbb{N}_0$ of real numbers that would imply the existence of a non-negative Borel measure $d\rho(x, y)$ on the real plane $\mathbb{R}^2$, for which

$$s_{m,n} = \int_{\mathbb{R}^2} x^m y^n \, d\rho(x, y), \qquad m, n \in \mathbb{N}_0. \tag{8.4.4}$$

Some variant of the solution of this problem is given by the theorem.

Theorem 8.4.2 *If a given two-indexed sequence of real numbers $(s_{m,n})$, $m, n \in \mathbb{N}_0$ has a representation (8.4.4), then the sequence is positive definite, i.e,*

$$\sum_{j,k,m,n\in\mathbb{N}_0}^{\infty} f_{j,k}\bar{f}_{m,n}s_{j+m,k+n} \geq 0, \tag{8.4.5}$$

for all finite sequences of complex numbers $\{f_{j,k}\}_{j,k=0}^{\infty}$, $f_{j,k} \in \mathbb{C}$.

The representation (8.4.4), for a given sequence of real numbers $(s_{m,n})$, $m, n \in \mathbb{N}_0$ exists and is unique if it is positive defined and

$$\sum_{p=1}^{\infty} \frac{1}{\sqrt[p]{\sum_{k=0}^{p} C_p^k \sqrt{s_{4p-4k,4k}}}} = \infty. \tag{8.4.6}$$

Recall that the condition (8.4.5) is the necessary for the representation (8.4.4). The conditions defined in (8.4.5) and (8.4.6) together guarantee not only the existence, but also the uniqueness of the representation (8.4.4) for the given sequence $(s_{m,n})$, $m, n \in \mathbb{N}_0$.

Proof The necessity of the condition(8.4.5) is proved simply. Actually, if the sequence $(s_{m,n})$, $m, n \in \mathbb{N}_0$ has the representation (8.4.4), then for an arbitrary finite sequence $f = (f_{m,n})_{m,n=0}^{\infty}$, $f_{m,n} \in \mathbb{C}$, we have

$$\sum_{j,k,m,n=0}^{\infty} f_{j,k}\bar{f}_{m,n}s_{j+m,k+n} = \int_{\mathbb{R}^2} \left| \sum_{m,n=0}^{\infty} f_{m,n}x^m y^n \right|^2 d\rho(x, y) \geq 0. \tag{8.4.7}$$

Let us denote by l the linear space of sequences $f = (f_{m,n})_{m,n=0}^{\infty}$, $f_{m,n} \in \mathbb{C}$, and we denote by l_{fin} its linear subspace consisting of finite sequences $f = (f_{m,n})_{m,n=0}^{\infty}$, i.e., sequences, which contain $f_{m,n} \neq 0$ for only finite numbers n and m. Let $\delta_{m,n}$, $m, n \in \mathbb{N}_0$ be the δ-sequence such that each $f \in l_{\text{fin}}$ has a representation $f = \sum_{n,m=0}^{\infty} f_{m,n}\delta_{m,n}$.

Let us consider two linear operators

$$(J_A f)_{j,k} = f_{j-1,k}, \; (J_B f)_{j,k} = f_{j,k-1}, \quad j, k \in \mathbb{N}_0, \quad f \in l_{\text{fin}}, \tag{8.4.8}$$

where always $f_{j,-1} = f_{-1,k} \equiv 0$. Operators J_A and J_B are the "creation" type operators. For the δ-sequence we get

$$J_A\delta_{j,k} = \delta_{j+1,k}, \quad J_B\delta_{j,k} = \delta_{j,k+1}. \tag{8.4.9}$$

Operators J_A and J_B are symmetric on finite vector with respect to the (quasi)scalar product

$$(f, g)_S = \sum_{j,k,m,n=0}^{\infty} f_{j,k}\bar{g}_{m,n}s_{j+m,k+n}, \quad f, g \in l_{\text{fin}}. \tag{8.4.10}$$

Indeed,

$$\begin{aligned}
(J_A f, g)_S &= \sum_{j,k,m,n=0}^{\infty} (J_A f)_{j,k}\bar{g}_{m,n}s_{j+m,k+n} \\
&= \sum_{j,k,m,n=0}^{\infty} f_{j-1,k}\bar{g}_{m,n}s_{j+m,k+n} \\
&= \sum_{j,k,m,n=0}^{\infty} f_{j,k}\bar{g}_{m,n}s_{j+m+1,k+n} \\
&= \sum_{j,k,m,n=0}^{\infty} f_{j,k}\bar{g}_{m-1,n}s_{j+m,k+n} \\
&= \sum_{j,k,m,n=0}^{\infty} f_{j,k}\overline{(J_A g)}_{m,n}s_{j+m,k+n} = (f, J_A g)_S;
\end{aligned}$$

$$\begin{aligned}(J_B f, g)_S &= \sum_{j,k,m,n=0}^{\infty} (J_B f)_{j,k} \bar{g}_{m,n} s_{j+m,k+n} \\ &= \sum_{j,k,m,n=0}^{\infty} f_{j,k-1} \bar{g}_{m,n} s_{j+m,k+n} \\ &= \sum_{j,k,m,n=0}^{\infty} f_{j,k} \bar{g}_{m,n} s_{j+n+1,k+m} \\ &= \sum_{j,k,m,n=0}^{\infty} f_{j,k} \bar{g}_{m-1,n} s_{j+n,k+m} \\ &= \sum_{j,k,m,n=0}^{\infty} f_{j,k} \overline{(J_B g)}_{m,n} s_{j+m,k+n} = (f, J_B g)_S.\end{aligned}$$

The operator J_A commutes with J_B on l_{fin}, i.e.,

$$(J_B J_A f)_{j,k} = f_{j-1,k-1} = (J_A J_B f)_{j,k}.$$

Let S be the Hilbert space obtained as the completion of the factor space

$$\dot{l}_{\text{fin}} = l_{\text{fin}} / \{h \in l_{\text{fin}} \mid (h, h)_S = 0\},$$

if (8.4.10) is a the quasi-scalar product.

The element f from S is a representative from the class $\dot{f}$ of the set of equivalent elements $\dot{l}_{\text{fin}}$. Hence, operators $\dot{J}_A$ and $\dot{J}_B$ are correctly defined in S. This fact is described detail in the literature. Similarly, in this case we get

$$\begin{aligned}\dot{J}_A \dot{f} &= (J_A f)^{\cdot}, \ f \in \operatorname{Dom}(\dot{J}_A) = \dot{l}_{\text{fin}}; \\ \dot{J}_B \dot{f} &= (J_B f)^{\cdot}, \ f \in \operatorname{Dom}(\dot{J}_B) = \dot{l}_{\text{fin}}.\end{aligned} \tag{8.4.11}$$

For further considerations, we denote by A and B the closures "$\sim$" of operators $\dot{J}_A$ and $\dot{J}_B$ in the space S. For simplicity, we suppose that the given sequence $(s_{m,n})$ is non-degenerate, i.e., if $(f, f)_S = 0$ for $f \in l_{\text{fin}}$, then $f = 0$, and now $\dot{f} = f$ and $\tilde{\dot{J}}_A = A$, $\tilde{\dot{J}}_A = B$. The investigation in general case is more complicated.

Let us assume also for a moment that operators A and B are self-adjoint. Later we will prove that A and B are self-adjoint and commuting under the condition (8.4.6). In general, conditions for Hermitian operators that have commuting extensions are known.

Let us construct the equipment of the space S, namely,

$$(l_2(p))_{-,S} \supset S \supset l_2(p) \supset l_{\text{fin}}, \tag{8.4.12}$$

where $l_2(p)$ is a weighted l_2-space with a weight $p = (p_{m,n})_{m,n=0}^{\infty}$, $p_n \geq 1$. The norm in $l_2(p)$ is given by the expression $\|f\|_{l_2(p)}^2 = \sum_{m,n=0}^{\infty} |f_{m,n}|^2 p_{m,n}$, where $(l_2(p))_{-,S} = \mathcal{H}_-$ is a negative space with respect to the positive space $l_2(p) = \mathcal{H}_+$ and the zero space $S = \mathcal{H}$. □

Lemma 8.4.3 *For the space S there exists enough fast increasing sequence $p_{m,n}$ such that the imbedding $l_2(p) \hookrightarrow S$ is quasi-nuclear.*

Proof The inequality (8.4.5) means also that the multimatrix $(K_{j,k;m,n})_{j,k,m,n=0}^{\infty}$ with coefficients $K_{j,k;m,n} = s_{j+m,k+n}$ is non-negative definite and, therefore,

$$|s_{j+n,k+m}|^2 = |K_{j,k;m,n}|^2 \leq K_{j,k;j,k} K_{m,n;m,n} = s_{j+k,j+k} s_{n+m,n+m}, \tag{8.4.13}$$

for $j, k, m, n \in \mathbb{N}_0$. Let the weight $q = (q_{j,k})_{j,k=0}^{\infty}$, $q_{j,k} \geq 1$, be such that $\sum_{j,k=0}^{\infty} s_{j+k,j+k} q_{j,k}^{-1} < \infty$. Then from (8.4.13), we get

$$\|f\|_S^2 = \sum_{j,k,m,n=0}^{\infty} f_{j,k} \bar{f}_{m,n} s_{j+m,k+n} \leq \left(\sum_{j,k=0}^{\infty} \frac{s_{j+k,j+k}}{q_{j,k}} \right) \|f\|_{l_2(q)}^2, \quad f \in l_{\text{fin}}.$$

Therefore, the imbedding $l_2(q) \hookrightarrow S$ is topological. And hence, if the inequality $\sum_{j,k=0}^{\infty} q_{j,k} p_{j,k}^{-1} < \infty$ holds true, then $l_2(p) \hookrightarrow l_2(q)$ is quasi-nuclear. The composition $l_2(p) \hookrightarrow S$ of the quasi-nuclear and topological imbedding is also quasi-nuclear one.

In the next step, we will use the chain (8.4.12) to construct generalized eigenvectors. The inner structure of the space $(l_2(p))_{-,S}$ is complicated, because of the complicated structure of S. This is a reason to introduce new auxiliary rigging

$$l = (l_{\text{fin}})' \supset (l_2(p^{-1})) \supset l_2 \supset l_2(p) \supset l_{\text{fin}}, \tag{8.4.14}$$

where $l_2(p^{-1})$, $p^{-1} = (p_{m,n}^{-1})_{m,n=0}^{\infty}$ is the negative space with respect to the positive space $l_2(p)$ and the zero space l_2. Chains (8.4.12) and (8.4.14) have the same positive space $l_2(p)$. The general Lemma 2.7.5 establishes that the space $(l_2(p))_{-,S}$ is isometric to the space $l_2(p^{-1})$.

Chains (8.4.12) and (8.4.14) will play the role of riggings (2.7.35).

Obviously that operators A and B are standardly connected with the chain (8.4.14),and the vector $\Omega = \delta_{0,0} \in l_{\text{fin}}$ is strong cyclic for operators of A and B, hence we can apply Theorem 8.4.1. Let $\xi_{x,y} \in (l_2(p))_{-,S}$ be a generalized eigenvector of operators A and B. Thus, in this case due to Theorem 8.4.1, we have

$$\begin{aligned} (\xi_{x,y}, Af)_S &= x(\xi_{x,y}, f)_S, \\ (\xi_{x,y}, Bf)_S &= y(\xi_{x,y}, f)_S, \quad (x, y) \in \mathbb{R}^2, \ f \in l_{\text{fin}}. \end{aligned} \tag{8.4.15}$$

Denote

$$P(x, y) = U\xi_{x,y} \in l_2(p^{-1}) \subset l, \quad P(x, y) = (P_{m,n}(x, y))_{m,n=0}^{\infty}.$$

By using (2.7.36), we can rewrite (8.4.15) in the form

$$(P(x,y), Af)_{l_2} = x(P(x,y), f)_{l_2},$$
$$(P(x,y), Bf)_{l_2} = y(P(x,y), f)_{l_2}, \quad (x,y) \in \mathbb{R}^2, \; f \in l_{\text{fin}}. \tag{8.4.16}$$

The corresponding Fourier transform has the form

$$S \supset l_{\text{fin}} \ni f \to (Ff)(x,y) = \hat{f}(x,y) = (f, P(x,y))_{l_2} \in L_2(\mathbb{R}^2, d\rho(x,y)). \tag{8.4.17}$$

Let us calculate $P(x,y)$. The operator A is generated by the rule (8.4.8), hence (8.4.16) gives

$$\sum_{m,n=0}^{\infty} x P_{m,n}(x,y)\bar{f}_{m,n} = x(P(x,y), f)_{l_2} = (P(x,y), Af)_{l_2}$$
$$= \sum_{m,n=0}^{\infty} P_{m+1,n}(x,y)\bar{f}_{m,n}, \;\forall f \in l_{\text{fin}}. \tag{8.4.18}$$

Similarly, using (8.4.8) and (8.4.16), we have

$$\sum_{m,n=0}^{\infty} y P_{m,n}(x,y)\bar{f}_{m,n} = y(P(x,y), f)_{l_2} = (P(x,y), Bf)_{l_2}$$
$$= \sum_{m,n=0}^{\infty} P_{m,n+1}(x,y)\bar{f}_{m,n}, \;\forall f \in l_{\text{fin}}. \tag{8.4.19}$$

Hence, we have

$$x P_{m,n}(x,y) = P_{m+1,n}(x,y), \quad y P_{m,n}(x,y) = P_{m,n+1}(x,y), \quad m,n \in \mathbb{N}_0.$$

Without loss of generality, we can take $P_{0,0}(x,y) = 1, \; (x,y) \in \mathbb{R}^2$. Then last two equalities give

$$P_{m,n}(x,y) = x^m y^n, \quad m,n \in \mathbb{N}_0. \tag{8.4.20}$$

Thus, the Fourier transform(8.4.17) has the form

$$S \supset l_{\text{fin}} \ni f \to (Ff)(x,y) = \hat{f}(x,y) = \sum_{m,n=0}^{\infty} f_{m,n} x^m y^n \in L_2(\mathbb{R}^2, d\rho(x,y)), \tag{8.4.21}$$

and

$$(f, g)_S = \int_{\mathbb{R}^2} \hat{f}(x, y)\overline{\hat{g}(x, y)}\, d\rho(x, y), \quad f, g \in l_{\text{fin}}. \tag{8.4.22}$$

To construct the Fourier transform (8.4.17) and use formulas (8.4.18)–(8.4.22) there is still necessary to check that $\Omega = \delta_{0,0} \in l_{\text{fin}}$ is the strong cyclic vector in the sense of the chain(8.4.12) for our operators A and B. But it is evidently true, since due to (8.4.9) we have: $A^p B^q \Omega = J_A^p J_B^q \delta_{0,0} = \delta_{p,q}$.

The Parseval equality (8.4.22) immediately leads to the representation (8.4.4) according to (8.4.20), (8.4.21): $\hat{\delta}_{m,n} = x^m y^n$ and $\hat{\delta}_{0,0} = 1$; from (8.4.10), for $m, n \in \mathbb{N}_0$, we get:

$$s_{m,n} = (\delta_{m,n}, \delta_{0,0})_S = (\hat{\delta}_{m,n}, \hat{\delta}_{0,0})_{L_2(\mathbb{R}^2, d\rho(x,y))} = \int_{\mathbb{R}^2} x^m y^n\, d\rho(x, y). \tag{8.4.23}$$

The uniqueness of the representation (8.4.4) follows from the self-adjointness and the commutativity of operators A and B in the strong resolvent sense. Thus, for the finish of the proof of Theorem 8.4.1 it is necessary only to check that the condition (8.4.6) provides the self-adjointness and the commutativity of A and B. In the next step we use Theorem 2.6.11. But for this (see Theorems 2.6.8 and 2.6.11) we must only check that the operator $A^2 + B^2$ has a total set $\mathcal{D}$ of quasi-analytic vectors.

Due to (8.4.8), Theorem 2.6.8, Lemma 2.6.7, and, in particular, (2.6.21), the operator $\mathcal{A} = A^2 + B^2$ acts on $\delta_{m,n} \in \mathcal{D}$ by the rule

$$\mathcal{A}\delta_{m,n} = (A^2 + B^2)\delta_{m,n} = \delta_{m+2,n} + \delta_{m,n+2}. \tag{8.4.24}$$

It is obvious that $\mathcal{A} \geq 0$. For $p \geq 1$ we have: $\mathcal{A}^p \delta_{m,n} = \sum_{k=0}^{p} C_p^k \delta_{m+2p-k,n+2k}$. According to (8.4.10), $\|f\|_S = \sqrt{(f, f)_S}$ is the norm in S. Hence, $\forall \delta_{m,n} \in \mathcal{D}$ we get

$$\begin{aligned} \|\mathcal{A}^p \delta_{m,n}\|_S &= \Big\|\sum_{k=0}^{p} C_p^k \delta_{m+2(p-k),n+2k}\Big\|_S \\ &\leq \sum_{k=0}^{p} C_p^k \|\delta_{m+2(p-k),n+2k}\| \\ &= \sum_{k=0}^{p} C_p^k \sqrt{s_{2m+4p-4k,2n+4k}}. \end{aligned} \tag{8.4.25}$$

Since

$$\sum_{p=1}^{\infty} \frac{1}{\sqrt[p]{\|\mathcal{A}^p \delta_{m,n}\|}} \geq \sum_{p=1}^{\infty} \frac{1}{\sqrt[p]{\sum_{k=0}^{p} C_p^k \sqrt{s_{2m+4p-4k,2n+4k}}}} = \infty, \ m, n \in \mathbb{N}_0,$$

then we conclude that the quasi-analyticity of the class $\mathrm{C}\{\|\mathcal{A}^p \delta_{m,n}\|\}$ follows from the quasi-analyticity of the class

$$\mathrm{C}\{\sqrt{\sum_{k=0}^{p} C_p^k \sqrt{s_{2m+4p-4k,2n+4k}}}\}$$

due to the quasi-analyticity properties. And this, in turn, is equivalent to the quasi-analyticity of the class

$$\mathrm{C}\{\sqrt{\sum_{k=0}^{p} C_p^k \sqrt{s_{4p-4k,4k}}}\}.$$

But this quasi-analyticity gives the condition (8.4.6), taking into account (8.4.25). This completes the proof of Theorem 8.4.2. □

Remark 8.4.4 The condition (8.4.5) is only necessary in Theorem 8.4.2. For the clearness we propose the simple contr-example. Let A and B be two Hermitian operators commuting on a linear set $\mathcal{D}$ which is dense in the Hilbert space $\mathcal{H}$ and it is invariant with respect to the action of A and B. Suppose that A and B are essentially self-adjoint on $\mathcal{D}$, i.e., $A = (A \upharpoonright \mathcal{D})^{\sim}$, $B = (B \upharpoonright \mathcal{D})^{\sim}$, but A and B are not commutative in the strong resolvent sense. The existence of such sort operators guarantees the Nelson example. By using such sort operators we can construct a sequence $s_{m,n} = (A^m B^n \delta_{0,0}, \delta_{0,0})$ that has no representation (8.4.4).

Remark 8.4.5 The condition (8.4.6) is new, but there are known many other conditions like (8.4.6), that guarantee the unfitness of the measure in Theorem 8.4.2.

Remark 8.4.6 To obtain the representation (8.4.4) we use also the Beresansky Yu. M. method of generalized eigenfunctions expansion. One say that we recovered the measure $d\rho(x, y)$. But we don't write any distribution function, that all would like to have. Actually, we need the measure for calculations of integrals with such a measure. And the important expression (8.4.23) gives us such a possibility. If we have to calculate the integral as in (8.4.22) for some functions $\hat{f}(x, y)$ and $\hat{g}(x, y)$, then we must to decompose $\hat{f}(x, y)$ and $\hat{g}(x, y)$ into series of polynomials of variables $x^n y^m$ and using given $(s_{m,n})$ and (8.4.23) to write(8.4.22) as a corresponding series. The correctness of the last conclusion is guaranteed by the Beresansky Yu. M. method.

The last remark is also relevant to all previous considered versions of the moment problem.

8.5 Examples of Matrices Corresponding to the Two-Dimensional Real Moment Problem

Let us find a condition that guarantees that the matrix J_A commutes with the matrix J_B. Multiplying these matrices, we get

$$J_A J_B = \begin{bmatrix} b_0 w_0 + c_0 v_0 & b_0 v_0 + c_0 w_1 & c_0 v_1 & 0 & 0 & \cdots \\ a_0 w_0 + b_1 u_0 & a_0 v_0 + b_1 w_1 + c_1 u_1 & b_1 v_1 + c_1 w_2 & c_1 v_2 & 0 & \cdots \\ a_1 u_0 & a_1 w_1 + b_2 u_1 & a_1 v_1 + b_2 w_2 + c_2 u_2 & b_2 v_2 + c_2 w_3 & c_2 v_3 & \cdots \\ \vdots & \vdots & \vdots & \vdots & \vdots & \ddots \end{bmatrix}. \tag{8.5.1}$$

The expression for $J_A J_B$ is obtained similarly to (8.5.1) with the change of a_n, b_n, and c_n to u_n, w_n, and v_n, respectively, and vice versa. Comparing the expressions for $J_A J_B$ and $J_B J_A$, we find out that the equality $J_A J_B = J_B J_A$ is equivalent to the fulfillment of such equalities (take into account that b_0 is a scalar, and put $w_0 = b_0$)

$$\begin{aligned} & c_0 u_0 = v_0 a_0; \\ & c_n v_{n+1} = v_n c_{n+1}, \\ & b_n v_n + c_n w_{n+1} = w_n c_n + v_n b_{n+1}, \\ & a_n v_n + b_{n+1} w_{n+1} + c_{n+1} u_{n+1} = u_n c_n + w_{n+1} b_{n+1} + v_{n+1} a_{n+1}, \quad n \in \mathbb{N}_0. \end{aligned} \tag{8.5.2}$$

Equalities

$$a_n w_n + b_{n+1} u_n = u_n b_n + w_{n+1} a_n, \quad a_{n+1} u_n = u_{n+1} a_n, \quad n \in \mathbb{N}_0$$

follow from third and second equalities of (8.5.2) writing their in adjoint form.

Thus, conditions (8.5.2) are necessary and sufficient for the matrix equality $J_A J_B = J_B J_A$. When norms of operators a_n, b_n, c_n and u_n, w_n, v_n are uniformly bounded with respect to $n \in \mathbb{N}_0$, then operators J_A and J_B are bounded, self-adjoint on $\mathbf{l}_2$ and conditions (8.5.2) gives the mutual commutativity of these operators.

Taking matrices a_0, b_0, c_0 as initial we can finding from (8.5.2) step by step a_1, b_1, c_1; a_2, b_2, c_2; ... etc. (but in non-uniquely manner).

The finding of matrices a_n, b_n, c_n and u_n, w_n, v_n, $n \in \mathbb{N}_0$, which are solutions of Eq. (8.5.2) such that a_n, c_n, and u_n, v_n have the form (8.1.26) and (8.1.44), respectively, is an enough complicated problem and we investigate here only some special cases.

Namely, we assume that for all matrices $w_n = b_n$, $n \in \mathbb{N}_0$. Then conditions (8.5.2) possess the form

$$\begin{aligned} &c_0 u_0 = v_0 a_0; \\ &c_n v_{n+1} = v_n c_{n+1}, \\ &b_n(v_n - c_n) = (v_n - c_n) b_{n+1}, \\ &a_n v_n + c_{n+1} u_{n+1} = u_n c_n + v_{n+1} a_{n+1}, \quad n \in \mathbb{N}_0. \end{aligned} \tag{8.5.3}$$

And if, additionally, put $w_n = b_n = 0$, then

$$\begin{aligned} &c_0 u_0 = v_0 a_0; \\ &c_n v_{n+1} = v_n c_{n+1}, \\ &a_n v_n + c_{n+1} u_{n+1} = u_n c_n + v_{n+1} a_{n+1}, \quad n \in \mathbb{N}_0. \end{aligned} \tag{8.5.4}$$

Furthermore, we assume that all matrices a_n, c_n, and u_n, v_n, $n \in \mathbb{N}_0$ have the form (8.1.26) and (8.1.44), where $a_{n;1,0}$, $a_{n;2,1}$, $\dots$, $a_{n;n+1,n}$, $c_{n;0,1}$, $c_{n;1,2}$, $\dots$, $c_{n;n,n+1}$, $u_{n;0,0}$, $u_{n;1,1}$, $\dots$, $u_{n;n,n}$, $v_{n;0,0}$, $v_{n;1,1}$, $\dots$, $v_{n;n,n}$, $\forall n \in \mathbb{N}_0$ are positive and all another elements of these matrices are equal to zero. Thus, for all $n \in \mathbb{N}_0$), blocks of our matrices have the form

$$u_n = \underbrace{\left.\begin{bmatrix} u_{n;0,0} & 0 & \dots & 0 \\ 0 & u_{n;1,1} & \dots & 0 \\ \vdots & \vdots & \ddots & \vdots \\ 0 & 0 & \dots & u_{n;n,n} \\ 0 & 0 & \dots & 0 \end{bmatrix}\right\} n+2}_{n+1}, \quad a_n = \underbrace{\left.\begin{bmatrix} 0 & 0 & \dots & 0 \\ a_{n;1,0} & 0 & \dots & 0 \\ 0 & a_{n;2,1} & \dots & 0 \\ \vdots & \vdots & \ddots & \vdots \\ 0 & 0 & \dots & a_{n;n+1,n} \end{bmatrix}\right\} n+1}_{n+2},$$

$$v_n = \underbrace{\left.\begin{bmatrix} v_{n;0,0} & 0 & \dots & 0 & 0 \\ 0 & v_{n;1,1} & \dots & 0 & 0 \\ \vdots & \vdots & \ddots & \vdots & \vdots \\ 0 & 0 & \dots & v_{n;n,n} & 0 \end{bmatrix}\right\} n+1}_{n+2}, \quad c_n = \underbrace{\left.\begin{bmatrix} 0 & c_{n;0,1} & 0 & \dots & 0 \\ 0 & 0 & c_{n;1,2} & \dots & 0 \\ 0 & \vdots & \vdots & \ddots & \vdots \\ 0 & 0 & 0 & \dots & c_{n;n,n+1} \end{bmatrix}\right\} n+2}_{n+1},$$

(8.5.5)

Multiplying matrices of the type (8.5.5) we can rewrite the first, second and fourth equality of (8.5.4) in the form of corresponding scalar equalities for elements of these matrices

$$
\begin{aligned}
v_{0;0,0}c_{1;0,1} &= v_{1;1,1}c_{0;0,1},\\
v_{1;0,0}c_{2;0,1} &= v_{2;1,1}c_{1;0,1},\\
v_{1;1,1}c_{2;1,2} &= v_{2;2,2}c_{1;1,2},\\
v_{2;0,0}c_{3;0,1} &= v_{3;1,1}c_{2;0,1},\\
v_{2;1,1}c_{3;1,2} &= v_{3;2,2}c_{2;1,2},\\
v_{2;2,2}c_{3;2,3} &= v_{3;3,3}c_{2;2,3},\\
&\dots\\
v_{n;0,0}c_{n+1;0,1} &= v_{n+1;1,1}c_{n;0,1},\\
v_{n;1,1}c_{n+1;1,2} &= v_{n+1;2,2}c_{n;1,2},\\
&\dots\\
v_{n;n,n}c_{n+1;n,n+1} &= v_{n+1;n+1,n+1}c_{n;n,n+1},
\end{aligned}
\tag{8.5.6}
$$

$$
\begin{aligned}
v_{0;0,0}c_{0;0,1} &= v_{1;1,1}c_{1;0,1},\\
v_{1;0,0}c_{1;0,1} &= v_{2;1,1}c_{2;0,1},\\
v_{1;1,1}c_{1;1,2} &= v_{2;2,2}c_{2;1,2},\\
v_{2;0,0}c_{2;0,1} &= v_{3;1,1}c_{3;0,1},\\
v_{2;1,1}c_{2;1,2} &= v_{3;2,2}c_{3;1,2},\\
v_{2;2,2}c_{2;2,3} &= v_{3;3,3}c_{3;2,3},\\
&\dots\\
v_{n;0,0}c_{n;0,1} &= v_{n+1;1,1}c_{n+1;0,1},\\
v_{n;1,1}c_{n;1,2} &= v_{n+1;2,2}c_{n+1;1,2},\\
&\dots\\
v_{n;n,n}c_{n;n,n+1} &= v_{n+1;n+1,n+1}c_{n+1;n,n+1},
\end{aligned}
\tag{8.5.7}
$$

where $v_{n;k,k} = u_{n;k,k}$ and $c_{n;k,k+1} = a_{n;k+1,k}$, $k = 0, 1, \dots, n$, $\forall n \in \mathbb{N}_0$. The system of equalities (8.5.6), (8.5.7) is equivalent to the system (8.5.3) for the case when $b_n = w_n = 0$, $n \in \mathbb{N}_0$.

Each system, consisting of two equations in each line in (8.5.6), (8.5.7), using the positivity of $v_{n;k,k}$ and $c_{n;k,k+1}$, $k = 0, 1, \dots, n$, $\forall n \in \mathbb{N}_0$, gives

$$\begin{array}{rlll} c_{1;0,1} = c_{0;0,1}, & c_{2;0,1} = c_{1;0,1}, & \dots, & c_{n+1;0,1} = c_{n;0,1}; \\ & c_{2;1,2} = c_{1;1,2}, & \dots, & c_{n+1;1,2} = c_{n;1,2}; \\ & & \dots, & \dots \\ & & \dots, & c_{n+1;n,n+1} = c_{n;n,n+1}. \end{array} \tag{8.5.8}$$

Let us combine the latest research into such an assertion.

Lemma 8.5.1 *Matrices J_A and J_B of the form* (8.1.26) *and* (8.1.44) *with elements a_n, c_n and u_n, v_n as in* (8.5.5) *and $b_n = w_n = 0$, $n \in \mathbb{N}_0$, are symmetric and commuting if for an arbitrary $v_{n;k,k} = u_{n;k,k} < c < \infty$, $k = 0, 1, \dots, n$ numbers $c_{n;k,k+1} = c_{n+1;k,k+1}$, $k = 0, 1, \dots, n$, $\forall n \in \mathbb{N}_0$ satisfy equalities* (8.5.8).

Proof For the construction of J_A and J_B there is suffices to choose the bounded sequence (δ_n), $|\delta_n| < c_1 < \infty$, $n \in \mathbb{N}_0$ and to put $c_{n;k,k+1} := \delta_n$, $k = 0, 1, \dots, n$, $\forall n \in \mathbb{N}_0$. □

Example 8.5.2 Take $b_n = w_n = 0$ and for (8.5.5) $v_{n;k,k} = u_{n;k,k} = c_{n;k,k+1} = a_{n;k+1,k} = 1$, $k = 0, 1, \dots, n$, $\forall n \in \mathbb{N}_0$. By the obvious observation we obtain matrices of the type J_A and J_B, that satisfy all conditions described in Theorems 8.1.5 and 8.1.9. Let us denote such matrices by J_{A1} and J_{B1}.

Example 8.5.3 By using previous example let us put $J_{A11} := J_{A1} + J_{B1}$. This matrix: (1) is obviously symmetric (as a sum of two symmetric ones); (2) is a bounded operator (as a sum of two bounded operators); (3) commutes with J_{B1}, since J_{A1} commutes with J_{B1}, and J_{B1} commutes with it self.

Example 8.5.4 Consider the degenerate two-dimensional moment problem in the case, i.e., the double integral decomposes into the product of two one-dimensional integrals:

$$s_{m,n} = \int\limits_{\mathbb{R}^2} x^m y^n \, d\rho(x, y) = a_m b_n = \int\limits_{\mathbb{R}} x^m d\rho_1(x) \int\limits_{\mathbb{R}} y^n \, d\rho_2(y), \qquad m, n \in \mathbb{N}_0.$$

Hence, we get two classical moment problems of Hamburger, each of which is associated with the usual Jacobi matrix:

$$J_a = \begin{bmatrix} \beta_0 & \alpha_0 & 0 & 0 & 0 & \dots \\ \alpha_0 & \beta_1 & \alpha_1 & 0 & 0 & \dots \\ 0 & \alpha_1 & \beta_2 & \alpha_2 & 0 & \dots \\ 0 & 0 & \alpha_2 & \beta_3 & \alpha_3 & \dots \\ \vdots & \vdots & \vdots & \vdots & \vdots & \ddots \end{bmatrix}, \quad J_b = \begin{bmatrix} \delta_0 & \gamma_0 & 0 & 0 & 0 & \dots \\ \gamma_0 & \delta_1 & \gamma_1 & 0 & 0 & \dots \\ 0 & \gamma_1 & \delta_2 & \gamma_2 & 0 & \dots \\ 0 & 0 & \gamma_2 & \delta_3 & \gamma_3 & \dots \\ \vdots & \vdots & \vdots & \vdots & \vdots & \ddots \end{bmatrix}. \tag{8.5.9}$$

In this case, matrices (8.1.26) and (8.1.44) now have the form

$$
J_A = \left[\begin{array}{c|cc|ccc|cccc|ccccc}
\beta_0 & 0 & \alpha_0 & & & & & & & & & & & & \\
0 & \beta_0 & 0 & 0 & \alpha_0 & 0 & & & & & & & & & \\
\alpha_0 & 0 & \beta_1 & 0 & 0 & \alpha_1 & & & & 0 & & & & & \\
 & 0 & 0 & \beta_0 & 0 & 0 & 0 & \alpha_0 & 0 & 0 & & & & & \\
 & \alpha_0 & 0 & 0 & \beta_1 & 0 & 0 & 0 & \alpha_1 & 0 & & & & & \\
 & 0 & \alpha_1 & 0 & 0 & \beta_2 & 0 & 0 & 0 & \alpha_2 & & & & & \\
 & & & 0 & 0 & 0 & \beta_0 & 0 & 0 & 0 & 0 & \alpha_0 & 0 & 0 & 0 \\
 & & & \alpha_0 & 0 & 0 & 0 & \beta_1 & 0 & 0 & 0 & 0 & \alpha_1 & 0 & 0 \\
 & 0 & & 0 & \alpha_1 & 0 & 0 & 0 & \beta_2 & 0 & 0 & 0 & 0 & \alpha_2 & 0 \\
 & & & 0 & 0 & \alpha_2 & 0 & 0 & 0 & \beta_3 & 0 & 0 & 0 & 0 & \alpha_3 \\
 & & & & \ddots & & & & \ddots & & & & \ddots & &
\end{array}\right],
$$

$$
J_B = \left[\begin{array}{c|cc|ccc|cccc|ccccc}
\delta_0 & \gamma_0 & 0 & & & & & & & & & & & & \\
\gamma_0 & \delta_1 & 0 & \gamma_1 & 0 & 0 & & & & & & & & & \\
0 & 0 & \delta_0 & 0 & \gamma_0 & 0 & & & & 0 & & & & & \\
 & \gamma_1 & 0 & \delta_2 & 0 & 0 & \gamma_2 & 0 & 0 & 0 & & & & & \\
 & 0 & \gamma_0 & 0 & \delta_1 & 0 & 0 & \gamma_1 & 0 & 0 & & & & & \\
 & 0 & 0 & 0 & 0 & \delta_0 & 0 & 0 & \gamma_0 & 0 & & & & & \\
 & & & \gamma_2 & 0 & 0 & \delta_3 & 0 & 0 & 0 & \gamma_3 & 0 & 0 & 0 & 0 \\
 & & & 0 & \gamma_2 & 0 & 0 & \delta_2 & 0 & 0 & 0 & \gamma_2 & 0 & 0 & 0 \\
 & 0 & & 0 & 0 & \gamma_0 & 0 & 0 & \delta_1 & 0 & 0 & 0 & \gamma_1 & 0 & 0 \\
 & & & 0 & 0 & 0 & 0 & 0 & 0 & \delta_0 & 0 & 0 & 0 & \gamma_0 & 0 \\
 & & & & \ddots & & & & \ddots & & & & \ddots & &
\end{array}\right].
$$

Note that polynomials orthogonal on the real plane are also decomposed into polynomials of separated variables x and y.

Bibliographical Notes

The two-dimensional real power moment problem is a direct generalization of the one-dimensional one, detailed in the works of Akhiezer N. I. [2, 3]. To obtain the integral representation of the two-dimensional moment problem, the method of the generalized eigenvectors expansion for a normal operator (namely the pair of commutative self-adjoint operators) is used, which was developed by Berezansky Yu. M. in [17, 18], and the spectral theory of sets of self-adjoint commutative operators Samoilenko Yu. S. [250]. The main content of the chapter is in works Dudkin M. E., Kozak V. [87, 88] and [89–92, 172]. Historically, the matrices corresponding to the real two-dimensional power moment problem were obtained by Gekhtman M. I.,

Kalyuzhnyĭ A. A. [107]. The two-dimensional moment problem was also considered by Èskin G. I. [101], Zarhina R. B. [299]. In the case of an indeterminate moment problem, there are issues of commutative extensions of symmetric operators, which are considered in the work of Bokhonov Yu. E. [57], Vainerman L. I. [287]. Orthogonal polynomials are generally known from works Geronimus Ya. L. [117], Szegö G. [278], Suetin P. K. [273, 274]. Deficiency indices of symmetric operators generated by block matrices were studied by Dyukarev Yu. M. [97]. The theory of equipped spaces can use works Koshmanenko V. D. and Dudkin M. E. [171]. By considering the issues of uniqueness of the integral representation, there used works Carleman T. [64], Nelson E. [225], Nussbaum A. E. [231, 232]. The Weyl function for the case of a pair of commuting self-adjoint operators is obtained as a generalization from the work of Brasche J. F., Malamud M. M., Neidhardt H. [59], where the Weyl function in the case of a single operator was considered; also in Gesztesy F., Simon B. [120]. The integral representation of the moment problem closely related to the two-dimensional moment problem is considered by Devinatz A. [81]. The commutativity of unbounded self-adjoint operators was considered by Schmüdgen K. [254]. He, in particular, considers two- and multi-dimensional moment problem [255], and also Kostyučenko A. G., Mityagin B. S. [168] . The construction of Jacobi type block matrices corresponding to the strong two-dimensional real moment problem can be found in the work of Kozak V. I. [172]. Difference equations and polynomials corresponding to multi-dimensional real moment problems are discussed by Xu Y. [296, 297]. Multi-dimensional moment problems have also been investigated in papers Haviland E. K. [135, 136], Fuglede B. [105], Petersen L. C. [233].

Chapter 9
Applications of the Spectral Theory of Jacobi Matrices and Their Generalizations to the Integration of Nonlinear Equations

In this chapter, the difference analogue of the well-known procedure for finding solutions of the Cauchy problem for some nonlinear partial differential equations, in particular the Korteweg-de-Vries (KdV) equation, is presented, using the inverse spectral problem for the Sturm-Liouville equation on the half-axis. More precisely, we consider the KdV, i.e., the partial differential equation

$$q_t(x,t) - 6q(x,t)q_x(x,t) + q_{xxx}(x,t) = 0, \quad (x,t) \in [0,\infty) \times [0,\infty)$$

with an unknown function $q(x,t)$, where

$$q_x = \frac{\partial q}{\partial x}, \quad q_t = \frac{\partial q}{\partial t}.$$

The Cauchy problem for this equation is as follows: to find the solution $q(x,t)$ for given initial data $q(x,0)$, $x \in [0,\infty)$. It is known that, according to a certain procedure, the inverse spectral problem for the Sturm-Liouville equation on the semi-axis $[0,\infty)$ with the potential $q(x,t)$ for fixed t, makes it possible to construct this solution of the Cauchy problem for the KdV equation.

The difference analog of this procedure is that, by solving the inverse spectral problem for the Jacobi matrix (that is, for the difference equation), it is possible to construct the solution of the Toda chain as analog of the KdV. This chapter is devoted to this problem and its generalizations.

Note that we only touch upon the spectral theory of Jacobi and block Jacobi matrices. The scattering-type problem for the difference equations is not used, although it gives interesting new results (see bibliographical notes to the chapter), in particular, in Introduction.

Y. M. Berezansky, M. E. Dudkin, *Jacobi Matrices and the Moment Problem*,
Operator Theory: Advances and Applications 294,
https://doi.org/10.1007/978-3-031-46387-7_9

Toda Chains The classical Toda chain on the semi-axis $[0, T)$, $T \leq \infty$ ("the semi-finite chain") has the form

$$\begin{aligned} \dot{\alpha}_n(t) &= \tfrac{1}{2}\alpha_n(t)(\beta_{n+1}(t) - \beta_n(t)), \\ \dot{\beta}_n(t) &= \alpha_n^2(t) - \alpha_{n-1}^2(t), \qquad n \in \mathbb{N}_0, \ \alpha_{-1}(t) = 0, \end{aligned} \tag{9.0.1}$$

where $\alpha_n(t)$, $\beta_n(t)$ are real continuously differentiable functions of $t \in [0, T)$, "$\cdot$"$= d/dt$. The expression (9.0.1) is a system of differential-difference equations. For (9.0.1) the Cauchy problem is considered: given initial conditions $\alpha_n(0)$, $\beta_n(0)$, $n \in \mathbb{N}_0$, it is necessary to find a solution $\alpha_n(t)$, $\beta_n(t)$, $n \in \mathbb{N}_0$ for $t \in [0, T)$, $T \leq \infty$. It turns out that it can be found essentially in an explicit form, applying the spectral theory of classical Jacobi matrices, which is presented in Sects. 2.1–2.9.

In the case when n in (9.0.1) changes along the whole $\mathbb{Z} = \ldots, -1, 0, 1, \ldots$ ("the double-infinite chain") the situation worsens: the solution of this problem is reduced to the integration of a system of certain linear equations but with known variable coefficients. These results will be presented in Sects. 9.2–9.5. At the end of Sect. 9.5 we will discuss some problems related to (9.0.1). Finally, we present some generalizations of these results to broader systems of equations of the type (9.0.1), as well as cases where it is possible to make calculations.

Note also that the Toda chain on the axis $\mathbb{Z}$ is often written in a slightly different form, namely,

$$\begin{aligned} \dot{a}_n(t) &= a_n(t)(b_n(t) - b_{n+1}(t)), \\ \dot{b}_n(t) &= 2(a_{n-1}^2(t) - a_n^2(t), \qquad n \in \mathbb{Z}, \ t \in [0, T), \ T \leq \infty, \end{aligned} \tag{9.0.2}$$

of variables $a_n(t)$, $b_n(t)$. The form (9.0.1) turns into the form (9.0.2), with $n \in \mathbb{Z}$ by the transformation

$$a_n(t) = \frac{\alpha_n(t)}{\sqrt{2}}, \quad b_n(t) = -\beta(t).$$

9.1 The Integration of the Toda Chain on the Semi-axis Using the Spectral Theory of Jacobi Matrices

We will assume that functions $\alpha_n(t)$, $\beta_n(t)$ in (9.0.1) are uniformly bounded with respect to $n \in \mathbb{N}_0$ and $a_n(t)$ are positive, i.e., $\exists C \in [0, \infty)$ such that

$$a_n(t), |b_n(t)| < C, \quad \alpha_n(t) > 0, \quad t \in [0, T), \ n \in \mathbb{N}_0. \tag{9.1.1}$$

First of all, let us rewrite the Toda chain in the so-called form of the Lax equation. To do this, we introduce the Jacobi matrix $J(t)$ by $\alpha_n(t)$, $\beta_n(t)$ depending on t, namely,

$$J(t) = \begin{bmatrix} \beta_0(t) & \alpha_0(t) & 0 & 0 & 0 & \dots \\ \alpha_0(t) & \beta_1(t) & \alpha_1(t) & 0 & 0 & \dots \\ 0 & \alpha_1(t) & \beta_2(t) & \alpha_2(t) & 0 & \dots \\ \vdots & \vdots & \vdots & \vdots & \vdots & \ddots \end{bmatrix}, \quad t \in [0, T). \tag{9.1.2}$$

The matrix (9.1.2) generates in a classical way the Hermitian operator $L(t)$ in the space l_2 (see Sects. 2.1–2.3), which we will denote by $\mathbf{J}(t)$, i.e., $\mathbf{J}(t)$ is generated by the matrix $J(t)$ (9.1.2) for each fixed $t \in [0, T)$. The boundedness of $\alpha_n(t)$, $\beta_n(t)$ and their differentiability ensure that this operator $\mathbf{J}(t)$ is bounded self-adjoint and weakly differentiable with respect to $t \in [0, T)$.

Let us introduce the matrix

$$A(t) = \frac{1}{2} \begin{bmatrix} 0 & -\alpha_0(t) & 0 & 0 & 0 & \dots \\ \alpha_0(t) & 0 & -\alpha_1(t) & 0 & 0 & \dots \\ 0 & \alpha_1(t) & 0 & -\alpha_2(t) & 0 & \dots \\ \vdots & \vdots & \vdots & \vdots & \vdots & \ddots \end{bmatrix}, \quad t \in [0, T), \tag{9.1.3}$$

which generates in l_2, similar to $J(t)$, an antisymmetric bounded operator $\mathbf{A}(t)$ depending on $t \in [0, T)$.

The following classical result is well known. *Toda system* (9.0.1) *is equivalent to the corresponding Lax equation*

$$\dot{J}(t) = J(t)A(t) - A(t)J(t) \text{ or } \dot{\mathbf{J}}(t) = \mathbf{J}(t)\mathbf{A}(t) - \mathbf{A}(t)\mathbf{J}(t), \; t \in [0, T). \tag{9.1.4}$$

The prof of this fact is very simple. Let us calculate $J(t)A(t) - A(t)J(t)$ using (9.1.2) and (9.1.3). Then the first equality in (9.1.4) is the same as (9.0.1). For $\alpha_n = \alpha_n(t)$ and $\beta_n = \beta_n(t)$, $n \in \mathbb{N}_0$ the product of matrices (9.1.2) and (9.1.3) has the form

$$JA = \frac{1}{2} \begin{bmatrix} \alpha_0^2 & -\beta_0\alpha_0 & -\alpha_0\alpha_1 & 0 & 0 & 0 & \dots \\ \beta_1\alpha_0 & -\alpha_0^2 + \alpha_1^2 & -\beta_1\alpha_1 & -\alpha_1\alpha_2 & & 0 & \dots \\ \alpha_0\alpha_1 & \beta_2\alpha_1 & -\alpha_1^2 + \alpha_2^2 & -\beta_2\alpha_2 & -\alpha_2\alpha_3 & 0 & \dots \\ 0 & \alpha_1\alpha_2 & \beta_3\alpha_2 & -\alpha_2^2 + \alpha_3^2 & -\beta_3\alpha_3 & -\alpha_3\alpha_4 & \dots \\ \vdots & \vdots & \vdots & \vdots & \vdots & \vdots & \ddots \end{bmatrix},$$

where, the $(n+3)$-th row, starting from the $n+1$-th column, has elements

$$\left[\alpha_n\alpha_{n+1}\ \beta_{n+2}\alpha_{n+1}\ -\alpha_{n+1}^2+\alpha_{n+2}^2\ -\beta_{n+2}\alpha_{n+2}\ -\alpha_{n+2}\alpha_{n+3}\right],$$

which are non-zero identically. Similarly, the product of matrices (9.1.3) and (9.1.2) has the form

$$AJ=\frac{1}{2}\begin{bmatrix} -\alpha_0^2 & -\alpha_0\beta_1 & -\alpha_0\alpha_1 & 0 & 0 & 0 & \dots \\ \alpha_0\beta_0 & \alpha_0^2-\alpha_1^2 & -\alpha_1\beta_2 & -\alpha_1\alpha_2 & & 0 & \dots \\ \alpha_0\alpha_1 & \alpha_1\beta_1 & \alpha_1^2-\alpha_2^2 & -\alpha_2\beta_3 & -\alpha_2\alpha_3 & 0 & \dots \\ 0 & \alpha_1\alpha_2 & \alpha_2\beta_2 & \alpha_2^2-\alpha_3^2 & -\alpha_3\beta_4 & -\alpha_3\alpha_4 & \dots \\ \vdots & \vdots & \vdots & \vdots & \vdots & \vdots & \ddots \end{bmatrix},$$

where the $(n+3)$-th row, starting from the $(n+1)$-th column, has elements

$$\left[\alpha_n\alpha_{n+1}\ \alpha_{n+1}\beta_{n+1}\ \alpha_{n+1}^2-\alpha_{n+2}^2\ -\alpha_{n+2}\beta_{n+3}\ -\alpha_{n+2}\alpha_{n+3}\right],$$

which are non-zero identically. Then the commutator $[JA]$ has the form

$$JA-AJ=\frac{1}{2}\begin{bmatrix} 2\alpha_0^2 & \alpha_0(\beta_1-\beta_0) & 0 & 0 & \dots \\ \alpha_0(\beta_1-\beta_0) & 2(\alpha_1^2-\alpha_0^2) & \alpha_1(\beta_2-\beta_1) & 0 & \dots \\ 0 & \alpha_1(\beta_2-\beta_1) & 2(\alpha_2^2-\alpha_1^2) & \alpha_2(\beta_3-\beta_2) & \dots \\ \vdots & \vdots & \vdots & \vdots & \ddots \end{bmatrix},\tag{9.1.5}$$

where the $(n+3)$-th row, starting from the $(n+2)$-th column, has elements

$$\left[\alpha_{n+1}(\beta_{n+2}-\beta_{n+1})\ 2(\alpha_{n+2}^2-\alpha_{n+1}^2)\ \alpha_{n+2}(\beta_{n+3}-\beta_{n+2})\right],$$

which are non-zero identically. Comparing (9.1.5) and elements from (9.1.2) written with the sign of the derivative, we get (9.0.1).

Here, we will use other linear operators $\mathbf{A}, \mathbf{B}, \mathbf{C}, \dots$ in the space l_2; denote their matrices by $A, B, C, \dots$

Let us also introduce some more notations.

In the space l_2, consider a standard orthonormal basis of the form $\delta_n = (0, \dots, 0, 1, 0, 0, \dots)$, where "1" stands at the n-th position ($n \in \mathbb{N}_0$). Let $\mathbf{B} : l_2 \to l_2$ be some bounded operator, then the matrix $B = (b_{j,k})_{j,k=0}^{\infty}$ corresponds to it, where $b_{j,k} = (\mathbf{B}\delta_k, \delta_j)_{l_2} \in \mathbb{C}$, and

$$(\mathbf{B}f)_j = \sum_{k=0}^{\infty} B_{j,k} f_k, \quad f = (f_n)_{n=0}^{\infty} \in l_2, \ j \in \mathbb{N}_0. \tag{9.1.6}$$

And conversely, each matrix $(b_{j,k})_{j,k=0}^{\infty}$ with uniformly bounded $b_{j,k} \in \mathbb{C}$ corresponds, according to (9.1.6), to a bounded operator $\mathbf{B} : l_2 \to l_2$.

Consider the resolvent $\mathbf{R}_z(t) = (\mathbf{J}(t) - z\mathbf{1})^{-1}$, $z \in \mathbb{C} \setminus \mathbb{R}$ of the bounded Hermitian operator $\mathbf{J}(t)$ generated by the matrix (9.1.2) in the space l_2 (see also the condition (9.1.1)). According to (9.1.1),

$$R_z(t) = (R_{z;j,k}(t))_{j,k=0}^{\infty}, \; R_{z;j,k}(t) = (\mathbf{R}_z\delta_k, \delta_j)_{l_2}. \tag{9.1.7}$$

Our immediate goal will be the derivation of the differential equation for the zero matrix element of the matrix (9.1.7), that is, for the so-called Weyl function

$$m(z;t) = R_{z;0,0}(t), \quad z \in \mathbb{C} \setminus \mathbb{R}, \; t \in [0, T), \; m(\bar{z};t) = \overline{m(z;t)}. \tag{9.1.8}$$

Theorem 9.1.1 *The Weyl function $m(z;t)$ satisfies the differential equation*

$$\dot{m}(z;t) = (z - \beta_0(t))m(z;t) + 1, \quad z \in \mathbb{C} \setminus \mathbb{R}, \; t \in [0, T). \tag{9.1.9}$$

Proof We will use the well-known formula for the derivative by t of the inverse operator-valued function $[0, T) \ni t \to \mathbf{C}(t)$:

$$(\mathbf{C}^{-1}(t))^{\cdot} = -\mathbf{C}^{-1}(t)\dot{\mathbf{C}}(t)\mathbf{C}^{-1}(t), \quad t \in [0, T). \tag{9.1.10}$$

According to (9.1.4), we get

$$\begin{aligned}(\mathbf{J}(t) - z\mathbf{1})^{\cdot} &= \dot{\mathbf{J}}(t) = \mathbf{J}(t)\mathbf{A}(t) - \mathbf{A}(t)\mathbf{J}(t)\\ &= (\mathbf{J}(t) - z\mathbf{1})\mathbf{A}(t) - \mathbf{A}(t)(\mathbf{J}(t) - z\mathbf{1}), \quad \forall z \in \mathbb{C} \setminus \mathbb{R},\end{aligned} \tag{9.1.11}$$

hence, due to (9.1.10) and (9.1.11), we get

$$\begin{aligned}\dot{\mathbf{R}}_z(t) &= ((\mathbf{J}(t) - z\mathbf{1})^{-1})^{\cdot} = -\mathbf{R}_z(t)(\mathbf{J}(t) - z\mathbf{1})^{\cdot}\mathbf{R}_z(t)\\ &= -\mathbf{R}_z(t)((\mathbf{J}(t) - z\mathbf{1})\mathbf{A}(t) - \mathbf{A}(t)(\mathbf{J}(t) - z\mathbf{1}))\mathbf{R}_z(t)\\ &= -((\mathbf{A}(t) - \mathbf{A}(t)(\mathbf{J}(t) - z\mathbf{1}))\mathbf{R}_z(t) = \mathbf{R}_z(t)\mathbf{A}(t) - \mathbf{A}(t)\mathbf{R}_z(t)\end{aligned} \tag{9.1.12}$$

thus

$$\dot{\mathbf{R}}_z(t) = [\mathbf{R}_z(t), \mathbf{A}(t)], \quad t \in [0, T)$$

(we use the standard notation for the commutator of two operators or theirs corresponding matrices $[B, C] = BC - CB$).

Let us calculate $\dot{m}(z;t)$. Note, that $\mathbf{A}^*(t) = -\mathbf{A}(t)$, i.e., we have an equality $A^*(t) = -A(t)$, hence from (9.1.8) and (9.1.12), we get

$$\begin{aligned}\dot{m}(z;t) &= (R_{z;0,0})(t) = (\dot{\mathbf{R}}(t)\delta_0, \delta_0)_{l_2} = ((\mathbf{R}_z(t)\mathbf{A}(t) - \mathbf{A}(t)\mathbf{R}_z(t))\delta_0, \delta_0)_{l_2} \\ &= (\mathbf{R}_z\mathbf{A}(t)\delta_0, \delta_0)_{l_2} + (\mathbf{R}_z\delta_0, \mathbf{A}(t)\delta_0)_{l_2}.\end{aligned} \tag{9.1.13}$$

From (9.1.3), it follows

$$\mathbf{A}(t)\delta_0 = (0, \frac{1}{2}\alpha_0(t), 0, 0, \ldots) = \frac{1}{2}\alpha_0(t)\delta_1. \tag{9.1.14}$$

Equalities (9.1.13) and (9.1.14) give

$$\begin{aligned}\dot{m}(z;t) &= (\mathbf{R}_z(t)\mathbf{A}(t)\delta_0, \delta_0)_{l_2} + (\mathbf{R}_z(t)\delta_0, \mathbf{A}(t)\delta_0)_{l_2} \\ &= \frac{1}{2}\alpha_0(t)((\mathbf{R}_z(t)\delta_1, \delta_0)_{l_2} + (\mathbf{R}_z(t)\delta_0, \delta_1)_{l_2}), \ z \in \mathbb{C} \setminus \mathbb{R}, \ t \in [0, T).\end{aligned} \tag{9.1.15}$$

But $\mathbf{1} = (\mathbf{J}(t) - z\mathbf{1})\mathbf{R}_z(t)$, then due to the form (9.1.2) of the matrix $J(t)$ and (9.1.8), we have

$$1 = \sum_{k=0}^{\infty}(J(t) - z1)_{0,k}(R_z(t))_{k,0} = (J(t) - z1)_{0,0}m(z;t) + (J(t))_{0,1}(R_z(t))_{1,0}$$

$$= (\beta_0(t) - z)m(z;t) + \alpha_0(t)(\mathbf{R}_z(t))_{1,0}.$$

And, hence,

$$\begin{aligned}(\mathbf{R}_z(t))_{1,0} &= \alpha_0^{-1}(t)(1 + (z - \beta_0(t))m(z;t)) \\ &= (\mathbf{R}_z(t)\delta_1, \delta_0)_{l_2}, \ z \in \mathbb{C} \setminus \mathbb{R}, \ t \in [0, T).\end{aligned} \tag{9.1.16}$$

To calculate the second term in (9.1.15), using the relation (9.1.16), we obtain the expression

$$\begin{aligned}(\mathbf{R}_z(t)\delta_0, \delta_1)_{l_2} &= (\delta_0, \mathbf{R}_z^*(t)\delta_1)_{l_2} = (\delta_0, \mathbf{R}_{\bar{z}}(t)\delta_1)_{l_2} = \overline{(\mathbf{R}_{\bar{z}}(t)\delta_1, \delta_0)_{l_2}} = \\ &= \alpha_0^{-1}(t)(1 + (z - \beta_0(t))\overline{m(\bar{z};t)}) = (\mathbf{R}_z(t)\delta_1, \delta_0)_{l_2}.\end{aligned}$$

That is, these two terms are the same.

Thus, (9.1.15) can be rewritten, taking into account the equality (9.1.16), in the form

$$\dot{m}(z;t) = 1 + (z - \beta_0(t))m(z;t), \quad z \in \mathbb{C} \setminus \mathbb{R},\ t \in [0,T).$$

The equality (9.1.9) is proved. □

As already mentioned, the solution of the Cauchy problem for the Toda chain (9.0.1) will be obtained using the spectral problem for the Jacobi matrix (9.1.2), *where* $t \in [0,T)$ *is now fixed.* Let us recall the facts we need, set out in Sects. 2.1–2.7.

Due to (9.1.1), the operator $\mathbf{J}(t)$, which is generated by the matrix in the space l_2, is bounded and self-adjoint. Its spectral measure $\mathfrak{B}(\mathbb{R}) \ni \Delta \to \rho(\Delta;t) \geq 0$ is finite and given on Borel sets $\Delta \in \mathbb{R}$ (we will also denote it by $d\rho(\lambda;t)$). And it is probability: $\rho(\mathbb{R};t) = 1$. The corresponding resolution of the identity $\mathfrak{B}(\mathbb{R}) \ni \Delta \to E(\Delta;t)$ is defined by the spectral integral

$$E(\Delta;t) = \int_{\Delta} P(\lambda;t)\, d\rho(\alpha;t), \quad \Delta \in \mathfrak{B}(\mathbb{R}), \tag{9.1.17}$$

where $P(\lambda;t)$ is the generalized projection operator, which acts from the space $l_2(p_j(t))$ into $l_2(p_j^{-1}(t))$, where for each $t \in [0,T)$, $p_j(t) > 0$ and $\sum_{j=0}^{\infty} p_j^{-1}(t) < \infty$.

According to Sects. 2.1–2.5, the Jacobi matrix (9.1.2) is associated with its polynomials of the first kind $P_j(z;t)$, which are solutions of the difference equation

$$(J(t)u)_j = \alpha_{j-1}(t)u_{j-1} + \beta_j(t)u_j + \alpha_j(t)u_{j+1} = zu_j, \quad j \in \mathbb{N}_0 \tag{9.1.18}$$

with initial conditions $u_{-1} = 0$ and $u_0 = 1$. Polynomials of the first kind $P_j(\lambda;t)$ of the variable $\lambda \in \mathbb{R}$ form complete orthonormal system in the space $L_2(\mathbb{R}, d\rho(\lambda;t))$, namely,

$$\int_{\mathbb{R}} P_j(\lambda;t)P_k(\lambda;t)\, d\rho(\lambda;t) = \delta_{j,k};\ j,k \in \mathbb{N}_0, \tag{9.1.19}$$

where $\delta_{j,k}$ denotes the Kronecker symbol.

The Fourier transform (expansion) by the (generalized) eigenvectors of the operator $\mathbf{J}(t)$ has the form

$$l_{2,0} \ni u = (u_j)_{j=0}^{\infty} \mapsto \tilde{u}(\lambda;t) = \sum_{j=0}^{\infty} u_j P_j(\lambda;t) \in L_2(\mathbb{R}, d\rho(\lambda;t)). \tag{9.1.20}$$

During the transformation (9.1.20), the action of the operator $\mathbf{J}(t)$ on the vectors $l_{2,0}$ of the space l_2 turns into the multiplication by λ operator in the space

$L_2(\mathbb{R}, d\rho(\lambda; t))$. The operator (9.1.20) is isometric after the extension by continuity and acts between the spaces l_2 and $L_2(\mathbb{R}, d\rho(\lambda; t))$.

Additionally, we note that the formula (9.1.20) can be written for a subspace l_2 that is wider than $l_{2,0}$, namely for vectors from $l_2(p; t) \supset l_{2,0}$ with the previously specified conditions on the sequence $(p_j(t))_{j=0}^{\infty}$.

After these reminders from the spectral theory of Jacobi matrices $J(t)$ with fixed $t \in [0, T)$, we note another fact related to the dependence of the Jacobi matrices on t.

Due to the condition (9.1.1), the corresponding operator $J(t)$, $t \in [0, T)$ is bounded self-adjoint in the space l_2 and depends continuously on t in the operator norm.

From here it is easy to deduce that its spectral measure $\rho(\Delta; t)$ for each $\Delta \in \mathfrak{B}(\mathbb{R})$ is a continuously integral function of $t \in [0, T)$, and $\rho(\mathbb{R}) = 1, \forall t \in [0, T)$.

Consider the measure $d\sigma(\lambda)$ which is given by the formula

$$\mathfrak{B}(\mathbb{R}) \ni \Delta \mapsto \sigma(\Delta) = \int_0^T a(t)\rho(\Delta; t)\, dt \geq 0, \tag{9.1.21}$$

where $a(t)$ is a fixed positive continuous function for which there exists the integral from 0 to T with respect to Lebesgue measure dt and this integral is equal one. Obviously, each measure $\rho(\Delta; t)$ at $t \in [0, T)$ is absolutely continuous with respect to $d\sigma(\lambda)$. Consider its derivative

$$\begin{gathered} r(\lambda; t) = \frac{d\rho(\lambda; t)}{d\rho(\lambda)}, \quad \lambda \in \mathbb{R}, \quad \text{i.e.,} \\ \rho(\Delta; t) = \int_\Delta r(\lambda; t)\, d\sigma(\lambda), \quad \Delta \in \mathfrak{B}(\mathbb{R}),\ t \in [0, T). \end{gathered} \tag{9.1.22}$$

Our immediate goal is to rewrite equation (9.1.9) in terms of the function $r(\lambda; t)$.

Theorem 9.1.2 *For $d\rho(\lambda)$-almost all $\lambda \in \mathbb{R}$, the function $r(\lambda; t)$ is a continuously differentiable with respect to $t \in [0, T)$ solution of the differential equation*

$$\dot{r}(\lambda; t) = (\lambda - \beta_0(t))r(\lambda; t), \quad t \in [0, T). \tag{9.1.23}$$

Proof First of all, we note that from the representation of the resolvent $\mathbf{R}_z(t)$ of the operator $\mathbf{J}(t)$, due to its resolution of the identity, i.e. the equality

$$\mathbf{R}_z(t) = \int_{\mathbb{R}} \frac{1}{\lambda - z}\, dE(\lambda; t), \quad z \in \mathbb{C} \setminus \mathbb{R},\ t \in [0, T), \tag{9.1.24}$$

and (9.1.8), it follows (see (9.1.22)), that

$$m(z;t) = R_{z;0,0}(t) = \int_{\mathbb{R}} \frac{1}{\lambda - z}\, d\rho(\lambda;t)$$
$$= \int_{\mathbb{R}} \frac{1}{\lambda - z} r(\lambda;t)\, d\sigma(\lambda), \quad z \in \mathbb{C}\setminus\mathbb{R},\ t \in [0,T). \tag{9.1.25}$$

□

By the definition (9.1.22) of the function $r(\lambda;t)$, one cannot directly assert that this function is continuous with respect to t. Therefore, it will now be necessary to use the concept of a generalized solution of the Eq. (9.1.23) (see, for example, [48], chapter 16).

Consider the space $L_2([0,T), dt)$ relative to the Lebesgue measure dt on $[0,T)$ and the corresponding scalar product $(u,v)_{L_2}$. For an arbitrary infinitely differentiable function $[0,T) \ni t \mapsto u(t) \in \mathbb{C}$, which vanishes in some neighborhoods of 0 and T, according to (9.1.22) and (9.1.25), (9.1.9), we have

$$
\begin{aligned}
&-\int_{\mathbb{R}} \frac{1}{\lambda - z} (r(\lambda;\cdot), \dot{u}(\cdot))_{L_2}\, d\sigma(\lambda) \\
&\quad = -\int_{\mathbb{R}} \frac{1}{\lambda - z}\, d(\rho(\lambda;\cdot), \dot{u}(\cdot))_{L_2} \\
&\quad = -\,(m(z;\cdot), \dot{u}(\cdot))_{L_2} = (\dot{m}(z;\cdot), u(\cdot))_{L_2} \\
&\quad = (((z - \beta_0(\cdot))m(z;\cdot) + 1), u(\cdot))_{L_2} \\
&\quad = ((z - \beta_0(\cdot)) \int_{\mathbb{R}} \frac{1}{\lambda - z}\, d\rho(\lambda;\cdot) + \int_{\mathbb{R}} \frac{\lambda - z}{\lambda - z}\, d\rho(\lambda;\cdot), u(\cdot))_{L_2} \\
&\quad = ((\lambda - \beta_0(\cdot))m(z;\cdot), u(\cdot))_{L_2} \\
&\quad = \int_{\mathbb{R}} \frac{1}{\lambda - z} ((\lambda - \beta_0(\cdot))\, d\rho(\lambda;\cdot), u(\cdot))_{L_2}, \quad \forall z \in \mathbb{C}\setminus\mathbb{R}.
\end{aligned}
\tag{9.1.26}
$$

We will use a well-known fact. If $d\omega(\lambda)$ is some charge on $\mathfrak{B}(\mathbb{R})$ with a bounded support, then the condition

$$\int_{\mathbb{R}} (\lambda - z)^{-1}\, d\omega(\lambda) = 0, \quad z \in \mathbb{C}\setminus\mathbb{R}$$

is equivalent to the equality $d\omega(\lambda) = 0$. By using this fact, we deduce from equality (9.1.26) that for $d\rho(\lambda)$-almost all $\lambda \in \mathbb{R}$, we have

$$-(r(\lambda;\cdot), \dot{u}(\cdot))_{L_2} = ((\lambda - \beta_0(\cdot))r(\lambda;\cdot), u(\cdot))_{L_2}. \tag{9.1.27}$$

The relation (9.1.27) means that $r(\lambda; t)$ is a generalized solution of the equation (9.1.23). But for ordinary linear differential equations, every generalized solution is a smooth solution closely to the boundary, i.e., to 0. Thus, $r(\lambda; t)$ is a smooth solution of the equation (9.1.23) on $[0, T)$. □

For the equation (9.1.23) with an arbitrary fixed $\lambda \in \mathbb{R}$, it is easy to write the solution of the corresponding Cauchy problem

$$r(\lambda; t) = e^{\int_0^t (\lambda - \beta_0(s))\, ds} r(\lambda; 0) = e^{\lambda t} e^{-\int_0^t \beta_0(s)\, ds} r(\lambda; 0), \tag{9.1.28}$$

with initial conditions $r(\lambda, 0)$, where $t \in [0, T)$, $\lambda \in \mathbb{R}$.

Using (9.1.22) and (9.1.28) for the spectral function $\rho(\lambda; t) = \rho((-\infty, \lambda); t)$, $\lambda \in \mathbb{R}$ of the operator $\mathbf{J}(t)$, $t \in [0, T)$, we calculate the expression

$$\rho(\lambda; t) = \int_{-\infty}^{\lambda} r(\mu; t)\, d\sigma(\mu) = e^{-\int_0^t \beta_0(s)\, ds} \int_{-\infty}^{\lambda} e^{\mu t} r(\mu; 0)\, d\sigma(\mu)$$

$$= e^{-\int_0^t \beta_0(s)\, ds} \int_{-\infty}^{\lambda} e^{\mu t}\, d\rho(\mu; 0),$$

i.e.,

$$d\rho(\lambda; t) = e^{-\int_0^t \beta_0(s) ds} e^{\lambda}\, d\rho(\lambda; 0), \quad \lambda \in \mathbb{R},\ t \in [0, T). \tag{9.1.29}$$

This formula gives a rule for calculating the spectral measure of the operator $\mathbf{J}(t)$ by the spectral measure of the given initial operator $\mathbf{J}(0)$.

The function $\beta_0(t)$ cannot be arbitrary, since the measure $\rho(\mathbb{R}; t) = 1$, $\forall t \in [0, T)$. Therefore, it follows from (9.1.29) that

$$e^{\int_0^t \beta_0(s)\, ds} = \int_{\mathbb{R}} e^{\lambda t}\, d\rho(\lambda; 0), \quad t \in [0, T).$$

Differentiating this expression by t, we get

$$\beta_0(t) = \left(\int_{\mathbb{R}} e^{\lambda t} \, d\rho(\lambda; 0) \right)^{-1} \int_{\mathbb{R}} \lambda e^{\lambda t} \, d\rho(\lambda; 0), \quad t \in [0, T). \tag{9.1.30}$$

Let us now formulate the main theorem of this section, which is a consequence of Theorem 9.1.2, formulas (9.1.29), (9.1.30) and the Gram-Schmidt orthogonalization procedure applied to the sequence $1, \lambda, \lambda^2, \ldots$ with respect to the measure (9.1.1).

Theorem 9.1.3 *Consider the Cauchy problem for a semi-infinite Toda chain* (9.0.1)*. We need to find the solution* $\alpha_n(t) > 0$*,* $\beta_n(t)$*,* $t \in [0, T)$*,* $T \leq +\infty$*,* $n \in \mathbb{N}_0$ *of the system* (9.0.1) *according to the initial data* $\alpha_n(0)$*,* $\beta_n(0)$*,* $n \in \mathbb{N}_0$*.*

The procedure for finding this solution is the following. Using the initial data, we construct the Jacobi matrix $J(0)$ *according to the formula* (9.1.2) *and find its spectral measure* $d\rho(\lambda; 0)$*. Using the formulas* (9.1.29), (9.1.30)*, we find the spectral measure* $d\rho(\lambda; t)$ *of the matrix* $J(t)$ *at* $t \in [0, T)$ *and the solution of the Cauchy problem for* (9.0.1) *by formulas*

$$\begin{aligned} \alpha_n(t) &= \int_{\mathbb{R}} \lambda P_n(\lambda; t) P_{n+1}(\lambda; t) \, d\rho(\lambda; t), \\ \beta_n(t) &= \int_{\mathbb{R}} \lambda P_n^2(\lambda; t) \, d\rho(\lambda; t), \quad n \in \mathbb{N}_0, \ t \in [0, T). \end{aligned} \tag{9.1.31}$$

In (9.1.31)*,* $P_n(\lambda; t)$*,* $n \in \mathbb{N}_0$ *are polynomials of the first kind constructed by the orthogonalization of functions of the variable* $\lambda \in \mathbb{R}$*:* $\{1, \lambda, \lambda^2, \ldots\}$ *in the space* $L_2(\mathbb{R}, d\rho(\lambda; t))$*, where* $d\rho(\lambda; t)$ *is found by the formula* (9.1.29)*,* $t \in [0, T)$*.*

We also note that the function $\beta_0(t)$*,* $t \in [0, T)$*, in* (9.1.31)*, can be calculated directly by the formula* (9.1.30)*.*

It is clear that finding the solution of the Cauchy problem for the Toda chain (9.0.1) according to this theorem is a complicated procedure. But it is possible to find such initial data when this procedure becomes real. We will not consider these issues here.

Note also that similarly to Theorem 9.1.2, one can find an equation of the type (9.1.23) for the spectral function of the operator $\mathbf{J}(t)$, that is, for the function $[0, T) \ni t \mapsto \rho((-\infty, \lambda; t) \in \mathbb{R}$ with each fixed $\lambda \in \mathbb{R}$.

9.2 The Doubly Infinite Toda Chain and Its Equivalent Lax Equation for Block Jacobi Type Matrices

Instead of the usual space ℓ_2 over $\mathbb{N}_0$ used in the previous section, we will consider:

$$\mathbf{l}_2 = \mathcal{H}_0 \oplus \mathcal{H}_1 \oplus \mathcal{H}_2 \oplus \cdots, \quad \mathcal{H}_0 = \mathbb{C}^1, \quad \mathcal{H}_1 = \mathcal{H}_2 = \cdots = \mathbb{C}^2 = \mathbb{C}^1 \oplus \mathbb{C}^1;$$
$$\mathbf{l}_2 \ni f = (f_n)_{n=0}^{\infty}, \quad f_0 =: f_{0,0} \in \mathbb{C}^1, \quad f_n = (f_{n,0}, f_{n,1}) \in \mathbb{C}^2, \quad n \in \mathbb{N}. \tag{9.2.1}$$

The standard orthonormal basis $\mathbf{l}_2$ is formed by sequences of coordinate vectors from one space $\mathbb{C}^1$ and from infinite number of $\mathbb{C}^2$, namely,

$$\begin{aligned} \delta_0 &= (1, (0,0), \ldots) =: \delta_{0,0}, \\ \delta_{n,0} &= (0, \underbrace{(0,0), \ldots, (0,0)}_{n-1}, (1,0), (0,0), \ldots), \\ \delta_{n,1} &= (0, \underbrace{(0,0), \ldots, (0,0)}_{n-1}, (0,1), (0,0), \ldots), \quad n \in \mathbb{N}. \end{aligned} \tag{9.2.2}$$

The system (9.2.2) of vectors from $\mathbf{l}_2$ is orthonormal and each vector $f \in \mathbf{l}_2$ is a limit in $\mathbf{l}_2$ of the corresponding linear sequence of vectors (9.2.2). We can say that each vector in $\mathbf{l}_2$ is a linear combination (finite or infinite) of vectors (9.2.2).

Each bounded operator $\mathbf{A} \colon \mathbf{l}_2 \to \mathbf{l}_2$ can be written in the basis (9.2.2) as a block infinite matrix

$$A = \begin{bmatrix} A_{0;0} & A_{0;1} & A_{0;2} & \cdots \\ A_{1;0} & A_{1;1} & A_{1;2} & \cdots \\ A_{2;0} & A_{2;1} & A_{2;2} & \cdots \\ \vdots & \vdots & \vdots & \ddots \end{bmatrix}, \tag{9.2.3}$$

where blocks $A_{j;k}$ of (9.2.3) act as operators in the way

$$\begin{aligned} A_{0;0} &\colon \mathbb{C}^1 \to \mathbb{C}^1; \\ A_{0;k} &\colon \mathbb{C}^2 \to \mathbb{C}^1, \quad k \in \mathbb{N}; \\ A_{j;0} &\colon \mathbb{C}^1 \to \mathbb{C}^2, \quad j \in \mathbb{N}; \\ A_{j;k} &\colon \mathbb{C}^2 \to \mathbb{C}^2, \quad j, k \in \mathbb{N}. \end{aligned} \tag{9.2.4}$$

These operators $A_{j;k}$ have the form of numerical matrices

$$\begin{aligned}
&A_{0;0} = [A_{0,0;0,0}] : \mathbb{C}^1 \to \mathbb{C}^1;\\
&A_{0;k} = \left[A_{0,0;k,0}\ A_{0,0;k,1}\right] : \mathbb{C}^2 \to \mathbb{C}^1, \quad k \in \mathbb{N};\\
&A_{j;0} = \begin{bmatrix} A_{j,0;0,0} \\ A_{j,1;0,0} \end{bmatrix} : \mathbb{C}^1 \to \mathbb{C}^2; \quad j \in \mathbb{N};\\
&A_{j;k} = \begin{bmatrix} A_{j,0;k,0} & A_{j,0;k,1} \\ A_{j,1;k,0} & A_{j,1;k,1} \end{bmatrix} : \mathbb{C}^2 \to \mathbb{C}^2; \quad j,k \in \mathbb{N};\\
&A_{j,k;v_j,v_k} = (\mathbf{A}\delta_{k,v_k}, \delta_{j,v_j})_{\mathbf{l}_2}, \quad v_j, v_k \in \{0,1\} \quad j,k \in \mathbb{N}_0.
\end{aligned} \tag{9.2.5}$$

The action of the operator $\mathbf{A}$ specified in (9.2.3)–(9.2.5) is written in the form

$$\begin{aligned}
(\mathbf{A}f)_0 &= \sum_{k=0}^{\infty} A_{0;k} f_k = \left[A_{0,0;0,0}\right] f_0 + \sum_{k=1}^{\infty} \left[A_{0,0;k,0}\ A_{0,0;k,1}\right] (f_{k,0}, f_{k,1});\\
(\mathbf{A}f)_j &= \sum_{k=0}^{\infty} A_{j;k} f_k = \sum_{k=0}^{\infty} \begin{bmatrix} A_{j,0;k,0} & A_{j,0;k,1} \\ A_{j,1;k,0} & A_{j,1;k,1} \end{bmatrix} (f_{k,0}, f_{k,1}), \quad j \in \mathbb{N}.
\end{aligned} \tag{9.2.6}$$

Let us repeat that the orthonormal basis in the space $\mathbf{l}_2$ is (9.2.2). The matrix (9.2.3) has a special structure associated with the zero vector of this basis.

Note that for the adjoint operator $\mathbf{A}^* : \mathbf{l}_2 \to \mathbf{l}_2$, we have the relation

$$(\mathbf{A}^*\delta_{k,v_k}, \delta_{j,v_j})_{\mathbf{l}_2} = (\delta_{k,v_k}, \mathbf{A}\delta_{j,v_j})_{\mathbf{l}_2} = \overline{(\mathbf{A}\delta_{j,v_j}, \delta_{k,v_k})_{\mathbf{l}_2}},$$

that is, its matrix A^* is also of the type (9.2.3), namely, it is a block matrix adjoint to (9.2.3) (matrices $A^*_{k;j}$ are located at the corresponding positions j, k).

Let us move on to the double-infinite Toda chain. Such a chain has the form

$$\dot{\alpha}_n(t) = \frac{1}{2}\alpha_n(t)(\beta_{n+1}(t) - \beta_n(t)), \quad \dot{\beta}_n(t) = \alpha_n^2(t) - \alpha_{n-1}^2(t), \quad n \in \mathbb{Z}, \tag{9.2.7}$$

where $\alpha_n(t)$, $\beta_n(t)$ are real continuous differentiable functions of $t \in [0, T)$. The expression (9.2.7) is the system of differential-difference nonlinear equations and for (9.2.7) one can consider the Cauchy problem: for a given initial data $\alpha_n(0)$, $\beta_n(0)$, $n \in \mathbb{Z}$ to find a solution $\alpha_n(t)$, $\beta_n(t)$, $n \in \mathbb{Z}$ with $t \in [0, T)$.

Assume that all functions $\alpha_n(t)$, $\beta_n(t)$ are uniformly bounded with respect to n and

$$\alpha_n(t) > 0, \quad t \in [0, T), \quad n \in \mathbb{Z}. \tag{9.2.8}$$

Now we will directly prove that the Cauchy problem (9.2.7), (9.2.8) can be expressed in Lax form for block matrices in space $\mathbf{l}_2$.

By using functions $\alpha_n(t)$, $\beta_n(t)$, $n \in \mathbb{Z}$, $t \in [0, T)$, we construct block matrices for $\forall t \in [0, T)$ of the form (9.2.3) acting in the space $\mathbf{l}_2$ (9.2.1), namely,

$$J(t) = \begin{bmatrix} b_0(t) & a_0^*(t) & 0 & 0 & 0 \ldots \\ a_0(t) & b_1(t) & a_1(t) & 0 & 0 \ldots \\ 0 & a_1(t) & b_2(t) & a_2(t) & 0 \ldots \\ \vdots & \vdots & \vdots & \vdots & \vdots \ddots \end{bmatrix};$$

$$\begin{aligned} b_0(t) &= [\beta_0(t)] \colon \mathbb{C}^1 \to \mathbb{C}^1, \\ a_0(t) &= \begin{bmatrix} \alpha_0(t) \\ \alpha_{-1}(t) \end{bmatrix} \colon \mathbb{C}^1 \to \mathbb{C}^2, \\ a_0^*(t) &= [\alpha_0(t)\ \alpha_{-1}(t)] \colon \mathbb{C}^2 \to \mathbb{C}^1; \\ a_n(t) &= \begin{bmatrix} \alpha_n(t) & 0 \\ 0 & \alpha_{-n-1}(t) \end{bmatrix} \colon \mathbb{C}^2 \to \mathbb{C}^2, \\ b_n(t) &= \begin{bmatrix} \beta_n(t) & 0 \\ 0 & \beta_{-n}(t) \end{bmatrix} \colon \mathbb{C}^2 \to \mathbb{C}^2, \quad n \in \mathbb{N}, \quad \text{i.e.,} \end{aligned} \tag{9.2.9}$$

The matrix $J(t)$ in ordinary scalar form has the form

$$J(t) = \left[\begin{array}{c|cc|cc|cc|cc|c} \beta_0 & \alpha_0 & \alpha_{-1} & 0 & 0 & 0 & 0 & 0 & 0 & \ldots \\ \hline \alpha_0 & \beta_1 & 0 & \alpha_1 & 0 & 0 & 0 & 0 & 0 & \ldots \\ \alpha_{-1} & 0 & \beta_{-1} & 0 & \alpha_{-2} & 0 & 0 & 0 & 0 & \ldots \\ \hline 0 & \alpha_1 & 0 & \beta_2 & 0 & \alpha_2 & 0 & 0 & 0 & \ldots \\ 0 & 0 & \alpha_{-2} & 0 & \beta_{-2} & 0 & \alpha_{-3} & 0 & 0 & \ldots \\ \hline 0 & 0 & 0 & \alpha_2 & 0 & \beta_3 & 0 & \alpha_3 & 0 & \ldots \\ 0 & 0 & 0 & 0 & \alpha_{-3} & 0 & \beta_{-3} & 0 & \alpha_{-4} & \ldots \\ \hline \vdots & \vdots & \vdots & \vdots & \vdots & \vdots & \vdots & \vdots & \vdots & \ddots \end{array}\right].$$

The action of the matrix (9.2.9), on vectors $f = (f_n)_{n=0}^{\infty} \in \mathbf{l}_2$, has three components, namely,

$$\begin{aligned} (J(t)f)_n &= a_{n-1}(t)f_{n-1} + b_n(t)f_n + a_n(t)f_{n+1}, \quad n \in \mathbb{N}, \\ (J(t)f)_0 &= b_0(t)f_0 + a_0^*(t)f_1, \quad f = (f_n)_{n=0}^{\infty} \in \mathbf{l}_2, \quad \forall t \in [0, T). \end{aligned} \tag{9.2.10}$$

The condition (9.2.8) on the coefficients of the matrix $J(t)$ ensures that the corresponding operator $\mathbf{J}(t)$ is bounded, real, and self-adjoint in the space $\mathbf{l}_2$.

The Lax equation for the matrix $J(t)$ (9.2.9) has the form

$$\begin{aligned} \dot{J}(t) &= J(t)A(t) - A(t)J(t) = [J(t), A(t)] \quad \text{or} \\ \dot{\mathbf{J}}(t) &= \mathbf{J}(t)\mathbf{A}(t) - \mathbf{A}(t)\mathbf{J}(t), \quad t \in [0, T), \end{aligned} \tag{9.2.11}$$

where $A(t)$ is also a block matrix of type (9.2.3) (with coefficients depending on t); $\mathbf{J}(t), \mathbf{A}(t)$ denote the bounded operators in $\mathbf{l}_2$ according to (9.2.9) and $A(t)$. A further goal is to show the equivalence of Eqs. (9.2.7) and (9.2.11) for a matrix $A(t)$ of a certain form.

Let us present the matrix $A(t)$ with an internal structure as (9.2.9), but asymmetric, i.e.,

$$A(t) = \begin{bmatrix} \widetilde{b}_0 & \widetilde{c}_0 & 0 & 0 & \cdots \\ \widetilde{a}_0 & \widetilde{b}_1 & \widetilde{c}_1 & 0 & \cdots \\ 0 & \widetilde{a}_1 & \widetilde{b}_2 & \widetilde{c}_2 & \cdots \\ 0 & 0 & \widetilde{a}_2 & \widetilde{b}_3 & \cdots \\ \vdots & \vdots & \vdots & \vdots & \ddots \end{bmatrix}$$

$$:= \frac{1}{2} \left[\begin{array}{c|cc|cc|cc|c} 0 & -\alpha_0 & \alpha_{-1} & 0 & 0 & 0 & 0 & \cdots \\ \hline \alpha_0 & 0 & 0 & -\alpha_1 & 0 & 0 & 0 & \cdots \\ -\alpha_{-1} & 0 & 0 & 0 & \alpha_{-2} & 0 & 0 & \cdots \\ \hline 0 & \alpha_1 & 0 & 0 & 0 & -\alpha_2 & 0 & \cdots \\ 0 & 0 & -\alpha_{-2} & 0 & 0 & 0 & \alpha_{-3} & \cdots \\ \hline 0 & 0 & 0 & \alpha_2 & 0 & 0 & 0 & \cdots \\ 0 & 0 & 0 & 0 & -\alpha_{-3} & 0 & 0 & \cdots \\ \hline \vdots & \vdots & \vdots & \vdots & \vdots & \vdots & \vdots & \ddots \end{array} \right], \tag{9.2.12}$$

where elements $\alpha_n, n \in \mathbb{Z}$ of matrices $\widetilde{a}_n, \widetilde{c}_n, n \in \mathbb{N}_0$ depend on t, $\forall t \in [0, T)$,.

Theorem 9.2.1 *The Toda equation* (9.2.7) *with conditions* (9.2.8) *and the Lax equation* (9.2.11) *with tri-diagonal block matrices* $J(t)$ *of the type* (9.2.9) *and* $A(t)$ *of the type* (9.2.12), *are equivalent, i.e., have the same solution.*

Proof For block matrices A, B of the type (9.2.3), we have the usual rule for the calculating coefficients during their multiplication

$$(AB)_{j;k} = \sum_{l=0}^{\infty} A_{j;l} B_{l;k}, \quad j, k \in \mathbb{N}_0.$$

Using this rule, one can simply calculate elements of the product of matrices J and A, (9.2.9), (9.2.12). Taking into account that $c_0 = a_0^*$, $c_n = a_n$, $n \in \mathbb{N}$ for each $n \in \mathbb{N}_0$, we have

$$\begin{aligned}
(JA)_{n-2;n} &= c_{n-2}\widetilde{c}_{n-1}, \\
(JA)_{n-1;n} &= b_{n-1}\widetilde{c}_{n-1} + c_{n-1}\widetilde{b}_n, \\
(JA)_{n;n} &= a_{n-1}\widetilde{c}_{n-1} + b_n\widetilde{b}_n + c_n\widetilde{a}_n, \\
(JA)_{n+1;n} &= a_n\widetilde{b}_n + b_{n+1}\widetilde{a}_n, \\
(JA)_{n+2;n} &= a_{n+1}\widetilde{a}_n,
\end{aligned} \tag{9.2.13}$$

where elements with negative indices are considered equal to zero.

Expressions similar to (9.2.13) are performed for AJ if a_n, b_n, c_n are replaced with $\widetilde{a}_n, \widetilde{b}_n, \widetilde{c}_n$ and $\widetilde{a}_n, \widetilde{b}_n, \widetilde{c}_n$ on a_n, b_n, c_n vice versa.

We compare elements on the left and right of the equality (9.2.11) and, using the formulas (9.2.13) and similar formulas for AJ, $\forall n \in \mathbb{N}$, we obtain equations

$$\begin{aligned}
0 &= c_n\widetilde{c}_{n+1} - \widetilde{c}_n c_{n+1}, \\
\dot{c}_n &= b_n\widetilde{c}_n + c_n\widetilde{b}_{n+1} - \widetilde{b}_n c_n - \widetilde{c}_n b_{n+1}, \\
\dot{b}_n &= a_{n-1}\widetilde{c}_{n-1} + b_n\widetilde{b}_n + c_n\widetilde{a}_n - \widetilde{a}_{n-1}c_{n-1} - \widetilde{b}_n b_n - \widetilde{c}_n a_n, \\
\dot{a}_n &= a_n\widetilde{b}_n + b_{n+1}\widetilde{a}_n - \widetilde{a}_n b_n - \widetilde{b}_{n+1}a_n, \\
0 &= a_{n+1}\widetilde{a}_n - \widetilde{a}_{n+1}a_n
\end{aligned} \tag{9.2.14}$$

(as before, elements of (9.2.14) with negative indices are considered equal to zero).

According to (9.2.12), we have

$$\begin{aligned}
&\widetilde{a}_0 = \frac{1}{2}\begin{bmatrix} \alpha_0 \\ -\alpha_{-1} \end{bmatrix}, \quad \widetilde{a}_n = \frac{1}{2}\begin{bmatrix} \alpha_n & 0 \\ 0 & -\alpha_{-n-1} \end{bmatrix}, \quad n \in \mathbb{N}; \\
&\widetilde{c}_0 = \frac{1}{2}\begin{bmatrix} -\alpha_0 & \alpha_{-1} \end{bmatrix}, \quad \widetilde{c}_n = \frac{1}{2}\begin{bmatrix} -\alpha_n & 0 \\ 0 & \alpha_{-n-1} \end{bmatrix}, \quad n \in \mathbb{N}; \\
&\widetilde{b}_n = 0, \quad n \in \mathbb{N}_0.
\end{aligned} \tag{9.2.15}$$

The second equality and (9.2.9), (9.2.15) give

$$\dot{c}_0 = \dot{a}_0^* = \begin{bmatrix} \dot{\alpha}_0 & \dot{\alpha}_{-1} \end{bmatrix} = \frac{1}{2}\beta_0\begin{bmatrix} -\alpha_0 & \alpha_{-1} \end{bmatrix} - \frac{1}{2}\begin{bmatrix} -\alpha_0 & \alpha_{-1} \end{bmatrix}\begin{bmatrix} \beta_1 & 0 \\ 0 & \beta_{-1} \end{bmatrix},$$

i.e., $\dot{\alpha}_0 = \frac{1}{2}\alpha_0(\beta_1 - \beta_0)$, $\dot{\alpha}_{-1} = \frac{1}{2}\alpha_{-1}(\beta_0 - \beta_{-1})$;

$$\dot{c}_n = \dot{a}_n = \begin{bmatrix} \dot{\alpha}_n & 0 \\ 0 & \dot{\alpha}_{-n-1} \end{bmatrix} = \frac{1}{2}\begin{bmatrix} \beta_n & 0 \\ 0 & \beta_{-n} \end{bmatrix}\begin{bmatrix} -\alpha_n & 0 \\ 0 & \alpha_{-n-1} \end{bmatrix}$$
$$-\frac{1}{2}\begin{bmatrix} -\alpha_n & 0 \\ 0 & \alpha_{-n-1} \end{bmatrix}\begin{bmatrix} \beta_{n+1} & 0 \\ 0 & \beta_{-n-1} \end{bmatrix}.$$

Therefore,

$$\dot{\alpha}_n = \frac{1}{2}\alpha_n(\beta_{n+1} - \beta_n), \quad \dot{\alpha}_{-n-1} = \frac{1}{2}\alpha_{-n-1}(\beta_{-n} - \beta_{-n-1}), \quad n \in \mathbb{N}.$$

Thus, it is shown that the second equality in (9.2.14) is equivalent to the first equality in (9.2.7).

The third equality in (9.2.14) and (9.2.9), (9.2.15) give

$$\dot{b}_0 = [\dot{\beta}_0] = c_0\widetilde{a}_0 - \widetilde{c}_0 a_0 = \frac{1}{2}[\alpha_0 \quad \alpha_{-1}]\begin{bmatrix} \alpha_0 \\ -\alpha_{-1} \end{bmatrix} - \frac{1}{2}[-\alpha_0 \quad \alpha_{-1}]\begin{bmatrix} \alpha_0 \\ \alpha_{-1} \end{bmatrix},$$

i.e., $\dot{\beta}_0 = \alpha_0^2 - \alpha_{-1}^2$;

$$\dot{b}_1 = \begin{bmatrix} \dot{\beta}_1 & 0 \\ 0 & \dot{\beta}_{-1} \end{bmatrix} = \frac{1}{2}\begin{bmatrix} \alpha_0 \\ \alpha_{-1} \end{bmatrix}[-\alpha_0 \quad \alpha_{-1}] + \frac{1}{2}\begin{bmatrix} \alpha_1 & 0 \\ 0 & \alpha_{-2} \end{bmatrix}\begin{bmatrix} \alpha_1 & 0 \\ 0 & -\alpha_{-2} \end{bmatrix}$$
$$-\frac{1}{2}\begin{bmatrix} \alpha_0 \\ -\alpha_{-1} \end{bmatrix}\begin{bmatrix} \alpha_0 \ \alpha_{-1} \end{bmatrix} - \frac{1}{2}\begin{bmatrix} -\alpha_1 & 0 \\ 0 & \alpha_{-2} \end{bmatrix}\begin{bmatrix} \alpha_1 & 0 \\ 0 & \alpha_{-2} \end{bmatrix},$$

i.e., $\dot{\beta}_1 = \alpha_1^2 - \alpha_0^2, \quad \dot{\beta}_{-1} = \alpha_{-1}^2 - \alpha_{-2}^2$;

$$\dot{b}_n = \begin{bmatrix} \dot{\beta}_n & 0 \\ 0 & \dot{\beta}_{-n} \end{bmatrix} = \frac{1}{2}\begin{bmatrix} \alpha_{n-1} & 0 \\ 0 & \alpha_{-n} \end{bmatrix}\begin{bmatrix} -\alpha_{n-1} & 0 \\ 0 & \alpha_{-n} \end{bmatrix}$$
$$+\frac{1}{2}\begin{bmatrix} \alpha_n & 0 \\ 0 & \alpha_{-n-1} \end{bmatrix}\begin{bmatrix} \alpha_n & 0 \\ 0 & -\alpha_{-n-1} \end{bmatrix} - \frac{1}{2}\begin{bmatrix} \alpha_{n-1} & 0 \\ 0 & -\alpha_{-n} \end{bmatrix}\begin{bmatrix} \alpha_{n-1} & 0 \\ 0 & \alpha_{-n} \end{bmatrix}$$
$$-\frac{1}{2}\begin{bmatrix} -\alpha_n & 0 \\ 0 & \alpha_{-n-1} \end{bmatrix}\begin{bmatrix} \alpha_n & 0 \\ 0 & \alpha_{-n-1} \end{bmatrix},$$

i.e., $\dot{\beta}_n = \alpha_n^2 - \alpha_{n-1}^2, \quad \dot{\beta}_{-n} = \alpha_{-n}^2, \quad n = 2, 3, \ldots$.

Recent calculations show that the third equation in (9.2.14) is equivalent to the second equation in (9.2.7).

It is also easy to calculate that the fourth equation in (9.2.14) for considered matrices is equivalent to the first equation in (9.2.7). Thus, we have

$$\dot{a}_0 = \begin{bmatrix} \dot{\alpha}_0 \\ \dot{\alpha}_{-1} \end{bmatrix} = \frac{1}{2}\begin{bmatrix} \beta_1 & 0 \\ 0 & \beta_{-1} \end{bmatrix}\begin{bmatrix} \alpha_0 \\ -\alpha_{-1} \end{bmatrix} - \frac{1}{2}\begin{bmatrix} \alpha_0 \\ -\alpha_{-1} \end{bmatrix}[\beta_0],$$

i.e., $\dot{\alpha}_0 = \frac{1}{2}\alpha_0(\beta_1 - \beta_0), \quad \dot{\alpha}_{-1} = \frac{1}{2}\alpha_{-1}(\beta_0 - \beta_{-1})$;

$$\dot{a}_n = \begin{bmatrix} \dot{\alpha}_n & 0 \\ 0 & \dot{\alpha}_{-n-1} \end{bmatrix} = \frac{1}{2}\begin{bmatrix} \beta_{n+1} & 0 \\ 0 & \beta_{-n-1} \end{bmatrix}\begin{bmatrix} \alpha_n & 0 \\ 0 & -\alpha_{-n-1} \end{bmatrix}$$
$$- \frac{1}{2}\begin{bmatrix} \alpha_n & 0 \\ 0 & -\alpha_{-n-1} \end{bmatrix}\begin{bmatrix} \beta_n & 0 \\ 0 & \beta_{-n} \end{bmatrix}.$$

Therefore, $\dot{\alpha}_n = \frac{1}{2}\alpha_n(\beta_{n+1} - \beta_n)$, $\dot{\alpha}_{-n-1} = \frac{1}{2}\alpha_{-n-1}(\beta_{-n} - \beta_{-n-1})$, $n \in \mathbb{N}$.

The first and last equalities in (9.2.14) for considered matrices are performed automatically. Thus, for the first equality we have

$$c_n\widetilde{c}_{n+1} - \widetilde{c}_n c_{n+1} = \frac{1}{2}\begin{bmatrix} \alpha_n & 0 \\ 0 & \alpha_{-n-1} \end{bmatrix}\begin{bmatrix} -\alpha_{n+1} & 0 \\ 0 & \alpha_{-n-2} \end{bmatrix}$$
$$- \frac{1}{2}\begin{bmatrix} -\alpha_n & 0 \\ 0 & \alpha_{-n-1} \end{bmatrix}\begin{bmatrix} \alpha_{n+1} & 0 \\ 0 & \alpha_{-n-2} \end{bmatrix} = 0, \qquad n \in \mathbb{N}.$$

And for the last equality we have

$$a_1\widetilde{a}_0 - \widetilde{a}_1 a_0 = \frac{1}{2}\begin{bmatrix} \alpha_1 & 0 \\ 0 & \alpha_{-2} \end{bmatrix}\begin{bmatrix} \alpha_0 \\ -\alpha_{-1} \end{bmatrix} - \frac{1}{2}\begin{bmatrix} \alpha_1 & 0 \\ 0 & -\alpha_{-2} \end{bmatrix}\begin{bmatrix} \alpha_0 \\ \alpha_{-1} \end{bmatrix} = 0;$$

$$a_{n+1}\widetilde{a}_n - \widetilde{a}_{n+1}a_n = \frac{1}{2}\begin{bmatrix} \alpha_{n+1} & 0 \\ 0 & \alpha_{-n-2} \end{bmatrix}\begin{bmatrix} \alpha_n & 0 \\ 0 & -\alpha_{-n-1} \end{bmatrix}$$
$$- \frac{1}{2}\begin{bmatrix} \alpha_{n+1} & 0 \\ 0 & -\alpha_{-n-2} \end{bmatrix}\begin{bmatrix} \alpha_n & 0 \\ 0 & \alpha_{-n-1} \end{bmatrix} = 0.$$

As a result, it is proved that the Toda equation (9.2.7) and the Lax equation (9.2.11) are equivalent. □

9.3 The Spectral Theory of Corresponding to Toda Chains Block Jacobi Type Matrices

Throughout the section, we will consider block Jacobi matrices (9.2.9) for a fixed $t \in [0, T)$, on which elements of the matrix J depend (we will denote elements of the matrix J by a_n, α_m, b_n, β_m, where $n \in \mathbb{N}_0$, $m \in \mathbb{Z}$). The matrix J generates the self-adjoint operator **J** in the space $\mathbf{l}_2$ (9.2.1) in the usual way. We denote by $\mathbf{l}_{\text{fin}}$ the space of finite vectors with $\mathbf{l}_2$. Then our operator **J** is the closure of the operator $\mathbf{l}_{\text{fin}} \ni f \to Jf$ in the space $\mathbf{l}_2$.

For us, it is important to construct the generalized eigenvectors expansion of the operator **J** using the construction of Chap. 3. For this purpose, we will construct a quasi-nuclear equipment of the space $\mathbf{l}_2$, namely,

$$\mathbf{l} = (\mathbf{l}_{\text{fin}})' \supset \mathbf{l}_2(p^{-1}) \supset \mathbf{l}_2 \supset \mathbf{l}_2(p) \supset \mathbf{l}_{\text{fin}}. \tag{9.3.1}$$

Here $\mathbf{l}_2(p)$ denotes the space $\mathbf{l}_2$ with a weight $p = (p_n)_{n=0}^{\infty}$ such that $p \geq 1$, $\sum_{n=0}^{\infty} p_n^{-1} < \infty$, $\|f\|^2_{\mathbf{l}_2(p)} = \|f_0\|^2_{\mathbb{C}^1} p_0 + \|f_1\|^2_{\mathbb{C}^2} p_1 + \|f_2\|^2_{\mathbb{C}^2} p_2 + \ldots$. The space $\mathbf{l}_2(p^{-1})$ is the similar to $\mathbf{l}_2(p)$ but with the change of p on $p^{-1} = (p_n^{-1})_{n=0}^{\infty}$; $\mathbf{l}$ is the space of all sequences $f = (f_n)_{n=0}^{\infty}$, $f_0 \in \mathbb{C}^1$, $\forall n \in \mathbb{N}$; $f_n \in \mathbb{C}^2$. For $\alpha \in \mathbf{l}$ and $\xi \in \mathbf{l}_{\text{fin}}$, there exists a bilinear form (α, ξ) generated by the scalar product $\mathbf{l}_2$.

Then for our operator **J** and its resolution of the identity $E(\triangle)$ there exists a Borel probability measure $d\rho_b(\lambda)$ on $\mathbb{R}$ (the base spectral measure), for which

$$E(\triangle)f = \int_{\triangle} \mathbf{\Phi}(\lambda)\, d\rho_b(\lambda) f, \quad \mathbf{J}f = \int_{\mathbb{R}} \lambda \mathbf{\Phi}(\lambda)\, d\rho_b(\lambda), \quad f \in \mathbf{l}_2(p). \tag{9.3.2}$$

Here $\triangle$ denotes an arbitrary Borel set from $\mathbb{R}$, i.e., $\triangle \in \mathfrak{B}(\mathbb{R})$. For each $\lambda \in \mathbb{R}$, $\mathbf{\Phi}(\lambda)$ is bounded operator acting from $\mathbf{l}_2(p)$ into $\mathbf{l}_2(p^{-1})$ (the generalized projection operator): $\forall f \in \mathbf{l}_2(p)$, $\mathbf{\Phi}(\lambda)f$ is a generalized eigenvector of the operator **J**. The operator $\mathbf{\Phi}(\lambda)$, for all $\lambda \in \mathbb{R}$ is positive (more precisely, non-negative), i.e., $\forall f \in \mathbf{l}_2(p)$, $(\mathbf{\Phi}(\lambda)f, f) \geq 0$.

For each $f \in \mathbf{l}_2(p)$, $\varphi(\lambda) = \mathbf{\Phi}(\lambda)f \in \mathbf{l}_2(p) \supset \mathbf{l}_2$ is generalized eigenvector of the operator **J** with corresponding eigenvalue λ, hence, $(\varphi(\lambda), \mathbf{J}u)_{\mathbf{l}_2} = \lambda(\varphi(\lambda), u)_{\mathbf{l}_2}$ for arbitrary $u \in \mathbf{l}_{\text{fin}}$, and, hence, we get $J\varphi(\lambda) = \lambda\varphi(\lambda)$. By using (9.2.10) and (9.2.9), we have

$$\begin{aligned} &\beta_0\varphi_{0,0}(\lambda) + \alpha_0\varphi_{1,0}(\lambda) + \alpha_{-1}\varphi_{1,1}(\lambda) = b_0\varphi_0(\lambda) + a_0^*\varphi_1(\lambda) = \lambda\varphi_0(\lambda), \\ &a_{n-1}\varphi_{n-1}(\lambda) + b_n\varphi_n(\lambda) + a_n\varphi_{n+1}(\lambda) = \lambda\varphi_n(\lambda), \quad n \in \mathbb{N}, \quad \forall\lambda \in \mathbb{R}, \end{aligned} \tag{9.3.3}$$

where $\varphi(\lambda) = (\varphi_n(\lambda))_{n=0}^{\infty} \in \mathbf{l}_2(p^{-1}) \subset \mathbf{l}$. Let us emphasize that for $n \in \mathbb{N}$ we have $\varphi_n(\lambda) = (\varphi_{n,0}(\lambda), \varphi_{n,1}(\lambda))$, in particular $\varphi_0(\lambda) = \varphi_{0,0}(\lambda)$.

Let us consider the equality (9.3.3) with a fixed $\lambda \in \mathbb{R}$. Let $\varphi_{0,0}$ and $\varphi_{1,0}$ be given, namely $\varphi_{0,0}(\lambda) = c_0$, $\varphi_{1,0}(\lambda) = c_1$, where $c_0, c_1 \in \mathbb{R}$ are initial conditions.

Since $\alpha_{-1} > 0$, then the first equality of (9.3.3) gives $\varphi_{1,1}(\lambda) = \bar{\alpha}_{-1}^{1}((\lambda - \beta_0)c_0 - \alpha_0 c_1)$. By using (9.3.3) for $n = 1, 2, \ldots$, we can find step by step the solution $\varphi(\lambda) = (\varphi_n(\lambda))_{n=0}^{\infty}$ (9.3.3).

Denote by $\theta^{(\alpha)}(\lambda) = (\theta_n^{(\alpha)})_{n=0}^{\infty} = (\theta_{n,v_n}^{(\alpha)}(\lambda))_{n=0,v_n=0,1}^{\infty}$, $\alpha = 0, 1$, two solutions of the Eq. (9.3.3) with initial data

$$\theta_{0,0}^{(0)}(\lambda) = 1, \quad \theta_{1,0}^{(0)}(\lambda) = 0; \quad \theta_{0,0}^{(1)}(\lambda) = 0, \quad \theta_{1,0}^{(1)}(\lambda) = 1. \tag{9.3.4}$$

Equation (9.3.3) are linear, differential of the second order with a fixed λ, therefore, the solution is uniquely determined by the values at the two points $(0, 0)$ and $(1, 0)$ from the domain of the argument $(n, 0)$, $(n, 1)$, $n \in \mathbb{N}$, $0 = (0, 0)$. Each solution $\varphi(\lambda) = (\varphi_n(\lambda))_{n=0}^{\infty}$ of (9.3.3) can be obtained as a linear combination of the fundamental system of solutions (9.3.4). For example, the value written above for $\varphi_{1,1}(\lambda)$ is equal to $\varphi_{0,0}(\lambda)\theta^{(0)}(\lambda) + \varphi_{1,0}(\lambda)\theta^{(1)}(\lambda)$ at the point $(1, 1)$. By using solutions $\theta^{(0)}(\lambda)$ and $\theta_{1,0}^{(1)}(\lambda)$, we can write

$$\varphi(\lambda) = \varphi_{0,0}(\lambda)\theta^{(0)}(\lambda) + \varphi_{1,0}(\lambda)\theta_{1,0}^{(1)}(\lambda), \quad \forall\lambda \in \mathbb{R}, \quad \text{i.e.,} \tag{9.3.5}$$

$$\varphi_{n,v_n}(\lambda) = \varphi_{0,0}(\lambda)\theta_{n,v_n}^{(0)}(\lambda) + \varphi_{1,0}(\lambda)\theta_{n,v_n}^{(1)}(\lambda),$$
$$n \in \mathbb{N}, \ \nu \in \{0, 1\}; \quad n = 0, \ v_0 = 0.$$

It is important for us to obtain the representation (9.3.5) for the operator matrix $\mathbf{\Phi}(\lambda)$, which is now regarded as the operator $\mathbf{l}_{\text{fin}} \to \mathbf{l}$.

First, we note that each linear operator $\mathbf{A} : \mathbf{l}_{\text{fin}} \to \mathbf{l}$ can be represented by a block matrix of the type (9.2.3) with scalar elements $A_{j,k;v_j,v_k}$, where $j, k \in \mathbb{N}$, and the second set of indices v_j, v_k takes the value "0" or "1 ", but for indices $j, k = 0, k$ we use only the second index (for $j, k = j, 0$—only the first $j \in \mathbb{N}_0$).

We denote matrices of this type by

$$(A_{j,v_j;k,v_k})_{j,k=0;v_j,v_k=0,1}^{\infty}; \tag{9.3.6}$$

for the solutions (9.3.3), as mentioned, a similar rule is also used (if $n = 0$, then $v_0 = 0$).

Therefore, a matrix of the type (9.3.6) for our operator $\mathbf{\Phi}(\lambda) : \mathbf{l}_{\text{fin}} \to \mathbf{l}$ has a form

$$\begin{aligned} &\Phi(\lambda) = (\Phi_{j,v_j;k,v_k}(\lambda))_{j,k=0;v_j,v_k=0,1}^{\infty}, \\ &\Phi_{j,v_j;k,v_k}(\lambda) = (\mathbf{\Phi}(\lambda)\delta_{k,v_k}, \delta_{j,v_j})_{\mathbf{l}_2}, \quad j, k \in \mathbb{N}_0, \ v_j, v_k \in \{0, 1\}. \end{aligned} \tag{9.3.7}$$

It is essential for us to obtain a representation of the type (9.3.5) for the elements of the matrix (9.3.7).

Lemma 9.3.1 *Elements of the matrix* $\Phi(\lambda)$ *in* (9.3.7) *have a representation*

$$\begin{aligned}\Phi_{j,v_j;k,v_k}(\lambda) &= \Phi_{0,0;0,0}(\lambda)\theta^{(0)}_{j,v_j}(\lambda)\theta^{(0)}_{k,v_k}(\lambda) + \Phi_{0,0;1,0}(\lambda)\theta^{(1)}_{j,v_j}(\lambda)\theta^{(0)}_{k,v_k}(\lambda)\\ &+ \Phi_{1,0;0,0}(\lambda)\theta^{(0)}_{j,v_j}(\lambda)\theta^{(1)}_{k,v_k}(\lambda) + \Phi_{1,0;1,0}(\lambda)\theta^{(1)}_{j,v_j}(\lambda)\theta^{(1)}_{k,v_k}(\lambda),\\ &\qquad j,k\in\mathbb{N}_0,\quad v_j,v_k=0,1;\quad \lambda\in\mathbb{R}.\end{aligned} \tag{9.3.8}$$

Proof Let us first consider the case of $j,k\in\mathbb{N}$. For fixed k, v_k, we have

$$\Phi_{j,v_j;k,v_k}(\lambda) = (\boldsymbol{\Phi}(\lambda)\delta_{k,v_k}, \delta_{j,v_j})_{\mathbf{l}_2} = (\boldsymbol{\Phi}(\lambda)\delta_{k,v_k})_{j,v_j}$$

is a vector from $\mathbf{l}$ which is generalized eigenvector of the operator $\mathbf{J}$ with an eigenvalue λ. Thus, it is a solution of the difference equation (9.3.3) with initial data $\Phi_{0,0;k,v_k}(\lambda)$ and $\Phi_{1,0;k,v_k}(\lambda)$ and, according to (9.3.5), one can write

$$\Phi_{j,v_j;k,v_k}(\lambda) = \Phi_{0,0;k,v_k}(\lambda)\theta^{(0)}_{j,v_j}(\lambda) + \Phi_{1,0;k,v_k}(\lambda)\theta^{(1)}_{j,v_j}(\lambda),\quad j\in\mathbb{N},\ v_j=0,1. \tag{9.3.9}$$

In general, the generalized projection operator $\boldsymbol{\Phi}(\lambda)$ is formally Hermitian, therefore, our matrix (9.3.7) is Hermitian. Since the operator $\mathbf{J}$ is real, then $\boldsymbol{\Phi}(\lambda)$ is also real and this fact gives that its matrix (9.3.7) is real. Thus, as a result, the matrix (9.3.7) is real and symmetric. Applying this fact to (9.3.9), we get

$$\Phi_{0,0;k,v_k}(\lambda) = \Phi_{k,v_k;0,0}(\lambda) = \Phi_{0,0;0,0}(\lambda)\theta^{(0)}_{k,v_k}(\lambda) + \Phi_{1,0;0,0}(\lambda)\theta^{(1)}_{k,v_k}(\lambda),$$

$$\Phi_{1,0;k,v_k}(\lambda) = \Phi_{k,v_k;1,0}(\lambda) = \Phi_{0,0;1,0}(\lambda)\theta^{(0)}_{k,v_k}(\lambda) + \Phi_{1,0;1,0}(\lambda)\theta^{(1)}_{k,v_k}(\lambda).$$

Substituting these expressions into (9.3.9), we get (9.3.8).

Let $j=0$, $k\in\mathbb{N}$. According to our rule, we use only the index $k\in\mathbb{N}_0$ in the matrix (9.3.6), thus, the equality (9.3.9) is fulfilled. The situation is similar in the case of $k=0$, $j\in\mathbb{N}$. To obtain the complete proof, it is necessary to repeat the proof for the case $j=0$, $k\in\mathbb{N}$. □

Let us return to the first equality in (9.3.2). For $f,g\in\mathbf{l}_{\text{fin}}$ and $\Delta=\mathbb{R}$, matrices (9.3.7), we have

$$\begin{aligned}(f,g)_{\mathbf{l}_2} &= \int_{\mathbb{R}} (\boldsymbol{\Phi}(\lambda)f,g)_{\mathbf{l}_2}\,d\rho_b(\lambda)\\ &= \int_{\mathbb{R}} \Big[\sum_{\substack{j,k=0,\\ v_j,v_k=0,1}}^{\infty} \Phi_{j,v_j;k,v_k}(\lambda) f_{k,v_k}\overline{g}_{j,v_j}\Big]\,d\rho_b(\lambda).\end{aligned} \tag{9.3.10}$$

Let us rewrite this equality in a convenient form. To do this, we introduce the Fourier transform (on the generalized eigenvectors of the operator $\mathbf{J}$): for $f = (f_n)_{n=0}^{\infty} \in \mathbf{l}_{\text{fin}}$, we put

$$\widehat{f}(\lambda) = (\widehat{f_0}(\lambda), \widehat{f_1}(\lambda)) \in \mathbb{C}^2,$$

$$\widehat{f_\alpha}(\lambda) = \sum_{\substack{n=0,\\ \upsilon_n=0,1}}^{\infty} \theta_{n,\upsilon_n}^{(\alpha)}(\lambda) f_{n,\upsilon_n}, \quad \alpha = 0, 1, \quad \forall \lambda \in \mathbb{R}. \tag{9.3.11}$$

By using the representation (9.3.8) and (9.3.11), we get

$$\begin{aligned}\sum_{\substack{j,k=0,\\ \upsilon_j,\upsilon_k=0,1}}^{\infty} \Phi_{j,\upsilon_j;k,\upsilon_k}(\lambda) f_{k,\upsilon_k} \overline{g}_{j,\upsilon_j} &= \Phi_{0,0;0,0}(\lambda)\widehat{f_0}(\lambda)\overline{\widehat{g_0}(\lambda)} \\ &+ \Phi_{0,0;1,0}(\lambda)\widehat{f_1}(\lambda)\overline{\widehat{g_0}(\lambda)} + \Phi_{1,0;0,0}(\lambda)\widehat{f_0}(\lambda)\overline{\widehat{g_1}(\lambda)} \\ &+ \Phi_{1,0;1,0}(\lambda)\widehat{f_1}(\lambda)\overline{\widehat{g_1}(\lambda)} = (C(\lambda)\widehat{f}(\lambda), \widehat{g}(\lambda))_{\mathbb{C}^2};\end{aligned} \tag{9.3.12}$$

$$C(\lambda) := \begin{bmatrix} \Phi_{0,0;0,0}(\lambda) & \Phi_{0,0;1,0}(\lambda) \\ \Phi_{1,0;0,0}(\lambda) & \Phi_{1,0;1,0}(\lambda) \end{bmatrix}.$$

Thus, (9.3.10) can be rewritten in the form

$$(f, g)_{\mathbf{l}_2} = \int_{\mathbb{R}} (C(\lambda)\widehat{f}(\lambda), \widehat{g}(\lambda))_{\mathbb{C}^2}\, d\rho_b(\lambda) = \int_{\mathbb{R}} (d\rho_M(\lambda)\widehat{f}(\lambda), \widehat{g}(\lambda))_{\mathbb{C}^2}, \ \forall f, g \in \mathbf{l}_{\text{fin}},$$

$$\mathfrak{B}(\mathbb{R}) \ni \Delta \mapsto \rho_M(\Delta) = \int_{\Delta} C(\lambda)\, d\rho_b(\lambda) = \int_{\Delta} \begin{bmatrix} \Phi_{0,0;0,0}(\lambda) & \Phi_{0,0;1,0}(\lambda) \\ \Phi_{1,0;0,0}(\lambda) & \Phi_{1,0;1,0}(\lambda) \end{bmatrix} d\rho_b(\lambda). \tag{9.3.13}$$

This 2×2–matrix measure $d\rho_M(\lambda) = (d\rho_{\mu,\nu}(\lambda))_{\mu,\nu=0,1}$ is called the matrix (standard) spectral measure of the operator $\mathbf{J}$; the first equality in (9.3.13) is the corresponding Parseval equality. This measure has the properties described by the lemma.

Lemma 9.3.2 *For each $\Delta \in \mathfrak{B}(\mathbb{R})$ the matrix $\rho_M(\lambda)$ is real, symmetric and positive definite, that is, the matrix spectral measure $d\rho_M(\lambda)$ is real, non-negative. It is probabilistic, i.e., $\rho_M(\mathbb{R}) = 1$.*

Proof As above, we mentioned that, the matrix $\Phi(\lambda)$ in (9.3.7) is real and symmetric, therefore, the matrix $C(\lambda)$ with (9.3.12) is also real and symmetrical. The definition (9.3.13) gives that these properties pass to $\rho_M(\Delta)$ as well.

The non-negativity of the measure $d\rho_M(\lambda)$ follows from the positivity of the operator $\mathbf{\Phi}(\lambda)$. Namely, we construct a vector

$$f = (x_0, (x_1, 0), (0, 0), (0, 0), \dots) = x_0\delta_{0,0} + x_1\delta_{1,0} \in \mathbf{l}_{\text{fin}}, \quad \forall x = (x_0, x_1) \in \mathbb{C}^2.$$

Thus, we have

$$\begin{aligned}(C(\lambda)x, x)_{\mathbb{C}^2} &= \Phi_{0,0;0,0}(\lambda)x_0\bar{x}_0 + \Phi_{0,0;1,0}(\lambda)x_1\bar{x}_0 \\ &+ \Phi_{1,0;0,0}(\lambda)x_0\bar{x}_1 + \Phi_{1,0;1,0}(\lambda)x_1\bar{x}_1 \\ &= (\mathbf{\Phi}(\lambda)\delta_{0,0}, \delta_{0,0})_{\mathbf{l}_2}x_0\bar{x}_0 + (\mathbf{\Phi}(\lambda)\delta_{1,0}, \delta_{0,0})_{\mathbf{l}_2}x_1\bar{x}_0 \\ &+ (\mathbf{\Phi}(\lambda)\delta_{0,0}, \delta_{1,0})_{\mathbf{l}_2}x_0\bar{x}_1 + (\mathbf{\Phi}(\lambda)\delta_{1,0}, \delta_{1,0})_{\mathbf{l}_2}x_1\bar{x}_1 \\ &= (\mathbf{\Phi}(\lambda)f, f)_{\mathbf{l}_2} \geq 0.\end{aligned}$$

It remains to prove that the measure $d\rho_M(\lambda)$ is probabilistic. Let us use vectors $\delta_{0,0}$ and $\delta_{1,0}$, their Fourier transform according to (9.3.11), (9.3.4) is of the form $\widehat{\delta_{0,0}}(\lambda) = (1, 0)$, $\widehat{\delta_{1,0}}(\lambda) = (0, 1)$. Therefore, the Parseval equality (9.3.13) gives

$$\begin{aligned}1 &= (\delta_{0,0}, \delta_{0,0})_{\mathbf{l}_2} = \rho_{M;0,0}(\mathbb{R}), \\ 0 &= (\delta_{0,0}, \delta_{1,0})_{\mathbf{l}_2} = \rho_{M;1,0}(\mathbb{R}), \\ 1 &= (\delta_{1,0}, \delta_{1,0})_{\mathbf{l}_2} = \rho_{M;1,1}(\mathbb{R}), \\ &\text{i.e.,} \quad \rho_M(\mathbb{R}) = 1 : \mathbb{C}^2 \to \mathbb{C}^2.\end{aligned}$$

□

Now it is essential for us to rewrite the expressions $\theta^{(0)}(\lambda), \theta^{(1)}(\lambda)$ as matrix solutions. Thus, let us introduce the matrix-valued function of the variable $\lambda \in \mathbb{R}$ (see (9.3.4)), namely,

$$\begin{aligned}P_n(\lambda) &= \begin{bmatrix} P_{n;0,0}(\lambda) & P_{n;0,1}(\lambda) \\ P_{n;1,0}(\lambda) & P_{n;1,1}(\lambda) \end{bmatrix} = \begin{bmatrix} \theta^{(0)}_{n,0}(\lambda) & \theta^{(1)}_{n,0}(\lambda) \\ \theta^{(0)}_{n,1}(\lambda) & \theta^{(1)}_{n,1}(\lambda) \end{bmatrix}, \quad n \in \mathbb{N}, \\ P_0(\lambda) &= \begin{bmatrix} \theta^{(0)}_{0,0}(\lambda) & \theta^{(1)}_{0,0}(\lambda) \end{bmatrix} = \begin{bmatrix} 1 & 0 \end{bmatrix}.\end{aligned} \tag{9.3.14}$$

The following result holds true.

Lemma 9.3.3 *Matrices $P_n(\lambda)$, $n \in \mathbb{N}_0$ satisfy equations*

$$\begin{aligned}&b_0 P_0(\lambda) + a_0^* P_1(\lambda) = \lambda P_0(\lambda), \\ &a_{n-1}P_{n-1}(\lambda) + b_n P_n(\lambda) + a_n P_{n+1}(\lambda) = \lambda P_n(\lambda), \quad n \in \mathbb{N}, \quad \forall \lambda \in \mathbb{R}.\end{aligned} \tag{9.3.15}$$

Conversely, the solution $P(\lambda) = (P_n(\lambda))_{n=0}^{\infty}$ *of the Eq.* (9.3.15) *with initial data*

$$P_{0;0,0}(\lambda) = 1, \quad P_{0;0,1}(\lambda) = 0; \quad P_{1;0,0}(\lambda) = 0, \quad P_{1;0,1}(\lambda) = 1 \tag{9.3.16}$$

has the form (9.3.14).

Proof Let us apply the left-hand side of the second equation in (9.3.15) to the vector $(1, 0) \in \mathbb{C}^2$. By using (9.3.14), we get

$$a_{n-1}(\theta_{n-1,0}^{(0)}(\lambda), \theta_{n-1,1}^{(0)}(\lambda)) + b_n(\theta_{n,0}^{(0)}(\lambda), \theta_{n,1}^{(0)}(\lambda)) + a_n(\theta_{n+1,0}^{(0)}(\lambda), \theta_{n+1,1}^{(0)}(\lambda)), \tag{9.3.17}$$

that is, the left-hand side of the second equality in (9.3.3) for the solution $\varphi(\lambda) = \theta^{(0)}(\lambda)$. Thus (9.3.17), according to (9.3.15), equals

$$\lambda(\theta_{n,0}^{(0)}(\lambda), \theta_{n,1}^{(0)}(\lambda)) = \lambda P_n(\lambda)(1, 0).$$

As a result, the second equality in (9.3.15), after these applications to the vector (0, 1) is fulfilled.

Similarly, it is proved that the second equality in (9.3.15), after the substitution of the vector (0, 1), also holds true (now, $\varphi(\lambda) = \theta^{(1)}(\lambda)$). Since the vectors (1, 0), (0, 1) are basic in $\mathbb{C}^2$, then the second equality in (9.3.15) is fulfilled.

The first equality is also satisfied: the applying its right-hand side to the vector (1, 0) gives

$$b_0(\theta_{0,0}^{(0)}(\lambda), 0) + a_0^*(\theta_{1,0}^{(0)}(\lambda), \theta_{1,1}^{(0)}(\lambda)),$$

that is, the left side of the equation (9.3.3) for $\varphi(\lambda) = \theta^{(0)}(\lambda)$. This application of the right-hand side gives λ, which is necessary.

Conversely, the lemma is obvious, Since the Eqs. (9.3.15) and (9.3.3) are equivalent, then the solution

$$\varphi(\lambda) = (P_{0;0,0}(\lambda), (P_{1;0,0}(\lambda), P_{1;1,0}(\lambda)), (P_{2;0,0}(\lambda), P_{2;1,0}(\lambda)), \ldots)$$

of Eqs. (9.3.3) has the first initial data (9.3.4), and

$$\varphi(\lambda) = (P_{0;0,1}(\lambda), (P_{1;0,1}(\lambda), P_{1;1,1}(\lambda)), (P_{2;0,1}(\lambda), P_{2;1,1}(\lambda)), \ldots)$$

has the second initial data. This representation implies (9.3.14). □

We formulate the results obtained above in the form of a theorem on the direct spectral problem for the operator **J**.

Theorem 9.3.4 *Consider the Jacobi matrix J of the form* (9.2.9) *with conditions on its coefficients, formulated at the beginning of the section. This matrix generates* *a self-adjoint bounded operator* $\mathbf{J}$ *in the space* $\mathbf{l}_2$ *in* (9.2.1). *The equipment* (9.3.1) *is associated with* $\mathbf{J}$.

The direct spectral problem for $\mathbf{J}$ *has the following form. The complete system of generalized eigenvectors of the operator* $\mathbf{J}$ *has a form*

$$\varphi_\alpha(\lambda) = \big((P_{n;0,\alpha}(\lambda), P_{n;1,\alpha}(\lambda))_{n=0}^{\infty}\big) \in \mathbf{l}, \quad \alpha = 0, 1, \quad \forall\lambda \in \mathbb{R}, \tag{9.3.18}$$

where "matrix polynomials"

$$P_n(\lambda) = \begin{bmatrix} P_{n;0,0}(\lambda) & P_{n;0,1}(\lambda) \\ P_{n;1,0}(\lambda) & P_{n;1,1}(\lambda) \end{bmatrix} \tag{9.3.19}$$

are solutions of the system (9.3.15) *with two initial conditions* (9.3.16) (*the last index in* $P_{n;\alpha,\beta}(\lambda)$ *is a number of a solution* (9.3.19)).

This system is complete in the sense that if we introduce $\forall f \in \mathbf{l}_{\text{fin}}$ *and* $\lambda \in \mathbb{R}$ ($\mathbb{C}^2$ *is a vectors value*) *the Fourier transform*

$$\begin{aligned} \widehat{f}(\lambda) &= (\widehat{f}_0(\lambda), \widehat{f}_1(\lambda)) \in \mathbb{C}^2, \\ \widehat{f}(\lambda) &= \sum_{n=0}^{\infty} P_n^*(\lambda) f_n = \\ &= f_0 + \sum_{n=1}^{\infty} (P_{n;0,0}(\lambda) f_{n,0} + P_{n;1,0}(\lambda) f_{n,1}) \\ &\quad + \sum_{n=1}^{\infty} (P_{n;0,1}(\lambda) f_{n,0} + P_{n;1,1}(\lambda) f_{n,1}), \end{aligned} \tag{9.3.20}$$

then we get the Parseval equality

$$(f, g)_{\mathbf{l}_2} = \int_{\mathbb{R}} (d\rho_M(\lambda) \widehat{f}(\lambda), \widehat{g}(\lambda))_{\mathbb{C}^2}, \quad f, g \in \mathbf{l}_{\text{fin}}. \tag{9.3.21}$$

Here, $d\rho_M(\lambda) = (d\rho_{\alpha,\beta}(\lambda))_{\alpha,\beta=0}^{1}$ *is a* (2×2)*-matrix of the non-negative spectral measure of the operator* $\mathbf{J}$*; it is the probability measure, i.e.,* $\rho_M(\mathbb{R}) = 1$*. The imbedding* $\mathbf{l}_{\text{fin}} \ni f \mapsto \widehat{f}(\lambda)$ *after the extension by continuity is the isometric operator between spaces* $\mathbf{l}_2$ *and* $L_2(\mathbb{C}^2, \mathbb{R}, d\rho_M(\lambda))$.

Note that, here $L_2 := L_2(\mathbb{C}, \mathbb{R}, d\rho_M(\lambda))$ denotes the Hilbert space of functions $\mathbb{R} \ni \lambda \mapsto F(\lambda) \in \mathbb{C}^2$, constructed by the scalar product

$$(F, G)_{L_2} = \int_{\mathbb{R}} (d\rho_M(\lambda) F(\lambda), G(\lambda))_{\mathbb{C}^2} \tag{9.3.22}$$

in the classical way: first the integral (9.3.22) is introduced for the simple functions $F(\lambda)$, $G(\lambda)$ (that is, finite combinations of $\mathbb{C}^2-$ characteristic functions) and then it is taken the completion (for more details, see Chapter 2).

Proof The first part of the theorem follows from Lemma 9.3.3. For the second part, it should first be noted that definitions (9.3.11) and (9.3.20) with (9.3.14), are the same. Then, it is necessary to apply (9.3.13). The density $\widehat{f}(\lambda)$, $f \in \mathbf{l}_2$ will be discussed later, when the inverse spectral problem for **J** will be considered. □

Let us pass to the inverse spectral problem for **J**, that is, to the restore the matrix J by its spectral measure $d\rho_M(\lambda)$.

First, we present a simple corollary from the Parseval equality (9.3.21), which we will understand as its extension for all $f, g \in \mathbf{l}_2$. Let in the expression (9.3.21), $g = \delta_{n,v_n}$, then according to (9.3.20), we have

$$(\widehat{\delta_{n,v_n}})(\lambda) = (P_{n;v_n,0}(\lambda), P_{n;v_n,1}(\lambda)) = P_n^*(\lambda)\delta_{n,v_n}, \quad n \in \mathbb{N}_0, \quad v_n = 0, 1. \tag{9.3.23}$$

Thus, we have the formula for restoring f by its Fourier image

$$\begin{aligned} f_{n,v_n} &= \int_{\mathbb{R}} (d\rho_M(\lambda)\widehat{f}(\lambda), (P_{n;v_n,0}(\lambda), P_{n;v_n,1}(\lambda)))_{\mathbb{C}^2} \\ &= \int_{\mathbb{R}} (P_n(\lambda)\, d\rho_M(\lambda)\widehat{f}(\lambda), \delta_{n,v_n})_{\mathbb{C}^2}. \end{aligned} \tag{9.3.24}$$

By setting $f = \delta_{j,v_j}$ in (9.3.14) and using (9.3.23), we obtain orthogonality conditions for $P_{n;\alpha,\beta}(\lambda)$, namely,

$$\begin{aligned} &\int_{\mathbb{R}} (d\rho_M(\lambda)(P_{j;\alpha,0}(\lambda), P_{j;\alpha,1}(\lambda)), (P_{k;\beta,0}(\lambda), P_{k;\beta,1}(\lambda)))_{\mathbb{C}^2} \\ &= (\delta_{j,\alpha})_{k,\beta} = \delta_{j,k}\delta_{\alpha,\beta}, \quad j, k \in \mathbb{N}_0, \quad \alpha, \beta = 0, 1 \end{aligned} \tag{9.3.25}$$

(at the end of the expression (9.3.25), $\delta_{m,n}$ denotes the usual Kronecker symbol).

It follows from (9.3.25) that the system of $\mathbb{R}^2$-valued functions

$$\begin{aligned} &(P_{0;0,0}(\lambda), P_{0;0,1}(\lambda)); (P_{1;0,0}(\lambda), P_{1;0,1}(\lambda)), (P_{1;1,0}(\lambda), P_{1;1,1}(\lambda)); \\ &(P_{2;0,0}(\lambda), P_{2;0,1}(\lambda)), (P_{2;1,0}(\lambda), P_{2;1,1}(\lambda)); \ldots, \quad \lambda \in \mathbb{R}, \end{aligned} \tag{9.3.26}$$

consists orthogonal elements.

Using equalities (9.3.15) and initial data (9.3.16), one can simply calculate $P_n(\lambda), n \in \mathbb{N}_0$ step by step. Thus, we have

$$P_0(\lambda) = [1 \quad 0] =: \begin{bmatrix} 1 & 0 \\ 0 & 0 \end{bmatrix}, \quad P_1(\lambda) = \begin{bmatrix} 0 & 1 \\ \frac{1}{\alpha_{-1}}(\lambda - \beta_0) & -\frac{\alpha_0}{\alpha_{-1}} \end{bmatrix}. \tag{9.3.27}$$

The matrix polynomials $P_2(\lambda), P_3(\lambda), \ldots$ have a more complicated distribution of λ by elements of the matrix $P_n(\lambda)$. It is significant that matrices $a_n, b_n, n \in \mathbb{N}$ are diagonal in (9.2.9) (it is possible to construct a similar theory with non-diagonal matrices, but it is more difficult). Hence, we have a lemma.

Lemma 9.3.5 *Assume that all matrices $a_n, b_n, n \in \mathbb{N}$ in (9.2.9) are diagonal. Then the matrix for $n \geq 2$ has polynomial elements $P_n(\lambda)$ and these polynomials have corresponding degree, i.e., $P_{n;0,0}(\lambda)$ has the degree $n-2$, $P_{n;0,1}(\lambda)$ and $P_{n;1,1}(\lambda)$ have the degree $n-1$, and $P_{n;1,0}(\lambda)$ has the degree n. Higher coefficients are non-zero.*

Proof First, we note that if c is a diagonal (2×2)-matrix with non-zero diagonal elements $c_{0,0}, c_{1,1} \in \mathbb{C}$, then their multiplication by the matrix $d = (d_{\alpha,\beta})_{\alpha,\beta=0}^1$ is a matrix with elements equal to the product of $d_{\alpha,\beta}$ by a scalar $c_{0,0}$ or $c_{1,1}$. Thus, the structure of the matrix d is preserved after multiplying it on the left by c.

Let $n = 2$. Then, from (9.3.15), we have

$$P_2(\lambda) = a_1^{-1}(\lambda P_1(\lambda) - b_1 P_1(\lambda) - a_0 P_0(\lambda))$$

and it has a structure specified in the lemma according to (9.3.27). For $\{n = 3, 4, \ldots\}$ we get from (9.3.15) step by step. The structure of the matrix $P_n(\lambda)$ is preserved. □

We denote by $R_m(\lambda)$ some ordinary polynomial of the degree $m \in \mathbb{N}_0$ with real coefficients and a non-zero (higher) leading coefficient. Then, from Lemma 9.3.5, we get that the sequence (9.3.26) of $\mathbb{R}^2$-polynomials has the form

$$\begin{aligned} &(1, 0); (0, R_0(\lambda)), (R_1(\lambda), R_0(\lambda)); \\ &(R_0(\lambda), R_1(\lambda)), (R_2(\lambda), R_1(\lambda)); \\ &(R_1(\lambda), R_2(\lambda)), (R_3(\lambda), R_2(\lambda)); \ldots; \\ &(R_n(\lambda), R_{n+1}(\lambda)), (R_{n+2}(\lambda), R_{n+1}(\lambda)); \ldots \end{aligned} \tag{9.3.28}$$

(with different polynomials $R_n(\lambda)$ in different places but with the same index n).

It is simple to construct $\mathbb{R}^2$-polynomials from linear combinations of given elementary $\mathbb{R}^2$-polynomials from (9.3.28).

Consider the sequence of basic $\mathbb{R}^2$-polynomials

$$(1, 0); (0, 1), (\lambda, 0); (0, \lambda), (\lambda^2, 0); \ldots; (0, \lambda^n), (\lambda^{n+1}, 0); \ldots. \tag{9.3.29}$$

Then all $\mathbb{R}^2$-polynomials $(P(\lambda), Q(\lambda))$, where $P(\lambda)$, $Q(\lambda)$ is some polynomial of degree $m, n \in \mathbb{N}_0$, respectively, is some linear combination of base polynomials (9.3.29). Namely, let

$$P(\lambda) = \sum_{j=0}^{m} p_j \lambda^j, \; Q(\lambda) = \sum_{k=0}^{n} q_k \lambda^k,$$

then it is obvious

$$(P(\lambda), Q(\lambda)) = \sum_{j=0}^{m} p_j (\lambda^j, 0) + \sum_{k=0}^{n} q_k (0, \lambda^k).$$

We can also say that if $m \in \mathbb{N}_0$ and $n = m + 1$, then the corresponding $\mathbb{R}^2$-polynomial $(P(\lambda), Q(\lambda))$ is a linear combination of the first $\mathbb{R}^2$-polynomials from (9.3.29) to $(0, \lambda^{m+1})$. If $m \in \mathbb{N}$ and $n = m - 1$, then for $(P(\lambda), Q(\lambda))$, it is necessary to take $(\lambda^m, 0)$ in (9.3.29).

To solve the inverse spectral problem, it is necessary to make the orthogonalization procedure for $\mathbb{R}^2$– polynomials (9.3.29) in the space

$$L_2(\mathbb{R}^2, \mathbb{R}, d\rho_M(\lambda)) \subset L_2(\mathbb{C}^2, \mathbb{R}, d\rho_M(\lambda)).$$

All $\mathbb{R}^2$-polynomials belong to the space $L_2(\mathbb{R}^2, \mathbb{R}, d\rho_M(\lambda))$ and they are linearly independent in this space, if the *measure* $d\rho_M(\lambda)$ *has a bounded support and is concentrated on a non-finite number of points in* $\mathbb{R}$ (the proof of this statement follows directly from a similar fact for ordinary polynomials and a scalar measure on $\mathbb{R}$).

We have the theorem follows from previous facts.

Theorem 9.3.6 *Let* $\mathbf{J}$ *be a bounded self-adjoint operator in the space* $\mathbf{l}_2$ *generated by the matrix* J *of the form* (9.2.9). *Its spectral measure* $d\rho_M(\lambda)$ *has a bounded support and is concentrated in an infinite number of points on* $\mathbb{R}$. *The corresponding to* $\mathbf{J}$ *generalized eigenvectors* (9.3.26) *can be obtained from the classical Gram-Schmidt orthogonalization procedure applied to the sequence* (9.3.29) *in the space* $L_2(\mathbb{R}^2, \mathbb{R}, d\rho_M(\lambda))$.

Proof First, let us explain the existence of the above-mentioned properties of the measure $d\rho_M(\lambda)$. Its support is bounded because the operator $\mathbf{J}$ is bounded. Assume that the measure $d\rho_M(\lambda)$ is concentrated in a finite number of points λ on $\mathbb{R}$. Then from (9.3.22), it follows that the space $L_2(\mathbb{C}^2, \mathbb{R}, d\rho_M(\lambda))$ must be finite-dimensional, but this contradicts of the isometry (9.3.21) between $\mathbf{l}_2$ and $L_2(\mathbb{C}^2, \mathbb{R}, d\rho_M(\lambda))$.

Consider the generalized eigenvectors (9.3.26) of the operator $\mathbf{J}$. The first and the second are $(1, 0)$, $(0, 1)$, that is, the first two vectors from (9.3.29). Then, we apply the Gram-Schmidt orthogonalization procedure in the space $L_2(\mathbb{R}^2, \mathbb{R}, d\rho_M(\lambda))$.

Lemma 9.3.5 and the orthonormality of vectors from (9.3.26) shows that the statement of the theorem holds true. □

Note also that the boundedness of the support $d\rho_M(\lambda)$ implies that the set of all $\mathbb{R}^2$-polynomials is dense in $L_2(\mathbb{R}^2, \mathbb{R}, d\rho_M(\lambda))$. Thus, Theorem 9.3.6 gives that the *set* $\{\widehat{f}(\lambda), f \in \mathbf{l}_{\text{fin}}\}$ *is dense in the space* $L_2(\mathbb{C}^2, \mathbb{R}, d\rho_M(\lambda))$ (this fact we use by the proof of Theorem 9.3.4).

The elements of the matrix J can be restored by the spectral measure $d\rho_M(\lambda)$ of the corresponding operator $\mathbf{J}$, similarly to classical Jacobi matrices. For this, it is convenient to introduce the integral of the form

$$\int_{\mathbb{R}} A(\lambda)\, d\rho_M(\lambda) B(\lambda) = C, \tag{9.3.30}$$

where $\mathbb{R} \ni \lambda \mapsto A(\lambda), B(\lambda)$ is a (2×2)-matrix-valued function. By definition, the matrix C is such that

$$(Cx, y)_{\mathbb{C}^2} = \int_{\mathbb{R}} (d\rho_M(\lambda) B(\lambda) x, A^*(\lambda) y)_{\mathbb{C}^2}, \quad \forall x, y \in \mathbb{C}^2.$$

By using the definitions (9.3.30) and (9.3.19), we can simply write the orthogonality conditions (9.3.25) in the form

$$\begin{aligned} &\int_{\mathbb{R}} P_j(\lambda)\, d\rho_M(\lambda) P_k^*(\lambda) = \delta_{j,k} 1, \quad j, k \in \mathbb{N}_0, \quad (j,k) \neq (0,0); \\ &\int_{\mathbb{R}} P_0(\lambda)\, d\rho_M(\lambda) P_0^*(\lambda) = \begin{bmatrix} 1 & 0 \\ 0 & 0 \end{bmatrix}, \end{aligned} \tag{9.3.31}$$

where, as in (9.3.25), $\delta_{j,k}$ denotes the Kronecker symbol.

Lemma 9.3.7 *Equalities*

$$\begin{aligned} a_n^* &= \int_{\mathbb{R}} \lambda P_n(\lambda)\, d\rho_M(\lambda) P_{n+1}^*(\lambda), \\ b_n &= \int_{\mathbb{R}} \lambda P_n(\lambda)\, d\rho_M(\lambda) P_n^*(\lambda), \quad n \in \mathbb{N}_0, \\ a_n^* &= a_n, \quad n \in \mathbb{N} \end{aligned} \tag{9.3.32}$$

hold true.

Proof Let $n \in \mathbb{N}$. Consider the second equality in (9.3.15), multiply it by $P^*_{n+1}(\lambda)$ and integrate over $\mathbb{R}$. The relation (9.3.31) gives the first formula in (9.3.32), since $a_n = a^*_n$.

To obtain the second formula in (9.3.32), it is necessary to repeat previous calculations with the replacement of $P^*_{n+1}(\lambda)$ with $P^*_n(\lambda)$.

Let us use the matrices (9.3.27) for $n = 0$ and the first equality in (9.3.15). Then, as before, we get the formula (9.3.32) for $n = 0$. □

The previous considerations lead to the theorem.

Theorem 9.3.8 *Let J in* (9.2.9) *be a block Jacobi matrix with the properties specified at the beginning of Sect.* 9.2 *with diagonal matrices $a_n, b_n, n \in \mathbb{N}$. Then elements of the matrix J are restored according to formulas* (9.3.32).

It is also possible to construct the matrix J using formulas (9.3.32) according to the previously specified (2×2)-matrix measure $d\rho_M(\lambda)$, which was discussed above. First, it is necessary to construct the matrix polynomials $P(\lambda), n \in \mathbb{N}_0$. For this construction, the Gram-Schmidt orthogonalization procedure is used in the space $L_2(\mathbb{R}^2, \mathbb{R}, d\rho_M(\lambda))$ for $\mathbb{R}^2$-polynomials (9.3.29) and the sequence (9.3.26) is obtained. Formulas (9.3.19) give $P_n(\lambda), n \in \mathbb{N}_0$.

Thus, it can be proved that the matrix J, constructed in this way, has the necessary properties and $a_n, b_n, n \in \mathbb{N}$ are diagonal matrices; the given measure $d\rho_M(\lambda)$ is the spectral measure of the corresponding operator $\mathbf{J}$. But we will not discuss this result in detail.

At the end of this section, we present additional constructions related to the direct spectral problem for the matrix J (9.2.9) and corresponding operators $\mathbf{J}$ (although they will not be used in the following text).

I. We recall that matrix polynomials (9.3.19) are analogues of polynomials $P_n(\lambda)$ of the first kind associated with the classical Jacobi matrices. Let us now construct an analogue of the polynomial $Q_n(\lambda)$ of the second kind (see Chapter 3, Sect. 3.4). The matrix J can be more general: $\forall n \in \mathbb{N}$, i.e., the matrices a_n, b_n are not necessarily diagonal.

From (9.3.2), (9.3.7), (9.3.8) and (9.3.11) it follows that $\forall z \in \mathbb{C} \setminus \mathbb{R}$ for the resolvent $\mathbf{R}_z$ of the operator $\mathbf{J}$, we have:

$$\widehat{(\mathbf{R}_z f)}(\lambda) = (\lambda - z)^{-1} \widehat{f}(\lambda), \quad \lambda \in \mathbb{R}, \quad f \in \mathbf{l}_2. \tag{9.3.33}$$

Therefore, (9.3.23) gives

$$\widehat{(\mathbf{R}_z \delta_{n,\nu_n})}(\lambda) = (\lambda - z)^{-1} (P_{n;\nu_n,0}(\lambda), P_{n;\nu_n,1}(\lambda)) = (\lambda - z)^{-1} P^*_n(\lambda) \delta_{n,\nu_n},$$
$$\nu_n = 0, 1, \quad \forall n \in \mathbb{N}_0, \quad z \in \mathbb{C} \setminus \mathbb{R}, \quad \lambda \in \mathbb{R}. \tag{9.3.34}$$

By using (9.3.24) with a change of n on m, from (9.3.34), $\forall z \in \mathbb{C} \setminus \mathbb{R}$, we get

$$(\mathbf{R}_z \delta_{m,\beta})_{n,\alpha} = \int_{\mathbb{R}} \frac{1}{\lambda - z} (P_n(\lambda)\, d\rho_M(\lambda) \widehat{\delta}_{m,\beta}(\lambda), \delta_{n,\alpha})_{\mathbb{C}^2}, \quad \alpha = 0, 1, \quad \text{i.e.,}$$

$$(\mathbf{R}_z \delta_{m,\beta})_n = \int_{\mathbb{R}} \frac{1}{\lambda - z} P_n(\lambda)\, d\rho_M(\lambda) \widehat{\delta}_{m,\beta}(\lambda), \quad m, n \in \mathbb{N}_0, \quad \beta = 0, 1. \tag{9.3.35}$$

By using the last equality, we get

$$\begin{aligned}(\mathbf{R}_z \delta_{m,\beta})_n &= \int_{\mathbb{R}} (\lambda - z)^{-1} (P_n(\lambda) - P_n(z))\, d\rho_M(\lambda) \widehat{\delta}_{m,\beta}(\lambda) \\ &\quad + P_n(z) \int_{\mathbb{R}} (\lambda - z)^{-1}\, d\rho_M(\lambda) \widehat{\delta}_{m,\beta}(\lambda), \\ &\quad m, n \in \mathbb{N}_0, \quad \beta = 0, 1, \quad z \in \mathbb{C} \setminus \mathbb{R}.\end{aligned} \tag{9.3.36}$$

Let $m = 0$, $\beta = 0$ and $m = 1$, $\beta = 0$ in (9.3.36). According to (9.3.23) and (9.3.27), we have

$$\begin{aligned}(\mathbf{R}_z \delta_{0,0})_n &= \int_{\mathbb{R}} (\lambda - z)^{-1} (P_n(\lambda) - P_n(z))\, d\rho_M(\lambda)(1, 0) \\ &\quad + P_n(z) \int_{\mathbb{R}} (\lambda - z)^{-1}\, d\rho_M(\lambda)(1, 0), \\ (\mathbf{R}_z \delta_{1,0})_n &= \int_{\mathbb{R}} (\lambda - z)^{-1} (P_n(\lambda) - P_n(z))\, d\rho_M(\lambda)(0, 1) \\ &\quad + P_n(z) \int_{\mathbb{R}} (\lambda - z)^{-1}\, d\rho_M(\lambda)(0, 1), \quad \forall n \in \mathbb{N}_0.\end{aligned} \tag{9.3.37}$$

Let $x = (x_0, x_1)$, $y = (y_0, y_1)$ be two arbitrary vectors from $\mathbb{C}^2$; consider (2×2)-matrix

$$\begin{bmatrix} x_0 & y_0 \\ x_1 & y_1 \end{bmatrix} =: [x \quad y]. \tag{9.3.38}$$

In such notations, two equalities (9.3.37) can be written in the form

$$\begin{aligned}\left[(\mathbf{R}_z\delta_{0,0})_n\ (\mathbf{R}_z\delta_{1,0})_n\right] &= \int_{\mathbb{R}} (\lambda - z)^{-1}(P_n(\lambda) - P_n(z))\, d\rho_M(\lambda) \\ &+ P_n(z)\int_{\mathbb{R}} (\lambda - z)^{-1}\, d\rho_M(\lambda), \quad n \in \mathbb{N}_0, \quad z \in \mathbb{C}\setminus\mathbb{R}\end{aligned} \tag{9.3.39}$$

(the symmetry of the matrix $d\rho_M(\lambda)$ is used here).

As for classical Jacobi matrices, we can say that

$$Q_n(z) = \int_{\mathbb{R}} (\lambda - z)^{-1}(P_n(\lambda) - P_n(z))\, d\rho_M(\lambda), \quad n \in \mathbb{N}_0, \quad z \in \mathbb{C}\setminus\mathbb{R} \tag{9.3.40}$$

is a matrix of polynomials of the second kind. The matrix-valued function

$$m(z) = \int_{\mathbb{R}} (\lambda - z)^{-1}\, d\rho_M(\lambda), \quad z \in \mathbb{C}\setminus\mathbb{R}, \tag{9.3.41}$$

is, by definition, a Weyl matrix function. The expression (9.3.39) has now a form

$$\left[(\mathbf{R}_z\delta_{0,0})_n\ (\mathbf{R}_z\delta_{1,0})_n\right] = Q_n(z) + P_n(z)m(z), \quad n \in \mathbb{N}_0, \quad z \in \mathbb{C}\setminus\mathbb{R}. \tag{9.3.42}$$

Thus, the important relation (9.3.42) is proved.

Now we will show that (as in the classical theory) the sequence $(Q_n(z))_{n=0}^{\infty}$ is a solution of the difference equations (9.3.15) for $n \in \mathbb{N}$ but with another initial data like $(P_n(\lambda))_{n=0}^{\infty}$.

Lemma 9.3.9 *The sequence* $Q(z) := (Q_n(z))_{n=0}^{\infty}$, $z \in \mathbb{C}$ *is correctly defined as a solution of difference equations* (9.3.15) *for* $n \in \mathbb{N}$ *with* λ *changed to* z *and an initial data*

$$Q_0(z) = 0, \quad Q_1(z) = \begin{bmatrix} 0 & 0 \\ \alpha_{-1}^{-1} & 0 \end{bmatrix}. \tag{9.3.43}$$

Proof First, note that all elements of the matrix $P_n(\lambda)$ are polynomials, thus the expression $(\lambda - z)^{-1}(P_n(\lambda) - P_n(z))$ is correctly defined for all $z \in \mathbb{C}$ and (9.3.40) holds true for all $z \in \mathbb{C}$. If we apply the difference expression $\mathcal{L}$ from the second equality on the left (9.3.15)–(9.3.40), we get

$$\begin{aligned}(\mathcal{L}Q(t))_n &= \int_{\mathbb{R}} (\lambda - z)^{-1}((\mathcal{L}P(\lambda))_n - (\mathcal{L}P(t))_n)\, d\rho_M(\lambda) \\ &= \int_{\mathbb{R}} (\lambda - z)^{-1}(\lambda P_n(\lambda) - zP_n(z))\, d\rho_M(\lambda)\end{aligned}$$

$$= z \int_{\mathbb{R}} (\lambda - z)^{-1}(P_n(\lambda) - P_n(z))\, d\rho_M(\lambda)$$

$$+ \int_{\mathbb{R}} P_n(\lambda)\, d\rho_M(\lambda) = zQ_n(z), \quad \forall n \in \mathbb{N},\ z \in \mathbb{C},$$

(the equality $\int_{\mathbb{R}} P_n(\lambda)\, d\rho(\lambda) = 0$ follows from (9.3.31) and $\rho_M(\mathbb{R}) = 1$).

The initial data (9.3.43) follows from (9.3.27) and $\rho_M(\mathbb{R}) = 1$. Note that the solution of the second equation in (9.3.15) can be found by replacing $P_n(\lambda)$ with $Q_n(\lambda)$ step by step for $n \in \mathbb{N}$, because the matrix element a_n has an inverse one.

□

Remark 9.3.10 It should be noted that the above spectral theory holds true also for general Hermitian matrices J of the form (9.2.9), when the corresponding operator $\mathbf{J}$ is not bounded in the space $\mathbf{l}_2$. But it is Hermitian with equal deficiency indices (since J is real and symmetric). Thus, it has self-adjoint extensions in the space $\mathbf{l}_2$. These extensions can be described, and the representation (9.3.42) plays an essential role in the description.

II. The construction of the previous section makes it possible to find the Weyl function $m(z)$ in a certain way (9.3.41) and, therefore also the spectral measure $d\rho_M(\lambda)$. Let us explain this approach. For the matrix J of the form (9.2.9) and its matrix polynomials $P_n(z)$ of the form (9.3.19), we consider the series

$$\sum_{n=0}^{\infty} P_n^*(z) P_n(z), \quad z \in \mathbb{C}. \tag{9.3.44}$$

Just understand that the *series* (9.3.44) *is divergent in the sense that*

$$\lim_{N\to\infty} \sum_{n=0}^{N} (P_n^*(z) P_n(z)x, x)_{\mathbb{C}^2} = \sum_{n=0}^{\infty} \| P_n(z)x \|^2_{\mathbb{C}^2} = \infty, \tag{9.3.45}$$

where $z \in \mathbb{C} \setminus \mathbb{R}$, $\forall x \in \mathbb{C}^2$, $x \neq 0$.

Indeed, it follows from (9.3.15) that in the case of a finite sum in (9.3.45), the vector $\varphi(z) = (P_n(z)x)_{n=0}^{\infty}$ belongs to $\mathbf{l}_2$ and is a non-zero eigenvector of the operator $\mathbf{J}$ with a complex eigenvalue z. And this is impossible, since $\mathbf{J}$ is self-adjoint.

Remark 9.3.11 Such a situation also exists for J with unbounded elements, but in the case when the corresponding operator $\mathbf{J}$ is self-adjoint in $\mathbf{l}_2$. Now we can formulate and prove a simple, but in some cases useful theorem.

Theorem 9.3.12 *Suppose, we can find a* (2×2)*-matrix function* $\mathbb{C} \setminus \mathbb{R} \ni z \mapsto f(z)$ *for which*

$$\sum_{n=0}^{\infty} \|(Q_n(z) + P_n(z) f(z))x\|^2_{\mathbb{C}^2} < \infty, \quad \forall x \in \mathbb{C}^2. \tag{9.3.46}$$

Then, $m(z) = f(z)$, $z \in \mathbb{C} \setminus \mathbb{R}$.

Proof First, we note that the condition (9.3.46) is fulfilled for $f(z) = m(z)$. Indeed, using (9.3.42) and (9.3.38), we get that

$$\begin{aligned}
&(Q_n(z) + P_n(z) m(z))x \\
&= ((\mathbf{R}_z \delta_{0,0})_{n,0} x_0 + (\mathbf{R}_z \delta_{1,0})_{n,0} x_1, (\mathbf{R}_z \delta_{0,0})_{n,1} x_0 + (\mathbf{R}_z \delta_{1,0})_{n,1} x_1)), \\
&\quad n \in \mathbb{N}_0, \quad \forall x = (x_0, x_1) \in \mathbb{C}^2
\end{aligned}$$

and the last sequence of vectors from $\mathbb{C}^2$ belongs to $\mathbf{l}_2$.

From (9.3.46), for $m(z)$ and $f(z)$, we get

$$\begin{aligned}
&\Big(\sum_{n=0}^{\infty} \|P_n(z)(m(z) - f(z))x\|^2_{\mathbb{C}^2}\Big)^{1/2} \\
&\quad \leq \Big(\sum_{n=0}^{\infty} \|(Q_n(z) + P_n(z) m(z))x\|^2_{\mathbb{C}^2}\Big)^{1/2} \\
&\quad + \Big(\sum_{n=0}^{\infty} \|(Q_n(z) + P_n(z) f(z))x\|^2_{\mathbb{C}^2}\|\Big)^{1/2} < \infty.
\end{aligned} \tag{9.3.47}$$

If $m(z) \neq f(z)$ for some $z \in \mathbb{C} \setminus \mathbb{R}$, then there exists $x \in \mathbb{C}^2$, for which $(m(z) - f(z))x \neq 0$ and the inequality (9.3.47) contradicts (9.3.45). □

9.4 Equations for the Weyl Function and Spectral Matrices

In Sect. 9.2 it is shown that the doubly infinite Toda equation can be written as the Lax equation (9.2.11) for the block Jacobi matrices $J(t)$, $A(t)$. The corresponding operator $\mathbf{J}(t)$ in the space $\mathbf{l}_2$ of the form (9.2.1) for each fixed $t \in [0, T)$, is a block Jacobi type operator that was studied in Sect. 9.3. Our goal now is to get the equation for the Weyl matrix function (9.3.41) from (9.2.11) and the derivatives of the spectral measure. The result is similar to Theorems 9.1.1, 9.1.2.

We start by deriving the corresponding differential equations for the Weyl matrix function $m(z;t)$ of the operator $\mathbf{J}(t)$ in the space $\mathbf{l}_2$. According to (9.3.41), this (2×2)-matrix function is

$$m(z;t) = \begin{bmatrix} m_{0,0}(z;t) & m_{0,1}(z;t) \\ m_{1,0}(z;t) & m_{1,1}(z;t) \end{bmatrix} = \int_{\mathbb{R}} \frac{1}{\lambda - z}\, d\rho_M(\lambda;t)$$

$$= \begin{bmatrix} (\mathbf{R}_z(t)\delta_{0,0}, \delta_{0,0})_{\mathbf{l}_2} & (\mathbf{R}_z(t)\delta_{1,0}, \delta_{0,0})_{\mathbf{l}_2} \\ (\mathbf{R}_z(t)\delta_{0,0}, \delta_{1,0})_{\mathbf{l}_2} & (\mathbf{R}_z(t)\delta_{1,0}, \delta_{1,0})_{\mathbf{l}_2} \end{bmatrix}, \quad z \in \mathbb{C} \setminus \mathbb{R}, \quad t \in [0, T), \tag{9.4.1}$$

where $d\rho_M(\lambda;t)$ is a matrix spectral measure of the operator $\mathbf{J}(t)$ and $\mathbf{R}_z(t)$ is its resolvent (see (9.2.2))

$$\delta_{0,0} = (1, (0,0), (0,0), \ldots), \quad \delta_{1,0} = (0, (1,0), (0,0), \ldots). \tag{9.4.2}$$

Note that the operator $\mathbf{J}(t)$ is self-adjoint and real. It gives that $\mathbf{R}_z^*(t) = \mathbf{R}_{\bar{z}}(t) = \overline{\mathbf{R}_z(t)}$, where the line over text "—" denotes the complex conjugation in the space $\mathbf{l}_2$. The last equality and reality of δ_{n,v_n} give

$$(\mathbf{R}_z(t)\delta_{k,v_k}, \delta_{j,v_j})_{\mathbf{l}_2} = (\delta_{k,v_k}, \overline{\mathbf{R}_z(t)\delta_{j,v_j}})_{\mathbf{l}_2} = (\mathbf{R}_z(t)\delta_{j,v_j}, \delta_{k,v_k})_{\mathbf{l}_2}.$$

Thus, we have

$$\begin{aligned} &(\mathbf{R}_z(t)\delta_{k,v_k}, \delta_{j,v_j})_{\mathbf{l}_2} = (\mathbf{R}_z(t)\delta_{j,v_j}, \delta_{k,v_k})_{\mathbf{l}_2}, \quad \forall j,k \in \mathbb{N}_0, \; v_j, v_k = 0, 1, \\ &m_{0,1}(z;t) = m_{1,0}(z;t), \quad z \in \mathbb{C} \setminus \mathbb{R}, \quad t \in [0, T). \end{aligned} \tag{9.4.3}$$

Our immediate goal is to construct a matrix analog of Theorem 9.1.1. To find a system of differential equations for t with respect to the functions $m_{\alpha,\beta}(z;t)$, we will use the idea of deriving the equation (9.1.9) from the Lax equation, but now this derivation is more complicated than in Sect. 9.1, since the block matrices $R_z(t)$ corresponding to the operators $\mathbf{R}_z(t)$ and $A(t)$ of the form (9.2.12) are cumbersome.

We apply the general construction (9.2.3)–(9.2.6) to our operator $\mathbf{R}_z(t)$. Thus, for the corresponding matrices R_z (for a fixed t) we have, using (9.4.3),

$$\begin{aligned} &R_z = (R_{z;j;k})_{j,k=0}^{\infty} = (R_{z;j,v_j;k,v_k})_{j,k=0,v_j,v_k=0,1}^{\infty}, \\ &R_{z;j,v_j;k,v_k} = (\mathbf{R}_z\delta_{k,v_k}, \delta_{j,v_j})_{\mathbf{l}_2} = (\mathbf{R}_z\delta_{j,v_j}, \delta_{k,v_k})_{\mathbf{l}_2}. \end{aligned} \tag{9.4.4}$$

In particular $\forall z \in \mathbb{C} \setminus \mathbb{R}$, $t \in [0, T)$

$$\begin{aligned} R_{z;0,0;0,0} &= m_{0,0}(z;t), \\ R_{z;0,0;1,0} &= m_{0,1}(z;t), \\ R_{z;1,0;0,0} &= m_{1,0}(z;t) = m_{0,1}(z;t), \\ R_{z;1,0;1,0} &= m_{1,1}(z;t). \end{aligned} \tag{9.4.5}$$

So, we start with the Lax equation (9.2.11) and make transformations as in Sect. 9.1. For the resolvent $\mathbf{R}_z(t)$ of the operator $\mathbf{J}(t)$ from the second equality in (9.2.11), we simply obtain

$$\begin{aligned} &\dot{\mathbf{R}}_z(t) = -\mathbf{R}_z(t)(\mathbf{J}(t) - z\mathbf{1})^{\cdot}\mathbf{R}_z(t) = -\mathbf{R}_z(t)\dot{\mathbf{J}}(t)\mathbf{R}_z(t) = [\mathbf{R}_z(t), \mathbf{A}(t)], \quad \text{or} \\ &\dot{R}_z(t) = [R_z(t), A(t)] = R_z(t)A(t) - A(t)R_z(t), \quad z \in \mathbb{C} \setminus \mathbb{R}, \quad t \in [0, T). \end{aligned} \tag{9.4.6}$$

Let us calculate $\dot{m}_{\mu,\nu}(z;t)$ for $\mu, \nu = 0.1$. If we notice that for the adjoint operator $\mathbf{A}^*(t) = -\mathbf{A}(t)$, then we get

$$\begin{aligned} \dot{m}_{0,0}(z;t) &= (\dot{\mathbf{R}}_z(t)\delta_{0,0}, \delta_{0,0})_{\mathbf{l}_2} = ((\mathbf{R}_z(t)\mathbf{A}(t) - \mathbf{A}(t)\mathbf{R}_z(t))\delta_{0,0}, \delta_{0,0})_{\mathbf{l}_2} \\ &= (\mathbf{R}_z(t)\mathbf{A}(t)\delta_{0,0}, \delta_{0,0})_{\mathbf{l}_2} + (\mathbf{R}_z(t)\delta_{0,0}, \mathbf{A}(t)\delta_{0,0})_{\mathbf{l}_2}. \end{aligned} \tag{9.4.7}$$

According to (9.2.12) and (9.2.6), (9.4.4), we have

$$\mathbf{A}(t)\delta_{0,0} = \Big(0, \Big(\frac{1}{2}\alpha_0, -\frac{1}{2}\alpha_{-1}\Big), 0, 0, \ldots\Big) = \frac{1}{2}\alpha_0\delta_{1,0} - \frac{1}{2}\alpha_{-1}\delta_{1,1}; \tag{9.4.8}$$

$$(\mathbf{R}_z(t)f)_j = \sum_{k=0}^{\infty} R_{z;j;k}(t)f_k, \quad R_{z;j,v_j;k,v_k}(t) = (\mathbf{R}_z(t)\delta_{k,v_k}, \delta_{j,v_j}), \tag{9.4.9}$$

$$v_j, v_k = 0, 1, \quad f \in \mathbf{l}_2.$$

Therefore,

$$\begin{aligned} &(\mathbf{R}_z(t)\mathbf{A}(t)\delta_{0,0}, \delta_{0,0})_{\mathbf{l}_2} = (\mathbf{R}_z(t)(\frac{1}{2}\alpha_0\delta_{1,0} - \frac{1}{2}\alpha_{-1}\delta_{1,1}), \delta_{0,0})_{\mathbf{l}_2} \\ &= \frac{1}{2}\alpha_0(\mathbf{R}_z(t)\delta_{1,0}, \delta_{0,0})_{\mathbf{l}_2} - \frac{1}{2}\alpha_{-1}(\mathbf{R}_z(t)\delta_{1,1}, \delta_{0,0})_{\mathbf{l}_2}; \\ &(\mathbf{R}_z(t)\delta_{0,0}, \mathbf{A}(t)\delta_{0,0})_{\mathbf{l}_2} = (\mathbf{R}_z(t)\delta_{0,0}, \frac{1}{2}\alpha_0\delta_{1,0} - \frac{1}{2}\alpha_{-1}\delta_{1,1})_{\mathbf{l}_2} \\ &= \frac{1}{2}\alpha_0(\mathbf{R}_z(t)\delta_{0,0}, \delta_{1,0})_{\mathbf{l}_2} - \frac{1}{2}\alpha_{-1}(\mathbf{R}_z(t)\delta_{0,0}, \delta_{1,1})_{\mathbf{l}_2}. \end{aligned}$$

From (9.4.7), the last two equations, and (9.4.3), we get

$$\begin{aligned}\dot{m}_{0,0}(z;t) &= \alpha_0 m_{0,1}(z;t) - \frac{1}{2}\alpha_{-1}(\mathbf{R}_z(t)\delta_{0,0},\delta_{1,1})_{\mathrm{l}_2} - \frac{1}{2}\alpha_{-1}(\mathbf{R}_z(t)\delta_{1,1},\delta_{0,0})_{\mathrm{l}_2}\\ &= \alpha_0 m_{0,1}(z;t) - \alpha_{-1}(\mathbf{R}_z(t)\delta_{0,0},\delta_{1,1})_{\mathrm{l}_2}, \quad z\in\mathbb{C}\setminus\mathbb{R}, \quad t\in[0,T).\end{aligned} \tag{9.4.10}$$

Consider $m_{0,1}(z;t)$. Similarly to (9.4.7) from (9.4.6), we have

$$\begin{aligned}\dot{m}_{0,1}(z;t) = (\dot{\mathbf{R}}_z(t)\delta_{1,0},\delta_{0,0})_{\mathrm{l}_2} &= (\mathbf{R}_z(t)\mathbf{A}(t)\delta_{1,0},\delta_{0,0})_{\mathrm{l}_2}\\ &+ (\mathbf{R}_z(t)\delta_{1,0},\mathbf{A}(t)\delta_{0,0})_{\mathrm{l}_2}.\end{aligned} \tag{9.4.11}$$

According to (9.2.12), (9.2.6), and (9.4.4), we get

$$\mathbf{A}(t)\delta_{1,0} = (-\frac{1}{2}\alpha_0,(0,0),(\frac{1}{2}\alpha_1,0),(0,0),\ldots) = -\frac{1}{2}\alpha_0\delta_{0,0} + \frac{1}{2}\alpha_1\delta_{2,0}; \tag{9.4.12}$$

$$\begin{aligned}(\mathbf{R}_z(t)\mathbf{A}(t)\delta_{1,0},\delta_{0,0})_{\mathrm{l}_2} &= (\mathbf{R}_z(t)(-\frac{1}{2}\alpha_0\delta_{0,0} + \frac{1}{2}\alpha_1\delta_{2,0}),\delta_{0,0})_{\mathrm{l}_2}\\ &= -\frac{1}{2}\alpha_0(\mathbf{R}_z(t)\delta_{0,0},\delta_{0,0})_{\mathrm{l}_2} + \frac{1}{2}\alpha_1(\mathbf{R}_z(t)\delta_{2,0},\delta_{0,0})_{\mathrm{l}_2}.\end{aligned} \tag{9.4.13}$$

For the last term in (9.4.11), using (9.4.9), (9.4.8), we have

$$\begin{aligned}(\mathbf{R}_z(t)\delta_{1,0},\mathbf{A}(t)\delta_{0,0})_{\mathrm{l}_2} &= (\mathbf{R}_z(t)\delta_{1,0},\frac{1}{2}\alpha_0\delta_{1,0} - \frac{1}{2}\alpha_{-1}\delta_{1,1})_{\mathrm{l}_2}\\ &= \frac{1}{2}\alpha_0(\mathbf{R}_z(t)\delta_{1,0},\delta_{1,0})_{\mathrm{l}_2} - \frac{1}{2}\alpha_{-1}(\mathbf{R}_z(t)\delta_{1,0},\delta_{1,1})_{\mathrm{l}_2}.\end{aligned} \tag{9.4.14}$$

From (9.4.11), (9.4.13), and (9.4.14), we get

$$\begin{aligned}\dot{m}_{0,1}(z;t) = &-\frac{1}{2}\alpha_0 m_{0,0}(z;t) + \frac{1}{2}\alpha_0 m_{1,1}(z;t) - \frac{1}{2}\alpha_{-1}(\mathbf{R}_z(t)\delta_{1,0},\delta_{1,1})_{\mathrm{l}_2}\\ &+ \frac{1}{2}\alpha_1(\mathbf{R}_z(t)\delta_{2,0},\delta_{0,0})_{\mathrm{l}_2}, \quad z\in\mathbb{C}\setminus\mathbb{R}, \quad t\in[0,T).\end{aligned} \tag{9.4.15}$$

Consider $m_{1,1}(z;t)$. Similarly to (9.4.7) and (9.4.11) from (9.4.6), we have

$$\begin{aligned}\dot{m}_{1,1}(z;t) = (\dot{\mathbf{R}}_z(t)\delta_{1,0},\delta_{1,0})_{\mathrm{l}_2} &= (\mathbf{R}_z(t)\mathbf{A}(t)\delta_{1,0},\delta_{1,0})_{\mathrm{l}_2}\\ &+ (\mathbf{R}_z(t)\delta_{1,0},\mathbf{A}(t)\delta_{1,0})_{\mathrm{l}_2}.\end{aligned} \tag{9.4.16}$$

Let us calculate the first term in this sum. Using the expression (9.4.12), we get

$$\begin{aligned}(\mathbf{R}_z(t)\mathbf{A}(t)\delta_{1,0},\delta_{1,0})_{\mathbf{l}_2} &= (\mathbf{R}_z(t)(-\frac{1}{2}\alpha_0\delta_{0,0}+\frac{1}{2}\alpha_1\delta_{2,0}),\delta_{1,0})_{\mathbf{l}_2}\\ &= -\frac{1}{2}\alpha_0(\mathbf{R}_z(t)\delta_{0,0},\delta_{1,0})_{\mathbf{l}_2}+\frac{1}{2}\alpha_1(\mathbf{R}_z(t)\delta_{2,0},\delta_{1,0})_{\mathbf{l}_2}.\end{aligned} \tag{9.4.17}$$

For the second term, we find from (9.4.12) that

$$\begin{aligned}(\mathbf{R}_z(t)\delta_{1,0},\mathbf{A}(t)\delta_{1,0})_{\mathbf{l}_2} &= (\mathbf{R}_z(t)\delta_{1,0},-\frac{1}{2}\alpha_0\delta_{0,0}+\frac{1}{2}\alpha_1\delta_{2,0})_{\mathbf{l}_2}\\ &= -\frac{1}{2}\alpha_0(\mathbf{R}_z(t)\delta_{1,0},\delta_{0,0})_{\mathbf{l}_2}+\frac{1}{2}\alpha_1(\mathbf{R}_z(t)\delta_{1,0},\delta_{2,0})_{\mathbf{l}_2}.\end{aligned} \tag{9.4.18}$$

From (9.4.16), (9.4.17), (9.4.18) and (9.4.3), we get

$$\begin{aligned}\dot{m}_{1,1}(z;t) &= -\alpha_0 m_{0,1}(z;t)+\frac{1}{2}\alpha_1(\mathbf{R}_z(t)\delta_{2,0},\delta_{1,0})_{\mathbf{l}_2}+\frac{1}{2}\alpha_1(\mathbf{R}_z(t)\delta_{1,0},\delta_{2,0})_{\mathbf{l}_2}\\ &= -\alpha_0 m_{0,1}(z;t)+\alpha_1(\mathbf{R}_z(t)\delta_{2,0},\delta_{1,0})_{\mathbf{l}_2}, \quad z\in\mathbb{C}\setminus\mathbb{R}, \quad t\in[0,T).\end{aligned} \tag{9.4.19}$$

As a result of calculations, we get expressions for $\dot{m}_{0,0}(z;t)$, $\dot{m}_{0,1}(z;t)$, $\dot{m}_{1,1}(z;t)$, i.e., (9.4.10), (9.4.15), (9.4.19). To calculate the system of differential equations for the unknowns $m_{0,0}(z;t)$, $m_{0,1}(z;t)$, $m_{1,1}(z;t)$ of the variable t, it is necessary to obtain expressions

$$(\mathbf{R}_z(t)\delta_{0,0},\delta_{1,1})_{\mathbf{l}_2},\ (\mathbf{R}_z(t)\delta_{1,0},\delta_{1,1})_{\mathbf{l}_2},\ (\mathbf{R}_z(t)\delta_{2,0},\delta_{0,0})_{\mathbf{l}_2},\ (\mathbf{R}_z(t)\delta_{2,0},\delta_{1,0})_{\mathbf{l}_2}$$

in (9.4.10), (9.4.15), (9.4.19) using unknown and known elements of matrices $J(t)$ and $A(t)$. The identity

$$\mathbf{1}+z\mathbf{R}_z(t)=\mathbf{R}_z(t)\mathbf{J}(t), \quad z\in\mathbb{C}\setminus\mathbb{R}, \quad t\in[0,T), \tag{9.4.20}$$

is used for this purpose.

First, let us transfer the relation (9.4.10) into the equation. From (9.4.20), we have

$$1+zm_{0,0}(z;t)=((\mathbf{1}+z\mathbf{R}_z(t))\delta_{0,0},\delta_{0,0})_{\mathbf{l}_2}=(\mathbf{R}_z(t)\mathbf{J}(t)\delta_{0,0},\delta_{0,0})_{\mathbf{l}_2}. \tag{9.4.21}$$

According to (9.2.9), we have

$$\mathbf{J}(t)\delta_{0,0}=\beta_0\delta_{0,0}+\alpha_0\delta_{1,0}+\alpha_{-1}\delta_{1,1}, \tag{9.4.22}$$

hence, (9.4.21), gives

$$\begin{aligned}1 + zm_{0,0}(z;t) &= (\mathbf{R}_z(t)(\beta_0\delta_{0,0} + \alpha_0\delta_{1,0} + \alpha_{-1}\delta_{1,1}), \delta_{0,0})_{\mathrm{l}_2}\\ &= \beta_0(\mathbf{R}_z(t)\delta_{0,0}, \delta_{0,0})_{\mathrm{l}_2} + \alpha_0(\mathbf{R}_z(t)\delta_{1,0}, \delta_{0,0})_{\mathrm{l}_2} + \alpha_{-1}(\mathbf{R}_z(t)\delta_{1,1}, \delta_{0,0})_{\mathrm{l}_2}\\ &= \beta_0 m_{0,0}(z;t) + \alpha_0 m_{0,1}(z;t) + \alpha_{-1}(\mathbf{R}_z(t)\delta_{1,1}, \delta_{0,0})_{\mathrm{l}_2}.\end{aligned}$$

Thus, we get

$$\alpha_{-1}(\mathbf{R}_z(t)\delta_{1,1}, \delta_{0,0})_{\mathrm{l}_2} = 1 + (z - \beta_0)m_{0,0}(z;t) - \alpha_0 m_{0,1}(z;t).$$

By using (9.4.3) and substituting this expression into (9.4.10), we get

$$\dot{m}_{0,0}(z;t) = 2\alpha_0 m_{0,1}(z;t) + (\beta_0 - z)m_{0,0}(z;t) - 1, \quad z \in \mathbb{C}\setminus\mathbb{R}, \quad t \in [0,T). \tag{9.4.23}$$

Substitute (9.4.15) into the equation. From (9.4.20) and (9.2.9), we have

$$\begin{aligned}&zm_{0,1}(z;t) = ((\mathbf{1} + z\mathbf{R}_z(t))\delta_{1,0}, \delta_{0,0})_{\mathrm{l}_2} = (\mathbf{R}_z(t)\mathbf{J}(t)\delta_{1,0}, \delta_{0,0})_{\mathrm{l}_2};\\ &\mathbf{J}(t)\delta_{1,0} = \alpha_0\delta_{0,0} + \beta_1\delta_{1,0} + \alpha_1\delta_{2,0}.\end{aligned} \tag{9.4.24}$$

Therefore,

$$\begin{aligned}zm_{0,1}(z;t) &= (\mathbf{R}_z(t)(\alpha_0\delta_{0,0} + \beta_1\delta_{1,0} + \alpha_1\delta_{2,0}), \delta_{0,0})_{\mathrm{l}_2}\\ &= \alpha_0(\mathbf{R}_z(t)\delta_{0,0}, \delta_{0,0})_{\mathrm{l}_2} + \beta_1(\mathbf{R}_z(t)\delta_{1,0}, \delta_{0,0})_{\mathrm{l}_2} + \alpha_1(\mathbf{R}_z(t)\delta_{2,0}, \delta_{0,0})_{\mathrm{l}_2}\\ &= \alpha_0 m_{0,0}(z;t) + \beta_1 m_{0,1}(z;t) + \alpha_1(\mathbf{R}_z(t)\delta_{2,0}, \delta_{0,0})_{\mathrm{l}_2}.\end{aligned}$$

Hence, we get

$$\alpha_1(\mathbf{R}_z(t)\delta_{2,0}, \delta_{0,0})_{\mathrm{l}_2} = (z - \beta_1)m_{0,1}(z;t) - \alpha_0 m_{0,0}(z;t). \tag{9.4.25}$$

Thus, the expression for the fourth term in (9.4.15) was found. Consider the third term. Let us rewrite the expression (9.4.24) (using (9.4.22)) in the form

$$\begin{aligned}zm_{0,1}(z;t) &= ((\mathbf{1} + z\mathbf{R}_z(t))\delta_{0,0}, \delta_{1,0})_{\mathrm{l}_2} = (\mathbf{R}_z(t)\mathbf{J}(t)\delta_{0,0}, \delta_{1,0})_{\mathrm{l}_2}\\ &= (\mathbf{R}_z(t)(\beta_0\delta_{0,0} + \alpha_0\delta_{1,0} + \alpha_{-1}\delta_{1,1}), \delta_{1,0})_{\mathrm{l}_2}\\ &= \beta_0(\mathbf{R}_z(t)\delta_{0,0}, \delta_{1,0})_{\mathrm{l}_2} + \alpha_0(\mathbf{R}_z(t)\delta_{1,0}, \delta_{1,0})_{\mathrm{l}_2} + \alpha_{-1}(\mathbf{R}_z(t)\delta_{1,1}, \delta_{1,0})_{\mathrm{l}_2}\\ &= \beta_0 m_{0,1}(z;t) + \alpha_0 m_{1,1}(z;t) + \alpha_{-1}(\mathbf{R}_z(t)\delta_{1,1}, \delta_{1,0})_{\mathrm{l}_2}.\end{aligned}$$

By using (9.4.3), we get

$$\begin{aligned}\alpha_{-1}(\mathbf{R}_z(t)\delta_{1,0},\delta_{1,1})_{l_2} &= \alpha_{-1}(\mathbf{R}_z(t)\delta_{1,1},\delta_{1,0})_{l_2}\\ &= (z-\beta_0)m_{0,1}(z;t)-\alpha_0 m_{1,1}(z;t).\end{aligned} \tag{9.4.26}$$

By using (9.4.26) and (9.4.25), we can write (9.4.15) in the form

$$\begin{aligned}\dot{m}_{0,1}(z;t) =& -\frac{1}{2}\alpha_0 m_{0,0}(z;t)+\frac{1}{2}\alpha_0 m_{1,1}(z;t)\\ &-\frac{1}{2}(z-\beta_0)m_{0,1}(z;t)+\frac{1}{2}\alpha_0 m_{1,1}(z;t)\\ &+\frac{1}{2}(z-\beta_1)m_{0,1}(z;t)-\frac{1}{2}\alpha_0 m_{0,0}(z;t)\\ =& -\alpha_0 m_{0,0}(z;t)+\alpha_0 m_{1,1}(z;t)+\frac{1}{2}(\beta_0-\beta_1)m_{0,1}(z;t).\end{aligned}$$

As a result, for $z\in\mathbb{C}\setminus\mathbb{R}$, $t\in[0,T)$, we get the equation

$$\dot{m}_{0,1}(z;t) = -\alpha_0 m_{0,0}(z;t)+\alpha_0 m_{1,1}(z;t)+\frac{1}{2}(\beta_0-\beta_1)m_{0,1}(z;t). \tag{9.4.27}$$

Substitute (9.4.19) into the equation. From (9.4.20) and (9.4.24), using (9.4.3), we have

$$\begin{aligned}1+zm_{1,1}(z;t) &= ((\mathbf{1}+z\mathbf{R}_z(t))\delta_{1,0},\delta_{1,0})_{l_2} = (\mathbf{R}_z(t)\mathbf{J}(t)\delta_{1,0},\delta_{1,0})_{l_2}\\ &= (\mathbf{R}_z(t)(\alpha_0\delta_{0,0}+\beta_1\delta_{1,0}+\alpha_1\delta_{2,0}),\delta_{1,0})_{l_2}\\ &= \alpha_0(\mathbf{R}_z(t)\delta_{0,0},\delta_{1,0})_{l_2}+\beta_1(\mathbf{R}_z(t)\delta_{1,0},\delta_{1,0})_{l_2}+\alpha_1(\mathbf{R}_z(t)\delta_{2,0},\delta_{1,0})_{l_2}\\ &= \alpha_0 m_{0,1}(z;t)+\beta_1 m_{1,1}(z;t)+\alpha_1(\mathbf{R}_z(t)\delta_{2,0},\delta_{1,0})_{l_2}.\end{aligned}$$

We get

$$\alpha_1(\mathbf{R}_z(t)\delta_{2,0},\delta_{1,0})_{l_2} = 1+(z-\beta_1)m_{1,1}(z;t)-\alpha_0 m_{0,1}(z;t).$$

Substituting this expression into (9.4.19), we get

$$\dot{m}_{1,1}(z;t) = -2\alpha_0 m_{0,1}(z;t)-(\beta_1-z)m_{1,1}(z;t)+1,\quad z\in\mathbb{C}\setminus\mathbb{R},\quad t\in[0,T). \tag{9.4.28}$$

As a result, we get three differential equations, i.e., (9.4.23), (9.4.27) and (9.4.28) for unknown functions $m_{0,0}(z;t)$, $m_{0,1}(z;t)$, $m_{1,1}(z;t)$. Thus, we proved the theorem.

Theorem 9.4.1 *Elements* $m_{0,0}(z;t)$, $m_{0,1}(z;t)$, *and* $m_{1,1}(z;t)$ *of the Weyl matrix* (9.4.1) *satisfy the system of differential equations*

$$\begin{aligned}
\dot{m}_{0,0}(z;t) &= (\beta_0(t) - z)m_{0,0}(z;t) + 2\alpha_0(t)m_{0,1}(z;t) - 1, \\
\dot{m}_{0,1}(z;t) &= -\alpha_0(t)m_{0,0}(z;t) + \frac{1}{2}(\beta_0(t) - \beta_1(t))m_{0,1}(z;t) + \alpha_0(t)m_{1,1}(z;t), \\
\dot{m}_{1,1}(z;t) &= -2\alpha_0(t)m_{0,1}(z;t) - (\beta_1(t) - z)m_{1,1}(z;t) + 1; \\
m_{1,0}(z;t) &= m_{0,1}(z;t), \qquad z \in \mathbb{C} \setminus \mathbb{R}, \quad t \in [0, T).
\end{aligned} \tag{9.4.29}$$

As in Sect. 9.1 it is possible to rewrite the system (9.4.29) in terms of measures (charges) $d\rho_{\alpha,\beta}(\lambda;t), \alpha, \beta = 0, 1$, elements of the (2×2)-matrix of the spectral measure $d\rho_M(\lambda;t)$. More precisely, let us rewrite (9.4.29) in terms of their derivatives. These charges are related to $m_{\alpha,\beta}(z;t)$ by equalities (9.4.1). For these measures, the standard notation is preserved, namely,

$$\mathfrak{B}(\mathbb{R}) \ni \Delta \mapsto \rho_{\alpha,\beta}(\Delta;t), \ \rho_M(\Delta;t); \quad \alpha, \beta = 0, 1, \quad \forall t \in [0, T). \tag{9.4.30}$$

It is easy to understand that there exists a measure $d\sigma(\lambda)$ independent of t for which all measures $d\sigma_{\alpha,\beta}(\lambda;t), t \in [0, T)$ are absolutely continuous with respect to $d\rho(\lambda)$. In fact, here it is necessary to repeat the construction related to formulas (9.1.21) and (9.1.22). Namely, we have the lemma.

Lemma 9.4.2 *Consider the mapping*

$$\mathfrak{B}(\mathbb{R}) \ni \Delta \mapsto \int_0^T a(t) \operatorname{Tr} \rho_M(\Delta;t)\, dt =: \sigma(\Delta) \geq 0, \tag{9.4.31}$$

where $a(t)$ *is a fixed continuous positive function integrable on the interval* $[0, T]$ *with respect to the Lebesgue measure and such an integral is equal to one.*

Then, the mapping (9.4.31) *is an integral with a finite measure on* $\mathbb{R}$ *and all measures* (9.4.30) *are absolutely continuous with respect to* $d\sigma(\lambda)$.

Proof Since

$$\mathfrak{B}(\mathbb{R}) \ni \Delta \mapsto \operatorname{Tr} \rho_M(\Delta;t)dt = \rho_{0,0}(\Delta;t) + \rho_{1,1}(\Delta;t) \leq 2, \quad \forall t \in [0, T)$$

is a non-negative measure, then by the passing to the limit under the integral sign, we obtain that (9.4.31) is a non-negative measure.

Let for $\Delta \in \mathfrak{B}(\mathbb{R})$ we have $\sigma(\Delta) = 0$, then from (9.4.31) we get that $\operatorname{Tr} \rho_M(\Delta;t) = 0$ for each $t \in [0, T)$. Since the matrix $\rho_M(\Delta;t)$ is non-negative, then it follows from the last equality that $\rho_M(\Delta;t) = 0$, i.e., the measure $\rho_M(\Delta;t) = 0$. □

We denote the Radon-Nikodym derivative $d\rho_M(\lambda;t)/d\sigma(\lambda)$ by

$$r_M(\lambda;t) = (r_{\alpha,\beta}(\lambda;t))_{\alpha,\beta=0}^1.$$

Thus, we have

$$\begin{aligned}\rho_M(\Delta;t) &= \int_\Delta r_M(\lambda;t)\,d\sigma(\lambda),\\ \rho_{\alpha,\beta}(\Delta;t) &= \int_\Delta r_{\alpha,\beta}(\lambda;t)\,d\sigma(\lambda),\\ \Delta\in \mathfrak{B}(\mathbb{R}),\quad & t\in[0,T),\quad \alpha,\beta=0,1.\end{aligned} \tag{9.4.32}$$

We use the spectral function (in the matrix and in the scalar forms):

$$\begin{aligned}\rho_M(\lambda;t) &= \rho_M((-\infty,\lambda);t),\\ \rho_{\alpha,\beta}(\lambda;t) &= \rho_{\alpha,\beta}((-\infty,\lambda);t);\\ \lambda\in\mathbb{R},\quad & t\in[0,T),\quad \alpha,\beta=0,1.\end{aligned} \tag{9.4.33}$$

As in Sect. 9.1, it is important to rewrite (9.4.29) as a system with unknowns $r_{\alpha,\beta}(\lambda;t)$, now a similar system is also useful. Thus, we get the theorem.

Theorem 9.4.3 *The system* (9.4.29) *is equivalent to the system of differential equations*

$$\begin{aligned}\dot r_{0,0}(\lambda;t) &= (\beta_0(t)-\lambda)r_{0,0}(\lambda;t) + 2\alpha_0(t)r_{0,1}(\lambda;t),\\ \dot r_{0,1}(\lambda;t) &= -\alpha_0(t)r_{0,0}(\lambda;t) + \frac{1}{2}(\beta_0(t)-\beta_1(t))r_{0,1}(\lambda;t) + \alpha_0(t)r_{1,1}(\lambda;t),\\ \dot r_{1,1}(\lambda;t) &= -2\alpha_0(t)r_{0,1}(\lambda;t) - (\beta_1(t)-\lambda)r_{1,1}(\lambda;t);\\ r_{1,0}(\lambda;t) &= r_{0,1}(\lambda;t),\qquad t\in[0,T).\end{aligned} \tag{9.4.34}$$

with unknowns $r_{\alpha,\beta}(\lambda;t)$*: for almost all* $\lambda\in\mathbb{R}$ *with respect to the measure* $d\sigma(\lambda)$.

Functions $r_{\alpha,\beta}(\lambda;t)$ *in* (9.4.34) *are once continuously differentiable on* $[0,T)$ *for* $d\sigma(\lambda)$*-almost of all* $\lambda\in\mathbb{R}$.

Proof The proof is similar to the proof of Theorem 9.1.2. Namely, let $(u,v)_{L_2}$ be the scalar product in the usual space $L_2([0,T),dt)\ni u,v$. By using the first equality in (9.4.29) and (9.4.32), $\forall z\in\mathbb{C}\setminus\mathbb{R}$ and for an arbitrary infinitely differentiable complex-valued function $u_{0,0}(t)$, $t \in [0, T)$, which vanishes in some neighborhood of points 0 and T, we get

$$\begin{aligned}
&-\int_{\mathbb{R}} \frac{1}{\lambda - z}(r_{0,0}(\lambda;\cdot), \dot{u}_{0,0}(\cdot))_{L_2}\, d\sigma(\lambda) \\
&\qquad = -\int_{\mathbb{R}} \frac{1}{\lambda - z}\, d(\rho_{0,0}(\lambda;\cdot), \dot{u}_{0,0}(\cdot))_{L_2} \\
&\qquad = -(m_{0,0}(z;\cdot), \dot{u}_{0,0}(\cdot))_{L_2} = (\dot{m}_{0,0}(z;\cdot), u_{0,0}(\cdot))_{L_2} \\
&\qquad = ((\beta_0(\cdot) - z)m_{0,0}(z;\cdot) + 2\alpha_0(\cdot)m_{0,1}(z;\cdot) - 1, u_{0,0}(\cdot))_{L_2} \\
&\qquad = \int_{\mathbb{R}} \frac{1}{\lambda - z}((\beta_0(\cdot) - \lambda)\, d\rho_{0,0}(\lambda;\cdot) + 2\alpha_0(\cdot)\, d\rho_{0,1}(\lambda;\cdot), u_{0,0}(\cdot))_{L_2} \\
&\qquad = \int_{\mathbb{R}} \frac{1}{\lambda - z}((\beta_0(\cdot) - \lambda)r_{0,0}(\lambda;\cdot) + 2\alpha_0(\cdot)r_{0,1}(\lambda;\cdot), u_{0,0}(\cdot))_{L_2} d\sigma(\lambda).
\end{aligned} \tag{9.4.35}$$

Here, we use the equality

$$\begin{aligned}
&(\beta_0(t) - z)m_{0,0}(z;t) + 2\alpha_0(t)m_{0,1}(z;t) - 1 \\
&= \int_{\mathbb{R}} \frac{1}{\lambda - z}((\beta_0(t) - \lambda)\, d\rho_{0,0}(\lambda;t) \\
&\quad + \int_{\mathbb{R}} \frac{1}{\lambda - z} 2\alpha_0(t)\, d\rho_{0,1}(\lambda;t), \quad z \in \mathbb{C} \setminus \mathbb{R}, \quad t \in [0, T).
\end{aligned}$$

This equality holds true, since $\forall t \in [0, T)$ $\rho_{0,0}(\mathbb{R}; t) = 1$, hence, we have

$$-1 = \int_{\mathbb{R}} (z - \lambda)(\lambda - z)^{-1}\, d\rho_{0,0}(\lambda; t).$$

Similarly, to the proof of Theorem 9.1.2, from (9.4.30) that for $d\sigma(\lambda)$-almost all $\lambda \in \mathbb{R}$, we get

$$-(r(\lambda,\cdot), \dot{u}_{0,0}(\cdot))_{L_2} = ((\beta_0(\cdot) - \lambda)r_{0,0}(\lambda;\cdot) + 2\alpha_0(\cdot)r_{0,1}(\lambda;\cdot), u_{0,0}(\cdot))_{L_2}. \tag{9.4.36}$$

Similar equalities can be obtained using the second and third equalities from (9.4.29) (now it is necessary to remind that $\rho_{1,1}(\mathbb{R};t) = 1, t \in [0,t))$. These equalities are possible

$$
\begin{aligned}
(-r_{0,1}(\lambda;\cdot), \dot u_{0,1}(\cdot))_{L_2} &= (-\alpha_0(\cdot)r_{0,0}(\lambda;\cdot) + \frac{1}{2}(\beta_0(\cdot)-\beta_1(\cdot))r_{0,1}(\lambda;\cdot) \\
&\quad + \alpha_0(\cdot)r_{1,1}(\lambda;\cdot), u_{0,1}(\cdot))_{L_2}, \\
(-r_{1,1}(\lambda;\cdot), \dot u_{1,1}(\cdot))_{L_2} &= (-2\alpha_0(\cdot)r_{0,1}(\lambda;\cdot) \\
&\quad - (\beta_1(\cdot)-\lambda)r_{1,1}(\lambda;\cdot), u_{1,1}(\cdot))_{L_2},
\end{aligned}
\tag{9.4.37}
$$

for $d\sigma(\lambda)$-almost all $\lambda \in \mathbb{R}$, where $u_{0,1}(t)$ and $u_{1,1}(t)$ are arbitrary functions of the type like $u_{0,0}(t)$.

Note that functions $u_{0,0}(t)$, $u_{0,1}(t)$ and $u_{1,1}(t)$ in (9.4.36), (9.4.37) are arbitrary that vanish in some neighborhood of points $0, T$. Thus, $r_{0,0}(\lambda;t)$, $r_{0,1}(\lambda;t)$ and $r_{1,1}(\lambda;t)$ (for $d\sigma(\lambda)$-almost all $\lambda \in \mathbb{R}$) is a generalized solution (in terms of the classical theory of generalized functions) of the system (9.4.34). But for a system of ordinary linear differential equations with smooth coefficients, it is known that, every generalized solution of such equations is actually a smooth solution. Thus, we can assume that the vector $(r_{0,0}(\lambda;t), r_{0,1}(\lambda;t), r_{1,1}(\lambda;t))$ is continuously differentiable once on $[0,T)$ and is a solution of the Cauchy problem for (9.4.34) with initial data $(r_{0,0}(\lambda;0), r_{0,1}(\lambda;0), r_{1,1}(\lambda;0))$. □

It is useful to rewrite the systems (9.4.29) and (9.4.34) using Toda chains (9.2.7). In these two systems, coefficients $\alpha_0(t)$, $\beta_0(t)$, $\beta_1(t)$ are the same as functions in (9.2.7). Thus we have

$$
\frac{1}{2}(\beta_0(t)-\beta_1(t)) = -\frac{\dot\alpha_0(t)}{\alpha_0(t)}, \quad \beta_1(t) = \beta_0(t) + 2\frac{\dot\alpha_0(t)}{\alpha_0(t)}, \quad t \in [0,T). \tag{9.4.38}
$$

By using these equalities, we can rewrite, for example, the equation (9.4.34) in the form for $d\sigma(\lambda)$-almost all $\lambda \in \mathbb{R}$, namely,

$$
\begin{aligned}
\dot r_{0,0}(\lambda;t) &= (\beta_0(t)-\lambda)r_{0,0}(\lambda;t) + 2\alpha_0(t)r_{0,1}(\lambda;t), \\
\dot r_{0,1}(\lambda;t) &= -\alpha_0(t)r_{0,0}(\lambda;t) - \frac{\dot\alpha_0(t)}{\alpha_0(t)}r_{0,1}(\lambda;t) + \alpha_0(t)r_{1,1}(\lambda;t), \\
\dot r_{1,1}(\lambda;t) &= -2\alpha_0(t)r_{0,1}(\lambda;t) - (\beta_0(t)-\lambda+2\frac{\dot\alpha_0(t)}{\alpha_0(t)})r_{1,1}(\lambda;t); \\
r_{1,0}(\lambda;t) &= r_{0,1}(\lambda;t), \quad t \in [0,T).
\end{aligned}
\tag{9.4.39}
$$

If we suppose hat $\alpha_0(t)$ is a constant $\alpha_0(t) = \alpha_0(0)$, $t \in [0, T)$, then the system (9.4.39) has a simplified form

$$\begin{aligned}
\dot{r}_{0,0}(\lambda; t) &= (\beta_0(t) - \lambda) r_{0,0}(\lambda; t) + 2\varkappa\, r_{0,1}(\lambda; t), \\
\dot{r}_{0,1}(\lambda; t) &= -\varkappa\, r_{0,0}(\lambda; t) + \varkappa\, r_{1,1}(\lambda; t), \\
\dot{r}_{1,1}(\lambda; t) &= -2\varkappa\, r_{0,1}(\lambda; t) - (\beta_0(t) - \lambda) r_{1,1}(\lambda; t); \\
r_{1,0}(\lambda; t) &= r_{0,1}(\lambda; t), \quad t \in [0, T), \quad \varkappa = \alpha_0(t).
\end{aligned} \tag{9.4.40}$$

Systems (9.4.29), (9.4.34), (9.4.39), and (9.4.40) are similar to Eqs. (9.1.8) and (9.1.23), but unfortunately, it is not possible to calculate in the general case the solution of the Cauchy problem for (9.4.34), (9.4.39) and even for (9.4.40), as in Sect. 9.1. Thus, we can only say that *if for some specific coefficients of the systems* (9.4.39), (9.4.40) *it is possible to calculate the solutions* $r_{0,0}(\lambda; t)$, $r_{0,1}(\lambda; t)$, $r_{1,1}(\lambda; t)$ *Cauchy problem with initial data* $r_{0,0}(\lambda; 0)$, $r_{0,1}(\lambda; 0)$, $r_{1,1}(\lambda; 0)$. *Then, using formulas* (9.4.32), (9.4.33), *we can find the spectral measure* $d\rho_M(\lambda; t)$ *of operators* $\mathbf{J}(t)$, $t \in [0, T)$.

Note that in the last case, one can apply the representation of the solution of the Cauchy problem for (9.4.39), (9.4.40) of the type (9.1.28), (9.1.29) and thus, one can avoid the measure $d\sigma(\lambda)$. The functions $\alpha_0(t)$, $\beta_0(t)$, $t \in [0, T)$ must be found under conditions that the matrix measure $d\rho_M(\lambda; t)$ is probabilistic for all $t \in [0, T)$.

Remark 9.4.4 One useful fact related to the matrix $r_M(\lambda; t)$ from (9.4.32) should be noted. Namely, let us denote by $D(\lambda; t)$ its determinant

$$D(\lambda; t) = \operatorname{Det} r_M(\lambda; t) = r_{0,0}(\lambda; t) r_{1,1}(\lambda; t) - r_{0,1}^2(\lambda; t), \quad \lambda \in \mathbb{R}, \quad t \in [0, T). \tag{9.4.41}$$

Then, for $d\sigma(\lambda)$-almost all $\lambda \in \mathbb{R}$, the equality

$$D(\lambda; t) = \frac{\alpha_0^2(0)}{\alpha_0^2(t)} D(\lambda; 0), \quad t \in [0, T), \tag{9.4.42}$$

holds true.

To prove (9.4.42), it is necessary to calculate $\dot{D}(\lambda; t)$. By using (9.4.41) and (9.4.39) after a number of simple transformations, we obtain, for $d\sigma(\lambda)$-almost all $\lambda \in \mathbb{R}$,

$$\dot{D}(\lambda; t) = -2 \frac{\dot{\alpha}_0(t)}{\alpha_0(t)} D(\lambda; t), \quad t \in [0, T). \tag{9.4.43}$$

The integrating of the equation (9.4.43) leads to (9.4.42).

If it follows from some additional information about the solution of the Cauchy problem for (9.2.7) that $D(\lambda; t)$, for λ, the set of positive measures $d\sigma(\lambda)$ does not depend on t, then from (9.4.42), we get that $\alpha_0(t) = \alpha_0(0)$, $t \in [0, T)$, and *the system (9.4.39) has a very simplified form (9.4.40).*

Note also that operators $\mathbf{J}(t)$ are isospectral, therefore, their spectrum is on the fixed interval $[a, b] \subset \mathbb{R}$.

9.5 The Basic Statement and Applications to Hamiltonian Systems

In this section, the procedure for finding the solution of the Cauchy problem for a doubly infinite Toda chain will be given, and the system of differential equations associated with Toda chains will also be considered. The construction is similar to the case of a semi-infinite Toda chain, see Sect. 9.2. Therefore, consider the doubly infinite Toda chain (9.2.7)

$$\begin{aligned} \dot{\alpha}_n(t) &= \frac{1}{2}\alpha_n(t)(\beta_{n+1}(t) - \beta_n(t)), \\ \dot{\beta}_n(t) &= \alpha_n^2(t) - \alpha_{n-1}^2(t), \quad n \in \mathbb{Z}, \quad t \in [0, T),\ T \leq \infty; \end{aligned} \tag{9.5.1}$$

where $\alpha_n(t) > 0$, $\beta_n(t)$ are real continuously differentiable functions on $[0, T)$. For (9.5.1), we formulate the Cauchy problem: find the solution $\alpha_n(t)$, $\beta_n(t)$ according to the initial data $\alpha_n(0)$, $\beta_n(0)$, $n \in \mathbb{Z}$, $t \in [0, T)$. The solution is formulated in the form of the theorem.

Theorem 9.5.1 *There is a uniformly bounded, with respect to $n \in \mathbb{Z}$ and $t \in [0, T)$, solution of the above Cauchy problem for* (9.5.1) *for a fixed $T < \infty$, we get this continuous solution by the following procedure.*

(1) By using the initial data, we construct a matrix of the type (9.2.9), *namely,*

$$J(0) = \left[\begin{array}{c|cc|cc|cc|c} \beta_0(0) & \alpha_0(0) & \alpha_{-1}(0) & 0 & 0 & 0 & 0 & \cdots \\ \hline \alpha_0(0) & \beta_1(0) & 0 & \alpha_1(0) & 0 & 0 & 0 & \cdots \\ \alpha_{-1}(0) & 0 & \beta_{-1}(0) & 0 & \alpha_{-2}(0) & 0 & 0 & \cdots \\ \hline 0 & \alpha_1(0) & 0 & \beta_2(0) & 0 & \alpha_2(0) & 0 & \cdots \\ 0 & 0 & \alpha_{-2}(0) & 0 & \beta_{-2}(0) & 0 & \alpha_{-3}(0) & \cdots \\ \hline 0 & 0 & 0 & \alpha_2(0) & 0 & \beta_3(0) & 0 & \cdots \\ 0 & 0 & 0 & 0 & \alpha_{-3}(0) & 0 & \beta_{-3}(0) & \cdots \\ \hline \vdots & \vdots & \vdots & \vdots & \vdots & \vdots & \vdots & \ddots \end{array}\right] \tag{9.5.2}$$

and we serch a (2×2)*-matrix spectral measure*

$$d\rho_M(\lambda; 0) = \begin{bmatrix} d\rho_{0,0}(\lambda; 0) & d\rho_{0,1}(\lambda; 0) \\ d\rho_{0,1}(\lambda; 0) & d\rho_{1,1}(\lambda; 0) \end{bmatrix}, \quad \lambda \in \mathbb{R}. \tag{9.5.3}$$

(2) *We consider the linear system* (9.4.39) *on* $[0, T)$ (*or* (9.4.40)) *of differential equations with respect to unknown continuously differentiable functions* $r_{\alpha,\beta}(\lambda; t)$, $\alpha, \beta = 0, 1$. *Here,* $\alpha_0(t) > 0$ *and* $\beta_0(t) \in \mathbb{R}$ *are fixed continuously differentiable functions,* $\lambda \in \mathbb{R}$, $t \in [0, T)$.

The relationship between $r_{\alpha,\beta}(\lambda; t)$ *and the spectral measure* $d\rho_M(\lambda; t)$ *and the function* $\rho_M(\lambda; t)$ *is given by formulas* (9.4.32) *and* (9.4.33), *where* $d\sigma(\lambda)$ *is a fixed measure on* $\mathbb{B}(\mathbb{R})$ *defined by a continuous positive function* $a(t)$ *on* $t \in [0, T)$ *according to the formula* (9.4.31).

We find the solution of the Cauchy problem (9.4.39) *and* (9.4.40) *with initial data* (9.5.3) *in terms of* $r_{\alpha,\beta}(\lambda; 0)$. *By using this solution* $\rho_{\alpha,\beta}(\lambda; t)$, *we search functions* $\alpha_0(t)$, $\beta_0(t)$ *according to conditions* $\rho_{\alpha_0,\beta_0}(\mathbb{R}; t) = 1$, $t \in [0, T)$. *Such a solution is denoted by* $\rho(\lambda; t)$ *in terms of* $\rho_M(\lambda; t)$.

(3) *For a fixed* $t \in [0, T)$ *in the Hilbert space* $L_2(\mathbb{C}^2, \mathbb{R}, d\rho_M(\lambda; t))$ *of the type* (9.3.22), *we consider a sequence of* $\mathbb{C}^2$*-vector-valued functions of the variable* λ (*see* (9.3.29)), *namely,*

$$(1, 0); ((0, 1), (\lambda, 0)); ((0, \lambda), (\lambda^2, 0)); \ldots; ((0, \lambda^n), (\lambda^{n+1}, 0)); \ldots . \tag{9.5.4}$$

We apply the Gram-Schmidt orthogonalization procedure to the sequence (9.5.4). *As result we get a sequence of vectors from* $L_2(\mathbb{C}^2, \mathbb{R}, d\rho_M(\lambda; t))$ (*see* (9.3.26)), *namely,*

$$\begin{gathered}(P_{0;0,0}(\lambda; t), P_{0;0,1}(\lambda; t)), (P_{1;0,0}(\lambda; t), P_{1;0,1}(\lambda; t)), (P_{1;1,0}(\lambda; t), P_{1;1,1}(\lambda; t));\\ (P_{2;0,0}(\lambda; t), P_{2;0,1}(\lambda; t)), (P_{2;1,0}(\lambda; t), P_{2;1,1}(\lambda; t)); \ldots .\end{gathered} \tag{9.5.5}$$

(4) *The required solution of the Cauchy problem is found using formulas* (9.3.32)*:*

$$\begin{bmatrix} \alpha_n(t) & 0 \\ 0 & \alpha_{-n-1}(t) \end{bmatrix} = \int_{\mathbb{R}} \lambda P_n(\lambda; t)\, d\rho_M(\lambda; t) P_{n+1}^*(\lambda; t), \quad n \in \mathbb{N},$$

$$\begin{bmatrix} \alpha_0(t) & \alpha_{-1}(t) \\ 0 & 0 \end{bmatrix} = \int_{\mathbb{R}} \lambda P_0(\lambda; t)\, d\rho_M(\lambda; t) P_1^*(\lambda; t);$$

$$\begin{bmatrix} \beta_n(t) & 0 \\ 0 & \beta_{-n}(t) \end{bmatrix} = \int_{\mathbb{R}} \lambda P_n(\lambda; t)\, d\rho_M(\lambda; t) P_n^*(\lambda; t), \quad n \in \mathbb{N}_0,$$

$$P_n(\lambda; t) = \begin{bmatrix} P_{n;0,0}(\lambda; t) & P_{n;0,1}(\lambda; t) \\ P_{n;0,1}(\lambda; t) & P_{n;1,1}(\lambda; t) \end{bmatrix}, \quad n \in \mathbb{N}_0,$$

$$P_{0;0,1}(\lambda; t) = P_{0;1,1}(\lambda; t) = 0, \quad \forall t \in [0, T), \quad \lambda \in \mathbb{R}.$$

Proof The proof follows from results of Sects. 9.2–9.4. □

The Toda Chain Represents a Hamiltonian System It describes the dynamics of an infinite chain of particles $q_n(t), n \in \mathbb{Z}$ along the whole axis $\mathbb{R}$ with an exponential interaction. The corresponding Hamiltonian has the form

$$H(p, q) = \frac{1}{2} \sum_{n=-\infty}^{\infty} p_n^2 + \sum_{n=-\infty}^{\infty} e^{q_n - q_{n+1}}, \tag{9.5.6}$$

where $q_n = q_n(t)$ is the n-th coordinate of a particle, and $p_n = p_n(t)$ is its moment, $t \in [0, T)$. The Hamilton equation

$$\dot{q}_n = \frac{\partial H}{\partial p_n} = p_n, \quad \dot{p}_n = -\frac{\partial H}{\partial q_n} = -e^{q_n - q_{n+1}} + e^{q_{n-1} - q_n}, \quad n \in \mathbb{Z}$$

can be rewritten as a system

$$\ddot{x}_n(t) = e^{x_{n-1}(t) - x_n(t)} - e^{x_n(t) - x_{n+1}(t)}, \quad n \in \mathbb{Z}, \quad t \in [0, T). \tag{9.5.7}$$

Here $x_n(t) \in \mathbb{R}$ is the n-th coordinate of the point $q_n(t)$ and $\dot{x}_n(t)$ is its moment $p_n(t)$.

Thus, we have a system (9.5.7) of nonlinear differential equations of the second order. For this system, we can set the Cauchy problem: for the given initial conditions $x_n(0)$, $\dot{x}_n(0)$, $n \in \mathbb{Z}$, it is necessary to find the solution $x_n(t)$, $n \in \mathbb{Z}$, $t \in [0, T)$.

Let us introduce Flaschka variables

$$\alpha_n(t) = e^{\frac{1}{2}(x_n(t) - x_{n+1}(t))}, \quad \beta_n(t) = -\dot{x}_n(t), \quad n \in \mathbb{Z}, \quad t \in [0, T). \tag{9.5.8}$$

Then the system (9.5.7) passes into a doubly infinite Toda chain (9.5.1) with the corresponding Cauchy problem. Now we can apply Theorem 9.5.1 to our case, taking into account that our conditions about the uniform boundedness of $\alpha_n(t)$, $\beta_n(t)$ mean in terms of $x_n(t)$ the boundedness

$$|x_n(t) - x_{n+1}(t)|, \quad |\dot{x}_n(t)|, \quad n \in \mathbb{Z}, \quad t \in [0, T) \tag{9.5.9}$$

with a fixed $T < \infty$. Therefore, we can formulate a conclusion.

Corollary 9.5.2 *The previously formulated Cauchy problem* (9.5.7) *with a uniform boundedness* (9.5.9) *turns at* (9.5.8) *into Theorem* 9.5.1. *The procedure of Theorem* 9.5.1 *makes it possible to find a solution of the Cauchy problem* (9.5.7).

Let us now turn to the problem of the shock wave. Consider a system (9.5.7), but for $n \in \mathbb{N}_0$,

$$\ddot{x}_n(t) = e^{x_{n-1}(t)-x_n(t)} - e^{x_n(t)-x_{n+1}(t)}, \quad n \in \mathbb{N}_0, \quad t \in [0, T). \tag{9.5.10}$$

For this system, it is possible to formulate a shock wave problem, that is, the following mixed problem: it is necessary to find the solution (9.5.10) x_n, $n \in \mathbb{N}_0$, $t \in [0, T)$ if initial data are $x_n(0)$, $\dot{x}_n(0)$, $n \in \mathbb{N}_0$ and boundary data are $x_{-1}(t) = f(t)$, $t \in [0, T)$, where $f(t)$ is a given real continuous function.

In the next text, we will consider a variant of this problem ("type of shock wave problem", in which it is necessary to find a bounded solution (9.5.10), if $x_n(0)$, $\dot{x}_n(0)$ are known, $n \in \mathbb{N}_0$, and $x_{-1}(t) - x_0(t) = f(t)$, $t \in [0, T)$ (instead of $x_{-1}(t)$).

Now using the Flaschka variables (9.5.8), but for $n \in \mathbb{N}_0$, from (9.5.8), (9.5.7), $\forall t \in [0, T)$, we get

$$\begin{aligned} &\dot{\alpha}_n(t) = \frac{1}{2}\alpha_n(t)(\beta_{n+1}(t) - \beta_n(t)), \quad n \in \mathbb{N}_0; \\ &\dot{\beta}_n(t) = -\ddot{x}_n(t) = \alpha_n^2(t) - \alpha_{n-1}^2(t), \quad n \in \mathbb{N}; \\ &\dot{\beta}_0(t) = -\ddot{x}_0(t) = \alpha_0^2(t) - e^{x_{-1}(t)-x_0(t)} = \alpha_0^2(t) + \varphi(t), \quad \varphi(t) := -e^{f(t)}. \end{aligned} \tag{9.5.11}$$

Let us rewrite the Eq. (9.5.11) in the form of the Lax equation with usual Jacobi matrices (9.1.2) and (9.1.3). Consider the matrix $J(t)$ of the form (9.1.2), elements of which are solutions of the system (9.5.11). Using the constructions of Sect. 9.2, it is not difficult to understand that the system (9.5.11) is equivalent to the inhomogeneous Lax equation

$$\begin{aligned} &\dot{J}(t) = J(t)A(t) - A(t)J(t) + \varphi(t)B, \quad \forall t \in [0, T), \\ &B = (b_{\alpha,\beta})_{\alpha,\beta=0}^{\infty}, \quad b_{0,0} = 1, \quad b_{\alpha,\beta} = 0, \quad \text{if} \quad \alpha + \beta > 0. \end{aligned} \tag{9.5.12}$$

Let $\mathbf{J}(t)$ be a bounded self-adjoint operator in ℓ_2 generated by $J(t)$, and $\mathbf{R}_z(t)$ be its resolvent; $\mathbf{A}(t)$, $\mathbf{B} = \mathbf{B}^2$ are operators generated by $A(t)$ and B. Then, according to (9.1.10), (9.5.12) and (9.1.12), we have

$$\begin{aligned} \dot{\mathbf{R}}_z(t) &= ((\mathbf{J}(t) - z\mathbf{1})^{-1})^{\cdot} = -\mathbf{R}_z(t)(\mathbf{J}(t) - z\mathbf{1})^{\cdot}\mathbf{R}_z(t) \\ &= -\mathbf{R}_z(t)\dot{\mathbf{J}}(t)\mathbf{R}_z(t) = -\mathbf{R}_z(t)([\mathbf{J}(t), \mathbf{A}(t)] + \varphi(t)\mathbf{B})\mathbf{R}_z(t) \\ &= -\mathbf{R}_z(t)[\mathbf{J}(t) - z\mathbf{1}, \mathbf{A}(t)]\mathbf{R}_z(t) - \varphi(t)\mathbf{R}_z(t)\mathbf{B}\mathbf{R}_z(t) \\ &= [\mathbf{R}_z(t), \mathbf{A}(t)] - \varphi(t)\mathbf{R}_z(t)\mathbf{B}\mathbf{R}_z(t), \quad z \in \mathbb{C} \setminus \mathbb{R}, \quad t \in [0, T). \end{aligned} \tag{9.5.13}$$

By using (9.5.13) instead of (9.1.12), the analogue of the equation (9.4.29) for $m(z;t)$ in our case is derived. Let $\delta_0 = (1, 0, 0, \ldots)$, then from (9.5.13), we get

$$\begin{aligned}\dot{m}(z;t) &= (\dot{\mathbf{R}}_z(t)\delta_0, \delta_0)_{\ell_2}\\ &= ([\mathbf{R}_z(t), \mathbf{A}(t)]\delta_0, \delta_0)_{\ell_2} - \varphi(t)(\mathbf{R}_z(t)\mathbf{B}\mathbf{R}_z(t)\delta_0, \delta_0)_{\ell_2}, \\ &\forall z \in \mathbb{C} \setminus \mathbb{R}, t \in [0, T).\end{aligned} \tag{9.5.14}$$

For the first term on the right-hand side in (9.5.14), we can repeat the transformation (9.1.13)–(9.1.16) and obtain $(z - \beta_0(t))m(z;t) + 1$.

For the second term we have

$$\begin{aligned}(\mathbf{R}_z(t)\mathbf{B}\mathbf{R}_z(t)\delta_0, \delta_0)_{\ell_2} &= (\mathbf{B}\mathbf{R}_z(t)\delta_0, \mathbf{R}_z^*(t)\delta_0)_{\ell_2}\\ &= (\mathbf{B}\mathbf{R}_z(t)\delta_0, \mathbf{B}\mathbf{R}_z^*(t)\delta_0)_{\ell_2}\\ &= (\mathbf{R}_z(t)\delta_0, \delta_0)_{\ell_2}\overline{(\mathbf{R}_z^*(t)\delta_0, \delta_0)}_{\ell_2}\\ &= (\mathbf{R}_z(t)\delta_0, \delta_0)_{\ell_2}^2 = m^2(z;t).\end{aligned}$$

Thus, (9.5.14) gives the Riccati type equation for $m(z;t)$, namely,

$$\dot{m}(z;t) = (z - \beta_0(t))m(z;t) + 1 - \varphi(t)m^2(z;t), \quad z \in \mathbb{C} \setminus \mathbb{R}, \quad t \in [0, T). \tag{9.5.15}$$

The procedure for integrating the equation (9.5.15) can be given. For this, let us note that we have

$$m(z;t) = \int_{\mathbb{R}} \frac{1}{\lambda - z}\, d\rho(\lambda;t), \quad z \in \mathbb{Z} \setminus \mathbb{R}, \quad t \in [0, T), \tag{9.5.16}$$

where $d\rho(\lambda;t)$ is the spectral measure of the operator $\mathbf{J}(t)$. Since the norms of the operators are uniformly bounded over t, the support $d\rho(\lambda;t)$ for all $t \in [0, T)$ belongs to some ball $\{\lambda \in \mathbb{R} | |\lambda| \le R\}$. Thus, for $|z| > R$, $t \in [0, T)$, we can write

$$\begin{aligned}m(z;t) &= \int_{\mathbb{R}} \frac{1}{\lambda - z}\, d\rho(\lambda;t) = -\frac{1}{z}\int_{\mathbb{R}} \frac{1}{1 - \lambda/z}\, d\rho(\lambda;t)\\ &= -\frac{1}{z}\int_{\mathbb{R}} \left(\sum_{n=0}^{\infty}\left(\frac{\lambda}{z}\right)^n\right) d\rho(\lambda;t) = -\sum_{n=0}^{\infty} \frac{s_n(t)}{z^{n+1}},\end{aligned} \tag{9.5.17}$$

$$s_n(t) = \int_{\mathbb{R}} \lambda^n\, d\rho(\lambda;t).$$

By using the approach of Sect. 9.1, it is possible to prove that elements $s_n(t)$, $n \in \mathbb{N}_0$ are continuously differentiable and

$$\dot{s}_n(t) = \int_{\mathbb{R}} \lambda^n \, d\dot{\rho}(\lambda; t), \quad n \in \mathbb{N}_0, \quad \forall t \in [0, T),$$
$$\dot{m}(z; t) = -\sum_{n=0}^{\infty} \frac{\dot{s}_n(t)}{z^{n+1}}, \quad |z| > R. \tag{9.5.18}$$

According to (9.5.17), (9.5.18), the Eq. (9.5.15) can be written in the form of an expression containing moments.

Namely, let us first note that the Eq. (9.5.15) is equivalent to

$$\dot{m}(z; t) = \int_{\mathbb{R}} \frac{\lambda - \beta_0(t)}{\lambda - z} \, d\rho(\lambda; t) - \varphi(t) m^2(z; t), \quad \forall z \in \mathbb{C} \setminus \mathbb{R}, \ t \in [0, T). \tag{9.5.19}$$

By using (9.5.16)–(9.5.18) we have, instead of (9.5.19),

$$\sum_{n=0}^{\infty} \frac{\dot{s}_n(t)}{z^{n+1}} = \sum_{n=0}^{\infty} \frac{1}{z^{n+1}} (s_{n+1}(t) - \beta_0(t) s_n(t)) + \varphi(t) \sum_{j,k=0}^{\infty} \frac{s_j(t) s_k(t)}{z^{j+k+2}}, \quad |z| > R. \tag{9.5.20}$$

We compare the coefficients in (9.5.20) near powers of $\frac{1}{z^{n+1}}$ for $n = 0, 1, \ldots$. As a result, using the equation $\rho(\mathbb{R}; t) = 1$, we get

$$s_1(t) = \beta_0(t), \ \dot{s}_1(t) = s_2(t) - \beta_0(t) s_1(t) + \varphi(t),$$
$$\dot{s}_2(t) = s_3(t) - \beta_0(t) s_2(t) + 2\varphi(t) s_1(t), \ldots,$$
$$\dot{s}_n(t) = s_{n+1}(t) - \beta_0(t) s_n(t) + \varphi(t) \sum_{j,k=0;\, j+k=n-1}^{\infty} s_j(t) s_k(t), \tag{9.5.21}$$
$$n \in \mathbb{N}, \ t \in [0, T).$$

The formula (9.5.21) gives a method for calculating moments $s_n(t)$ of the spectral measure $d\rho(\lambda; t)$ of the Jacobi operator $\mathbf{J}(t)$. Thus, with the help of classical formulas for the expression of elements of the Jacobi matrix $J(t)$ by moments of its spectral measure, it is possible to obtain formulas for the solution of the generalized shock wave problem. Therefore, we can formulate the theorem.

Theorem 9.5.3 *Consider the generalized Toda shock wave type problem: it is necessary to find the solution of the system* (9.5.10)*, if* $x_n(0)$, $\dot{x}_n(0)$, $n \in \mathbb{N}_0$, *and* $x_{-1}(t) - x_0(t) = f(t)$, $t \in [0, T)$. *For this solution, it is necessary to go to Flaschka variables*

$$\alpha_n(t) = e^{\frac{1}{2}(x_n(t) - x_{n+1}(t))}, \quad n \in \mathbb{N}_0,$$

$$\beta_n(t) = -\dot{x}_n(t), \quad n \in \mathbb{N},$$

$$\beta_0(t) = \alpha_0^2(t) - e^{f(t)}; \quad t \in [0, T).$$

After that, it is necessary to calculate $\forall t \in [0, T)$ *moments* $s_n(t), n \in \mathbb{N}$ *of the spectral measure* $d\rho(\lambda; t)$ *of the Jacobi matrix* $J(t)(s_0(t) = 1)$ *using formulas* (9.5.21)*, where* $\varphi(t) = e^{-f(t)}$. *Then the solution of the problem with Flaschka variables* $\alpha_n(t)$, $\beta_n(t)$, $n \in \mathbb{N}_0$ *is obtained by classical formulas that give the calculation of elements of the Jacobi matrix by its moments.*

Remark 9.5.4 By using Theorem 9.5.3, it is possible to find the solution of the Cauchy problem for the semi-infinite Toda chain in a slightly different way than that specified in Sect. 9.2. In this case, the above functions $\varphi(t)$ are equal to zero, that is, $f(t) = -\infty$.

Remark 9.5.5 It is possible for the main problem (the Cauchy problem for a doubly infinite Toda chain) to apply the way of the previous remark. But for this reason, it is necessary first to develop the theory of moments of the "classical type", but related to the Jacobi matrices of Sect. 9.3.

Bibliographical Notes

The beginning of the researched topic is, of course, the works of Toda M. [282], Flaschka H. [103, 104]. Materials of the section appeared under the influence of the works of Berezanskiĭ Yu. M. [24, 25, 27, 28, 32] and works together with Gekhtman M. I., Shmoĭsh M. E., Mokhon'ko A. A. [41, 42, 46]. The research was carried out under the influence of the works of Sakhnovich L. A. [249], Moser J. [219]. The integration of non-Abelian chains can be found in work of Gekhtman M. I. [106], Bruschi M., Manakov S. V., Ragnisco O., Levi D. [60, 61]. A group approach to the integration of Toda chains was studied by Daletskiĭ A. Yu., Podkolzin G. B. [71]. Integration in the class of Hilbert-Schmidt operators is in the work of Zhernakov N. V. [300]. Integration of nonlinear equations by the methods of algebraic geometry is in the work of Krichever I. M. [191]. Non-isospectral problems were considered by Shmoĭsh M. E. [259], Mokhon'ko O. A. [217, 218], Berezanskiĭ Yu. M. [29]. A separate novelty of the section is the study of doubly infinite Toda chains, which are developed in the works of Berezansky Yu. [33]. Further development of the Toda chains subject is in works Deift P.,

Kriecherbauer T., Venakides S. [73], Deift P., Li L. C., Tomei C. [74] Egorova I., Michor J., Teschl G. [99], Faybusovich L. Gekhtman M. [102], Kaup D. J., Neuberger D. H. [154, 155]. In connection with the indefinite metric, the Toda problem was considered in works of Kodama Y., Ye J. [166, 167]. The Toda shock wave we can see, for example, by Venakides S., Deift P., Oba R. [288, 289], Holian B. L., Straub G. K. [139]. Periodic and special solutions related to Toda chains can be found in works of Kac M. and van Moerbeke P. [147, 148]. Some new class of solutions is in works Gesztesy F., Holden H., Michor J., Teschl G., Regner W. [118, 119], in particular, unbounded in works Szczepanik M. [277]. Finally, we note the monograph by Teschl G. [281], where many problems related to the integration of nonlinear chains are considered.

Some other results on Toda chains related to the inverse spectral problem are in [120]. The application of this approach to Schur flows is in [126].

References

1. Adams, D. R., & Hedberg, L. I. (1996). *Function spaces and potential theory. Grundlehren der mathematischen Wissenschaften [Fundamental Principles of Mathematical Sciences]* (vol. 314, xii+366pp.). Springer.
2. Akhiezer, N. I. (1941). Infinite matrices Jacobi and the problem of moments (in Russian). *Uspekhi Matematicheskikh Nauk, 9*, 126–156.
3. Akhiezer, N. I. (1965). *The classical moment problem* (vol. X, 252pp.).
4. Akhiezer, N. I., & Glazman, I. M. (2013). *Theory of linear operators in hilbert space. Dover books on mathematics* (378pp.). Courier Corporation.
5. Albeverio, S., Brasche, J. F., & Röckner, M. (1989). Dirichlet forms and generalized Schrödinger operators. In *Schrödinger operators (Sønderborg, 1988). Lecture notes in physics* (vol. 345, pp. 1–42). Springer.
6. Albeverio, S., & Kurasov, P. (2000). Singular perturbations of differential operators. In *Solvable Schrödinger type operators. London Mathematical Society Lecture Note Series* (vol. 271, xiv+429pp.). Cambridge University Press.
7. Aleksandryan, R. A., & Mirzakhanyan, E. A. (1979). *General topology* (in Russian). (336pp.). Higher School.
8. Antonevich, A. B., Knjazev, P. N., & Radyno, J. V. (1978). Tasks and exercises in functional analysis. In S. G. Krein (Ed.). *Problems and exercises in functional analysis* (in Russian). (205pp.). Vysh. Shkola.
9. Antonevich, A. B., & Radyno, Y. V. (2003). *Functional analysis and integral equations* (Russian) (430 pp.). BGU.
10. Arkhangel'skiĭ, A. V., & Ponomarev, V. I. (1974). *Fundamentals of general topology in tasks and exercises* (in Russian). (424pp.). Nauka.
11. Arlinskiĭ, Y. M. (1986). On commuting and normal extensions of Hermitian operators (Russian). *Izvestiya Vysshikh Uchebnykh Zavedenii. Matematika, 12*, 3–10, 76.
12. Arlinskiĭ, Y., Golinskiĭ, L., & Tsekanovskiĭ, E. (2008). Contractions with rank one defect operators and truncated CMV matrices. *Journal of Functional Analysis, 254*(1), 154–195.
13. Atzmon, A. (1975). A moment problem for positive measures on the unit disc. *Pacific Journal of Mathematics, 59*(2), 317–325.
14. Azizov, T. Y., & Iokhvidov, I. S. (1989). *Linear operators in spaces with an indefinite metric. Pure and applied mathematics* (Trans. from Russian by E. R. Dawson). (xii+304pp.). Wiley.
15. Banach, S. (1993). *Théorie des opérations linéaires* (in French). (iv+128pp.). Éditions Jacques Gabay. Reprint of the 1932 original.

Y. M. Berezansky, M. E. Dudkin, *Jacobi Matrices and the Moment Problem*,
Operator Theory: Advances and Applications 294,
https://doi.org/10.1007/978-3-031-46387-7

16. Barut, A., & Rączka, R. (1986). *Theory of group representations and applications* (2nd ed., xx+717pp.). World Scientific.
17. Berezanskiĭ, Y. M. (1956). Expansion according to eigenfunction of a partial difference equation of order two (in Russian). *Trudy Moskovskogo Matematicheskogo Obshchestva, 5*, 203–268.
18. Berezanskiĭ, Y. M. (1956). On expansion according to eigenfunctions of general self-adjoint differential operators in (Russian). *Doklady Akademii Nauk SSSR (N.S.), 108*, 379–382.
19. Berezanskiĭ, Y. M. (1956). Generalization of Bochner's theorem to expansions according to eigenfunctions of partial differential equations. (Russian) *Doklady Akademii Nauk SSSR (N.S.), 110* 893–896.
20. Berezanskii, Y. M. (1959). Representation of positive definite kernels by eigenfunctions of differential equations (in Russian). *Matematicheskii Sbornik (N.S.), 47*(89), 145–176.
21. Berezanskiĭ, J. M. (1968). *Expansions in eigenfunctions of selfadjoint operators. Translations of mathematical monographs* (vol. 17, ix+809pp.). American Mathematical Society. Translated from the Russian by R. Bolstein, J. M. Danskin, J. Rovnyak and L. Shulman.
22. Berezanskiĭ, Y. M. (1980). Bilinear forms and Hilbert riggings (in Russian). In *Spectral analysis of differential operators. Akademii Nauk Ukrainskoj SSR* (vol. 134, pp. 83–106).
23. Berezanskiĭ, Y. M. (1986). *Selfadjoint operators in spaces of functions of infinitely many variables. Translations of Mathematical Monographs* (vol. 63, xiv+383pp.). American Mathematical Society. Translated from the Russian by H. H. McFaden. Translation edited by Ben Silver.
24. Berezanskiĭ, Y. M. (1985). Integration of nonlinear difference equations by the method of the inverse spectral problem (Russian). *Doklady Akademii Nauk SSSR, 281*(1), 16–19.
25. Berezanskiĭ, Y. M. (1985). A remark on a loaded Toda lattice (in Russian). *Ukrainskii Matematicheskii Zhurnal, 37*(3), 352–355, 405.
26. Berezanskiĭ, Y. M. (2003). The generalized moment problem associated with correlation measures (in Russian) Funktsional. *Funktsional'nyi Analiz i ego Prilozheniya, 37*(4), 86–91; translation in Funct. Anal. Appl. 37 (2003), no. 4, 311–315.
27. Berezanski, Y. M. (1986). The integration of semi-infinite Toda chain by means of inverse spectral problem. *Reports on Mathematical Physics, 24*(1), 21–47.
28. Berezansky, J. M. (1984). *The integration of semi-infinite Toda chain by means of inverse spectral problem.* Akademii Nauk Ukrainskoj SSR (84–79, 42 pp.).
29. Berezansky, Y. M. (1996). Integration of nonlinear nonisospectral difference-differential equations by means of the inverse spectral problem. In *Nonlinear Physics, Theory and Experiment, Proceedings of the First Workshop (Galliopoli (Lecce), Italy, June 29–July 7, 1995)* (pp. 11–20). World Scientific.
30. Berezansky, Y. M. (2000). Spectral theory of commutative Jacobi fields: Direct and inverse problems. In *Operator theory and its applications (Winnipeg, MB, 1998). Fields institute communications* (vol. 25, pp. 211–224). American Mathematical Society.
31. Berezansky, Y. M. (2002). Some generalizations of the classical moment problem. *Integral Equations Operator Theory, 44*(3), 255–289.
32. Berezansky, Y. M. (2008). Integration of the modified double-infinite Toda lattice with the help of inverse spectral problem. *Ukrainskii Matematicheskii Zhurnal, 60*(4), 453–469; reprinted in Ukrainian Math. J. 60 (2008), no. 4, 521–539.
33. Berezansky, Y. (2009). The integration of double-infinite Toda lattice by means of inverse spectral problem and related questions. *Methods of Functional Analysis and Topology, 15*(2), 101–136.
34. Berezansky, Y. M. (2009). Spectral theory of the infinite block Jacobi type normal matrices, orthogonal polynomials on a complex domain, and the complex moment problem. In *Modern Analysis and Applications. The Mark Krein Centenary Conference. Vol. 2: Differential Operators and Mechanics. Operator theory: Advances and applications* (vol. 191, pp. 37–50). Birkhöuser.
35. Berezansky, Y. M., & Dudkin, M. E. (2004). The complex moment problem in the exponential form. *Methods of Functional Analysis and Topology, 10*(4), 1–10.

36. Berezansky, Y. M., & Dudkin, M. E. (2005). The direct and inverse spectral problems for the block Jacobi type unitary matrices. *Methods of Functional Analysis and Topology, 11*(4), 327–345.
37. Berezansky, Y. M., & Dudkin, M. E. (2006). The complex moment problem and direct and inverse spectral problems for the block Jacobi type bounded normal matrices. *Methods of Functional Analysis and Topology, 12*(1), 1–31.
38. Berezansky, Y. M., & Dudkin, M. E. (2006). The complex moment problem and direct and inverse spectral problems for the block Jacobi type bounded normal matrices. In *International Workshop "Nonlinear Physics and Mathematics" (NLPM 2006) Kiev, May 25–27* (pp. 4–5).
39. Berezansky, Y. M., & Dudkin, M. E. (2007). On the complex moment problem. *Mathematische Nachrichten, 280*(1–2), 60–73.
40. Berezansky, Y. M., & Dudkin, M. E. (2010). The strong Hamburger moment problem and related direct and inverse spectral problems for block Jacobi-Laurent matrices. *Methods of Functional Analysis and Topology, 16*(3), 203–241.
41. Berezanskiĭ, Y. M., Gekhtman, M. I., & Shmoĭsh, M. E. (1986). Integration of certain chains of nonlinear difference equations by the method of the inverse spectral problem (in Russian). *Ukrainskii Matematicheskii Zhurnal, 38*(1), 84–89, 134.
42. Berezanskiĭ, Y. M., & Gekhtman, M. I. (1990). Inverse problem of spectral analysis and nonabelian chains of nonlinear equations (in Russian). *Ukrainskii Matematicheskii Zhurnal, 42*(6), 730–747; translation in Ukrainian Math. J. 42 (1990), no. 6, 645–658.
43. Berezansky, Y. M., & Kondratiev, Y. G. (1995). *Spectral methods in infinite-dimensional analysis. Mathematical physics and applied mathematics* (vol. 12/1, xviii+576pp.). Kluwer Academic Publishers. Translated from the 1988 Russian original by P. V. Malyshev and D. V. Malyshev and revised by the authors.
44. Berezansky, Y. M., & Kondratiev, Y. G. (1995). *Spectral methods in infinite-dimensional analysis. Mathematical physics and applied mathematics* (vol. 1–2. viii+432pp.). Kluwer Academic Publishers.
45. Berezansky, Y. M., & Mierzejewski, D. A. (2007). The investigation of a generalized moment problem associated with correlation measures. *Methods of Functional Analysis and Topology, 13*(2), 124–151.
46. Berezanskiĭ, Y. M., & Mokhon'ko, A. A. (2008). Integration of some nonlinear differential-difference equations using the spectral theory of normal block-Jacobi matrices (in Russian). *Funktsional'nyi Analiz i ego Prilozheniya, 42*(1), 1–21, 95; translation in Funct. Anal. Appl. 42 (2008), no. 1, 1–18.
47. Berezansky, Y. M., Sheftel, Z. G., & Us, G. F. (1996). *Functional analysis, vol. I. Operator theory: Advances and applications*. Translated from the 1990 Russian original by Peter V. Malyshev (vol. 85, xx+423pp.). Birkhäuser.
48. Berezansky, Y. M., Sheftel, Z. G., & Us, G. F. (1996) *Functional analysis. Vol. II. Operator theory: Advances and applications*. Translated from the 1990 Russian original by Peter V. Malyshev (vol. 86, xvi+293pp.). Birkhäuser.
49. Berezansky, Y., & Shmoish, M. (1994). Nonisospectral flows on semi-infinite Jacobi matrices. *Journal of Nonlinear Mathematical Physics, 1*(2), 116–146.
50. Berezanskiĭ, Y. M., Zhernakov, N. V., & Us, G. F. (1987). Operator stochastic integrals (in Russian). *Ukrainskii Matematicheskii Zhurnal, 39*(2), 144–149, 270.
51. Berezin, F. A., & Shubin, M. A. (1991). *The Schrödinger equation. Mathematics and its Applications (Soviet Series)* (vol. 66, xviii+555pp.). Kluwer Academic Publishers. Translated from the 1983 Russian edition by Yu. Rajabov, D. A. Leĭtes and N. A. Sakharova and revised by Shubin. With contributions by G. L. Litvinov and Leĭtes.
52. Berg, C., Christensen, J. P., & Ressel, P. (1984). *Harmonic analysis on semigroups. Theory of positive definite and related functions. Graduate Texts in Mathematics* (vol. 100, x+289pp.). Springer.
53. Biriuk, G., & Coddington, E. A. (1964). Normal extensions of unbounded formally normal operators. *Journal of Mathematics and Mechanics, 13*, 617–637.

54. Birman, M. S., & Solomjak, M. Z. (1987). *Spectral theory of selfadjoint operators in Hilbert space. Mathematics and its Applications (Soviet Series)* (xv+301pp.). D. Reidel. Translated from the 1980 Russian original by S. Khrushchëv and V. Peller.
55. Bisgaard, T. M. (1989). The two-sided complex moment problem. *Arkiv för Matematik, 27*(1), 23–28.
56. Bisgaard, T. M. (2002). A note on factoring of positive definite functions on semigroups. *Mathematische Nachrichten, 236*, 31–46.
57. Bokhonov, Y. E. (1990). Selfadjoint extensions of commuting Hermitian operators (in Russian). *Ukrainskii Matematicheskii Zhurnal, 42*(5), 695–697; translation in Ukrainian Math. J. 42 (1990), no. 5, 614–616 (1991).
58. Brasche, J. F., Koshmanenko, V., & Neidhardt, H. (1994). New aspects of Krein's extension theory. *Ukrainskii Matematicheskii Zhurnal, 46*(1–2), 37–54; reprinted in Ukrainian Math. J. 46 (1994), no. 1–2, 34–53.
59. Brasche, J. F., Malamud, M. M., & Neidhardt, H. (2002). Weyl function and spectral properties of self-adjoint extensions. *Integral Equations Operator Theory, 43*(3), 264–289.
60. Bruschi, M., Manakov, S. V., Ragnisco, O., & Levi, D. (1980). The nonabelian Toda lattice-discrete analogue of the matrix Schrödinger spectral problem. *Journal of Mathematical Physics, 21*(12), 2749–2753.
61. Bruschi, M., Ragnisco, O., & Levi, D. (1981). Evolution equations associated with the discrete analog of the matrix Schrödinger spectral problem solvable by the inverse spectral transform. *Journal of Mathematical Physics, 22*(11), 2463–2471.
62. Cantero, M. J., Moral, L., & Velázquez, L. (2003). Five-diagonal matrices and zeros of orthogonal polynomials on the unit circle. *Linear Algebra and its Applications, 362*, 29–56.
63. Cantero, M. J., Moral, L., & Velá zquez, L. (2005). Minimal representations of unitary operators and orthogonal polynomials on the unit circle. *Linear Algebra and its Applications, 408*, 40–65.
64. Carleman, T. (1926). *Les fonctions quasi analytiques* (116pp.).
65. Coddington, E. A. (1960). Normal extensions of formally normal operators. *Pacific Journal of Mathematics, 10*, 1203–1209.
66. Coddington, E. A. (1965). Formally normal operators having no normal extensions. *Canadian Journal of Mathematics, 17*, 1030–1040.
67. Coddington, E. A., & Levinson, N. (1955). *Theory of ordinary differential equations* (xii+429pp.). McGraw-Hill.
68. Collatz, L. (1966). *Functional analysis and numerical mathematics* (xx+473pp.). Academic Press. Translated from the German by Hansjörg Oser.
69. Conway, J. B. (1990). *A course in functional analysis. Graduate texts in mathematics* (vol. 96, 2nd edn., xvi+399pp.). Springer.
70. Curto, R. E., & Fialkow, L. A. (1991). Recursiveness, positivity, and truncated moment problems. *Houston Journal of Mathematics, 17*(4), 603–635.
71. Daletskiĭ, A. Y., & Podkolzin, G. B. (1989). Group approach to the integration of an infinite Toda lattice (in Russian). *Ukrainskii Matematicheskii Zhurnal, 40*(4), 518–521, 544; translation in Ukrainian Math. J. 40 (1988), no. 4, 445–447 (1989).
72. Day, M. M. (1973). *Normed linear spaces. Ergebnisse der Mathematik und ihrer Grenzgebiete, Band* (vol. 21, 3rd edn., viii+211pp.). Springer.
73. Deift, P., Kriecherbauer, T., & Venakides, S. (1995). Forced lattice vibrations. I, II. *Communications on Pure and Applied Mathematics, 48* (1995), no. 11, 1187–1249, 1251–1298.
74. Deift, P., Li, L. C., & Tomei, C. (1985). Toda flows with infinitely many variables. *Journal of Functional Analysis, 64*(3), 358–402.
75. Derkach, V., & Kovalyov, I. (2015). On a class of generalized Stieltjes continued fractions. *Methods of Functional Analysis and Topology, 21*(4), 315–335.
76. Derkach, V. A., & Malamud, M. M. (1991). Generalized resolvents and the boundary value problems for Hermitian operators with gaps. *Journal of Functional Analysis, 95*(1), 1–95.
77. Derkach, V. A., & Malamud, M. M. (1995). The extension theory of Hermitian operators and the moment problem. *Journal of Mathematical Sciences, 73*(2), 141–242.

78. Derevyagin, M., & Derkach, V. (2004). Spectral problems for generalized Jacobi matrices. *Linear Algebra and its Applications, 382*, 1–24.
79. Derevyagin, M. S., & Derkach, V. A. (2007). On the convergence of Padé approximants of generalized Nevanlinna functions (in Russian). *Trudy Moskovskogo Matematicheskogo Obshchestva, 68*, 133–182. ISBN: 978-5-382-00135-7; translation in Trans. Moscow Math. Soc. 2007, 119–162.
80. Derkach, V. A., & Malamud, M. M. (2017). The theory of extensions of symmetric operators (in Russian). In *Proceedings of the Institute of Mathematics of the National Academy of Sciences of Ukraine* (vol. 104, 573pp.). Institute of Mathematics of the National Academy of Sciences of Ukraine.
81. Devinatz, A. (1954). Integral representations of positive definite functions. II. *Transactions of the American Mathematical Society, 77*, 455–480.
82. Dieudonné, J. (1963). *Fondements de l'analyse moderne* (in French). Traduit de l'anglais par D. Huet. Avant-propos de G. Julia. Cahiers Scientifiques, Fasc. XXVIII Gauthier-Villars (xviii+372pp.). Éditeur.
83. Dudkin, M. E. (2007). About normal extensions of formally normal operators. Modern Analysis and Applications (MAA2007) dedicated to the centenary of Mark Krein, Odessa, Ukraine, April 9–14, p. 43.
84. Dudkin, M. E. (2008). An exact inner structure of the block Jacobi-type unitary matrices connected with the corresponding direct and inverse spectral problems. *Methods of Functional Analysis and Topology, 14*(2), 168–176.
85. Dudkin, M. E. (2012). The complex moment problem in the exponential form with direct and inverse spectral problems for the block Jacobi type correspondence matrices. *Methods of Functional Analysis and Topology, 18*(2), 111–139.
86. Dudkin, M. E. (2013). The inner structure of the Jacobi-Laurent matrix related to the strong Hamburger moment problem. *Methods of Functional Analysis and Topology, 19*(2), 97–107.
87. Dudkin, M. E., & Kozak V. (2014). Direct and inverse spectral problems for block Jacobi type bounded symmetric matrices related to the two dimensional real moment problem. *Methods of Functional Analysis and Topology, 20*(3), 219–251.
88. Dudkin, M. E., & Kozak, V. I. (2014). A direct problem for block matrices of Jacobi type, corresponding to the two-dimensional real moment problem. *Scientific News of NTUU "KPI"*(4), 41–47.
89. Dudkin, M. E., & Kozak, V. I. (2015). Polynomials of the second kind in the two-dimensional moment problem (in Ukrainian). *Scientific News of NTUU "KPI"*(4), 41–46.
90. Dudkin, M. E., & Kozak, V. I. (2016). Jacobi-type block matrices that correspond to the two-dimensional moment problem: Polynomials of the second kind and the Weyl function (in Ukrainian). *Ukraïnskii Matematicheskii Zhurnal, 68*(4), 495–505; translation in Ukrainian Math. J. 68 (2016), no. 4, 557–569.
91. Dudkin, M. E., & Kozak V. I. (2016). Conditions for the unity of the measure corresponding to the two-dimensional problem of moments (Ukrainian). *Scientific News of NTUU KPI*(4), 32–37.
92. Dudkin, M. E., & Kozak, V. I. (2016). A direct spectral problem with Jacobi-type block matrices corresponding to a strong two-dimensional moment problem. *NaUKMA, Physical and Mathematical Sciences, 178*, 16–22.
93. Dudkin, M. E., & Nizhnik, L. P. (20101). Singularly perturbed normal operators. *Methods of Functional Analysis and Topology, 16*(4), 298–303.
94. Dunford, N., & Schwartz, J. T. (1958). *Linear operators. I. General theory. Pure and applied mathematics*. With the assistance of W. G. Bade and R. G. Bartle (vol. 7, xiv+858pp.). Interscience Publishers.
95. Dunford, N., & Schwartz, J. T. (1963). *Linear operators. Part II: Spectral theory. Self adjoint operators in Hilbert space* (ix+pp. 859-1923+7 pp.). Interscience Publishers. With the assistance of William G. Bade and Robert G. Bartle.

96. Dunford, N., & Schwartz, J. T. (1971). *Linear operators. Part III: Spectral operators. Pure and applied mathematics*. (vol. VII, i–xx and 1925–2592). Interscience Publishers. With the assistance of William G. Bade and Robert G. Bartle.
97. Dyukarev, Y. M. (2006). Deficiency numbers of symmetric operators generated by block Jacobi matrices (in Russian). *Matematicheskii Sbornik, 197*(8), 73–100; translation in Sb. Math. 197 (2006), no. 7–8, 1177–1203.
98. Edwards, R. E. (1965). *Functional analysis. Theory and applications* (xiii+781pp.). Holt, Rinehart and Winston.
99. Egorova, I., Michor, J., & Teschl, G. (2009). Soliton solutions of the Toda hierarchy on quasi-periodic backgrounds revisited. *Mathematische Nachrichten, 282*(4), 526–539.
100. Erdélyi, A., Magnus, W., Oberhettinger, F., & Tricomi, F. G. (1953). *Higher transcendental functions*. Vols. I, II. Based, in part, on notes left by Harry Bateman (xxvi+302, xvii+396pp.). McGraw.
101. Èskin, G. I. (1960). A sufficient condition for the solvability of a multi-dimensional problem of moments (in Russian). *Doklady Akademii Nauk SSSR, 133*, 540–543; translated as Soviet Math. Dokl. 1 1960, 895–898.
102. Faybusovich, L., & Gekhtman, M. (2000). Elementary Toda orbits and integrable lattices. *Journal of Mathematical Physics, 41*(5), 2905–2921.
103. Flaschka, H. (1974). The Toda lattice. I. Existence of integrals. *Physical Review B, 9*, 1924–1925.
104. Flaschka, H. (1974). On the Toda lattice. II. Inverse-scattering solution. *Progress of Theoretical Physics, 51*, 703–716.
105. Fuglede, B. (1983). The multidimensional moment problem. *Expositiones Mathematicae, 1*(1), 47–65.
106. Gekhtman, M. I. (1990). Integration of nonabelian chains of Toda type (in Russian). *Funktsional'nyi Analiz i ego Prilozheniya, 24*(3), 72–73; translation in Funct. Anal. Appl. 24 (1990), no. 3, 231–233 (1991).
107. Gekhtman, M. I., & Kalyuzhnyĭ, A. Å. (1991). Spectral theory of orthogonal polynomials of several variables (in Russian). *Ukrainskii Matematicheskii Zhurnal, 43*(10), 1437–1440; translation in Ukrainian Math. J. 43 (1991), no. 10, 1334–1337 (1992).
108. Gelbaum, B. R., & Olmsted, J. M. H. (1964). *Counterexamples in analysis* (xxiv+194pp.). The Mathesis Series Holden-Day.
109. Gel'fand, I. M., Graev, M. I., & Vilenkin, N. Y. (2016). *Generalized functions. Vol. 5. Integral geometry and representation theory* (xvii+449pp.). AMS Chelsea Publishing. Translated from the 1962 Russian original by Eugene Saletan. Reprint of the 1966 English translation.
110. Gel'fand, I. M., Graev, M. I., & Pyatetskii-Shapiro, I. I. (2016). *Generalized functions. Vol. 6. Representation theory and automorphic functions* (xvi+426pp.). AMS Chelsea Publishing. Translated from the 1966 Russian original by K. A. Hirsch. Reprint of the 1969 English translation.
111. Gel'fand, I. M., & Kostyucenko, A. G. (1955). Expansion in eigenfunctions of differential and other operators (in Russian). *Doklady Akademii Nauk SSSR (N.S.), 103*, 349–352.
112. Gel'fand, I. M., Raikow, D. A., & Schilow, G. E. (1964). *Kommutative normierte Algebren* (in German). *Übersetzung und wissenschaftliche Redaktion von Helmut Boseck. Mathematische Forschungsberichte* (xi+332pp.). Deutscher Verlag der Wissenschaften.
113. Gel'fand, I. M., & Shilov, G. E. (2016).*Generalized functions. Vol. 1. Properties and operations* (xviii+423 pp.). AMS Chelsea Publishing. Translated from the 1958 Russian original by Eugene Saletan. Reprint of the 1964 English translation.
114. Gel'fand, I. M., & Shilov, G. E. (2016). *Generalized functions. Vol. 2. Spaces of fundamental and generalized functions* (2016. x+261pp.). AMS Chelsea Publishing. Translated from the 1958 Russian original by Morris D. Friedman, Amiel Feinstein and Christian P. Peltzer. Reprint of the 1968 English translation.
115. Gel'fand, I. M., & Shilov, G. E. (2016) *Generalized functions. Vol. 3. Theory of differential equations* (x+222 pp.). AMS Chelsea Publishing. Translated from the 1958 Russian original [MR0106410] by Meinhard E. Mayer. Reprint of the 1967 English translation.

116. Gel'fand, I. M., & Vilenkin, N. Y. (2016). *Generalized functions. Vol. 4. Applications of harmonic analysis* (xiv+384pp.). AMS Chelsea Publishing. Translated from the 1961 Russian original by Amiel Feinstein. Reprint of the 1964 English translation.
117. Geronimus, L. Y. (1961). *Orthogonal polynomials: Estimates, asymptotic formulas, and series of polynomials orthogonal on the unit circle and on an interval*. Authorized translation from the Russian Consultants Bureau (vi+242pp.).
118. Gesztesy, F., Holden, H., Michor, J., & Teschl, G. (2008). *Soliton equations and their algebro-geometric solutions. Vol. II. (1+1)-dimensional discrete models. Cambridge studies in advanced mathematics* (vol. 114, x+438pp.). Cambridge University Press.
119. Gesztesy, F., & Renger, W. (1997). New classes of Toda soliton solutions. *Communications in Mathematical Physics, 184*(1), 27–50 (1997).
120. Gesztesy, F., & Simon, B. (1997). m-functions and inverse spectral analysis for finite and semi-infinite Jacobi matrices. *Journal d'Analyse Mathématique, 73*, 267–297.
121. Gikhman, I. I., & Skorokhod, A. V. (1996). *Introduction to the theory of random processes* (xiv+516pp.). Dover Publications. Translated from the 1965 Russian original. Reprint of the 1969 English translation. With a preface by Warren M. Hirsch.
122. Glazman, I. M. (1966). *Direct methods of qualitative spectral analysis of singular differential operators* (ix+234pp.). Daniel Davey. Translated from the Russian by the IPST staff. Israel Program for Scientific Translations, Jerusalem.
123. Glazman, I. M., & Ljubič, J. I. (1974). *Finite-dimensional linear analysis: A systematic presentation in problem form* (xvi+520pp.). The M.I.T. Press. Translated from the Russian and edited by G. P. Barker and G. Kuerti.
124. Glazman, I. M., & Naĭman, P. B. (1955). On the convex hull of orthogonal spectral functions (in Russian). *Doklady Akademii Nauk SSSR (N.S.), 102*, 445–448.
125. Gohberg, I. C., & Kreĭn, M. G. (1969). *Introduction to the theory of linear nonselfadjoint operators* (vol. 18, xv+378pp.). American Mathematical Society. Translated from the Russian by A. Feinstein Translations of Mathematical Monographs.
126. Golinskiĭ, L. B. (2006). Schur flows and orthogonal polynomials on the unit circle (in Russian). *Matematicheskii Sbornik, 197*(8), 41–62; translation in Sb. Math. 197 (2006), no. 7–8, 1145–1165.
127. Golinskii, L., & Totik, V. (2007). *Orthogonal polynomials: From Jacobi to Simon*. Spectral theory and mathematical physics: A Festschrift in honor of Barry Simon's 60th birthday. Proceedings of symposia in pure mathematics (vol. 76, Part 2, pp. 821–874). American Mathematical Society.
128. Gorbachuk, V. I., & Gorbachuk, M. L. (1991). *Boundary value problems for operator differential equations. Mathematics and its Applications (Soviet Series)* (vol. 48, xii+347 pp.). Kluwer Academic Publishers. Translated and revised from the 1984 Russian original.
129. Gorbachuk, M. L., & Gorbachuk, V. I. (1994). The theory of selfadjoint extensions of symmetric operators; entire operators and boundary value problems (in Ukrainian). *Ukraïnskii Matematicheskii Zhurnal, 46*(1–2), 55–62; translation in Ukrainian Math. J. 46 (1994), no. 1–2, 54–61.
130. Gorbačuk, M. L., & Mihaĭlec, V. A. (1976). Semibounded selfadjoint extensions of symmetric operators (in Russian). *Doklady Akademii Nauk SSSR, 226*(4), 765–767.
131. Gårding, L. (1953). *Eigenfunction expansions connected with elliptic differential operators*. Tolfte Skandinaviska Matematikerkongressen, Lund (pp. 44–55). Lunds Universitets Matematiska Institution, Lund, (1954).
132. Gårding, L. (1954). *Applications of the theory of direct integrals of Hilbert spaces to some integral and differential operators. The Institute for Fluid Dynamics and Applied Mathematics, Lecture series, no. 11* (23pp.). University of Maryland.
133. Halmos, P. R. (1950). *Measure theory* (xi+304pp.). D. Van Nostrand.
134. Halmos, P. R. (1982). *A Hilbert space problem book. Second edition. Encyclopedia of mathematics and its applications, 17. Graduate Texts in Mathematics* (vol. 19, xvii+369pp.). Springer.

135. Haviland, E. K. (1935). On the momentum problem for distribution functions in more than one dimension. *American Journal of Mathematics, 57*(3), 562–568.
136. Haviland, E. K. (1936). On the momentum problem for distribution functions in more than one dimension. II. *American Journal of Mathematics, 58*(1), 164–168.
137. Hendriksen, E., & Nijhuis, C. (2000). Laurent-Jacobi matrices and the strong Hamburger moment problem. *Acta Applicandae Mathematicae, 61*(1–3), 119–132. Proceedings of the International Conference on Rational Approximation, ICRA99 (Antwerp).
138. Hille, E., & Phillips, R. S. (1974). *Functional analysis and semi-groups* (vol. 31, xii+808pp.). American Mathematical Society. Third printing of the revised edition of 1957.
139. Holian, B. L., & Straub, G. K. (1978). Molecular dynamics of shock waves in one-dimensional chains. *Physical Review B, 18*(4), 1593–1608.
140. Hutson, V., Pym, J. S., & Cloud, M. J. (2005). *Applications of functional analysis and operator theory. Mathematics in science and engineering* (vol. 200, 2nd edn., xiv+426pp.). Elsevier.
141. Ivasiuk, I. Y. (2009). Direct spectral problem for the generalized Jacobi Hermitian matrices. *Methods of Functional Analysis and Topology, 15*(1), 3–14.
142. Jones, W. B., & Njåstad, O. (1999). Orthogonal Laurent polynomials and strong moment theory: A survey. Continued fractions and geometric function theory (CONFUN) (Trondheim, 1997). *Journal of Computational and Applied Mathematics, 105*(1–2), 51–91.
143. Jones, W. B., Thron, W. J., & Njåstad, O. (1984). Orthogonal Laurent polynomials and the strong Hamburger moment problem. *Journal of Mathematical Analysis and Applications, 98*(2), 528–554.
144. Jones, W. B., Thron, W. J., & Waadeland, H. (1980). A strong Stieltjes moment problem. *Transactions of the American Mathematical Society, 261*(2), 503–528.
145. Jørgensen, P. E. T. (1981). A uniqueness theorem for the Heisenberg-Weyl commutation relations with nonselfadjoint position operator. *American Journal of Mathematics, 103*(2), 273–287.
146. Kac, G. I. (1958). Expansion in characteristic functions of self-adjoint operators (in Russian). *Doklady Akademii Nauk SSSR (N.S.), 119*, 19–22.
147. Kac, M., & van Moerbeke, P. (1975). On some periodic Toda lattices. *Proceedings of the National Academy of Sciences of the United States of America, 72*, 1627–1629.
148. Kac, M., & van Moerbeke, P. (1975). A complete solution of the periodic Toda problem. *Proceedings of the National Academy of Sciences of the United States of America, 72*(8), 2879–2880.
149. Kamke, E. (1959). *Differentialgleichungen. LÖsungsmethoden und Lösungen. Teil I: Gewöhnliche Differentialgleichungen. 6. Aufl.; Teil II: Partielle Differentialgleichungen erster Ordnung für eine gesuchte Funktion. 4. Aufl* (in German). Mathematik und ihre Anwendungen in Physik und Technik, Reihe A, Band 18 Akademische Verlagsgesellschaft Geest and Portig K.-G., Leipzig, 1959 I. (xxvi+666pp., II. xv+243pp.).
150. Kantorovich, L. V., & Akilov, G. P. (1964). *Functional analysis in normed spaces. International series of monographs in pure and applied mathematics* (vol. 46, xiii+771pp.). The Macmillan Company. Translated from the Russian by D. E. Brown Edited by A. P. Robertson.
151. Kato, T. (1955). Quadratic forms in Hilbert spaces and asymptotic perturbation series. University of California, Department of Mathematics, Berkeley (ii+145pp.).
152. Kato, T. (1966). *Perturbation theory for linear operators. Die Grundlehren der mathematischen Wissenschaften* (xix+592pp.). Springer.
153. Kats, I. S., & Nudel'man, A. A. (1996). The Stieltjes strong moment problem (in Russian). *Algebra i Analiz, 8*(6), 26–56; translation in St. Petersburg Math. J. 8 (1997), no. 6, 931–950.
154. Kaup, D. J. (1984). The forced Toda lattice: An example of an almost integrable system. *Journal of Mathematical Physics, 25*(2), 277–281.
155. Kaup, D. J., & Neuberger, D. H. (1984). The soliton birth rate in the forced Toda lattice. *Journal of Mathematical Physics, 25*(2), 282–284.
156. Killip, R., & Nenciu, I. (2007). CMV: The unitary analogue of Jacobi matrices. *Communications on Pure and Applied Mathematics, 60*(8), 1148–1188.

157. Kilpi, Y. (1953). Über lineare normale Transformation im Hilbertschen Raum (in German). *Annales Academiae Scientiarum Fennicae Sereries A. I. Mathematical Physics, 1953*(154), 38pp.
158. Kilpi, Y. (1954). Lineare normale Transformationen mit einem einfachen Spektrum und das komplexe Momentenproblem (in German). *Annales Academiae Scientiarum Fennicae Sereries A. I. Mathematical Physics 1954*(176), 34pp.
159. Kilpi, Y. (1955). Zur Theorie des mehrdimensionalen Momentenproblems (in German). *Annales Academiae Scientiarum Fennicae Sereries A. I., 1955*(205), 13pp.
160. Kilpi, Y. (1957). Über das komplexe Momentenproblem (in German). *Annales Academiae Scientiarum Fennicae Sereries A. I., 236*, 32pp.
161. Kirillov, A. A. (1976). *Elements of the theory of representations. Grundlehren der Mathematischen Wissenschaften* (Band 220, xi+315pp.). Springer.
162. Kirillov, A. A., & Gvishiani, A. D. (1982). *Theorems and problems in functional analysis. Problem books in mathematics* (ix+347pp.). Springer. Translated from the Russian by Harold H. McFaden.
163. Kolmogorov, A. N., & Fomin, S. V. (1975). *Introductory real analysis.* (xii+403pp.). Springer. Translated from the second Russian edition and edited by Richard A. Silverman. Corrected reprinting.
164. Kočubeĭ, A. N. (1979). Symmetric operators commuting with a family of unitary operators (in Russian). *Funktsional'nyi Analiz i ego Prilozheniya, 13*(4), 77–78.
165. Kochubeĭ, A. N. (1982). Elliptic operators with boundary conditions on a subset of measure zero (in Russian). *Funktsional'nyi Analiz i ego Prilozheniya, 16*(2), 74–75.
166. Kodama, Y., & Ye, J. (1996). Toda hierarchy with indefinite metric. *Physics D, 91*(4), 321–339.
167. Kodama, Y., & Ye, J. (1998). Toda lattices with indefinite metric. II. Topology of the ISO-spectral manifolds. *Physics D, 121*(1–2), 89–108.
168. Kostyučenko, A. G., & Mityagin, B. S. (1960). The multi-dimensional problem of moments (in Russian). *Doklady Akademii Nauk SSSR, 131*, 1249–1252; translated as Soviet Math. Dokl. 1 1960 415–419.
169. Kostjucenko, A. G., & Sargsjan, I. S. (1979). *Distribution of eigenvalues. Selfadjoint ordinary differential operators* (in Russian) (400pp.). Nauka.
170. Koshmanenko, V. (1999). *Singular quadratic forms in perturbation theory. Mathematics and its applications* (vol. 474, viii+308pp.). Kluwer Academic Publishers. Translated from the 1993 Russian original by P. V. Malyshev and D. V. Malyshev.
171. Koshmanenko, V., & Dudkin, M. (2016). *The method of rigged spaces in singular perturbation theory of self-adjoint operators. Operator theory: Advances and applications* (vol. 253, xx+237pp.). Birkäuser/Springer.
172. Kozak, V. I. (2015). Construction of Jacobi-type block matrices corresponding to a stong two-dimensional real moment problem (in Ukrainian). *Scientific Notes of NaUKMA, 21*(165), 19–25.
173. Krasnosel'skij, M. A., Lifshits, J. A., & Sobolev, A. V. (1989). *Positive linear systems. The method of positive operators. Sigma Series in Applied Mathematics* (vol. 5, viii+354pp.). Heldermann Verlag. Translated from the Russian by Jürgen Appell.
174. Krasnosel'skiĭ, M. A., Vaĭnikko, G. M., Zabreĭko, P. P., Rutitskii, Y. B., & Stetsenko, V. Y. (1972). *Approximate solution of operator equations* (xii+484pp.). Wolters-Noordhoff Publishing. Translated from the Russian by D. Louvish.
175. Krasnosel'skiĭ, M. A., & Zabreĭko, P. P. (1984). *Geometrical methods of nonlinear analysis. Grundlehren der mathematischen Wissenschaften [Fundamental Principles of Mathematical Sciences]* (vol. 263, xix+409pp.). Springer. Translated from the Russian by Christian C. Fenske.
176. Krasnosel'skiĭ, M. A., Zabreĭko, P. P., Pustyl'nik, E. I., & Sobolevskiĭ, P. E. (1976). *Integral operators in spaces of summable functions. Monographs and Textbooks on Mechanics of Solids and Fluids, Mechanics: Analysis* (xv+520 pp.). Noordhoff International Publishing. Translated from the Russian by T. Ando.

177. Krein, M. (1944). On Hermitian operators whose deficiency indices are 1. *C. R. (Doklady) Academy of Sciences URSS (N.S.), 43*, 323–326.
178. Krein, M. (1944). On Hermitian operators with deficiency indices equal to one. II. *C. R. (Doklady) Academy of Sciences URSS (N. S.), 44*, 131–134.
179. Krein, M. (1945). On a problem of extrapolation of A. N. Kolmogoroff. *C. R. (Doklady) Academy of Sciences URSS (N. S.), 46*, 306–309.
180. Krein, M. (1946). On a general method of decomposing Hermite-positive nuclei into elementary products. *C. R. (Doklady) Academy of Sciences URSS (N.S.), 53*, 3–6.
181. Krein, M. (1947). The theory of self-adjoint extensions of semi-bounded Hermitian transformations and its applications. I (in Russian). *Recreational Mathematics [Matematicheskii Sbornik] N.S., 20*(62), 431–495.
182. Krein, M. G. (1967). The description of all solutions of the truncated power moment problem and some problems of operator theory (in Russian). *Istoriko-Matematicheskie Issledovaniya 2*, 2, 114–132.
183. Krein, S. G. (1982). *Linear equations in Banach spaces* (xii+102 pp.). Birkhäuser. Translated from the Russian by A. Iacob. With an introduction by I. Gohberg.
184. Krein, S. G. (Ed.). Vilenkin, N. Y., Gorin, E. A., Kostyuchenko, A. G., Krasnosel'skiĭ, S. G., Kreĭn, S. G., Maslov, V. P., Mityagin, B. S., Petunin, Y. I., Rutitskii, Y. B., Sobolev, V. I., Stetsenko, V. Y., Faddeev, L. D., Tsitlanadze, E. S. (1972). *Functional analysis* (xv+380pp.). Wolters-Noordhoff Publishing. Translated from the Russian by Richard E. Flaherty. English edition edited by George F. Votruba with the collaboration of Leo F. Boron.
185. Kreĭn, M. G. (1948). On Hermitian operators with directed functionals (in Ukrainian). *Akad. Nauk Ukrain. RSR. Zbirnik Prac' Inst. Mat. 1948*(10), 83–106.
186. Kreĭn, M. G. (1949). The fundamental propositions of the theory of representations of Hermitian operators with deficiency index (m,m) (in Russian). *Ukrainskii Matematicheskii Zhurnal, 1*(2), 3–66.
187. Kreĭn, M. (1949). Infinite J-matrices and a matrix-moment problem (in Russian). *Doklady Akademii Nauk SSSR (N.S.), 69*, 125–128.
188. Kreĭn, M. G., & Krasnosel'skiĭ, M. A. (1947). Fundamental theorems on the extension of Hermitian operators and certain of their applications to the theory of orthogonal polynomials and the problem of moments (in Russian). *Uspehi Matematicheskik Nauk (N. S.), 2*(3), 60–106.
189. Kreĭn, M. G., & Nudel'man, A. A. (1977). *The Markov moment problem and extremal problems. Ideas and problems of P. L. Čebyšev and A. A. Markov and their further development. Translations of Mathematical Monographs* (vol. 50, v+417pp.). American Mathematical Society. Translated from the Russian by D. Louvish.
190. Kreĭn, S. G. (1971). *Linear differential equations in Banach space. Translations of Mathematical Monographs* (vol. 29, v+390pp.). American Mathematical Society. Translated from the Russian by J. M. Danskin.
191. Krichever, I. M. (1977). Integration of nonlinear equations by the methods of algebraic geometry (in Russian). *Funkcional. Anal. i Priložen. 11*(1), 15–31, 96.
192. Kutateladze, S. S. (1996). *Fundamentals of functional analysis. Kluwer texts in the mathematical sciences* (vol. 12, xiv+276pp.). Kluwer Academic Publishers. Translated from the second (1995) edition.
193. Kuzhel', A. V. (1981). An analogue of M. G. Kreĭn's formula for resolvents of nonselfadjoint extensions of a Hermitian operator (in Russian). *Teor. Funktsiĭ Funktsional'nyi Analiz i ego Prilozheniya, 36*, 49–55, 127.
194. Kuzhel, A. V., & Kuzhel, S. A. (1998). Regular extensions of Hermitian operators. Translated from the Russian by P. Malyshev and D. Malyshev. (xii+273pp.). VSP.
195. Ladyzhenskaya, O. A. (1985). *The boundary value problems of mathematical physics. Applied mathematical sciences* (vol. 49, xxx+322pp.). Springer. Translated from the Russian by Jack Lohwater [Arthur J. Lohwater].
196. Lax, P. D., & Milgram, A. N. (1954). *Parabolic equations. Contributions to the theory of partial differential equations. Annals of mathematics studies* (vol. 33, pp. 167–190). Princeton University Press.

197. Levitan, B. M. (1987). *Inverse Sturm-Liouville problems* (x+240pp.). VSP. Translated from the Russian by O. Efimov.
198. Levitan, B. M., & Sargsjan I. S. (1975). *Introduction to spectral theory: Selfadjoint ordinary differential operators. Translations of mathematical monographs* (vol. 39, xi+525pp.). American Mathematical Society. Translated from the Russian by Amiel Feinstein.
199. Lions, J.-L., & Magenes, E. (1968). *Problèmes aux limites non homogènes et applications* (in French). *Travaux et Recherches Mathématiques* (vol. 17 xx+372pp.). Dunod.
200. Lions, J.-L. (1961). Équations différentielles opérationnelles et problémes aux limites (in French). Die Grundlehren der mathematischen Wissenschaften (Band 111, ix+292pp.). Springer.
201. Liusternik, L. A., & Sobolev, V. J. (1961). *Elements of functional analysis. Russian monographs and texts on advanced mathematics and physics* (vol. 5, x+411pp.). Hindustan Publishing.
202. Loomis, L. H. (1953). *An introduction to abstract harmonic analysis* (x+190pp.). D. Van Nostrand.
203. Lyantse, V. È., & Storozh, O. G. (1983). *Methods of the theory of unbounded operators* (in Russian) (211pp.). Naukova Dumka.
204. Malamud, M. M., & Mogilevskii, V. I. (2002). Krein type formula for canonical resolvents of dual pairs of linear relations. *Methods of Functional Analysis and Topology, 8*(4), 72–100.
205. Manakov, S. V. (1975). Complete integrability and stochastization of discrete dynamical systems. *Soviet Physics JETP, 40*(2), 269–274; translated from Ž. Èksper. Teoret. Fiz. 67 (1974), no. 2, 543–555.
206. Mandelbrojt, S. (1952). *Séries adhérentes. régularisation des suites. Applications* (in French) (xiv+277pp.). Gauthier-Villars.
207. Marchenko, V. A. (1986). *Sturm-Liouville operators and applications. Operator theory: Advances and applications* (vol. 22, xii+367pp.). Birkhäuser. Translated from the Russian by A. Iacob.
208. Markuševič, A. I. (1967). *Theory of analytic functions. Vol. I. Fundamentals of the theory* (in Russian) (2nd edn., 486pp.). Nauka.
209. Markuševič, A. I. (1968). *Theory of analytic functions. Vol. II. Further construction of the theory* (in Russian). (Second revised and augmented edition, 624pp.). Nauka.
210. Maurin, K. (1958). Entwicklung positiv definiter Kerne nach Eigendistributionen. Differenzierbarkeit der Spektralfunktion eines hypoelliptischen Operators (in German). *Bulletin L'Académie Polonaise des Science, Série des Sciences Mathématiques, Astronomiques et Physiques, 6*, 149–155.
211. Maurin, K. (1959). *Methods of Hilbert space* (in Polish). Polska Akademia Nauk. Monografie Matematyczne, Tom 36 Panstwowe Wydawnictwo Naukowe, Warsaw (363pp.).
212. Maurin, K. (1960). Eine Bemerkung zur allgemeinen Eigenfunktionsentwicklungen für vertauschbare Operatorensysteme beliebiger Mächtigkeit. *Bulletin L'Académie Polonaise des Science, Série des Sciences Mathématiques, Astronomiques et Physiques, 8*, 381–384.
213. Maurin, K. (1968). *General eigenfunction expansions and unitary representations of topological groups. Monografie Matematyczne, Tom 48 PWN* (367pp.). Polish Scientific Publishers.
214. Maz'ja, V. G. (1985). *Sobolev spaces. Springer series in soviet mathematics.* Translated from the Russian by T. O. Shaposhnikova (xix+486pp.). Springer.
215. Mihlin, S. G. (1960). *Linear integral equations. Russian monographs and texts on advanced mathematics and physics* (vol. I, vii+223pp.). Gordon and Breach Publishers.
216. Mikhaĭlov, V. P. (1978). *Partial differential equations* (397pp.). Translated from the Russian by P. C. Sinha; distributed by Imported Publications.
217. Mokhon'ko, O. A. (2005). Some solvable classes of nonlinear nonisospectral difference equations (in Ukrainian). *Ukrainskii Matematicheskii Zhurnal, 57*(3), 356–365; translation in Ukrainian Math. J. 57 (2005), no. 3, 427–439.
218. Mokhon'ko, O. A. (2008). Nonisospectral flows on semiinfinite unitary block Jacobi matrices. *Ukrainskii Matematicheskii Zhurnal, 60*(4), 521–544; reprinted in Ukrainian Math. J. 60 (2008), no. 4, 598–622.

219. Moser, J. (1975). Three integrable Hamiltonian systems connected with isospectral deformations. *Advances in Mathematics, 16*, 197–220.
220. Mynbaev, K. T., & Otelbaev, M. O. (1988). *Weighted function spaces and the spectrum of differential operators* (in Russian) (286pp.). Nauka. With an English summary.
221. Naimark, M. A. (1964). *Normed rings*. Revised edition Translated from the first Russian edition by Leo F. Boron P. Noordhoff N. V., Groningen (xvi+572pp.).
222. Naimark, M. A. (1967). *Linear differential operators. Part I: Elementary theory of linear differential operators* (xiii+144pp.). Frederick Ungar Publishing.
Naimark, M. A. (1968). *Linear differential operators. Part II: linear differential operators in Hilbert space* (xv+352pp.). With additional material by the author, and a supplement by V. È. Ljance. Translated from the Russian by E. R. Dawson. English translation edited by W. N. Everitt Frederick.
223. Naĭmark, M., & Stern, A. (1979). *Théorie des représentations des groupes* (in French) (608 pp.). Translated from the Russian by A. Sossinski.
224. Natanson, I. P. (1981). *Theorie der Funktionen einer reellen Veränderlichen* (German). *Mathematische Lehrbücher und Monographien, I. Abteilung: Mathematische Lehrbücher [Mathematical Textbooks and Monographs, Part I: Mathematical Textbooks], VI* (5th edn., xii+590 pp.). Akademie-Verlag. Translated from the 1957 Russian edition. Edited by Karl Bögel.
225. Nelson, E. (1959). Analytic vectors. *Annals of Mathematics, 70*, 572–615.
226. Nelson, E. (1964). Interaction of nonrelativistic particles with a quantized scalar field. *Journal of Mathematical Physics, 5*, 1190–1197.
227. Neumark, M. A. (1947). Extremal spectral functions of a symmetric operator (in Russian). *Bulletin of the Academy of Sciences of the USSR, Mathematics Series [Izvestia Akad. Nauk SSSR] 11*, 327–344.
228. Njåstad, O. (1996). Solutions of the strong Hamburger moment problem. *Journal of Mathematical Analysis and Applications, 197*(1), 227–248.
229. Nikol'skiĭ, S. M. (1975). *Approximation of functions of several variables and imbedding theorems* (Band, 205, viii+418pp.). Springer. Translated from the Russian by John M. Danskin, Jr. Die Grundlehren der mathematischen Wissenschaften.
230. Nudel'man, A. A. (1953). On the application of completely and absolutely monotone sequences to the problem of moments (in Russian). *Uspekhi Matematicheskikh Nauk (N.S.), 8*(6), 119–124.
231. Nussbaum, A. E. (1965). Quasi-analytic vectors. *Arkiv för Matematik, 6*, 179–191.
232. Nussbaum, A. E. (1969). A note on quasi-analytic vectors. *Studia Mathematica, 33*, 305–309.
233. Petersen, L. C. (1982). On the relation between the multidimensional moment problem and the one-dimensional moment problem. *Mathematica Scandinavica, 51*(2), 361–366.
234. Petrovsky, I. G. (1991). *Lectures on partial differential equations* (x+245pp.). Dover Publications. Translated from the Russian byA. Shenitzer. Reprint of the 1964 English translation.
235. Petrovskiĭ, I. G. (1957). *Lectures on the theory of integral equations* (vi+97pp.). Graylock Press. Translated from the second revised Russian edition by H. Kamel and H. Komm.
236. Plesner, A. I. (1969). *Spectral theory of linear operators. Vols. I, II* (vol. I: xii+235pp. vol. II: viii+271pp.). Frederick Ungar Publishing. Translated from the Russian by Merlynd K. Nestell and Alan G. Gibbs Frederick.
237. Pontryagin, L. S. (1988). *Selected scientific works. Vol. III. Continuous groups* (343pp.). Nauka.
238. Pulemyotov, A. D. (2003). Support of a joint resolution of identity and the projection spectral theorem. *Infinite Dimensional Analysis, Quantum Probability and Related Topics, 6*(4), 549–561.
239. Reed, M., & Simon, B. (1975). *Methods of modern mathematical physics. II. Fourier analysis, self-adjointness* (xv+361pp.). Academic Press.
240. Reed, M., & Simon, B. (1978). *Methods of modern mathematical physics. IV. Analysis of operators* (xv+396pp.). Academic Press.

241. Reed, M., & Simon, B. (1979). *Methods of modern mathematical physics. III. Scattering theory* (xv+463pp.). Academic Press.
242. Reed, M., & Simon, B. (1980). *Methods of modern mathematical physics. I. Functional analysis* (2nd edn., xv+400pp.). Academic Press.
243. Richtmyer, R. D. (1978). *Principles of advanced mathematical physics. Texts and Monographs in Physics* (vol. I., xv+422pp.). Springer.
244. Richtmyer, R. D. (1981). *Principles of advanced mathematical physics. Texts and monographs in physics* (vol. II, ix+322pp.). Springer.
245. Riesz, F., & Sz.-Nagy, B. (1990). *Functional analysis. Dover books on advanced mathematics* (xii+504pp.). Dover Publications. Translated from the second French edition by Leo F. Boron. Reprint of the 1955 original.
246. Robertson, A. P., & Robertson, W. (1980). *Topological vector spaces. Cambridge tracts in mathematics* (vol. 53, viii+172pp.). Cambridge University Press. Reprint of the second edition.
247. Rudin, W. (1973). *Functional analysis*. McGraw-Hill.
248. Sadovnichiĭ, V. A. (1991). *Theory of operators. Contemporary Soviet Mathematics* (xii+396pp.). Consultants Bureau. Translated from the second Russian edition by Roger Cooke.
249. Sakhnovich, L. A. (1989). On a semi-infinite Toda chain (in Russian). *Teoreticeskaa i Matematiceskaa Fizika, 81*(1), 12–23; translation in Theoret. and Math. Phys. 81 (1989), no. 1, 1018–1026 (1990).
250. Samoĭlenko, Y. S. (1991). *Spectral theory of families of selfadjoint operators. Mathematics and its applications (Soviet Series)* (vol. 57, xvi+293pp.). Kluwer Academic Publishers. Translated from the Russian by E. V. Tisjachnij.
251. Schaefer, H. H., & Wolff, M. P. (1999). *Topological vector spaces*. Graduate texts in mathematics (vol. 3., 2nd edn., xii+346pp.). Springer.
252. Schmüdgen, K. (1979). An example of a positive polynomial which is not a sum of squares of polynomials. A positive, but not strongly positive functional. *Mathematische Nachrichten, 88*, 385–390.
253. Schmüdgen, K. (1985). A formally normal operator having no normal extension. *Proceedings of the American Mathematical Society, 95*(3), 503–504.
254. Schmüdgen, K. (1986). On commuting unbounded selfadjoint operators. IV. *Mathematische Nachrichten, 125*, 83–102.
255. Schmüdgen, K. (2017). *The moment problem. Graduate Texts in Mathematics* (vol. 277, xii+535pp.). Springer.
256. Shilov, G. E. (1973). *Mathematical analysis. Vol. 1: Elementary real and complex analysis* (xi+516pp.). The M.I.T. Press. Revised English edition translated from the Russian and edited by Richard A. Silverman.
257. Shilov, G. E. (1996). *Elementary real and complex analysis* (xii+516 pp.). Dover Publications. Revised English edition translated from the Russian and edited by Richard A. Silverman. Corrected reprint of the 1973 English edition.
258. Shilov, G. E., & Gurevich, B. L. (1977). *Integral, measure and derivative: A unified approach. Dover books on advanced mathematics* (xiv+233pp.). Dover Publications. Translated from the Russian and edited by Richard A. Silverman. Revised English edition.
259. Shmoĭsh, M. E. (1989). Nonisospectral deformations of Jacobi matrices, and nonlinear difference equations (in Russian). *Ukrainskii Matematicheskii Zhurnal, 41*(2), 282–285, 289; translation in Ukrainian Math. J. 41 (1989), no. 2, 254–257.
260. Shohat, J. A., & Tamarkin, J. D. (1943). *The problem of moments. American mathematical society mathematical surveys* (vol. I, xiv+140pp.). American Mathematical Society.
261. Simonov, K. K. (2006). Orthogonal matrix Laurent polynomials (in Russian). *Matematicheskie Zametki, 79*(2), 316–320; translation in Math. Notes 79 (2006), no. 1–2, 291–295
262. Simon, B. (2004). Orthogonal polynomials on the unit circle: New results. *International Mathematics Research Notices, 53*, 2837–2880.

263. Simon, B. (2005). *Orthogonal polynomials on the unit circle. Part 1. Classical theory. American mathematical society colloquium publications* (vol. 54, part 1, xxvi+466pp.). American Mathematical Society.
264. Simon, B. (2005). *Orthogonal polynomials on the unit circle. Part 2. Spectral theory. American mathematical society colloquium publications* (vol. 54, part 2, pp. i–xxii and 467–1044). American Mathematical Society.
265. Simon, B. (2005). OPUC on one foot. *Bulletin of the American Mathematical Society (N.S.), 42*(4), 431–460.
266. Simon, B. (2007). CMV matrices: Five years after. *Journal of Computational and Applied Mathematics, 208*(1), 120–154.
267. Simonov, K. K. (2006). Strong matrix moment problem of Hamburger. *Methods of Functional Analysis and Topology, 12*(2), 183–196.
268. Sobolev, S. L., & Vaskevich, V. L. (1997). *The theory of cubature formulas. Mathematics and its applications* (vol. 415, xxii+416pp.). Kluwer Academic Publishers. Translated from the 1996 Russian original and with a foreword by S. S. Kutateladze. Revised by Vaskevich.
269. Stochel, J., & Szafraniec, F. H. (1985). On normal extensions of unbounded operators. *I. J. Operator Theory, 14*(1), 31–55.
270. Stochel, J., & Szafraniec, F. H. (1989). On normal extensions of unbounded operators. II. *Acta Scientiarum Mathematicarum (Szeged), 53*(1–2), 153–177.
271. Stochel, J., & Szafraniec, F. H. (1994). Algebraic operators and moments on algebraic sets. *Portugaliae Mathematica, 51*(1), 25–45.
272. Stochel, J., & Szafraniec, F. H. (1998). The complex moment problem and subnormality: A polar decomposition approach. *Journal of Functional Analysis, 159*(2), 432–491.
273. Suetin, P. K. (1966). Fundamental properties of polynomials orthogonal on a contour (in Russian). *Uspekhi Matematicheskikh Nauk, 21*(2), 41–88.
274. Suetin, P. K. (1999). *Orthogonal polynomials in two variables*. Analytical methods and special functions (vol. 3, xx+348pp.). Gordon and Breach Science Publishers. Translated from the 1988 Russian original by E. V. Pankratiev [E. V. Pankrat'ev].
275. Szafraniec, F. H. (1981). Boundedness of the shift operator related to positive definite forms: An application to moment problems. *Arkiv för Matematik, 19*(2), 251–259.
276. Szafraniec, F. H. (2006). On normal extensions of unbounded operators. IV. A matrix construction. In *Operator theory and indefinite inner product spaces. Operator theory: Advances and applications* (vol. 163, pp. 337–350).
277. Szczepanik, M. (1997). Integration of semi-infinite Toda chain in some class of unbounded solutions. *Reports on Mathematical Physics, 39*(2), 263–273.
278. Szegö, G. (1967). *Orthogonal polynomials. American mathematical society colloquium publications* (vol. 23, 3rd edn., xiii+423pp.). American Mathematical Society.
279. Sz. Nagy, B. (1942). *Spektraldarstellung linearer Transformationen des Hilbertschen Raumes* (in German). *Ergebnisse der Mathematik und ihrer Grenzgebiete* (vol. 5, no. 5., iv+80pp.). Springer.
280. Telyakovsky, S. A. (1980). *Collection of problems on the theory of functions of a real variable* (in Russian). Nauka, Glav. red. fiz.- mat. lit. (112pp.).
281. Teschl, G. (2000). *Jacobi operators and completely integrable nonlinear lattices. Mathematical surveys and monographs* (vol. 72, xvii+351pp.). American Mathematical Society.
282. Toda, M. (1981). *Theory of nonlinear lattices. Springer series in solid-state sciences*. Translated from the Japanese by the author (vol. 20, x+205 pp.). Springer.
283. Trenogin, V. A. (1980). *Functional analysis* (in Russian) (496pp.). Nauka.
284. Trenogin, V. A., Pisarevskiĭ V. M., & Soboleva T. S. (1984). *Tasks and exercises in functional analysis* (in Russian). Nauka, Glav. red. fiz.- mat. lit. (256pp.).
285. Triebel, H. (1995). *Interpolation theory, function spaces, differential operators* (2nd edn., 532pp.). Johann Ambrosius Barth.
286. Triebel, H. (1997). *Fractals and spectra. Related to Fourier analysis and function spaces. Monographs in mathematics* (vol. 91, viii+271pp.). Birkhäuser Verlag.

287. Vainerman, L. I. (1980). Extensions of closed operators in Hilbert space (in Russian). *Matematicheskie Zametki, 28*(6), 833–842, 960.
288. Venakides, S. (1991). The Toda shock problem. In *Inverse scattering and applications (Amherst, MA, 1990)* (vol. 122, pp. 119–133). American Mathematical Society.
289. Venakides, S., Deift, P., & Oba, R. (1991). The Toda shock problem. *Communications on Pure and Applied Mathematics, 44*(8–9), 1171–1242.
290. Višik, M. I. (1952). On general boundary problems for elliptic differential equations (in Russian). *Trudy Moskovskogo Matematicheskogo Obshchestva, 1*, 187–246.
291. Vladimirov, V. S. (1971). *Equations of mathematical physics. Pure and applied mathematics* (vol. 3, vi+418pp.). Marcel Dekker. Translated from the Russian by Audrey Littlewood Edited by Alan Jeffrey.
292. Vladimirov, V. S. (2002). *Methods of the theory of generalized functions. Analytical methods and special functions* (vol. 6, xiv+311 pp.). Taylor and Francis.
293. Vladimirov, V. S., Mikhailov, V. P., Vasharin, A. A., Karimova, K. K., Sidorov, Y. V., & Shabunin, M. I. (1986). *A collection of problems on the equations of mathematical physics* (288pp.). Springer. Translated from the Russian by Eugene Yankovsky [E. Yankovskiĭ].
294. Vulih, B. Z. (1967). *Vvedenie v funktsional'nyi analiz* (in Russian) *[Introduction to functional analysis]*. Second revised and augmented edition Izdat (415pp.). Nauka.
295. Wouk, A. (1953). Difference equations and J-matrices. *Duke Mathematical Journal, 20*, 141–159.
296. Xu, Y. (1994). Block Jacobi matrices and zeros of multivariate orthogonal polynomials. *Transactions of the American Mathematical Society, 342*(2), 855–866.
297. Xu, Y. (1997). *On orthogonal polynomials in several variables. Special functions, q-series and related topics (Toronto, ON, 1995). Fields institute communications* (vol. 14, pp. 247–270). American Mathematical Society.
298. Yosida, K. (1965). *Functional analysis. Die Grundlehren der mathematischen Wissenschaften* (Band 123, xi+458pp.). Academic Press.
299. Zarhina, R. B. (1959). On the two-dimensional problem of moments (in Russian). *Doklady Akademii Nauk SSSR, 124*, 743–746.
300. Zhernakov, N. V. (1987). Integration of Toda lattices in the class of Hilbert-Schmidt operators (in Russian). *Ukrainskii Matematicheskii Zhurnal, 39*(5), 645–648, 680.

Subject Index

C
Changing variables in the integral, 35
Chain of Toda, 414
Circle weil-Hamburger, 143, 146, 222
Completion by two norm, 87
Criterion
  of commutativity, 62
  of self-adjointness, 23
    hyperbolic, 55
    parabolic, 56
    quasi-analytic, 57
    Schrödinger, 50

D
Defect number, 22
Deficiency indices, 23
Deficiency subspace, 22
Differentiation
  an operator-valued measure, 94
  resolution of the identity, 99
Direct product of measures, 35
Domain of an operator, 17

E
Equality (identity)
  of Hilbert, 24
  of Parseval, 112
Equipment
  of the Hilbert space, 64
  nuclear, 84
  by projection limits, 73
  of the quasi-nuclear space, 68
Extension of an operator, 17
  Friedrichs, 92

F
Form
  bilinear, 90, 85
    closed, 91
    Hermitian, 90
    positive, 91
  semi-bounded, 91
  sesquilinear, 90
Form-sum operators, 93
Formulae
  Neumann, 29
  Stone, 42
Function of an operator, 38

G
Generalized
  eigenvalue, 102, 106
  eigenvector, 102, 106
  projection, 100
  spectrum, 102

I
Infinitesimal generator, 45

L
Limit
  inductive, 74
  projective, 72, 80

Y. M. Berezansky, M. E. Dudkin, *Jacobi Matrices and the Moment Problem*,
Operator Theory: Advances and Applications 294,
https://doi.org/10.1007/978-3-031-46387-7

M
Matrix
  CMV, a unitary operator, 310
  Jacobi, 122
  Jacobi-Laurent, 183
  of a normal operators, 256
Moment problem
  complex, 275
    in the exponential form, 361
  Hamburger, 148
  strong, 183
  trigonometric, 316
  two-dimensional real, 400

N
Norm of Hilbert-Schmidt, 96, 184

O
Operator
  adjoint, 20
  closable, 19
  closed, 19
  essentially self-adjoint, 23
  graph, 18
  Hermitian, 23
  isometric, 25
  linear, 17
  invertible, 18
  non-negative, 24
  normal, 35, 43
  self-adjoint, 23
  semi-bounded, 24
  standardly connected, 105, 277
  unitary, 25

P
Prerigging, 89
Product of resolutions of the identity, 35
Positive definite function, 48
Point
  regular, 21
  regular type, 21
Polynomial
  first kind, 123
  Laurent, 192
    first kind, 215
    second kind, 216
  second kind, 138

Q
Quasi-analytic class, 58

R
Range of an operator, 18
Resolvent, 24
Resolution of the identity, 30, 35, 38, 100
Restriction of an operator, 17

S
Space
  Hilbert, 17
  positive, 63
  negative, 64
  rigged, 63
  topological, 70
    linear, 71
Spectral expansion
  of a bounded operator, 36
  unitary operator, 39
Spectral integral
  of bounded function, 33
  direct of Hilbert spaces, 112
  of unbounded function, 33
Spectrum, 24
Support of a measure, 38

T
Theorem
  Banach on closed graph, 21
  Denjoy-Càrleman, 58
  Friedrichs, 92
  KLMN, 93
  Krasnoselsky-Krein, 22
  projection spectral
    for a finite set of operators, 105
    for a normal operator, 106
    for a self-adjoint operator, 102
  spectral for a normal operator, 40
  Stone, 45
Tensor product
  of chains, 79
  of operators, 78
Trace
  measure, 95
  of an operator, 94
Transform
  Fourier, 112
  Cayley, 25
    inverse, 26

V
Vector
analytic, 60
eigenvector, 22
common of a set of operators, 108
cyclic, 110
strong, 277
quasi-analytic, 58
Stieltjes vector, 60
entire, 60

Notation Index

A^*, 20
$\tilde{A}$, 19
A^{-1}, 18
$\mathcal{A}$, 36
$\mathbb{C}$, 17
C_0^∞, 45
$C\{m_n\}$, 57
c.l.s., 77
$\mathfrak{D}(A)$, dom (A), 17
dim, 22
$E(\alpha)$, 30
$g(A)$, 102
$\mathcal{H}$, 17
$\mathbb{I}$, 17
ind lim, 74
Γ_A, 18
Ker$A$, 18
K_z, 146, 222
L_∞, 33
L_0, 33
$(\mathbf{l}_{\text{fin}})'$, 205
$\mathbf{l}_{\text{fin}}$, 205
$\mathbf{l}_2$, 191, 193, 248, 327
$\mathcal{L}(\mathcal{H}_+, \mathcal{H}_-)$, 69
l.s., 77
N_z, 22
(m, n), 23
pr lim, 72
$\mathbb{R}$, 1
R_z, 24
$\mathfrak{R}(A)$, ran (A), 18
$S(A)$, 24
$\mathcal{S}(\mathbb{R}^N)$, 86
supp, 38
$\langle R, \mathfrak{R}\rangle$, 30
$\chi_\alpha(\lambda)$, 31
$\mathcal{I}_F$, 34
$\mathbb{T}$, 39
Tr$(\,\cdot\,)$, 94
$U(t)$, 44
$\|\cdot\|_{\mathcal{H}_-}$, 64
$\|\cdot\|_{\mathcal{H}_+}$, 63
$\|\cdot\|_{\mathcal{H}_0}$, 63
$\oplus$, 18
$\otimes$, 76
$|\cdot|$, 79, 96
$\hookrightarrow$, 189
$\restriction$, 17
$\sim$, 20, 21

Y. M. Berezansky, M. E. Dudkin, *Jacobi Matrices and the Moment Problem*,
Operator Theory: Advances and Applications 294,
https://doi.org/10.1007/978-3-031-46387-7